GENERAL CHEMISTRY

GENERAL CHEMISTRY

JOHN B. RUSSELL

Professor of Chemistry
Humboldt State University

McGraw-Hill Book Company

New York St. Louis San Francisco Auckland Bogotá Hamburg
Johannesburg London Madrid Mexico Montreal New Delhi Panama
Paris São Paulo Singapore Sydney Tokyo Toronto

GENERAL CHEMISTRY

Copyright © 1980 by McGraw-Hill, Inc. All rights reserved. Printed in the United States of America. No part of this publication may be reproduced, stored in a retrieval system, or transmitted, in any form or by any means, electronic, mechanical, photocopying, recording, or otherwise, without the prior written permission of the publisher.

1234567890   DODO   89876543210

This book was set in Times Roman by York Graphic Services, Inc. The editors were Jay Ricci, Anne T. Vinnicombe, and Sibyl Golden; the designer was Hermann Strohbach; the production supervisor was Charles Hess. The drawings were done by J & R Services, Inc.
R. R. Donnelley & Sons Company was printer and binder.

Library of Congress Cataloging in Publication Data

Russell, John Blair, date
    General chemistry.

    Includes index.
    1. Chemistry. I. Title.
QD31.2.R87   540   79-20059
ISBN 0-07-054310-0

Cover: Photomicrograph of a crystal of sodium chloride grown by evaporation of an aqueous solution of table salt to dryness. The color results from an optical technique known as Nomarski differential interference contrast.
Magnification: x270. (*Photomicrograph by Eric V. Gravé.*)

To My Mother and Father

CONTENTS

ix   CONTENTS

x CONTENTS

PREFACE

This text is intended for use in a full-year introductory chemistry course designed for students majoring in chemistry and other sciences, for students majoring in science-related disciplines, and for students who for any number of reasons simply wish to learn about chemistry. Among the students who take "freshman chemistry" today the spectrum of vocational goals, educational backgrounds, and motivations seems to be broader than ever before, and throughout the book I have kept this in mind. The effectiveness of this book depends upon the student's ability to read with comprehension at the college freshman level and to handle simple quantitative problems. To be specific, the student who has in high school followed a balanced, college-preparatory program which included two years of algebra should be able to handle this book. I have not assumed that the reader has studied any calculus, physics, or chemistry. But I have assumed that the text will be read and studied with considerable care.

Those of us who teach chemistry are aware that chemistry looms as an awesome challenge to many students. I believe an important reason is that chemical terms and concepts often seem so strange and abstract that the very language of chemistry tends to intimidate the student. Accordingly, I have tried to compensate for the natural human tendency to be wary of the unfamiliar by presenting chemistry in a way which is straightforward, clear, and logical. Above all, I have tried to help students see that chemical concepts are *reasonable* concepts. Those students who do their part should become almost as comfortable talking about antibonding orbitals and octahedral complexes as about apples and oranges.

The first twelve chapters of this book emphasize the structures of chemical systems and the dependence of properties on structure. With a few exceptions the progression of topics leads from the microscopic to the macroscopic. Atomic structure and bonding are presented from a descriptive quantum-mechanical perspective, and four chapters are devoted to these topics. Stoichiometry and ideal-gas behavior are introduced early, partly to allow students to cut their teeth early on chemical calculations, and partly to assist instructors in planning concomitant laboratory work. The properties of *ideal* gases, liquids, and solids are thoroughly discussed before *real* matter, and deviations from ideal behavior and perfect structure are considered. Except for the early coverage of mixtures of gases, solutions are not discussed until the properties of pure matter have been thoroughly described.

Chapters 13 to 24 cover aspects of chemical change and descriptive chemistry. One chapter is devoted to the special characteristics of aqueous-solution reactions. The topic of chemical kinetics is discussed in considerable detail, but without the use of calculus. Equilibrium is covered in three chapters, two of which are devoted to aqueous-solution equilibria. Chemical thermodynamics is discussed in a separate chapter and then used to introduce the topic of electrochemistry. Three chapters are devoted to descriptive inorganic chemistry and one to organic chemistry. The final chapter describes some important properties of nuclear change.

This text incorporates a number of features which should prove useful to the student. Each chapter is prefaced by a short section entitled *To the student* in which the purpose of the chapter is informally described. *Added comments* have been set off at many places which always seem to give difficulty to some students. These are brief digressions which point out potential trouble spots or provide alternative descriptions of relationships or concepts. About 200 *examples* of calculations are included within the chapters, and approximately 1200 *questions and problems* are provided at the chapter ends. Each chapter also concludes with a *summary* and a list of *key terms*. The *appendixes* are extensive; perhaps the most important of these is Appendix A, a *glossary of important terms* used throughout the book. *Answers to selected numerical problems* are given in Appendix J.

Supplements to this book which are available from the publisher include a *Student/Instructor Solution Supplement* by Roger Weiss, a *Study Guide* by Norman Eatough, and an *Instructor's Manual.* Vincent Sollimo has prepared an accompanying laboratory manual, *General Chemistry in the Laboratory.*

Prefaces are intended to be read first but are usually written last. As I reach the conclusion of what has turned out to be a three-year task, I am more aware than ever of the debt I owe to my teachers. I am fortunate to have been able to study and work under people such as Werner Bromund, Arthur Campbell, Paul Flory, Clyde Mason, Michell Sienko, and Luke Steiner. Their influence on my teaching has been considerable, and I wish to express my appreciation to them at this time. I also owe a debt of gratitude to another chemist, my father, who was the first to teach me what chemistry is really all about.

In writing this book I have relied heavily on the advice of my colleagues at Humboldt State University. I am greatly indebted to Tom Borgers, Greg Bowman, Thomas Clark, Clyde Davis, Mervin Hanson, Richard Paselk, M. G. Suryaraman, Robert Wallace, Roger Weiss, and William Wood, all of the Department of Chemistry, and to Frederick Cranston, of the Department of Physics and Physical Science, each of whom more than once kept me from going astray. I also wish to thank the reviewers Edwin H. Abbott, David L. Adams, John L. Burmeister, Gregory J. Exarhos, Norman Eatough, William A. Johnson, Fred H. Redmore, and Joseph R. Wiebush whose critical evaluations of the manuscript resulted in many large and small improvements. Especially valuable were the comments of David Adams whose ability to solve pedagogical problems was phenomenal. Lastly, I would like to express my unbounded gratitude to the staff at McGraw-Hill, and especially to Anne T. Vinnicombe, for whom the word indefatigable seems to have been coined.

John B. Russell

GENERAL CHEMISTRY

1

PRELIMINARIES AND PREMISES

TO THE STUDENT

This chapter is an introductory one. It begins with a discussion of chemistry as a science and follows with a brief outline of the history of chemistry. This historical perspective in itself is not of overwhelming importance, but some knowledge of how chemical ideas were developed gives greater significance to the ideas themselves. Following that is a short discussion of scientific methodology and definitions and explanations of some important terms. Knowing exactly how scientific terms are used is essential for accurate, efficient communication of ideas. Finally, numerical calculations are considered. Emphasis is placed on keeping track of significant figures, on using numbers written in exponential form, and on the SI system of metric units. Chemistry is a quantitative science, and it is therefore important for you to learn how to perform calculations correctly and to express your results properly.

**1-1
Chemistry: what, why, and how?**

What is chemistry?

At one time it was easy to define chemistry. The traditional definition goes something like this: Chemistry is the study of the nature, properties, and composition of matter, and how these undergo changes. That served as a perfectly adequate definition as late as the 1930s, when natural science (the systematic knowledge of nature) seemed quite clearly divisible into the physical and biological sciences, with the former being comprised of physics, chemistry, geology, and astronomy and the latter consisting of botany and zoology. This classification is still used, but the emergence of important fields of study such as oceanography, paleobotany, meteorology, and biochemistry, for example, have made it increasingly clear that the dividing lines between the sciences are no longer at all sharp. Chemistry, for instance, now overlaps so much with geology (thus we have *geochemistry*), astronomy (*astrochemistry*), and physics (*physical chemistry*) that it is probably impossible to devise a really good modern definition of chemistry, except, perhaps, to fall back on the operational definition: chemistry is what chemists do. (And what chemists do is what this book is all about!)

1

Why study chemistry?

Chemistry plays an important part in all of the other natural sciences, basic and applied. Plant growth and metabolism, the formation of igneous rocks, the role played by ozone in the upper atmosphere, the degradation of environmental pollutants, the properties of lunar soil, and the medical action of drugs: none of these can be understood without the knowledge and perspective provided by chemistry. Indeed, many people study chemistry so that they can apply it to their own particular fields of interest.

Of course, chemistry itself is the field of interest for many people, too. Many study chemistry not to apply it to another field, but simply to learn more about the physical world and the behavior of matter from a chemical viewpoint. Some simply like "what chemists do" and so decide to "do it" themselves.

How to study chemistry

Giving advice on studying is risky, but the following list of suggestions may be useful:

1 *Read this book slowly.* Sometimes too much emphasis is placed on speed reading. Many persons can read a light novel at a rate of 100 pages per hour and understand all that needs to be understood. Reading a scientific or technical book is another matter, however. A student with a fast reading rate may be able to read this text at an average of 20 pages per hour. But if you find yourself reading at a much slower rate, especially in chapters that are more difficult for you, do not feel that anything is wrong. Just keep going slowly and steadily—and learn!

2 Use the *questions and problems* at the end of each chapter. They were prepared to help you learn. They should not be regarded as extra appendages, but as integral parts of the chapters, intended not only to help you check your understanding of the chapter material but also to help you actually *learn* the material. Much chemistry needs to be *done,* not just read about, if a real understanding is to be reached.

3 *Study critically.* As you read statements in this book, question them. *Check cross-references* always. If the meaning of a technical word is not clear, or if you have forgotten it, look it up. The index is a complete one and should be used frequently. Be smart about the way you study: study critically.

4 Do not skip over the *examples of numerical calculations* which you will frequently find in this book. Follow each one step by step. Do not hurry. These sample calculations will teach you more than just how to get the answer: they will help you learn, but only if you follow them through.

5 Do not be intimidated by names and concepts which seem terribly strange to you. *Strange* just means "unfamiliar," so give yourself a chance to become familiar with new ideas. Your being less familiar with orbitals, chelates, and entropy than with apples, elephants, and television should not prevent you from thinking and reasoning about these new things.

6 Keep an open mind!

1-2 The way it all started

The origins of chemistry are almost as old as mankind. The first "prechemical" activities undoubtedly consisted of little more than chance observations. The Stone Age creature probably marveled at the way a piece of wood changed as it

The earliest "chemical" activities

was consumed by fire, at the changes in appearance, smell, and taste of a piece of meat as it rotted, and at the changes in color and hardness of a piece of clay as it baked near a fire. Simple observations such as these were the first steps and eventually led to more significant accomplishments. Copper metal, for example, was undoubtedly known in the Stone Age, but because of its scarcity it was probably considered to be a mere curiosity. But copper *ore* was much more abundant, and early man must have accidentally learned that if some pretty blue stones were heated in a fire they would change to copper metal. (The big breakthrough came in about 4000 B.C., when it was realized that heating copper ore would *always* produce copper metal.) Bronze, an *alloy* (metallic combination) of copper and tin, was probably first made accidentally, when copper and tin ores were simultaneously heated in a fire. Copper was too soft for armor and spear tips, but bronze was not. By 3000 B.C. the Bronze Age had begun and early civilization had been changed.

To make a very long story short, copper and bronze gave way to iron and steel, the latter being an iron-carbon alloy. Metals were very important in early civilization and the practice of metallurgy provided a wealth of chemical information. Egyptians, for example, learned how to obtain many different metals from their ores, and according to some experts the word *chemistry* is derived from an ancient word *khemeia,* which may refer to the Egyptians' name for their own country, *Kham.* However, some experts believe *chemistry* came from the Greek word *chyma,* which means "to melt or cast a metal."

But the "chemical" accomplishments of the ancients were not limited to metallurgy. By 3000 B.C. the art of making glass was flourishing in Egypt, as were the techniques for manufacturing dyes, pigments, artificial gems, and intoxicating beverages. Gunpowder was made and used by the early inhabitants of what are now China and India.

It was not until around 600 B.C. that the beginnings of chemical theory emerged. Thales, a Greek philosopher, proposed that all chemical change was merely a change in the aspect of one fundamental material or *element.* Later, Empedocles, who lived about 450 B.C. proposed that there were four elements: earth, air, fire, and water. This idea was amplified by the great Greek philospher, Aristotle (384–322 B.C.), who considered that each element resulted from the combination of two of the four fundamental qualities: hot, cold, wet, and dry. Only four of the six possible combinations were permitted according to Aristotle. Thus, hot and dry could combine to form fire, hot and wet to form air, cold and dry to form earth, and cold and wet to form water.

The Greek philosophers raised another important question: Was matter continuous or discontinuous? If matter were *continuous,* or jellylike, in nature, then any piece of it could be broken into smaller and smaller fragments, and this division and subdivision could occur without limit. If, on the other hand, matter were *discontinuous,* or granular, then successive subdivision of any substance could occur only until the smallest indivisible granules were obtained. Two Greek philosophers, Leucippus and Democritus, who lived about 400 B.C., were the first of the advocates of discontinuity. Democritus named the ultimately small, indivisible granules *atomos* ("indivisible") and our word *atom* is derived directly. Thus, the view that matter is not indefinitely subdivisible is known as *atomism.*

The alchemists   The art of *khemeia* flourished in the early Egyptian and Greek civilizations. During these periods chemical reactions seemed so fascinating and mysterious that they were thought to have religious, mystical, or occult implications. One of the greatest challenges faced by *khemeia* was that of *transmutation,* a word applied to the much sought for process which would change one element into another—cheap and plentiful lead into rare and costly gold, for instance. After the peak of Greek civilization, *khemeia* and the search for a successful recipe for transmutation were kept alive by the Romans, the Persians, and later by the Indians, Chinese, and Arabs. The Arabs kept *khemeia* alive until about A.D. 1100. The word *khemeia* became *al-kimiya* in Arabic (the prefix *al* means "the" in Arabic) and from this word the English word *alchemy* was later derived.

The period of alchemy is sometimes designated as spanning the years from 300 B.C. to A.D. 1500. Arabic alchemy was important during the interval from about A.D. 600 to A.D. 1100. The most famous of the Arab alchemists was a man known today as "Geber," although his real name was Jabir ibn-Hayyan, who lived about A.D. 800. Geber made many efforts to create gold, finally becoming convinced that if mercury, which is a metal, and sulfur, which is golden-colored, were mixed properly, gold would result. The Greeks thought that a transmutation-assisting powder might be necessary to make gold, and they called such a powder *xerion* from the Greek word meaning "dry." In Arabic this word became *al-iksir,* from which our word *elixir* is derived. Needless to say Geber never found *al-iksir,* but he spent much of his life searching for it.

Following the Christian crusades, which started in 1096, contact between the East and West became more frequent and alchemical knowledge began to infiltrate Western Europe. Medieval alchemy was closely interwoven with numerology, astrology, mysticism, and the "black arts," and its thrust was toward finding a method for "manufacturing" gold. Alchemists were obsessed in their search for Geber's elixir, which they eventually renamed the *philosopher's stone*. (Natural science was once called *natural philosophy*.)

Paracelsus and   The forces of alchemy underwent a change shortly after 1500, largely through the
iatrochemistry   work of a Swiss, Phillipus Aureolus Theophrastus Bombastus von Hohenheim (1493–1541). He is better known to us by his self-selected pseudonym, Paracelsus, which means "better than Celsus." (Celsus was an early Roman who wrote about medicines and the medical arts.) Paracelsus used alchemical methods in order to find medicines which would cure disease. He searched for the philosopher's stone, but primarily because he thought it could function as the "elixir of life," a substance which would prolong life and good health indefinitely. During his lifetime Paracelsus acquired quite a reputation as a drug manufacturer and physician throughout much of Europe. Many of his remedies for sickness seem by today's standards to be harsh and even deadly. On the other hand, he apparently learned (through trial and error) that very small doses of poisonous substances were sometimes beneficial in curing certain diseases, a fact which is well accepted today. Thus, for example, Paracelsus used substances containing lead, mercury, and arsenic in his preparations. Paracelsus believed that all matter was composed of three "principles," sulfur, mercury, and salt, which were themselves each composed of varying proportions of the four elements of Aristotle: earth, air, fire, and water. (To Paracelsus the words *sulfur, mercury,* and *salt* had broader

meanings than they do today.) Paracelsus died in 1541, likely a victim of accidental self-poisoning. The period from roughly 1500 to 1650 is sometimes known as the era of *iatrochemistry,* or medical chemistry.

Boyle and the phlogiston period

During the seventeenth century attentions were turned to the behavior and properties of gases. One Irish chemist, Robert Boyle, discovered the simple relationship which exists between the volume of a gas and its pressure. This relationship is known in much of the world as *Boyle's law* (Sec. 4-2). He also showed that sound cannot be transmitted in a vacuum and that air is necessary for animal life. His work with different gases and liquids led to the development of many ingenious new experimental techniques which entailed measurements of high precision. Boyle's interest in chemical theory eventually prompted him to write his most famous work, *The Sceptical Chymist,* a book published in 1661. (The prefix in *alchemy* was dropped by Boyle, and the field was known as *chemistry* from this time on.)

Interest in gases during the seventeenth and early eighteenth centuries led to an important, though erroneous, theory regarding the nature of burning and of corrosion. This was the *phlogiston* theory. In 1669 the German, J. J. Becher, expanded on the Greek notion that when combustion took place something escaped, and another German, G. E. Stahl, later called the "something" phlogiston. Stahl believed that combustible substances were rich in phlogiston, and that combustion was simply the loss of this mysterious substance. Further, corrosion or rusting of metals was also believed to be a slow loss of phlogiston from the metal. Stahl never solved the problem of the gain in weight which metals exhibit when they corrode. Some chemists actually suggested that phlogiston had a negative weight, but this only caused trouble when attempting to explain why, for example, a piece of wood lost weight when it was burned.

Lavoisier and the modern era

In 1743 the French chemist A. L. Lavoisier was born. He, too, became interested in the process of combustion, but unlike most of his predecessors, he carefully designed his experiments so that he could precisely *weigh* both the combustibles and their products. Lavoisier proceeded to burn everything he could get his hands on, even a diamond, and was able to show that when a substance corrodes in a sealed container, the resulting gain in weight is compensated by a corresponding loss in weight of the air in the container. Thus, reasoned Lavoisier, when a metal corrodes, something in the air enters, or combines with, the metal. Lavoisier also correctly explained that the apparent loss of weight accompanying the burning of a substance such as wood is merely the result of the products of combustion being gaseous. Lavoisier maintained that if the weights of all substances involved in a chemical reaction are considered there is no overall loss or gain in weight. This generalization is really a preliminary version of one of the foundations of chemistry, the law of conservation of mass (Sec. 2-2).

Lavoisier is generally said to have ushered in the modern era of chemistry. In 1789 he published an important textbook, *Elementary Treatise on Chemistry.* Soon thereafter most of the uncertainties about elements, compounds, atoms, and chemical change became at least tentatively resolved, and the march forward to new experimental and theoretical discoveries accelerated greatly.

ADDED COMMENT

This has been a brief survey of the history of chemistry up to the modern era. You may wish to read more of the story of how chemistry came to be. If so, check your library; some of the chapters in the story will amaze and fascinate you.

**1-3
Science and its
methods**

Viewed from an historical perspective, it is clear that scientific knowledge has been obtained and that therefore science has "advanced" in a series of fairly logical steps. On the other hand, counterparts to these steps are difficult to identify in the day-to-day professional activities of a scientist. In order to understand the reasons for this difference, we should first examine what is commonly called the *scientific method*.

*Observations and
laws*

It can be considered that science begins with *observations*. These observations may be chance ones (the cave man noticed that the meat of the saber-toothed tiger tasted better when the tiger had been killed in a forest fire, than when he killed the animal himself), or they may be observations made under closely controlled experimental conditions. The records kept of these observations are called scientific *data,* and often examination of sets of data reveals regularities, similarities, or consistencies. Sometimes these regularities can be concisely stated in a summary statement or generalization known as a *law*.[1]

Consider an example: Suppose that you have been studying chemical reactions between gases and that you have been able to measure the volumes of gases used up and formed in a number of reactions. Consider, for instance, the reaction between hydrogen gas and oxygen gas to form gaseous water (steam):

$$\text{Hydrogen} + \text{oxygen} \longrightarrow \text{water}$$

If all of your volumes were measured at the same temperature and pressure, then your data would look something like this:

Experiment	Volume of hydrogen used, liters	Volume of oxygen used, liters	Volume of water formed, liters
1	3.4	1.7	3.4
2	7.2	3.6	7.2
3	9.0	4.5	9.0

It can be seen that in each experiment the hydrogen/oxygen/water ratio is $2:1:2$, that is, that 2 volumes of hydrogen combine with 1 volume of oxygen to form 2 volumes of water vapor.

If you repeated your experiments with other gas reactions, you would observe similar results. If, for example, you studied the reaction

$$\text{Hydrogen} + \text{nitrogen} \longrightarrow \text{ammonia}$$

you would find the hydrogen/nitrogen/ammonia volume ratio to be $3:1:2$. In

---

[1] This kind of law is a *natural law* and should be distinguished from a statutory law, one passed by a parliament, congress, or legislature.

fact, for each such gas reaction you would find that *the volumes of all gases involved can be expressed in ratios of small whole numbers, if all measurements are made at the same temperature and pressure.* This generalization is a *law,* and if you had discovered it for yourself, you would have followed about the same path as that followed by the French chemist Joseph Gay-Lussac in 1808.

The process of arriving at a law from observation and data is an example of *inductive reasoning* in which recognition of consistencies and regularities in observed facts suggests inferences which can be stated as concise generalizations. These generalizations are laws.

Theories   Curiosity is an important human characteristic, and a natural law often provokes that curiosity. We wish to know *why* something behaves the way it does, *why* combining volumes of gases are expressible as ratios of small whole numbers, for instance. Proposed answers to such "why" questions are called *theories,* or, in the case of tentative answers or explanations, *hypotheses.*

Theories are the keys to scientific progress. They are explanations of behavior expressed in terms of the known behavior of more familiar objects. Thus, a theory explains observed behavior by using a *model* which is more easily understood. We shall see in Sec. 4-5, for instance, that the behavior of gases, including that expressed in Gay-Lussac's law of combining volumes, can be rather neatly explained by picturing gases as being composed of molecules which are like tiny billiard balls in constant motion, colliding and rebounding off each other and off the walls of the container. This is part of what is known as kinetic-molecular theory, a theory which uses tiny imaginary "billiard balls" as a model to help account for the observed behavior of gases. Sometimes the model upon which a theory is based is a physical one; that is, it is based upon familiar physical objects. Some theories, however, use mathematical models. It should be noted that neither a physical nor a mathematical model will explain anything to someone who is not already familiar with the physical objects or mathematical relationships used in the model.[2]

The power of a successful theory is not merely found in its success in explaining observed phenomena but in its ability to make *predictions* which can be *tested* by designing and carrying out new experiments. If the new observations and data obtained from these experiments agree with prediction, then the theory is further substantiated. If they do not, then it is necessary to modify the theory, or perhaps to junk it entirely and search for a new one. The point is that the scientific process begins to repeat itself at this stage; that is, new observations, data, and laws tend to lead to new theories, which lead to new experiments, observations, data, and so on. Thus, the method of science is just a logical approach to learning about nature. It is summarized in Fig. 1-1.

Remember that a theory is only a model; it is not reality itself. Thus, gases are not really composed of billiard balls. Gases *act* as if they were composed of particles which have some of the properties of billiard balls. We will never know the ultimate nature of a gas, that is, what a gas really *is,* but kinetic-molecular

[2]We have used the word *explain.* In a sense it is not possible for us ever to provide an ultimate explanation for anything, because after each explanation, we may again say "Why?" In the context of a theory, *to explain* just means "to make seem reasonable."

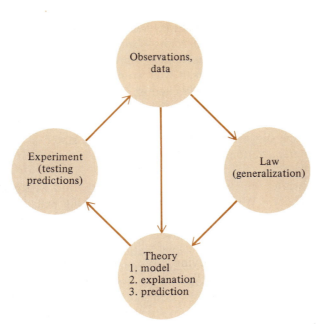

Figure 1-1
The advance of science: a repeating cycle. Science can be described in terms of a series of repeating cycles. Observations and data (and laws) lead to the proposal of theories that, in turn, suggest predictions which can be tested by designing new experiments, and the whole process starts all over again.

theory helps us understand gas behavior. The distinction between what is actually observed as a fact and what is used as a model is often lost, even by scientists. This is especially true when a theory becomes accepted for many years. Then the tendency is to treat the model as if it were real itself. In reality no one knows with *absolute* certainty whether or not molecules really exist, but there is such overwhelming evidence in their favor, that we now speak of them as real entities. The careful scientist, however, keeps in mind that the molecule is only a concept invented by people (a *construct*) to help them understand the universe.

The scientist   A scientist is sometimes depicted as someone who works in a laboratory, an office, or the field, using the scientific method, that is, the cyclic sequence of steps outlined above. This description is misleading for a number of reasons. First, the progress of science is often slow, and scientists may spend their entire professional lives collecting data, working on a theory, or designing experiments. Thus, there may not be time for a given scientist to pass through even one of the stages of the scientific method. Second, many scientists have varying amounts of freedom to follow a single research project through to its ultimate conclusion. There is nothing undemocratic here; in most cases it is simply a case of a corporate employer feeling that it would be to the company's best interests for a scientist-employee to switch to another research project. Last, a scientist does not work by mechanically following a prescribed sequence of steps. He or she certainly does not enter the laboratory in the morning thinking, "Well, yesterday I stated a law, so today I'll devise a theory"! Instead, the scientist often uses a combination of educated guesses, hunches, and flashes of inspiration to supplement all the knowledge provided by education and training. The problem faced may be difficult, and so progress toward the solution is often slow, tortuous, and frustrating. Finding the solution generally requires much creative scientific inspiration and even more plain hard work.

1-4
Words and definitions

Most of us tend to use words rather carelessly, and although this is not a good practice, it is usually not as disastrous in everyday life as it is in science. Thus, it is important to understand the meanings of new words and expressions, so that they will not be hidden sources of confusion later. This is especially true for certain scientific words which have different meanings when used in everyday conversation. Such words include *energy, work,* and *mass,* to name three. When words are defined or explained, take the time to be sure you understand their meanings. Learning to use a chemical vocabulary will make your study of chemistry proceed much more smoothly. Remember, too, that if you forget the meaning of a term, a glossary can be found in App. A.

Some fundamental terms

Chemistry is a way of studying matter. What is matter? As is true with many of those words which are really basic to science, *matter* is hard to define. It is often said that matter is anything which has mass and occupies space. But then what are "mass" and "space"? Although we can define these, the process yields little real insight into what matter is. So let us just say that matter is anything which has real, physical existence; in a word matter is just *stuff.* Iron, air, wood, gold, milk, aspirin, monkeys, rubber, and pizza—these are all matter. Some things which are not matter are heat, cold, colors, dreams, hopes, ideas, sunlight, beauty, fear, and x rays. None of these is "stuff"; none is matter.

A sample of matter can be either a pure substance or a mixture. A *pure substance* has a fixed, characteristic composition and a fixed, definite set of properties. Pure substances include copper, salt, diamond, water, table sugar, oxygen, mercury, vitamin C, and ozone. We will see in Sec. 2-2 that a pure substance may be a single *element,* such as copper or oxygen, or a *compound* of two or more elements in a *fixed* ratio, such as salt (39.34 percent sodium and 60.66 percent chlorine) or table sugar (42.11 percent carbon, 6.48 percent hydrogen, and 51.41 percent oxygen).

A *mixture* is a collection of pure substances simply mixed together. Its composition is *variable,* as are its properties. Examples of mixtures are milk, wood, concrete, saltwater, air, granite, motor oil, chocolate, and elephants.

A pure substance can be a solid, a liquid, or a gas; these are the three *states of matter.* A *solid* maintains its volume and shape; a *liquid,* its volume only; and a *gas,* neither. Solids tend to be hard and unyielding; liquids maintain their volumes and flow to adopt the shapes of their containers; gases also flow and in addition expand to occupy all of their containers. The ability to flow is called *fluidity,* and so gases and liquids are called *fluids.*

One of the goals of chemistry is to be able to describe the properties of matter in terms of its internal structure, the arrangement and interrelationship of its parts. This word, *structure,* sometimes refers to the physical arrangement of particles, such as atoms or molecules, in space. At other times it is used to indicate some other arrangement, such as the arrangement of energy levels of an electron in an atom.

The structure of matter determines its properties. Properties can be classed as either physical or chemical. A *physical property* of a substance can be characterized without specific reference to any other substance and usually describes the response of the substance to some external influence, such as heat, light, force, electricity, etc. Physical properties include boiling point, melting point, thermal (heat) conductivity, color, refractive index, reflectivity, hardness, tensile strength, fluidity, and electrical conductivity.

A *chemical property,* on the other hand, describes a *chemical change*: the interaction of one substance with another, or the change of one substance into another. Iron rusts in a moist environment, unrefrigerated milk turns sour, wood burns in air, dynamite explodes—each of these is a chemical property because each involves chemical change. During chemical changes, substances are actually changed into other substances. The simultaneous disappearance of some substances (called the *reactants*) and appearance of others (the *products*) is characteristic of chemical change (chemical reaction). Chemical changes are generally characterized by pronounced internal structural rearrangements.

Physical changes are not characterized by the transformation of one substance into another, but rather by the change of the form of a given substance. The bending of a piece of copper wire fails to change the copper into another substance; crushing a block of ice leaves only crushed ice; melting an iron nail yields a substance still called iron: These are all usually accepted as physical changes.

Properties of matter may also be categorized as either macroscopic or microscopic. A *macroscopic property* describes characteristics or behavior of a sample which is large enough to see, handle, manipulate, weigh, etc. A *microscopic property* describes the behavior of a much smaller sample of matter, an atom or molecule, for instance. Macroscopic and microscopic properties are often different. A banana is yellow, but we do not use color to describe an atom. Some properties, on the other hand, can be either microscopic or macroscopic; *mass* is one of these.

Another word we will often use is *system*. A system is a portion of the universe which we wish to observe or consider. The size of the portion is usually small and a system may be a real one (in a test tube or flask, for example), or an imaginary one which we are just talking about. Often we will place some restriction on a system. There will be occasions, for instance, when we will wish to discuss a system whose temperature is not allowed to change.

1-5 Numbers: their use and misuse

Chemistry is a quantitative science. There are important relationships in chemistry which can be expressed satisfactorily only in the language of mathematics. Furthermore, in many cases a real understanding and appreciation of these relationships comes only with the working of numerical problems based on the relationships. In solving these you will need to use numbers which are written in exponential form and to pay proper attention to significant figures in your calculations.

Exponential notation

Some numbers are either so very large or so very small that using our ordinary decimal system proves to be inconvenient and cumbersome. For instance, not only is it a waste of zeros to write 0.0000000000000000000472, but it also gives only a vague feeling about how small this number is. (Did you actually count the zeros?) How much more satisfactory it is simply to write 4.72×10^{-20}. A number written this way is said to be written in *exponential,* or *scientific, notation.* The notation itself should not be new to you, but a few facts about numbers written this way should be pointed out:

1 Each number is written as a product of two numbers. (There is nothing tricky about this; we could write 8 as 4×2, if we wished.) The two numbers are (1) a coefficient and (2) 10 raised to some power. The power (exponent) should be a positive or negative integer, or zero.

2 When two numbers written exponentially are to be multiplied (or divided), first multiply (or divide) the two coefficients and then add (or subtract) the exponents, as shown in the following examples:

Multiplication:

$$(2 \times 10^3) \times (4 \times 10^6) = (2 \times 4) \times (10^3 \times 10^6) = 8 \times 10^9$$

$$(3 \times 10^3) \times (2 \times 10^{-6}) = (3 \times 2) \times (10^3 \times 10^{-6}) = 6 \times 10^{-3}$$

Division:

$$\frac{4 \times 10^6}{2 \times 10^2} = \frac{4}{2} \times \frac{10^6}{10^2} = 2 \times 10^4$$

$$\frac{8 \times 10^{16}}{4 \times 10^{-3}} = \frac{8}{4} \times \frac{10^{16}}{10^{-3}} = 2 \times 10^{19}$$

3 Numbers provided as data or resulting from a calculation are usually, though not always, expressed with one digit to the left of the decimal place in the coefficient. Thus,

$$(4.0 \times 10^{-8}) \times (7.0 \times 10^{-12}) = 28 \times 10^{-20} = 2.8 \times 10^{-19}$$

$$\frac{4 \times 10^{-8}}{5 \times 10^{-12}} = 0.8 \times 10^4 = 8 \times 10^3$$

4 Adding or subtracting numbers is accomplished by first changing the way they are written, if necessary, so that the powers of 10 are the same. Then the coefficients can be added conventionally:

$$
\begin{array}{rcl}
6.0 \ \times 10^8 & \longrightarrow & 0.60 \times 10^9 \\
+8.14 \times 10^9 & \longrightarrow & +8.14 \times 10^9 \\
\hline
& & 8.74 \times 10^9
\end{array}
$$

$$
\begin{array}{rcl}
8.37 \times 10^{-4} & \longrightarrow & 8.37 \times 10^{-4} \\
-3.0 \ \times 10^{-5} & \longrightarrow & -0.30 \times 10^{-4} \\
\hline
& & 8.07 \times 10^{-4}
\end{array}
$$

5 When taking the square or cube root of a number, first rewrite it, if necessary, so that the exponent is divisible by a 2 or 3, respectively:[3]

---

[3] The symbol $\approx$, which follows, means "is approximately equal to."

$$\sqrt{4 \times 10^{12}} = \sqrt{4} \times \sqrt{10^{12}} = 2 \times 10^6$$

$$\sqrt{4 \times 10^{13}} = \sqrt{40 \times 10^{12}} = \sqrt{40} \times \sqrt{10^{12}} \approx 6 \times 10^6$$

$$\sqrt[3]{4 \times 10^{-13}} = \sqrt[3]{400 \times 10^{-15}} = \sqrt[3]{400} \times \sqrt[3]{10^{-15}} \approx 7 \times 10^{-5}$$

Measurement, accuracy, and precision

Most of the numbers that we encounter in everyday life (and in science, as well) are not exact numbers but are instead the results of measurements; hence, they are only approximate. When you buy "five gallons" of gasoline for your car, you get only approximately 5 gal, and a "one-pound" package of butter contains only approximately 1 lb. Usually we do not worry much about the accuracy of such quantities; we assume that they are close enough for our purposes. *Accuracy* refers to the truthfulness of the number in an absolute sense. The term *precision* refers *not* to the absolute truthfulness of the number but to the degree of effort and care which was taken in obtaining the number.

In order to see the contrast between the terms *accuracy* and *precision* look at an example: Consider the "five gallons" of gasoline you put in your car. The dials on the pump at the service station can easily be read to the nearest one-tenth of a gallon. (The lines or graduations on the right-hand dial are one-tenth of a gallon apart.) Thus, without any great effort the amount of gasoline delivered can be seen, *apparently,* to be 5.0 gal, not 5.1 or 4.9 gal. The measurement of the volume of gasoline delivered into your tank can be said to be *precise* to the nearest 0.1 gal; that is, the number 5.0 has a precision of 0.1. With a little care the space between two adjacent graduations on the "one-tenth gallon," or right-hand, dial on the pump can be divided up into 10 imaginary spaces, and the flow of gasoline can be stopped so that a little more than 5.0 gal has been delivered. Estimate (using our imaginary subdivisions on the dial) that 5.02 gal have been delivered. Reading the dial this carefully has allowed us to specify the volume of gasoline delivered to the nearest 0.01 gal, an increase in the precision of the measurement. Does high precision always mean high accuracy? Not always. Suppose that the metering device in the gasoline pump were defective so that the pump delivered only about 4 gal of gas when the meter read 5.02 gal. In this (unlikely) event the precision of the last measurement would still be very high (one-hundredth of a gallon), but the accuracy would be very low (only about 1 gal).

Significant figures

Since each number we encounter (with a few exceptions, to be discussed below) is either directly or indirectly the result of a measurement, and since each is therefore limited in its precision, it is useful to know that precision. There are several different ways for expressing the precision inherent in a number, but the simplest and most commonly used is the method whereby the number of significant figures in the number shows the relative precision of the number itself. Compare, for example, the numbers 5.32 and 5.3. The number 5.32 is expressed to the nearest 0.01; its precision is 0.01 part in 5.32 parts, or $(0.01/5.32) \times 100 = 0.2$ percent. On the other hand, the number 5.3 is expressed only to the nearest 0.1; its precision is therefore only 0.1 part in 5.3 parts or $(0.1/5.3) \times 100 = 2$ percent. (Note that the position of the decimal point has no effect on the determination of precision.) *Significant figures,* or *significant digits,* as they are often called, are digits which serve to establish the value of the number; zeros shown merely to locate the decimal point are *not* significant figures. Thus, the number 4.56 has *three* significant figures, and so does the number 0.00456. In each of these cases the 4, the 5, and the 6 are the significant figures; in the second example the two zeros just to the right of the decimal point are only place markers. (The zero on the far left is not really necessary at all; it is often written, however, for the purpose of alerting the reader that a decimal point is lurking around somewhere.)

When a number is written correctly a rule usually followed is that the *last* significant figure (on the right) is the one which may be somewhat uncertain or doubtful. This means, for example, that in the number 4.56 the 4 and the 5 are known with certainty, but there is some doubt about the 6. The 6 is still specified, however, and what this really means is that the *most probable* value of the last digit is 6.

To find the number of significant figures in a numerical quantity simply count all the digits starting at the first nonzero digit at the left. If a number is written in exponential notation, count only the significant figures in the coefficient. (The exponential part is considered an exact multiplier.) In each example below, the number of significant figures is specified for each number:

Number	Significant figures
7	1
7.4	2
7.0	2
7.40	3
7.4321	5
7.0000	5
0.0007	1
0.07	1
0.0700	3
700.07	5
7.0007×10^2	5
7×10^{14}	1
7.40×10^{-21}	3

Note that in these examples the location of the decimal point bears no relation to the number of significant figures. A common error is to say that the number of significant figures is the number of places to the right of the decimal point. *This is wrong!* Thus, 0.0432 and 4.32 each have three significant figures, not four or two.

Write each number to as many significant figures as can be justified. Watch out especially for zeros at the end. If, for example, a volume of 5 ml of water is measured to the nearest milliliter, then it is correct to write the volume as 5 ml. But if the volume is measured more precisely, say to the nearest 0.01 ml, then the volume should be expressed as 5.00 ml.

A kind of number which occasionally crops up is the *exact number*. Examples of this kind of number are the number of players on a basketball team (exactly 5), the number of corners formed by the intersection of two roads (exactly 4), and the number of sides of a triangle (exactly 3). Each of these numbers may be considered to be written with an infinite number of zeros (all of which are significant figures) to the right of the decimal point.

A special problem arises with one kind of number. Most of us are in the habit of expressing large numbers by using zeros at the right; for example, we might write "twenty-seven thousand" as 27,000. The question is, are those zeros significant or not? Unfortunately there is no way of knowing from the number itself. If 27,000 is being expressed to the nearest thousand, then it has two significant figures; if, on the other hand, it is being expressed to the nearest unit, then it has five significant figures. Just looking at "27,000" gives us no clue as to how precise the number is. The way out of this dilemma is to write the number using exponential notation. Then we can easily see that 2.7×10^4 has two significant figures, while 2.7000×10^4 has five.

In carrying out calculations based on experimentally determined numbers it is important that each result is expressed to the proper number of significant figures. To be specific, a calculated result must not express higher or lower precision than can be justified by the numbers on which the calculation is based. In order to ensure that this is the case, two rules for arithmetic calculations are used. The first rule states that *in addition and subtraction the number of digits to the right of the decimal point in a calculated result should be the same as the number of such digits in the term which has the fewest digits to the right of the decimal point.*

An example should help here. Suppose that you have just weighed out 12.47 oz of sugar and placed it in a jar previously weighed at 7 oz. What is the total weight? The answer is found as follows:

Weight of sugar	12.47 oz
Weight of jar	7. oz
Total weight	19. oz

Since the "7" has no digits to the right of the decimal point, the answer must be rounded off so as to be expressed similarly. (Because the weight of the jar is known only to the nearest ounce, the sum must not be expressed to any greater precision than this.)

In practice the rounding off may take place before or after the actual addition or subtraction. Rounding off before addition is done like this:

$$
\begin{array}{rcl}
1.627 & \longrightarrow & 1.6 \\
23.1 & \longrightarrow & 23.1 \\
4.06 & \longrightarrow & 4.1 \\
106.91 & \longrightarrow & \underline{106.9} \\
& & 135.7
\end{array}
$$

Rounding off after addition is done this way:

$$
\begin{array}{l}
1.627 \\
23.1 \\
4.06 \\
\underline{106.91} \\
135.697 \longrightarrow 135.7
\end{array}
$$

The two rounding-off procedures generally yield similar, if not identical, results.

In *multiplication* and *division* a different rule is used. Here it is necessary to count significant figures and to be sure that *the number of significant figures in the result is the same as the least number of significant figures in any of the multiplied or divided terms.* Some examples of how this rule is used are

$$\frac{1.473}{2.6} = 0.57$$

$$3.94 \times 2.12345 = 8.37$$

$$9 \times 0.00043 = 0.004$$

$$(4.1 \times 10^6) \times (9.653 \times 10^3) = 3.9 \times 10^{10}$$

$$\frac{6.734 \times 10^3}{7.41 \times 10^8} = 9.09 \times 10^{-6}$$

$$\frac{3.6(7.431 \times 10^8)}{1.49(6.67 \times 10^4)} = 2.7 \times 10^4$$

Note that in each of the above examples the result has been rounded off so that its number of significant figures is the same as the least number of significant figures in any of the numbers used in the calculation.

**1-6
The metric system
of units**

All of us use various units of measurement in our everyday lives. These include the familiar quarts, gallons, inches, miles, ounces, tons, seconds, years, as well as many others. In addition, the following units are used in specialized applications: furlongs, drams, carats, cables, cords, chains, nautical miles, sidereal days, and many dozens more! For years, however, the units used in science have almost exclusively been *metric units*. Furthermore, most of the world now uses the metric system, and even the United States seems to be gradually phasing in metric units.

SI units

In 1960 an international organization, the General Conference of Weights and Measures, agreed upon a unified version of the metric system which is gradually being adopted by most of the world. The units in this system are known as *SI units*. (SI stands for *Système International d'Unités*, and is the symbol used in all languages to denote these units.) The system is actually a descendant of the meter-kilogram-second (mks) system, a metric system of units used by scientists for years.

Seven *base units* constitute the foundation of the SI system. These units and the quantities which they measure are listed in Table 1-1. In this book we will have occasion to use each of these except for the candela, the unit of light intensity.

**Table 1-1
The seven SI base units**

Physical quantity	Name of unit	Symbol
Mass	Kilogram	kg
Length	Meter	m
Time	Second	s
Electric current	Ampere	A
Temperature	Kelvin	K
Luminous intensity	Candela	cd
Quantity of substance	Mole	mol

The SI base unit of *mass* is the *kilogram* (kg). It is a convenient unit of intermediate size; 1 kg is about 2.2 pounds. (Mass and weight are not the same; we will look at the difference in Sec. 2-1.)

The SI base unit of *length* is the *meter* (m). It is also of intermediate size, being a little larger than a yard, about 39 inches.

The SI base unit of *time* is the *second* (s). The rest of the SI base units are listed in App. B.

Closely related to the SI base units are the *derived* units. These include units of volume, area, force, energy, pressure, power, and others. Some of the derived units have special names. The *newton* (N), for example, is the SI unit of *force*. One newton is the force necessary to cause a mass of one kilogram to accelerate one meter per second per second. In other words

$$1\,N = 1\,\frac{kg\,m}{s^2}$$

The SI unit of *energy* is the *joule* (J). This is defined as the energy expended (or the work done, which is equivalent) when a force of one newton moves an object one meter in the direction in which the force is applied. Thus

$$1\,J = 1\,N\,m = 1\,\frac{kg\,m^2}{s^2}$$

Some SI derived units have no special names. These include the unit of *area*, the *square meter* (m^2), and of *volume,* the cubic *meter* (m^3).

Multiple and submultiple metric units

It is often convenient to use units which are larger or smaller than the SI base or derived units. One of the big advantages of the metric system, including the SI version, is that it is a *decimal* system. This means that different units which measure a given quantity (say, length) but which are of different size are related by powers of 10. The power of 10 is indicated by a prefix in the name of the unit which relates its size to that of the base unit. For instance, the prefix *kilo-* means 10^3 (or 1000). Thus, one kilometer is 10^3 meters, or

$$1\,km = 1 \times 10^3\,m$$

Sometimes the prefix in the name of a unit indicates that the unit is smaller than the related base unit and is hence a submultiple of it. The prefix *milli-* means 10^{-3} (or $\frac{1}{1000}$), so one millimeter is 10^{-3} meter, or

$$1\,mm = 1 \times 10^{-3}\,m$$

Table 1-2 Common prefixes in the metric system

Prefix	Abbreviation	Meaning	Examples
Kilo	k	10^3	1 kilogram = 10^3 grams (1 kg = 10^3 g)
Deci	d	10^{-1}	1 decimeter = 10^{-1} meter (1 dm = 10^{-1} m)
Centi	c	10^{-2}	1 centimeter = 10^{-2} meter (1 cm = 10^{-2} m)
Milli	m	10^{-3}	1 milligram = 10^{-3} gram (1 mg = 10^{-3} g)
Micro	μ*	10^{-6}	1 microgram = 10^{-6} gram (1 μg = 10^{-6} g)
Nano	n	10^{-9}	1 nanometer = 10^{-9} meter (1 nm = 10^{-9} m)

*μ is the Greek letter *mu,* pronounced "mew."

Table 1-3
Conversion equations

Quantity	Metric-English	English-metric
Length	1 km = 0.621 mi	1 mi = 1.61 km
	1 m = 1.09 yd	1 yd = 0.914 m
	1 m = 39.4 in	1 in = 2.54×10^{-2} m
	1 cm = 0.394 in	1 in = 2.54 cm
Mass	1 kg = 2.20 lb	1 lb = 0.454 kg
	1 g = 3.53×10^{-2} oz	1 oz = 28.3 g
Volume or capacity	1 liter = 1.06 qt	1 qt = 0.946 liter
	1 ml = 6.10×10^{-2} in^3	1 in^3 = 16.4 ml
	1 cm^3 = 6.10×10^{-2} in^3	1 in^3 = 16.4 cm^3

The metric prefixes most commonly used in chemistry are shown in Table 1-2. (Others are shown in App. B.)

Current practice in chemistry is to use SI units in most cases. Sometimes, however, other units are found to be especially convenient and so are still used. The SI derived unit of *pressure* is the *pascal* (Pa), which is defined as one newton of force per square meter of area, or

$$1 \text{ Pa} = 1 \frac{\text{N}}{\text{m}^2} = 1 \frac{\text{kg}}{\text{m}^2 \text{ s}^2}$$

But chemists have been slow to adopt the pascal; the *standard atmosphere* (atm) and *millimeter of mercury* (mmHg) will probably remain in common use for some time. We will discuss these non-SI units in Sec. 4-1.

Two units of *volume* which are metric but non-SI units are the *liter* and its submultiple the milliliter (ml). Although the liter was once defined as the volume occupied by one kilogram of water at 3.98°C, it was redefined in 1964 as being exactly equal to one cubic decimeter. (The change is a very small one.) The milliliter is thus the same as the cubic centimeter. We will use milliliters and liters frequently in this text, since they are of a convenient size for many chemical purposes. Remember that

$$1 \text{ liter} = 1 \text{ dm}^3 \qquad 1 \text{ ml} = 1 \text{ cm}^3$$

(Note the above abbreviations for cubic decimeter and cubic centimeter. The latter has also been abbreviated cc.)

In this book we will have little occasion to convert back and forth between metric and other units. If you are more familiar with English-system units, ounces, inches, and quarts, for instance, than with metric units, such as grams,

Table 1-4
Metric measurements

Distance		
	Thickness of a common pin	0.5 mm
	Length of a teaspoon	15 cm
	Height of a basketball player	2.1 m
	Distance of a marathon race	42 km
Mass	Common pin	50 mg
	Penny	3 g
	Apple	200 g
	Football player	100 kg
Volume	Teaspoon	5 ml
	Water glass	240 ml
	Basketball	7.2 liters

centimeters, and liters, then you should try to get a feeling for the magnitudes of the more common of the latter units. In Table 1-3 is a list of metric-English and English-metric conversion equations. Look also at Table 1-4, in which measurements of some common objects are given. You will soon find that metric units are easy to use and visualize.

1-7
Solving numerical
problems

Dimensional analysis

Whenever appropriate we will use a problem-solving approach which makes use of *dimensional analysis,* a technique in which one keeps track of the dimensions (units) of each quantity throughout a calculation. With this technique dimensions such as inches, pounds, centimeters, and grams are treated (multiplied, divided, etc.) much as if they were algebraic quantities. Suppose, for instance, that we wish to convert an English-unit measurement, 38.8 in, into centimeters. From Table 1-3 we see that

$$1 \text{ in} = 2.54 \text{ cm}$$

Therefore, we can write

$$\frac{2.54 \text{ cm}}{1 \text{ in}} = 1$$

because any fraction whose numerator and denominator are equal or equivalent is equal to unity. Furthermore, any quantity multiplied by unity remains unchanged, so we can multiply 38.8 in by the above fraction and obtain a quantity which is a measure of the same length:

$$38.8 \text{ in} \frac{2.54 \text{ cm}}{1 \text{ in}} = 98.6 \text{ cm}$$

Note that the unit in has been canceled, leaving cm as the final unit.

The quantity 2.54 cm/1 in is known as a *unit factor,* because it is a factor which is equal to unity (1). It is also often called a *unit conversion factor,* because with it one can convert from inch units to centimeter units. The quantity 1 in/2.54 cm is also a unit factor which we could use to convert from centimeters to inches.

● **EXAMPLE 1-1**

Problem   If you are 147 cm tall, what is your height in inches?

Solution   Since 2.54 cm = 1 in, the unit factor which converts centimeters to inches is

$$\frac{1 \text{ in}}{2.54 \text{ cm}} = 1$$

Therefore, 147 cm is

$$147 \text{ cm} \frac{1 \text{ in}}{2.54 \text{ cm}} = 57.9 \text{ in} \quad ●$$

In the above example how did we know that we should multiply by the unit factor 1 in/2.54 cm and not 2.54 cm/1 in? If we had multiplied by the latter, the units would have come out wrong:

$$147 \text{ cm} \frac{2.54 \text{ cm}}{1 \text{ in}} = 373 \frac{\text{cm}^2}{\text{in}}$$

Since cm^2/in is not a unit of length, we know that we have made an error. *Choose the appropriate unit factor so that you obtain the desired units in your answer.*

Next consider an example in which two unit factors are needed in order to make two consecutive conversions.

● **EXAMPLE 1-2**

Problem   Jet airliners frequently travel at an altitude of 8.0 mi. How many meters is this?

Solution   From Table 1-3 we see that

$$1 \text{ mi} = 1.61 \text{ km}$$

Thus we have the unit factor

$$\frac{1.61 \text{ km}}{1 \text{ mi}} = 1$$

which we can use to convert from miles to kilometers

$$8.0 \text{ mi} \frac{1.61 \text{ km}}{1 \text{ mi}} = 13 \text{ km}$$

Now, in order to convert from kilometers to meters we need another conversion equation. From Table 1-2 we have

$$1 \text{ km} = 10^3 \text{ m}$$

so that

$$\frac{10^3 \text{ m}}{1 \text{ km}} = 1$$

Now we can make the last conversion.

$$13 \text{ km} \frac{10^3 \text{ m}}{1 \text{ km}} = 13 \times 10^3 \text{ m} = 1.3 \times 10^4 \text{ m}$$

The two conversions could have been written down more concisely as

$$8.0 \text{ mi} \frac{1.61 \text{ km}}{1 \text{ mi}} \frac{10^3 \text{ m}}{1 \text{ km}} = 1.3 \times 10^4 \text{ m}$$

It can be seen that the *mile* and *kilometer* units cancel, leaving *meters*. ●

● **EXAMPLE 1-3**

Problem   You are traveling in a car at a speed of 55 miles per hour (mi/h). How fast is this in meters per second?

Solution   Here we wish to convert miles to meters and hours to seconds. The conversion equations and corresponding unit factors are

Conversion equation	Unit factor
1 mi = 1.61 km (Table 1-3)	$\frac{1.61 \text{ km}}{1 \text{ mi}} = 1$
1 km = 10^3 m (Table 1-2)	$\frac{10^3 \text{ m}}{1 \text{ km}} = 1$
1 h = 60 min	$\frac{1 \text{ h}}{60 \text{ min}} = 1$
1 min = 60 s	$\frac{1 \text{ min}}{60 \text{ s}} = 1$

Putting it all together we have

$$55 \, \frac{\text{mi}}{\text{h}} \left(\frac{1.61 \text{ km}}{1 \text{ mi}}\right)\left(\frac{10^3 \text{ m}}{1 \text{ km}}\right)\left(\frac{1 \text{ h}}{60 \text{ min}}\right)\left(\frac{1 \text{ min}}{60 \text{ s}}\right) = 25 \, \frac{\text{m}}{\text{s}} \quad \bullet$$

Unit factors may be *squared* or *cubed* when necessary, since $1^2 = 1^3 = 1$. This is illustrated in the following example.

● **EXAMPLE 1-4**

Problem   A piece of wood has a volume of 4.5 in³. What is its volume in cubic centimeters?

Solution   Since 1 in = 2.54 cm,

$$\frac{2.54 \text{ cm}}{1 \text{ in}} = 1$$

so that

$$\left(\frac{2.54 \text{ cm}}{1 \text{ in}}\right)^3 = 1 \qquad \text{or} \qquad \frac{16.4 \text{ cm}^3}{1 \text{ in}^3} = 1$$

Thus, 4.5 in³ is

$$4.5 \text{ in}^3 \, \frac{16.4 \text{ cm}^3}{1 \text{ in}^3} = 74 \text{ cm}^3 \quad \bullet$$

Finally, we will consider a more complex problem.

● **EXAMPLE 1-5**

Problem   Two cubic feet of gasoline weighs 84 lb. How many cubic centimeters does 25 g of gasoline occupy (1 in³ = 16.4 ml)?

Solution   We start with 25 g of gasoline. In kilograms this is

$$25 \text{ g} \, \frac{1 \text{ kg}}{10^3 \text{ g}}$$

Since 1 kg = 2.20 lb (Table 1-3), the weight (actually, mass) of the gasoline in pounds is

$$25 \text{ g} \left(\frac{1 \text{ kg}}{10^3 \text{ g}}\right)\left(\frac{2.20 \text{ lb}}{1 \text{ kg}}\right)$$

Since 2.0 ft³ of gasoline weighs 84 lb, each of these two quantities measures the *same amount of gasoline,* so we can say that they are equivalent quantities. Using the symbol $\sim$ to mean *is equivalent to,* can say that *for gasoline*

$$2.0 \text{ ft}^3 \sim 84 \text{ lb}$$

From this we obtain the unit factor (valid for gasoline only)

$$\frac{2.0 \text{ ft}^3}{84 \text{ lb}} = 1$$

Thus, the volume in cubic feet occupied by the 25 g of gasoline is

$$25 \text{ g} \left(\frac{1 \text{ kg}}{10^3 \text{ g}}\right)\left(\frac{2.20 \text{ lb}}{1 \text{ kg}}\right)\left(\frac{2.0 \text{ ft}^3}{84 \text{ lb}}\right)$$

Since 1 ft = 12 in, the volume of the gasoline expressed in cubic inches is

$$25 \text{ g} \left(\frac{1 \text{ kg}}{10^3 \text{ g}}\right)\left(\frac{2.20 \text{ lb}}{1 \text{ kg}}\right)\left(\frac{2.0 \text{ ft}^3}{84 \text{ kg}}\right)\left(\frac{12 \text{ in}}{1 \text{ ft}}\right)^3$$

In milliliters this is

$$25 \text{ g} \left(\frac{1 \text{ kg}}{10^3 \text{ g}}\right)\left(\frac{2.20 \text{ lb}}{1 \text{ kg}}\right)\left(\frac{2.0 \text{ ft}^3}{84 \text{ kg}}\right)\left(\frac{12 \text{ in}}{1 \text{ ft}}\right)^3 \left(\frac{16.4 \text{ ml}}{1 \text{ in}^3}\right)$$

and in cubic centimeters this is

$$25 \text{ g} \left(\frac{1 \text{ kg}}{10^3 \text{ g}}\right)\left(\frac{2.20 \text{ lb}}{1 \text{ kg}}\right)\left(\frac{2.0 \text{ ft}^3}{84 \text{ kg}}\right)\left(\frac{12 \text{ in}}{1 \text{ ft}}\right)^3 \left(\frac{16.4 \text{ ml}}{1 \text{ in}^3}\right)\left(\frac{1 \text{ cm}^3}{1 \text{ ml}}\right)$$

Calculating this out we find that the 25 g of gasoline occupies 37 cm³. ●

The above example illustrates the value of the dimensional analysis approach in a not-so-simple calculation. If we had set up one of the unit factors incorrectly, the final units would not have come out to be cubic centimeters, and we would have been alerted to the fact that at least one step was wrong.

ADDED COMMENT

Using the dimensional analysis approach to problem solving does not guarantee that you will not make a mistake. Even if the units come out right you may still have made an arithmetic error. On the other hand, if the units do *not* come out right, you have been given a red flag which should warn you that your method was wrong.

SUMMARY

In this chapter we started out by talking a little about chemistry: what it is, why it is important, and how to go about learning it. Then we moved to a description of the *historical roots of chemistry,* in which we followed chemistry's development from the ancient Egyptians through the periods of *alchemy, iatrochemistry,* and *phlogiston* to the work of *Lavoisier* and the beginnings of chemistry as we know it today.

Chemistry is a science, and so we considered the way science, in general, advances. Special emphasis was placed on the roles played by *natural laws* and *scientific theories.* In addition, the role of the scientist and the nature of individual scientific endeavor were mentioned.

Some important terms which are basic to chemical language were next introduced. The nature of *matter* was briefly discussed, as were two kinds of matter: *pure substances* and *mixtures. Solids, liquids,* and *gases* were provisionally defined. The relation between *structure* and *properties* was mentioned, and *physical* and *chemical properties* and *changes* were contrasted, as were *microscopic* and *macroscopic properties.*

The last sections of the chapter treated various aspects of numerical relationships and calculations. *Exponential notation* was briefly described, as was the proper handling of *significant figures,* or *digits,* in calculations. The *metric system* was then summarized, and the universal *SI units* were introduced. Lastly, numerical problem solving was discussed from the standpoint of *dimensional analysis.* Several examples of the *unit factor* method of solving problems were given.

KEY TERMS

Accuracy (Sec. 1-5)
Alchemy (Sec. 1-2)
Atom (Sec. 1-2)
Chemical property (Sec. 1-4)
Compound (Sec. 1-4)
Continuous (Sec. 1-2)
Dimensional analysis (Sec. 1-7)
Discontinuous (Sec. 1-2)
Element (Sec. 1-2)
Exponential notation (Sec. 1-5)
Gas (Sec. 1-4)
Hypothesis (Sec. 1-3)
Iatrochemistry (Sec. 1-2)
Khemeia (Sec. 1-2)
Law (Sec. 1-3)
Liquid (Sec. 1-4)
Macroscopic property (Sec. 1-4)

Matter (Sec. 1-4)
Metric system (Sec. 1-6)
Microscopic property (Sec. 1-4)
Mixture (Sec. 1-4)
Phlogiston (Sec. 1-2)
Physical property (Sec. 1-4)
Precision (Sec. 1-5)
Pure substance (Sec. 1-4)
SI units (Sec. 1-6)
Scientific method (Sec. 1-3)
Significant figures (digits) (Sec. 1-5)
Solid (Sec. 1-4)
Structure (Sec. 1-4)
System (Sec. 1-4)
Theory (Sec. 1-3)
Unit factor (Sec. 1-7)

QUESTIONS AND PROBLEMS

Chemical history

1-1 It is sometimes said that early Greek philosophers tried to learn abut the structure of matter by using pure logic but *no* experimental data. Criticize this statement.

1-2 Comment on the contributions made to the early development of chemistry by **(a)** the Egyptians **(b)** the Greeks **(c)** the Arabs.

1-3 What was *alchemy?* Why was the period of alchemy important?

1-4 Identify each of the following early "chemists": Paracelsus, Geber, Lavoisier, Boyle.

1-5 What was *phlogiston* supposed to be? What properties was it supposed to possess?

The scientific method

1-6 Discuss the role of each of the following in the advance of science: **(a)** observation **(b)** data **(c)** laws **(d)** hypotheses **(e)** theories **(f)** experiments.

1-7 It has been said that a well-established and well-tested theory eventually becomes a law. Is this wrong? Explain.

1-8 In what way do theories make use of models?

1-9 What is meant by the statement, "A good scientist is a good artist"?

Matter

1-10 Give five examples of the things which are **(a)** matter **(b)** not matter.

1-11 Why is *matter* hard to define?

1-12 Table salt is a compound of sodium and chlorine. How do we know it is not a mixture?

1-13 Explain how you could distinguish between the two members of each of the following pairs, and in each case state the property used to make the distinction: **(a)** liquid water and ice **(b)** pure water and saltwater **(c)** aluminum and copper **(d)** pure water and pure alcohol **(e)** pure gaseous oxygen and pure gaseous nitrogen **(f)** white paraffin wax and white polyethylene plastic **(g)** cotton fibers and wool fibers.

1-14 Tell which of the following is a mixture and which is a pure substance: iron, water, salt, saltwater, granite, wood, gasoline, beer, chocolate, vinegar (look at the bottle label), dusty air, clean air.

1-15 Name a physical property exhibited by each of the following: water, sugar, mercury metal, copper metal, steel, maple syrup, oxygen gas, glass.

1-16 Name a chemical property exhibited by each of the following: water, alcohol, paraffin wax, potatoes, copper, nitroglycerin, milk.

Exponential notation and significant figures

1-17 Express each of the following as a number written in exponential notation with one digit to the left of the decimal point in the coefficient: **(a)** 100.4 **(b)** 0.0043 **(c)** 1.0 billion **(d)** 0.0000400 **(e)** 156,000 to three significant figures **(f)** 156,000 to four significant figures.

1-18 Tell how many significant figures there are in each of the following: **(a)** 4.96 **(b)** 162.9 **(c)** 100.01 **(d)** 100.00 **(e)** 0.123 **(f)** 0.003 **(g)** 0.0030 **(h)** 1.67×10^{16} **(i)** 4.1×10^{-23} **(j)** 4.0×10^2 **(k)** 404

1-19 A beaker weighing 45.3261 g is filled with each of the following, successively. (Nothing is removed.) Tell how much the total weighs after each addition: **(a)** 0.0031 g of salt **(b)** 1.197 g of water **(c)** 27.45 g of sugar **(d)** 38 g of milk **(e)** 88 g of maple syrup.

1-20 Each member of a class was told to measure the temperature of the air in a room in °C. The results submitted were as follows: 24.9, 24.8, 24.8, 25.0, 22.1, 24.9, 22.0, 22.2, 22.4, 25.0, 22.5, 24.7, 24.8, 22.2, 24.7. Discuss what can be concluded about the above data. Include in your discussion mention of the accuracy and precision of the measurements.

1-21 Perform the indicated arithmetic operations assuming that each number is the result of an experimental measurement:

(a) $146 + 4.12$

(b) $12.641 - 1.4$

(c) $1.42 + 11.196 - 3.8$

(d) $146.3 - 145.9$

(e) $(26.92 - 1.07)(4.33 + 5.0)$

(f) $\dfrac{1.090 + 436}{2.0}$

(g) $(4.7 \times 10^6)(1.4 \times 10^9)$

(h) $(4.7 \times 10^6) - (3.1 \times 10^5)$

(i) $(6.88 \times 10^{-8}) + (3.36 \times 10^{-10})$

(j) $(1.91 \times 10^{-4}) + (3.42 \times 10^{-3})$

(k) $(3.4 \times 10^6)(1.21 \times 10^{-4})$

(l) $(3.91 \times 10^{-2})(9.1 \times 10^{-3})$

(m) $\dfrac{6.67 \times 10^{-6}}{4.3 \times 10^{-4}}$

(n) $\dfrac{3.19 \times 10^{-4}}{2.02 \times 10^9}$

(o) $\dfrac{(3.146 \times 10^5)(2.04 \times 10^{-3})}{1.1 \times 10^{-9}}$

1-22 Using long division show that when you divide 7 by 304, the result *can be* expressed to only one significant figure. (Watch the zeros.)

The metric system

1-23 Indicate whether each of the following units is a measure of length, mass, volume, or time: **(a)** ml **(b)** kg **(c)** μs **(d)** cm^3 **(e)** mg **(f)** m^3 **(g)** μg **(h)** nm.

1-24 What does each of the following metric prefixes mean? **(a)** kilo **(b)** centi **(c)** milli **(d)** micro **(e)** nano.

1-25 For each of the following tell which is larger and by what factor: **(a)** 1 mm or 1 m **(b)** 1 mg or 1 kg **(c)** 1 ml or $1\ cm^3$ **(d)** 1 kl or $1\ mm^3$.

1-26 Convert each of the following to *grams:* **(a)** $2.65 \times 10^{-4}\ kg$ **(b)** $3.6 \times 10^4\ mg$ **(c)** $8.14 \times 10^3\ kg$ **(d)** $1.16 \times 10^9\ \mu g$.

1-27 Suggest some reasons for and against converting to a decimal *time* system.

Dimensional analysis and problem solving

1-28 Write the unit factor for converting: **(a)** inches to feet **(b)** feet to inches **(c)** millimeters to centimeters **(d)** centimeters to millimeters.

1-29 Write the unit factor for converting: **(a)** liters to quarts **(b)** meters to yards **(c)** miles to kilometers **(d)** cubic centimeters to milliliters **(e)** cubic meters to cubic millimeters **(f)** square centimeters to square meters.

1-30 Convert 4.76 cm to **(a)** meters **(b)** millimeters **(c)** inches **(d)** feet.

1-31 Convert 126 g to **(a)** kilograms **(b)** milligrams **(c)** ounces **(d)** pounds.

1-32 Convert $14.7\ g/cm^3$ to **(a)** g/ml **(b)** g/liter **(c)** g/m^3 **(d)** kg/m^3.

1-33 If tomatoes cost $0.69 per pound and each tomato weighs 4.5 oz, what is the average cost of one tomato?

1-34 If onions cost $0.29 per pound and $4.00 will buy 126 onions, how many ounces does an average onion weigh?

1-35 If a brick measuring 1.5 in $\times$ 3.0 in $\times$ 6 in weighs 38 oz, what total volume in liters would be occupied by 45 kg of bricks?

2 MATTER AND ENERGY

TO THE STUDENT

All *matter* is composed of atoms. In this chapter we take our first look at these little, invisible particles. Why should we study atoms at all? One good reason is that atoms interact and combine with each other, and the way they do this largely determines the characteristics of matter. Thus, for example, the odor, color, flavor, and texture of a ripe, red apple are ultimate consequences of atomic interactions in the apple. Studying atoms is indeed essential if we are to understand why matter behaves the way it does.

Energy, the ability to do work, is another one of those very basic ideas which is used in all of science. The concept of energy is extremely useful, and is probably already familiar to you. You will need to consider, furthermore, only a few different kinds of energy. Energy seems to come in fewer forms than does matter.

Heat is an important form of energy. It represents one of the ways by which energy can be added to or removed from a system. A word to the wise: The difference between heat and temperature is important. In Sec. 2-5 we talk about this difference, so read carefully! (The two terms are often confused with each other.)

**2-1
Matter: what is it?**

In Sec. 1-4 we said that matter is stuff which (1) *has mass* and (2) *occupies space*. The concept of something occupying space should not cause much difficulty, but what is mass? And how does mass differ from *weight*?

Mass, inertia, and weight

The mass of an object is a direct measure of the amount of matter in the object. Thus, an automobile has a larger mass than a bicycle, a chicken egg has more mass than a hummingbird egg, and a gallon of water has a higher mass than a gallon of air. (Solids, liquids, and gases all have mass, because all are matter.)

It is the mass of an object which determines two other important properties of that object. These are inertia and weight. *Inertia* is the resistance shown by an object to any attempt to change its state of motion. Any object at rest tends to remain at rest, and any moving object tends to remain moving at a constant speed and in a constant direction.[1] The state of motion of an object can be changed, of course, by applying a force; this can cause an object at rest to begin moving or a moving object to change its velocity. But it is difficult to change the state of motion of an object which has a high mass, while it is easy in the case of an object of low mass. Thus, a golf ball thrown at a window is not likely to be stopped by the glass, but a Ping-Pong ball might indeed bounce right off. In this example the inertia (tendency to keep moving) of the golf ball is clearly greater than that of the Ping-Pong ball; the reason is that the mass of the golf ball is higher than that of the Ping-Pong ball.

In addition to inertia another property is determined by an object's mass, i.e., its *weight*. On earth *weight is the force of gravitational attraction between an object and the earth*.[2] Unlike inertia, weight is determined by two factors in addition to the mass of the object. These are (1) the mass of the earth and (2) the distance between the object and the center (actually the "center of mass") of the earth. Changes in the mass of the earth are small, however, and so can be ignored, but since the earth is not perfectly spherical, being somewhat flattened at the poles, the weight of an object does indeed vary with location on the earth's surface. The weight of a given object is therefore greater at the North or South Pole than it is at the equator. Note that this is true not because the mass (quantity of matter) is any different but because the earth's poles are closer to its center than is any spot on its equator. Furthermore, measurements of the weight of an object yield different results when they are made on a mountaintop, at sea level, or high in a balloon. For chemical purposes we need to be able to make a direct measurement of the quantity of matter in any sample. Because of its location dependency, weight is not a useful quantity in chemistry. On the other hand, mass is a consistent and unchanging quantity, gives a direct measure of amount of matter, and is therefore much more useful than weight. In this book the unit of mass we will most commonly use is the *gram*.

One more observation: *To weigh* can mean either "to determine mass" or "to determine weight." The English language has need for a new verb!

ADDED COMMENT

Some students have trouble with "mass," because in common use the word can mean large physical size ("a mass of snow") or even large numbers ("a mass of flowers"). Note that the scientific use of the word is different from these and must not be confused with them.

Density   One of the properties used to characterize a substance is its density. *Density* is defined as the mass of a unit volume of a substance, or

---

[1] This is a statement of Newton's first law of motion. Isaac Newton (1642–1727) created much of the foundation of early physics with three laws of motion, statements which point out consistencies in the behavior of moving objects.

[2] On the moon, Mars, or Jupiter, weight is the gravitational attraction between an object and that body.

$$\text{Density} = \frac{\text{mass}}{\text{volume}}$$

What is the significance of this quantity? Density expresses how much matter is compacted into a given unit of volume. It is imprecise to say that "lead weighs more than aluminum," because a volume has not been specified. (One *cubic foot* of aluminum weighs more than one *cubic inch* of lead, for example.) When we say that lead has a higher density than aluminum, however, a constant unit of volume is implied. In any volume of lead there is more matter than in the same volume of aluminum. Sometimes a substance may be identified by measuring its density.

Using metric units the densities of solids are commonly expressed in units of *grams per cubic centimeter* (g/cm³), and those of gases in *grams per liter* (g/liter).

● **EXAMPLE 2-1**

Problem   A piece of redwood weighing 238.3 g is found to occupy a volume of 545 cm³. Calculate its density in grams per cubic centimeter.

Solution
$$\text{Density} = \frac{\text{mass}}{\text{volume}}$$
$$= \frac{238.3 \text{ g}}{545 \text{ cm}^3} = 0.437 \text{ g/cm}^3 \ ●$$

At this point we will introduce a symbolism which has been recommended for complex units. Instead of

$$\text{g/cm}^3 \quad \text{or} \quad \frac{\text{g}}{\text{cm}^3}$$

the suggested symbol is

$$\text{g cm}^{-3}$$

Similarly,

$$\text{g/liter} \quad \text{and} \quad \frac{\text{g}}{\text{liter}}$$

become

$$\text{g liter}^{-1}$$

Thus, the horizontal or slanted line representing *per* or *divided by* in such units disappears and is replaced by a negative exponent. (*Note:* g cm⁻³ is still read "grams per cubic centimeter.")

● **EXAMPLE 2-2**

Problem   Gem-quality ruby has a density of 4.10 g cm⁻³. What is the volume of a ruby which weighs 6.7 g?

Solution   Method I:
$$\text{Density} = \frac{\text{mass}}{\text{volume}}$$

Therefore

$$\text{Volume} = \frac{\text{mass}}{\text{density}}$$

$$= \frac{6.7 \text{ g}}{4.10 \text{ g cm}^{-3}} = 1.6 \text{ cm}^3$$

Method II: The stated density tells us that 1 cm^3 of ruby weighs 4.10 g. In other words, 1 cm^3 and 4.10 g are equivalent measures of the same quantity of ruby. Using the $\sim$ symbol as before to express equivalence, we have

$$1 \text{ cm} \sim 4.10 \text{ g}$$

Thus, we see that density can be used as a unit factor:

$$\frac{4.10 \text{ g}}{1 \text{ cm}^3} = 1$$

Therefore,

$$\frac{1 \text{ cm}^3}{4.10 \text{ g}} = 1$$

Thus, 6.7 g of ruby is

$$6.7 \text{ g} \frac{1 \text{ cm}^3}{4.10 \text{ g}} = 1.6 \text{ cm}^3 \quad \bullet$$

● **EXAMPLE 2-3**

Problem   What is the mass of a ruby (density = 4.10 g cm^{-3}) which has a volume of 0.421 cm^3?

Solution   Here we use the unit factor which changes the unit cm^3 to the unit g.

$$\text{Mass of ruby} = 0.421 \text{ cm}^3 \frac{4.10 \text{ g}}{1 \text{ cm}^3} = 1.73 \text{ g} \quad \bullet$$

Numerical values of densities vary widely. Table 2-1 shows densities in both

Table 2-1 Densities of some common substances (25°C, 1 atm)				
Substance	**State**	**Density, g cm^{-3}**	**Density, g liter^{-1}**	
Aluminum	Solid	2.70	2.70×10^3	
Balsa wood	Solid	0.12	1.2×10^2	
Beeswax	Solid	0.96	9.6×10^2	
Chlorine	Gas	2.90×10^{-3}	2.90	
Diamond	Solid	3.5	3.5×10^3	
Ebony wood	Solid	1.2	1.2×10^3	
Ethyl alcohol	Liquid	0.789	7.89×10^2	
Gold	Solid	19.3	1.93×10^4	
Lead	Solid	11.3	1.13×10^4	
Mercury	Liquid	13.6	1.36×10^4	
Milk	Liquid	1.03	1.03×10^3	
Nitrogen	Gas	1.31×10^{-3}	1.31	
Silver	Solid	10.5	1.05×10^4	
Water	Liquid	0.997	9.97×10^2	
Zinc	Solid	7.1	7.1×10^3	

g cm^{-3} and g liter^{-1} for a number of common substances. Note that the unit g liter^{-1} is a convenient one for gases.[3]

Levels of organization   Although the detailed structure of any piece of matter is very complex, the problem of describing that structure is greatly simplified because matter is organized in a hierarchy of structural levels. Consider, for example, a teaspoonful of sugar. At a very basic level we find that submicroscopic particles, protons, neutrons, and electrons, are organized into three kinds of larger particles which we call *carbon atoms, hydrogen atoms,* and *oxygen atoms.* At the next level we see these atoms grouped into units called *sugar molecules,* each of which is composed of 12 carbon atoms, 22 hydrogen atoms, and 11 oxygen atoms. Going to the next level we find the sugar molecules packed together in a beautifully symmetrical array called a *crystal lattice.* This structure is in turn responsible for the strikingly geometrical sugar crystal which it composes. Lastly, the collection of these individual sugar crystals make up the mound in the bowl of the spoon.

Structures and properties at any level of organization in a given kind of matter provide insights into the overall nature of that matter. In this book we do not concentrate on any one level but rather describe structures at each level with special emphasis on how a structure relates to the levels "above" and "below." At all times we stress the structure and organization within a level and how these influence the observed properties of the substance.

Living matter   The most extensive hierarchy of organizational levels is found in what we call living matter. What does *living* mean? What is "life"? These questions have perplexed philosophers and scientists alike for centuries. At one time it was thought that a mystical entity called the *vital force* was possessed by all living organisms and by all substances out of which they are composed. Although this idea gets resurrected occasionally, all evidence indicates that no vital force really exists. Instead, the unusual characteristics of living matter are a result of the monumental and enormously complex hierarchy of levels. Descriptions of structure and organization at all levels are necessary to provide a complete picture of the living organism, a challenging task for the biologist, biochemist, and biophysicist.

2-2 Kinds of matter   Figure 2-1 shows a useful way of classifying matter.

Pure substances and mixtures   A *pure substance* is essentially what its name implies: a single, uncontaminated substance. Thus, pure water is water and nothing else. It is a characteristic of a pure substance that the temperature at which it undergoes a change of state (melting, boiling, etc.) remains constant for the duration of the change. Thus, the

---

[3]Closely related to density is *specific gravity,* which is the ratio of the density of a substance to that of water. Since alcohol and water have densities of 0.80 g cm^{-3} and 1.0 g cm^{-3}, respectively, the specific gravity of ethanol is 0.80 g cm^{-3}/1.0 g cm^{-3} = 0.80. *Numerically* specific gravity is thus the same as density, but it has no units.

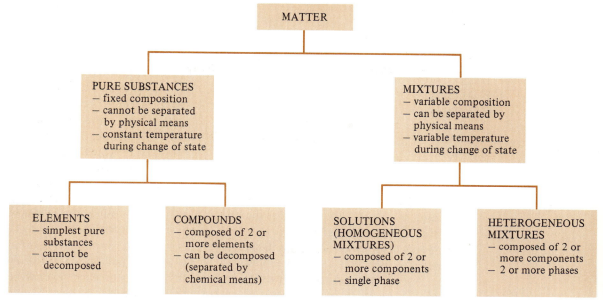

Figure 2-1 Classification of matter.

freezing point of 100 g of water remains at a constant 0°C from the onset of freezing of the first gram to the completion of freezing of the last.

A *mixture* is a physical combination or collection of two or more pure substances. Sometimes a mixture can be so identified by merely looking at it. A piece of granite can be seen to be a mixture of small grains of different components: white quartz, black mica, pink feldspar, etc. Some mixtures seem to disguise their nature, however. Saltwater can not be visually identified as a mixture; it looks just like pure water. Unlike the change-of-state temperature of a pure substance, that of a mixture does not remain constant as the transformation takes place. Rather, it changes, gradually in some cases, abruptly in others. We will discuss this further shortly.

Elements and compounds   There are two kinds of pure substances: *elements* and *compounds*. (Refer again to Fig. 2-1.) Elements are simple, fundamental substances. Compounds are composed of ("compounded of") elements and thus are more complex substances. A compound may be separated, or *decomposed,* into its constituent elements, but elements themselves resist all such attempts to be fragmented into simpler, more fundamental substances.

Although it is true that the smallest particles of an element (its atoms) may be broken up into still smaller fragments, a collection of one kind of fragment can not be assembled into a substance which can be observed and measured. Thus, we see that elements are the simplest pure substances which can be stored, observed, and manipulated in macroscopic-sized quantities.

The separation of a compound into its constituent elements is a chemical process. On the other hand, separation of a mixture into its components can be accomplished physically. The different mineral grains in a lump of granite can be

Table 2-2
Some elements and
their symbols

Element (English name)	Symbol	Source of symbol
Hydrogen	H	English
Oxygen	O	English
Carbon	C	English
Magnesium	Mg	English
Uranium	U	English
Chlorine	Cl	English
Chromium	Cr	English
Sodium	Na	Latin (natrium)
Potassium	K	Latin (kalium)
Tungsten	W	German (Wolfram)

separated by crushing the lump and then sorting through the rubble, perhaps with the aid of a magnifying glass. Saltwater can be physically separated by boiling it. (The water boils away, leaving the salt.)

There are over 100 known elements, of which about 90 occur naturally on earth. An element is often represented by an abbreviation known as a *chemical symbol.* This consists of one or two letters taken from the name of the element, usually in English, but occasionally in another language, usually Latin. The first letter of a chemical symbol is always capitalized. Examples of some symbols and the elements which they represent are given in Table 2-2. A complete list of all the known elements and their symbols is given inside the front cover of this book.

It is important that you recognize that compounds are not mixtures. Compounds can not be separated by physical means. In addition, they have fixed, characteristic compositions, while the compositions of mixtures are variable.[4] Thus the compound sodium chloride always has the composition 39.34 percent sodium (Na) and 60.66 percent chlorine (Cl) by mass. On the other hand, the composition of a mixture can vary over a wide range.

A compound is often represented by a chemical shorthand composed of the symbols of its elements. Thus NaCl represents sodium chloride, and H_2O represents water. These are called *chemical formulas;* we will consider them in Chap. 3.

Phases and states In a given sample of matter, pure or otherwise, a *phase* is defined as a region within which all the chemical and physical properties are uniformly the same. In a system consisting only of pure liquid water there is only one phase. A system which is composed of liquid water and one or more ice cubes consists of two phases. (The *number* of ice cubes is not important; 12 ice cubes, for example, constitute only one phase: ice.) When lubricating oil is added to water, the two liquids do not mix appreciably, and so the resulting system is a two-phase system, but alcohol and water mix completely yielding a single phase. A given phase may be *solid, liquid,* or *gaseous.* Almost any combination of these is possible, except that not more than one gaseous phase can be present in a system; all gases mix completely with each other yielding but a single phase. Figure 2-2 shows examples of several systems with various numbers of phases.

[4]There is, however, a class of compounds in which composition is variable. These are called *nonstoichiometric compounds* and will be mentioned again in Sec. 11-3.

System

Phases

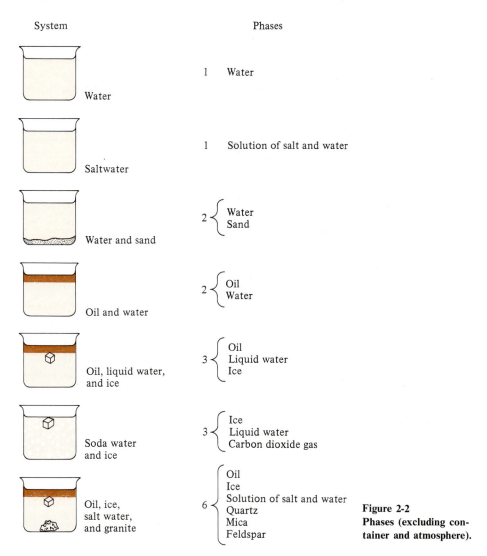

1   Water

Water

1   Solution of salt and water

Saltwater

$2\begin{cases} \text{Water} \\ \text{Sand} \end{cases}$

Water and sand

$2\begin{cases} \text{Oil} \\ \text{Water} \end{cases}$

Oil and water

$3\begin{cases} \text{Oil} \\ \text{Liquid water} \\ \text{Ice} \end{cases}$

Oil, liquid water, and ice

$3\begin{cases} \text{Ice} \\ \text{Liquid water} \\ \text{Carbon dioxide gas} \end{cases}$

Soda water and ice

$6\begin{cases} \text{Oil} \\ \text{Ice} \\ \text{Solution of salt and water} \\ \text{Quartz} \\ \text{Mica} \\ \text{Feldspar} \end{cases}$

Oil, ice, salt water, and granite

Figure 2-2
Phases (excluding container and atmosphere).

Mixtures: heterogeneous and homogeneous (solutions)

As is shown in Fig. 2-1, mixtures can be classified as either homogeneous or heterogeneous. A *homogeneous mixture* is usually called a *solution*. It consists of only *one phase*. Solutions include saltwater, gasoline, vinegar, vodka, and air. A solution may be solid, liquid, or gaseous, and its components may be either elements or compounds.

A *heterogeneous mixture* consists of two or more phases. Examples include mixtures of sand and salt, sand and water, gasoline and water, air bubbles and water, and dust and air.

How can we distinguish a solution (homogeneous mixture) from a pure substance (which is also homogeneous)? In the first place, a solution can be separated into its constituents by physical means, while a compound cannot. (And an element has only one constituent.) Secondly, a solution has a variable composition, while that of a compound is fixed.

In the laboratory a solution can usually be identified as such by observing the temperature at which it changes state. When saltwater is boiled, not only does the onset of boiling occur at a temperature above the boiling point of pure water, but the temperature slowly *rises,* as the water is boiled away and the concentration of salt in the remaining solution accordingly increases. With only a comparatively few exceptions, all solutions exhibit this behavior during changes of state.

ADDED COMMENT

At this point many of you may still not be clear in your minds about the difference between compounds and solutions, since both are more complex than elements. One added observation may be helpful: When you make a compound (from its elements or from simpler compounds) the process is often more noteworthy in some way than is the formation of a solution from its components. There may be a sharp change in appearance (color, texture, form, etc.) or considerable energy (often heat) given off or absorbed when a compound is prepared. But when a solution is formed, the process is commonly a more gentle one with little heat being involved and no startling changes in appearance. Unfortunately, these statements have many exceptions. Nevertheless, keep in mind that the formation of a compound often involves a more fundamental change in properties and structure (on the atomic and molecular level) than does the formation of a solution.

Laws of chemical change

Observation of the behavior of matter undergoing chemical change has led to the statement of several *laws of chemical change,* the first of which was stated by Lavoisier in 1774 and is now called the *law of conservation of mass.* This law states that during chemical change, no gain or loss of mass can be measured.[5] This is reasonable because chemical reactions do not result in any net measurable creation or destruction of matter.

● **EXAMPLE 2-4**

Problem

When the compound limestone (calcium carbonate) is heated, it decomposes to form quicklime (calcium oxide) and carbon dioxide gas. Suppose 40.0 g of limestone is decomposed, leaving 22.4 g of quicklime. How much carbon dioxide is released?

Solution

The law of conservation of mass tells us that in the chemical reaction

$$\text{Limestone} \longrightarrow \text{quicklime} + \text{carbon dioxide}$$

no change in total mass occurs. This means that the mass of the limestone decomposed equals the sum of the masses of the two products. Thus we can write

---

[5] The emphasis on measurement is made for two reasons. First, the law is a statement summarizing observations, that is, measurements, none of which has shown gain or loss in mass. Second, modern physics views mass and energy to be different but interconvertible (Sec. 24-5). Conversion of mass into energy and vice versa can be observed in nuclear processes, for example. The amount of energy which is equivalent to even a very small mass is, however, enormous, but in the case of chemical reactions the energy absorbed or released is not great enough to cause a *measurable* gain or loss in mass.

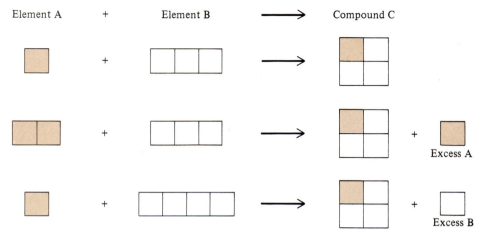

Element A   +   Element B   ⟶   Compound C

Excess A

Excess B

Figure 2-3
The laws of conservation of mass and definite composition: a hypothetical example. Different proportions of elements A (colored squares) and B (open squares) are mixed for reaction to form compound C, as shown. C is composed of 25 percent A and 75 percent B. Each square represents a certain mass, for example, 1 g.

$$\text{Mass}_{\text{carbon dioxide}} = \text{mass}_{\text{limestone}} - \text{mass}_{\text{quicklime}}$$
$$= 40.0 \text{ g} - 22.4 \text{ g} = 17.6 \text{ g}$$

A second law of chemical change is the *law of definite composition,* also known as the *law of constant composition* or the *law of definite proportions.* This law states that each compound has a fixed characteristic composition by mass. The laws of conservation of mass and definite composition are illustrated in Fig. 2-3. Notice that in each part of the figure the composition of compound C is constant, being 25 percent A and 75 percent B.

● **EXAMPLE 2-5**

Problem   The elements magnesium (Mg) and bromine (Br) combine directly with each other to form the compound magnesium bromide. In one experiment 6.00 g of Mg was mixed with 35.0 g of Br, and after reaction, it was found that although all the Br had reacted, 0.70 g of Mg remained in excess. What is the percentage composition of magnesium bromide by mass?

Solution
$$\text{Mass of bromine used} = 35.0 \text{ g}$$
$$\text{Mass of magnesium used} = 6.00 \text{ g} - 0.70 \text{ g} = 5.30 \text{ g}$$
$$\text{Mass of compound formed} = 35.0 \text{ g} + 5.30 \text{ g} = 40.3 \text{ g}$$

$$\% \text{ Mg} = \frac{\text{mass of Mg}}{\text{mass of compound}} \times 100$$
$$= \frac{5.30 \text{ g}}{40.3 \text{ g}} \times 100 = 13.2\%$$

$$\% \text{ Br} = \frac{35.0 \text{ g}}{40.3 \text{ g}} \times 100 = 86.8\%$$

● **EXAMPLE 2-6**

Problem   Iron (Fe) and oxygen (O) combine to form a compound which consists of 69.9 percent Fe and 30.1 percent O by mass. If 50.0 g of Fe and 40.0 g of O are reacted, how many grams of compound will be formed?

Solution   The above percentages tell us that a 100-g sample of the compound will consist of 69.9 g Fe and 30.1 g O. Since the ratio of Fe to O is fixed (law of definite composition), these masses of Fe and O can be said to be chemically equivalent:

$$69.9 \text{ g Fe} \sim 30.1 \text{ g O}$$

or

$$\frac{30.1 \text{ g O}}{69.9 \text{ g Fe}} = 1$$

(This means that the ratio of grams of O to grams of Fe in the compound is fixed at 30.1:69.9.) We can use this unit factor to find the number of grams of O needed to combine with 50.0 g of Fe. (We must not assume that all 40.0 g of O reacts.)

$$\text{No. of grams of O required} = 50.0 \text{ g Fe} \frac{30.1 \text{ g O}}{69.9 \text{ g Fe}} = 21.5 \text{ g O}$$

Since more than this amount of O (40.0 g) is present at the start, we conclude that (1) *all* the Fe will be used, and (2) only 21.5 g of O will be used. Therefore,

$$\text{Mass of compound formed} = (\text{mass of Fe used}) + (\text{mass of O used})$$

$$= 50.0 \text{ g} + 21.5 \text{ g} = 71.5 \text{ g}$$

and 40.0 g − 21.5 g = 18.5 g of O are left over in excess.

We see that the amount of compound formed is *limited* by the amount of Fe present. What if we had started the other way around, that is, by finding out how much Fe will combine with the original 40.0 g of O?

$$69.9 \text{ g Fe} \sim 30.1 \text{ g O} \qquad \text{or} \qquad \frac{69.9 \text{ g Fe}}{30.1 \text{ g O}} = 1$$

$$\text{No. of grams of Fe required (for 40.0 g O)} = 40.0 \text{ g O} \frac{69.9 \text{ g Fe}}{30.1 \text{ g O}} = 92.9 \text{ g Fe}$$

Since this is more than the iron provided, we know that the reaction is limited by the available iron and so we proceed as before. ●

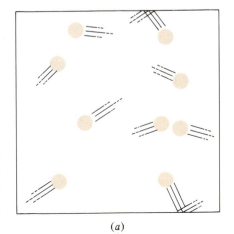
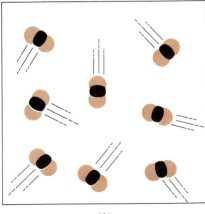

Figure 2-4
Two different gases:
(a) Monatomic (neon)
and (b) triatomic
(carbon dioxide).

(a)   (b)

Figure 2-5
Two different liquids:
(*a*) monatomic (neon)
and (*b*) triatomic (car-
bon dioxide).

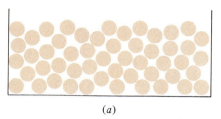

(*a*)

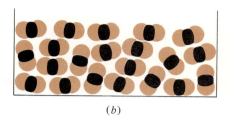

(*b*)

2-3
The microstructure
of matter

The notion that matter is composed of atoms traces back to the early Greeks (Sec. 1-2). Today we have much experimental evidence which supports our conception of the nature of atoms and of their role as building blocks of matter. A preliminary look at atoms is therefore in order at this point.

Aggregates of atoms

Atoms are not usually free and independent of each other. Under normal conditions they are found associated in aggregates or clusters of anywhere from two atoms up to numbers too large to count. When the cluster is small and tightly held together, it is called a *molecule* and acts as a single composite particle. Thus a molecule is a particle which consists of two or more atoms bonded together.

Gases consist of atoms or molecules which are essentially free and independent of each other. Figure 2-4 represents the molecules of two gases: neon and carbon dioxide. The particles of neon are single atoms, while those of carbon dioxide are *triatomic* molecules, each being composed of a carbon atom flanked by two oxygen atoms. In each drawing the molecules are shown to be quite far apart, rapidly moving and occasionally colliding with each other. The neon atoms undergo *translational* motion only. This is motion in which the molecule as a whole travels through space. The molecules of carbon dioxide undergo, in addition, a *rotational* motion; they tumble and spin between collisions. Although it is not represented in the drawing, the carbon and oxygen atoms can also oscillate or quiver against each other (*vibrational* motion).

Molecules are much closer to each other in liquids than in gases. Imagine that we have liquefied, or condensed, our samples of neon and carbon dioxide. Figure 2-5 shows the resulting structures in which the neon atoms and carbon dioxide molecules are shown much closer to each other. In each case motion still occurs but is severely restricted by the congestion. Each particle is "hemmed in" by neighboring molecules.

Freezing neon and carbon dioxide produces the structures represented in Fig. 2-6. In Fig. 2-6*a* the neon atoms are shown in a regular array which is actually

Figure 2-6
Two different solids: (*a*) monatomic (neon) and (*b*) triatomic (carbon dioxide). A single layer of atoms (neon) and molecules (carbon dioxide) is shown.

(*a*)

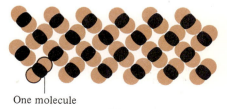

One molecule

(*b*)

Zn

S

Figure 2-7
Zinc sulfide (wurtzite). A single layer of atoms is shown.

ordered in three dimensions. They are packed together in the closest possible way. Carbon dioxide molecules, shown in Fig. 2-6b, are also packed together in a regular three-dimensional array, but they cannot be as closely packed as the neon atoms, as they are not spherical in shape.

In the structure of solid carbon dioxide individual molecules are identifiable, each being a more or less independent particle consisting of one carbon atom and two oxygen atoms. Such is not the case for all solids, however. Some solid substances are not molecular, that is, their atoms are not clustered in discrete[6] molecules. These solids still consist of atoms, of course, but the type of aggregation is different. One such substance is the form of zinc sulfide known as the mineral *wurtzite*. The structure of a portion of a crystal of wurtzite is shown schematically in Fig. 2-7. Here the smaller atoms are zinc, and the larger are sulfur. These atoms are linked together in a complex, three-dimensional array of which Fig. 2-7 is only a two-dimensional slice. Note that there are no discrete clusters of zinc and sulfur atoms which could be called molecules. There is only the regular array which repeats and repeats until the surface of the crystal is reached, many trillions of atoms away. Zinc sulfide does not consist of molecules.

**2-4
Energy: what is it?**

Energy is a fundamental concept which is used commonly but is hard to define rigorously. Defined in words, energy is the *ability* or *capacity to do work*. But what is work? *Work* is done if an object is moved against an opposing force. When, for example, we lift a book up off the table, we do work on the book, because we have *moved* the book against the *opposing force* of gravity. If something has the capacity to do work as defined above, we say it has energy.

Mechanical energy

Mechanical energy is energy which an object possesses because of either its motion or its position. Energy of motion is called *kinetic energy*. The kinetic energy possessed by a moving object depends upon two properties of that object: its mass (m) and its speed (s). These are related to kinetic energy (E_k) by the formula

$$E_k = \tfrac{1}{2}ms^2$$

which tells us that the faster an object is moving or the greater its mass, the greater is its kinetic energy. (Think: Which would you rather have hit you—a fast-moving golf ball or a slow-moving one? A golf ball or a cannonball moving at the same speed?)

Potential energy is energy possessed by an object not because of its motion, but

[6]*Discrete* means "separate and individually distinct." (Do not confuse this word with *discreet*.)

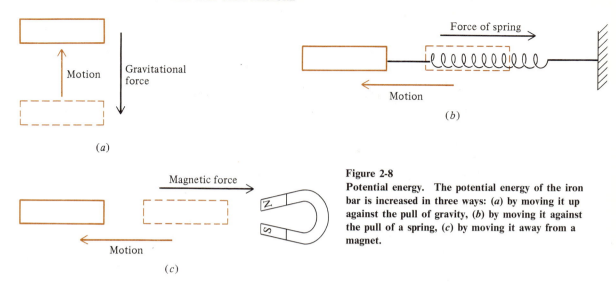

Figure 2-8
Potential energy. The potential energy of the iron bar is increased in three ways: (a) by moving it up against the pull of gravity, (b) by moving it against the pull of a spring, (c) by moving it away from a magnet.

because of its *position.* Objects "get" potential energy when they have work done on them. An iron bar, for example, can be given potential energy by lifting it up against the force of gravity (Fig. 2-8*a*), by moving it against the force exerted by an attached spring (Fig. 2-8*b*), or by moving it away from a magnet (Fig. 2-8*c*). In each case the potential energy of the bar increases, the farther it is moved against the opposing force. The potential energy (E_p) given to an object depends upon the distance (d) the object is moved or displaced and upon the force (F) against which it is moved:

$$E_p = Fd$$

Energy can be easily transformed from one form into another, but it can not be created or destroyed (law of conservation of energy). The potential energy of an object held above the earth can be gradually transformed into kinetic energy by releasing the object and hence allowing it to fall. As it falls, its potential energy decreases (as its distance from the earth decreases), and its kinetic energy increases (as its speed increases).

There are many forms of energy in addition to mechanical energy. These include electrical energy, heat energy, radiant (electromagnetic) energy, nuclear energy, chemical energy, and mass energy. Each of these is the capacity to do work. These forms of energy can also be interconverted.

2-5
Heat and temperature

The distinction between heat and temperature is one which is sometimes poorly understood but is very important. We use the word *heat,* or *heat energy,* to specify energy which is directly transferred from one object to another. Such energy is not in the form of heat before or after the transfer, only *during* the transfer. In other words, heat is energy *in transit.* Thus, after a quantity of heat energy has been transferred to an object, one should *not* say that the object "contains more heat." In such a case the increased energy should not be called heat, because it is no longer in transit.

What happens to the heat energy which has been absorbed by an object? Clearly the *total* energy of the object must somehow increase. This may occur in three ways: either the kinetic energies of the particles within the object increase, or their potential energies increase, or both increase simultaneously.

Various quantities of heat are either released or absorbed when chemical reactions occur. The quantity depends on the reaction and on the quantities of substance involved. A reaction which *releases heat* is called an *exothermic reaction;* one which *absorbs heat* from the surroundings is said to be *endothermic*.

The *temperature* of an object measures the *average kinetic energy* of the particles in the object. When energy such as heat is added to an object, its component particles (atoms, molecules, etc.) move more rapidly on the average, and so their average kinetic energy increases. We observe this increase as an increase in temperature. If two objects at different temperatures are placed in contact with each other, the temperature of the warmer object is seen to decrease and that of the cooler object to increase, until finally the two temperatures become the same. What really is happening here is that the warmer object (higher average kinetic energy per particle) loses heat energy to the cooler object (lower average kinetic energy per particle).

Sometimes when heat is added to an object, no temperature increase is observed. This means that average kinetic energy is not increasing. What becomes of the added heat energy? It increases the average *potential* energy of the particles which compose the object. This happens when a substance undergoes a change of state. Adding heat to ice at 0°C, for instance, causes no increase in temperature (the average kinetic energy of the molecules stays constant). The ice melts, however, forming liquid water, still at 0°C. In liquid water at 0°C the average molecular potential energy is greater than that in ice at the same temperature, even though the average kinetic energies are the same.

ADDED COMMENT

A word about the adjective *hot* may be in order: *Hot* pertains to temperature, not heat; a hot object is one whose temperature is high.

Note, also, that we have been using *heat* as a noun. As a verb this word is commonly used to mean two things: (1) "to raise the temperature of," as in "heat the water from 10 to 20°C"; and (2) "to add heat energy to," as in "heat the water at its boiling point for 5 min." The two meanings sometimes overlap, so some caution should be used with the verb *to heat*.

Units of energy and temperature

In this text we will use almost exclusively the SI energy unit, the joule (J) (Sec. 1-6). One joule is a moderate amount of energy. It is enough to lift a penny (about 3 g) a distance of 34 m (about 37 yd) or an "average human" a distance of 1 mm (about 0.04 in) against the force of gravity. For many chemical purposes the joule is less useful than its larger relative, the kilojoule (kJ), which is 10^3 J. When an ordinary wooden match burns, it releases about 1 kJ of heat, which is enough to raise the temperature of 500 g of water (a little more than a pint) by 0.5°C (0.8°F).

A common unit of energy is the calorie (cal). Although it is not an SI unit, it is familiar to most scientists and is likely to be used for some time. A calorie of energy is *larger* than a joule; it is equal to 4.18 joules. One calorie will raise the temperature of 1.00 gram of water 1.00 degree Celsius (1.00°C), which is the original basis for the definition of a calorie. Sometimes encountered is the nutritional Calorie (note the capitalization), which is really a kilocalorie (kcal), 10^3 cal.

There are several *temperature scales* in common use. The most important of these is the one based on the SI unit of temperature, called the *kelvin* (K).[7] The name of the scale is the *Kelvin scale* and it is defined as follows: The zero point on the Kelvin scale is the so-called absolute zero of temperature, the lowest temperature theoretically attainable. (More about this in Sec. 4-3.) The second fixed point on this scale is the temperature at which solid, liquid, and gaseous water can coexist indefinitely. This temperature, called the *triple-point temperature* of water (Chap. 11), is assigned the value of 273.1600 K. The Kelvin scale is one of several scales which are called *absolute scales,* because of the use of absolute zero as one fixed point.

The Celsius[8] scale is also in common use. Its units are called *degrees Celsius* (°C) and are identical in size to the kelvin. On the Celsius scale the triple-point temperature of water is 0.01°C, and the freezing point (under 1 atm of air pressure) is 0.00°C, while the boiling point (at 1 atm pressure) is 100.00°C. Thus, to convert from the Kelvin scale to the Celsius scale, it is merely necessary to add 273.15:

[7] Until recently the kelvin was called the *degree kelvin* (°K). The former is the SI name, however.

[8] A closely related scale is the Centigrade scale. It is essentially the same as the Celsius scale, so we need not define it.

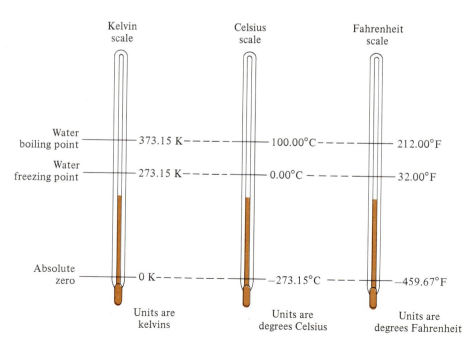

Figure 2-9
Temperature scales.

$$\text{Degrees Celsius} = \text{kelvins} + 273.15$$

or
$$°C = K + 273.15$$

Another scale which is familiar in North America is the Fahrenheit scale, on which the normal freezing and boiling points of water are 32 and 212°F, respectively. Conversion between Celsius and Fahrenheit temperatures may be accomplished by the following relationship:

$$\text{Degrees Celsius} = \tfrac{5}{9} \, (\text{degrees Fahrenheit} - 32)$$

or
$$°C = \tfrac{5}{9} \, (°F - 32)$$

The Kelvin, Celsius, and Fahrenheit scales are compared in Fig. 2-9.

SUMMARY

In this chapter we have described some of the properties of matter and energy. An important property of matter is its mass, which is related to both *inertia* and *weight*. Mass is a useful quantity because it directly measures the quantity of matter in an object. Also useful is *density*, which is *mass per unit volume*.

A sample of matter may be classified as either a *pure substance* or a *mixture*. A pure substance is either an *element* or a *compound*. A mixture may be *homogeneous* (composed of only one *phase*) or *heterogeneous* (composed of two or more phases). Homogeneous mixtures are called *solutions*.

Although reactants are consumed and products are formed during a chemical reaction, the total mass of the system does not measurably change. This is the *law of conservation of mass*. Another law, the *law of definite composition* states that a chemical compound has a characteristic fixed composition by mass. These are two *laws of chemical change*.

Matter is composed of building blocks called *atoms*. These may be bonded together in small aggregates, *molecules,* which are capable of acting as separate particles. Some matter, however, is not composed of molecules.

Energy is the capacity to do work. *Mechanical energy* is but one form of energy. It is either *kinetic energy* (energy of motion) or *potential energy* (energy of position). The *law of conservation of energy* states that energy may be transformed into different forms but may not be created or destroyed. *Heat* is a form of energy which is transferred from one system to another. Such a transfer may change the *temperature* of a system, where temperature is a measure of the average molecular kinetic energy of the system. Energy is commonly measured in joules (J) or calories (cal); temperature in kelvins (K), degrees Celsius (°C), or degrees Fahrenheit (°F).

KEY TERMS

Chemical change (Sec. 2-2)
Chemical formula (Sec. 2-2)
Chemical symbol (Sec. 2-2)
Compound (Sec. 2-2)
Conservation of energy (Sec. 2-4)
Conservation of mass (Sec. 2-2)
Definite composition (Sec. 2-2)
Density (Sec. 2-1)
Element (Sec. 2-2)
Energy (Sec. 2-4)
Heat (Sec. 2-5)
Heterogeneous (Sec. 2-2)
Homogeneous (Sec. 2-2)
Inertia (Sec. 2-1)
Kelvin scale (Sec. 2-5)
Kinetic energy (Sec. 2-4)

Mass (Sec. 2-1)
Matter (Sec. 2-1)
Mixture (Sec. 2-2)
Molecule (Sec. 2-3)
Phase (Sec. 2-2)
Potential energy (Sec. 2-4)
Pure substance (Sec. 2-2)
Rotational motion (Sec. 2-3)
Solution (Sec. 2-2)
State (Sec. 2-2)
Translational motion (Sec. 2-3)
Temperature (Sec. 2-5)
Vibrational motion (Sec. 2-3)
Weight (Sec. 2-1)
Work (Sec. 2-4)

QUESTIONS AND PROBLEMS

Mass and density

2-1 Explain the essential difference between **(a)** mass and weight **(b)** mass and density.

2-2 Consider an object in a space station, far away from any planets or stars. Does this object have weight? mass? Justify your answers.

2-3 Two objects A and B have the same mass, but the density of A is twice that of B. How do their volumes compare?

2-4 What is the density of **(a)** a solid, 225 g of which occupy a volume of 329 cm³? **(b)** a liquid, 28 cm³ of which weigh 26.4 g? **(c)** a gas, 22.4 liters of which weigh 79.0 g?

2-5 The density of a gas is usually less than the density of a liquid. Why is this so?

2-6 It is often said that magnesium is a light metal. How could this be stated more precisely?

2-7 A cube of redwood 3.00 cm on an edge weighs 11.8 g. What is the density of redwood?

2-8 A metal cube 5.00 cm on an edge is found to weigh 394 g. Chemical analysis of a small sample filed off the cube shows the metal to be aluminum (density = 2.70 g cm⁻³). What can you say about the cube?

2-9 A beaker containing 4.00×10^2 cm³ of a liquid having a density of 1.85 g cm⁻³ is found to weigh 884 g. What is the mass of the empty beaker?

2-10 On the atomic level what two factors determine the density of a substance?

2-11 Why do the densities of most liquids decrease with increasing temperature?

Kinds of matter

2-12 Describe the melting point of **(a)** a pure substance **(b)** a homogeneous mixture **(c)** a heterogeneous mixture. Account for any differences.

2-13 Define or explain each of the following terms: homogeneous, heterogeneous, phase, element, compound, solution.

2-14 In terms of their microstructures what are the essential differences between a solid, a liquid, and a gas?

2-15 If an excess of sodium chloride is added to 100 g of water at 25°C, the amount of the salt which dissolves is always a fixed 36.1 g. Can the saltwater combination be classified as a compound? Explain.

2-16 How many phases are present in each of the following well-mixed systems? **(a)** pure quartz sand **(b)** sand + salt **(c)** sand + salt + sugar **(d)** sand + salt + sugar + liquid water **(e)** sand + salt + sugar + water + gasoline **(f)** sand + salt + sugar + water + gasoline + copper metal **(g)** all of the previous plus ice **(h)** all of the previous plus beer.

2-17 Suppose that you were given a sample of a homogeneous liquid. What would you do in order to determine whether it is a solution or a pure substance?

2-18 Two pure substances A and B are mixed to form a homogeneous product with no visible A or B left over. There are several possibilities for the nature of the product. What are they?

Elements and compounds

2-19 When marble is heated strongly, it decomposes into a new solid and a gas. **(a)** Is marble an element or a compound? **(b)** If the solid and gas are compounds consisting in each case of two elements, what can you say about the number of elements of which marble is composed?

2-20 Compounds and solutions both represent combinations of simpler substances. What is the essential difference between them? How can they be distinguished in the laboratory?

2-21 A certain compound is analyzed and found to consist of 38.0 percent element A and 62.0 percent element B by mass. If A and B react directly to form the compound, how much compound could be prepared by mixing **(a)** 38.0 g of A with 90.0 g of B? **(b)** 80.0 g of A with 40.0 g of B? **(c)** 9.50 g of A with 14.0 g of B?

2-22 The compound calcium bromide consists of 20.0 percent calcium (Ca) and 80.0 percent bromine (Br) by mass. Water is composed of 88.9 percent oxygen and 11.1 percent hydrogen by mass. Suppose 10.0 g of calcium bromide is dissolved in 45.0 g of water. Calculate the percentage of each element in the resulting solution.

2-23 A sample of a certain compound contains 1.40 g of element A, 2.60 g of element B, and 3.80 g of element C. How many grams of this compound could be made from **(a)** 1.00 g of A, 1.00 g of B, and 1.00 g of C? **(b)** 1.00 g of A, 2.00 g of B, and 3.00 g of C?

2-24 Carbon dioxide is a compound consisting of 27.3 percent carbon and 72.7 percent oxygen by mass. **(a)** If 25.0 g of carbon is burned, how much carbon dioxide is formed? **(b)** If 25.0 g of oxygen is used in burning some carbon, how much carbon dioxide is formed?

2-25 Sulfur dioxide is a compound consisting of molecules, each of which is composed of one sulfur atom and two oxygen atoms. If a sulfur atom weighs twice as much as an oxygen atom, what is the percentage composition by mass of sulfur dioxide?

2-26 Using the relative masses of a sulfur atom and an oxygen atom given in the previous question, determine the percentage composition of sulfur trioxide, in which each molecule has three oxygen atoms and one sulfur atom.

Energy, heat, and temperature

2-27 What is heat? How is it different from other forms of energy? Why are heat and temperature often confused with each other?

2-28 Describe what happens on the molecular level when each of the following substances, initially at room temperature, is heated 10°C: **(a)** iron   **(b)** water   **(c)** air.

2-29 Describe what happens to the potential and kinetic energy of a bullet which is fired into the air and then falls to earth.

2-30 Compare and contrast kinetic and potential energy.

2-31 The burning of wood is a chemical reaction which *liberates* considerable energy. Why do you suppose energy is *required* in order to start a piece of wood burning?

2-32 Assuming that the potential energy of an object is zero at the surface of the earth, which of each of the following pairs has the larger gravitational potential energy: **(a)** a golf ball 1 m or a golf ball 2 m above the earth's surface?   **(b)** a cannonball or a golf ball each 1 m above the earth?   **(c)** a golf ball at the earth's surface or a golf ball 1 m below (in a hole)?   **(d)** a golf ball 1 m above the earth moving upward or a golf ball at the same location moving downward?

2-33 Compare the kinetic energy of   **(a)** a 2-ton automobile with a 1-ton automobile traveling at the same speed   **(b)** an automobile traveling at 50 mi h^{-1} with an identical automobile traveling at 25 mi h^{-1}.

2-34 Conversion of any temperature on the Kelvin scale to its Celsius equivalent always yields a different number, yet the size of the degree Celsius is identical to that of the kelvin. Explain.

2-35 Convert each of the following temperatures to kelvins: **(a)** 25°C   **(b)** −25°C   **(c)** 101°C   **(d)** −201°C.

2-36 Convert each of the following temperatures to degrees Celsius: **(a)** 75°F   **(b)** −15°F   **(c)** 325°F   **(d)** 325 K.

2-37 Suppose 4.90 kg of water is heated from 15 to 48°C. How much heat is absorbed by the water? Express your answer in   **(a)** kilocalories and   **(b)** kilojoules.

3 FORMULAS, EQUATIONS, AND STOICHIOMETRY

TO THE STUDENT

You may have already been familiar with the concepts of matter, energy, atoms, molecules, and so on; if so, then the preceding chapters were mostly a review for you. The ideas in those chapters are absolutely essential, however, because they provide the foundation for this chapter in which we are primarily concerned with two things: formulas and equations. We also introduce the *mole,* a central concept in chemistry. When you get to the section on moles (Sec. 3-3), read it carefully; it is not difficult, and it *is* important.

Stoichiometry, part of the title of this chapter, may be a new word for you. It comes from the Greek *stoicheon,* meaning "the measurement of elements." Today it is usually more broadly defined as the study of (1) relative quantities of elements combined in compounds and (2) relative quantities of elements and compounds involved in chemical reactions.

**3-1
Combinations and
collections of
atoms**

In Sec. 2-3 we described some of the ways in which atoms can be clustered to form larger groups. Just as we find it useful to use chemical symbols to denote individual atoms, so we also find it useful to use *chemical formulas* to denote specific groups or collections of atoms.

*The molecule and
molecular formulas*

A group of atoms bonded together to form a discrete, independent particle is called a *molecule.* We can write a *formula* to represent a molecule, using a symbol for each atom in the group and using subscript numbers when more than one atom of an element is present. This kind of formula is called a *molecular formula.* A molecule of water is composed of one oxygen atom and two hydrogen atoms, so the molecular formula for water is H_2O. A molecule of sucrose (cane sugar) consists of 12 C atoms, 22 H atoms, and 11 O atoms, so its molecular formula is $C_{12}H_{22}O_{11}$. Remember that this kind of formula gives the actual number of atoms of each element which are present in one molecule of a substance. Several molecules and their formulas are given in Fig. 3-1.

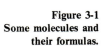

**Figure 3-1
Some molecules and
their formulas.**

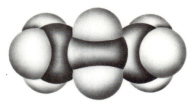

H$_2$O (water)   CO$_2$ (carbon dioxide)   C$_3$H$_8$ (propane)

Molecular formulas can be determined from the results of chemical analyses, as we will see in Sec. 3-4, provided that it is known that the substance is composed of molecules. This latter information can come from a variety of sources. In the case of solids, for example, a powerful technique known as x-ray diffraction (Sec. 10-2) provides information about the positions of atoms in the solid, and can therefore be used to decide whether or not discrete molecules exist.

Carbon dioxide is an example of a substance for which it is possible to write a molecular formula: CO$_2$ (see Fig. 3-1). Discrete molecules exist in the gas, liquid, and solid states. Solid carbon dioxide, known as *dry ice,* has the arrangement of CO$_2$ molecules shown in Fig. 3-2. Both *ball-and-stick* and *space-filling* drawings are shown. It is easy to pick out the discrete CO$_2$ molecules in this structure.

Empirical formulas   Another kind of formula, the *empirical formula,* gives the *relative* numbers of atoms of different elements in a compound.[1] The glucose (grape sugar) molecule consists of 6 C atoms, 12 H atoms, and 6 O atoms. The empirical formula of glucose is CH$_2$O and expresses only the *ratio* of C atoms to H atoms to O atoms in glucose. Since this ratio is always expressed in simplest form, an empirical formula is also often called a *simplest formula.* In the case of glucose the molecular formula (C$_6$H$_{12}$O$_6$) is an integral multiple of the empirical formula (CH$_2$O). However, the molecular and empirical formulas of sucrose (C$_{12}$H$_{22}$O$_{11}$) are the same, since the subscripts are not all divisible by any integer other than 1.

Empirical formulas are also used to represent substances which are not composed of molecules. One such substance is solid silicon carbide, also known as *carborundum.* In this compound each silicon atom is surrounded by (and bonded to) four adjacent carbon atoms, and each carbon atom is similarly surrounded by four silicon atoms, as shown in Fig. 3-3. This linkage of carbon and silicon atoms extends in three dimensions, ending only when the physical limits of the crystal of silicon carbide are reached. In such a structure no discrete molecules are identifiable. Perhaps the whole crystal could be called a *molecule,* but using this word to mean something so large is not ordinarily done.

In a sample of silicon carbide the total number of silicon atoms is equal to the total number of carbon atoms. In other words, the ratio of silicon atoms to carbon atoms is 1:1, so we write the empirical formula as SiC. Examples of empirical formulas are shown in Table 3-1. Note that although molecular formulas can be

---

[1]The word *empirical* means "based only on observation and measurement." In this case it is used because, as we will see in Sec. 3-4, an empirical formula can be determined from the percentage composition, that is, the *analysis* of the compound in terms of its component elements. No knowledge of molecules is required.

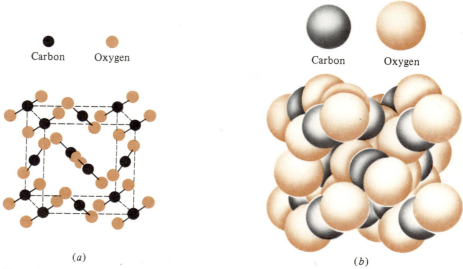

● Carbon   ● Oxygen   ⬤ Carbon   ⬤ Oxygen

(a)   (b)

Figure 3-2
Solid carbon dioxide (dry ice): two representations. (a) Ball-and-stick. (b) Space-filling. These are drawings of two commonly used types of three-dimensional models. Each type has its strengths and weaknesses: the ball-and-stick model more clearly shows geometrical relationships, while the space-filling model more accurately represents atomic sizes and molecular packing in the solid. In each model, however, discrete CO_2 molecules can be readily identified.

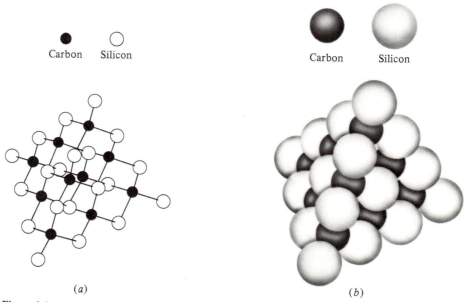

● Carbon   ○ Silicon   ⬤ Carbon   ⬤ Silicon

(a)   (b)

Figure 3-3
Silicon carbide (carborundum): two representations. (a) Ball-and-stick. (b) Space-filling. In each of these models it can be seen that there are no discrete SiC molecules, only a network of Si and C atoms extending in three dimensions.

Table 3-1
Molecular and empirical
formulas

Substance	Molecular formula	Empirical (simplest) formula
Water	H_2O	H_2O
Ammonia	NH_3	NH_3
Carbon dioxide	CO_2	CO_2
Sucrose (cane sugar)	$C_{12}H_{22}O_{11}$	$C_{12}H_{22}O_{11}$
Oxygen	O_2	O
Sulfur	S_8	S
Acetylene	C_2H_2	CH
Benzene	C_6H_6	CH
Silicon carbide		SiC
Sodium chloride		NaCl
Calcium chloride		$CaCl_2$
Sodium nitrate		$NaNO_3$

written only for molecular substances, empirical formulas can be written for all (pure) substances.

ADDED COMMENT

Did you notice the compounds acetylene and benzene in Table 3-1? They have identical empirical formulas but different molecular formulas. This is not unusual and shows that an empirical formula does not always uniquely specify a compound.

Structural formulas   Another kind of formula is the *structural formula*. For a molecular substance the structural formula not only gives the number of atoms of each element in the molecule but also indicates how they are bonded together. Figure 3-4 shows structural formulas of several molecules. Note that the last two, ethanol and dimethyl ether, have the same molecular formulas, C_2H_6O, but different structural formulas. This is an example of *isomerism*. *Isomers* are compounds which have the same molecular formula but differ in the arrangement of atoms.

3-2
Atomic weights and other masses

An important property of an atom is its *mass*. We might, of course, express the mass of an atom in any one of many units. The mass of one oxygen atom, for example, is 2.7×10^{-23} g, or 9.4×10^{-25} oz, or 4.1×10^{-22} grains, or 2.9×10^{-29} tons, or 1.3×10^{-22} carats. But each of these numbers is small. What we need is

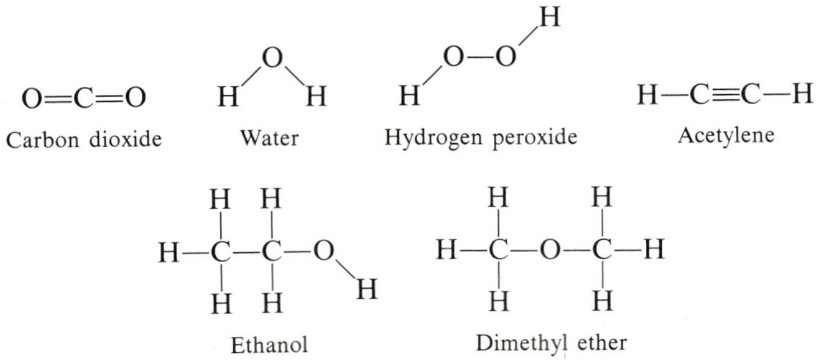

Figure 3-4
Structural formulas.

a unit of mass which is itself so small that masses of atoms turn out to be numerically greater than 1.

Masses of atoms: atomic weights

The mass of an atom is usually expressed using an extremely small unit called the *atomic mass unit* (amu). To the nearest integer the mass of one oxygen atom turns out to be 16 amu. Similarly, the mass of a fluorine atom is 19 amu, that of a chromium atom is 52 amu, and that of a gold atom is 197 amu. The lightest atom is hydrogen, mass 1 amu.

Expressed in atomic mass units, the masses of atoms are called *atomic weights.* A table of these atomic weights is given inside the front cover of this book. Each atomic weight in the table is given to at least four significant figures and represents the average mass of the atoms of a given element. Why *average?* Not all atoms of a given element have exactly the same mass. Different varieties of atoms of the same element, called *isotopes,* exist in many cases. Isotopes of an element have different masses; so for each element which does have naturally occurring isotopes the mass of an "average atom" is given as the atomic weight. One atomic mass unit is defined as one-twelfth of the mass of the most common isotope of carbon. (A more complete discussion of isotopes and of the determination of atomic weights will be found in Sec. 5-3.)

Specifying *atomic weight* implies that the units are atomic mass units. In this book, for instance, an atomic weight might be provided in a problem as "Atomic weight: N = 14.0." Although no units are explicitly stated here, this means that the mass of one nitrogen atom is 14.0 amu. Also, most tables of atomic weights do not explicitly state that the listed values are expressed in atomic mass units.

An important point about atomic weights is that they are *masses,* not weights. Why, then, are they called *weights?* Would not the term *atomic mass* be a better one? Perhaps, but the International Union of Pure and Applied Chemistry, or IUPAC (pronounced "you-pack" or "eye-you-pack"), has officially approved the term *atomic weight,* as have other international scientific groups.

Masses of molecules: molecular weights

The mass of a molecule is also usually expressed in atomic mass units. Because a molecule is a cluster of atoms, its mass is the sum of the masses of all the atoms in the cluster. Stated differently, the sum of the atomic weights of all the atoms in a molecule equals the *molecular weight.* (Here, again, we are speaking of a *mass.*) The atomic weights of carbon and hydrogen are 12.0 amu and 1.0 amu, respectively; thus, the molecular weight of ethylene, C_2H_4, is 2(12.0 amu) + 4(1.0 amu) = 28.0 amu.

Strictly speaking, the term *molecular weight* is used correctly only when expressing the mass of a *molecule* and is not appropriate in the case of nonmolecular substances.

Masses of formula units: formula weights

What, then, do we get when we add the weights of the atoms shown in an empirical formula? In this case the sum is called a *formula weight.* A formula weight is the mass of one *formula unit* in atomic mass units, where a formula unit is simply those atoms which are shown in the formula. The empirical formula of calcium chloride is $CaCl_2$, and so one formula unit of calcium chloride consists of one Ca atom and two Cl atoms. Since the atomic weight of Ca is 40.1 amu and that of Cl is 35.5 amu, the formula weight of $CaCl_2$ is 40.1 amu + 2(35.5 amu), or 111.1 amu.

(This quantity is occasionally called a molecular weight, but since $CaCl_2$ does not consist of molecules, the term *formula weight* is better.) In the case of a molecular formula, molecular weight and formula weight are equivalent names for the same thing.

● EXAMPLE 3-1

Problem Parathion is a toxic compound which has been used as an insecticide. Its molecular formula is $C_{10}H_{14}O_5NSP$. What is the molecular weight of parathion? (Atomic weights: C = 12.0; H = 1.0; O = 16.0; N = 14.0; S = 32.1; P = 31.0.)

Solution

10 C atoms: mass = 10(12.0 amu) = 120 amu
14 H atoms: mass = 14(1.0 amu) = 14 amu
 5 O atoms: mass = 5(16.0 amu) = 80.0 amu
 1 N atom: mass = 1(14.0 amu) = 14.0 amu
 1 S atom: mass = 1(32.1 amu) = 32.1 amu
 1 P atom: mass = 1(31.0 amu) = 31.0 amu
          Molecular weight of parathion = 291 amu   ●

● EXAMPLE 3-2

Problem The empirical formula for aluminum sulfate is $Al_2(SO_4)_3$. What is the formula weight of aluminum sulfate? (Atomic weights: Al = 27.0; S = 32.1; O = 16.0.)

Solution The formula $Al_2(SO_4)_3$ tells us that two Al atoms are combined with three SO_4 groups (called *sulfate* groups); each SO_4 group is in turn composed of one S atom and four O atoms. Thus, the total number of S atoms shown is 3, and of O atoms is 3×4, or 12. In other words, $Al_2(SO_4)_3$ might be alternatively written as $Al_2S_3O_{12}$.

 2 Al atoms: mass = 2(27.0 amu) = 54.0 amu
 3 S atoms: mass = 3(32.1 amu) = 96.3 amu
12 O atoms: mass = 12(16.0 amu) = 192 amu
          Formula weight of $Al_2(SO_4)_3$ = 342 amu   ●

3-3
The mole

At times it is useful to consider the properties and behavior of a single atom or, perhaps, a small group of atoms. In the laboratory, however, we must of necessity work with much larger quantities. For this reason we often keep track of and count atoms in large groups.

Avogadro's number and moles of atoms

Just as eggs are counted by the dozen, atoms are counted by the *mole,* a very large number, 6.02×10^{23}.[2] The number itself is called *Avogadro's number.*[3] To repeat, one mole of atoms is Avogadro's number of atoms, or 6.02×10^{23} atoms.

Why do we count atoms in "packages" of 6.02×10^{23}? To find out, look at the table of atomic weights (inside front cover). The atomic weight of oxygen is 16.0 amu. It turns out that if we assemble a group of 6.02×10^{23} oxygen atoms,

---

[2] A more precise value is 6.022045×10^{23}, but three significant figures will usually be sufficient for the purposes of this book.

[3] Lorenzo Romano Amedeo Carlo Avogadro di Quaregua e di Cerreto (1776–1856) was an Italian lawyer and physicist. He was one of the first to distinguish between atoms and molecules.

the entire collection weighs 16.0 g. Thus, we can say a *single* oxygen atom weighs 16.0 amu, and 1 mol of oxygen atoms (6.02×10^{23} atoms) weighs 16.0 g. (*Mol* is the official SI abbreviation for mole, not much of a saving in printer's ink.)

Next, consider atoms of another element, sulfur. From the table we see that the atomic weight of sulfur is 32.1 amu, and so each atom of sulfur weighs *almost exactly twice as much* as each atom of oxygen. Imagine, now, that we have two collections of atoms, one of oxygen and the other of sulfur, and that *the number of atoms in each collection is the same.* Under this condition the collection of sulfur atoms must weigh *almost exactly twice as much* as the collection of oxygen atoms.

Now consider one collection of 6.02×10^{23} sulfur atoms (1 mol) and another collection consisting of 6.02×10^{23} oxygen atoms (also 1 mol). Since a single sulfur atom weighs almost exactly twice as much as a single oxygen atom, the sulfur collection weighs almost twice as much as the oxygen collection. Furthermore, since the collection of oxygen atoms weighs 16.0 g, the collection of sulfur atoms must weigh almost exactly twice this, or 32.1 g. To summarize, one sulfur atom weighs 32.1 amu, and one oxygen atom weighs 16.0 amu; 1 mol of sulfur atoms weighs 32.1 g, and 1 mol of oxygen atoms weighs 16.0 g.

The relationship between atoms and moles of atoms is perfectly general, e.g., it applies to all atoms. In other words, the ratio of the masses of 1 mol of A atoms to 1 mol of B atoms is the same as the ratio of the mass of a single A atom to a single B atom. Furthermore, the mass of 1 mol of atoms of *any* element is *numerically* equal to the mass of a single atom, provided that the mass of the single atom is expressed in atomic mass units while that of 1 mol is expressed in grams.

ADDED COMMENT

You should now be able to see that a table of atomic weights, such as the one inside the front cover of this book, gives two separate but related pieces of information: (1) the mass of *one atom* of each element *expressed in atomic mass units* and (2) the mass of *1 mol of atoms* of each element *expressed in grams.*

We now know two things about the mole: One mole of atoms (1) consists of 6.02×10^{23} atoms and (2) has a mass in grams which is numerically equal to the atomic weight. Let us consider some applications of these relationships.

● **EXAMPLE 3-3**

Problem   A certain sample of nitrogen gas consists of 4.63×10^{22} N atoms. How many moles of N atoms is this?

Solution   Since there are 6.02×10^{23} N atoms in 1 mol of N atoms, we can write the equivalency,

$$6.02 \times 10^{23} \text{ N atoms} \sim 1 \text{ mol N atoms}$$

Therefore, 4.63×10^{22} N atoms are

$$4.63 \times 10^{22} \text{ N atoms} \frac{1 \text{ mol N atoms}}{6.02 \times 10^{23} \text{ N atoms}} = 0.0769 \text{ mol N atoms} \quad ●$$

● EXAMPLE 3-4

Problem   How many iron atoms are present in a piece of iron consisting of 1.00×10^{-4} mol of Fe atoms?

Solution   One mole of Fe atoms consists of 6.02×10^{23} Fe atoms; so 1.00×10^{-4} mol of Fe atoms consists of

$$1.00 \times 10^{-4} \text{ mol Fe atoms } \frac{6.02 \times 10^{23} \text{ Fe atoms}}{1 \text{ mol Fe atoms}} = 6.02 \times 10^{19} \text{ Fe atoms} \quad ●$$

● EXAMPLE 3-5

Problem   How many moles of Cu atoms are there in 3.05 g of copper? (Atomic weight: Cu = 63.5.)

Solution   One Cu atom weighs 63.5 amu; so 1 mol of Cu atoms weighs 63.5 g. Therefore, we can write the equivalency

$$1 \text{ mol Cu atoms} \sim 63.5 \text{ g Cu}$$

Therefore, 3.05 g of Cu is

$$3.05 \text{ g Cu } \frac{1 \text{ mol Cu atoms}}{63.5 \text{ g Cu}} = 0.0480 \text{ mol Cu atoms} \quad ●$$

● EXAMPLE 3-6

Problem   How much does 4.98×10^{-6} mol of Cu atoms weigh?

Solution   One mole of Cu atoms weighs 63.5 g (see Example 3-5). Thus, 4.98×10^{-6} mol of Cu atoms weighs

$$4.98 \times 10^{-6} \text{ mol Cu atoms } \frac{63.5 \text{ g Cu}}{1 \text{ mol Cu atoms}} = 3.16 \times 10^{-4} \text{ g Cu} \quad ●$$

● EXAMPLE 3-7

Problem   How many atoms are present in a lump of sulfur weighing 10.0 g? (Atomic weight: S = 32.1)

Solution   One S atom weighs 32.1 amu; so 1 mol of S atoms weighs 32.1 g. Therefore, 10.0 g of S is

$$10.0 \text{ g S } \frac{1 \text{ mol S atoms}}{32.1 \text{ g S}} = 0.312 \text{ mol S atoms}$$

Since 1 mol of S atoms consists of 6.02×10^{23} S atoms, 0.312 mol of S atoms is

$$0.312 \text{ mol S atoms } \frac{6.02 \times 10^{23} \text{ S atoms}}{1 \text{ mol S atoms}} = 1.88 \times 10^{23} \text{ S atoms}$$

Alternatively, we can set this calculation up in one composite step. (Notice how the units "cancel.")

$$10.0 \text{ g S } \left(\frac{1 \text{ mol S atoms}}{32.1 \text{ g S}} \right) \left(\frac{6.02 \times 10^{23} \text{ S atoms}}{1 \text{ mol S atoms}} \right) = 1.88 \times 10^{23} \text{ S atoms} \quad ●$$

● EXAMPLE 3-8

Problem   How many grams do 8.46×10^{24} atoms of fluorine weigh? (Atomic weight: F = 19.0.)

Solution 8.46×10^{24} atoms of F are

$$8.46 \times 10^{24} \text{ F atoms} \left(\frac{1 \text{ mol F atoms}}{6.02 \times 10^{23} \text{ F atoms}} \right) \left(\frac{19.0 \text{ g F}}{1 \text{ mol F atoms}} \right) = 267 \text{ g F} \quad \bullet$$

Moles of other things

We often use moles to count other things besides atoms—molecules, for instance. We may thus refer to a mole of water molecules, a mole of carbon dioxide molecules, a mole of sugar molecules, etc. In each case 1 mol consists of 6.02×10^{23} molecules, that is, Avogadro's number of molecules.

How much does a mole of molecules weigh? It has a mass *in grams* which is *numerically* equal to the molecular weight (the mass of one molecule in atomic mass units). This means that when we add atomic weights to get a molecular weight (in atomic mass units), we are simultaneously finding the mass of 1 mol of molecules (in grams).

● **EXAMPLE 3-9**

Problem   What is the mass of 1.00 mol of sulfur dioxide molecules if the molecular formula for sulfur dioxide is SO_2? (Atomic weights: S = 32.1; O = 16.0.)

Solution   The mass of a single SO_2 molecule equals 32.1 amu + 2(16.0 amu), or, 64.1 amu. Since one SO_2 molecule weighs 64.1 amu, 1 mol of SO_2 molecules weighs 64.1 g. ●

We can also count *formula units* by the mole. One mole of silicon carbide, SiC, formula units is thus 6.02×10^{23} formula units, each of which consists of one Si atom and one C atom. The mass of 1 mol of formula units is *numerically* equal to the formula weight, but expressed in grams.

● **EXAMPLE 3-10**

Problem   Calculate the mass of 1 mol of formula units of potassium nitrate, KNO_3. (Atomic weights: K = 39.1; N = 14.0; O = 16.0.)

Solution   The mass of one formula unit of KNO_3 is 39.1 amu + 14.0 amu + 3(16.0 amu), or, 101.1 amu. Therefore, the mass of 1 mol of KNO_3 formula units is 101.1 g. ●

We can count anything by the mole. Thus, we might speak of a mole of buckshot, a mole of eggs, or a mole of elephants, but for obvious reasons this is not often done.[4]

3-4 Formula stoichiometry

Each chemical formula has three meanings or interpretations: (1) a *qualitative* meaning, (2) a *microscopic quantitative* meaning, and (3) a *macroscopic quantitative* meaning.

Qualitatively a formula simply represents a substance. For example, H_2O represents *water,* NaCl represents *table salt,* and $C_9H_8O_4$ can be written instead of *aspirin.* In other words, a formula simply stands for the name of a substance.

---

[4]One full-grown elephant weighs about 2.7×10^{30} amu. How much does a mole of elephants weigh? How does this compare with the mass of the earth, 6.6×10^{21} tons (9.1×10^2 kg = 1 ton)?

Quantitative meanings of chemical formulas

On the *microscopic* (or *atomic*) *scale,* a *molecular formula* indicates the number of atoms of each element in one molecule. Thus, $C_{10}H_{14}N_2$ represents *one molecule* of the compound nicotine and tells us that this molecule consists of 10 carbon atoms, 14 hydrogen atoms, and 2 nitrogen atoms. On this scale the meaning of an *empirical formula* is similar, but describes the composition of a formula unit. For example, K_2SO_4 represents *one formula unit* of potassium sulfate and tells us that this formula unit consists of two potassium atoms, one sulfur atom, and four oxygen atoms. The empirical formula K_2SO_4 tells us that in this compound the *ratio* of K atoms to S atoms to O atoms is 2:1:4.

Any microscopic interpretation of a chemical formula is stated in terms of atoms, but can be translated into a *macroscopic* (or *laboratory*) scale interpretation stated in terms of moles of atoms. Thus, a molecular formula tells us the number of moles of atoms of each element in 1 mol of molecules of the compound. The molecular formula of nicotine, $C_{10}H_{14}N_2$, tells us that 1 mol of nicotine molecules consists of 10 mol of carbon atoms, 14 mol of hydrogen atoms, and 2 mol of nitrogen atoms. In the case of potassium sulfate, K_2SO_4, this empirical formula tells us that 1 mol of K_2SO_4 formula units consists of 2 mol of potassium atoms, 1 mol of sulfur atoms, and 4 mol of oxygen atoms.

● EXAMPLE 3-11

Problem The empirical formula for lithium carbonate is Li_2CO_3. One formula unit of Li_2CO_3 consists of how many of each kind of atom?

Solution The formula tells us that one formula unit of Li_2CO_3 consists of two Li atoms, one C atom, and three O atoms. ●

● EXAMPLE 3-12

Problem In 38 formula units of Li_2CO_3, how many of each kind of atom are present?

Solution In 38 formula units of Li_2CO_3 there are

$$38 \text{ formula units } \frac{2 \text{ Li atoms}}{1 \text{ formula unit}} = 76 \text{ Li atoms}$$

$$38 \text{ formula units } \frac{1 \text{ C atom}}{1 \text{ formula unit}} = 38 \text{ C atoms}$$

$$38 \text{ formula units } \frac{3 \text{ O atoms}}{1 \text{ formula unit}} = 114 \text{ O atoms} \quad ●$$

● EXAMPLE 3-13

Problem In 1.00 mol of Li_2CO_3 formula units how many moles of Li, C, and O atoms are present?

Solution The formula tells us that 1.00 mol of Li_2CO_3 formula units consists of 2.00 mol of Li atoms, 1.00 mol of C atoms, and 3.00 mol of O atoms. ●

● EXAMPLE 3-14

Problem How many moles of Li, C, and O atoms are present in 4.31×10^{-2} mol of Li_2CO_3 formula units?

Solution In 4.31×10^{-2} mol of Li_2CO_3 formula units there are

$$4.31 \times 10^{-2} \text{ mol Li}_2\text{CO}_3 \text{ formula units } \frac{2 \text{ mol Li atoms}}{1 \text{ mol Li}_2\text{CO}_3 \text{ formula units}}$$
$$= 8.62 \times 10^{-2} \text{ mol Li atoms}$$

$$4.31 \times 10^{-2} \text{ mol Li}_2\text{CO}_3 \text{ formula units } \frac{1 \text{ mol C atoms}}{1 \text{ mol Li}_2\text{CO}_3 \text{ formula units}}$$
$$= 4.31 \times 10^{-2} \text{ mol C atoms}$$

$$4.31 \times 10^{-2} \text{ mol Li}_2\text{CO}_3 \text{ formula units } \frac{3 \text{ mol O atoms}}{1 \text{ mol Li}_2\text{CO}_3 \text{ formula units}}$$
$$= 1.29 \times 10^{-1} \text{ mol O atoms} \bullet$$

● EXAMPLE 3-15

Problem   The molecular formula for caffeine is $C_8H_{10}O_2N_4$. In a sample consisting of 0.150 mol of caffeine molecules how many moles of C, H, O, and N atoms are present?

Solution   In 0.150 mol of $C_8H_{10}O_2N_4$ molecules there are

$$0.150 \text{ mol C}_8\text{H}_{10}\text{O}_2\text{N}_4 \text{ molecules } \frac{8 \text{ mol C atoms}}{1 \text{ mol C}_8\text{H}_{10}\text{O}_2\text{N}_4 \text{ molecules}} = 1.20 \text{ mol C atoms}$$

$$0.150 \text{ mol C}_8\text{H}_{10}\text{O}_2\text{N}_4 \text{ molecules } \frac{10 \text{ mol H atoms}}{1 \text{ mol C}_8\text{H}_{10}\text{O}_2\text{N}_4 \text{ molecules}} = 1.50 \text{ mol H atoms}$$

$$0.150 \text{ mol C}_8\text{H}_{10}\text{O}_2\text{N}_4 \text{ molecules } \frac{2 \text{ mol O atoms}}{1 \text{ mol C}_8\text{H}_{10}\text{O}_2\text{N}_4\text{molecules}} = 0.300 \text{ mol O atoms}$$

$$0.150 \text{ mol C}_8\text{H}_{10}\text{O}_2\text{N}_4 \text{ molecules } \frac{4 \text{ mol N atoms}}{1 \text{ mol C}_8\text{H}_{10}\text{O}_2\text{N}_4 \text{ molecules}} = 0.600 \text{ mol N atoms} \bullet$$

Elemental analysis   Given either the empirical or molecular formula we can easily determine the
from a formula   elemental analysis, or percentage composition, of a compound.

● EXAMPLE 3-16

Problem   Butyric acid, a compound produced when butter goes rotten, has the formula $C_4H_8O_2$. (Is this formula molecular or empirical?) What is the elemental analysis of butyric acid?

Solution   We start by finding the number of moles of C, H, and O atoms in 1 mol of $C_4H_8O_2$. (Actually, although 1 mol is convenient, any number of moles of $C_4H_8O_2$ will do.) One mole of $C_4H_8O_2$ molecules consists of 4 mol of C atoms, 8 mol of H atoms, and 2 mol of O atoms.

   Now we need to find how much these quantities weigh. From the table inside the front cover we obtain the following atomic weights: C = 12.0, H = 1.01, and O = 16.0. From these data we know that 1 mol of C atoms weighs 12.0 g, 1 mol of H atoms weighs 1.01 g, and 1 mol of O atoms weighs 16.0 grams. Therefore,

4 mol of C atoms weighs: $$4 \text{ mol C atoms } \frac{12.0 \text{ g C}}{1 \text{ mol C atoms}} = 48.0 \text{ g C}$$

8 mol of H atoms weighs: $$8 \text{ mol H atoms } \frac{1.01 \text{ g H}}{1 \text{ mol H atoms}} = 8.08 \text{ g H}$$

2 mol of O atoms weighs: $$2 \text{ mol O atoms } \frac{16.0 \text{ g O}}{1 \text{ mol O atoms}} = 32.0 \text{ g O}$$

The mass of 1 mol of $C_4H_8O_2$ is thus 48.0 g + 8.08 g + 32.0 g, or 88.1 g. Now all we need to do is take the fraction of 88.1 g of butyric acid which consists of each element and multiply by 100 to get its percent by mass:

$$\% \text{ C} = \frac{48.0 \text{ g}}{88.1 \text{ g}} \times 100 = 54.5\%$$

$$\% \text{ H} = \frac{8.08 \text{ g}}{88.1 \text{ g}} \times 100 = 9.17\%$$

$$\% \text{ O} = \frac{32.0 \text{ g}}{88.1 \text{ g}} \times 100 = 36.3\%$$

Empirical formula from elemental analysis

An empirical formula expresses the ratios among the numbers of moles of atoms in one mole of formula units. This provides us with a way of finding the empirical formula of a compound from its analysis.

EXAMPLE 3-17

Problem

Peroxyacetylnitrate (PAN) is believed to play a role in the formation of photochemical smog. It has the following percentage composition by mass: 19.8 percent C, 2.5 percent H, 66.1 percent O, and 11.6 percent N. What is the empirical formula of PAN? (Atomic weights: C = 12.0, H = 1.01, O = 16.0, N = 14.0.)

Solution

We start by taking some quantity of PAN and finding how many grams of C, H, O, and N are present in that quantity. Actually, *any* quantity will do because we are only after a ratio, but it is convenient to choose exactly 100 g. Then the number of grams of an element is numerically equal to the percentage of that element (19.8 percent of 100 g is 19.8 g, for example). In 100 g of PAN there are, therefore, 19.8 g of C, 2.5 g of H, 66.1 g of O, and 11.6 g of N.

Next we need to find the number of moles of each kind of atom. Since 1 mol of each has a mass which is numerically equal to its atomic weight, but expressed in grams,

$$19.8 \text{ g C} \frac{1 \text{ mol C atoms}}{12.0 \text{ g C}} = 1.65 \text{ mol C atoms}$$

$$2.5 \text{ g H} \frac{1 \text{ mol H atoms}}{1.01 \text{ g H}} = 2.5 \text{ mol H atoms}$$

$$66.1 \text{ g O} \frac{1 \text{ mol O atoms}}{16.0 \text{ g O}} = 4.13 \text{ mol O atoms}$$

$$11.6 \text{ g N} \frac{1 \text{ mol N atoms}}{14.0 \text{ g N}} = 0.829 \text{ mol N atoms}$$

What we are looking for is a ratio, several ratios in fact, because there are four elements. That ratio expresses the relative numbers of moles of atoms of each element and could be written (for C to H to O to N) as 1.65:2.5:4.13:0.829. But that leads to the empirical formula

$$C_{1.65}H_{2.5}O_{4.13}N_{0.829}$$

In order to convert the ratio into one expressed in integers we *divide each of the numbers by the smallest of them.* [*Note:* dividing the numbers in a ratio by the same (nonzero) number never changes the ratio.]

Element	Ratio of moles (original form)	Ratio of moles (revised form)
C	1.65	$\dfrac{1.65}{0.829} = 1.99$
	to	to
H	2.5	$\dfrac{2.5}{0.829} = 3.0$
	to	to
O	4.13	$\dfrac{4.13}{0.829} = 4.98$
	to	to
N	0.829	$\dfrac{0.829}{0.829} = 1.00$

These numbers can be rounded off giving us 2 to 3 to 5 to 1, and so the empirical formula is $C_2H_3O_5N$. ●

Molecular formulas If the empirical formula of a compound is known, then only one more piece of information is needed in order to determine the molecular formula, provided, of course, that the compound is composed of molecules. That information is the molecular weight.

● **EXAMPLE 3-18**

Problem Ethane is a molecular compound having the empirical formula CH_3. The molecular weight of ethane has been experimentally determined to be 30.1. Find its molecular formula.

Solution The ratio shown in the molecular formula must be equivalent to the ratio shown in the empirical formula, that is, $1:3$. That does not help much, however, because the molecular formula might be CH_3, or C_2H_6, or C_3H_9, or C_4H_{12}, etc. But for each of these possibilities the molecule would have a different mass, that is a different molecular weight. We can add up the atomic weights in order to find the formula weight corresponding to each of the possible formulas until we find one which agrees with the known molecular weight. But this type of trial-and-error comparison is not really necessary. Whatever the molecular formula is, it must be some multiple of the empirical formula. Thus, the molecular weight must be the *same* multiple of the empirical formula weight. So all we need do is divide the known molecular weight, 30.1 amu/molecule by the formula weight of CH_3, 15.03 amu/formula unit:

$$\frac{30.1 \text{ amu/molecule}}{15.03 \text{ amu/formula unit}} = 2 \frac{\text{formula units}}{\text{molecule}}$$

The molecular formula of ethane is, consequently, $(CH_3)_2$, or C_2H_6. ●

If we are provided with data on percentage composition and molecular weight and we need the molecular formula only, we can ignore the empirical formula completely.

● **EXAMPLE 3-19**

Problem The molecular weight of dioxan is 88.1 amu, and its analysis is 54.5 percent C, 9.15 percent

H, and 36.3 percent O by mass. What is the molecular formula of dioxan? (Atomic weights: C = 12.0, H = 1.01, O = 16.0.)

Solution   This time we want more than just a ratio. Since the actual number of atoms (as opposed to a ratio of atoms) in one dioxan molecule is the same as the number of moles of C, H, and O atoms, respectively, in 1 mol of dioxan molecules, all we need to find is the number of moles of C, H, and O atoms in 88.1 g, which is 1 mol, of dioxan.

We start by finding the number of grams of C, H, and O in 88.1 g of dioxan:

The number of grams of C is 54.5% of 88.1 g, or 0.545(88.1 g) = 48.1 g.
The number of grams of H is 9.15% of 88.1 g, or 0.0915(88.1 g) = 8.06 g.
The number of grams of O is 36.3% of 88.1 g, or 0.363(88.1 g) = 32.0 g.

How many moles of C, H, and O atoms is this?

$$48.1 \text{ g C } \frac{1 \text{ mol C atoms}}{12.0 \text{ g C}} = 4.01 \text{ mol C atoms}$$

$$8.06 \text{ g H } \frac{1 \text{ mol H atoms}}{1.01 \text{ g H}} = 7.98 \text{ mol H atoms}$$

$$32.0 \text{ g O } \frac{1 \text{ mol O atoms}}{16.0 \text{ g H}} = 2.00 \text{ mol O atoms}$$

These numbers round off to 4, 8, and 2, respectively, and represent the number of moles of C, H, and O atoms, respectively, in 1 mol (88.1 g) of dioxan molecules. Consequently, the molecular formula of dioxan is $C_4H_8O_2$. ●

ADDED COMMENT

It is true that we have given both analytical data (percentage composition) and molecular weights in these examples, without providing any hint as to how these data can be obtained. In Sec. 3-6 there is an example of how elemental analyses are actually obtained in the laboratory. The method of experimentally determining a molecular weight depends largely upon the nature of the compound in question. For gases a simple method exists, and it is considered in Sec. 4-4.

3-5
Chemical equations

Carbon burns in excess oxygen, and the product of the reaction is carbon dioxide. The *chemical equation* for this reaction is written

$$C(s) + O_2(g) \longrightarrow CO_2(g)$$

On the left we see the *reactants,* carbon and oxygen, and on the right, the sole *product,* carbon dioxide. "O_2" has been written for oxygen (rather than "O"), because it represents a molecule of oxygen. The equation may be read "carbon reacts (or combines) with oxygen to form carbon dioxide." The arrow can be read "forms," "yields," "reacts to form," etc. The notations (*s*) and (*g*) are not absolutely necessary but serve to supplement the basic equation by giving information about the states of the reactants and products. (*s*) means *solid,* (*l*) means *liquid,* and (*g*) means *gas.* Sometimes (*aq*) is used to indicate that a species is dissolved in aqueous (water) solution.

The symbols and formulas in the equation represent not only the names of the various substances but also atoms and molecules. Thus, the above equation may also be read "one carbon atom reacts (or combines) with one oxygen molecule to form one carbon dioxide molecule."

The above equation is *balanced*. A balanced equation must show, among other things, that atoms are conserved; all of the atoms in the reactants must be accounted for in the products.

Another equation

$$CH_4(g) + 2\,O_2(g) \longrightarrow CO_2(g) + 2H_2O(g)$$

represents the burning of methane (natural gas) in oxygen to form carbon dioxide and gaseous water (steam). This equation has the coefficient 2 in front of both the O_2 and the H_2O in order to make the equation balance. (A 1 is assumed to be in front of the other formulas.) Without these coefficients the hydrogen and oxygen atoms would not balance; that is, there would be unequal numbers on opposite sides of the equation.

Balancing equations by inspection   Many of the simpler equations can be balanced *by inspection*.

● **EXAMPLE 3-20**

Problem   Balance the equation for the burning of butane, C_4H_{10}, in oxygen to form carbon dioxide and water:

$$C_4H_{10} + O_2 \longrightarrow CO_2 + H_2O$$

Solution   Examine the equation and pick one element to balance first. A good procedure is to start with the formula that has the largest number of atoms or the largest number of different elements. In this case that is the C_4H_{10}. First balance the carbon. Noting that there are four carbons on the left side of the equation and only one on the right, start by putting a 4 in front of CO_2:

$$C_4H_{10} + O_2 \longrightarrow 4CO_2 + H_2O$$

Step 2: Now look at the other element, H, in the C_4H_{10}. In order to balance the H atoms we need to write a 5 in front of the H_2O:

$$C_4H_{10} + O_2 \longrightarrow 4CO_2 + 5H_2O$$

Step 3: The oxygen atoms are the only ones left unbalanced; so put the correct coefficient in front of the O_2. That coefficient must be a fraction, however, because there are a total of 13 (count them) atoms on the right. Therefore we write $\frac{13}{2}$ in front of the O_2:

$$C_4H_{10} + \tfrac{13}{2}O_2 \longrightarrow 4CO_2 + 5H_2O$$

Step 4: Now we can clear fractions by multiplying through by 2:

$$2C_4H_{10} + 13\,O_2 \longrightarrow 8CO_2 + 10H_2O$$

Step 5: Check by adding up the atoms of each element on each side of the equation:

Atom	Left side	Right side
C	$2 \times 4 = 8$	8
H	$2 \times 10 = 20$	$10 \times 2 = 20$
O	$13 \times 2 = 26$	$(8 \times 2) + 10 = 26$

One very important rule to follow when balancing an equation is this: *Never change the formula of a reactant or product during the balancing process.*

● **EXAMPLE 3-21**

Problem   Balance the equation for the burning of hydrogen in oxygen to form water:

$$H_2 + O_2 \longrightarrow H_2O$$

Incorrect solution   An incorrect way to balance this equation is to put a subscript 2 after O in H_2O.

$$H_2 + O_2 \longrightarrow H_2O_2$$

This is incorrect because it changes the equation so that it no longer describes the same reaction. The new equation shows hydrogen peroxide, H_2O_2, as the product, not water.

Correct solution   The equation should be balanced by putting a $\frac{1}{2}$ in front of the O_2:

$$H_2 + \tfrac{1}{2}O_2 \longrightarrow H_2O$$

and then multiplying through by 2 to clear of fractions (if desired):

$$2H_2 + O_2 \longrightarrow 2H_2O \ ●$$

ADDED COMMENT

Sometimes students get the mistaken idea that they ought to be able to complete an equation when they are given only the reactants, somehow figuring out what the products ought to be. This is not possible without more chemical knowledge than you have at this stage. For example, the product of the reaction of N_2 with O_2 might be NO_2, N_2O_2, N_2O_4, N_2O, or something else. At this stage of your chemistry education, it is not possible for you to choose the correct product.

3-6
Reaction
stoichiometry

The meanings of a chemical equation are analogous to the meanings of a chemical formula (Sec. 3-4). Each equation has (1) a qualitative meaning and both (2) microscopic and (3) macroscopic quantitative meanings.

Qualitatively a chemical equation simply describes what the reactants and products of a reaction are. Thus, the equation

$$4Fe(s) + 3\,O_2(g) \longrightarrow 2Fe_2O_3(s)$$

indicates that iron (Fe) reacts with oxygen (O_2) to form the compound ferric oxide (Fe_2O_3).

Quantitative
meanings of
chemical equations

On the *microscopic* scale a *balanced* equation indicates numerical relationships among units (atoms, molecules, formula units, etc.) used up or formed in a reaction.

$$4Fe(s) + \quad 3\,O_2(g) \quad \longrightarrow \quad 2Fe_2O_3(s)$$

4	+	3	$\longrightarrow$	2
iron		oxygen		ferric oxide
atoms		molecules		formula units

The coefficients in a balanced equation describe *fixed ratios* among these units. In this case the ratios among the numbers of Fe atoms and O_2 molecules used up and Fe_2O_3 formula units formed is 4:3:2. (Do not worry if you did not know that Fe_2O_3 is not composed of molecules. It is not important for now.)

A balanced equation establishes a chemical equivalency between reactants and products. Using $\sim$ as we have before to indicate this equivalence, we can say that

$$4 \text{ Fe atoms} \sim 3 \text{ } O_2 \text{ molecules} \sim 2 \text{ } Fe_2O_3 \text{ formula units}$$

from which we can write the following unit factors:

$$\frac{4 \text{ Fe atoms}}{3 \text{ } O_2 \text{ molecules}} = 1$$

$$\frac{4 \text{ Fe atoms}}{2 \text{ } Fe_2O_3 \text{ formula units}} = 1$$

$$\frac{3 \text{ } O_2 \text{ molecules}}{2 \text{ } Fe_2O_3 \text{ formula units}} = 1$$

(The reciprocal of each left-hand term is of course equal to unity, too.)

● **EXAMPLE 3-22**

Problem   Nitrogen gas and hydrogen gas combine at a moderate temperature and pressure to form the compound ammonia (NH_3) according to the reaction:

$$N_2(g) + 3H_2(g) \longrightarrow 2NH_3(g)$$

How many (a) H_2 molecules are used up and (b) NH_3 molecules are formed when 4.20×10^{21} N_2 molecules react?

Solution   (a) The equation shows that one N_2 molecule combines with three H_2 molecules. Thus, the number of H_2 molecules used up is

$$4.20 \times 10^{21} \text{ } N_2 \text{ molecules} \frac{3 \text{ } H_2 \text{ molecules}}{1 \text{ } N_2 \text{ molecule}} = 1.26 \times 10^{22} \text{ } H_2 \text{ molecules}$$

(b) The equation shows that one N_2 molecule reacts to form two NH_3 molecules. Thus, the number of NH_3 molecules formed is

$$4.20 \times 10^{21} \text{ } N_2 \text{ molecules} \frac{2 \text{ } NH_3 \text{ molecules}}{1 \text{ } N_2 \text{ molecule}} = 8.40 \times 10^{21} \text{ } NH_3 \text{ molecules} ●$$

● **EXAMPLE 3-23**

Problem   In the reaction of ammonia (NH_3) with oxygen (O_2) to form nitric oxide (NO) and water how many molecules of NO can be made from 3.60×10^{21} molecules of O_2?

Solution   First we must write the balanced equation.

Step 1: $NH_3 + O_2 \longrightarrow NO + H_2O$

Step 2: $NH_3 + O_2 \longrightarrow NO + \frac{3}{2}H_2O$ (H balanced)

Step 3: $NH_3 + \frac{5}{4}O_2 \longrightarrow NO + \frac{3}{2}H_2O$ (O balanced)

Step 4: $4NH_3 + 5 O_2 \longrightarrow 4NO + 6H_2O$ (fractions cleared)

Next, look at the balanced equation and see what the O_2/NO ratio is. The coefficients are 5 (for O_2) and 4 (for NO); so the ratio is 5:4. This means that 5 molecules of O_2 will form 4 molecules of NO. Therefore, the number of molecules of NO formed from 3.60×10^{21} O_2 molecules is

$$3.60 \times 10^{21} \text{ } O_2 \text{ molecules} \frac{4 \text{ NO molecules}}{5 \text{ } O_2 \text{ molecules}} = 2.88 \times 10^{21} \text{ NO molecules} \bullet$$

Each chemical equation also has a *macroscopic* meaning. It indicates numerical relationships among moles of atoms, molecules, formula units, etc., which are consumed or formed in a reaction.

$$4Fe(s) \quad + \quad 3\,O_2(g) \quad \longrightarrow \quad 2Fe_2O_3(s)$$

4	+	3	⟶	2
mol of iron atoms		mol of oxygen molecules		mol of ferric oxide formula units

The coefficients in a balanced equation describe *fixed ratios* among moles of the units involved in the reaction. For this reaction the ratio of moles of Fe atoms used up to moles of O_2 molecules used up to moles of Fe_2O_3 formula units formed is 4:3:2.

From this equation we can write the following chemical equivalencies: 4 mol Fe atoms $\sim$ 3 mol O_2 molecules $\sim$ 2 mol Fe_2O_3 formula units. We thus have the following unit factors for this reaction:

$$\frac{4 \text{ mol Fe atoms}}{3 \text{ mol } O_2 \text{ molecules}} = 1$$

$$\frac{4 \text{ mol Fe atoms}}{2 \text{ } Fe_2O_3 \text{ formula units}} = 1$$

$$\frac{3 \text{ mol } O_2 \text{ molecules}}{2 \text{ mol } Fe_2O_3 \text{ formula units}} = 1$$

● **EXAMPLE 3-24**

Problem   For the reaction

$$N_2(g) + 3H_2(g) \longrightarrow 2NH_3(g)$$

how many (a) moles of H_2 molecules are consumed and (b) moles of NH_3 molecules are formed when 1.38 mol of N_2 molecules react?

Solution   (a) The equation shows that 1 mol of N_2 molecules combine with 3 mol of H_2 molecules. Thus, the number of moles of H_2 molecules used up is

$$1.38 \text{ mol } N_2 \text{ molecules} \frac{3 \text{ mol } H_2 \text{ molecules}}{1 \text{ mol } N_2 \text{ molecules}} = 4.14 \text{ mol } H_2 \text{ molecules}$$

(b) The equation shows that 1 mol of N_2 molecules react to form 2 mol of NH_3 molecules. Thus the number of moles of NH_3 molecules formed is

$$1.38 \text{ mol } N_2 \text{ molecules} \frac{2 \text{ mol } NH_3 \text{ molecules}}{1 \text{ mol } N_2 \text{ molecules}} = 2.76 \text{ mol } NH_3 \text{ molecules} \bullet$$

● **EXAMPLE 3-25**

Problem   How many moles of NO molecules can be made from 6.98 mol of O_2 molecules in the reaction

$$4NH_3(g) + 5\,O_2(g) \longrightarrow 4NO(g) + 6H_2O(g)$$

Solution   Five moles of O_2 molecules form 4 mol of NO molecules; so the number of moles of NO molecules formed from 6.98 mol of O_2 molecules is

$$6.98 \text{ mol } O_2 \text{ molecules} \frac{4 \text{ mol NO molecules}}{5 \text{ mol } O_2 \text{ molecules}} = 5.58 \text{ mol NO molecules} \;\bullet$$

Since we know how much 1 mol weighs (the atomic or formula weight in grams) and how many units it is composed of (Avogadro's number), we can find both the numbers of units and the masses of substances used up and formed in chemical reactions.

● **EXAMPLE 3-26**

Problem   When lead sulfide, PbS, and lead oxide, PbO, are heated together the products are lead (metal) and sulfur dioxide, SO_2:

$$PbS(s) + 2PbO(s) \longrightarrow 3Pb(l) + SO_2(g)$$

If 14.0 g of lead oxide react according to the above equation, how many (a) moles of lead atoms, (b) grams of lead, (c) lead atoms, and (d) grams of sulfur dioxide are formed? (Atomic weights: Pb = 207; S = 32.1; O = 16.0.)

Solution   For each one of these four parts we must use the balanced equation, and since that equation states a quantitative relationship in terms of moles, we must first find out how many moles of lead oxide we are starting with. Adding the atomic weights of lead (207) and oxygen (16.0), we find the formula weight of lead oxide (PbO) to be 223. Thus, 1 mol of lead oxide formula units weighs 223 g. In 14.0 g of PbO there are, therefore,

$$14.0 \text{ g PbO} \frac{1 \text{ mol PbO formula units}}{223 \text{ g PbO}} = 6.28 \times 10^{-2} \text{ mol PbO formula units}$$

(a) The balanced equation shows that 2 mol of PbO formula units form 3 mol of Pb atoms. Therefore, 6.28×10^{-2} mol of PbO formula units form

$$6.28 \times 10^{-2} \text{ mol PbO formula units} \frac{3 \text{ mol Pb atoms}}{2 \text{ mol PbO formula units}}$$
$$= 9.42 \times 10^{-2} \text{ mol Pb atoms}$$

(b) The atomic weight of lead is 207; so we know that 1 mol of Pb atoms weighs 207 g. Thus, 9.42×10^{-2} mol of Pb atoms weighs

$$9.42 \times 10^{-2} \text{ mol Pb atoms} \frac{207 \text{ g Pb}}{1 \text{ mol Pb atoms}} = 19.5 \text{ g Pb}$$

(c) 9.42×10^{-2} mol of Pb atoms is

$$9.42 \times 10^{-2} \text{ mol Pb atoms} \frac{6.02 \times 10^{23} \text{ Pb atoms}}{1 \text{ mol Pb atoms}} = 5.67 \times 10^{22} \text{ Pb atoms}$$

(d) The balanced equation shows that 2 mol of PbO formula units form 1 mol of SO_2 molecules. Thus, 6.28×10^{-2} mol of PbO formula units will form

$$6.28 \times 10^{-2} \text{ mol PbO formula units} \frac{1 \text{ mol } SO_2 \text{ molecules}}{2 \text{ mol PbO formula units}}$$
$$= 3.14 \times 10^{-2} \text{ mol } SO_2 \text{ molecules}$$

The molecular weight of SO_2 is $32.1 + 2(16.0) = 64.1$; so 1 mol of SO_2 molecules weighs 64.1 g. Thus, 3.14×10^{-2} mol of SO_2 molecules weighs

$$3.14 \times 10^{-2} \text{ mol } SO_2 \text{ molecules } \frac{64.1 \text{ g } SO_2}{1 \text{ mol } SO_2 \text{ molecules}} = 2.01 \text{ g } SO_2 \text{ molecules} \quad \bullet$$

● **EXAMPLE 3-27**

Problem Thermite is a powdered mixture of aluminum and ferric oxide (Fe_2O_3). When raised to a high enough temperature, it reacts with a spectacular pyrotechnic display to form *molten* iron and aluminum oxide (Al_2O_3). If a mixture of 10.0 g of Al and 50.0 g of Fe_2O_3 are reacted, how many grams of iron are produced? (Atomic weights: Al = 27.0; Fe = 55.8; O = 16.0.)

Solution The balanced equation for the reaction is

$$2Al(s) + Fe_2O_3(s) \longrightarrow Al_2O_3(s) + 2Fe(l)$$

The problem here is to find out which reactant, Al or Fe_2O_3, limits the amount of iron formed and which reactant is present in excess. The number of moles of Al atoms present is

$$10.0 \text{ g Al} \frac{1 \text{ mol Al atoms}}{27.0 \text{ g Al}} = 0.370 \text{ mol Al atoms}$$

From the balanced equation we see that 2 mol of Al atoms require 1 mol of Fe_2O_3 formula units. Thus, the number of moles of Fe_2O_3 formula units required to react with 0.370 mol of Al atoms is

$$0.370 \text{ mol Al atoms} \frac{1 \text{ mol } Fe_2O_3 \text{ formula units}}{2 \text{ mol Al atoms}} = 0.185 \text{ mol } Fe_2O_3 \text{ formula units}$$

This number of formula units of Fe_2O_3 is needed; how many are supplied?
Fifty grams of Fe_2O_3 is

$$50.0 \text{ g } Fe_2O_3 \frac{1 \text{ mol } Fe_2O_3 \text{ formula units}}{159.6 \text{ g } Fe_2O_3} = 0.313 \text{ mol } Fe_2O_3 \text{ formula units}$$

This is more than the quantity of Fe_2O_3 required to react with all the aluminum, so we conclude that the Fe_2O_3 is present in excess and the Al limits (and determines) the amount of Fe formed.
Now, 0.370 mol of Al atoms react to form

$$0.370 \text{ mol Al atoms} \frac{1 \text{ mol Fe atoms}}{1 \text{ mol Al atoms}} = 0.370 \text{ mol Fe atoms}$$

These weigh

$$0.370 \text{ mol Fe atoms} \frac{55.8 \text{ g Fe}}{1 \text{ mol Fe atoms}} = 20.6 \text{ g Fe} \quad \bullet$$

The following is an example of how elemental analyses (percentage compositions) of compounds can be determined experimentally.

● **EXAMPLE 3-28**

Problem Xylene is a compound composed of carbon and hydrogen only. Its elemental analysis and empirical formula can be determined by a technique known as combustion analysis, in which a sample is burned in excess oxygen, and the products, carbon dioxide and water vapor in this case, are separated and weighed. If the combustion of a sample of xylene produces 33.4 g of CO_2 and 8.55 g of water, determine (a) the percentage composition and (b) the empirical formula of xylene. (Atomic weights: $C = 12.0$; $H = 1.01$; $O = 16.0$.)

Solution   (a) First we must find the molecular weights of CO_2 and H_2O.

$$CO_2: \quad 12.0 + 2(16.0) = 44.0$$

$$H_2O: \quad 2(1.01) + 16.0 = 18.0$$

Next we calculate the number of moles of CO_2 and H_2O molecules formed.

$$33.4 \text{ g } CO_2 \frac{1 \text{ mol } CO_2 \text{ molecules}}{44.0 \text{ g } CO_2} = 0.759 \text{ mol } CO_2 \text{ molecules}$$

$$8.55 \text{ g } H_2O \frac{1 \text{ mol } H_2O \text{ molecules}}{18.0 \text{ g } H_2O} = 0.475 \text{ mol } H_2O \text{ molecules}$$

Since each mole of CO_2 molecules contains 1 mol of C atoms, and since each mole of H_2O molecules contains 2 mol of H atoms, the numbers of moles of C and H atoms present in the CO_2 and H_2O, and *present also in the original xylene,* are

$$0.759 \text{ mol } CO_2 \text{ molecules} \frac{1 \text{ mol C atoms}}{1 \text{ mol } CO_2 \text{ molecules}} = 0.759 \text{ mol C atoms}$$

$$0.475 \text{ mol } H_2O \text{ molecules} \frac{2 \text{ mol H atoms}}{1 \text{ mol } H_2O \text{ molecules}} = 0.950 \text{ mol H atoms}$$

And these weigh

$$\text{Mass of C} = 0.759 \text{ mol C atoms} \frac{12.0 \text{ g C}}{1 \text{ mol C atoms}} = 9.11 \text{ g C}$$

$$\text{Mass of H} = 0.950 \text{ mol H atoms} \frac{1.01 \text{ g H}}{1 \text{ mol H atoms}} = 0.960 \text{ g H}$$

Since no other element is present in xylene, the total mass of the original sample was $9.11 \text{ g} + 0.960 \text{ g} = 10.07 \text{ g}$, and so the analysis of the compound is

$$\% \text{ C} = \frac{\text{mass of carbon}}{\text{mass of xylene}} \times 100$$

$$= \frac{9.11 \text{ g}}{10.07 \text{ g}} \times 100 = 90.5\%$$

$$\% \text{ H} = \frac{\text{mass of hydrogen}}{\text{mass of xylene}} \times 100$$

$$= \frac{0.960 \text{ g}}{10.07 \text{ g}} \times 100 = 9.53\%$$

(b) To find the empirical formula we need the ratio of moles of C atoms to moles of H atoms, each of which we have already found. This ratio is $0.759:0.950$ (C to H). If we divide each number by 0.759, the ratio becomes $1.00:1.25$, which is $1:\frac{5}{4}$, or $4:5$. Thus, the empirical formula is C_4H_5. ●

3-7
The mole: final
comments

The mole concept is a simple one, but by this time it should be clear how important it is. Up until now we have been careful to specify in words the kind of unit being counted by the mole. Thus, we have spoken of 4.14 mol of hydrogen *molecules,*

6.28×10^{-2} mol of lead oxide *formula units,* 4 mol of iron *atoms,* etc. At this time an added simplification in terminology can be made. If we always write the formula or symbol of the unit being counted by the mole, we can safely leave off the name of the unit; that is, we can say 4.14 mol of H_2, 6.28×10^{-2} mol of PbO, and 4 mol of Fe, to use the previous examples. It is important, however, to write the formula of a substance, not its name. Suppose you were to say (or write), "one mole of hydrogen." Do you see how ambiguous that would be? There is no way of telling whether you mean 1 mol of H atoms or 1 mol of H_2 molecules—the second amount of hydrogen is twice the first! So when you use the mole, be sure to specify just *what* it is that you are expressing moles of. Unless the context makes it clear, it is wise to specify "moles of H_2," or "moles of hydrogen molecules," not just "moles of hydrogen."

How many is a mole?   The word *mole* and the word *molecule* both have their origins in the Latin word *moles,* meaning "pile," or "mass," or "burden." It is important to gain a notion of about how much matter is present in a mole of something. A mole of liquid H_2O, for example, is a small mouthful; a mole of gaseous N_2 will inflate a round balloon to a diameter of about 14 in; a mole of $C_{12}H_{22}O_{11}$ (sugar) is about three-quarters of a pound. And a mole of each consists of 6.02×10^{23} molecules.

Note that Avogadro's number is an incredibly large number, too large to really comprehend. It is larger, for example, than the number of grains of sand on all the beaches in the world. If you had Avogadro's number of pennies, how much do you think they would weigh? The answer is about 2×10^{18} tons, which is about one three-thousandth of the mass of the earth! Rice paper can be made very thin, about one one-thousandth of an inch thick. Suppose you had Avogadro's number of sheets of thin rice paper in a single stack. How tall do you think the stack would be? Would you guess that it would be 100 million times higher than the distance from the earth to the sun? Yes, Avogadro's number is *huge!*

SUMMARY

In this chapter we started out by considering different kinds of *chemical formulas* and their meanings. A formula can be used *qualitatively* simply to represent a substance. In addition it can provide *quantitative* information in terms of ratios of atoms (*microscopic meaning*) and moles of atoms (*macroscopic meaning*) in a compound. A *molecular formula* indicates actual numbers of atoms in a molecule (or moles of atoms in a mole of molecules). An *empirical formula* specifies only ratios of atoms or of moles of atoms in a compound. An empirical formula can also be said to specify a formula unit and its atomic composition. *Structural formulas* indicate the bonding in a molecule.

Masses of atoms, *atomic weights,* are specified in *atomic mass units* (amu), where 1 amu is defined as one-twelfth of the mass of the most common isotope of carbon. Addition of the masses (in amu) of all the

atoms in a molecule gives the *molecular weight,* also in atomic mass units. The sum of the masses of the atoms indicated in any formula is called a *formula weight,* but it is called a molecular weight only if the formula is molecular.

One *mole* of objects is *Avogadro's number* of objects, where Avogadro's number is 6.02×10^{23}. Thus, 1 mol of atoms is 6.02×10^{23} atoms, 1 mol of molecules is 6.02×10^{23} molecules, etc. The mass of 1 mol of atoms (or molecules or formula units) equals the atomic (or molecular or formula) weight expressed in grams.

The results of a chemical analysis of a compound can be used to calculate ratios of moles of elements in the compound and from this the empirical formula. Molecular weight information then permits determination of the molecular formula.

Chemical equations describe chemical changes *quali-*

tatively and also *quantitatively* in terms of atoms, molecules, etc. (*microscopic meaning*), or of moles of atoms, molecules, etc. (*macroscopic meaning*). A *balanced* equation shows conservation of atoms (and hence of mass) by indicating equal numbers of atoms of each element on the two sides of the equation. A balanced chemical equation is useful for relating quantities of reactants and products involved in chemical reactions. The key relationship here is the one stated in terms of moles of reactants and products, as indicated by the coefficients in the balanced equation.

KEY TERMS

Atomic mass unit (Sec. 3-2)
Atomic weight (Sec. 3-2)
Avogadro's number (Sec. 3-3)
Balanced equation (Sec. 3-5)
Empirical formula (Sec. 3-1)
Formula unit (Sec. 3-2)
Formula weight (Sec. 3-2)
Isotope (Sec. 3-2)

Macroscopic scale (Sec. 3-4)
Microscopic scale (Sec. 3-4)
Mole (Sec. 3-3)
Molecule (Sec. 3-1)
Molecular formula (Sec. 3-1)
Molecular weight (Sec. 3-2)
Structural formula (Sec. 3-1)

QUESTIONS AND PROBLEMS

Note: For answering questions in this and later chapters, make use of the table of atomic weights inside the front cover of the book whenever you need it.

Atoms and moles of atoms

3-1 Express to three significant figures the mass of each of the following in atomic mass units: **(a)** one atom of chlorine **(b)** one atom of silver **(c)** 200 atoms of sulfur **(d)** 6.02×10^{23} atoms of calcium.

3-2 Calculate the atomic weight of element X given that 3.74×10^6 atoms of X weigh 2.20×10^8 amu.

3-3 Calculate the atomic weight of element Y given that 3.48×10^{-3} mol of Y atoms weighs 7.03×10^{-2} g.

3-4 **(a)** If a housefly weighs 1.0×10^{-2} g, what is the mass in grams of 1.0 mol of houseflies? **(b)** What is the mass in atomic mass units of one housefly?

3-5 **(a)** What is the mass in grams of one fluorine atom? **(b)** What is the mass in grams of 0.0349 mol of F atoms? **(c)** How many moles of F atoms are present in 0.150 g? **(d)** How many F atoms are present in 0.150 g?

3-6 Determine the number of atoms in each of the following: **(a)** 14.0 g of N **(b)** 50.0 g of O **(c)** 10.0 g of Fe **(d)** 5.00 g of B.

Molecules, formula units, and moles

3-7 Determine the mass in grams of each of the following to three significant figures: **(a)** 1 mol of Cl atoms **(b)** 2 mol of Cu atoms **(c)** 0.200 mol of Si atoms **(d)** 8.75×10^{-3} mol of K atoms **(e)** 1 mol of Cl_2 molecules **(f)** 0.300 mol of $CaCl_2$ formula units.

3-8 How many moles are present in each of the following? **(a)** 55.85 g of Fe **(b)** 46.0 g of NO_2 **(c)** 1.00 g of NH_3 **(d)** 324 g of $C_{12}H_{22}O_{11}$.

3-9 How many grams does each of the following weigh? **(a)** 0.255 mol of CO_2 **(b)** 4.67×10^{22} CO_2 molecules **(c)** one CO_2 molecule.

3-10 How many moles of oxygen atoms are present in each of the following? **(a)** 11.5 g of O **(b)** 4.62×10^{24} atoms of O **(c)** 9.20×10^{22} molecules of SO_3 **(d)** 4.20×10^{-2} mol of Na_2O formula units **(e)** 3.93×10^{-3} mol of P_4O_{10} molecules.

3-11 How many moles of nitrogen atoms are present in each of the following? **(a)** 2.00 g of N **(b)** 3.00×10^{20} atoms of N **(c)** 4.00×10^{24} molecules of NH_3 **(d)** 5.00×10^{23} molecules of N_2H_4 **(e)** 6.00×10^{-3} mol of NO_2 molecules **(f)** 7.00×10^{-2} mol of $Ca(NO_3)_2$ formula units.

3-12 Express the mass of 0.400 mol of carbon tetrachloride, CCl_4, in **(a)** grams **(b)** amu.

3-13 If a molecule of X weighs 7.89×10^{-23} g, what is the molecular weight (in amu) of X?

Formula stoichiometry

3-14 Each of the following is a correctly written molecular formula. In each case write the empirical formula for the substance: H_2O_2, P_4O_{10}, N_2O_4, C_4H_{10}, O_3.

3-15 A sample of a certain compound contains

7.0×10^{22} mol of sodium atoms, 3.5×10^{22} mol of sulfur atoms, and 1.4×10^{23} mol of oxygen atoms. What is the empirical formula of the compound?

3-16 A sample consisting of 7.5×10^{20} molecules of cyclohexane contains 4.5×10^{21} carbon atoms and 9.0×10^{21} hydrogen atoms. What is **(a)** the empirical (simplest) formula and **(b)** the molecular formula of cyclohexane?

3-17 A sample of acetaldehyde, C_3H_6O, consists of 1.6 moles of acetaldehyde molecules. How many moles of **(a)** C atoms, **(b)** H atoms, and **(c)** O atoms are present?

3-18 Ferrocene is a compound consisting of iron, carbon, and hydrogen. The molecule contains one iron atom and equal numbers of carbon and hydrogen atoms. A sample of ferrocene was found to contain 7.4×10^{-2} mol of iron and 7.4×10^{-1} mol of carbon. What is the molecular formula of ferrocene?

3-19 How many moles of hydrogen atoms are necessary to combine with 7.0×10^{-4} mol of nitrogen atoms to form the compound ammonia, NH_3?

3-20 Given eight atoms of oxygen, tell how many molecules of each of the following compounds it would be possible to make **(a)** N_2O **(b)** NO **(c)** NO_2 **(d)** N_2O_4.

3-21 Given 0.12 mol of carbon atoms, tell how many moles of each of the following molecules it would be possible to prepare **(a)** CH_4 **(b)** C_4H_8 **(c)** C_4H_{10} **(d)** $C_4H_{12}O_6$.

3-22 Given 25.0 g of carbon, tell how many grams of each of the following could be made: **(a)** C_2H_6 **(b)** C_6H_6 **(c)** Na_2CO_3 **(d)** C_3O_2.

3-23 Mercuric cyanate ("fulminate of mercury") is used as a primer in small-caliber ammunition. Its analysis is 70.48 percent Hg, 8.44 percent C, 9.84 percent N, and 11.24 percent O by mass. What is the empirical formula of mercuric cyanate?

3-24 Ammonium perchlorate has been used in some solid rocket-fuel formulations. Its percentage composition is 11.92 percent N, 3.43 percent H, 30.18 percent Cl, and 54.47 percent O. What is the empirical formula of ammonium perchlorate?

3-25 Progesterone is a common component of birth-control pills. If the molecular formula for progesterone is $C_{21}H_{30}O_2$, what is its elemental analysis (percentage composition by mass)?

3-26 Ascorbic acid, vitamin C, is $C_6H_8O_6$. What is its elemental analysis (percentage composition by mass)?

3-27 Aluminum sulfate, $Al_2(SO_4)_3$ is used in large quantities in the manufacture of paper. What is the elemental analysis (percentage composition by mass) of aluminum sulfate?

3-28 The elemental analysis of acetylsalicylic acid, or aspirin, is 60.0 percent C, 4.48 percent H, and 35.5 percent O. If the molecular weight of aspirin is 180.2, what is its molecular formula?

3-29 The compound *para*-dichlorobenzene has been often used as mothballs. It analyzes: 49.02 percent C, 2.743 percent H, and 48.24 percent Cl, and its molecular weight is 147.0. What is its molecular formula?

3-30 Butadiene is a compound which is used in manufacturing some kinds of synthetic rubber. It is composed of carbon and hydrogen only, and when it is burned in excess oxygen the only products are carbon dioxide and water. When a certain sample of butadiene is burned, 0.325 g of CO_2 and 0.0998 g of H_2O are formed. **(a)** What is the empirical formula of butadiene? **(b)** How many grams of butadiene were burned? **(c)** If the molecular weight of butadiene is 54.1, what is its molecular formula?

3-31 Methanol (wood alcohol) is a compound of carbon, hydrogen, and oxygen only. When 0.375 g of methanol is burned, 0.516 g of carbon dioxide and 0.421 g of water are formed. There are no other products. **(a)** What is the empirical formula of methanol? **(b)** If its molecular weight is 32.0, what is the molecular formula of methanol?

Chemical equations

3-32 Balance each of the following equations using integral coefficients:
(a) $C_2H_4 + O_2 \longrightarrow CO_2 + H_2O$
(b) $C_6H_6 + O_2 \longrightarrow CO_2 + H_2O$
(c) $C_6H_{14} + O_2 \longrightarrow CO_2 + H_2O$
(d) $C_2H_6 + H_2O \longrightarrow CO + H_2$
(e) $C_3H_8 + H_2O \longrightarrow CO + H_2$
(f) $C_2H_6 + O_2 \longrightarrow CO + H_2O$

3-33 Balance each of the following equations using integral coefficients:
(a) $Fe_2O_3 + C \longrightarrow Fe + CO$
(b) $Fe_2O_3 + CO \longrightarrow Fe + CO_2$
(c) $Fe_2O_3 + C \longrightarrow Fe + CO_2$
(d) $Fe_2O_3 + CO \longrightarrow FeO + CO_2$
(e) $Fe_2O_3 + H_2 \longrightarrow Fe + H_2O$
(f) $Fe_3O_4 + H_2 \longrightarrow Fe + H_2O$

3-34 Balance each of the following equations using integral coefficients:
(a) $KClO_3 + S \longrightarrow KCl + SO_2$
(b) $P_4 + Cl_2 \longrightarrow PCl_3$
(c) $KClO_3 \longrightarrow KCl + O_2$
(d) $KClO_3 \longrightarrow KClO_4 + KCl$
(e) $Al + O_2 \longrightarrow Al_2O_3$
(f) $NH_3 + O_2 \longrightarrow NO + H_2O$

(g) $Hg_2CrO_4 \longrightarrow Hg + Cr_2O_3 + O_2$
(h) $HI + H_2SO_4 \longrightarrow I_2 + H_2S + H_2O$
(i) $KClO_3 + HCl \longrightarrow KCl + ClO_2 + Cl_2 + H_2O$
(j) $B + KOH \longrightarrow K_3BO_3 + H_2$
(k) $C_8H_{18}O_4 + O_2 \longrightarrow CO_2 + H_2O$

Equation stoichiometry

3-35 Sulfur dioxide, SO_2, reacts with oxygen gas to form sulfur trioxide, SO_3:

$$2SO_2(g) + O_2(g) \longrightarrow 2SO_3(g)$$

How many SO_3 molecules could be made via this reaction from **(a)** 2 SO_2 molecules **(b)** 28 SO_2 molecules **(c)** 1 O_2 molecule **(d)** 38 O_2 molecules **(e)** 2 SO_2 and 1 O_2 molecules **(f)** 16 SO_2 and 7 O_2 molecules **(g)** 38 SO_2 and 21 O_2 molecules.

3-36 Refer to the reaction of Prob. 3-35. Tell how many *moles* of SO_3 can be made via this reaction from **(a)** 0.100 mol of SO_2 **(b)** 0.100 mol of O_2 **(c)** 22.2 g of SO_2 **(d)** 14.6 g of O_2.

3-37 Refer to the reaction of Prob. 3-35. How many *grams* of SO_3 can be made via this reaction from **(a)** 0.275 mol of SO_2 **(b)** 1.43 mol of O_2 **(c)** 10.0 g of O_2 **(d)** 4.63×10^{22} O_2 molecules.

3-38 Under appropriate conditions nitrogen will react with hydrogen to form ammonia:

$$N_2(g) + 3H_2(g) \longrightarrow 2NH_3(g)$$

From each of the following specified quantities of reactants determine the maximum number of NH_3 molecules which could be made via this reaction:
(a) 100 N_2 molecules and 400 H_2 molecules
(b) 400 N_2 molecules and 300 H_2 molecules
(c) 145 N_2 molecules and 381 H_2 molecules
(d) 137 N_2 molecules and 438 H_2 molecules.

3-39 Refer to the reaction of Prob. 3-38. From each of the following specified quantities of reactants determine the maximum number of *grams* of NH_3 which could be produced via this reaction: **(a)** 0.300 mol of N_2 and 0.300 mol of H_2 **(b)** 0.300 g of N_2 and 0.300 g of H_2 **(c)** 1.82 g of N_2 and 0.364 g of H_2.

3-40 Suppose 0.26 mol of iron atoms is reacted with 0.40 mol of oxygen atoms to form the product ferric oxide, Fe_2O_3. Which element is left over in excess, and by how much?

3-41 Hydrogen gas reacts with ferric oxide, Fe_2O_3, at elevated temperatures to form water vapor and iron. In order to produce 825 g of iron by this reaction, how many grams of **(a)** iron oxide and **(b)** hydrogen are needed?

3-42 When heated to a very high temperature, limestone (calcium carbonate, $CaCO_3$) decomposes to form quicklime (solid calcium oxide, CaO) and gaseous carbon dioxide. A crucible containing some limestone weighs 30.695 g. It is heated strongly to decompose all the limestone. After cooling to room temperature, it is found to weigh 30.141 g. What is the mass of the crucible?

3-43 Calcium carbonate ($CaCO_3$) and magnesium carbonate ($MgCO_3$) each decompose when strongly heated to form calcium oxide [CaO(*s*)] and magnesium oxide [MgO(*s*)], respectively. In each case the only other product is gaseous carbon dioxide. A mixture of the two carbonates weighing a total of 15.22 g is strongly heated, cooled, and then found to weigh 8.29 g. What is the percent $CaCO_3$ in the mixture?

4 IDEAL GASES

TO THE STUDENT

This chapter focuses on one of the three states of matter—gases. Why do we consider gases first? The main reason is that gases are in many respects much simpler to study than are solids or liquids. For instance, gases are remarkably uniform in their behavior, so uniform that we find it useful to imagine an *ideal gas* and then to compare the properties of gases which really exist, *real gases,* with those of the hypothetical ideal gas. Most real gases, it turns out, show approximately ideal-gas behavior, especially if the pressure is not too high and the temperature not too low. This chapter is about such gases.

This chapter provides a good illustration of how a theory is used to account for observed behavior. *Kinetic-molecular theory* is not complicated, can be used to describe gas behavior both qualitatively and quantitatively, and uses a simple physical model to make gas behavior seem quite reasonable.

**4-1
Observing gas
behavior**

For a sample of gas consisting of a certain number of moles of molecules there are three measurable quantities, or *variables,* which are mathematically related to each other. These are *volume, pressure,* and *temperature.* Before developing a relationship among these variables, we need to consider the meaning of each of these terms.

Volume

Because a gas spontaneously expands to fill its container completely, the volume of a gas, the volume *occupied by* a gas, is the capacity of the container enclosing it. The SI base unit of length (Sec. 1-6) is the meter (m), and so the fundamental unit of volume is the *cubic meter* (m^3), a rather large unit. A *cubic decimeter* (dm^3), based on the decimeter (one-tenth of a meter), is a more convenient unit of volume for our purposes. It is usually called by another name: *liter.*[1] For smaller

---

[1]Today a liter is defined as being exactly equal to a cubic decimeter. (Prior to 1964, 1 liter equaled 1.000028 dm^3.)

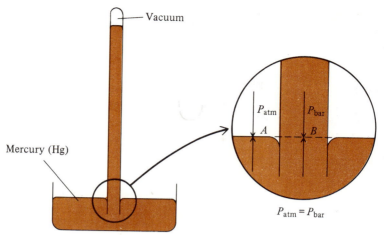

Figure 4-1

The barometer. At point A the atmospheric pressure (arrow down) is equal to the pressure ex-erted up by the mercury (arrow up). (If it were not, the mercury would move.) At point B the pressure exerted downward by the column of mercury (arrow down) is similarly equal to the up-ward pressure (arrow up). But the pressure upward at point B must equal that upward at point A, because A and B are at the same level. Therefore atmospheric pressure P_{atm} equals the pressure produced by the column of mercury in the barometer, P_{bar}.

volumes we will use the *cubic centimeter* (cm³), which is the same as the *milliliter* (ml).

Pressure   Pressure is defined as *force per unit area,* that is, the total force on a surface divided by the area of that surface. The fundmental SI derived unit of pressure is the *pascal* (Pa), which is one newton of force per square meter of area.[2] For the purpose of expressing gas pressures, however, the pascal is not used much in chemistry. More common are the *standard atmosphere* and the *millimeter of mercury*. The standard atmosphere, or, more simply, atmosphere (atm), is the pressure exerted at 0 degrees Celsius by liquid mercury at a depth of 760 millime-ters (76 cm) below its surface. This pressure is called *one atmosphere* because it is roughly equal to atmospheric pressure at sea level. Atmospheric pressure con-stantly changes, but when it is exactly 1 atm, the height of a column of mercury in a *barometer* at 0°C is exactly 760 mm. A mercurial barometer is shown in Fig. 4-1.

We will also express gas pressures in *millimeters of mercury* (mmHg), where 1 millimeter of mercury is defined as $\frac{1}{760}$ of an atmosphere. One millimeter of mercury is also known as one *torr*.[3]

Closely related to the barometer is the *manometer,* a device used to measure the pressure of a gas. One type of simple manometer is the *open-end* mercury

---

[2] Blaise Pascal was a French physicist and mathematician who lived from 1623 to 1662.

[3] The unit torr is named after Evangelista Torricelli (1608–1647), the Italian physicist who invented the barometer. Not being SI-approved units, the atmosphere and mmHg (torr) may eventually be phased out.

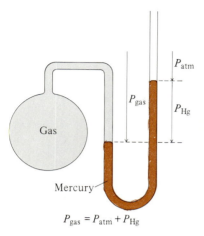

$$P_{gas} = P_{atm} + P_{Hg}$$

Figure 4-2

An open-end manometer. At the level of the lower broken line the pressure exerted downward in the right tube is $P_{atm} + P_{Hg}$, where P_{Hg} is simply the length of the column of mercury (mmHg) between the broken lines. In the left tube the gas exerts the same pressure downward, providing the basis for the stated equality.

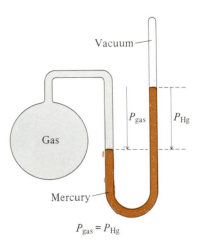

$$P_{gas} = P_{Hg}$$

Figure 4-3

A closed-end manometer. The pressure of the gas, P_{gas}, is equal to the pressure P_{Hg} exerted downward by the mercury in the right column between the broken lines. This pressure is expressed as the height of this column, that is, in millimeters of mercury.

manometer, shown in Fig. 4-2. With such a manometer the pressure of the gas equals atmospheric pressure in millimeters of mercury plus the difference, in millimeters, between the heights of the two liquid levels. In the drawing the mercury level is higher in the right-hand tube than in the left, because the pressure of the gas is *greater than* that of the atmosphere. At the level of the lower broken line the pressure exerted downward in the right-hand tube is $P_{atm} + P_{Hg}$, where P_{Hg} is the pressure produced by the column of mercury above that point as measured by the length of that column. In the left-hand tube the gas exerts the same pressure downward (at the same level); so we can write

$$P_{gas} = P_{atm} + P_{Hg}$$

If the mercury level in the right-hand tube of the manometer in Fig. 4-2 were *lower* than that in the left, this would mean that the pressure of the gas is *less* than that of the atmosphere. As before, the difference between the levels of the mercury in the two tubes is the difference between P_{gas} and P_{atm}, and so again

$$P_{gas} = P_{atm} + P_{Hg}$$

where P_{Hg}, the difference between the mercury levels, is a negative number this time. (*Note:* P_{Hg} is not a pressure; it is the *difference* between two pressures.)

Another type of manometer (Fig. 4-3) is the *closed-end* manometer. In this manometer the end of the tube is closed and the space above the mercury is evacuated. Thus, the atmosphere does not add its pressure to that exerted by the mercury in the right-hand column. Actually, in the evacuated space at the top of this manometer and at the top of a barometer, Fig. 4-1, the pressure is not quite zero because of the (extremely small) tendency of mercury to evaporate. (See Secs. 4-4 and 10-6.)

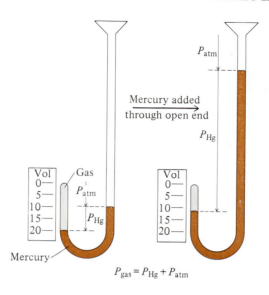

Figure 4-4
Boyle's-law apparatus. The pressure on the gas trapped in the closed end of the tube can be changed by adding more mercury through the open end. As the pressure on the gas increases, its volume decreases.

Temperature   In Sec. 2-5 we described the Celsius and Kelvin temperature scales. The measurement of temperature depends upon the measurement of any physical property which changes predictably with changing temperature. A conventional thermometer's usefulness depends on the fact that the liquid in the tube *expands* with increasing temperature. Because the length of the column of liquid in the capillary depends upon the total volume of the liquid, the stem of the thermometer can be graduated in temperature units, and the temperature read directly. Other physical properties used to measure temperature are the resistance of an electrical conductor, the electrical voltage generated at the junction of two dissimilar pieces of metal (a thermocouple), and the magnetic properties of certain materials.

**4-2
Pressure-volume
relationship:
Boyle's law**   Robert Boyle (Sec. 1-2) is best known today in connection with what is known as *Boyle's law.* In the latter half of the seventeenth century Boyle and Edme Mariotte, a French physicist, independently studied the way the volume occupied by a gas at a fixed temperature changes as the pressure on the gas is changed. Each used an apparatus similar to the one pictured in Fig. 4-4, in which a sample of gas is trapped in the closed end of a calibrated manometer. The pressure on the gas can be increased by adding mercury to the open end of the manometer, and the volume occupied by the gas is measured at each pressure.

   The results of a series of pressure-volume measurements made on hydrogen at about room temperature are summarized in Table 4-1. Two things are apparent from these data: The first is that as the pressure on the hydrogen goes up, its volume decreases. The second is that the increase in pressure and decrease in volume occur in such a way that the *product* of the pressure and volume remains constant. This constancy of the PV product was what attracted the attentions of Boyle and Mariotte. Their results can be summarized:

$$PV = k$$

where P represents the gas pressure, V is its volume, and k is an unchanging

Measurement	Pressure, mmHg	Volume, ml	Pressure × volume, mmHg ml
I	700	25.0	1.75×10^4
II	830	21.1	1.75×10^4
III	890	19.7	1.75×10^4
IV	1060	16.5	1.75×10^4
V	1240	14.1	1.75×10^4
VI	1510	11.6	1.75×10^4

number (a constant) and has a fixed value for any given (fixed) temperature and amount (number of moles) of gas. Letting n represent the number of moles of gas and T, the temperature, we can write

$$PV = k \qquad \text{(at constant } T, n)$$

This relationship is known as *Boyle's law*.[4]

Boyle's law can be stated in an alternative way. Dividing both sides of the above equality by P, we obtain

$$V = k \frac{1}{P}$$

The symbol k is a constant of proportionality, and so in words this means that the volume of a fixed quantity of a gas is *inversely proportional* to its pressure at constant temperature. (Pressure and volume are inversely proportional to each other.)

ADDED COMMENT

The concept of inverse proportionality should not cause you any grief. Consider a fixed quantity (fixed number of moles) of gas at a constant temperature:

If we *double* the pressure, the volume will be *halved*.

If we *triple* the pressure, the volume will change to *one-third its original value*.

If we *quadruple* (4×) the pressure, the volume will decrease to *one-fourth its starting value*.

If the pressure is *increased to* $\frac{7}{4}$ its original value, the volume will *decrease to* $\frac{4}{7}$ what it was at the start.

Decreasing the pressure by a factor of $\frac{700}{760}$ will result in an *increase* in the volume by a factor of $\frac{760}{700}$.

All of these statements are implied by the simple *inverse* proportionality between pressure and volume.

Boyle's law and the ideal gas

Very precise measurements show that the PV product for hydrogen at 20°C is *not quite* constant. (Hydrogen does not follow Boyle's law "exactly.") Other gases

---

[4] In much of continental Europe it is known as *Mariotte's law*.

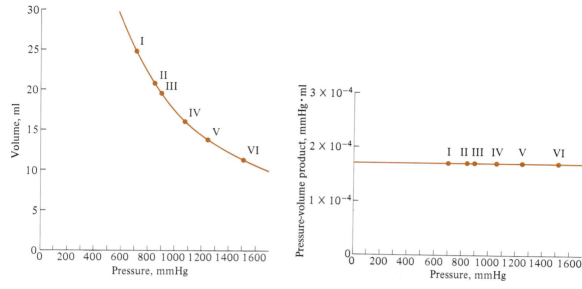

Figure 4-5
Boyle's law: *P* versus *V*.

Figure 4-6
Boyle's law: *PV* versus *P*.

deviate slightly from Boyle's law in their PV behavior, too. Further, it can be shown that the extent of deviation from Boyle's law behavior is larger at low temperatures and high pressures. Since *all gases approximate Boyle's law behavior at low pressures and high temperatures,* we find it convenient to speak of the *ideal gas,* a hypothetical gas which obeys Boyle's law *exactly* at all temperatures and pressures. This allows us to say that the *PV* behavior of a *real gas* (one which really exists) *approaches that of the ideal gas as the temperature is raised or as the pressure is lowered.*

Boyle's law behavior is but part of the definition of the ideal gas; other parts of that definition will follow shortly. For many gases near room temperature and normal atmospheric pressure the deviations from ideal-gas behavior are not large. We will consider the nature and causes of these deviations in Sec. 11-1.

Boyle's law: graphical representations

The data from Table 4-1 can be plotted on a graph such as the one shown in Fig. 4-5. For each of the six measurements the pressure is plotted on the horizontal axis (the abscissa) and the corresponding observed volume on the vertical axis (the ordinate). The line connecting the points is a curve known as a *hyperbola,* a fact which we could have predicted from the form of the equation $PV = k$.[5]

Another graphical representation of Boyle's-law behavior is shown in Fig. 4-6, where the pressure-volume product, PV, is plotted on the ordinate against the pressure on the abscissa. A straight line with no slope is the result, because the PV product is constant.

---

[5] Figure 4-5 actually shows only one branch of the hyperbola. The other branch is not physically significant, corresponding to negative pressures and volumes.

What does "at constant temperature" signify? It means that even if there should be a tendency for a gas to become cooler or warmer as its pressure and volume change, it is prevented from doing so by being enclosed in a constant-temperature device, often a water bath. For small changes in pressure the air of the room may serve to hold the temperature of the apparatus essentially constant.

Boyle's-law calculations

● **EXAMPLE 4-1**

Problem   An ideal gas is enclosed in a Boyle's-law apparatus. Its volume is 247 ml at a pressure of 625 mmHg. If the pressure is increased to 825 mmHg, what will be the new volume occupied by the gas if the temperature is held constant?

Solution   Method 1: According to Boyle's law, the product of the pressure of a gas times its volume remains unchanged even though the pressure and volume individually change. Using subscripts 1 and 2 to represent the *initial* and *final* states, respectively, we can write

$$P_1 V_1 = P_2 V_2$$

or, solving for V_2, the final volume

$$V_2 = \frac{P_1 V_1}{P_2}$$

$$= \frac{(625 \text{ mmHg})(247 \text{ ml})}{825 \text{ mmHg}} = 187 \text{ ml}$$

Method 2: The pressure of the gas increases from 625 to 825 mmHg, or in other words, by a factor of $\frac{825}{625}$. Because volume and pressure are *inversely* proportional, the volume must *decrease* by a factor of $\frac{625}{825}$. Defining V_1 and V_2 as in Method 1 above, we can write

$$V_2 = V_1 \times \text{(ratio of pressures)}$$

$$= 247 \text{ ml} \times \frac{625 \text{ mmHg}}{825 \text{ mmHg}} = 187 \text{ ml}$$

What we have done here is to multiply the original volume (247 ml) by the *ratio* of the two pressures. In this case we know ahead of time that the pressure *increase* will cause a volume *decrease,* so we make the ratio of pressures a fraction *less than* 1, that is, we put the *smaller* number (625) in the *numerator* and the *larger* (825) in the *denominator*. This makes the final calculated volume come out less than 247 ml. ●

● **EXAMPLE 4-2**

Problem   Suppose 4.63 liters of an ideal gas at 1.23 atm is expanded at constant temperature until the pressure is 4.14×10^{-2} atm. What is the final volume of the gas?

Solution   Method 1: $$P_1 V_1 = P_2 V_2$$

$$V_2 = \frac{P_1 V_1}{P_2}$$

$$= \frac{(1.23 \text{ atm})(4.63 \text{ liters})}{4.14 \times 10^{-2} \text{ atm}} = 138 \text{ liters}$$

Method 2: The pressure *decreases* by a factor of $4.14 \times 10^{-2}/1.23$. (Note that we do not need to convert from atmospheres to any other unit; we do need the ratio of the two pressures expressed in the *same* units, however.) The volume accordingly *increases* by a factor of $1.23/4.14 \times 10^{-2}$.

$$V_2 = 4.63 \text{ liters} \times \frac{1.23 \text{ atm}}{4.14 \times 10^{-2} \text{ atm}} = 138 \text{ liters}$$

Notice that the ratio of pressures is a fraction *greater* than 1, ensuring that our final volume will turn out to be greater than 4.63 liters. ●

● **EXAMPLE 4-3**

Problem   Suppose 10.9 ml of an ideal gas at 765 mmHg is expanded at constant temperature until its volume is 38.1 ml. What is the final pressure?

Solution   Method 1:

$$P_1 V_1 = P_2 V_2$$

$$P_2 = \frac{P_1 V_1}{V_2}$$

$$= \frac{(765 \text{ mmHg})(10.9 \text{ ml})}{38.1 \text{ ml}} = 219 \text{ mmHg}$$

Method 2: The volume *increases* by a factor of $38.1/10.9$. Because we know that the pressure and volume are inversely proportional, we predict that the pressure will *decrease*, and by a factor of $10.9/38.1$.

$$P_2 = 765 \text{ mmHg} \times \frac{10.9 \text{ ml}}{38.1 \text{ ml}} = 219 \text{ mmHg}$$

(Note that $10.9/38.1$ is a fraction *less* than 1.) ●

4-3
Temperature effects:
Charles' law

In 1787 a French physicist, Jacques Charles, investigated the change in the volumes of oxygen, hydrogen, carbon dioxide, and air caused by temperature changes. Charles found that each of these gases expands the same relative amount when heated from 0 to 80°C at constant pressure. From 1802 to 1808, Joseph Gay-Lussac, also French, showed that many more gases could be added to Charles' list. More importantly, he showed that for each degree Celsius increase each gas expands $\frac{1}{273}$ of its volume at 0°C, provided that the pressure is kept constant.

Results of three experiments Charles or Gay-Lussac might have performed are shown in Fig. 4-7. Here for three different samples of hydrogen gas are plotted a series of points, each of which represents the volume of the sample at a given temperature. For each sample of gas the pressure is kept constant. Each line can be extrapolated to the point where it crosses the horizontal temperature axis, that is, where $V = 0$. The graph seems to indicate that in each case if the sample of hydrogen gas could be cooled to that temperature, -273°C, the volume of the gas would be reduced to zero, meaning, presumably, that the gas would cease to exist.

Similar experiments with *different* gases yield data which, when plotted, fall on straight lines intersecting the temperature axis at this same temperature, -273°C. Since any volume less than zero is impossible, this low temperature is

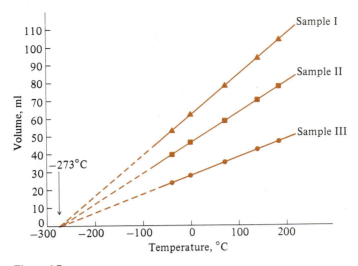

Figure 4-7
Charles' law: *V* versus *T*. Each set of points represents a series of measurements of the volume of a sample of hydrogen gas at different temperatures, pressure being held constant. Data for three samples of different mass are plotted. For each sample the points lie on a straight line, which, when extrapolated, intersects the temperature axis at −273°C.

called *absolute zero,* the lowest possible temperature. (To five significant figures absolute zero is −273.15°C.)

The equation for each straight line shown in Fig. 4-7 can be written

$$V = a(t + 273) \qquad \text{(at constant } P, n\text{)}$$

where *V* is the volume of the gas, *t* is the Celsius temperature, and *a* is the slope of the straight line. Since the temperature on the Celsius scale is related to that on the Kelvin scale by *T* (kelvins) = *t* (degrees Celsius) + 273 (Sec. 2-5), we can write[6]

$$V = aT \qquad \text{(at constant } P, n\text{)}$$

This simple relationship tells us that the volume of a fixed quantity (number of moles) of gas is *directly proportional* to its temperature on the Kelvin scale if the pressure is held constant. As was pointed out in Sec. 2-5 the Kelvin scale is a so-called *absolute temperature* scale, because its lower fixed point is the absolute zero of temperature. (Other absolute scales exist which use the same lower fixed point, but a different-sized degree.) Thus, we can say more concisely that the volume and absolute temperature of a (fixed quantity of) gas are directly proportional (at constant pressure). This is a statement of *Charles' law,* also frequently called *Gay-Lussac's law.*

Charles' law and the ideal gas   As was the case with Boyle's law, real gases show typical Charles' law behavior only at higher temperatures and lower pressures. Charles' law is an exact de-

---

[6] In this book we will consistently use a (capital) *T* to represent temperature on the Kelvin scale and a (lowercase) *t* for temperature on the Celsius scale.

scription only of the behavior of the hypothetical *ideal gas,* but understanding this behavior gives us something with which to compare the behavior of real gases (Sec. 11-1). As a real gas is cooled at constant pressure starting at a temperature well above its condensation point, its volume at first decreases linearly as is shown in Fig. 4-7. As the temperature approaches the condensation point, the line begins to curve (downward, under most conditions), that is, the gas does not behave ideally when its temperature is close to its condensation point. Of cource the gas ceases to be a gas when it condenses, and as a liquid its behavior is nothing at all like that described by Charles' law. Finally, the complete disappearance of a gas at absolute zero, as predicted by Charles' law, is something we need not concern ourselves about, since all gases condense at temperatures above absolute zero. Absolute zero is just the temperature at which the volume of a gas *would* become zero, *if* it did not condense and *if* it behaved ideally down to that temperature.

Charles' law calculations ● **EXAMPLE 4-4**

Problem   An ideal gas occupies a volume of 1.28 liters at 25°C. If the temperature is raised to 50°C, what is the new volume of the gas if the pressure remains constant?

Solution   Method 1: For a given quantity of gas at constant pressure,

$$V = aT$$

Therefore,

$$\frac{V}{T} = a$$

Since a is a constant, this means that the V/T ratio remains constant, even though V and T individually change. If, as before, we let the subscripts 1 and 2 represent *initial* and *final* states, respectively, we have

$$\frac{V_1}{T_1} = \frac{V_2}{T_2}$$

Solving for V_2,

$$V_2 = \frac{V_1 T_2}{T_1}$$

The *absolute* temperatures are $T_1 = 25 + 273 = 298$ K and $T_2 = 50 + 273 = 323$ K. Thus, we have

$$V_2 = \frac{(1.28 \text{ liters})(323 \text{ K})}{298 \text{ K}} = 1.39 \text{ liters}$$

Method 2: Volume and *absolute* temperature are directly proportional, so we first find the Kelvin temperatures: 25°C is 25 + 273, or 298 K, and 50°C is 50 + 273, or 323 K. Temperature *increases* by a factor of $\frac{323}{298}$. Volume and temperature are *directly* proportional; so the volume will also *increase* by a factor of $\frac{323}{298}$. (The gas *expands* when it is heated.)

$$V_2 = 1.28 \text{ liters} \times \frac{323 \text{ K}}{298 \text{ K}} = 1.39 \text{ liters}$$

Here we have multiplied the initial volume (1.28 liters) by the ratio of temperatures, putting the *larger* number in the *numerator* of the ratio. ●

● **EXAMPLE 4-5**

Problem   A sample of ideal gas has a volume of 128 cm³ at −27°C. To what temperature must the gas be heated at constant pressure if the final volume is to be 214 cm³?

Solution   Method 1: First we convert the given temperature to kelvins: −27°C is (−27 + 273), or 246 K. In Method 1 of Example 4-4 we showed that in an ideal-gas expansion at constant pressure the V/T ratio remains constant. Thus we write, as before:

$$\frac{V_1}{T_1} = \frac{V_2}{T_2}$$

$$T_2 = \frac{V_2 T_1}{V_1}$$

$$= \frac{(214 \text{ cm}^3)(246 \text{ K})}{128 \text{ cm}^3} = 411 \text{ K}$$

If we wish to express the final temperature in degrees Celsius,

$$t_2 = 411 - 273 = 138°C$$

Method 2: $T_1 = -27 + 273 = 246$ K. At constant pressure the only way to *expand* a gas is to *raise* its temperature. We thus multiply the initial temperature by a ratio of volumes which is greater than one, that is, by $\frac{214}{128}$.

$$T_2 = 246 \text{ K} \times \frac{214 \text{ cm}^3}{128 \text{ cm}^3} = 411 \text{ K}$$

$$t_2 = 411 - 273 = 138°C \quad ●$$

***Combined
calculations***   The same approach can be used to calculate a change in volume accompanying a change of pressure *and* temperature or, for that matter, a change in any one of the three variables brought about by changes in the other two.

● **EXAMPLE 4-6**

Problem   Suppose 2.65 liters of an ideal gas at 25°C and 1.00 atm is simultaneously warmed and compressed until the final temperature is 75°C and the final pressure is 2.00 atm. What is the final volume?

Solution   Method 1: The inverse proportionality between volume and pressure expressed by Boyle's law can be written

$$V \propto \frac{1}{P} \qquad \text{(at constant } T, n)$$

where $\propto$ means *is proportional to*. The direct proportionality between volume and temperature expressed by Charles' law can be similarly written:

$$V \propto T \qquad \text{(at constant } P, n)$$

Combining these two statements of proportionality we have

$$V \propto \frac{T}{P} \qquad \text{(at constant } n)$$

or $$V = c \frac{T}{P}$$

where c is a constant of proportionality. The last equation rearranges to

$$\frac{PV}{T} = c$$

which is sometimes referred to as the *combined gas law.* Thus, when pressure, volume, and temperature *all* change, the ratio of PV to T remains a constant, and so we can write

$$\frac{P_1 V_1}{T_1} = \frac{P_2 V_2}{T_2}$$

In this problem the initial and final pressures, temperatures, and volumes are summarized below:

	P	T	V
Initial	1.00 atm	25 + 273 = 298 K	2.65 liters
Final	2.00 atm	75 + 273 = 348 K	?

Solving the above equation for V_2 we have

$$V_2 = \frac{P_1 V_1 T_2}{P_2 T_1}$$

$$= \frac{(1.00 \text{ atm})(2.65 \text{ liters})(348 \text{ K})}{(2.00 \text{ atm})(298 \text{ K})} = 1.55 \text{ liters}$$

Method 2: The final volume will be the same no matter how the temperature and pressure are increased to their final values. In other words, it makes no difference whether the temperature changes first, the pressure changes first, or they both change simultaneously. In each case the final volume is the same. As long as that is the case, we will imagine that first the temperature rises, and then the pressure is increased. Following the logic and procedure of Method 2 in each of the previous examples we reason that an *increase* in temperature alone will cause an *increase* in volume:

$$V = 2.65 \text{ liters} \times \frac{348 \text{ K}}{298 \text{ K}} = 3.09 \text{ liters}$$

Now what happens to the 3.09 liters when the pressure is increased? A pressure *increase* will cause the volume to *decrease:*

$$V_2 = 3.09 \text{ liters} \times \frac{1.00 \text{ atm}}{2.00 \text{ atm}} = 1.55 \text{ liters}$$

This calculation can be set up more conveniently in a single step

$$V_2 = \underbrace{2.65 \text{ liters}}_{\substack{\text{Initial} \\ \text{volume}}} \times \underbrace{\frac{348 \text{ K}}{298 \text{ K}}} \times \frac{1.00 \text{ atm}}{2.00 \text{ atm}} = 1.55 \text{ liters}$$

Volume after
temperature change only

(This is the final volume after temperature *and* volume change) ●

● **EXAMPLE 4-7**

Problem   A sample of an ideal gas occupies a volume of 68.1 ml at 945 torr and 18°C. What volume would it occupy at 118°C and 745 torr?

Solution

	P	T	V
Initial	945 torr	18 + 273 = 291 K	68.1 ml
Final	745 torr	118 + 273 = 391 K	?

Method 1:

$$\frac{P_1 V_1}{T_1} = \frac{P_2 V_2}{T_2}$$

$$V_2 = \frac{P_1 V_1 T_2}{P_2 T_1}$$

$$= \frac{(945 \text{ torr})(68.1 \text{ ml})(391 \text{ K})}{(745 \text{ torr})(291 \text{ K})} = 116 \text{ ml}$$

Method 2: Temperature *increases* by a factor of $\frac{391}{291}$; so volume will also *increase* by a factor of $\frac{391}{291}$. Pressure *decreases* by a factor of $\frac{745}{945}$; so volume will *increase* by a factor of $\frac{945}{745}$.

$$V_2 = 68.1 \text{ ml} \times \frac{391 \text{ K}}{291 \text{ K}} \times \frac{945 \text{ torr}}{745 \text{ torr}} = 116 \text{ ml} \quad \bullet$$

Using similar techniques we can solve problems in which the temperature or pressure is sought.

4-4
Ideal-gas behavior

Avogadro's principle

When hydrogen and oxygen gases react to form gaseous water, a simple relationship exists among the volumes of the reactants and of the product, if these volumes are all measured at the same temperature and pressure:

$$\text{Two volumes of hydrogen} + \text{one volume of oxygen} \longrightarrow \text{two volumes of water}$$

Compare this with the balanced equation for the reaction:

$$2H_2(g) + O_2(g) \longrightarrow 2H_2O(g)$$

The relationship among the volumes is an example of *Gay-Lussac's law of combining volumes,* which states: *When measured under the same conditions of temperature and pressure, the volumes of gaseous reactants and products of a reaction are in ratios of small integers.* A second example of this law is found in the reaction by which ammonia gas is synthesized from nitrogen and hydrogen gases. The relative volumes (all measured at the same temperature and pressure) are

$$\text{One volume of nitrogen} + \text{three volumes of hydrogen} \longrightarrow \text{two volumes of ammonia}$$

and the equation for this reaction is

$$N_2(g) + 3H_2(g) \longrightarrow 2NH_3(g)$$

Why is the relationship among the *volumes* exactly the same as the relationship among the *numbers of molecules* of the different substances, as shown by the coefficients in the balanced equation? The answer is that *equal volumes of different gases contain equal numbers of molecules when measured at the same pressure and temperature.* This suggestion was first made by Avogadro in 1811 and is now known as *Avogadro's principle.*[7]

By means of Avogadro's principle we can see that the ratio of the number of molecules in two volumes of gas (measured at the same temperature and pressure) is the same as the ratio of the volumes themselves. In other words at constant temperature and pressure the volume of a sample of gas is proportional to the number of molecules in the sample. Furthermore, since the molecules may be counted by the *mole,* we can say that at constant temperature and pressure the volume of a gas sample is proportional to the number of moles present, or

$$V \propto n \qquad \text{(at constant } T, P)$$

The ideal-gas law Boyle's law, Charles' law, and Avogadro's principle are all statements of proportionality which describe ideal gases. We can summarize these as

Boyle's law: $V \propto \dfrac{1}{P}$ (at constant T, n)

Charles' law: $V \propto T$ (at constant P, n)

Avogadro's principle: $V \propto n$ (at constant T, P)

By combining the three proportionalities, we get

$$V \propto \frac{1}{P} Tn$$

That this proportionality is equivalent to the previous three can be seen by holding constant any *two* of the quantities, P, T, and n, and then seeing how the volume depends on the remaining variable.

If we now rewrite the above proportionality as an *equality*, we can see that

$$V = R \frac{1}{P} Tn$$

where R is a constant of proportionality. This equality is usually written in the form

$$PV = nRT$$

and is known as the *ideal-gas law* or *perfect-gas law.*[8]

Determination of R In order to use the ideal-gas law we need to know the numerical value of R. To determine it we might take PVT data for any gas and calculate R from

$$R = \frac{PV}{nT}$$

---

[7] This has also been called *Avogadro's law* or *Avogadro's hypothesis.* The use of the term *law* is not appropriate here, however, since Avogadro's statement was not a summary of observed facts but was rather a suggested explanation for Gay-Lussac's law of combining volumes.

[8] It is also known as the *equation of state for an ideal gas.* An equation of state is an equation which relates variables which can be used to define or describe the state of a system. (Here we are using *state* to mean the exact condition of a system, not to indicate whether it is solid, liquid, or gas.) Specifying all but one of the variables in an equation of state uniquely fixes the last.

Table 4-2
Calculation of R for different gases at 0°C and 1 atm

Gas	n, mol	V, liters	$\dfrac{PV}{nT}$, liter atm K^{-1} mol^{-1}
Helium	0.13641	3.0591	0.082101
Oxygen	0.12394	2.7755	0.081984
Methane	0.10912	2.4399	0.081859
Carbon dioxide	0.12722	2.8441	0.081844
Ethylene	0.11398	2.5350	0.081423

Values of R calculated this way for several gases at 0°C and 1 atm pressure are shown in Table 4-2. Note that the calculated values of R are not all the same. The reason for this is that these gases differ slightly in PVT behavior, and none is truly ideal.

How do we find an accurate value of R? If we really had an ideal gas to work with, we could calculate R from PVT measurements on it. Unfortunately, the ideal gas is imaginary. The way of this dilemma is to make a series of PVT measurements on a real gas at progressively lower pressures. We then find that PV/nT approaches the value 0.082057 liter atm K^{-1} mol^{-1}, as the pressure approaches zero. Furthermore, we obtain the same result for *all* gases, showing that real-gas behavior approaches that of an ideal gas as the pressure becomes progressively lower.

The symbol R is known as the *ideal-*, or *universal-*, *gas constant*. We will normally need to use it expressed to only three significant figures:

$$R = 0.0821 \text{ liter atm K}^{-1}\text{mol}^{-1}$$

R can also be expressed in other units, some of which are given in App. B.

ADDED COMMENT

The above units of R may seem strange at first, but you should soon get used to them. As long as pressure is expressed in atmospheres, volume in liters, and temperatures in kelvins, the units of R come out to be liter-atmospheres per kelvin-mole:

$$R = \frac{PV}{nT} = \frac{(\text{atmospheres})(\text{liters})}{(\text{moles})(\text{kelvins})}$$
$$= \frac{\text{liter atm}}{\text{K mol}} = \text{liter atm K}^{-1}\text{mol}^{-1}$$

Ideal-gas law calculations

The relationship between P, V, n, and T expressed by the ideal-gas law can be used to calculate any one of these variables from the other three.

● **EXAMPLE 4-8**

Problem   Suppose 0.176 mol of an ideal gas occupies 8.64 liters at a pressure of 0.432 atm. What is the temperature of the gas in degrees Celsius?

Solution $$PV = nRT$$
$$T = \frac{PV}{nR}$$

$$= \frac{(0.432 \text{ atm})(8.64 \text{ liters})}{(0.176 \text{ mol})(0.0821 \text{ liter atm K}^{-1}\text{mol}^{-1})} = 258 \text{ K}$$

To convert to degrees Celsius we need only subtract 273 from the above result:

$$t = 258 - 273 = -15°C \quad \bullet$$

● EXAMPLE 4-9

Problem   Suppose 5.00 g of oxygen gas, O_2, at 35°C is enclosed in a container having a capacity of 6.00 liters. Assuming ideal-gas behavior, calculate the pressure of the oxygen in millimeters of mercury. (Atomic weight: $O = 16.0$.)

Solution   One mole of O_2 weighs $2(16.0) = 32.0$ g. 5.00 g of O_2 is, therefore, $5.00 \text{ g}/32.0 \text{ g mol}^{-1}$, or 0.156 mol. 35°C is $35 + 273 = 308$ K.

$$PV = nRT$$

$$P = \frac{nRT}{V}$$

$$= \frac{(0.156 \text{ mol})(0.0821 \text{ liter atm K}^{-1}\text{mol}^{-1})(308 \text{ K})}{6.00 \text{ liters}} = 0.659 \text{ atm}$$

$$0.659 \text{ atm} \frac{760 \text{ mmHg}}{1 \text{ atm}} = 500 \text{ mmHg} \quad \bullet$$

Molar volume of an ideal gas at standard temperature and pressure   Using the ideal-gas law we can calculate what volume would be occupied by 1 mol of an ideal gas at any temperature and pressure. A reference condition which is commonly used for describing gas properties is 0°C (273.15 K) and 1.0000 atm (760.00 mmHg), called *standard temperature and pressure,* or more briefly, *STP.* The volume occupied by one mole, or *molar volume,* of an ideal gas at STP is

$$V = \frac{nRT}{P}$$

$$= \frac{(1.0000 \text{ mol})(0.082057 \text{ liter atm K}^{-1}\text{mol}^{-1})(273.15 \text{ K})}{1.0000 \text{ atm}} = 22.414 \text{ liters}$$

It is handy to remember this value to three significant figures. *Remember: One mole of an ideal gas occupies 22.4 liters at STP.* The molar volumes of real gases at STP are not far from this value, as is shown in Table 4-3.

Molecular weight from gas density   The ideal-gas law provides a convenient way of finding the molecular weight of a gas from its density.

Table 4-3
Molar volume of gases at STP

Gas	Molar volume, liter mol^{-1}
Hydrogen	22.428
Helium	22.426
Oxygen	22.394
Carbon dioxide	22.256
Ammonia	22.094
Ideal gas	22.414

84   IDEAL GASES

● **EXAMPLE 4-10**

The density of phosphine gas is 1.26 g liter^{-1} at 50°C and 747 mmHg. Assume that phosphine behaves ideally, and calculate its molecular weight.

$$50°C = 50 + 273 = 323 \text{ K}$$

$$747 \text{ mmHg} = 747 \text{ mmHg} \frac{1 \text{ atm}}{760 \text{ mmHg}} = 0.983 \text{ atm}$$

Since the density is given as 1.26 g liter^{-1}, we will use the ideal-gas law to find how many *moles* are present in 1.00 liter. Then we can find how much 1 mol weighs.

$$PV = nRT$$

$$n = \frac{PV}{RT}$$

$$= \frac{(0.983 \text{ atm})(1.00 \text{ liter})}{(0.0821 \text{ liter atm K}^{-1}\text{mol}^{-1})(323 \text{ K})} = 0.0371 \text{ mol}$$

Now we know that 1.00 liter of phosphine *weighs* 1.26 g and *consists of* 0.0371 mol. In other words, 0.0371 mol of phosphine weighs 1.26 g. Thus, we can conclude that 1 mol of phosphine weighs

$$\frac{1.26 \text{ g}}{0.0371 \text{ mol}} = 34.0 \text{ g mol}^{-1}$$

The molecular weight of phosphine is therefore 34.0. ●

ADDED COMMENT

If the last division was unexpected, look at it this way: *Suppose* it had been true that "2 mol of phosphine weigh 10 g." Then you would have divided 10 by 2 in order to get the mass of 1 mol. That is the reasoning behind the last step in the solution to the problem.

The gas-density method is but one way of experimentally determining molecular weights. It provides an experimental basis for finding molecular formulas as well, as was discussed in Sec. 3-4.

Mixtures of gases: Dalton's law of partial pressures

In 1801 John Dalton observed that different gases in a mixture seem to exert pressure on the inside of the container walls independently of each other. Thus, the measured pressure of a mixture of gases is the sum of the pressures the gases would exert if each were alone in the container. *Dalton's law of partial pressures* states that *the total pressure exerted by a mixture of gases is equal to the sum of the partial pressures of the individual gases.* A *partial pressure* is defined as the pressure a gas would exert if it were the only gas in the container. As an illustration consider the mixture of hydrogen and helium gases shown in Fig. 4-8. The pressure gauge in the illustration indicates a total pressure. Because each gas exerts a pressure independently of the other, the total pressure, 400 mmHg, is equal to the sum of the partial pressure of helium, 100 mmHg, and that of hydrogen, 300 mmHg. Dalton's law of partial pressures is obeyed closely by most mixtures of gases, provided that the gases do not react.

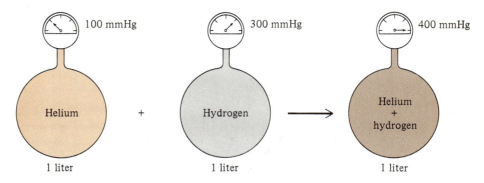

$$P_{He} + P_{H_2} = P_{total} \quad \text{(at constant T)}$$

Figure 4-8
Dalton's law of partial pressures. One liter each of helium (at 100 mmHg) and hydrogen (at 300 mmHg) are transferred together to a third 1-liter container. If the temperatures of each gas and of the mixture are all the same, the pressure of the mixture is 400 mmHg.

● **EXAMPLE 4-11**

Problem   Suppose 1.00 g each of H_2, O_2, and N_2 gases are placed together in a 10.0-liter container at 125°C. Assume ideal behavior, and calculate the total pressure in atmospheres. (Atomic weights: H = 1.01; O = 16.0; N = 14.0.)

Solution   First, calculate the number of moles of each gas:

$$n_{H_2} = 1.00 \text{ g} \times \frac{1 \text{ mol}}{2(1.01) \text{ g}} = 0.495 \text{ mol}$$

$$n_{O_2} = 1.00 \text{ g} \times \frac{1 \text{ mol}}{2(16.0) \text{ g}} = 0.0313 \text{ mol}$$

$$n_{N_2} = 1.00 \text{ g} \times \frac{1 \text{ mol}}{2(14.0) \text{ g}} = 0.0357 \text{ mol}$$

Second, calculate *each* partial pressure

$$T = 125 + 273 = 398 \text{ K}$$

For each gas $PV = nRT$; so

$$P_{H_2} = \frac{n_{H_2}RT}{V}$$

$$= \frac{(0.495 \text{ mol})(0.0821 \text{ liter atm K}^{-1}\text{mol}^{-1})(398 \text{ K})}{10.0 \text{ liters}} = 1.62 \text{ atm}$$

$$P_{O_2} = \frac{n_{O_2}RT}{V}$$

$$= \frac{(0.0313 \text{ mol})(0.0821 \text{ liter atm K}^{-1}\text{mol}^{-1})(398 \text{ K})}{10.0 \text{ liters}} = 0.102 \text{ atm}$$

$$P_{N_2} = \frac{n_{N_2}RT}{V}$$

$$= \frac{(0.0357 \text{ mol})(0.0821 \text{ liter atm K}^{-1}\text{mol}^{-1})(398 \text{ K})}{10.0 \text{ liters}} = 0.117 \text{ atm}$$

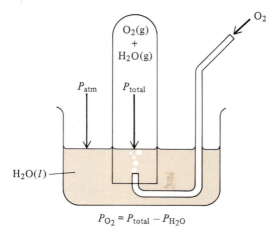

$$P_{O_2} = P_{total} - P_{H_2O}$$

Figure 4-9
Oxygen gas collected by displacement of water. When the level inside the gas-collection container is even with the level outside, $P_{total} = P_{atm}$, which can be measured using a barometer. If the temperature is known, P_{H_2O}, the vapor pressure of water, can be obtained, and thus P_{O_2} can be calculated by subtraction.

Third, add the partial pressures:

$$P_{total} = P_{H_2} + P_{O_2} + P_{N_2}$$
$$= 1.62 \text{ atm} + 0.102 \text{ atm} + 0.117 \text{ atm} = 1.84 \text{ atm}$$

Notice that the total pressure depends in actuality upon the total number of moles of molecules in the container, *irrespective of what they are*. In other words we could have set this problem up as follows:

$$n_{total} = n_{H_2} + n_{O_2} + n_{N_2}$$
$$= 0.495 \text{ mol} + 0.0313 \text{ mol} + 0.0357 \text{ mol} = 0.562 \text{ mol}$$

$$P_{total} = \frac{n_{total}RT}{V}$$

$$= \frac{(0.562 \text{ mol})(0.0821 \text{ liter atm K}^{-1}\text{mol}^{-1})(398 \text{ K})}{10.0 \text{ liters}} = 1.84 \text{ atm}$$

Gases collected over water

It is common practice in the laboratory to collect a gas by displacement of water. Figure 4-9 shows a sample of oxygen gas which has been collected this way. The gas is said to be "wet" because it is not pure, consisting of a mixture of oxygen and water molecules. When *dry* oxygen (or any other gas) is allowed to come in contact with liquid water, some of the water *evaporates;* that is, water molecules

Table 4-4
Vapor pressure of water

Temperature, °C	Vapor pressure, mmHg	Vapor pressure, atm
0	4.58	6.03×10^{-3}
5	6.54	8.61×10^{-3}
10	9.21	1.21×10^{-2}
15	12.79	1.68×10^{-2}
20	17.54	2.31×10^{-2}
25	23.76	3.13×10^{-2}
30	31.82	4.19×10^{-2}
35	42.18	5.55×10^{-2}
40	55.32	7.28×10^{-2}
45	71.88	9.46×10^{-2}
50	92.51	1.22×10^{-1}

leave the liquid and mix with the oxygen molecules. (We will examine the process of evaporation more closely in Sec. 10-6.) The water molecules continue to leave the liquid until the partial pressure of the water vapor (gas) has reached a maximum value. This value *depends only on the temperature,* is usually reached quickly, and is called the *vapor pressure* of water. The total pressure of the mixture of oxygen and water vapor equals the sum of the two partial pressures. The partial pressure of the oxygen is therefore equal to the total pressure *less* the partial pressure (the vapor pressure) of the water, or $P_{O_2} = P_{total} - P_{H_2O}$. Values of the vapor pressure of water at different temperatures are tabulated in Table 4-4, and, more extensively, in App. F.

● **EXAMPLE 4-12**

Problem   Suppose 0.157 g of a certain gas collected over water occupies a volume of 135 ml at 25°C and 745 mmHg. Assuming ideal behavior, determine the molecular weight of the gas.

Solution

$$T = 25 + 273 = 298 \text{ K}$$

At 25°C the vapor pressure of water is 23.76 mmHg (Table 4-4). Since this is the partial pressure of water in the mixture,

$$P_{gas} = P_{total} - P_{H_2O} = 745 \text{ mmHg} - 23.76 \text{ mmHg} = 721 \text{ mmHg}$$

In atmospheres, this pressure is

$$P_{gas} = 721 \text{ mmHg} \frac{1 \text{ atm}}{760 \text{ mmHg}} = 0.949 \text{ atm}$$

$$V = 135 \text{ ml} \frac{1 \text{ liter}}{1000 \text{ ml}} = 0.135 \text{ liter}$$

(This last conversion, from milliliters to liters, amounts to just moving the decimal point three places to the left, thus dividing by 1000.)

The number of moles of gas is

$$n = \frac{PV}{RT}$$

$$= \frac{(0.949 \text{ atm})(0.135 \text{ liter})}{(0.0821 \text{ liter atm K}^{-1}\text{mol}^{-1})(298 \text{ K})} = 5.24 \times 10^{-3} \text{ mol}$$

Since 5.24×10^{-3} mol of gas weighs 0.157 g, 1 mol weighs

$$\frac{0.157 \text{ g}}{5.24 \times 10^{-3} \text{ mol}} = 30.0 \text{ g mol}^{-1}$$

Therefore, the molecular weight of the gas is 30.0. ●

Graham's laws of diffusion and effusion   In 1829 Thomas Graham, an English chemist, reported the results of his observations on the rates of diffusion of various gases. *Diffusion* is the term given to the passage of a substance throughout another medium. (For example, when some foods are cooked, they evolve gases which soon *diffuse* through the air and are detected by our noses.) Graham found that *the rate of diffusion of one gas through another is inversely proportional to the square root of the density of the gas.* If we let *d* represent density, we can write

$$\text{Rate} = \text{constant} \times \frac{1}{\sqrt{d}}$$

or, for gases A and B,

$$\frac{\text{Rate}_A}{\text{Rate}_B} = \frac{\sqrt{d_B}}{\sqrt{d_A}}$$

These are statements of *Graham's law of diffusion.*

At any given temperature and pressure the density and molecular weight of an ideal gas are directly proportional, as can be shown algebraically as follows, representing density by *d* and mass by *m*:

$$d = \frac{m}{V} = \frac{m}{nRT/P} = \frac{mP}{nRT}$$

Since $n = m/M$, where M is the molecular weight,

$$d = \frac{mP}{(m/M)RT} = \frac{P}{RT}M \quad \text{or} \quad d \propto M$$

Since density and molecular weight are proportional, we can write Graham's law of diffusion as

$$\frac{\text{Rate}_A}{\text{Rate}_B} = \frac{\sqrt{M_B}}{\sqrt{M_A}}$$

In 1846 Graham reported a second observation, one relating to the effusion of gases. *Effusion* of a gas is its passage through a pinhole opening or orifice. *Graham's law of effusion* is analogous to his law of diffusion: *The rate of effusion of a gas is inversely proportional to the square root of its density, or of its molecular weight.*

Graham's laws provide another way of determining molecular weights from experimental measurements.

● EXAMPLE 4-13

Problem   The rate of effusion of an unknown gas (X) through a pinhole is found to be only 0.279 times the rate of effusion of hydrogen (H_2) gas through the same pinhole, if both gases are at STP. What is the molecular weight of the unknown gas? (Atomic weight: H = 1.01.)

Solution

$$\frac{\text{Rate}_X}{\text{Rate}_{H_2}} = \frac{\sqrt{M_{H_2}}}{\sqrt{M_X}}$$

$$\sqrt{M_X} = \frac{\text{Rate}_{H_2}}{\text{Rate}_X} \times \sqrt{M_{H_2}}$$

$$= \frac{1}{0.279} \times \sqrt{2(1.01)} = 5.09$$

$$M_X = 26.0$$

The molecular weight of the unknown gas is 26.0. ●

ADDED COMMENT

Be sure you are aware of the *inverse* proportionality. It means that the *larger* the molecular weight is, the *slower* will be the rate of diffusion or effusion of the gas. Also, note that *rate* and *time* are inversely proportional to each other. (The *higher* a car's speed, the *smaller* is its travel time to a certain destination, for example.) Thus, if the molecular weight of A is 9 times that of B, it will take 3 times ($\sqrt{9} = 3$) as long for a given number of moles of A to effuse or diffuse than for the same number of moles of B under the same conditions.

4-5
Kinetic-molecular
theory

We have at this stage discussed the more important properties of ideal gases. These properties are summarized in the concise statements known collectively as *the gas laws,* that is, Boyle's law, Charles' law, Dalton's law, Graham's laws, etc. But we have not, until now, asked the question, "Why?" In Sec. 1-3 we talked about the theory and its role in scientific methodology. A theory attempts to make observed behavior seem more reasonable, that is, to explain it in terms of some kind of model. Kinetic-molecular theory, described below, provides a good example of a theory which makes use of a physical model, sometimes called the *billiard-ball model,* of the structure of the gaseous state.

Kinetic-molecular theory, also called, more simply, *kinetic theory,* rests upon the following assumptions:

1 A gas is composed of a large number of tiny particles, molecules, which are so small that their sizes are negligible compared to the average distances between them and compared to the size of their container. Most of the measured volume of a gas is just empty space.
2 The molecules of a gas are in constant, rapid, random, straight-line motion. Because of this translational motion (Sec. 2-3), the molecules collide frequently with each other and with the walls of the container. All collisions are *elastic;* that is, there is no *net* loss (or gain) of kinetic energy at each collision. Although one molecule may lose energy to another during a collision, the *total* energy of the colliding pair remains unchanged.
3 Except during collisions, the molecules in a gas are completely independent of each other; there are no forces of attraction or repulsion between the molecules of a gas.
4 At any instant there is a wide range of molecular speeds, with some molecules moving very rapidly while others are briefly motionless. This gives rise to a similar wide range of molecular kinetic energies. The *average kinetic energy* of all the molecules is, however, *proportional to the absolute temperature.*

According to this model an instantaneous picture of a gas taken through an ultra-high-power microscope would show something like that pictured in Fig. 4-10. Here we see that a gas is indeed mostly empty space, but space which is being rapidly transversed by the tiny molecules in their chaotic motion.

Analysis of the
assumptions

How well do the above assumptions correlate with observed gas properties? Assumption 1, that most of a gas is empty space, is justified by the observation that

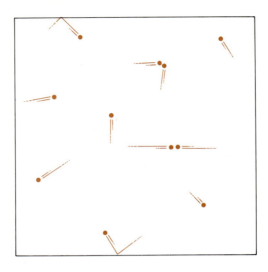

Figure 4-10
Kinetic-molecular model of a gas.

gases are very compressible. It is necessary only to double the pressure of a gas in order to halve its volume. (The small pressure increase required to halve the volume of a gas supports the notion that compression consists of merely forcing the molecules closer together by reducing the volume of space available to them.)

Assumption 2 is supported by a number of observations. First, any gas tends to expand spontaneously to fill its container completely. This could not occur, if the molecules were stationary. Second, the collisions must be elastic. If they were not, that is, if kinetic energy were lost by the entire collection of molecules as a result of collisions, the molecules would gradually slow down. This would mean a continuous decrease in temperature, and presumably the molecules would eventually settle down and sit on the bottom of the container. But a gas in a well-insulated container does not "run down" in this way; even if it did, where would the energy go? One can imagine some possibilities, such as conversion to potential, or even radiant, energy, but none of these is consistent with observations. A third support of assumption 2 comes from the observation of *brownian motion,* first reported by Robert Brown, a Scottish botanist, who in 1827 observed that tiny pollen grains suspended in water undergo a "tremulous motion," observable under the microscope. Later the same motion was observed with small particles, such as particles of smoke, suspended in air. The smaller the particle the more violent is the jerky, zigzag movement. The phenomenon of brownian motion directly supports the notion of moving molecules, as is shown in Fig. 4-11. If at one instant a small particle, shown greatly enlarged in Fig. 4-11*a*, is bombarded by a larger number of rapidly moving molecules from the left than from the right, it will be seen to jump suddenly to the right. If an instant later it is hit by more from the right, it will jump to the left, as is shown in Fig. 4-11*b*. The rapid, zigzags of brownian motion are a result of the unequal bombardment of the small particle by molecules coming from opposite directions.

Like assumption 2, assumption 3 is supported by the fact that a gas tends to expand spontaneously. Gases at both high and low pressures expand to fill their containers completely.

Let us take a close look at assumption 4: There is a distribution of molecular

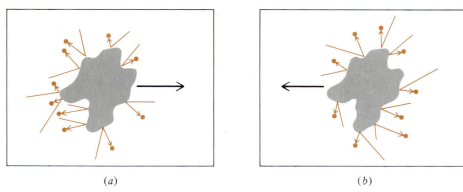

(a)   (b)

Figure 4-11
Brownian motion. (a) More molecules bombard the particle from the left than from the right; the particle moves to the right. (b) An instant later the situation is reversed.

kinetic energies in a gas. That distribution is represented graphically in Fig. 4-12 for a gas at three different temperatures. Notice how even at low temperatures there is a wide distribution of kinetic energies. As the temperature of the gas is raised, the added heat goes to increase the fraction of the molecules with high kinetic energies, while the fraction with low kinetic energies accordingly drops, thus raising the *average* kinetic energy of the molecules.

That the average kinetic energy is directly proportional to the absolute temperature is supported by Graham's laws of effusion and diffusion as can be shown as follows: Two different gases, A and B, at the same temperature would have, according to the theory, the same average molecular kinetic energy. As we mentioned in Sec. 2-4, the kinetic energy of a single molecule is given by

$$E_k = \tfrac{1}{2}ms^2$$

Figure 4-12
Molecular kinetic energy distributions. Average kinetic energies are indicated by the broken lines. Notice how the distribution "smears out" as the temperature is raised and that the *average* molecular kinetic energy becomes higher. (More molecules move faster as the temperature is raised.)

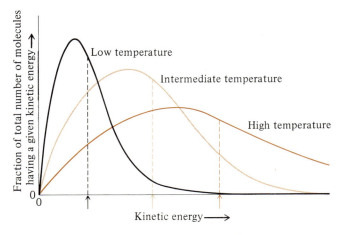

where m is the mass of the molecule and s is its speed. For a large collection of identical molecules the average kinetic energy is

$$\overline{E_k} = \tfrac{1}{2}m\overline{s^2}$$

where a bar is written over a symbol to indicate an average quantity. Note that $\overline{s^2}$ means the *average* of the *squared* speeds. (Each of the individual molecular speeds is squared, and then the average of these is taken.) Since the average molecular kinetic energies of the two gases A and B are equal, we can write

$$\tfrac{1}{2}m_A\overline{s_A^2} = \tfrac{1}{2}m_B\overline{s_B^2}$$

or, rearranging,

$$\frac{\overline{s_A^2}}{\overline{s_B^2}} = \frac{m_B}{m_A}$$

Now, taking the square root of each side we get

$$\frac{\sqrt{\overline{s_A^2}}}{\sqrt{\overline{s_B^2}}} = \frac{\sqrt{m_B}}{\sqrt{m_A}}$$

The quantity $\sqrt{\overline{s^2}}$ is called the *root-mean-square* speed. (The name means the square *root* of the average, or *mean,* of the *squares* of the speeds.)

Using arguments beyond the scope of this book, it is possible to show that the root-mean-square speed is proportional to the average speed, that is,

$$\sqrt{\overline{s^2}} \propto \overline{s}$$

Therefore

$$\frac{\overline{s_A}}{\overline{s_B}} = \frac{\sqrt{m_B}}{\sqrt{m_A}}$$

If we express the masses of molecules A and B in atomic mass units, these masses are molecular weights. Also, since rate of diffusion or effusion depends on average molecular speed, Graham's laws immediately follow:

$$\frac{\text{Rate}_A}{\text{Rate}_B} = \frac{\sqrt{M_B}}{\sqrt{M_A}}$$

The fact that it is possible to *derive* Graham's laws from assumption 4 gives enormous support to that assumption and, moreover, to the whole theory.

Success of the theory   Kinetic-molecular theory proves to be successful in accounting for much observed gas behavior.

BOYLE'S LAW. Figure 4-13 shows two samples of gas containing the same number of molecules at the same temperature. The volume of the gas in Fig. 4-13b is one-half that of the gas in Fig. 4-13a. As a result the molecules have less space in b in which to move. They collide more frequently, therefore, with the walls of the container. The average speed of the molecules is no greater in b than in a, because the gases are at the same temperature, but the increased number of collisions per

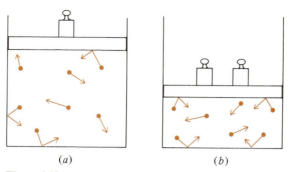

Figure 4-13
Boyle's law and kinetic-molecular theory. Each sample of gas is shown enclosed by a cylinder fitted with a sliding piston. The samples contain equal numbers of molecules at the same temperature. Because of the smaller volume available in (*b*), the molecules collide more frequently with the walls, producing a higher pressure than in (*a*).

second with the walls produces the effect we observe as higher pressure. The theory thus makes Boyle's law seem qualitatively reasonable, at least.

CHARLES' LAW. Figure 4-14 pictures a gas which undergoes thermal expansion in two stages. Initially (Fig. 4-14*a*) the gas occupies a certain volume at some temperature and pressure. Then (Fig. 4-14*b*) the temperature is raised, but the volume is held constant, causing a pressure increase. In Fig. 4-14*c* the temperature is the same as in *b*, but the pressure has been allowed to drop down to its initial value by allowing the volume to increase. Thus, in Fig. 4-14*c* we see the gas at its original pressure but at a higher volume and temperature. The pressure in *b* has risen from its initial value, because at the higher temperature the average kinetic energy of the molecules is greater, which means they have higher average speed. Higher speed means that the molecules collide with the container walls

Figure 4-14
Charles' law and kinetic-molecular theory. (*a*) Low *T*, low *P*, low *V*. (*b*) High *T*, high *P*, low *V*. (*c*) High *T*, original (low) *P*, high *V*. In (*a*) the gas at low temperature is at a low pressure, equal to that provided by the "weight" on the sliding piston. In (*b*) the gas has been heated and so the average molecular speed is higher; but expansion has been prevented, and so the pressure is higher. When the pressure is allowed to fall to its original value (*c*), volume increases. (The temperature is maintained constant in the last step.)

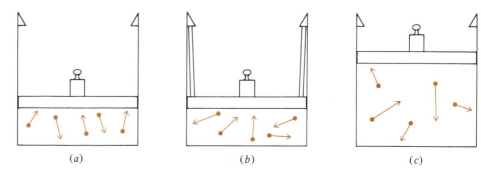

(1) *more often* and (2) *with greater violence,* on the average. We observe this as a higher pressure. In order to reduce the pressure to its original value (as we must, if we are going to account for Charles' law), we must allow the gas to expand. Thus, at the higher temperature the gas occupies a larger volume if pressure is the same as it was at the start. Charles' law is therefore also accounted for, at least qualitatively.

DALTON'S LAW. Each molecule of a gas behaves essentially independently of the others, except during collisions. Thus, no molecule is influenced by others in its collisions with the container walls. Molecule A contributes toward the total pressure just as if molecule B were not there, for example. Thus in a mixture of gases each will exert its own partial pressure independently of the presence of other gases, and the total pressure will be the sum of the partial pressures.

IDEAL-GAS LAW. The quantitative relationships expressed in Boyle's law, Charles' law, and Avogadro's principle and combined in the ideal-gas law can be derived by (1) using the assumptions of kinetic-molecular theory and (2) applying the principles of physics to a system composed of moving, colliding molecules. This derivation is given in App. E. That it is possible to start with this theory and actually derive $PV = nRT$ is an outstanding example of the power of the theory.

OTHER SUCCESSES. We have already mentioned that kinetic-molecular theory successfully accounts for brownian motion and Graham's laws. In addition many other gas properties, for example, thermal conductivity and viscosity, can be accounted for in terms of this theory of moving molecules. Some of the power of this theory is indicated by the fact that it can be extended to account for the behavior of real gases and, to some extent, of liquids and solids, as well.

4-6
Gas stoichiometry

We have seen (Sec. 4-4) that the volumes of gases consumed or formed in a chemical reaction are related by the ratios of the coefficients of these substances in the balanced equation, if the volumes are measured at the same pressure and temperature.

● **EXAMPLE 4-14**

Problem

When ethane, C_2H_6, burns in oxygen, the products are carbon dioxide and water. If 1.26 liters of ethane are burned in 4.50 liters of oxygen, how many liters of carbon dioxide and of water vapor are formed, if all volumes are measured at 400°C and 4.00 atm pressure? Assume ideal-gas behavior.

Solution

First, we write the equation and balance it by inspection (Sec. 3-5):

$$2C_2H_6(g) + 7\,O_2(g) \longrightarrow 4CO_2(g) + 6H_2O(g)$$

As long as the gas volumes are all measured at the same temperature and pressure, we can compare these volumes immediately. From the balanced equation we see that 2 vol of C_2H_6 require 7 vol of O_2, or, since volumes are expressed in liters in this problem:

$$2 \text{ liters } C_2H_6 \sim 7 \text{ liters } O_2$$

Next we see which gas is present in excess, and which is used up. If all of the 1.26 liters of C_2H_6 is used up, the number of liters of O_2 required is

$$1.26 \text{ liters } C_2H_6 \frac{7 \text{ liters } O_2}{2 \text{ liters } C_2H_6} = 4.41 \text{ liters } O_2$$

Since we have more than this volume of O_2 present, we conclude that O_2 is present in excess and the amounts of the products formed are limited by the amount of C_2H_6 burned. Thus, the volume of CO_2 formed is

$$1.26 \text{ liters } C_2H_6 \frac{4 \text{ liters } CO_2}{2 \text{ liters } C_2H_6} = 2.52 \text{ liters } CO_2$$

Similarly, the volume of H_2O formed is

$$1.26 \text{ liters } C_2H_6 \frac{6 \text{ liters } H_2O}{2 \text{ liters } C_2H_6} = 3.78 \text{ liters } H_2O$$

In Example 4-14, we were able to solve the problem quickly because all volumes were measured at the same temperature and pressure. It is a little more complicated when volumes are measured at different pressures and/or temperatures.

● **EXAMPLE 4-15**

Problem   Propane, C_3H_8, reacts with (burns in) oxygen to form CO_2 and H_2O. If 1.00 liter of propane at 21°C and 8.44 atm is burned in excess oxygen, how many liters of carbon dioxide measured at 925°C and 1.00 atm are formed?

Solution   The balanced equation is found to be

$$C_3H_8(g) + 5 O_2(g) \longrightarrow 3CO_2(g) + 4H_2O(g)$$

Method 1: First we find the volume which would be occupied by the propane at 925°C and 1.00 atm.

$$21°C = 273 + 21 = 294 \text{ K}$$

$$925°C = 925 + 273 = 1198 \text{ K}$$

$$V = 1.00 \text{ liter} \times \frac{1198 \text{ K}}{294 \text{ K}} \times \frac{8.44 \text{ atm}}{1.00 \text{ atm}} = 34.4 \text{ liters}$$

Now we find the volume occupied by the CO_2 formed.

$$V_{CO_2} = 34.4 \text{ liters } C_3H_8 \frac{3 \text{ liters } CO_2}{1 \text{ liter } C_3H_8} = 103 \text{ liters } CO_2$$

Method 2: First we find the number of moles of C_3H_8 from the ideal-gas law:

$$n = \frac{PV}{RT}$$

$$= \frac{(8.44 \text{ atm})(1.00 \text{ liter})}{(0.0821 \text{ liter atm K}^{-1}\text{mol}^{-1})(294 \text{ K})} = 0.350 \text{ mol}$$

This will form

$$0.350 \text{ mol } C_3H_8 \frac{3 \text{ mol } CO_2}{1 \text{ mol } C_3H_8} = 1.05 \text{ mol } CO_2$$

The volume of the CO_2 at 1.00 atm and 1198 K is

$$V = \frac{nRT}{P}$$

$$= \frac{(1.05 \text{ mol})(0.0821 \text{ liter atm K}^{-1}\text{mol}^{-1})(1198 \text{ K})}{1.00 \text{ atm}} = 103 \text{ liters } CO_2 \;\bullet$$

Putting stoichiometry and the gas laws all together we can solve problems like the following one.

● **EXAMPLE 4-16**

Problem   Suppose 1.00 liter of ethane, C_2H_6, measured at 25°C and 745 mmHg is placed in a strong 5.00-liter container along with 6.00 g of oxygen gas. The combustion reaction (explosion, in this case) is started by means of a spark. Then the *strong* container is cooled to 150°C. What is the total pressure (in mmHg) inside the container? (Atomic weight: O = 16.0.)

Solution   First we find the temperature in kelvins and pressure in atmospheres.

$$25°C = 25 + 273 = 298 \text{ K}$$

$$745 \text{ mmHg} = 745 \text{ mmHg} \frac{1 \text{ atm}}{760 \text{ mmHg}} = 0.980 \text{ atm}$$

Next we find the number of moles of C_2H_6

$$n_{C_2H_6} = \frac{PV}{RT}$$

$$= \frac{(0.980 \text{ atm})(1.00 \text{ liter})}{(0.0821 \text{ liter atm K}^{-1}\text{mol}^{-1})(298 \text{ K})} = 0.0401 \text{ mol}$$

Now, for the number of moles of O_2,

$$n_{O_2} = 6.00 \text{ g } O_2 \frac{1 \text{ mol } O_2}{2(16) \text{ g } O_2} = 0.188 \text{ mol } O_2$$

The balanced equation is

$$2C_2H_6(g) + 7 O_2(g) \longrightarrow 4CO_2(g) + 6H_2O(g)$$

From this we see that the number of moles of O_2 required for reaction with 0.0401 mol of C_2H_6 is

$$0.0401 \text{ mol } C_2H_6 \frac{7 \text{ mol } O_2}{2 \text{ mol } C_2H_6} = 0.140 \text{ mol } O_2$$

Since we actually have 0.188 mol of O_2, we conclude that all of the 0.0401 mol of C_2H_6 is used up, and 0.188 − 0.140, or 0.048, mol of O_2 is left over in excess.

Number of moles of CO_2 formed $= 0.0401 \text{ mol } C_2H_6 \frac{4 \text{ mol } CO_2}{2 \text{ mol } C_2H_6} = 0.0802 \text{ mol } CO_2$

Number of moles of H_2O formed $= 0.0401 \text{ mol } C_2H_6 \frac{6 \text{ mol } H_2O}{2 \text{ mol } C_2H_6} = 0.120 \text{ mol } H_2O$

Now let us summarize what we have found out so far

$$2C_2H_6(g) + 7\,O_2(g) \longrightarrow 4CO_2(g) + 6H_2O(g)$$

No. of moles at start	0.0401	0.188	0	0
No. of moles at end	0	0.048	0.0802	0.120

Now that we know how many moles of each gas (O_2, CO_2, and H_2O) are left after the reaction, we could use the ideal-gas law to find the partial pressure of each, and then add the partial pressures to get the total pressure. But it is faster to find the *total* number of moles and use the ideal-gas law just once to find the total pressure.

$$n_{total} = n_{O_2} + n_{CO_2} + n_{H_2O}$$

$$= 0.048 \text{ mol} + 0.0802 \text{ mol} + 0.120 \text{ mol} = 0.248 \text{ mol}$$

$$150°C = 150 + 273 = 423 \text{ K}$$

$$P_{total} = \frac{n_{total}RT}{V} = \frac{(0.248 \text{ mol})(0.0821 \text{ liter atm K}^{-1}\text{mol}^{-1})(423 \text{ K})}{5.00 \text{ liters}}$$

$$= 1.72 \text{ atm}$$

In millimeters of mercury this is

$$1.72 \text{ atm}\, \frac{760 \text{ mmHg}}{1 \text{ atm}} = 1.31 \times 10^3 \text{ mmHg}$$

SUMMARY

The most important concept in this chapter is that of the hypothetical *ideal gas*. The behavior of a *real gas* approaches that of the ideal gas as the pressure becomes lower or as the temperature becomes higher. Boyle's and Charles' laws describe proportionalities among the pressure, volume, and temperature of a given quantity (number of moles) of an ideal gas. *Boyle's law* describes an inverse proportionality between volume and pressure at constant temperature. *Charles' law* describes a direct proportionality between volume and absolute temperature at constant pressure.

Avogadro's principle states that equal volumes of different ideal gases at the same pressure and temperature consist of equal numbers of molecules. When this principle is combined with Boyle's and Charles' laws, the result is a highly useful generalization, the *ideal-gas law*. Using this relationship, $PV = nRT$, any of the four variables P, V, n, or T can be determined for a gas if the other three are known (assuming ideal behavior). Using the ideal-gas law, one can show that the *molar volume of an ideal gas at standard temperature and pressure* (STP; 0°C and 1 atm) is 22.4 liter mol^{-1}.

In gas mixtures the measured total pressure is the sum of the partial pressures of the individual gases in the mixture. This is *Dalton's law of partial pressures*. (A partial pressure is the pressure which a gas would exert, if it were alone in the given container.) The total pressure of a gas in contact with water equals the sum of the partial pressure of the gas itself plus the *vapor pressure* of water.

Diffusion of one gas through another and effusion of a gas through a pinhole are described by *Graham's laws*, which can be used as a basis for the experimental determination of molecular weight. The rate of diffusion (or effusion) is inversely proportional to the square root of the molecular weight of the gas.

Observed behavior of gases can be accounted for both qualitatively and quantitatively in terms of *kinetic-molecular theory*. This theory provides a description of a gas in terms of tiny, rapidly moving molecules which collide elastically with each other and the walls of the container. Except during collisions the molecules are independent of each other and their average kinetic energy is proportional to the absolute temperature. Using this model, we can justify observed gas behavior as described by the gas laws.

Avogadro's principle provides the basis for *Gay-Lussac's law of combining volumes*: ratios of volumes of reacting gases at the same pressure and temperature can be expressed with small integers. This provides us with a stoichiometric relationship among volumes of reacting gases which is identical to the relationship among their numbers of moles.

KEY TERMS

Absolute zero (Sec. 4-3)
Avogadro's principle (Sec. 4-4)
Barometer (Sec. 4-1)
Boyle's law (Sec. 4-2)
Charles' law (Sec. 4-3)
Dalton's law (Sec. 4-4)
Diffusion (Sec. 4-4)
Effusion (Sec. 4-4)
Gay-Lussac's law (Sec. 4-4)
Graham's law (Sec. 4-4)
Ideal gas (Secs. 4-2 and 4-4)

Ideal-gas constant (Sec. 4-4)
Ideal-gas law (Sec. 4-4)
Kinetic-molecular theory (Sec. 4-5)
Molar volume (Sec. 4-4)
Manometer (Sec. 4-1)
Millimeter of mercury (Sec. 4-1)
Partial pressure (Sec. 4-4)
STP (Sec. 4-4)
Standard atmosphere (Sec. 4-1)
Vapor pressure (Sec. 4-4)

QUESTIONS AND PROBLEMS

Note: *In answering each of the following assume that each gas behaves as an ideal gas.*

Barometer

4-1 Why is the height of the mercury column in a barometer independent of its diameter?

4-2 At 25°C the density of mercury is 13.5 g ml^{-1} and that of water is 1.00 g ml^{-1}. On a day when the barometeric pressure is 1.00 atm and the temperature is 25°C, how high would the column in a water-filled barometer be? (Neglect the vapor pressure of water.)

Boyle's law

4-3 Suppose 10.0 liters of hydrogen at STP are compressed to a volume of 1.88 liters at constant temperature. What is the final pressure?

4-4 Suppose 2.00 liters of oxygen at STP are expanded until the pressure has dropped to 2.05×10^{-3} atm. What is the final volume?

Charles' law

4-5 Assume 168 ml of carbon dioxide at 760 mmHg and 290°C are cooled to 0°C at constant pressure. What volume does the gas now occupy?

4-6 The volume of a sample of nitrogen is 1.41 cm^3 at 140°C and 1.46 atm. What volume would the gas occupy at 190°C at the same pressure?

4-7 A 1.40-liter container is filled with methane at 645 mmHg and 25°C, and the container is then sealed. If the flask is heated to 225°C, what is the final pressure? (Ignore any expansion of the flask itself.)

Diffusion and effusion

4-8 What is the difference between diffusion and effusion? How can measurements of these properties be used to obtain molecular weights of gases?

4-9 How do Graham's laws support the basic assumptions of kinetic-molecular theory?

4-10 Assuming comparable conditions compare the rates of diffusion of each of the following pairs of gases **(a)** H_2 and O_2 **(b)** N_2O and N_2 **(c)** UF_6 and H_2.

4-11 A certain gas, X, effuses through a pinhole at a rate of 4.73×10^{-4} mol s^{-1}. If methane, CH_4, effuses through the same pinhole under comparable conditions at a rate of 1.43×10^{-3} mol s^{-1}, what is the molecular weight of X?

Molar volume

4-12 What is the molar volume of nitrogen (N_2) gas at 225°C and 1.00 atm?

4-13 What is the molar volume of ethane (C_2H_6) at 0°C and 975 mmHg?

Combined gas laws

4-14 Suppose 40.0 ml of nitrogen at 98°C and 0.384 atm are compressed until the final pressure is 1.76 atm at 45°C. What is the final volume?

4-15 A sample of carbon dioxide occupies a volume of 34.0 liters at 28°C and 946 mmHg. What volume would the gas occupy at STP?

4-16 How many liters would 1.00 g of oxygen (O_2) occupy **(a)** at STP **(b)** at 145°C and 2.00 atm?

4-17 Determine how many grams each of the following weighs: **(a)** 1.00 liter of C_2H_6 at STP **(b)** 425 ml of C_2H_6 at 1.40 atm and 68°C.

4-18 Determine how many liters each of the following occupies: **(a)** 14.0 g of propane, C_3H_8, at STP **(b)** 14.0 g of butane, C_4H_{10}, at 64°C and 629 mmHg.

4-19 Determine the pressure in atmospheres of each of the following: **(a)** 4.50×10^{22} molecules of O_2 in a 4.50-liter container at 25°C **(b)** 3.92×10^{22} molecules of CH_4 *and* 7.21×10^{22} molecules of CO_2 in a 6.75-liter container at 425 K.

4-20 A 5.00-liter flask is filled with carbon dioxide gas at 1.50 atm and -15°C. How much does the CO_2 weigh?

4-21 The lowest pressure ever achieved in the laboratory was estimated to be about 1×10^{-15} mmHg, a pressure lower than that in outer space. Calculate how many molecules of gas would be present in a 1-liter container at this pressure at 0°C.

Dalton's law

4-22 Suppose 40.0 ml of hydrogen and 60.0 ml of nitrogen, each at STP, are both transferred to the same 125-ml container. What is the pressure of the mixture at 0°C?

4-23 Suppose 25.0 ml of oxygen at 25°C and 705 mmHg are added to a 30.0-ml container which already contains carbon dioxide at 35°C and 735 mmHg. If the temperature of the mixture is brought to 28°C, what is its pressure?

4-24 Suppose 0.500 g of oxygen (O_2) gas is collected at 30°C. Calculate the volume in milliliters occupied by the gas if it is collected **(a)** dry at 735 mmHg **(b)** over water at a (total) pressure of 735 mmHg.

4-25 A given sample of CO_2 gas weighs 10.0 g. Calculate its volume when collected over water at 50°C and 0.952 atm.

4-26 Oxygen weighing 1.00 g is placed in a container which holds an unknown liquid. The oxygen does not dissolve appreciably in the liquid, and the volume of the gas phase is 0.850 liter. If the pressure of the gas mixture is 739 mmHg and its temperature is 25°C, what is the vapor pressure of the unknown liquid at 25°C?

Gas density and molecular weight

4-27 Find the density of ethane gas (C_2H_6) in g liter^{-1} at 0.480 atm and 25°C.

4-28 The density of nitrogen gas is 1.52 g liter^{-1} at 177°C and 2.00 atm. How many N_2 molecules are present in a 5.00-liter flask filled with nitrogen under these conditions?

4-29 The density of oxygen gas is 1.43 g liter^{-1} at STP. What is its density at 400°C and 2.50 atm?

4-30 What is the molecular weight of a gas, 0.224 g of which occupies 238 cm^3 at 728 mmHg and 99°C?

4-31 What is the molecular weight of a gas if its density is 3.79 g liter^{-1} at STP?

4-32 What is the molecular weight of a gas if its density is 3.39 g liter^{-1} at 722 mmHg and 135°C?

Kinetic-molecular theory

4-33 Give a theoretical explanation for each of the following: **(a)** heating increases the pressure of a gas enclosed in a container of fixed volume **(b)** tiny dust particles in the air exhibit brownian motion **(c)** a gas expands very rapidly to fill the volume of a container **(d)** a gas-filled balloon gets larger as its temperature is raised **(e)** doubling the number of grams of nitrogen in a given container results in a doubling of the pressure of the gas, if the temperature stays constant.

4-34 Two containers of identical capacity are initially evacuated and maintained at the same constant temperature. A certain quantity of H_2 gas is put in the first container, and exactly the same mass of CH_4 gas is put in the second. Compare the two gas samples *quantitatively* with regard to the **(a)** number of molecules in the container **(b)** pressure **(c)** average molecular speed.

4-35 Two containers of identical capacity are initially evacuated. Container 1 is maintained at a constant 400 K and container 2 at a constant 200 K. Container 1 is then filled with oxygen gas and container 2 with hydrogen gas, each at 740 mmHg. Compare the two samples of gas *quantitatively* with regard to the **(a)** number of molecules **(b)** average molecular kinetic energy **(c)** mass **(d)** average molecular speed.

4-36 Suppose that you are given samples of two different gases A and B. The molecular weight of A is *twice* that of B. The average speed of the A molecules is *twice* that of the B molecules. If both samples contain the same number of molecules per liter, and if the pressure of B is 3.0 atm, what is the pressure of A?

Molecular formula

4-37 The empirical formula of a certain compound is CH_2. If its density as a gas at STP is 2.50 g liter^{-1}, what is its molecular formula?

4-38 The elemental analysis of a certain compound is 24.3 percent C, 4.1 percent H, and 71.6 percent Cl by mass. If 0.132 g of this compound occupies 41.4 ml at 741 mmHg and 96°C, what is the molecular formula of the compound?

4-39 The density of a certain gas is 1.28 g liter^{-1} at 56°C and 454 mmHg. Its percentage composition is 62.0 percent C, 10.4 percent H, and 27.6 percent O by mass. What is the molecular formula of the gas?

Gas stoichiometry

4-40 Suppose 1.64 liters of H_2 gas, measured at 38°C and 2.40 atm, are burned in excess oxygen to form water. How many liters of oxygen, measured at 38°C and 1.20 atm, are consumed?

4-41 Diborane, B_2H_6, can be burned in oxygen gas to form boric oxide, B_2O_3, and water. How many liters of O_2 at 1.25 atm and 25°C are needed to produce 10.0 g of B_2O_3 by this reaction?

4-42 Suppose 0.560 g of O_2 is placed in a container and irradiated with ultraviolet light. This converts 0.100 g of the O_2 to O_3, ozone, the only product. Calculate the volume of the mixture at 740 mmHg and 91°C.

4-43 When potassium nitrate, KNO_3, undergoes thermal decomposition, the products are potassium nitrite, KNO_2, and O_2 gas. If 5.00 g of KNO_3 are decomposed, how many liters of oxygen gas are formed, if the O_2 is collected over water at 15°C and 748 mmHg?

4-44 Ammonia, NH_3, reacts with cupric oxide, CuO, to form copper metal, water, and nitrogen gas. Suppose 25.0 g of CuO are consumed in this reaction. **(a)** How many liters of NH_3 at 31°C and 0.900 atmospheres are also consumed? **(b)** How many N_2 molecules are formed? **(c)** If the H_2O and N_2 are trapped in a 12.0-liter container, what is the total pressure at 300°C?

4-45 When nitroglycerin, $C_3H_5N_3O_9$, explodes, the products are all gaseous: CO_2, N_2, NO, and H_2O. If 1.00 g of nitroglycerin is exploded, what is the volume of the mixture of products at 1.00 atmosphere and 2.00×10^3°C?

4-46 Propellant power for certain rockets has been provided by the reaction of liquid hydrazine, N_2H_4, with liquid dinitrogen tetraoxide, N_2O_4. The products of the reaction are gaseous N_2 and H_2O. When 25.0 g of hydrazine react, **(a)** how many liters of N_2 are formed at STP? **(b)** how many liters of H_2O are formed at 975°C and 1.00 atm?

5 THE ATOM

TO THE STUDENT

Although the Greek philosophers Democritus and Leucippus (Sec. 1-2) guessed correctly about the existence of atoms, they could not conceive of anything as subtle as the *structure* of a single atom. But atoms are *not* homogeneous, structureless little pellets. They are composed of smaller, lighter, more fundamental particles, and our study of chemistry now takes us to a consideration of how these particles are assembled into atoms.

Why do we need to study atomic structure? It turns out that the physical and chemical behavior of matter depends upon the ways in which atoms bond and otherwise interact with each other. These interactions are thus very important, and to understand them we must first look closely at the atom's component particles.

5-1
The divisible atom

The notion that atoms might have an internal structure, that they might themselves be composed of smaller particles, was not taken very seriously until the latter part of the nineteenth century. Even Dalton's theory (below), the first atom-based theory to account satisfactorily for many aspects of chemical behavior, assumes that atoms are indivisible and structureless.

The Dalton
indivisible atom

John Dalton, an English science teacher, pointed out in about 1803 that most of the chemical observations made through the eighteenth century could be accounted for by simply assuming that matter is composed of atoms. Dalton proposed that

1 All matter is composed of ultimate particles, atoms.
2 Atoms are permanent and indivisible and can be neither created nor destroyed.
3 All atoms of a given element are identical in all their properties, and atoms of different elements have different properties. (In other words, different elements have different properties because their atoms are different.)

101

4 Chemical change consists of a combination, separation, or rearrangement of atoms.

5 Compounds are composed of atoms of different elements in fixed ratios.

Using these simple ideas, Dalton made most chemical observations of his day seem very reasonable indeed. His theory was successful, for example, in explaining why *mass is conserved* during a chemical reaction. (If each atom has its own characteristic mass, and if atoms are rearranged but otherwise left unchanged during a chemical reaction, then the total mass of the atoms of the reactants must be the same as that of the atoms of the products.) The law of *definite composition* is also accounted for. (Dalton pointed out that this law follows naturally if each compound is characterized by having fixed ratios among the numbers of atoms of its different component elements. For example, since the compound carbon dioxide is composed of carbon and oxygen atoms in a ratio of $1:2$, respectively, and since the masses of carbon and oxygen atoms are fixed, it follows that the composition of carbon dioxide by mass must also be definitely fixed.)

It is interesting to note that Dalton's theory put an end to the search for a method of transmuting base metals into gold. (See Sec. 1-2.) Most metals were recognized by Dalton as being elements, and since an atom of lead was considered to be permanent, there could be no way to change it into an atom of gold. Dalton was confused about a couple of points, the distinction between a molecule and an atom, for example. Also, he arrived at incorrect formulas for certain compounds. Nevertheless, his contributions were of tremendous importance. He started the scientific world thinking seriously about atoms and thereby set the stage for the next great question to be asked, "How and of what are atoms constructed?"

Electrolysis experiments

The first indication that matter and electricity might be related came as a result of some experiments with electrolysis. *Electrolysis* is the use of an electric current to cause a chemical reaction to take place. (The suffix *-lysis* comes from a Greek word meaning "to split or break up." Electrolysis is therefore the splitting up of something by electrical means.) In 1800 two English chemists, William Nicholson and Anthony Carlisle, first demonstrated the decomposition of water into hydrogen and oxygen gases by electrolysis. Later two other English chemists, Humphrey Davy and his assistant Michael Faraday, studied electrolysis intensively. In 1832 Faraday was able to show that there is a quantitative relationship between the amount of electricity used and the amount of products formed during electrolysis. He also showed that for a given amount of electricity used the amount of product formed depends upon the nature of that product, that is, upon what substance it is. These observations are usually summarized more precisely in statements known as *Faraday's laws* (see Sec. 19-2).

Discharge-tube experiments

Figure 5-1a shows a device known as a *gas discharge tube*. It consists of a cylindrical glass tube closed at both ends and fitted with two flat-disk electrodes. It is attached to a vacuum pump by means of the small side tube shown, and the electrodes are connected to a source of high voltage, say, 20,000 V. Such discharge tubes are often called *Crookes tubes,* after William Crookes, who used them during the latter half of the nineteenth century to investigate the effects of electrical discharges in gases at low pressure.

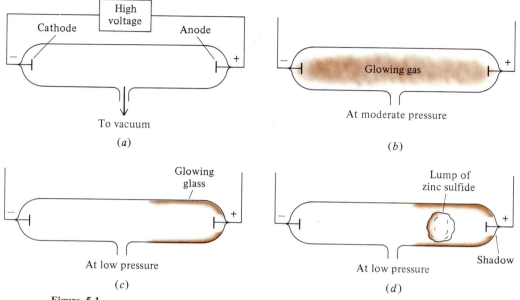

Figure 5-1
Discharge-tube experiments.

When a high voltage is applied to a discharge tube at atmospheric pressure, little is observed. If the vacuum pump is started, however, the gas inside the tube soon begins to emit a soft glow, as shown in Fig. 5-1*b*. As the pressure is further reduced, the glowing region moves toward the positively charged electrode, the *anode*. At still lower pressures the glow from within the tube fades out, and instead the glass at the anode end begins to emit a greenish glow, Fig. 5-1*c*.

If the above experiment is performed with a special tube into which a lump of zinc sulfide has been inserted, as in Fig. 5-1*d*, the side of the zinc sulfide facing the negatively charged electrode, the *cathode,* emits a bright glow (phosphorescence). This glow, when examined under a low-power microscope, is revealed actually to be composed of countless small, bright flashes of light. In addition, the lump of zinc sulfide casts a sharp-edged shadow on the anode end of the tube.

The results described above can be obtained using *any* metal for the electrodes. Furthermore, starting the experiment with various gases in the tube produces no change in results, except that the color of the glow from within the tube changes from gas to gas.

These gas-discharge-tube experiments can be interpreted as follows: At lower pressure something leaves the cathode and moves toward the anode. That something was first named a *cathode ray,* and the name is still used. The cathode ray is not radiant energy, such as light. The flashes of light from the zinc sulfide indicate that the cathode ray is composed of a stream of particles, each flash being caused by a cathode-ray particle colliding with the zinc sulfide. The shadow cast by the lump of zinc sulfide has sharp edges, and so particles must travel in straight lines away from the cathode. (If the particles could travel in curved paths, the shadow would be absent or have fuzzy edges.) The glow from the inside of the tube results from collisions of the particles with the molecules of gas in the tube.

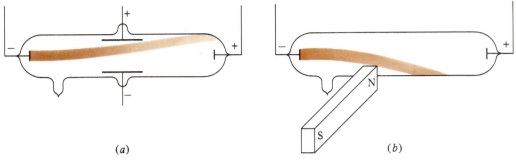

Figure 5-2
Deflection of cathode rays by applied fields: (*a*) electric field, (*b*) magnetic field.

At very low pressure most of the particles reach the anode end of the tube without colliding with any gas molecules, where they cause the glass to glow.

Can we learn more from cathode-ray discharge tubes? Yes; cathode rays can be bent by applying an *electric field*. Figure 5-2*a* shows a discharge tube which has been modified by adding a pair of metal plates. When these plates are given opposite electrical charges, the beam of particles undergoes a deflection, as illustrated. A similar experiment was first successfully performed by J. J. Thomson in 1897 and indicates that the particles in the cathode ray carry a *negative* electrical charge.

The stream of particles constituting the cathode ray can also be deflected by a *magnetic field*, as is illustrated in Fig. 5-2*b*. The direction of deflection agrees with that predicted by the laws of physics, *if* the particles are assumed to be *negatively* charged.

The particles of which cathode rays are composed always have the same electrical charge and the same mass. Because their properties are independent of the cathode material, it can be concluded *that they are present in all matter*. These particles are called *electrons*.

In an elegant experiment Thomson simultaneously applied an electric and a magnetic field to cathode rays and from his results was able to calculate the ratio of the charge of the electron to its mass (*e/m*). The present-day value of this charge-to-mass ratio is

$$\frac{e}{m} = -1.76 \times 10^8 \, \text{C g}^{-1}$$

where C represents *coulomb*, a unit of electrical charge. (See App. B.)

In 1908 Robert A. Millikan performed a classic experiment in which he determined the charge on the electron. He sprayed a fine mist of oil droplets between two metal plates which could be electrically charged. A single droplet could be observed by means of a microscope, and such a droplet could be observed to fall through the air under the influence of gravity. Millikan then irradiated the space between the plates with x rays. These knocked electrons out of some of the molecules of the air, and some of these electrons were caught by oil droplets. By electrically charging the upper plate positive and the lower negative, Millikan could stop the fall of an oil droplet, as the upward electrostatic force on the droplet was balanced against the downward gravitational force. From the amount of charge on the plates and the mass of the droplet (determined from its rate of fall through air when the plates were uncharged), the charge on the droplet was determined.

Millikan found that the droplets were always charged with a multiple of -1.6×10^{-19} C. Reasoning that a droplet could pick up any *integral* number of electrons, he concluded that the charge on an electron is -1.6×10^{-19} C.

Combining the value of the charge on the electron obtained by Millikan with the charge-to-mass ratio of Thomson, we find the mass of the electron to be

$$m = \frac{e}{e/m} = \frac{-1.6 \times 10^{-19} \text{ C}}{-1.76 \times 10^{8} \text{ C g}^{-1}} = 9.1 \times 10^{-28} \text{ g}$$

ADDED COMMENT

All electrons are alike in charge and mass. They are present in all matter and are *one* of the subatomic building blocks.

Cathode-ray discharge-tube experiments tell us that matter contains negatively charged particles, electrons. Can we learn still more? Yes, by using a modified discharge tube, one in which the cathode has had a hole drilled in it, as shown in Fig. 5-3. Such a tube was first used by E. Goldstein in 1886. When Goldstein evacuated the tube and applied a high voltage across the electrodes, he observed a glowing streamer emerging from the hole in the cathode on the opposite side from the anode, as is shown in the diagram. Goldstein called this streamer a *canal ray*. A canal ray has a color which depends upon the nature of the residual gas in the tube. Also, because it is possible to deflect the path of a canal ray by using an electric or magnetic field, it can be shown that the ray consists of a mixture of different particles, even if only a single, pure gas is present (at low pressure) in the tube. The particles are all *positively* charged, but if we determine the magnitude of these charges, we find that they are not all the same. Rather, they are multiples of 1.6×10^{-19} C; that is, we find particles with charges of 1.6×10^{-19} C, 3.2×10^{-19} C, 4.8×10^{-19} C, and so on.

All this can be accounted for as follows: The cathode ray, as we have seen, consists of a stream of electrons which leave the cathode and move toward the anode. Some of these electrons collide with gas molecules and do so vigorously enough to knock out other electrons. Each of these molecules now lacks one or more electrons and consequently carries a *positive charge*. The anode and cathode create an electric field within the discharge tube, and this causes the positively charged molecules to accelerate toward the cathode. Most of them strike the cathode, but a few coast through the hole and can be detected by the glow they produce on the other side. These charged molecules carry various amounts of

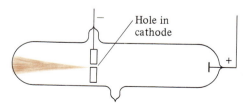

Figure 5-3
Goldstein's canal-ray tube.

positive charge depending on how many electrons they have lost. Their masses are always much larger than that of an electron and, in addition, depend upon the identity of the residual gas in the tube.

Why is it true that if an electron, which is *negatively* charged, is removed from a molecule, the molecule acquires a *positive* charge? It can be shown experimentally that molecules carry no net electrical charge. If removal of an electron causes a molecule to have a positive charge, it must be that molecules are composed of positively charged particles as well as electrons. When the total positive charge equals the total negative charge (due to the electrons), the *net* charge is zero. But if one or more electrons are lost, the positive charge now outweighs the negative charge; so the molecule is said to carry a *net* positive charge.

Atoms or molecules which have gained or lost electrons are called *negative* or *positive ions,* respectively. Each ion is represented by a symbol or formula followed by a superscript indication of the number of unit charges it carries. Thus, Na^+ represents a sodium atom which has *lost* one electron; S^{2-}, a sulfur atom which has *gained* two electrons; and O_2^+, an oxygen molecule which has lost an electron.

The Thomson atom   What is present in an atom besides electrons? What is the atom's structure? In 1898 J. J. Thomson proposed answers to these questions with a model for the structure of the atom. Thomson reasoned as follows: Electrons can be knocked out of an atom leaving a positive ion which has a much higher mass than an electron. Perhaps, therefore, each atom is composed of a large, massive positive part plus a number of smaller, lighter electrons. Specifically, the Thomson model of the atom was a sphere of positively charged electricity in which some electrons were imbedded. The positive portion contained most of the mass of the atom. (Later, Thomson postulated that the electrons were arranged in rings and were moving in circular orbits through the positively charged sphere.) There was something intuitively satisfying about this model, and it was accepted by physicists and chemists, until the experiments by Rutherford, Geiger, and Marsden, described below, showed that the Thomson model could not be correct.

In 1896 W. Röntgen was experimenting with a cathode-ray discharge tube which was totally enclosed by pieces of black cardboard. Nearby happened to be lying a piece of paper coated with a compound which, like zinc sulfide, would emit light when struck by cathode rays. Working in a darkened room, Röntgen noticed that the paper glowed. Later he found that photographic plates (the forerunners of photographic film) in unopened boxes had unexplainably been fogged (exposed). He eventually decided that unknown radiations, or "x rays," had passed through glass, paper, and cardboard, causing the strange phenomenon. X rays are formed when high-energy electrons (cathode rays) strike a metal or other target. They are electromagnetic radiation much like light, but higher in frequency (Sec. 5-3).

Later in the same year the French physicist H. Becquerel showed that compounds of the element uranium would expose a photographic plate, even if the plate were completely protected from light by opaque paper. His work was continued by Marie and Pierre Curie and others who found that several elements released radiations which were similar to x rays. Eventually it was recognized that such "radioactive" substances contain atoms which undergo decay or disintegration. (Atoms are not indestructible!) Three kinds of radiation are observed resulting from the disintegration of naturally radioactive elements. These are called α (alpha) particles, β (beta) particles, and γ (gamma) rays. It turns out that *α particles* are rapidly moving nuclei (see below) of helium atoms, *β particles* are rapidly moving electrons, and γ *rays* are essentially high-energy x rays.

The nuclear atom   Shortly after the start of the twentieth century, E. Rutherford, H. Geiger, and E. Marsden studied the paths of α particles shot at thin sheets of various materials such as paper, mica, and gold. They observed that the paths of some of the particles were deflected as they passed through these substances, that is, the particles were *scattered*. Rutherford was intrigued by this scattering and asked Geiger and Marsden to make a more detailed study of the scattering of α particles by thin (0.01 mm) gold foil. Figure 5-4 schematically shows the apparatus they used. In this apparatus α particles were detected by means of a small movable screen coated with zinc sulfide.

The results of the experiment were startling at the time. Rutherford, thinking in terms of the Thomson model of the atom, reasoned that nowhere during its passage through the foil would an α particle encounter any great concentration of charge, and therefore, at no time would any great forces be exerted on the particle. He expected most α particles, therefore, to pass through the foil un-scattered, with a few, perhaps, showing scattering at small angles, as shown in Fig. 5-5a. Although most of the α particles did indeed go through the foil with little or no scattering, an appreciable number were scattered at moderate and large angles. For some the scattering angle was greater than 90°; this means that these α particles actually bounced back from the foil, emerging from the same side that they had just entered. Rutherford realized that only very strong repulsions could cause the α particles to reverse their directions, and since the Thomson atom gave no clue as to any such repulsion, Rutherford concluded that the Thomson model must be wrong.

At this point Rutherford recalled an idea which has been proposed in 1904 by a Japanese physicist, H. Nagaoka, who suggested that an atom might be composed of a very small positively charged nucleus (at the center of the atom) surrounded by a comparatively large region containing the electrons. Rutherford correctly

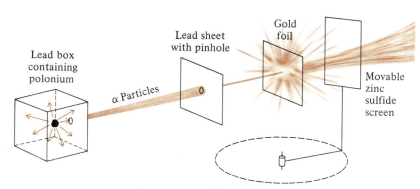

Figure 5-4
The Rutherford-Geiger-Marsden experiment.

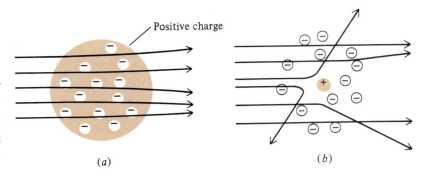

Figure 5-5
Expected deflections of
α particles: (a) Thom-
son atom, small deflec-
tions only; (b) Ruther-
ford atom, large and
small deflections.

(a)   (b)

reasoned that only if an α particle came close to the compact, positively charged, and massive[1] nucleus would any deflection occur, but close-approaching α particles would indeed experience large deflections, as shown in Fig. 5-5b.

Rutherford recognized that the Geiger-Marsden results could not be used to distinguish between a positively charged nucleus and a negatively charged one, since a massive negative nucleus would cause large deflections also. He assumed that the nucleus was positive, however, because he knew that electrons have relatively low masses, and if a nucleus consisted of a collection of electrons, then an α particle would be more likely to "flip" the nucleus out of an atom than to be deflected itself. Since deflection of α particles was observed to occur, Rutherford proposed that the *nonelectronic,* that is, *positive* and *massive,* part of the atom was associated with the nucleus.

The modern atom   Our present-day model of the atom is a direct descendant of the Rutherford atom. We believe the atom to be composed of two regions: (1) the tiny *nucleus,* comprising all the positive charge and practically all the mass of the atom and (2) the *extranuclear region* (everything else), composed of electrons. The positive charge on the nucleus is often expressed in units which are the *magnitude* of the charge on the electron and is called the *atomic number, Z.* For example, an atom of lithium has a nucleus carrying a charge of $+4.8 \times 10^{-19}$ C and three electrons each having a charge of -1.6×10^{-19} C. In electronic-charge units, the nucleus of a lithium atom has a 3+ charge, and each of the three electrons a 1− charge. The atomic number (Z) of lithium is 3.

ADDED COMMENT

It is important to realize how extremely small the nucleus actually is. Although an atom is a tiny particle, its nucleus is only about one one-hundred-thousandth as large. This means that if an atom could be enlarged so that its nucleus was the size of a tennis ball, then the entire atom would be about 4 mi across!

It is the atomic number, that is, the charge on the nucleus, which is related to an element's identity. Each atom of sulfur has an atomic number of 16, and each

[1] *Massive* here, and throughout this book, is used to mean "having a high mass," *not* "physically large."

iron atom has an atomic number of 26. Thus, to identify an atom of an element, or the element itself, it is sufficient to specify the atomic number.

There is abundant experimental evidence that the nucleus generally consists of two kinds of component particles, or *nucleons*. The first of these is the *proton*, which has a charge (in electronic-charge units) of $+1$. The second of these is the *neutron*, so-named because it is neutral; that is, it carries no charge at all. Since a whole atom carries no net electrical charge, the number of protons in the nucleus must be equal to the number of electrons in the extranuclear region. Also, *the number of protons in the nucleus equals the atomic number.* Except for the ordinary hydrogen nucleus all nuclei contain one or more neutrons. The protons and neutrons in the nucleus of an atom are almost entirely responsible for the atom's mass, because electrons are so very light.

Since the mass of an atom is almost exclusively determined by the total number of nucleons, we call that number the *mass number* and represent it by A. If a particular atom has 47 protons and 60 neutrons, for example, its mass number A is 107.

ADDED COMMENT

Be careful! The mass number A is *not a mass*. It is an integer representing the total number of nucleons.

Any specific atom can be designated using the following shorthand symbolism: Immediately preceding the chemical symbol for the element is written the atomic number Z as a subscript and the mass number A as a superscript. Therefore,

$$_{Z}^{A}X$$

indicates an atom of element X having an atomic number Z and a mass number A.[2] Since the atomic number is uniquely tied to the identity of the element, it is redundant and can be omitted. Thus, for example, we can write either $_{8}^{16}O$, or ^{16}O (each read as "oxygen 16" or "O 16") to represent this particular atom of oxygen.

All atoms of a given element have the same atomic number and therefore both the same nuclear charge and the same number of electrons in the extranuclear region. Can atoms of a given element differ from each other in any way? Yes, their mass numbers (and hence their masses) can be different. Atoms which have the same atomic number but different mass numbers are called *isotopes*, from Greek words meaning "same place." (If the elements are listed in order of their increasing atomic numbers, all the isotopes of an element occupy the same place in the list.) For example, there are three naturally occurring isotopes of oxygen: $_{8}^{16}O$, $_{8}^{17}O$, and $_{8}^{18}O$. These isotopes differ in mass because different mass numbers mean different numbers of neutrons $(A - Z)$, in the nucleus. Thus isotopes are sometimes defined as atoms of the same element having different numbers of neutrons. The numbers of electrons and nucleons in several different atoms are indicated in Table 5-1.

[2]In some older books you may find the mass number occupying a different position, a superscript written *after* the symbol: $_{Z}X^{A}$.

Table 5-1
Numbers of protons,
neutrons, and electrons
in various atoms

| Atom | Population of nucleus | | Population of extranuclear region |
	Protons	Neutrons	Electrons
$^{1}_{1}H$	1	0	1
$^{2}_{1}H$	1	1	1
$^{3}_{1}H$	1	2	1
$^{16}_{8}O$	8	8	8
$^{17}_{8}O$	8	9	8
$^{18}_{8}O$	8	10	8
$^{238}_{92}U$	92	146	92

**5-2
Atomic weights**

We have seen (Sec. 3-2) that atomic weights are customarily expressed in atomic mass units (amu). *One atomic mass unit* (1 amu) is defined as being exactly one-twelfth of the mass of a $^{12}_{6}C$ atom. This is equivalent to assigning the value 12 amu as the mass of one $^{12}_{6}C$ atom. On the atomic-mass-unit scale, the mass of each atom is expressed relative to the value 12 amu for a $^{12}_{6}C$ atom.[3]

Isotopic abundances

Most elements are found as a mixture of isotopes. Boron, for example, occurs naturally as a mixture of $^{10}_{5}B$ (19.78 percent) and $^{11}_{5}B$ (80.22 percent). The percentages quoted are *number percentages,* that is, out of every 100 boron atoms 19.78 are $^{10}_{5}B$ and 80.22 are $^{11}_{5}B$. (If you dislike the fractional atoms, consider that out of every 10,000 B atoms, 1978 are $^{10}_{5}B$ and 8022 are $^{11}_{5}B$.) The relative abundances of the isotopes of an element vary slightly depending upon the origin of the sample, but such variations are ordinarily extremely small.

Isotopic masses and relative isotopic abundances are usually determined by means of a technique known as *mass spectrometry.* The *mass spectrometer* is a descendant of the device used by J. J. Thomson to determine the charge-to-mass ratio of the electron. In one modern version of this instrument atoms are first converted to positive ions, *ionized,* by being bombarded by energetic electrons. The ions are then accelerated by means of an electric field, and the resulting beam is bent by passing it through a magnetic field. The amount the beam is bent depends on the charge-to-mass ratio of the particles in the beam, and so a beam which originally consisted of a mixture of isotopes is separated into a series of beams, each one containing ions of a specific charge-to-mass ratio. The beams can then be individually detected, either by means of a photographic film (in a *mass spectrograph*) or an ion detector (in a *mass spectrometer*).

Figure 5-6a schematically shows a mass spectrometer in the process of separating the three isotopes of naturally occurring neon. In the diagram only neon atoms which have lost *one* electron are considered; hence all the resulting ions have the same charge (1+) and differ only in mass. Figure 5-6b shows the resulting mass spectrum. From the heights of the peaks relative isotopic abundances can be determined.

Atomic weights

The atomic weight of an element is expressed in atomic mass units and consists of the *average mass* of the atoms of that element. Such an average must be *weighted;*

---

[3]Other definitions of the atomic mass unit have been used in the not-so-distant past, but the practical differences among these definitions are so small that they need not concern us here.

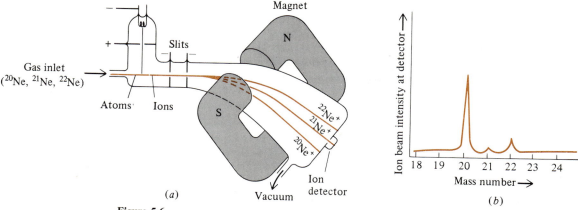

Figure 5-6

The mass spectrometer. (*a*) **Schematic diagram:** The separation of three isotopes of neon is shown. Neon atoms become ionized upon impact with electrons. These ions are then accelerated and pass through slits in two metal plates which are negatively charged, the second more highly than the first. Then they enter the magnetic field. The path of the ^{20}Ne$^+$ ions is bent the most by the magnetic field, that of the ^{22}Ne$^+$ ions, the least. The voltage on the slits is slowly raised to cause the Ne$^+$ ion beams one by one to sweep across the ion detector, which is connected to a strip-chart recorder. (*b*) **Mass spectrum of neon:** The tree isotopes of neon leave their peaks on the graph drawn by the strip-chart recorder. From the heights of these peaks the relative abundances can be calculated: 90.92 percent ^{20}Ne; 0.257 percent ^{21}Ne; and 8.82 percent ^{22}Ne.

that is, it must take account of the relative abundances of the various isotopes of the element.

● EXAMPLE 5-1

Problem   Copper occurs on earth as an isotopic mixture of 69.09 percent ^{63}Cu (mass = 62.93 amu per atom) and 30.91 percent ^{65}Cu (mass = 64.93 amu per atom). What is the atomic weight of copper?

Solution   We solve this problem by finding the average mass of some number, say 100, of copper atoms. Since 100 Cu atoms consist of 69.09 ^{63}Cu atoms and 30.91 ^{65}Cu atoms, the total mass of these 100 atoms is

$$(69.09 \text{ atoms})(62.93 \text{ amu atom}^{-1}) + (30.91 \text{ atoms})(64.93 \text{ amu atom}^{-1}) = 6355 \text{ amu}$$

The average mass of a copper atom is therefore

$$\frac{6355 \text{ amu}}{100 \text{ atoms}} = 63.55 \text{ amu atom}^{-1} \quad ●$$

It is important to remember that the atomic weight scale is a *relative* scale. The atomic weight of an element is the average mass of the atoms of that element expressed relative to the mass of a $^{12}_{6}$C atom, which is set equal to 12. Consider, for instance, the atomic weight of magnesium, which is known to be 24.3. This means that one ("average") magnesium atom has a mass of 24.3 amu. Since the mass of a $^{12}_{6}$C atom is 12 amu, one Mg atom weighs a little more than twice as much as one C atom. Similarly, chlorine, atomic weight 35.5, weighs a little less than three times as much as a $^{12}_{6}C$ atom.

ADDED COMMENT

Sometimes the following question is asked: Why is it that atomic weights are not integers (whole numbers)? First of all, most atomic weights are *average* masses, as we have just pointed out. Second, even though mass numbers must be integers (because a nucleus cannot contain a fractional number of nucleons), neither the mass of a proton (1.007276 amu) nor that of a neutron (1.008665 amu) is an integer. (By comparison the mass of an electron is 0.000549 amu.) Third, incredible though it may seem, the mass of an atom is actually a little *less* than the sum of the masses of its component protons, neutrons, and electrons. (We explore this paradox in Sec. 24-5.) There is really no reason for atomic weights to be whole numbers.

Be sure not to confuse mass number with atomic weight. A mass number must always be an integer; it is simply the number of particles in a nucleus. But the atomic weight of an element is the weighted average of the masses of all the isotopes of the element. It is expressed in atomic mass units and indicates the mass of the "average atom" compared to a ^{12}C atom, which has a mass equal to 12 amu.

5-3
Electrons in atoms

No sooner had Rutherford, Geiger, and Marsden shown that the nuclear model of the atom was correct, than much of the scientific world began asking, "What are the electrons *doing?*" Rutherford himself at first suggested that the atom had a *planetary* structure with the nucleus corresponding to the sun in our solar system, and the electrons to planets which travel through empty space in fixed orbits or trajectories. Such a model of atomic structure is certainly attractive. One can easily imagine, for instance, that the tendency for an electron to fly out of the atom (due to so-called centrifugal force) is exactly balanced by the electrical attraction between the negatively charged electron and the positively charged nucleus. But Rutherford and many others recognized that there was a fatal flaw in this simple planetary model.

The noncollapsing atom

Consider the situation from a logical standpoint. Imagine that you are looking at a hydrogen atom so greatly enlarged that you can see its nucleus and, far away from it, its one electron. Now there are only two possibilities for the state of motion of the electron: either it is moving or it is not. If it is stationary, then the electrostatic attraction which tends to draw unlike charges together should cause the electron to move toward the nucleus. It can be shown, moreover, that the electron should reach the nucleus within a small fraction of a second. It is not understood what would happen then, but that is really not important. What is important is that the atom would effectively collapse if its electron were sucked into the nucleus. (Remember that the size of the atom is really the size of the extranuclear region, that is, the region in which the electron is found.) The same reasoning can be applied to all other atoms in the universe, and so we reach the conclusion that at least according to this "stationary-electron" model the entire universe should collapse—within a small fraction of a second! Since the universe has not collapsed, we have no choice but to reject this model of the atom.

The stationary-electron model of the atom probably seems absurd to you, but it has been necessary to dispose of this possibility before considering the other alternative—the simple planetary model. This model assumes that the electron is moving rapidly around the nucleus. Its path or trajectory might be simple, such as a circular or elliptical orbit, or the electron might execute some more complex kind of movement, but in either case it must move roughly *around* the nucleus so that the centrifugal effect will prevent the electron from being drawn into the nucleus. The fatal flaw in this model is that it is inconsistent with the principles of classical physics. (Classical physics just means pre-twentieth-century physics, physics before the advent of quantum ideas.) These principles show that *any time a charged particle experiences an acceleration, it must emit radiant energy*. Acceleration of an object is defined as a change in its velocity, that is, a change in either its speed or its direction of motion. Furthermore, an object following a circular orbit experiences an acceleration toward the center of the circle; if it did not, it would follow a straight line and fly off into space. Thus we must conclude that a planetary electron, a charged particle, ought to radiate energy because of this acceleration. This is an unreasonable conclusion for two reasons: (1) such a radiation of energy is not observed, and (2) if the electron were to lose energy, it would have to slow down, and so the radius of its orbit would decrease. (Think of it this way: The centrifugal effect, which tends to throw it outward, would become less.) Thus, as the electron radiated energy it would spiral into the nucleus. As before, the atom would essentially collapse, and also as before, it can be shown that this collapse would take only a small fraction of a second. Because of this prediction of atomic collapse, and for other reasons as well, we must conclude that the planetary model should be rejected. It predicts an event (the immediate and almost instantaneous collapse of the universe!) which is not observed.

What a dilemma! Either (1) the electron is moving, or (2) it is not, and each alternative is inconsistent with observation. At this point the thought may have crossed your mind that something must be wrong with classical physics, and in a sense that is correct. It is not so much that classical physics is really wrong, as that it is inadequate to describe events occurring on the atomic scale. It was necessary to modify classical physics in order to resolve the dilemma of the noncollapsing (or collapsing, depending on how you look at it) atom.

The first important attempt to develop a nonclassical model of the atom was made by Niels Bohr. Although his model was not completely successful, it introduced some startling new concepts which led ultimately to the development of the modern model of atomic structure. Bohr felt that the clue to atomic structure was to be found in the nature of the light emitted by substances at high temperatures or under the influence of an electric discharge. More specifically, Bohr believed that this light was produced when electrons in atoms underwent energy changes. Before we see what Bohr proposed, however, we should examine some of the characteristics of light and other forms of radiant energy.

Radiant energy   *Radiant energy,* also called *electromagnetic energy,* travels at 3.00×10^8 m s^{-1} in a vacuum, for example. (This is about 186,000 mi s^{-1}, which by comparison makes

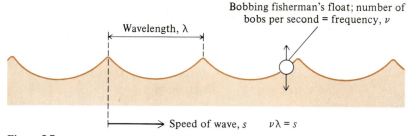

Bobbing fisherman's float; number of
bobs per second = frequency, ν

Wavelength, λ

Speed of wave, s $\nu\lambda = s$

Figure 5-7
A water wave. A fisherman's float will bob up and down as the water wave moves past. The distance between successive crests (or troughs) of the wave is called the *wavelength*. Since one wavelength passes by for each bob of the float, the product of the number of bobs per second (the *frequency*) times the wavelength equals the distance traveled by the wave in one second, the *speed* of the wave.

an interplanetary rocket look like a turtle.) Such energy exhibits what is known as *wave motion*.

What is wave motion? You are probably familiar with water waves, one of which is shown schematically in Fig. 5-7. As the wave moves from left to right, any floating object, such as the fisherman's float shown, bobs up and down as successive crests and troughs of the water wave pass it. When we speak of the wave, we mean the repetitive disturbance which travels across the surface of the water and causes the float to bob. The *frequency* of the wave is represented by the Greek letter nu, ν, and is the number of bobs made by the float per second.

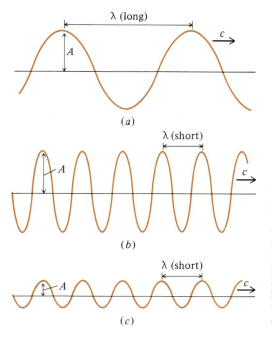

Figure 5-8
Electromagnetic waves. The frequency ν of each wave is the number of successive crests (or troughs) which pass a stationary point in a second. Thus frequency times wavelength (λ) equals the speed of the wave (a constant): $\nu\lambda = c$. Frequency and wavelength can be seen to be inversely proportional; if one is large, the other is small. On the other hand, the amplitude A of the wave can vary independently, as can be seen by comparing waves b and c.

The *wavelength* is represented by lambda, λ, and is the distance between successive corresponding points, such as crests. The product of these is the *speed s* at which the wave moves across the water, that is,

$$\nu\lambda = s$$

Similar to the water wave is the *electromagnetic wave,* by which radiant energy is transmitted. An electromagnetic wave is a combination of an electric and a magnetic field, both of which are oscillating and traveling through space. What does that mean? It means that a tiny, electrically charged particle will "bob" back and forth as the electromagnetic wave goes by, in much the same way as the fisherman's float bobs in response to a water wave. Also, a tiny magnet will be set into a similar oscillatory motion by the electromagnetic wave.

Figure 5-8 schematically illustrates three examples of electromagnetic waves. The shape of each wave is that of the familiar sine wave, and the wavelength, frequency, and speed are all related as they were for the water wave. Because the speed of all electromagnetic waves is a constant, at least in a vacuum, it is designated by *c*, so that we may write

$$\nu\lambda = c$$

This means that the frequency of a wave and its wavelength are inversely proportional; as one goes up the other goes down.

The range of observed electromagnetic frequencies (and of corresponding wavelengths) is depicted in Fig. 5-9. The so-called visible spectrum is the relatively narrow band of wavelengths and corresponding frequencies which our eyes (and brains) are able to detect. This is shown expanded in the diagram so that you

Figure 5-9
The electromagnetic spectrum. Energy increases from left to right in the diagram.

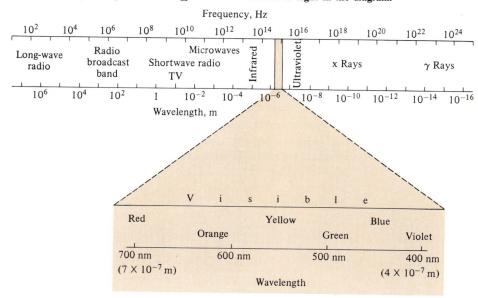

can correlate wavelength with color. Short-wavelength (high-frequency) waves we *see* as violet, for example. In Fig. 5-9 frequencies are indicated in *hertz* (Hz), a unit derived from the SI base units. It is the same as *cycles per second,* or s^{-1} (read "reciprocal seconds," "inverse seconds," or just "per second"). Wavelengths are indicated in the figure in meters (m), except for the visible spectrum where they are shown as nanometers (nm). (1 nm = 10^{-9} m.)

● EXAMPLE 5-2

Problem   A certain AM radio station broadcasts on a frequency assigned to it which is 980 kilohertz (kHz). What is the wavelength of the electromagnetic wave broadcast by the station?

Solution $$980 \text{ kHz} = 980 \times 10^3 \text{ Hz} = 980 \times 10^3 \text{ s}^{-1}$$

Rearranging the relationship $\nu\lambda = c$ to solve for λ, we get

$$\lambda = \frac{c}{\nu}$$

$$= \frac{3.00 \times 10^8 \text{ m s}^{-1}}{980 \times 10^3 \text{ s}^{-1}} = 306 \text{ m}$$

The wavelength of this radio wave is thus 306 m, or about 335 yd. ●

Atomic spectroscopy   White light, such as sunlight, is composed of a collection of electromagnetic waves with wavelengths ranging from about 400 nm to 700 nm. This mixture of waves can be separated by using an optical prism, which not only bends a ray of light (this is called *refraction*) but also bends light of different wavelengths by different amounts (*dispersion*). Figure 5-10 shows a ray of sunlight being refracted and dispersed by a prism. Notice how the white light produces a continuum of colors from red (long wavelength) to violet (short wavelength). Such a spectrum is called a *continuous spectrum* and can also be obtained from the light emitted by the incandescent filament of an ordinary light bulb.[4]

When substances are heated to a very high temperature by a flame or an electric arc or spark, light is given off, and the color depends upon the substance being investigated. Sodium and sodium-containing compounds give off bright yellow light, while potassium and potassium-containing compounds produce a weaker violet color. Such colors can be investigated spectroscopically in order to determine what wavelengths of light are emitted. When this is done with hydrogen gas the results are like those shown in Fig. 5-11. Instead of a continuous spectrum, a *line spectrum* is produced. Each line on the screen is produced by light of a single discrete wavelength and frequency, and the whole series of lines is called the *Balmer* series, after J. J. Balmer, who studied it in 1885. Balmer discovered a surprisingly simple relationship among the different wavelengths which produce the lines in the series. One form of this relationship is

$$\frac{1}{\lambda} = R\left(\frac{1}{2^2} - \frac{1}{n^2}\right)$$

---

[4]The sun's spectrum is actually not quite continuous. Certain discrete wavelengths are missing from sunlight, as evidenced by some dark lines, *Fraunhofer lines,* in its spectrum. Light of these wavelengths is absorbed by the outer mantle of the sun.

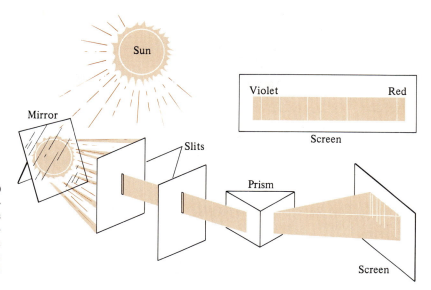

Figure 5-10
The spectrum of sunlight (Fraunhofer lines not shown). The colors on the screen form a continuous band from violet at one end to red at the other.

Here λ is the wavelength in nanometers, R is a constant now called the *Rydberg constant*, and n is an integer, 3 or larger. The value of R is $1.10 \times 10^{-2}\,\mathrm{nm}^{-1}$. Balmer found that this formula could be used to predict the wavelength of each of the lines within experimental error just by substituting integers for n. In other words each value of n (3 or larger) substituted in the Balmer formula yields the correct wavelength of one of the spectral lines.

Hydrogen produces other series of spectral lines, such as the *Lyman series* in the ultraviolet region of the spectrum, and the *Paschen series* in the infrared. Each series can be described by an equation which generates the wavelengths of all the lines in the series, as Balmer's equation did for the Balmer series. The different equations can be combined into a single relation, sometimes called the *Rydberg equation*

$$\frac{1}{\lambda} = R \left(\frac{1}{n_1{}^2} - \frac{1}{n_2{}^2} \right) \qquad (5\text{-}1)$$

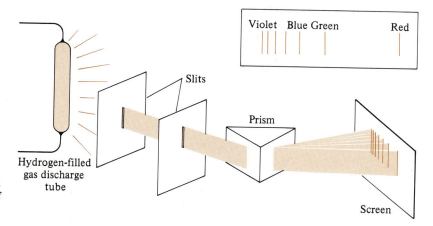

Figure 5-11
The line spectrum of hydrogen.

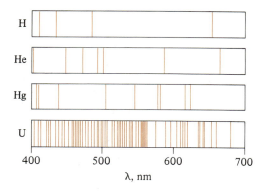

Figure 5-12
Line spectra of some elements.

This equation generates the wavelengths of all the lines in each series in the following way:

Lyman series: $n_1 = 1$ $n_2 = 2, 3, 4, 5, \ldots, \infty$
Balmer series: $n_1 = 2$ $n_2 = 3, 4, 5, 6, \ldots, \infty$
Paschen series: $n_1 = 3$ $n_2 = 4, 5, 6, 7, \ldots, \infty$

One more very important point should be noted: other elements besides hydrogen give line spectra, too. These spectra are more complex, consisting of more lines and of lines whose wavelengths are not related by anything as simple as the Rydberg equation. But the mere fact that elements in general exhibit *line* spectra proves to be very significant. The line spectra of several elements are shown in Fig. 5-12.

ADDED COMMENT

Why so much fuss about spectra? Mainly because the existence of line spectra provided the major experimental support for the Bohr model of the atom, which we are about to discuss. The Rutherford-Geiger-Marsden experiments showed in a general way where the electrons are in an atom, but they also pointed out the inadequacy of classical physics in justifying an atom's stability. In developing his model of the atom, Bohr took his cue from the countless line spectra which had been observed by spectroscopists during the nineteenth and early twentieth centuries.

The Bohr atom   In 1913 Neils Bohr, a Danish physicist, considered the dilemma of the noncollapsing atom. He said that there must be some as yet unknown physical principles which describe electrons in atoms. Although Bohr's theory was shown to have serious flaws, Bohr was brave enough to question classical physics, and his work encouraged others to find out why classical physics fails for small particles and to develop the new physics, quantum mechanics.

Bohr started out by assuming that when a substance is heated and emits light, it must do so because its atoms have absorbed energy from the flame or electric discharge. Bohr then suggested that it is actually the electrons which absorb energy and then reemit it as light. But, Bohr asked, why is the radiation limited to certain specific wavelengths, each of which produces a line in the element's spectrum? With what must have been a flash of inspiration he saw that there could be only one explanation for the discrete wavelengths. He reasoned that these would result if *the energy of an electron in an atom is quantized.* This means that an electron in an atom is not free to have just any amount of energy, but only *certain, specific, discrete amounts of energy.*

What is the connection between the energy of an electron and the wavelength of the light emitted from an atom? Why did Bohr see that quantized electronic energy levels account for line spectra? At the beginning of the twentieth century Max Planck and Albert Einstein, using entirely different approaches, demonstrated that all electromagnetic radiation behaves as if it were composed of little energy packets, now called *photons.* They showed that each photon has an *energy* which is proportional to the *frequency* of the light:

$$E_{photon} = h\nu \qquad (5\text{-}2)$$

where the proportionality constant, h, is called *Planck's constant* (and has the value 6.63×10^{-34} J s). We have already seen that frequency, wavelength, and speed of light are related by the equation

$$\nu\lambda = c \qquad \text{or} \qquad \nu = \frac{c}{\lambda}$$

so we can write Eq. (5-2) as

$$E_{photon} = \frac{hc}{\lambda} \qquad (5\text{-}3)$$

This equation shows that if electromagnetic radiation has a single, discrete wavelength, then the energy of its photons must have a single value (since h and c are both constants). Bohr pointed out that if an electron in an atom drops from one quantized energy level to another, it should be expected to emit light of a single energy and therefore of a single wavelength.

Bohr described the origin of line spectra as follows: Each atom of an element has a set of available quantized energies, or *energy levels,* for its electrons. Normally the atom is in its *ground state;* that is, all the electrons are in the lowest available levels. When the atom absorbs energy from a flame or electric discharge, its electrons are raised into higher energy levels. The atom is now said to be in an *excited state.* But now there are unoccupied lower energy levels and so each electron drops down from a higher quantized level E_2 to a lower one, E_1. For a given electron the *difference* between these energies is the amount of energy lost by the electron and equals the energy of the photon of electromagnetic energy (light) which is emitted. In other words,

$$E_2 - E_1 = E_{photon} = \frac{hc}{\lambda} \qquad (5\text{-}4)$$

Thus a discrete spectral line of wavelength λ is accounted for by the quantized electronic energies E_2 and E_1. In a given atom many such transitions from higher to lower level are possible. Each contributes toward a single discrete line in the spectrum of that element. (Of course, many such transitions in many atoms are necessary to produce a line which is visible.)

ADDED COMMENT

The concept of quantization of energy is important. Consider the analogy with a ladder. You can stand on each rung, but you cannot stand between rungs. There are, therefore, only certain, specific, discrete elevations above ground for your feet. (Let us ignore the brief periods during which you go from one rung to another.) Your elevation is quantized. In a similar way the energy of an electron in an atom is quantized, because it cannot have just any amount of energy, but only certain *permitted* amounts.

The Bohr theory proved to be most successful when applied to the hydrogen atom. Bohr showed that the total energy (kinetic plus potential) of the electron in a hydrogen atom should be quantized in the following way:

$$E = -A \frac{1}{n^2} \tag{5-5}$$

where E is the total energy, n is a positive integer, called a *quantum number,* and A is a constant. When a succession of positive integers is substituted for n in this relationship, a series of energies is obtained, as follows:

n	1	2	3	4	5	.	.	.	∞
E	$-A$	$-\dfrac{A}{4}$	$-\dfrac{A}{9}$	$-\dfrac{A}{16}$	$-\dfrac{A}{25}$	.	.	.	0

These energies are shown schematically in Fig. 5-13, where they look much like rungs on a ladder, except that the rungs get progressively closer together going up. Each energy level in the drawing is an allowed amount of energy; the electron cannot have "in between" amounts.

Do not be concerned about the fact that the energies shown in Fig. 5-13 are negative. This results from the fact that the highest energy level, the one with

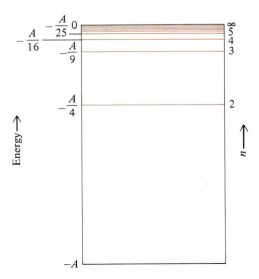

Figure 5-13
Energy levels in the hydrogen atom.

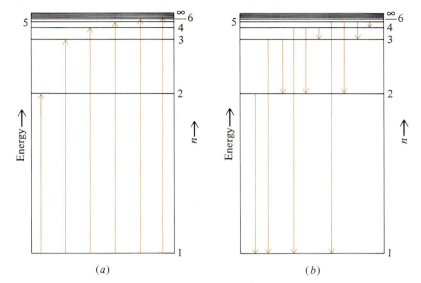

Figure 5-14
Electronic transitions
in a hydrogen atom.
(a) Some transitions
from ground state to
excited states. (*Energy*
***is absorbed.*) (b) Some**
transitions from excited
states to lower states.
(*Energy is emitted.*)

$n = \infty$, has been somewhat arbitrarily set at $E = 0$; so the others, being lower in energy, must be negative. Furthermore, the location of the zero point in the diagram is not important, since we are concerned mainly with *differences* in energy associated with transitions made by an electron from one level to another.

The electron in a ground-state hydrogen atom can absorb various discrete amounts of energy and can in this way be raised into any of the higher levels. Such transitions are shown schematically in Fig. 5-14a. The length of each upward-pointing arrow in the diagram is a measure of the amount of energy absorbed by the atom when it undergoes the transition. (Transitions to the $n = \infty$ level correspond to loss of the electron by the atom.)

According to the Bohr theory each line in the spectrum of hydrogen corresponds to a specific "downward" transition, that is, from an excited atomic state to some lower state. Some of these transitions are shown schematically in Fig. 5-14b.

Applying the Bohr expression for electronic energy [Eq. (5-5)] to a state of higher energy, E_2, we can write

$$E_2 = -A \frac{1}{n_2{}^2}$$

and to a lower energy state,

$$E_1 = -A \frac{1}{n_1{}^2}$$

Thus the energy of the photon of light emitted when the electron drops from E_2 to E_1 is obtained from Eq. (5-4).

$$E_{\text{photon}} = \frac{hc}{\lambda} = E_2 - E_1$$

$$= -A \frac{1}{n_2{}^2} - \left(-A \frac{1}{n_1{}^2}\right) = A \left(\frac{1}{n_1{}^2} - \frac{1}{n_2{}^2}\right)$$

From this we get after rearranging

$$\frac{1}{\lambda} = \frac{A}{hc}\left(\frac{1}{n_1{}^2} - \frac{1}{n_2{}^2}\right)$$

We can see that this equation has exactly the same form as the Rydberg equation [Eq. (5-1)], where A/hc is equal to R, the Rydberg constant. One of the great successes of the Bohr theory was the fact that by using it one could *derive* the Rydberg equation, as we have just done. Moreover, Bohr was able to calculate A and thereby obtain a value for the Rydberg constant which agreed with the experimentally obtained value to five significant figures! In 1913 such qualitative *and* quantitative agreement between theory and experiment seemed very exciting to the scientific world.

What did Bohr have to say about what electrons in atoms are *doing?* His model is a *modified* planetary model in which each quantized energy level corresponds to a specific, stable, circular electronic orbit with a quantized radius.[5] Orbits with large radii correspond to high energy levels. The planetary aspect of the Bohr model is, however, one of its weakest characteristics, for reasons which will become apparent in Chap. 6. For this reason we will speak no more of electrons following orbits or trajectories around the nucleus. The flurry of excitement over the Bohr theory subsided rather soon when it became apparent that it had several serious weaknesses. Some of these will be mentioned later, but for now it is enough to point out that the Bohr theory works well only for one atom—hydrogen. It can be adapted to work after a fashion for several other atoms, but for atoms of most elements it is a dismal failure, because the series of spectral lines predicted by the theory simply do not correspond to what is actually observed.

The Bohr model of the atom did, however, serve as an important stepping-stone on the path to present-day concepts of atomic structure. The notion of *quantization* of energy was a perplexing one in the early 1900s. The work of Planck, Einstein, and Bohr seemed revolutionary, because for years it had been believed that "Nature doesn't make a jump,"[6] or, more elegantly, "There are no finite discontinuities in nature"! But the concept of quantization has survived, even if much of the rest of the Bohr model has not.

---

[5] A. Sommerfeld and others extended the Bohr model to include *elliptical* orbits.

[6] A literal translation from the Latin, *natura non facit saltum.*

SUMMARY

In this chapter we have outlined the development of the concept of the atom from the Dalton indivisible model to the Bohr planetary-electron model. The Dalton model was based on the proposition that atoms were indivisible, unchangeable, and apparently structureless. However, *electrolysis* studies and *gas-discharge-tube* experiments showed that atoms had positive and negative parts. Rutherford and his associates showed that the atom was constructed with a tiny, massive, positively charged *nucleus* surrounded by negatively charged *electrons,* a model which is considered essentially correct today.

Atoms have nuclei which consist of protons and, except for $^1_1$H, neutrons. The number of protons in the nucleus is called the *atomic number Z,* and the sum of the numbers of protons and neutrons is the *mass number A.* Each element is characterized by a specific atomic number, but different atoms of a given element may have different numbers of neutrons, hence different mass numbers. Such atoms are called *isotopes*. The *atomic weight* of an element is a weighted average of the masses of its naturally occurring isotopes and is expressed on a scale on which the mass of one $_{6}^{12}C$ atom is set equal to exactly 12 amu.

The Rutherford nuclear model of the atom was more precisely defined by Bohr, who postulated that electrons in an atom had quantized orbits and energies. Although attractive in many respects, this model proved inadequate in accounting for observed atomic spectra other than that of hydrogen. Because of its ideas about energy quantization, the Bohr model did indeed make a useful contribution and served as a stepping-stone to the quantum-mechanical atom (Chap. 6).

KEY TERMS

Anode (Sec. 5-1)
Atomic number (Sec. 5-1)
Canal ray (Sec. 5-1)
Cathode (Sec. 5-1)
Cathode ray (Sec. 5-1)
Discharge tube (Sec. 5-1)
Electrolysis (Sec. 5-1)
Electron (Sec. 5-1)
Excited state (Sec. 5-3)
Extranuclear region (Sec. 5-1)
Frequency (Sec. 5-3)
Ground state (Sec. 5-3)

Ion (Sec. 5-1)
Isotope (Sec. 5-1)
Line spectrum (Sec. 5-3)
Mass number (Sec. 5-1)
Mass spectrometer (Sec. 5-2)
Neutron (Sec. 5-1)
Nucleon (Sec. 5-1)
Nucleus (Sec. 5-1)
Photon (Sec. 5-3)
Proton (Sec. 5-1)
Quantization (Sec. 5-3)
Wavelength (Sec. 5-3)

QUESTIONS AND PROBLEMS

Atomic models

5-1 Compare the Dalton, Thomson, Rutherford, and Bohr models of the atom.

5-2 Describe the Thomson model of the atom and show how the Rutherford-Geiger-Marsden experiments made it necessary to reject this model.

5-3 Outline the key elements of Dalton's atomic theory. Which of these are not consistent with the modern view of the atom?

5-4 How did the gas-discharge-tube experiments of Crookes and others show that the atom was composed of smaller particles?

Nuclear atom

5-5 Imagine for a moment that the nucleus of an atom carries a *negative* electrical charge, instead of a positive one. Sketch the path of an α particle as it approaches such a hypothetical nucleus. Could α particles be deflected back in the general direction from which they came, if atoms had negative nuclei?

5-6 What would have been the results of the Rutherford-Geiger-Marsden α-particle scattering experiment, if most of the mass of an atom were accounted for by its electrons, all other characteristics remaining the same?

5-7 What is the nature of cathode rays? of canal rays? How and why do canal rays change their character when the gas inside a discharge tube is changed?

5-8 Compare the characteristics of the three subatomic particles: proton, neutron, electron.

5-9 A certain atom has a radius of 0.15 nm. If its nucleus has a radius of 1.5×10^{-6} nm, compare the density of the nucleus with that of the whole atom.

5-10 How do discharge-tube experiments show that electrons are present in all matter?

5-11 When one or more electrons are removed from an atom, the resulting particle is a *positive* ion. Explain.

5-12 Tell how many **(a)** protons and **(b)** neutrons are present in the nucleus of each of the following atoms: ^{14}N, ^{15}N, ^{179}Ta, ^{234}U.

5-13 Give the total number of electrons present in each of the following atoms or ions: N, O, O^{2-}, Na^{+}, Sr^{2+}, Sn^{4+}.

5-14 (a) What is the charge in coulombs on a nucleus of an atom of iron? **(b)** What is its charge in units of the electronic charge?

5-15 The mass of a proton is 1.007 amu. What is the

mass in grams of **(a)** 1 mol of protons **(b)** one proton?

Atomic weight

5-16 Chlorine occurs naturally as a mixture of two isotopes: ^{35}Cl (mass 34.97 amu) and ^{37}Cl (mass 36.96 amu). If the relative isotopic abundance of ^{35}Cl is 75.35 percent, what is the atomic weight of Cl?

5-17 Magnesium occurs naturally as a mixture of three isotopes: 78.70 percent ^{24}Mg (mass 23.99 amu), 10.13 percent ^{25}Mg (mass 24.99 amu), and 11.17 percent ^{26}Mg (mass 25.98 amu). What is the atomic weight of Mg?

5-18 Boron occurs naturally as a mixture of the two isotopes ^{10}B (mass 10.01 amu) and ^{11}B (mass 11.01 amu). If the atomic weight of boron is 10.81, what are the relative abundances of the two isotopes?

5-19 Silver (atomic weight 107.87) occurs in nature as a mixture of 51.82 percent ^{107}Ag and 48.18 percent ^{109}Ag. If the isotopic weight of ^{107}Ag is 106.90 amu, what is that of ^{109}Ag?

Bohr atom

5-20 Describe the Bohr model of the atom. How does it differ from a planetary model based on classical physics?

5-21 How does the existence of line spectra provide support for the Bohr model of the atom?

5-22 What is a *photon?* How is the energy of a photon of electromagnetic energy related to **(a)** frequency **(b)** wavelength?

5-23 In terms of the Bohr theory of the structure of the atom why is it that electrons do not spiral into the nucleus?

5-24 A ripple travels across the otherwise smooth surface of a lake. If the wavelength is 8.0 in and the frequency is 2.0 s^{-1}, what is the speed of the ripple in inches per second?

5-25 What is the wavelength (in nanometers) of blue light which has a frequency of 6.60×10^{14} Hz $(6.60 \times 10^{14} \text{ s}^{-1})$?

5-26 What is the frequency (in megahertz) of a citizen's band (CB) radio wave which has a wavelength of 11.0 m?

5-27 If atomic energy levels were not quantized, but could vary within certain limits, what would atomic spectra look like?

5-28 Using the Rydberg equation calculate to three significant figures the wavelengths of the lines in the Lyman series for $n_2 = 2$ through 8. As n_2 approaches infinity what limiting value does λ approach?

6

ELECTRONS: ENERGIES, WAVES, AND PROBABILITIES

TO THE STUDENT

The notion that electrons in an atom are discrete little particles which fly around the nucleus in well-defined orbits is an appealing one. How many times have you seen pictures supposedly depicting electrons whizzing around a nucleus? They often show up in newspapers, "popular" scientific magazines, and on television. In fact, such pictures are so common that it is hard to believe that the Bohr model of atomic structure was laid to rest about 1925, a pretty long time ago!

In this chapter we leave the Bohr atom and move on to the present-day theory of atomic structure: quantum mechanics. In some ways the quantum-mechanical model of the atom is quite similar to the Bohr theory. Electronic energies, for example, are quantized according to both theories, although only quantum mechanics gives us a satisfactory reason. You will soon learn to use the quantum-mechanical model of the atom to account for observed atomic properties, one of the most important of which is the ability of atoms to form bonds. No orbiting-electron model of the atom has been able to account successfully for this ability.

**6-1
The quantum-mechanical model**

During the first part of the twentieth century, the science of physics experienced a revolution which eventually influenced all other sciences. This revolution occurred, at least originally, in *mechanics,* which is broadly defined as "the branch of physics which seeks to formulate general rules for predicting the behavior of a physical system under the influence of any type of interaction with its environment."[1]

The failure of classical mechanics

The new kind of physics is called *quantum mechanics,* or *wave mechanics.* The older kind of mechanics, based on Newton's laws of motion, is now called *classical*

---

[1] Daniel N. Lapedes (ed.), *McGraw-Hill Dictionary of Scientific and Technical Terms,* McGraw-Hill Book Co., New York, 1976.

mechanics, or *newtonian mechanics.* As Bohr concluded, classical mechanics is inadequate for predicting the behavior of small particles such as molecules, atoms, electrons, nuclei, etc. Quantum mechanics, on the other hand, is successful. Furthermore, it can be shown that the mathematical relationships of quantum mechanics can be simplified and reduced to those of classical mechanics when the objects considered are much larger than atoms or molecules. Thus classical mechanics can be viewed as a simplified version of quantum mechanics, which describes the behavior of objects which are large enough so that quantum effects are not significant.

Quantum effects include, first of all, the quantization of energy. As has already been mentioned (Sec. 5-3), Planck and Einstein showed that energy is "packaged" in little bundles, corpuscles, or *quanta.* The name *photon* is given to a quantum of electromagnetic energy. Thus electromagnetic *waves* can exhibit some of the properties of *particles.*

A second quantum effect is that *particles* can exhibit the properties of *waves.* This may seem strange at first because it means that quantities like wavelength and frequency can be used to describe particles such as electrons, atoms, fleas, and elephants! This surprising idea originated mostly with the work of a French physicist, Louis de Broglie (pronounced, approximately: duh Bruh-yee). De Broglie suggested in 1924 that electrons, previously considered typical particles, have wavelike properties.

De Broglie concluded that electrons might have wavelengths, by first combining two relationships, one derived by Einstein, the other by Planck. Einstein showed that the total energy E of any particle is proportional to its mass m, the proportionality constant being the square of the speed of light in a vacuum, c, or

$$E = mc^2 \qquad \text{(energy of a particle of mass } m)$$

The Planck relationship relating the energy of a wave to its frequency has already been mentioned.

$$E = h\nu \qquad \text{(energy of a wave of frequency } \nu)$$

De Broglie first pointed out that the effective mass of a photon of electromagnetic energy such as light could be found from its wavelength by combining the above two relationships.

$$mc^2 = h\nu$$

Since $\nu\lambda = c$ (Sec. 5-3),

$$mc^2 = h\left(\frac{c}{\lambda}\right) \qquad \text{and} \qquad m = \frac{h}{\lambda c}$$

De Broglie went further, however. Reasoning that if something as wavelike as light could be considered a stream of particles, he suggested that maybe something as particlelike as an electron could be thought of as a wave. Unlike light, which travels at a constant velocity, electrons travel at various velocities; so substituting v for the velocity of an electron in the above relationship, de Broglie obtained

$$m = \frac{h}{\lambda v} \qquad \text{or} \qquad \lambda = \frac{h}{mv}$$

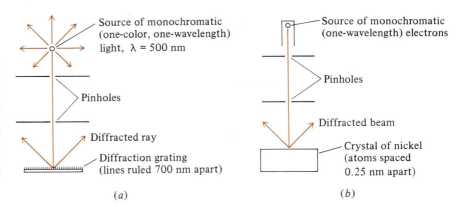

**Figure 6-1
Diffraction of light and
electrons.** (*a*) Mono-
chromatic light is dif-
fracted by a reflection
grating. (*b*) Mono-
chromatic electrons are
diffracted by a crystal.
(Davisson-Germer ex-
periment.)

(*a*)   (*b*)

This allows us to calculate the wavelength of an electron, indeed, of any particle, from the particle's mass and velocity.

Direct evidence supporting de Broglie's electron waves came from the experiments of two groups of physicists: C. Davisson and L. H. Germer in the United States and G. P. Thomson (the son of J. J. Thomson) and A. Reid in Scotland. Each of these two groups showed that it is possible to *diffract* electrons.[2]   It is important to know that diffraction is a phenomenon which is characteristic of waves; when diffraction occurs, waves must be present. As is shown in Fig. 6-1*a*, light can be diffracted, that is, bent and separated into several beams, by a *diffraction grating*, which is a series of closely and regularly spaced lines ruled on a flat surface. In the illustration the lines are ruled on a mirror, making a so-called reflection grating. Diffraction of the light occurs if its wavelength is about the same as the distance between the ruled lines.

Using the de Broglie relationship we can calculate the wavelength of an electron and thus the approximate spacing of lines which should diffract an electron beam. Planck's constant h is 6.63×10^{-34} J s, and since one joule is defined (App. B) as

$$1 \text{ J} = 1 \text{ kg m}^2 \text{ s}^{-2}$$

then $$h = 6.63 \times 10^{-34} \text{ kg m}^2 \text{ s}^{-1}$$

The mass of an electron is 9.1×10^{-31} kg. For an electron traveling at 4.0×10^4 m s^{-1} we have

$$\lambda = \frac{h}{mv}$$

$$= \frac{6.63 \times 10^{-34} \text{ kg m}^2 \text{ s}^{-1}}{(9.1 \times 10^{-31} \text{ kg})(4.0 \times 10^4 \text{ m s}^{-1})} = 1.8 \times 10^{-8} \text{ m}$$

In nanometers this is

$$(1.8 \times 10^{-8} \text{ m}) \frac{10^9 \text{ nm}}{1 \text{ m}} = 18 \text{ nm}$$

---

[2]Davisson and Germer, who were the first to demonstrate the effect, diffracted electrons off a metal surface. Thomson and Reid diffracted electrons by passing them through a thin foil.

This shows that electrons moving at a speed of 4.0×10^4 m s^{-1} have a wavelength of 18 nm. These electrons would be diffracted by a grating having lines about 18 nm (less than a millionth of an inch) apart, but ruling the lines this close together is impossible. Fortunately, suitable ready-made gratings are naturally available in the form of crystals. We will see in Chap. 10 that in a crystalline material closely spaced layers of atoms can serve as a diffraction grating. Davisson and Germer took a crystal of nickel (distance between atomic layers, 22 nm) and observed that diffraction of electrons occurred, as is shown in Fig. 6-1b. Electrons behave as waves, as well as particles!

According to the de Broglie relationship $\lambda = h/mv$, *all* particles should exhibit wavelike properties. The objects of our everyday world probably do possess the properties of waves, but these objects have such large masses (compared to the tiny value of h, Planck's constant) that the wavelength of even a flea, for instance, is negligibly small. The velocity term v, in the denominator of the de Broglie equation, could in principle be so small that even a flea's wavelength would be large enough to be measured. Unfortunately, the velocity would have to be of the order of 10^{-3} nm per year, and such motion is not detectable. Thus wavelike character is usually significant only for things about as small as electrons.

● EXAMPLE 6-1

Problem   Calculate the wavelength of a flea weighing 1.5 mg and jumping at a velocity of 2.0 m s^{-1}.

Solution m is 1.5 mg, which is

$$1.5 \text{ mg} \left(\frac{1 \text{ g}}{10^3 \text{ mg}} \right) \left(\frac{1 \text{ kg}}{10^3 \text{ g}} \right) = 1.5 \times 10^{-6} \text{ kg}$$

$$v = 2.0 \text{ m s}^{-1}$$

$$h = 6.63 \times 10^{-34} \text{ J s} = 6.63 \times 10^{-34} \text{ kg m}^2 \text{ s}^{-1}$$

$$\lambda = \frac{h}{mv}$$

$$= \frac{6.63 \times 10^{-34} \text{ kg m}^2 \text{ s}^{-1}}{(1.5 \times 10^{-6} \text{ kg})(2.0 \text{ m s}^{-1})} = 2.2 \times 10^{-28} \text{ m} = 2.2 \times 10^{-28} \text{ m} \frac{10^9 \text{ nm}}{1 \text{ m}}$$

$$= 2.2 \times 10^{-19} \text{ nm}$$

Since this is much smaller than the size of even a small atom (about 10 nm), we can safely ignore the wavelength of a flea. (A flea is more particle than wave!) ●

Electronic energies   In the quantum-mechanical atom an electron is considered to behave as much like a wave as like a particle. Although it is easy to imagine waves in water, we obviously need to learn a new kind of thinking for electron waves in atoms. Furthermore, our thinking must allow us to answer three questions: (1) What are the energies of the electrons in an atom? (2) Where is each electron? (3) What is it doing?

One last comment before we move forward. The electronic energy levels which we are about to describe are *hydrogenlike* energy levels; that is, they are similar to the levels obtained from quantum-mechanical calculations for a one-electron atom, hydrogen. Although this may seem like an oversimplification, in practice it works out very well.

SHELLS. In order to describe electronic energies in an atom, it is convenient to organize them into sets each of which is called a *shell*. A shell is therefore a collection of one or more quantized energy levels.

ADDED COMMENT

The word *shell* is unfortunate, being a carry-over from the Bohr theory. Do not think of a hollow shell like a nutshell when you see this word. In Sec. 6-4 we will give this word a geometric interpretation. Until then, try not to associate any geometric shape with it. For now, a shell is a set of energy levels.

Shells are named in two different ways: (1) by using the letters K, L, M, N, O, P, . . . , and (2) by specifying the number n of the shell, where n can be 1, 2, 3, 4, 5, 6, ... Electrons in the K shell ($n = 1$) have the lowest permitted energy.

The maximum electron population of a shell can be shown both theoretically and experimentally to be equal to $2n^2$. Thus the K shell ($n = 1$) can hold up to two electrons, the L shell ($n = 2$) up to eight electrons, and so on. The shells which consist of the lower electronic energies in an atom are schematically shown at the left of Fig. 6-2.

SUBSHELLS. Not all electrons in a given shell have the same energy. Each shell is thus considered to consist of, or contain, one or more *subshells*. (The number of subshells in any given shell is equal to n.) Within a given shell the subshells are named in order of *increasing energy*: *s, p, d, f, g, h, i,* . . . . (Only the first four, *s, p, d,* and *f,* are needed for describing atoms in their ground states.) A specific subshell is named by using the number which designates the shell, plus the letter which indicates the subshell, such as, for example, 2*p*, 4*d*, 5*f*, etc.

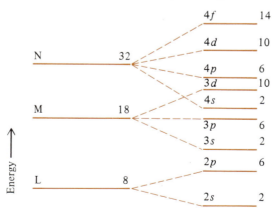

Figure 6-2
Shells and subshells.

As stated above the number of subshells into which a given shell is subdivided depends on the quantum number n. The K shell ($n = 1$) has but *one* subshell, called the 1s subshell. The L shell ($n = 2$) has *two:* 2s and 2p; the M shell ($n = 3$) *three:* 3s, 3p, and 3d; the N shell ($n = 4$) *four:* 4s, 4p, 4d, and 4f, etc.

The *maximum* electronic population of an individual subshell depends upon whether the subshell is s, p, d, or f. There can be no more than 2 electrons in any s subshell, 6 in any p subshell, 10 in any d subshell, and 14 in any f subshell.

Confirmation of the existence of subshells comes from studies of atomic spectra. Subdivision of shells into subshells occurs because electrons in an atom are not independent of each other but interact electrically. Even hydrogen atoms with their single electrons show this splitting of energy levels in the presence of an electric field. (This is called the *Stark effect.*)

Energy relationships among the subshells of the first four shells (K, L, M, and N) are shown at the right of Fig. 6-2. Note the energy overlap of the 4s subshell with the 3d. The 4s subshell actually is of *lower* energy than the 3d. This is partially a result of the bunching up of levels at higher energies (toward the top of the diagram).

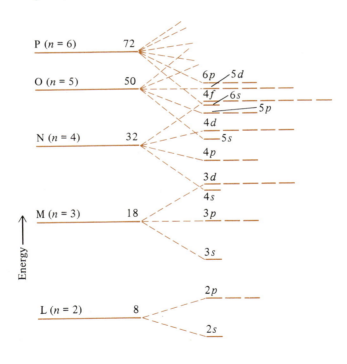

Figure 6-3
Filling diagram
(partial).

| Shell | Maximum population, e^- | Subshell |

ORBITALS. In an isolated atom all electrons in a given subshell have the same energy. Under perturbing influences, however, their energies may be further split into sublevels called *orbitals*. Such splitting occurs, for example, in the presence of a magnetic field. (This is called the *Zeeman effect*.)

The word *orbital*, is, like *shell*, somewhat unfortunate. It is derived from the word *orbit*, as used by Bohr, but it does not in any way signify an orbit. For now think of an orbital simply as an energy level. Later (Sec. 6-4) we will see that *orbital* can be taken to mean a region of space in which an electron of a certain energy is likely to be found.

The number of orbitals into which a subshell is subdivided or split depends on its identity. The number of orbitals in an *s*, *p*, *d*, or *f* subshell is 1, 3, 5, or 7, respectively. Comparing these numbers with the maximum population of each kind of subshell (2, 6, 10, or 14, respectively), we see that *each orbital has a maximum population of two electrons.*

For the lower energy levels the subdivision of subshells into orbitals is shown schematically in Fig. 6-3. In this diagram the orbitals of a given subshell are all shown at the same energy, as they would exist in an isolated atom not subject to external fields.

ELECTRON SPIN. Theoretical calculations and spectroscopic data show that each orbital may contain one or two electrons. We may also refer to an *empty*, or *unoccupied*, orbital, which is a potentially available energy level with no electrons in it. Furthermore, both theory and experiment support the idea that when two electrons occupy the same orbital, they *have different spins*. What does this mean? An interesting experiment performed in 1921 by Otto Stern and Walter Gerlach showed that an electron acts as if it could *spin* in either of two opposite directions. The Stern-Gerlach experiment, shown in Fig. 6-4, consists of passing a beam of metal atoms (Stern and Gerlach used silver, sodium, potassium, and others) through an inhomogeneous (nonuniform) magnetic field. With these metals the beam divides into two components, one being deflected up and the other down. The principles of physics tell us that a particle passing through such a magnetic field will be deflected (1) if the particle is *charged* and (2) if it is *spinning*. If a spinning electron were passed through such a magnetic field, we would expect it to be deflected either up or down, depending on the direction of its spin. But what about a sodium atom? A sodium atom has 11 electrons. Five of these spin in one direction, and six in the other. The five which spin in the first direction have their magnetic effects exactly canceled by five of those which spin

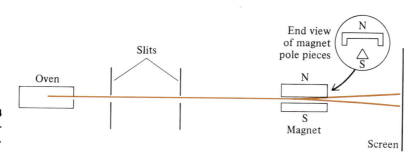

Figure 6-4
Stern-Gerlach experiment.

in the opposite direction. Thus the magnetic effects of ten electrons are self-canceling. But the last electron is uncompensated for and causes the entire atom to be deflected. Since half of the sodium atoms have their "eleventh" electrons spinning one way, and half the other, two beams of atoms emerge from the magnetic field. There are other experimental methods of showing that the electron has spin, but the Stern-Gerlach experiment is the most direct.

Each orbital in an atom can accommodate a maximum of two electrons, one spinning one way and the other spinning in the opposite way. We say that two such electrons are *paired* and have opposing, opposite, or *antiparallel* spins. The spins of electrons contribute to the magnetic behavior of matter. This is because any spinning or rotating object which is electrically charged generates a magnetic field, that is, acts as if it were a magnet. Substances which contain atoms with orbitals which are *half-filled* (contain only one electron) are weakly attracted into a magnetic field, that is, "toward a magnet." Such a weak attraction is called *paramagnetism* and is evidence of the existence of *unpaired* electrons in a substance.[3]

Electronic configurations

The specific way in which the orbitals of an atom are occupied by electrons is called the *electronic configuration* of that atom. We can predict the electronic configurations of many atoms by making use of the diagram of Fig. 6-3. This is usually called a *filling diagram,* because by filling it from the bottom up with electrons we can build up the electronic configuration of each atom. (Figure 6-3 is better called a filling diagram than an energy-level diagram, because the actual sequence of orbital energies is not exactly the same for all atoms.)

In order to determine ground-state configurations we follow a method known as the *Aufbau* (German "building-up") procedure: electrons are added to the filling diagram, starting at the bottom and working up. This ensures that for each atom all electrons are in the lowest energy levels available to them. We start by building up the configurations of the two simplest atoms: hydrogen ($Z = 1$) and helium ($Z = 2$). We are at present interested in considering only neutral atoms, not ions; so in each case the atomic number, Z, gives us the number of electrons. We indicate the electronic configurations of these two atoms as

$$\text{H } (Z = 1): \quad 1s^1 \qquad \text{He } (Z = 2): \quad 1s^2$$

In each case the superscript indicates the electron population of the $1s$ orbital.

An alternative way of showing orbital populations called the *orbital diagram* uses a little horizontal line (or sometimes a square or a circle) to represent an orbital. Electrons are represented by half arrows as follows:

$$\text{H } (Z = 1): \quad \underline{\uparrow}_{1s} \qquad \text{He } (Z = 2): \quad \underline{\uparrow\downarrow}_{1s}$$

The two electrons paired in the $1s$ orbital of helium have antiparallel spins, as indicated by the opposing half arrows.

[3] A much stronger attraction into a magnetic field, *ferromagnetism,* is shown by the elements iron, cobalt, nickel, as well as some other substances. It is commonly called, simply, "magnetism," occurs only in the solid state, and will be discussed in Sec. 22-6. In addition, the phenomenon of *diamagnetism,* a feeble *repulsion* out of a magnetic field is shown by all matter, but is so weak that it is masked by paramagnetism or ferromagnetism, when present.

Table 6-1 Electronic configurations: lithium through neon

Element	Atomic number, Z	Electronic configuration	Orbital diagram $1s$	$2s$	$2p$
Li	3	$1s^22s^1$	↑↓	↑	___ ___ ___
Be	4	$1s^22s^2$	↑↓	↑↓	___ ___ ___
B	5	$1s^22s^22p^1$	↑↓	↑↓	↑ ___ ___
C	6	$1s^22s^22p^2$	↑↓	↑↓	↑ ↑ ___
N	7	$1s^22s^22p^3$	↑↓	↑↓	↑ ↑ ↑
O	8	$1s^22s^22p^4$	↑↓	↑↓	↑↓ ↑ ↑
F	9	$1s^22s^22p^5$	↑↓	↑↓	↑↓ ↑↓ ↑
Ne	10	$1s^22s^22p^6$	↑↓	↑↓	↑↓ ↑↓ ↑↓

The atoms of the next eight elements, $Z = 3$ through 10, have the electronic configurations and corresponding orbital diagrams that are shown in Table 6-1. (Refer to the filling diagram, Fig. 6-3, as you follow the buildup of these atoms.) Note that the $2s$ orbital fills before the $2p$ because its energy is lower. Note also that when electrons are added to the $2p$ subshell, they spread out to occupy different orbitals. Thus the orbital diagram for carbon ($Z = 6$) in its *ground state* is

$$\underset{1s}{↑↓} \; \Big| \; \underset{2s}{↑↓} \quad \underset{2p}{↑ \quad ↑ \quad ___}$$

and *not*

$$\underset{1s}{↑↓} \; \Big| \; \underset{2s}{↑↓} \quad \underset{2p}{↑↓ \quad ___ \quad ___}$$

Note also that unpaired electrons in different orbitals have parallel spins. This is an illustration of an important generalization known as *Hund's rule: Electrons in a given subshell tend to remain unpaired (in separate orbitals) with parallel spins.* Hund's rule is illustrated in nitrogen ($Z = 7$) and oxygen ($Z = 8$), as well.

In each of the eight atoms, Li through Ne (Table 6-1), the K shell is complete ($1s^2$). Since this is the configuration of helium, the designation [He], referring to the *helium core,* is often used instead of $1s^2$. Thus the configuration of carbon can be represented as

$$C\ (Z = 6): \quad [He]2s^22p^2$$

Helium is an element known as a *noble gas* (Sec. 7-2).

The next eight atoms are built up much as the previous eight. Since the atom just completed is neon ($1s^22s^22p^6$), another noble gas, we will represent the *neon core* as [Ne] in the following atoms:

Na ($Z = 11$):	[Ne]$3s^1$	P ($Z = 15$):	[Ne]$3s^23p^3$
Mg ($Z = 12$):	[Ne]$3s^2$	S ($Z = 16$):	[Ne]$3s^23p^4$
Al ($Z = 13$):	[Ne]$3s^23p^1$	Cl ($Z = 17$):	[Ne]$3s^23p^5$
Si ($Z = 14$):	[Ne]$3s^23p^2$	Ar ($Z = 18$):	[Ne]$3s^23p^6$

**Table 6-2
Electronic
configurations
of the elements**

Element	Z	K (1) s	L (2) s	L (2) p	M (3) s	M (3) p	M (3) d	N (4) s	N (4) p	N (4) d	N (4) f	O (5) s	O (5) p	O (5) d	O (5) f	P (6) s	P (6) p	P (6) d	P (6) f	Q (7) s
H	1	1																		
He	2	2																		
Li	3	2	1																	
Be	4	2	2																	
B	5	2	2	1																
C	6	2	2	2																
N	7	2	2	3																
O	8	2	2	4																
F	9	2	2	5																
Ne	10	2	2	6																
Na	11	2	2	6	1															
Mg	12	2	2	6	2															
Al	13	2	2	6	2	1														
Si	14	2	2	6	2	2														
P	15	2	2	6	2	3														
S	16	2	2	6	2	4														
Cl	17	2	2	6	2	5														
Ar	18	2	2	6	2	6														
K	19	2	2	6	2	6		1												
Ca	20	2	2	6	2	6		2												
Sc	21	2	2	6	2	6	1	2												
Ti	22	2	2	6	2	6	2	2												
V	23	2	2	6	2	6	3	2												
Cr	24	2	2	6	2	6	5	1												
Mn	25	2	2	6	2	6	5	2												
Fe	26	2	2	6	2	6	6	2												
Co	27	2	2	6	2	6	7	2												
Ni	28	2	2	6	2	6	8	2												
Cu	29	2	2	6	2	6	10	1												
Zn	30	2	2	6	2	6	10	2												
Ga	31	2	2	6	2	6	10	2	1											
Ge	32	2	2	6	2	6	10	2	2											
As	33	2	2	6	2	6	10	2	3											
Se	34	2	2	6	2	6	10	2	4											
Br	35	2	2	6	2	6	10	2	5											
Kr	36	2	2	6	2	6	10	2	6											
Rb	37	2	2	6	2	6	10	2	6			1								
Sr	38	2	2	6	2	6	10	2	6			2								
Y	39	2	2	6	2	6	10	2	6	1		2								
Zr	40	2	2	6	2	6	10	2	6	2		2								
Nb	41	2	2	6	2	6	10	2	6	4		1								
Mo	42	2	2	6	2	6	10	2	6	5		1								
Tc	43	2	2	6	2	6	10	2	6	6		2								
Ru	44	2	2	6	2	6	10	2	6	7		1								
Rh	45	2	2	6	2	6	10	2	6	8		1								
Pd	46	2	2	6	2	6	10	2	6	10										
Ag	47	2	2	6	2	6	10	2	6	10		1								
Cd	48	2	2	6	2	6	10	2	6	10		2								
In	49	2	2	6	2	6	10	2	6	10		2	1							
Sn	50	2	2	6	2	6	10	2	6	10		2	2							
Sb	51	2	2	6	2	6	10	2	6	10		2	3							
Te	52	2	2	6	2	6	10	2	6	10		2	4							
I	53	2	2	6	2	6	10	2	6	10		2	5							
Xe	54	2	2	6	2	6	10	2	6	10		2	6							

[He] (Li–Ne)
[Ne] (Na–Ar)
[Ar] (K–Kr)
[Kr] (Rb–Xe)

Table 6-2
(Continued)

Element	Z	K (1) s	L (2) s	L (2) p	M (2) s	M (2) p	M (2) d	N (4) s	N (4) p	N (4) d	N (4) f	O (5) s	O (5) p	O (5) d	O (5) f	P (6) s	P (6) p	P (6) d	P (6) f	Q (7) s
Cs	55	2	2	6	2	6	10	2	6	10		2	6			1				
Ba	56	2	2	6	2	6	10	2	6	10		2	6			2				
La	57	2	2	6	2	6	10	2	6	10		2	6	1		2				
Ce	58	2	2	6	2	6	10	2	6	10	1	2	6	1		2				
Pr	59	2	2	6	2	6	10	2	6	10	3	2	6			2				
Nd	60	2	2	6	2	6	10	2	6	10	4	2	6			2				
Pm	61	2	2	6	2	6	10	2	6	10	5	2	6			2				
Sm	62	2	2	6	2	6	10	2	6	10	6	2	6			2				
Eu	63	2	2	6	2	6	10	2	6	10	7	2	6			2				
Gd	64	2	2	6	2	6	10	2	6	10	7	2	6	1		2				
Tb	65	2	2	6	2	6	10	2	6	10	9	2	6			2				
Dy	66	2	2	6	2	6	10	2	6	10	10	2	6			2				
Ho	67	2	2	6	2	6	10	2	6	10	11	2	6			2				
Er	68	2	2	6	2	6	10	2	6	10	12	2	6			2				
Tm	69	2	2	6	2	6	10	2	6	10	13	2	6			2				
Yb	70	2	2	6	2	6	10	2	6	10	14	2	6			2				
Lu	71	2	2	6	2	6	10	2	6	10	14	2	6	1		2				
Hf	72	2	2	6	2	6	10	2	6	10	14	2	6	2		2				
Ta	73	2	2	6	2	6	10	2	6	10	14	2	6	3		2				
W	74	2	2	6	2	6	10	2	6	10	14	2	6	4		2				
Re	75	2	2	6	2	6	10	2	6	10	14	2	6	5		2				
Os	76	2	2	6	2	6	10	2	6	10	14	2	6	6		2				
Ir	77	2	2	6	2	6	10	2	6	10	14	2	6	7		2				
Pt	78	2	2	6	2	6	10	2	6	10	14	2	6	9		1				
Au	79	2	2	6	2	6	10	2	6	10	14	2	6	10		1				
Hg	80	2	2	6	2	6	10	2	6	10	14	2	6	10		2				
Tl	81	2	2	6	2	6	10	2	6	10	14	2	6	10		2	1			
Pb	82	2	2	6	2	6	10	2	6	10	14	2	6	10		2	2			
Bi	83	2	2	6	2	6	10	2	6	10	14	2	6	10		2	3			
Po	84	2	2	6	2	6	10	2	6	10	14	2	6	10		2	4			
At	85	2	2	6	2	6	10	2	6	10	14	2	6	10		2	5			
Rn	86	2	2	6	2	6	10	2	6	10	14	2	6	10		2	6			
Fr	87	2	2	6	2	6	10	2	6	10	14	2	6	10		2	6			1
Ra	88	2	2	6	2	6	10	2	6	10	14	2	6	10		2	6			2
Ac	89	2	2	6	2	6	10	2	6	10	14	2	6	10		2	6	1		2
Th	90	2	2	6	2	6	10	2	6	10	14	2	6	10		2	6	2		2
Pa	91	2	2	6	2	6	10	2	6	10	14	2	6	10	2	2	6	1		2
U	92	2	2	6	2	6	10	2	6	10	14	2	6	10	3	2	6	1		2
Np	93	2	2	6	2	6	10	2	6	10	14	2	6	10	4	2	6	1		2
Pu	94	2	2	6	2	6	10	2	6	10	14	2	6	10	6	2	6			2
Am	95	2	2	6	2	6	10	2	6	10	14	2	6	10	7	2	6			2
Cm	96	2	2	6	2	6	10	2	6	10	14	2	6	10	7	2	6	1		2
Bk	97	2	2	6	2	6	10	2	6	10	14	2	6	10	9	2	6			2
Cf	98	2	2	6	2	6	10	2	6	10	14	2	6	10	10	2	6			2
Es	99	2	2	6	2	6	10	2	6	10	14	2	6	10	11	2	6			2
Fm	100	2	2	6	2	6	10	2	6	10	14	2	6	10	12	2	6			1
Md	101	2	2	6	2	6	10	2	6	10	14	2	6	10	13	2	6			2
No	102	2	2	6	2	6	10	2	6	10	14	2	6	10	14	2	6			2
Lr	103	2	2	6	2	6	10	2	6	10	14	2	6	10	14	2	6	1		2
Rf	104	2	2	6	2	6	10	2	6	10	14	2	6	10	14	2	6	2		2
Ha	105	2	2	6	2	6	10	2	6	10	14	2	6	10	14	2	6	3		2
	106	2	2	6	2	6	10	2	6	10	14	2	6	10	14	2	6	4		2

[Xe] (elements Cs 55 – Rn 86)

[Rn] (elements Fr 87 – 106)

With argon ($Z = 18$) we have again reached an ns^2np^6 last-shell configuration (n is the number of the shell).

Although the third, or M, shell has an empty d subshell at this point, the next electrons are added to the $4s$ subshell, because, as can be seen from the filling diagram, the $4s$ energy level fills before the $3d$. Representing the argon core by [Ar] we have for the next two elements:

$$\text{K }(Z = 19):\qquad [\text{Ar}]4s^1 \qquad \text{Ca }(Z = 20):\qquad [\text{Ar}]4s^2$$

And then we begin adding electrons to the $3d$ subshell. The filling obeys Hund's rule, as the electrons spread out to occupy separate d orbitals, when possible.

Sc ($Z = 21$):	$[\text{Ar}]3d^14s^2$	Fe ($Z = 26$):	$[\text{Ar}]3d^64s^2$
Ti ($Z = 22$):	$[\text{Ar}]3d^24s^2$	Co ($Z = 27$):	$[\text{Ar}]3d^74s^2$
V ($Z = 23$):	$[\text{Ar}]3d^34s^2$	Ni ($Z = 28$):	$[\text{Ar}]3d^84s^2$
Cr ($Z = 24$):	$[\text{Ar}]3d^54s^1$	Cu ($Z = 29$):	$[\text{Ar}]3d^{10}4s^1$
Mn ($Z = 25$):	$[\text{Ar}]3d^54s^2$	Zn ($Z = 30$):	$[\text{Ar}]3d^{10}4s^2$

Here we see electrons being added to the $3d$ subshell while the population of the $4s$ subshell remains almost constant; it contains two electrons except in the atoms of chromium ($Z = 24$) and copper ($Z = 29$). In each of these elements the $4s$ subshell contains but one electron, and the "missing" electron can be seen to be present in the $3d$ subshell. Thus the orbital diagrams of chromium and copper are, respectively:

This illustrates the general rule that an atom gains extra stability by either exactly *half-filling* (one electron per orbital) or *filling* (two electrons per orbital) a subshell composed of more than one orbital. In chromium and copper the $3d$ and $4s$ energies are close enough together for one of the $4s$ electrons to move to the $3d$ subshell in order to half-fill (in Cr) or fill (in Cu) this subshell.

The next six electrons add to the $4p$ subshell:

Ga ($Z = 31$):	$[\text{Ar}]3d^{10}4s^24p^1$	Se ($Z = 34$):	$[\text{Ar}]3d^{10}4s^24p^4$
Ge ($Z = 32$):	$[\text{Ar}]3d^{10}4s^24p^2$	Br ($Z = 35$):	$[\text{Ar}]3d^{10}4s^24p^5$
As ($Z = 33$):	$[\text{Ar}]3d^{10}4s^24p^3$	Kr ($Z = 36$):	$[\text{Ar}]3d^{10}4s^24p^6$

Krypton ($Z = 36$), having a last-shell ns^2np^6 configuration, is a noble gas.

Table 6-2 shows the electronic configurations of all the elements. Note that after each ns subshell (for $n = 4$, or larger) has been filled, the $(n - 1)d$ subshell is filled and is followed by the np subshell, the resulting ns^2np^6 configuration being a noble gas. Note also that for $n = 6$, or larger, the filling of the $(n - 2)f$ subshell intervenes between ns and $(n - 1)d$.

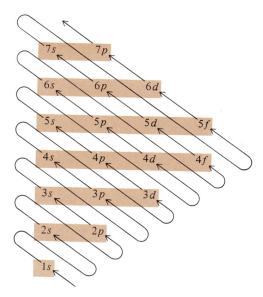

Figure 6-5
Subshell filling sequence. This diagram is helpful for remembering the order in which the subshells are filled. Note that it is constructed by placing on the same horizontal line all subshells with a given value of n. The filling sequence is found by following the diagonal arrows starting at the lower left, as shown.

The filling sequence is not exactly regular, but it can be summarized approximately by Fig. 6-5. Some of the deviations from a perfect filling order can be accounted for by the tendency to exactly half-fill or fill a subshell. The electronic configurations shown in Table 6-2 are those of gaseous, isolated atoms, uninfluenced by any external fields; those of bonded atoms are generally different from those shown. We will return to a consideration of electronic configurations in Chap. 7, when we will consider correlations between configurations and properties.

6-2
One-dimensional standing waves

Having considered the energies of electrons, we now turn our attentions to the questions of where the electrons in an atom are located and what they are doing. Because quantum mechanics considers electrons in atoms to have wavelike properties, we will first consider some of the characteristics of waves in general. The most familiar examples of standing waves are those produced in musical instruments, and of these, the easiest to describe are the waves produced in stringed instruments.

The vibrating string

Figure 6-6 shows a stretched guitar string being plucked. It is pulled back (*a*, in the figure) at its center and then released. The resulting oscillation or vibration is shown in successive stages (*b* through *n*) only fractions of a second apart. Figure 6-6*o* is a composite of all the stages *b* through *n*. The vibration of the string is called a *standing wave,* because it does not appear to move to the left or right along the string as a *running wave* would. The guitar string is fixed at its ends and cannot move at these points. Such points, at which no motion occurs, are called *nodes.*

Figure 6-6 shows only one of the many ways in which a stretched string may vibrate. If, as is shown in Fig. 6-7*a*, the string is held motionless at its center, and is plucked half-way from its center to one end, a different standing wave results.

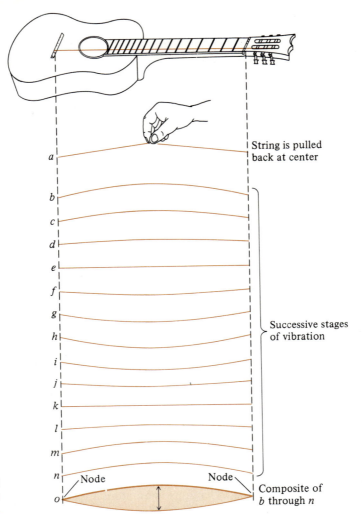

String is pulled
back at center

Successive stages
of vibration

Composite of
b through *n*

Node   Node

Figure 6-6
A standing wave in a
guitar string.

At the instant the string is released the stop at the center is withdrawn, and the string starts its vibration as shown in *b* through *n*. Figure 6-7*o* is a composite representation. Note that this mode of vibration has *three* nodes, one at each end as before, and a new one at the center. *Remember:* Nodes are places on the string where there is no displacement as the string vibrates. Closely related to nodes are *antinodes,* locations halfway between adjacent nodes, where the displacement of the string is at a maximum.

A stretched string may be caused to vibrate in many different modes. Some of these are shown in Fig. 6-8. Each mode has associated with it a characteristic number of nodes.

The modes of vibration shown in Fig. 6-8 are often said to be *allowed,* or *permitted,* which really means that they are the ones which are possible. Certain additional types of vibration may be imagined, but are *forbidden,* that is, they are impossible. Some of these are shown in Fig. 6-9. If you try to set a guitar string

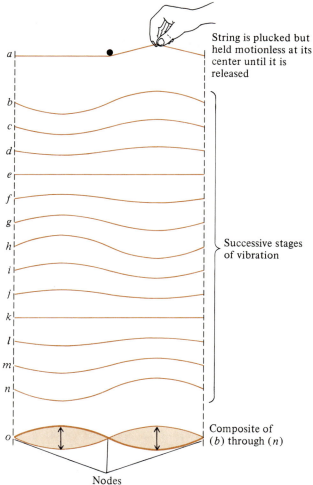

String is plucked but held motionless at its center until it is released

Successive stages of vibration

Composite of (b) through (n)

Nodes

Figure 6-7
Plucked string: a second mode of vibration.

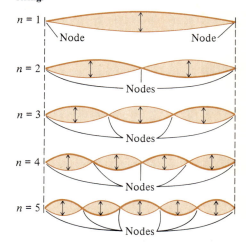

Figure 6-8
Some allowed modes of vibration of a stretched string.

$n = 1$ Node   Node

$n = 2$ Nodes

$n = 3$ Nodes

$n = 4$ Nodes

$n = 5$ Nodes

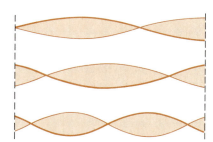

Figure 6-9
Some forbidden modes of vibration of a stretched string.

into vibration in one of these modes, the oscillation of the string very quickly becomes damped out, and almost no musical note results. A glance at the location of the nodes explains why this is true. The spacing of the nodes must be regular, and there must be nodes at the ends of the string, because here the string *cannot* move.

In summary, a stretched string may vibrate in a number of different modes, each with its own characteristic number of nodes (and antinodes), provided that for each mode of vibration a node exists at each end of the string. (The prescription that nodes be present at the ends is known as a *boundary condition*.)

Quantization   The vibrations of a stretched string can be said to be *quantized* because certain specific modes of vibration are allowed. We can label the allowed modes of vibration by assigning a *quantum number* to each. The allowed modes shown in

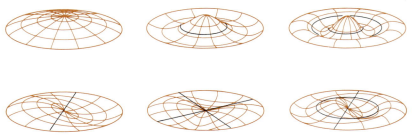

Figure 6-10
Vibrations of a drumhead. These are instantaneous representations of a drumhead vibrating in six different modes. Radial and angular lines (color lines) have been drawn on the drumhead, and the vertical displacement has been exaggerated for clarity. In each mode there is a node at the rim. Additional nodes are shown as black lines.

Fig. 6-8 could be designated $n = 1$, 2, 3, 4, and 5, respectively, where n is a quantum number and is one less than the number of nodes in each case. (Thus n equals the number of antinodes.) Lastly, each mode of vibration has associated with it a certain discrete energy, such that the larger the quantum number n, the higher is the energy of vibration.

6-3
Two-dimensional standing waves

It is easy to extend our ideas about standing waves to a two-dimensional case, the vibrating head of a drum.

The drumhead

The vibration of a drumhead is not always a simple up-and-down motion. A drum struck at various different points on its drumhead makes different sounds because different kinds, actually modes, of vibration of the drumhead are possible. Some of the ways in which a drumhead can vibrate are shown in Fig. 6-10. (In these drawings the vertical scale has been exaggerated to make the patterns more obvious.)

Radial vibrations

Figure 6-11 shows a drum being struck in the exact center of its drumhead. Depending upon how hard it is struck, the drumhead will vibrate in one (or a combination) of various modes of *radial* vibrations, the first three of which are shown. The first mode is one in which the drumhead makes a simple up-and-down displacement. This mode has only one node, but unlike the case of the stretched string, where the nodes are points, this node is a line, a circle around the rim of the drum. This node must exist for all modes of vibration because the boundary condition for a vibrating drumhead is that the displacement must always be zero around the periphery.

The second mode of radial vibration of a drumhead has a second circular node, as shown. When the drumhead moves up near the rim, it moves down near the center. At any point *on* the node there is no motion of the drumhead at all in either the upward or downward direction.

The third mode of radial vibration has one more circular node, for a total of

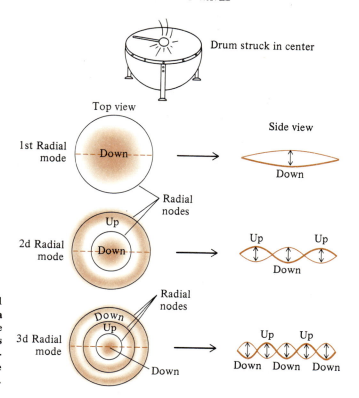

Figure 6-11
Radial vibrations of a drumhead. Each side view shown in the cross section along the dotted line shown in the corresponding top view.

three. In this case, as before, the modes separate regions of the drumhead which are moving in opposite directions at any instant.

Angular vibrations   When the drum is struck off center, modes of *angular* vibration are produced in the drumhead. Four of these are shown in Fig. 6-12. Each of the first two has, in addition to the ever-present radial node around the rim, one *angular node*, a straight line passing through the center of the drumhead. On one side of the angular node the drumhead moves up, while on the other it moves down. An instant later the locations of the sides are reversed, and the left is down while the right is up, the alternation continuing as the vibration goes on.

The third and fourth modes shown in Fig. 6-12 each have two angular nodes. As always, there is no upward or downward motion at any point on a node, and the nodes separate regions where the drumhead is moving in opposite directions.

Quantization   As in the case of the one-dimensional vibrating string the vibrations of a drumhead are quantized in energy, but this time, because a second dimension has been added, they are also quantized in *orientation* within the plane. In order to classify or describe such two-dimensional vibrations, we need not just one, but two quantum numbers, one to describe the frequency or energy of vibration and one to specify its orientation.[4]

[4]See, for example, J. Waser, K. N. Trueblood, and C. M. Knobler, *Chem One,* McGraw-Hill Book Co., New York, 1976, pp. 224, 225.

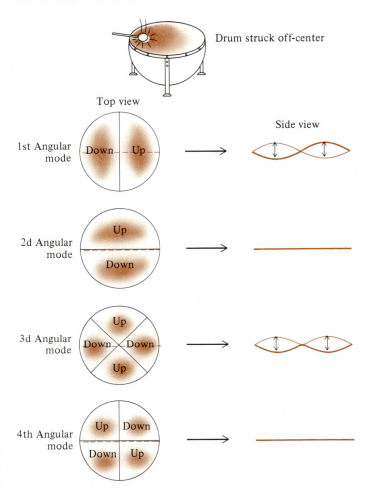

Drum struck off-center

Top view

Side view

1st Angular mode

Down | Up

2d Angular mode

Up

Down

3d Angular mode

Down | Down

Up

Up

4th Angular mode

Up | Down

Down | Up

Figure 6-12
Angular vibrations of a
drumhead.

6-4
Electrons: three-
dimensional waves

The previous discussion could easily be extended to similar standing-wave vibrations in three-dimensional space. For example, the placement of several loudspeakers in various locations in a closed room can be used to produce different standing waves in the vibration of the air in the room. As before, nodes and antinodes are present, but in this case the nodes are *surfaces*. Investigating these modes of vibration is interesting, but it is unnecessary here, and so we will simply observe that standing waves in three dimensions do exist.

The Heisenberg
uncertainty
principle

In 1927 Werner Heisenberg, a German physicist, developed an important relationship known as the *uncertainty principle*. This relationship puts a severe limitation on our ability to learn about the motion of a small particle. The uncertainty principle indicates that it is impossible to know simultaneously both the *momentum* and *position* of a small particle with any degree of certainty.

The crux of the uncertainty principle is that in order to learn anything about the momentum (product of mass times velocity) of a particle, we must somehow interact with that particle. Consider, for example, a small piece of goose-down fluff drifting slowly to the floor in a draft-free room. Suppose you wished to learn about the motion of this bit of fluff *in the dark!* If your sense of touch were sensitive enough and you were skillful enough, you could reach out with your hand, allow the fluff to brush by it, and get an idea about the fluff's position and motion. But the very act of touching the fluff would change its motion slightly, so that it would no longer fall as it would have, had you not touched it. Your crude measurement of the fluff's position and momentum would cause a change in the very quantities you wished to determine.

The situation is similar for any particle as tiny as an electron. No instrument can "feel" or "see" an electron *without severely influencing the electron's motion.* If, for example, a "supermicroscope" were devised for the purpose of measuring the position of an electron, it would have to use radiation with a much smaller wavelength than that of light. (In order for a small object to be seen in a microscope, the wavelength of the light used must be smaller than the size of the object.) Our imaginary supermicroscope would therefore need to be designed to use x rays, γ rays, or perhaps, electrons. But the energy of such radiation would be so large that it would change the momentum, and hence the velocity, of the electron by a large, uncertain amount.

The uncertainty principle may be paraphrased: the closer we attempt to look at a tiny particle, the more indistinct our view of it becomes. The problem is not important for large objects such as bits of dust, automobiles, or baseballs. Here the Heisenberg uncertainty associated with each measurement is negligible compared with the size of the measurement itself. But for an electron we are forced to conclude that any physical picture, any mental model, of the electronic structure of the atom must not precisely and simultaneously (1) locate the electron and (2) describe its motion. We are thus led to the conclusion that any picture of the atom as concrete as the Bohr picture with its well-defined electronic trajectories or orbits must be wrong, wrong in the sense that the uncertainty principle has been violated.[5]

ADDED COMMENT

At this point we are about to develop a description of the quantum-mechanical atom. We will need to refer to some higher mathematics, but we are not going to "do" any. The picture of the atom which we will develop is easy to accept; just keep an open mind, and do not be intimidated by a few Greek letters!

Wave equations   It is possible to describe any wavelike motion by using a kind of mathematical equation known as a *wave equation*. We have already mentioned de Broglie's suggestion (Sec. 6-1) that electrons can be considered to be waves. In 1926 Erwin Schrödinger, who with Heisenberg is considered to be co-founder of quantum

---

[5] The Heisenberg uncertainty principle may be stated mathematically as $\Delta p \, \Delta x \approx h$. Here Δp represents the uncertainty in the momentum (mass times velocity) of a particle, and Δx represents the uncertainty in its *position*. The product of these is shown by the above relationship to be of the order of h, Planck's constant. This means that if one uncertainty is low, the other must be high. In other words, the more we know about how a particle is moving, the less we know about where it is, and the more we know about where it is, the less we know about how it is moving.

mechanics, wrote a wave equation for the electron in a hydrogen atom. And thus quantum mechanics, also known as wave mechanics, was born. The solutions to the Schrödinger wave equation provide information about the various energy states for the hydrogen atom and can also be used to describe other atoms. The approach can even be used to describe bonding and the properties of molecules. Because quantum-mechanical calculations are difficult and tedious to perform with accuracy, many could not be done until the advent of the high-speed electronic computer, but today the practical power of the quantum-mechanical approach has finally been realized.

A wave equation is not much like an algebraic equation. It is of a type known as a *differential* equation. It is important to know that a differential equation normally has not just one, or two, but rather a whole series of solutions. When, for example, a wave equation is written and then solved for the case of a vibrating string, each solution corresponds to one mode of vibration of the string and can be used to find wavelength, frequency, energy, etc., for that mode. This is also true for wave equations written for electrons in atoms. From such a wave equation a series of separate solutions is obtained, each of which corresponds to a different energy state. Thus in quantum mechanics it is not necessary to *assume* quantization of energy, as it was in the case of the Bohr atom. Instead, the condition of energy quantization falls quite naturally out of the theory. It happens as a direct consequence of the electron's wavelike character.

Each solution to a wave equation is called a *wave function* and is represented by the Greek letter Ψ. (This should be pronounced "psee," but is very commonly mispronounced "psi" or "si," to rhyme with "eye.") The first solution to the wave equation for the hydrogen atom can be written as

$$\Psi_{1s} = Ae^{-Br} \tag{6-1}$$

Here A and B are constants (which can be evaluated from the theory). e is also a constant, being the base of the natural logarithms (2.71828 . . .), and r represents the distance from the nucleus. Ψ corresponds to the *amplitude* of the electron wave. (The amplitude of a wave in a vibrating string is the size of the displacement of the string at an antinode. See Fig. 5-8.) More important is the value of Ψ^2. This represents *the probability of finding the electron within some tiny region which is at a distance r from the nucleus.*

We now see how quantum mechanics meets the challenge posed by the uncertainty principle. Instead of providing a concrete picture of an electron's position and motion, it speaks in terms of probabilities of finding the electron at various places. Since Ψ^2 is a probability of finding an electron within some tiny region or volume, it is a probability per unit volume and is called, therefore, a *probability density* (recall that ordinary density is mass per unit volume).

The 1s orbital   Look at Eq. (6-1) again. It is the solution to the wave equation for an electron in a 1s orbital. The exponent is $-Br$, and because of the negative sign, Ψ will decrease (quite rapidly) as r increases. This means that the farther we go from the nucleus, the smaller will be the probability of finding the electron. This relation between probability density, Ψ^2, and distance from the nucleus, r, can be shown on a graph, Fig. 6-13, in which values of Ψ^2 are plotted against r. Looking at the graph [or at Eq. (6-1)] we see that the probability is highest at (immediately

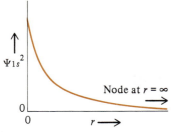

Figure 6-13
Probability-density plot for a 1s electron. This shows the probability of finding a 1s electron in a tiny volume at a distance r from the nucleus.

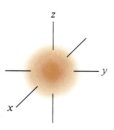

Figure 6-14
The electronic charge cloud of a 1s electron.

adjacent to) the nucleus. This does *not* mean that the electron *is* next to the nucleus, merely that the chances of finding it are highest there. The chance of finding the electron gets smaller the farther we go from the nucleus, but that chance never really gets to zero (except at infinity). It soon gets so low, however, that for practical purposes we need not worry about it.

Thinking in terms of probabilities like this may seem difficult at first. Perhaps the following discussion will make it easier. Imagine that the electron is rapidly moving in some way which is pretty much undescribable, but in which it spends more time near the nucleus and less and less time at distances farther and farther away. In this way the negative charge carried by the electron becomes, in effect, smeared out into a blob or cloud which is most concentrated or dense at the center and which becomes progressively thinner going away from the nucleus. This electron *charge cloud* takes the shape of a sphere (for a 1s electron) and would look, if we could see it, something like Fig. 6-14. Now suppose that we cut the sphere through the middle and look at a cross section. We would see something like Fig. 6-15. The intensity of shading is shown greatest at the nucleus; this indicates that the density or concentration of electronic charge is greatest there. In other words the probability of finding the electron is greatest at the nucleus, which is exactly what Eq. (6-1) and Fig. 6-13 say.

Another way of depicting the probability-density distribution is obtained by constructing a surface (in three dimensions), every point on which has the same value of Ψ^2. For the electron in a 1s orbital such a surface is a sphere, as shown in Fig. 6-16. Since the actual shape of the electron charge cloud is spherically symmetrical, any number of spheres could be so constructed, the larger ones

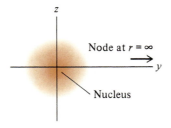

Figure 6-15
Probability density for a 1s electron. This is a cross sectional view of the charge cloud shown in Fig. 6-14.

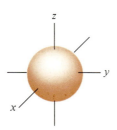

Figure 6-16
Boundary surface for a 1s orbital.

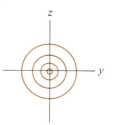

Figure 6-17
Contour diagram for a 1s electron.

containing a larger percentage of the total electronic charge. These surfaces of constant probability density, Ψ^2, are called *boundary surfaces* and are usually constructed large enough so that they enclose a large percentage, say 90 percent, of the electronic charge. Stated differently, the probability that the electron is *somewhere inside* the sphere is 90 percent.

A last way of showing the shape of a probability-density distribution is shown in Fig. 6-17. Each circle in the diagram is the cross-section of a boundary surface, the larger ones obviously enclosing more of the electronic charge than the smaller. Such a diagram is often called a *contour diagram* because each line represents a contour of constant probability density in much the same way as lines on a topographic map represent contours of constant elevation.

Ψ^2 tells us the probability of finding an electron within a tiny element of volume *in a specified location*. It is also common to show the probability of finding an electron *at a specified distance* from the nucleus, *irrespective of angular direction*. This is called the *radial probability* and is plotted for a 1s electron in Fig. 6-18. Each point on the curve, called the *radial probability distribution*, represents the total probability of finding an electron in all of the tiny volume elements which are at a distance r from the nucleus.

Notice that the radial probability distribution curve starts at 0 (at $r = 0$), increases, goes through a maximum, and then decreases. Why is this curve different from the probability-density curve shown in Fig. 6-13? The answer is found in the fact that the number of tiny volume elements at a distance r from the nucleus increases as r increases. This is true because all these volume elements are part of a thin spherical shell of radius r, and as r increases, the number of volume elements out of which the shell is composed effectively increases. At small distances from the nucleus the radial probability is low even though Ψ^2 is high, because the number of volume elements in the shell is low. As r increases, the

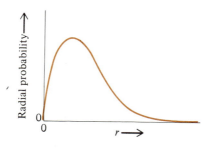

Figure 6-18
Radial probability distribution for a 1s electron.

number of volume elements increases while Ψ^2 decreases. This makes the radial probability, that is, the probability of finding the electron within the thin spherical shell, first increase, as the size of the shell increases while Ψ^2 is still fairly high, and then decrease, as Ψ^2 drops to very low values. (*Note: within the shell* means "in the shell," *not* enclosed by it. Think of the tiny volume elements as being like microscopic shreds of tomato skin, not like pieces of the interior of the tomato.)

The maximum in the 1s radial probability curve shown in Fig. 6-18 shows the distance from the nucleus at which the electron is most probably found. This means that it is most probably found somewhere in a thin, spherical shell of this radius. It is interesting to note that this distance is exactly that predicted by the Bohr theory for the radius of the orbit of a 1s electron.

The 2s and 3s orbitals

Each *s* orbital is spherically symmetrical. In addition, each has a spherical *node* at infinity. (This statement merely means that the probability of finding an electron approaches zero at a distance approaching infinity in any direction from the nucleus.) A 2s orbital has, in addition, another spherical node. This second node shows up as a place where the value of Ψ briefly reaches zero in a probability-density plot (Fig. 6-19). Also shown is a cross section of the electron charge-cloud representation of a 2s electron. The node is a spherical surface separating the central, high-density region from a shell-shaped high-density region farther out.

A 3s electron has the characteristics shown in Fig. 6-20. Notice that it has *two* spherical nodes at intermediate distances in addition, of course, to the node at infinity.

The 2p and 3p orbitals

The shape of a boundary surface is related to the shape of the three-dimensional electron standing wave but is more properly described as either (1) the shape of the region in which the electron is most probably found, (2) the shape of the electron

Figure 6-19
A 2s orbital. (*a*) **Probability-density plot (not drawn to scale).** (*b*) **Cross section of electronic charge cloud.**

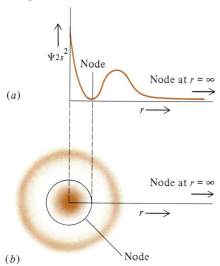

Figure 6-20
A 3s orbital. (*a*) **Probability-density plot (not drawn to scale).** (*b*) **Cross section of electronic charge cloud.**

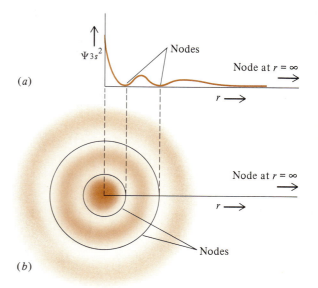

charge cloud, (3) the shape of the electronic probability-density distribution, or, most simply, (4) the *shape of the orbital*.

You should recall from Sec. 6-1 that the 2p subshell of an atom consists of *three* orbitals. Each of these is of the same energy, in the absence of external influences, at least. But each does not have the same orientation in space. Of the three, we will first consider the $2p_y$ orbital. The wave function for an electron in this orbital can be written as

$$\Psi_{2p_y} = Cye^{-Dr} \tag{6-2}$$

Here C and D are constants, and, as before, e is the base for the system of natural logarithms, also a constant. Compare this equation with Eq. (6-1), namely,

$$\Psi_{1s} = Ae^{-Br}$$

The two wave functions are similar, but the one for the $2p_y$ electron has an extra y in it. This y tells us that the probability-density distribution for the $2p_y$ electron cannot be spherically symmetrical. Considering the usual xyz coordinate system (usually called three-dimensional *cartesian* coordinates), it means that anywhere that the value of y is zero, the value of Ψ, and hence Ψ^2, must be zero. And where is y zero? Anywhere on the plane defined by the x and z axes, the so-called xz plane.

The electron charge-cloud distribution for a $2p_y$ electron is shown in Fig. 6-21. It clearly does not have the spherical symmetry of the 1s orbital. It is composed, instead, of two *lobes,* each of which is like a sphere which has been somewhat flattened. (Imagine that you are holding two spherical balloons, side by side, between your hands. Now imagine that you have pushed the balloons together slightly, while cupping your hands a little. The resulting form is close to the shape of a 2p orbital.)

The contour diagram for a $2p_y$ electron is shown in Fig. 6-22. Here, as before, several contours are shown, each of which is a cross section of a boundary surface which encloses a different amount of electronic charge.

The probability of finding a $2p_y$ electron on the xz plane is zero. That this is true can be seen from the perspective of Fig. 6-21. This plane is a *node* which

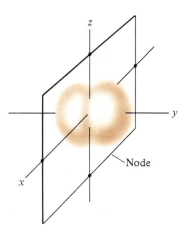

Figure 6-21
The electronic charge cloud of a $2p_y$ electron.

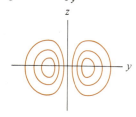

Figure 6-22
Contour diagram for a $2p_y$ orbital.

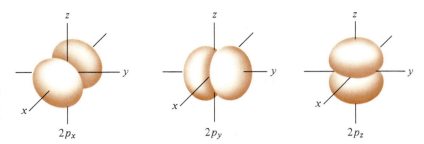

Figure 6-23
Boundary surface dia-
grams of the three 2p
orbitals.

separates the two lobes of the orbital. Everywhere on this nodal plane the probability of finding the electron is zero, even though there is a high-probability region on each side of the node. Thus the electron appears to spend considerable time on each side of the node, but is never on the node itself. This seems absurd until we remember that we are dealing with an electron which has wavelike properties. Do you recall the standing waves in a vibrating string or drumhead? These have nodes, too, and a lot of action can occur on each side of a node with none at the node itself.

The existence of nodes in probability-density distributions has prompted much conjecture about the answer to the question, "How does the electron get across the node?" One clever response goes, "Since the node is a geometrical plane, and since a plane has zero thickness, it should take zero time for the electron to pass through it!" Actually, we should not worry about the electron's crossing the node. Just as a vibrating string has nodes and antinodes, the electronic charge distribution can have similar nodes and antinodes. Thus the electron can be thought of as being smeared out into the two lobes of a p orbital with a node in between. Think of an electron as a *wave*.[6]

The set of three 2p orbitals is pictured in Fig. 6-23. The boundary surfaces of these orbitals are identical in shape, but their orientations differ. The $2p_x$ orbital has the x axis running through it, and the yz plane is a node. The y and z axes run through the $2p_y$ and $2p_z$ orbitals, respectively, and their respective nodes are the xz and xy planes.

The probability-density curve for a 2p electron is nearly always plotted so that values of Ψ^2 are shown at various distances from the nucleus *along that axis which runs through the lobes*. (It remains zero, of course, along each of the other two axes.) Such a plot is shown in Fig. 6-24. Note that the value of Ψ^2 starts at zero (at the node), increases, passes through a maximum, and finally approaches zero again as r approaches infinity (another node). Also shown in the diagram is the cross section of the electron charge cloud.

A 3p orbital is like a 2p orbital, except that it has an extra radial node. The charge distribution is shown in Fig. 6-25, in which both the probability density plot and the charge-cloud cross section are shown.

[6]In 1928 Paul A. M. Dirac developed an elegant version of quantum mechanics which incorporated aspects of the theory of relativity. The results of such relativistic quantum mechanics show that the probability density does not quite become zero at a node. Nodes are hence surfaces at which Ψ^2 is very low, almost zero.

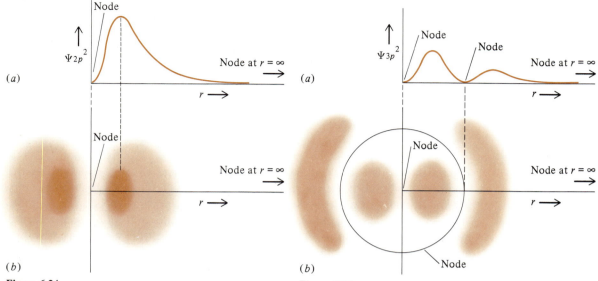

Figure 6-24
A **2p** orbital. (*a*) Probability-density plot. (*b*) Cross section of electronic charge cloud.

Figure 6-25
A **3p** orbital. (*a*) Probability-density plot. (*b*) Cross section of electronic charge cloud.

The 3d orbitals   The 3*d* subshell consists of five orbitals. These are equivalent in energy (in the unperturbed atom) and are shown in charge-cloud representations in Fig. 6-26. The $3d_{xy}$ orbital consists of four lobes, each of which is located between the *x* and *y* coordinate axes. The $3d_{xz}$ and $3d_{yz}$ orbitals have lobes which are similarly

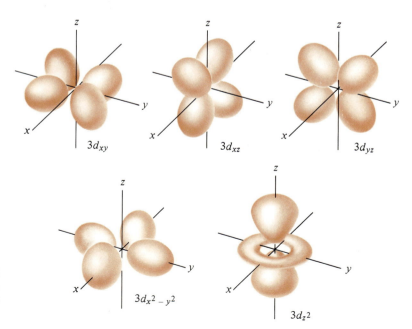

Figure 6-26
Boundary surface diagrams of the five 3d orbitals.

located between the x and z axes and between the y and z axes, respectively. The $3d_{x^2-y^2}$ orbital also has four lobes, but these are oriented so that the x and y axes pass right through the lobe centers. The $3d_{z^2}$ is different; it has a pair of lobes pointing along the z axis plus a "doughnut" around the middle, as is shown in the diagram. These $3d$ orbitals and the similar $4d$ and $5d$ orbitals will become important for us later in accounting for the bonding and other properties of the transition elements.

The f orbitals

The f orbitals are even more complex than the d orbitals, having up to eight lobes. The f orbitals are important in connection with lanthanoid and actinoid chemistry.

Multiple-electron distributions

We have seen that an orbital may accommodate either one or two electrons. What is the difference between the electronic charge-cloud density associated with *one* electron in a given orbital as compared with two electrons in the same orbital? The answer is that in any tiny volume of space the (total) probability of finding any one electron (*either* electron) is twice as great as that of finding one specific electron. Within any small volume the density of the total charge cloud is twice as great for two electrons as it is for only one electron in the same orbital.

Do the fuzzy charge clouds penetrate each other? Can two electrons thus occupy the same space? The answer is yes, provided that we think of the electrons either as waves, or as particles, both of which are rapidly traversing the same space over a period of time.

What is the shape of the *total* electronic charge-cloud distribution for the $2p_x^1 2p_y^1$ configurations? In this case the edges of the lobes of the $2p_x$ orbital overlap those of the $2p_y$ orbital to form a distribution which looks like a doughnut, shown in Fig. 6-27.

What is the shape of the total electronic charge-cloud distribution for the $2p_x^1 2p_y^1 2p_z^1$ configuration? The edges of the orbitals overlap to produce an overall *spherical* charge distribution, as shown in Fig. 6-28.

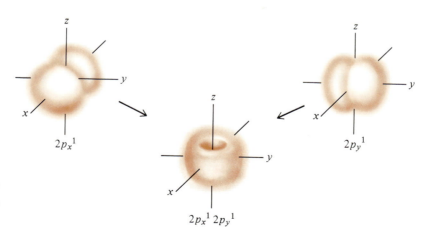

$2p_x^1$

$2p_y^1$

$2p_x^1 2p_y^1$

Figure 6-27
A $2p_x^1 2p_y^1$ charge cloud.

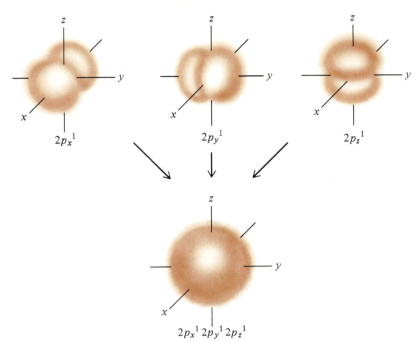

Figure 6-28
A $2p_x{}^1 2p_y{}^1 2p_z{}^1$ charge cloud.

6-5
Quantum numbers

In order to designate the subshell and orbital we have up until now used labels such as $2p_x$, $3d_{xy}$, etc. An alternate way of labeling an electron makes use of *four* quantum numbers.

The principal quantum number n

The first quantum number, represented by the letter n, is called the *principal quantum number*. This gives a rough indication of the *energy* of an electron and of its *average distance from the nucleus*. The allowed values for n are the *positive integers:* 1, 2, 3, 4, ...

The principal quantum number tells us the *shell* in which the electron is found. Thus $n = 1$ corresponds to the first, or K, shell; $n = 2$ to the second, or L, shell. This quantum number is essentially the same as the n of the Bohr theory.

The azimuthal quantum number l

The second quantum number, l, is usually called the *azimuthal quantum number*. It specifies the *subshell,* and hence the *orbital shape*.

The azimuthal quantum number l can have integral values ranging from zero up to $n - 1$. $l = 0$ signifies an s subshell, which means that the orbital must be spherically symmetrical. $l = 1$ corresponds to a p subshell, which means that the orbital must have the typical two-lobed shape of a p orbital. Similarly, $l = 2$ corresponds to a d subshell, and $l = 3$ to an f subshell.

The magnetic quantum number m_l

The third quantum number is the *magnetic quantum number, m_l*. (The term *magnetic* refers to the fact that the different orbitals of a given subshell have different quantized energies in the presence of a magnetic field.) m_l gives information about the *orientation* of an orbital in space.

The magnetic quantum number m_l can have integral values ranging from a minimum of $-l$ to a maximum of $+l$. (If $l = 3$, for instance, m_l may be -3, -2,

Table 6-3
The four quantum
numbers

Name	Symbol	Information provided	Possible values
Principal	n	Shell	1, 2, 3, 4, . . .
Azimuthal	l	Subshell	$0, 1, 2, \ldots, n-1$
Magnetic	m_l	Orbital	$-l, \ldots, 0, \ldots, l$
Spin	m_s	Spin	$+\frac{1}{2}, -\frac{1}{2}$

$-1, 0, +1, +2,$ or $+3$.) It would be nice if we could always associate a particular value of m_l with one of the orbitals we have described. Unfortunately that is not possible. Each p subshell, for example, consists of three independent orbitals. But each orbital corresponds to a certain wave function, a solution to the wave equation. Furthermore, there is more than one way of choosing a set of three "p-type" wave functions as independent solutions to the wave equation. One way gives us a set of three orbitals with m_l values of $-1, 0$, and $+1$, respectively. These are useful in describing the magnetic and spectral properties of atoms. Another way gives us the familiar p_x, p_y, and p_z orbitals. These are useful in accounting for other atomic behavior, such as bonding. It is the latter set which we will use the most in this book.[7]

The spin quantum number m_s

The last quantum number, the *spin quantum number,* is represented by m_s. It specifies the spin of the electron and has a value of either $+\frac{1}{2}$ or $-\frac{1}{2}$.

The four quantum numbers, the information they provide, and their possible values are summarized in Table 6-3.

The Pauli exclusion principle

No two electrons in a given atom can be in exactly the same state. In other words, all electrons in an atom are somehow different. This is a rough statement of the *Pauli exclusion principle,* which is usually worded: *No two electrons in an atom can have the same values for all four quantum numbers.*

Values of each of the quantum numbers for each of the electrons in a chlorine atom (Cl; $Z = 17$) are given as an example in Table 6-4.

Quantum numbers and nodes

In the case of the vibrating string (Sec. 6-2) we showed that a quantum number could be assigned to each mode of vibration. Also, we pointed out that the number of nodes in each vibration mode can be determined from the corresponding quantum number. A system of two quantum numbers can be used to specify the number of radial and angular nodes in a vibrating drumhead. Thus we may anticipate that the numbers of various kinds of nodes in an electronic charge cloud are related to some of its quantum numbers.

The principal quantum number n specifies the *total number of nodes* in the charge cloud. An s electron has only radial nodes, spherical surfaces on which the probability density is everywhere zero. Refer back to Figs. 6-15, 6-19, and 6-20. These show the cross section of the electronic charge cloud for a $1s$, a $2s$, and a $3s$ electron, respectively. The $1s$ distribution has a radial node only at infinity (∞). That is, if we go out from the nucleus *in any direction,* Ψ^2 becomes zero only when r equals infinity. The $2s$ charge cloud has this node at infinity but has, in addition,

---

[7] Each of the two sets of p orbitals does contain one element in common: the p orbital for which $m_l = 0$ is identical to the p_z orbital.

Table 6-4
Quantum numbers for each of the 17 electrons in a chlorine atom

Electron	n	l	m_l	m_s
1	1 ⎫ K	0 ⎫ s	0 ⎫ $1s$	$-\frac{1}{2}$
2	1 ⎭	0 ⎭	0 ⎭	$+\frac{1}{2}$
3	2	0 ⎫ s	0 ⎫ $2s$	$-\frac{1}{2}$
4	2	0 ⎭	0 ⎭	$+\frac{1}{2}$
5	2	1	-1	$-\frac{1}{2}$
6	2	1	-1	$+\frac{1}{2}$
7	2 ⎬ L	1	0	$-\frac{1}{2}$
8	2	1 ⎬ p	0 ⎬ $2p_x, 2p_y, 2p_z$	$+\frac{1}{2}$
9	2	1	$+1$	$-\frac{1}{2}$
10	2	1	$+1$	$+\frac{1}{2}$
11	3	0 ⎫ s	0 ⎫ $3s$	$-\frac{1}{2}$
12	3	0 ⎭	0 ⎭	$+\frac{1}{2}$
13	3	1	-1	$-\frac{1}{2}$
14	3 ⎬ M	1	-1	$+\frac{1}{2}$
15	3	1 ⎬ p	0 ⎬ $3p_x, 3p_y, 3p_z$	$-\frac{1}{2}$
16	3	1	0	$+\frac{1}{2}$
17	3	1	$+1$	$-\frac{1}{2}$

a second radial node at an intermediate distance. The $3s$ electron has *two* intermediate radial nodes plus the one at infinity.

The azimuthal quantum number l specifies the number of *angular nodes*. Angular nodes are surfaces of zero probability density which pass through the origin, that is, the nucleus. For the orbitals which we have considered these surfaces are either *planes* or *cones*. Refer again to Fig. 6-24 (for $2p$ electrons) and to Fig. 6-25 (for $3p$ electrons). For any p electron $l = 1$. Correspondingly, the $2p$ and $3p$ distributions each show one angular node, a plane passing through the nucleus perpendicular to the axis of the orbital.

Look at the $3d$ distributions shown in Fig. 6-26. For any d electron, $l = 2$; so we should look for *two* angular nodes in each of these charge clouds. For the d_{xy}, d_{yz}, d_{xz}, and $d_{x^2-y^2}$ distributions these are planes. For the d_{z^2} the nodes are, however, conical. Each cone separates one of the lobes from the "doughnut" of the charge around the middle.

SUMMARY

In this chapter we examined the extranuclear region of the atom. The *quantum-mechanical model* of the atom considers electrons to behave as *waves,* and the consequence of this is that their energies and distributions in space are both *quantized.* Direct evidence of the wavelike character of electrons comes from their diffraction by the regularly spaced atoms in a crystal.

The quantization of electronic energies can be described in terms of *shells, subshells,* and *orbitals,* which are discrete energy-level sets and subsets. Each orbital can accommodate up to two electrons with *antiparallel spins.* Orbitals are grouped into subshells (s, p, d, f, . . .), which are in turn grouped into shells. The detailed assignment of the electrons in an atom to these energy levels is called the *electronic configuration* of the atom. *Ground-state* configurations are configurations of lowest-energy states and can be predicted with considerable accuracy by following the *Aufbau* procedure and using a *filling diagram.*

By limiting the exactness with which position and

momentum of electrons can be specified, the *Heisenberg uncertainty principle* shows that we must reject the Bohr model of the atom in favor of a quantum-mechanical description. In this we describe regions of high probability of finding an electron; these regions are the spatial distributions which correspond to orbitals. The shape of the distribution, often called the *shape of the orbital,* depends upon the subshell: an *s* orbital is spherical, a *p* orbital is two-lobed, etc. Different orbitals in a given subshell have different orientations. Thus the three *2p* orbitals are identical in shape but are oriented at right angles to each other.

Each electron in an atom can be assigned values of the four *quantum numbers, n, l, m_l,* and *m_s.* Specifying these for an electron is equivalent to specifying the electron's shell, subshell, orbital, and spin. According to the *Pauli exclusion principle* no two electrons in an atom can have identical sets of quantum numbers.

KEY TERMS

Antinode (Sec. 6-2)
Antiparallel spin (Sec. 6-1)
Aufbau procedure (Sec. 6-1)
Azimuthal quantum number (Sec. 6-5)
Boundary surface (Sec. 6-4)
Classical mechanics (Sec. 6-1)
Contour diagram (Sec. 6-4)
Diffraction grating (Sec. 6-1)
Electron diffraction (Sec. 6-1)
Electron spin (Sec. 6-1)
Electronic charge cloud (Sec. 6-4)
Electronic configuration (Sec. 6-1)
Filling diagram (Sec. 6-1)
Heisenberg uncertainty principle (Sec. 6-4)
Hund's rule (Sec. 6-1)
Magnetic quantum number (Sec. 6-5)
Noble-gas core (Sec. 6-1)

Node (Sec. 6-2)
Orbital (Sec. 6-1)
Orbital diagram (Sec. 6-1)
Orbital lobe (Sec. 6-4)
Parallel spins (Sec. 6-1)
Paramagnetism (Sec. 6-1)
Principal quantum number (Sec. 6-5)
Probability density (Sec. 6-4)
Quantum mechanics (Sec. 6-1)
Quantum number (Secs. 6-1 and 6-5)
Shell (Sec. 6-1)
Subshell (Sec. 6-1)
Spin quantum number (Sec. 6-5)
Wave equation (Sec. 6-4)
Wave function (Sec. 6-4)
Wave mechanics (Sec. 6-1)

QUESTIONS AND PROBLEMS

Waves and electrons

6-1 How can it be shown experimentally that electrons have properties of waves?

6-2 Calculate the wavelength of an electron (mass = 9.1×10^{-31} kg) traveling at 5.0×10^6 m s^{-1}.

6-3 Calculate the wavelength of an elephant (mass = 4500 kg) traveling at 1.0 m s^{-1}. Comment on the wave nature of an elephant.

6-4 Show how the Bohr theory of the atom is inconsistent with the Heisenberg uncertainty principle.

6-5 If all particles have wavelike characteristics, why is it that we do not observe diffraction of large particles such as bullets and baseballs?

Electronic configurations

6-6 Using the filling diagram (Fig. 6-3), write the ground-state electronic configuration of each of the following atoms: **(a)** N ($Z = 7$) **(b)** Si ($Z = 14$) **(c)** V ($Z = 23$) **(d)** Se ($Z = 34$) **(e)** Kr ($Z = 36$).

6-7 Draw an orbital diagram for the ground state of each atom in Prob. 6-6.

6-8 Give the maximum population of each of the following: **(a)** the L shell **(b)** the N shell **(c)** the 4*p* subshell **(d)** the 5*f* subshell **(e)** the $3p_y$ orbital **(f)** the $4d_{xy}$ orbital.

6-9 Give the number of orbitals in **(a)** a *d* subshell **(b)** an *f* subshell **(c)** an M shell **(d)** an O shell.

6-10 Predict which of the following atoms should be paramagnetic in their ground states: **(a)** Li $(Z = 3)$ **(b)** Mg $(Z = 12)$ **(c)** S $(Z = 16)$ **(d)** Fe $(Z = 26)$ **(e)** Zn $(Z = 30)$ **(f)** Ge $(Z = 32)$ **(g)** Kr $(Z = 36)$.

6-11 Each of the following is the subshell configuration after the "last electron" has been added to it according to the *Aufbau* procedure. In each case write the symbol of the atom and its complete electronic configuration: **(a)** $2p^4$ **(b)** $3s^1$ **(c)** $3p^2$ **(d)** $3d^2$ **(e)** $3d^7$ **(f)** $3p^5$ **(g)** $4s^2$.

6-12 Account for the fact that the N-shell configuration in Cr $(Z = 24)$ and Cu $(Z = 29)$ is $4s^1$ and not $4s^2$.

6-13 Discuss two kinds of experimental evidence which show that the electron has the property of spin.

Electrons: spatial distributions

6-14 The probability of finding a $1s$ and a $2s$ electron is highest at the nucleus in each case. Considering that this is true, is it correct to say that the $2s$ subshell is farther from the nucleus than the $1s$ subshell? Explain what is meant by the latter statement.

6-15 Describe the differences in size and shape between each of the following pairs of orbitals: **(a)** $2s$ and $3s$ **(b)** $2s$ and $2p_x$ **(c)** $2p_x$ and $2p_y$ **(d)** $2p_x$ and $3p_x$.

6-16 Tell how many radial (spherical) and angular nodes each of the following orbitals has **(a)** $4s$ **(b)** $3d$ **(c)** $2p$ **(d)** $5p$ **(e)** $4f$.

6-17 Prepare a table showing values of the four quantum numbers for each electron in each of the following atoms: **(a)** N **(b)** S **(c)** Si **(d)** Cr.

6-18 A vibrating string always has nodes at its ends. Explain.

6-19 Every atomic orbital has a node at infinity. Explain.

6-20 Draw a contour diagram for each of the following orbitals: **(a)** $1s$ **(b)** $2s$ **(c)** $2p_x$ **(d)** $3d_{xy}$ **(e)** $3d_{z^2}$.

6-21 Draw a contour diagram for each of the following orbitals: **(a)** $n = 1$, $l = 0$ **(b)** $n = 2$, $l = 1$ **(c)** $n = 3$, $l = 1$.

6-22 Imagine an isolated ground-state N atom $(Z = 7)$. **(a)** What is its electronic configuration? **(b)** Compare the energies of its three p electrons.

(c) How would these energies change if two negative charges were to approach the atom, one from each direction along the z axis? **(d)** If four negative ions were to approach the atom along the z and y axes, how would the orbital energies change?

6-23 Using probability-density curves show that a $2p$ electron is less strongly bound to the nucleus than a $2s$ electron.

6-24 Consider one electron of one atom on the end of your nose. What are the chances of finding that electron on the planet Mars? Explain.

6-25 Describe the *total* electronic charge-cloud distribution of the atoms with each of the following atomic numbers: **(a)** $Z = 1$ **(b)** $Z = 2$ **(c)** $Z = 3$ **(d)** $Z = 4$ **(e)** $Z = 5$ **(f)** $Z = 6$.

6-26 Describe what is meant by a *node* in the case of **(a)** a vibrating string **(b)** a vibrating drumhead **(c)** an electron wave in an atom.

6-27 Compare the shapes of the orbitals in each of the following pairs: **(a)** $1s$ and $2s$ **(b)** $2s$ and $2p$ **(c)** $3p_x$ and $3p_y$ **(d)** $3p_x$ and $3d_{xy}$ **(e)** $3d_{xy}$ and $3d_{z^2}$.

6-28 State a general rule allowing you to predict the number and type of nodes in an orbital from its principal and azimuthal quantum numbers.

Quantum numbers

6-29 What are the four quantum numbers used for specifying electrons in atoms? What is the physical significance of each?

6-30 By considering the number of possible values which the magnetic quantum number, m_l, may have show that each d subshell can accommodate 10 electrons.

6-31 State the Pauli exclusion principle in terms of quantum numbers. Restate it in terms of shells, subshells, orbitals, and electron spin.

6-32 Which of the following sets of quantum numbers (listed in the order n, l, m_l, m_s) are impossible for an electron in an atom? **(a)** 4, 2, 0, $+\frac{1}{2}$ **(b)** 3, 3, -3, $-\frac{1}{2}$ **(c)** 2, 0, $+1$, $+\frac{1}{2}$ **(d)** 4, 3, 0, $+\frac{1}{2}$ **(e)** 3, 2, -2, -1.

7

CHEMICAL PERIODICITY

TO THE STUDENT

By the early 1800s enough elements had been discovered, and their properties and those of their compounds well enough characterized, that many similarities in physical and chemical properties had become apparent. Certain groups of elements could logically be called *families,* for instance. This led the chemists of the day to search for *numerical* correlations among elements, and since by that time atomic weights were known (or presumed to be known) for many elements, what could be more natural than to look for correlations between trends in observed properties and atomic weights? The first such correlations made were few in number and attracted little attention in the scientific world. However, these led ultimately to a powerful statement of the interrelationships among the properties of the elements: the *periodic law.* This chapter is about that law.

7-1
The discovery of the periodic law

In 1829 Johann W. Döbereiner, a German chemist, observed that certain groups of three elements had similar properties. Döbereiner's *triads,* as he called them, included chlorine, bromine, and iodine (part of a family now known as the *halogens*); calcium, strontium, and barium (of the *alkaline-earth metals*); and sulfur, selenium, and tellurium (of the *chalocogens*). Döbereiner pointed out that the middle member of each triad had an atomic weight which was very close to the arithmetic mean (average) of the atomic weights of the other two elements. Most chemists of the day were unimpressed. Döbereiner's relationship seemed to work for a comparatively few elements. He was apparently the first, however, to come up with a systematic relationship among some elements.

In 1859 the German chemist R. W. Bunsen (inventor of the Bunsen burner) and physicist G. R. Kirchhoff developed the spectroscope, which rapidly led to the discovery of many new elements. Then, in 1860, the Italian chemist Stanislav Cannizzaro clarified the distinction between atoms and molecules and showed that many previously determined "atomic weights" were really molecular weights.

157

This permitted the tabulation of a reasonably accurate and consistent set of atomic weights of the elements.

The next reported attempt to correlate properties with atomic weight did not occur until 1862, when a French geologist, Alexandre de Chancourtois, listed the elements then known on a line which spiraled around a cylinder from bottom to top. He divided the circumference of the cylinder into 16 subdivisions and proceeded to show that elements with similar properties ended up above each other on adjacent turns of the spiral. De Chancourtois called his representation a *telluric screw*. It also attracted little attention.

In 1864 the English chemist John Newlands, who was a music lover, reported that if the elements were listed in order of increasing atomic weight, there was a repetition in properties every eighth element.[1] Newlands believed that there was some mystical connection between music and chemistry and was laughed at by his fellow chemists for making the association. The ridicule apparently obscured the real significance of Newlands' work, which he called the *law of octaves,* and it was not publicly recognized until 23 years later. But long before this several versions of the periodic table had been published.

Meyer and Mendeleev

The concept of chemical periodicity owes its development more to the two chemists Lothar Meyer (German) and Dmitri Mendeleev (Russian) than to anyone else. Working independently they discovered the periodic law and published periodic tables of the elements. Meyer first published in 1864, but in 1869 extended his table to include over 50 elements. He showed that when various properties such as molar volume, boiling point, brittleness, etc. were plotted as functions of atomic weight, a periodically repeating curve was obtained in each case. In the same year Mendeleev published the results of his work and included his own version of a periodic table. Mendeleev continued his work, revising and improving the table. In 1871 he published the version shown in Table 7-1. With this table he was able to show that certain as-yet-undiscovered elements ought to exist, elements which would fill in the blanks of the table. In this way he not only predicted the existence of the elements gallium and germanium, but he also estimated their properties with remarkable accuracy. The value of the periodic table in organizing chemical knowledge had been demonstrated.

**7-2
The periodic law**

Both Meyer and Mendeleev listed the elements sequentially in order of increasing *atomic weight*. Today we know that periodicity in properties is best shown if the listing is done in order of increasing *atomic number*. This difference affects few elements, since an increase in atomic number generally means an accompanying increase in atomic weight. (There are a few exceptions, however; as examples see potassium, $Z = 19$, and iodine, $Z = 53$.)

The *periodic law* states that *if the elements are listed sequentially in order of increasing atomic number, a periodic repetition in properties is observed*. In order to see what this means, imagine that we have listed all the elements in such a

---

[1] Newlands' relationship seemed to work, because in his day the noble gases were as yet undiscovered.

Table 7-1
Mendeleev's periodic table with atomic weights

	Group I	Group II	Group III	Group IV	Group V	Group VI	Group VII	Group VIII		
1	H 1									
2	Li 7	Be 9.4	B 11	C 12	N 14	O 16	F 19			
3	Na 23	Mg 24	Al 27.3	Si 28	P 31	S 32	Cl 35.5			
4	K 39	Ca 40	— 44	Ti 48	V 51	Cr 52	Mn 55	Fe 56	Co 59	Ni 59
5	Cu 63	Zn 65	— 68	— 72	As 75	Se 78	Br 80			
6	Rb 85	Sr 87	Yt 88	Zr 90	Nb 94	Mo 96	— 100	Ru 104	Rh 104	Pd 106
7	Ag 108	Cd 112	In 113	Sn 118	Sb 122	Te 125	I 127			
8	Cs 133	Ba 137	Di 138?	Ce 140?						
9										
10			Er 178	La 180	Ta 182	W 184		Os 195	Ir 197	Pt 198
11	Au 199	Hg 200	Tl 204	Pb 207	Bi 208					
12				Th 231		U 240				

sequence and are considering their properties. One set of elements which would soon attract our attention is the *noble gases*. These are elements each of which is a *gas* at ordinary temperatures and pressures and is chemically quite unreactive, or *noble*. (This word has been used in this way ever since the precious metals of royalty, notably gold, silver, and platinum, all of which are quite unreactive, were called the *noble metals*.) The noble gases are

Helium	(He, $Z = 2$)	Krypton	(Kr, $Z = 36$)
Neon	(Ne, $Z = 10$)	Xenon	(Xe, $Z = 54$)
Argon	(Ar, $Z = 18$)	Radon	(Ra, $Z = 86$)

Thus we periodically encounter one of these noble gases in our list of elements.

Now comes the interesting part. Each element immediately *following* a noble gas is a metal; it is a special kind of metal, one which is so highly reactive that it even reacts, for instance, with water. This set of elements having similar chemical properties is called the *alkali metals*. They are

Lithium	(Li, $Z = 3$)	Rubidium	(Rb, $Z = 37$)
Sodium	(Na, $Z = 11$)	Cesium	(Cs, $Z = 55$)
Potassium	(K, $Z = 19$)	Francium	(Fr, $Z = 87$)

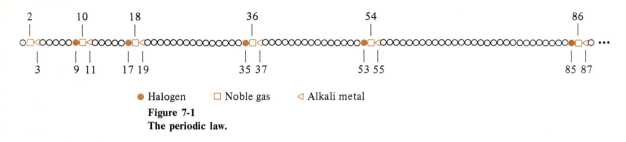

● Halogen   □ Noble gas   ◁ Alkali metal

Figure 7-1
The periodic law.

Furthermore, each element immediately *preceding* each of the noble gases except helium is a highly reactive nonmetallic element called a *halogen*. All the halogens have similar chemical properties. The halogens are

Fluorine	(F, $Z = 9$)	Iodine	(I, $Z = 53$)
Chlorine	(Cl, $Z = 17$)	Astatine	(At, $Z = 85$)
Bromine	(Br, $Z = 35$)		

Considering only these three families of elements (halogens, noble gases, and alkali metals) we find that they show up in our listed sequence of elements as shown in Fig. 7-1. Except for the fact that helium is not preceded by a halogen, the sequence, halogen–noble gas–alkali metal, *repeats periodically*. Other families of elements could be added to our sequence, as well.

The modern
periodic table

The repetition expressed in the periodic law is the basis for the modern *periodic table,* in which families of elements, such as halogens, noble gases, and alkali metals, are placed in vertical columns. Such a periodic table is shown in Fig. 7-2 and also inside the front cover of this book. The horizontal rows in the table are called *periods*. They are numbered from 1 through 7 using arabic numerals. Note that they vary greatly in length from the first period (two elements) to the sixth (32 elements) and seventh (potentially 32 elements) periods.

The vertical columns in the periodic table are called *groups*. Each of these is a family of elements. Some groups consist of five or six elements each. These are the so-called *main, representative,* or *A groups*. They are numbered from one through seven, using roman numerals and the letter A. Thus the halogens, for instance, are the elements of group VIIA (the A designation is sometimes omitted, and they are thus simply referred to as group VII). Also included with the main-group elements is group 0, the noble gases. The shorter groups are called *subgroups* or *B groups*. Thus, for example, copper, silver, and gold together constitute group IB, also known as the *copper subgroup*.

Notice these special characteristics of the periodic table (Fig. 7-2):

1 Hydrogen is placed all by itself. It is one of the two members of the first period, and its properties set it aside from all other elements. (Some versions of the periodic table place hydrogen above lithium in group IA and sometimes above fluorine in group VIIA, but it is clearly neither an alkali metal nor a halogen.)

Groups	IA	IIA												IIIA	IVA	VA	VIA	VIIA	0
								H 1											He 2
2	Li 3	Be 4												B 5	C 6	N 7	O 8	F 9	Ne 10
3	Na 11	Mg 12	IIIB	IVB	VB	VIB	VIIB		VIIIB		IB	IIB		Al 13	Si 14	P 15	S 16	Cl 17	Ar 18
4	K 19	Ca 20	Sc 21	Ti 22	V 23	Cr 24	Mn 25	Fe 26	Co 27	Ni 28	Cu 29	Zn 30	Ga 31	Ge 32	As 33	Se 34	Br 35	Kr 36
5	Rb 37	Sr 38	Y 39	Zr 40	Nb 41	Mo 42	Tc 43	Ru 44	Rh 45	Pd 46	Ag 47	Cd 48	In 49	Sn 50	Sb 51	Te 52	I 53	Xe 54
6	Cs 55	Ba 56	La 57	* 58-71	Hf 72	Ta 73	W 74	Re 75	Os 76	Ir 77	Pt 78	Au 79	Hg 80	Tl 81	Pb 82	Bi 83	Po 84	At 85	Rn 86
7	Fr 87	Ra 88	Ac 89	† 90-103	Rf 104	Ha 105	106												

*Lanthanoids	Ce 58	Pr 59	Nd 60	Pm 61	Sm 62	Eu 63	Gd 64	Tb 65	Dy 66	Ho 67	Er 68	Tm 69	Yb 70	Lu 71
†Actinoids	Th 90	Pa 91	U 92	Np 93	Pu 94	Am 95	Cm 96	Bk 97	Cf 98	Es 99	Fm 100	Md 101	No 102	Lr 103

Figure 7-2
The periodic table of the elements.

2 Group VIIIB (also sometimes called, simply, group VIII) consists of a total of nine elements.

3 The sixth period includes the 14 elements from $Z = 58$, cerium, through $Z = 71$, lutetium. These elements follow lanthanum ($Z = 57$) and are called the *lanthanoids*[2]. They would be shown in the appropriate location, except that this would make the table too wide. This sequence is shown separately and at the bottom of the main portion of the table.

4 Similarly, the seventh period includes the 14 *actinoids*[3] following actinium ($Z = 89$). Both the lanthanoids and actinoids are considered to be part of group IIIB, the scandium subgroup.

5 The most important thing to note about the periodic table as a whole is that *a new period, or cycle, starts when a repetition in properties begins in the sequence of elements*. Thus lithium ($Z = 3$) is an alkali metal, and when the next alkali metal, sodium ($Z = 11$), rolls around, it is placed under lithium, and so on. In this way each group is a family of elements having similar properties.

---

[2] They are also often called the *lanthanides* or the *rare earth elements*. The term *lanthanoid* has been recommended by IUPAC.

[3] Older names include *actinides* and *heavy rare-earth elements*.

At this point we will return to a further consideration of the building up (*Aufbau*) of the electronic configurations of the elements (Sec. 6-1).

When we compare electronic configurations (Table 6-2) with the periodic table we find that each period is started by adding an electron to a *new* (previously unoccupied) shell. Thus hydrogen and the group IA elements have an ns^1 configuration, where n is the principal quantum number of the "last," or outermost, shell. This shell is almost always referred to as the *valence shell,* and so from now on we will use this term. (Valence means *combining capacity,* and the electrons in the outermost shell are largely responsible for the way in which most atoms form bonds, or combine, with other atoms.) *Note:* The valence shell of an atom is the occupied shell with the highest principal quantum number n.

The main-group elements have valence-shell populations ranging from one through eight electrons, the respective configurations being ns^1 and ns^2 for groups IA and IIA and ns^2np^1 through ns^2np^6 for groups IIIA through 0. These elements are often called the *representative elements.* The valence-shell configuration ns^2np^6 is an especially stable one, as can be seen by the low reactivities of the noble gases. This configuration is often called an *octet,* and the generalization that it is especially stable is known as the *octet rule.*

The B-group elements in the periodic table are most commonly called the *transition elements.* Each horizontal series of transition elements corresponds to the belated filling of the d subshell of the $(n - 1)$ shell of these atoms. Since any d subshell can accommodate a total of 10 electrons, this filling builds up 10 transition elements in periods 4 and 5. Across the lanthanoid and actinoid series the subshell being filled in each case is the f subshell of the $(n - 2)$ shell. The f

Figure 7-3
The periodic table and electronic configurations.

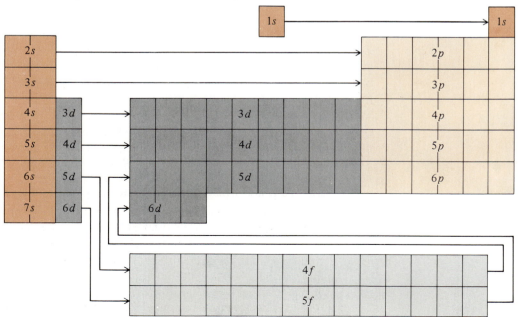

subshells have a maximum population of 14 electrons; so there are 14 lanthanoids and 14 actinoids.

The periodic table can be used to predict the electronic configurations of most atoms. Figure 7-3 summarizes the relationship between orbital filling and the periodic table. In each indicated region of the table the "last" electrons are added to the designated subshell: in groups IA and IIA to the ns subshell, in groups IIIA through 0 to the np subshell, in the transition elements to the $(n-1)d$ subshell, and in the lanthanoids and actinoids to the $(n-2)f$ subshell.

ADDED COMMENT

You should be able to predict the configurations of most atoms from their locations in the periodic table. As mentioned earlier, some atoms have slightly irregular configurations. These irregularities are basically a result of the very small differences between orbital energies in certain atoms; their effect on chemical behavior is minimal. Why should you bother to learn how to predict electronic configurations at all? One of the most exciting aspects of chemistry is the correlation of observed properties with theoretical structure, and that is just where we are heading. We will shortly see that substances which act like metals are located at the left side of the periodic table and hence have few valence-shell electrons. This relationship between number of valence-shell electrons and metallic character is but a simple (actually, somewhat oversimplified) example, but it illustrates how we will be able to use electronic configurations to help us account for *properties*.

7-3
Trends in atomic
properties

A number of measurable properties of atoms show periodic variation with atomic number. In this section we will describe the variation of three such properties: atomic radius, ionization energy, and electron affinity.

Atomic radius

How big is an atom? You could answer "fantastically small" or "infinitely large," and be right either way! An atom has no sharp boundary, no limit beyond which its electron charge cloud vanishes. It is true that we could arbitrarily define an atom's size as being that of a boundary surface which contains most, say 95 percent, of its total electronic charge, but an *experimental* measure of atomic size is much more desirable.

Measuring an atom, however, is not without problems. The difficulty, it turns out, lies not with the experimental techniques, but rather with the interpretations of the results. Good experimental methods exist, for example, for measuring the center-to-center distance between adjacent atoms in a molecule. (X-ray diffraction is one of these. We will discuss it in Sec. 10-2.) In the hydrogen, H_2, molecule the internuclear distance, or *bond length*, has been found to be 0.074 nm, and in carbon (diamond) the distance between centers of adjacent (bonded) atoms is 0.154 nm. Now, if we define half the distance between the centers of adjacent atoms as the *radius* of the atom, then the radius of an H atom comes out to be 0.037 nm and that of a C atom, 0.077 nm. But what about a compound in which C is bonded to H? If each atom maintained a constant, characteristic radius, then the C-H internuclear distance would be 0.037 nm + 0.077 nm, or 0.114 nm. It turns out that in the methane (CH_4) molecule the measured C-H bond length is

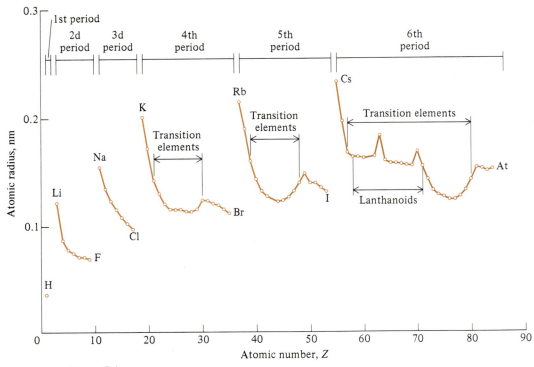

Figure 7-4
Atomic radii.

0.110 nm. Should we dismiss the discrepancy (0.004 nm) as experimental error? No, the methods used are far too accurate. The point is that the effective radius of an atom depends on how it is bonded. The contribution an atom makes to the total bond length depends on the nature of the bond, which in turn depends in part on the properties of the other atom.

In spite of these difficulties it is possible to assemble a set of approximate atomic radii derived from interatomic-distance measurements. These are made on various compounds chosen so that each of the atoms is involved in about the same kind of bonding. The results are shown graphically in Fig. 7-4. The periodic change in atomic radius shows clearly in this graph. Compare it with the periodic table, and note how each period starts (group IA) with a large atom, and how the radii then generally decrease across the remainder of the period, clearly illustrating the periodic law.

How can we account for this decrease in atomic size across a period? As we build up the atoms across a period we add electrons to the valence shell of the atom, while simultaneously increasing the nuclear charge. The increased nuclear charge attracts *all* the electrons more strongly, so they are all drawn in more closely to the nucleus. The effect is clearest in the second and third periods, which include no transition elements. Table 7-2 shows the variation in atomic radii of the second-period elements, as well as their electronic configurations.

But look at the transition elements in periods 4, 5, and 6. The above-mentioned

**Table 7-2
Atomic radii of the
second-period elements**

Atom	Nuclear charge	Electronic configuration	Radius, nm
Li	3+	$[He]2s^1$	0.123
Be	4+	$[He]2s^2$	0.089
B	5+	$[He]2s^22p^1$	0.080
C	6+	$[He]2s^22p^2$	0.077
N	7+	$[He]2s^22p^3$	0.074
O	8+	$[He]2s^22p^4$	0.074
F	9+	$[He]2s^22p^5$	0.072

decrease in size is interrupted when the transition elements intervene in the middle of a period. Why is this? Remember that when we proceed across a transition element series, we are adding electrons not to the valence shell of the atom but to the $(n - 1)d$ subshell. The average distance between the nucleus and these electrons is less than the average distance between the nucleus and the valence-shell electrons, so that the inner electrons partially screen the outer electrons from the attractive influence of the nucleus. (Saying this is equivalent to saying that the outer electrons experience a repulsion from the inner electrons, and that this repulsion partially compensates for the nuclear attraction. The screening effect of inner electrons always reduces the positive charge which an outer electron "feels" to a value well below the actual charge on the nucleus.) The shrinkage in atomic radii becomes less, then, across a transition-element series. Toward the end of each such series, the d subshell of the second shell from the outside approaches its maximum population, 10 electrons, and the density of electronic charge in this shell becomes very high, increasing the screening effect and reducing the *effective* nuclear charge. Thus the elements at the ends of these series actually get larger. Only after a transition element series has been completed by filling up the d subshell to its capacity of 10 electrons are additional electrons added to the valence shell, causing the resumption of the shrinkage across the period. [In Fig. 7-4 the radii of the noble-gas atoms (group 0) have not been plotted. These elements form few compounds, and data, where available, are not comparable.]

In the lanthanoids we see an even greater effectiveness of screening by inner electrons. Going across this series, we add electrons to the $(n - 2)$ shell, while simultaneously increasing the nuclear charge. The increased screening of the deeply buried f subshell almost completely cancels out the increase in nuclear charge, and there is, as a result, only a small shrinkage across the series. (The small upward spikes at europium, $Z = 63$, and ytterbium, $Z = 70$, are not significant. The bonding in the substances from which these radii were obtained is not comparable to that in the substances used for the other lanthanoids.)

Going from one lanthanoid to the next the effectiveness of screening by the inner electrons results in a very small increase in *effective* nuclear charge. Thus the shrinkage in radius averages only 0.001 nm from one atom to the next, but because there are 14 elements in the series, there is a *total* contraction of 0.013 nm across the whole series. This total contraction, the *lanthanoid contraction,* is significant, as we will see shortly.

How does atomic size vary among atoms within a group in the periodic table? Down any of the A-group elements, the radii tend to increase. Table 7-3 shows this increase for the alkali metals (group IA). For these elements we might jump to the conclusion that the increase in nuclear charge (from 3+ for Li to 55+ for

Table 7-3
Atomic radii of the
alkali metals (group IA)

Atom	Nuclear charge	Electronic configuration	Radius, nm
Li	3+	$[He]2s^1$	0.123
Na	11+	$[Ne]3s^1$	0.157
K	19+	$[Ar]4s^1$	0.203
Rb	37+	$[Kr]5s^1$	0.216
Cs	55+	$[Xe]6s^1$	0.235

Ca) ought to pull in the electrons progressively more strongly and cause a size decrease. Instead, the valence shell becomes progressively farther from the nucleus as the total number of shells increases. The increased screening by inner electrons keeps the effective nuclear charge from increasing, and that, plus the increase in the principal quantum number of the valence-shell electrons, causes the observed increase down a group.

The size variation down a B group is less spectacular, to say the least. Going from any fourth-period transition element to the element immediately below it in the fifth period, there is an appreciable increase in atomic radius. From the fifth to the sixth period, however, there is virtually no change. Consider, for example, the atomic radii of titanium ($Z = 22$), zirconium ($Z = 40$), and hafnium ($Z = 72$), as shown in Table 7-4. The expected increase in atomic radius from zirconium to hafnium is not observed. The situation is similar for all of the sixth-period transition elements; each atomic radius is very close to that of the corresponding fifth-period element. Why? The answer is found in the *lanthanoid contraction*. As has been pointed out, the belated filling of the $4f$ subshell results in lutetium, the last of the lanthanoids, being 0.013 nm smaller in radius than lanthanum. This means that the elements following lutetium have unusually small atoms, because their nuclear charges are higher than they would be if the lanthanoids had not been built up before them. The lanthanoid contraction almost exactly cancels out the effect of the added "last" shell in the sixth period, so the sixth- and fifth-period transition elements are close together in atomic size.

The similarity in size and electronic configuration between the fifth- and sixth-period transition elements accounts for their striking similarity in respective properties. Zirconium and hafnium, for example, not only form chemical compounds which have similar formulas, but these compounds also have similar properties (melting points, boiling points, solubilities, etc.), and so they are often quite difficult to separate. Zirconium and hafnium compounds are usually found together in nature.

Ionization energy

When energy is absorbed by an isolated atom in its ground state, an electron may be raised from one quantized energy level to another. If enough energy is supplied, the electron may be completely removed from the atom, leaving behind a *positive ion*. The electron most easily removed from an atom is the one least strongly bound to the atom. The process of forming a positive ion from an

Table 7-4
Atomic radii of the
elements of group IVB

Atom	Nuclear charge	Electronic configuration	Radius, nm
Ti	22+	$[Ar]3d^24s^2$	0.132
Zr	40+	$[Kr]4d^25s^2$	0.145
Hf	72+	$[Xe]5d^26s^2$	0.144

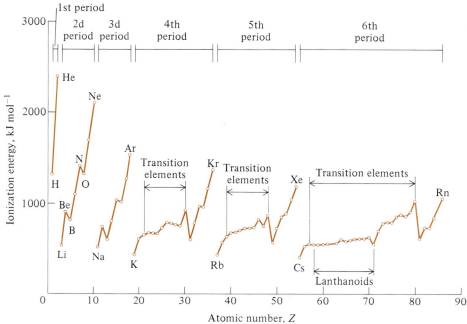

**Figure 7-5
First ionization
energies.**

isolated, neutral, ground-state atom by removing an electron from it is called *ionization,* and the minimum energy required to accomplish this is called the *ionization energy.* Since an isolated atom is essentially free from the perturbing influence of any nearby atom, the term implies an atom in the gaseous state. Thus ionization energy is the energy necessary to cause the following process to occur:

$$M(g) \longrightarrow M^+(g) + e^-$$

Since more than one electron may be removed from an atom, the energy required to cause the above process to occur is often called the *first* ionization energy. The *second* ionization energy is the energy required to remove a second electron, that is, to cause the following process to occur:

$$M^+(g) \longrightarrow M^{2+}(g) + e^-$$

Ionization energies are usually expressed in electronvolts[4] per atom, kilocalories per mole, or kilojoules per mole. We will use the SI unit, kJ mol^{-1}. (Conversion factors are given in App. B.)

The way in which ionization energies vary with atomic number provides another good illustration of the periodic law. Figure 7-5 shows this variation for the first six periods. In general the ionization energy starts low at the beginning of each period and then increases across the period. How can we account for this trend? As nuclear charge increases across a period the outer-shell electrons become more strongly bound to the nucleus. The decrease in atomic radius across a period also

---

[4]One electronvolt is the energy necessary to raise an electron through an electrical potential difference of one volt. Ionization energies are often called ionization *potentials,* but we will not use this term, as it is sometimes misleading.

168   CHEMICAL PERIODICITY

Table 7-5
First ionization energies of the second-period elements

Atom	Nuclear charge	Electronic configuration	Ionization energy, kJ mol^{-1}
Li	3+	$1s^22s^1$	520
Be	4+	$1s^22s^2$	899
B	5+	$1s^22s^22p_x^1$	801
C	6+	$1s^22s^22p_x^12p_y^1$	1086
N	7+	$1s^22s^22p_x^12p_y^12p_z^1$	1402
O	8+	$1s^22s^22p_x^22p_y^12p_z^1$	1314
F	9+	$1s^22s^22p_x^22p_y^22p_z^1$	1681
Ne	10+	$1s^22s^22p_x^22p_y^22p_z^2$	2081

tends to make these electrons more tightly bound. These two effects combine to produce an increase in ionization energy across a period.

Notice the high ionization energies for the noble gases. For all of these except helium the removal of an electron breaks up the *octet* of eight electrons in the outer shell. As we have previously mentioned, a valence-shell octet is an especially stable arrangement, and the high ionization energies of neon, argon, krypton, xenon, and radon are evidence of this stability. (Helium has only a K shell and can therefore accommodate only two electrons in its outer shell. These two electrons serve the same stability-producing function for helium that an octet does for the rest of the noble gases.)

Notice, however, that the trend across a period is not without some irregularities, even in periods 2 and 3, where there are no transition elements. Look at the values of the ionization energies for the second period, as shown in Table 7-5. The table and the graph of Fig. 7-5 both show that the ionization energy for boron ($Z = 5$) is actually *lower* than that for beryllium ($Z = 6$). The same is true for oxygen ($Z = 8$) as compared to nitrogen ($Z = 7$). These irregularities in the general trend for increasing ionization energy across a period may be accounted for as follows: In B the electron to be removed is in a $2p$ orbital, while in Be it is in a $2s$ orbital. Electrons in $2s$ orbitals have maxima in their probability-density curves which are closer to the nucleus. (Each s electron has a maximum which is next to the nucleus, in fact.) Thus a $2s$ electron (1) is more strongly bound to the nucleus and (2) can partially screen a $2p$ electron from the nucleus. This makes it easier to remove the $2p$ electron from B than the $2s$ electron from Be. The case of oxygen is slightly different. The ionization energy of O is lower than would otherwise be expected, because in this case the electron comes from a $2p$ orbital which has another electron in it. The two electrons occupying the same orbital (region of space) repel each other, making it easier to remove one. Similar irregularities (which can be similarly explained) can be found in periods 3, 4, 5, and 6. Note also the very slow increase in ionization energy across each transition-element series. This is caused by the gradual increase in nuclear charge which is almost compensated by the increased screening of the inner-shell electrons.

How does ionization energy vary down a group in the periodic table? Figure 7-5 shows that it decreases. The decrease is additionally shown for the alkali metals (group IA) in Table 7-6. The increased screening effect of the inner-shell electrons (recall that the number of shells increases going down any group) largely compensates for the increased nuclear charge, so that the main consideration is the

Table 7-6
First ionization energies of the alkali metals (group IA)

Atom	Nuclear charge	Electronic configuration	Ionization energy, kJ mol⁻¹
Li	3+	$[He]2s^1$	520
Na	11+	$[Ne]3s^1$	496
K	19+	$[Ar]4s^1$	419
Rb	37+	$[Kr]5s^1$	403
Cs	55+	$[Xe]6s^1$	376

increase in atomic radius. Thus ionization energy decreases; it becomes easier to remove an electron.

The second, third, fourth, and so on, ionization energies are the energies needed to remove a second, third, fourth, and so on, electron, respectively. In Fig. 7-6 are shown the first through the sixteenth ionization energies for sulfur ($Z = 16$). The pronounced increase in successive ionization energies is a result of the fact that each electron must leave an ion which is more highly charged than the ion left by the previous electron. Since the attractive force between the departing electron and the ion increases accordingly, ionization energy increases. Note, however, the large jump at the seventh ionization energy and again at the fifteenth. These can not be explained on the basis of nuclear charge increase alone. The seventh electron must come from a completed octet in the L shell, and this shell is closer to

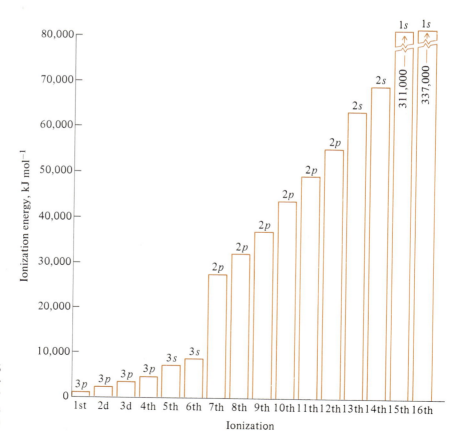

Figure 7-6
Ionization energies for sulfur. (The "origin" of each electron is shown.)

the nucleus than the M shell, which has supplied the first through the sixth electrons. Again at the fifteenth a sharp increase occurs, this time because the previously unused K shell must be broken into. Removal of an electron from an outer-shell noble-gas configuration (two electrons in the case of the K shell; eight electrons for all others) requires an extra amount of energy.

Electron affinity — An atom can pick up an electron, too. The resulting particle is a *negative* ion, and the process can be written

$$X(g) + e^- \longrightarrow X^-(g)$$

This process is normally accompanied by the *release* of energy, the amount of which measures how tightly the electron becomes bound to the atom. The *electron affinity* of an atom is accordingly defined as *the amount of energy released when a gaseous (isolated) ground-state atom gains an electron.*

Electron affinities are difficult to measure, and accurate values are not known for all elements. Values for the elements of the first three periods are shown plotted against atomic number in Fig. 7-7. (Not all of these have been experimentally obtained; a few are taken from theoretical calculations.) Some of the electron affinities are negative; this means that energy is *absorbed* when the electron is added.

Figure 7-7
Electron affinities of the elements of the first three periods. (A few values have been calculated theoretically.)

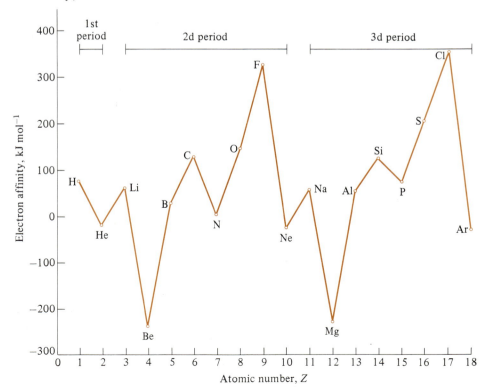

Table 7-7
Electron affinities of the
halogens (group VIIA)

Atom	Nuclear charge	Electronic configuration	Electron affinity, kJ mol^{-1}
F	9+	$[He]2s^2 2p^5$	333
Cl	17+	$[Ne]3s^2 3p^5$	348
Br	35+	$[Ar]4s^2 4p^5$	324
I	53+	$[Kr]5s^2 5p^5$	296

The periodicity in electron affinity shows clearly in Fig. 7-7. Across a period, the values generally increase as nuclear charge increases. Each alkali metal (Li, Na) has a small (positive) electron affinity. The next elements, the alkaline-earth metals (Be and Mg), have very negative values, however. This can be explained by the fact that these atoms must accept the added electron in the $2p$ and $3p$ subshell, respectively. Screening by the $2s$ and $3s$ subshells lowers the effective nuclear charge "felt" by these $2p$ and $3p$ electrons; so electron affinity is low.

After these elements the electron affinity increases across the period until it drops down suddenly at group VA (N and P in Fig. 7-7). This sudden decrease is accounted for by the fact that the added electron must enter an already half-full $2p$ (in the case of N) or $3p$ (in P) subshell. Repulsion between the two electrons in the same orbital lowers the electron affinity. From here on the increasing nuclear charge increases the electron affinity until it reaches a maximum in each period at a halogen (group VII). Addition of an electron to a halogen atom completes the octet in the valence shell. Finally, the electron affinities of the noble gases are low; each of these atoms shows little tendency to start a new shell.

Going down a group electron affinities undergo a general decrease. This is a result of the fact that the valence shell is progressively farther from the nucleus with progressively more inner shells screening the nuclear charge. Values for the halogens are shown in Table 7-7. Each of these elements has a comparatively high electron affinity; adding an electron completes the outer-shell octet and creates a noble-gas configuration. The value for fluorine is, however, out of line, being smaller than that of chlorine. As a matter of fact, all of the elements of the second period have what appear to be anomalously low electron affinities. (See Fig. 7-7.) This is apparently an indirect result of the small sizes of the atoms of these elements. Electron repulsion within the compact L shell seems to partially compensate for the high attraction from the nucleus, thus reducing electron affinity.

As is the case with ionization energies, electron affinities may be expressed in any convenient energy units. Electronvolts (per atom) are commonly used, but we will be consistent and use kilojoules per mole of atoms.

7-4
Trends in physical
properties

The physical properties of the elements can be used to demonstrate the periodic law. Properties such as melting point, boiling point, thermal conductivity, electrical conductivity, hardness, and density show periodic variations with atomic number. The variation is often not a very regular one, however, because the relationship between such properties and electronic configuration is often not very direct.

Figure 7-8 shows the densities of the elements at 25°C plotted against atomic number. (The density of each element which is normally a gas at that tempera-

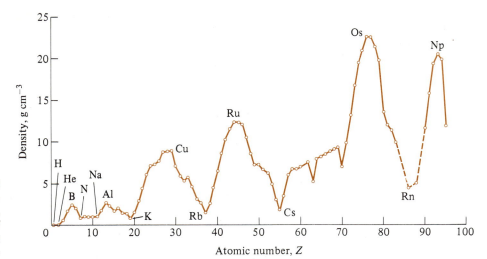

Figure 7-8
Densities of the ele-
ments, 25°C (for gases,
density is that of
liquid at boiling
point).

ture is that of the liquid at its boiling point.) The graph goes through a succession of maxima and minima, clearly illustrating the periodic law.

Figure 7-9 shows the periodicity in the melting points of the elements. Again we see a periodic succession of maxima and minima. A comparison of the graphs of Figs. 7-8 and 7-9 with the periodic table shows that corresponding parts of each

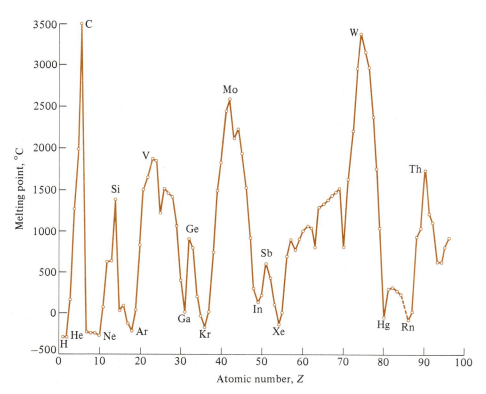

Figure 7-9
Melting points of the
elements.

Table 7-8
Atomic combining ratios
in chlorine compounds

Period	Group I	Group II	Group III	Group IV	Group V	Group VI	Group VII
2	LiCl	$BeCl_2$	BCl_3	CCl_4	NCl_3	OCl_2	FCl
3	NaCl	$MgCl_2$	$AlCl_3$	$SiCl_4$	PCl_3	SCl_2	(ClCl)
4	KCl	$CaCl_2$	$GaCl_3$	$GeCl_4$	$AsCl_3$	$SeCl_2$	BrCl

cycle (maxima or minima, for instance) tend to occur in atoms in the same region of the periodic table. The best examples of this are perhaps the minima in melting points shown by the noble gases.

7-5
Trends in chemical properties

The work of Mendeleev and Meyer was based to a large extent on periodicity in chemical properties. We will be discussing trends in chemical properties later in this book, but brief mention of a few is in order now.

Stoichiometry

Atomic combining ratios in compounds illustrate chemical periodicity in a rather spectacular way. Table 7-8 shows the formulas of a number of representative-element compounds of chlorine. The atomic combining ratio can be seen to increase from 1:1 in the group-IA compounds to 1:4 in the group-IVA compounds and then to decrease back down to 1:1 in the group-VIIA compounds. (Chlorine gas, Cl_2, is included in the table even though it is not a compound. Also, some of the elements listed form one or more additional chlorine compounds having other atomic combining ratios.) This periodicity in stoichiometry can be similarly illustrated with compounds other than chlorine compounds, oxygen and hydrogen compounds, for example.

Metallic properties

If we proceed across a period in the periodic table, we observe a gradual transition in character of the elements from *metallic* to *nonmetallic*. What is a metal? A metal is a substance whose physical properties include high electrical and thermal conductivity and a remarkable luster or shine. One chemical property of a metal is its ability to form a compound with hydrogen and oxygen, a *hydroxo* compound, which is known as a *base*. Nonmetals are just the opposite. They are poor or nonconductors of heat and electricity, have a dull luster, and form hydroxo compounds which are known as *acids*. (We will define acids and bases in Sec. 12-6 and 13-1. For now it is enough to know that they are classes of compounds.)

In the second period there is a gradual change from the typically metallic properties of lithium ($Z = 3$) to the typically nonmetallic properties of fluorine ($Z = 9$). The noble gas neon then intervenes between fluorine and the next metal, sodium ($Z = 11$), where the sequence from metal to nonmetal starts again. Elements between the metals on the left and the nonmetals on the right are intermediate in properties and are called *metalloids,* or *semimetals.*

In the long periods, 4 through 7, a much longer interval is needed for the gradation from metal to nonmetal to occur. In the fourth period, for instance, potassium ($Z = 19$) and calcium ($Z = 20$) are excellent metals, but the appearance of clearly nonmetallic properties does not occur until the last part of the period. The transition elements scandium ($Z = 21$) through zinc ($Z = 30$) are more metallic than nonmetallic, and are thus often called *transition metals.*

SUMMARY

This chapter has been concerned with the *periodic law,* which states that the properties of the elements are periodic functions of their atomic numbers. We followed the development of thinking about such periodic behavior from Döbereiner and his triads to Meyer and Mendeleev and their early periodic tables. The modern periodic table is a direct descendant of these.

Periodicity in the electronic configurations of atoms is a result of the quantum-mechanical arrangement of shells, subshells, and orbitals. This electronic periodicity is the ultimate cause for the observed periodicity in atomic, physical, and chemical properties.

Three important atomic properties were introduced in this chapter. These are *atomic radius, ionization energy,* and *electron affinity.* We will make use of each of these in later chapters.

KEY TERMS

A group (Sec. 7-2)
Actinoid (Sec. 7-2)
Alkali metal (Sec. 7-2)
Atomic radius (Sec. 7-3)
B group (Sec. 7-2)
Electron affinity (Sec. 7-3)
Group (Sec. 7-2)
Halogen (Sec. 7-2)
Ionization energy (Sec. 7-3)
Lanthanoid (Sec. 7-2)
Lanthanoid contraction (Sec. 7-3)
Main group (Sec. 7-2)
Metal (Sec. 7-5)

Metalloid (Sec. 7-5)
Noble gas (Sec. 7-2)
Nonmetal (Sec. 7-5)
Octet rule (Sec. 7-2)
Period (Sec. 7-2)
Periodic law (Sec. 7-2)
Periodic table (Sec. 7-2)
Representative element (Sec. 7-2)
Semimetal (Sec. 7-5)
Subgroup (Sec. 7-2)
Transition element (Sec. 7-2)
Transition metal (Sec. 7-5)
Valence shell (Sec. 7-2)

QUESTIONS AND PROBLEMS

Historical

7-1 What were Döbereiner's triads? Using the periodic table find some sets of elements (other than those listed in this chapter) which Döbereiner might have classified as triads.

7-2 Draw a representation of the telluric screw of de Chancourtois as it would look if it were unrolled off the cylinder and flattened into a plane.

7-3 Newlands attempted to show that there was a periodic similarity in properties every *eighth* element, yet today we see that for periods 2 and 3 it occurs every *ninth* element. Explain.

7-4 How did Mendeleev predict the properties of elements not yet discovered?

Periodic table

7-5 Define or explain each of the following: period, group, subgroup, representative element, transition element, lanthanoid.

7-6 Why does the number of elements in each period in the periodic table increase going down the table?

7-7 Which of the following are *not* transition elements? **(a)** Ni $(Z = 28)$ **(b)** W $(Z = 74)$ **(c)** Pb $(Z = 82)$ **(d)** Eu $(Z = 63)$ **(e)** Ra $(Z = 88)$.

7-8 Which element is **(a)** a halogen in the fifth period? **(b)** a noble gas in the third period? **(c)** an alkali metal with one more occupied shell than potassium? **(d)** a transition element with a $4d^3$ configuration?

Electronic configurations

7-9 Using only the periodic table give the electronic configuration of the ground state of each of the following: N $(Z = 7)$; Si $(Z = 14)$; V $(Z = 23)$; Se $(Z = 34)$; Sr $(Z = 38)$; Mg^{2+} $(Z = 12)$.

7-10 Using only the periodic table give the symbol of the atom having each of the following ground-state valence-shell configurations: **(a)** $2s^2$ **(b)** $3s^23p^1$ **(c)** $4s^2$ **(d)** $6s^26p^5$.

7-11 Using only the periodic table predict which of the following atoms ought to be paramagnetic in their ground states: **(a)** Li $(Z = 3)$ **(b)** Mg $(Z = 12)$ **(c)** S $(Z = 16)$ **(d)** Zn $(Z = 30)$ **(e)** Ba $(Z = 56)$ **(f)** Re $(Z = 75)$ **(g)** Cu^{2+} $(Z = 29)$ **(h)** Fe^{3+} $(Z = 26)$.

7-12 Give the symbols of all the elements which have **(a)** valence-shell $4s^1$ configurations **(b)** valence-shell $5s^2 5p^2$ configurations **(c)** exactly half-filled $3d$ subshells **(d)** exactly filled $3d$ subshells.

7-13 Which of the following atoms do *not* have spherical total electronic charge-cloud distributions: Na, O, Ca, Xe, Cr, Mn? (Use only the periodic table.)

Atomic properties

7-14 What is meant by the *size* of an atom? What are the problems associated with atomic size determinations?

7-15 Where in the periodic table do you find the elements with the highest first ionization energies? Explain.

7-16 The volume of a sample of solid C containing 1 mol of atoms is 5.0 cm^3, while that of a similar sample of N is 14 cm^3. Calculate the density of each of these two solids.

7-17 For each of the following pairs of atoms indicate which has the higher first ionization energy and briefly explain why: **(a)** S and P **(b)** Al and Mg **(c)** Sr and Rb **(d)** Cu and Zn **(e)** Rn and At **(f)** K and Rb.

7-18 How is the first ionization energy of the chloride ion Cl^- related to the electron affinity of the chlorine atom Cl?

7-19 Why is the second ionization energy of *any* atom greater than its first?

7-20 For each of the following pairs of atoms state which would be expected to have the higher electron affinity, and explain why: **(a)** Br and I **(b)** Li and F **(c)** F and Ne.

7-21 The following particles are *isoelectronic,* that is, they each have the same electronic configuration. Rank them in order of decreasing radius: Ne, F^-, Na^+, O^{2-}, Mg^{2+}.

7-22 Rank the following in order of decreasing ionic radius: Se^{2-}, S^{2-}, Te^{2-}, O^{2-}.

7-23 The first ionization energy of Na is 496 kJ mol^{-1}. The electron affinity of Cl is 348 kJ mol^{-1}. One mole of gaseous Na atoms reacts with 1 mol of gaseous Cl atoms to form 1 mol each of Na^+ and Cl^-. Does the overall process liberate or absorb energy? How much?

7-24 Calculate the frequency and wavelength of light necessary to ionize lithium atoms, given that the first ionization energy is 520 kJ mol^{-1}.

7-25 The first ionization energy of M is 367 kJ mol^{-1}. One mole of M atoms react with 1 mol of X atoms to form 1 mol each of M^+ and X^- ions, absorbing 255 kJ of heat in the process. If all reactants and products are gaseous, isolated particles, what is the electron affinity of X?

7-26 Which of the following would you expect to have the smaller ionic radius: Fe^{2+} or Fe^{3+}? Explain.

7-27 Which would you expect to have the higher electron affinity: C or N? Explain.

7-28 The first ionization energy of gold $(Z = 79)$ is *higher* than that of silver $(Z = 47)$, directly above it in the periodic table. Explain.

Physical and chemical properties

7-29 Why is it that the density of an element is not directly proportional to its atomic weight?

7-30 In what regions of the periodic table do you find **(a)** metals **(b)** nonmetals **(c)** metalloids?

7-31 State one chemical difference and one physical difference between a metal and a nonmetal.

7-32 Which of the following would be expected to be a metal? **(a)** H $(Z = 1)$ **(b)** Ca $(Z = 20)$ **(c)** Br $(Z = 35)$ **(d)** Si $(Z = 14)$ **(e)** Xe $(Z = 54)$ **(f)** Tc $(Z = 43)$ **(g)** Fr $(Z = 87)$

7-33 Which of the following hydroxo compounds would you expect to be acidic? **(a)** KOH **(b)** ClOH **(c)** $Ca(OH)_2$ **(d)** $SO_2(OH)_2$ **(e)** $Ti(OH)_2$ **(f)** TlOH **(g)** $Al(OH)_3$

8

CHEMICAL BONDING

TO THE STUDENT

One of the most intriguing aspects of chemistry is the study of forces between atoms. In this chapter we focus on two of the strongest interatomic forces: ionic bonds and covalent bonds. It is important to note that most bonds are not 100% ionic or 100% covalent. Instead, most have intermediate characteristics. But it is easiest to understand these intermediate bonds by relating them to the pure or ideal bond types.

8-1
Ionic bonding

The ionic bond is the electrostatic force which attracts particles with opposite electrical charges. (Any two particles or objects which carry opposite electrical charges tend to be attracted to each other. This statement is part of *Coulomb's law,* and the forces of attraction are known as *electrostatic,* or *coulombic, forces.*)

The formation of ions

An ion is formed when an atom gains or loses one or more electrons. Loss of an electron by an atom yields a positive ion, or *cation,*

$$M(g) \longrightarrow M^+(g) + e^-$$

while gain of an electron yields a negative ion, or *anion,*

$$X(g) + e^- \longrightarrow X^-(g)$$

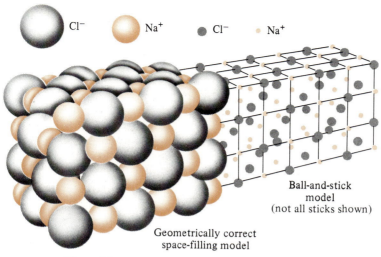

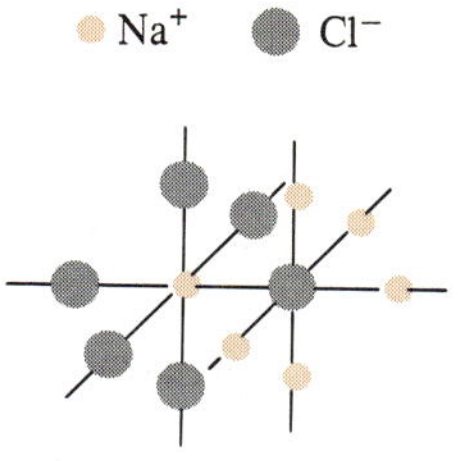

Figure 8-2
A small portion of the NaCl structure. This drawing shows that an Na^+ ion forms six ionic bonds with its six nearest-neighbor Cl^- ions. Similarly, a Cl^- ion forms six bonds with its six nearest-neighbor Na^+ ions.

Ball-and-stick model
(not all sticks shown)

Geometrically correct space-filling model

Figure 8-1
The sodium chloride structure.

If the electron lost by M is gained by X, then the overall process can be represented by

$$M(g) + X(g) \longrightarrow M^+(g) + X^-(g)$$

The M^+ and X^- ions thus formed are then attracted to each other because they have opposite electrical charges:

$$M^+(g) + X^-(g) \longrightarrow M^+X^-(g)$$

This attraction is called an *ionic* bond, or, sometimes, an *electrovalent* bond. It is the electrostatic attraction between ions of opposite charge. In the sequence of events just described, the end result is a gaseous M^+X^- unit which is called an *ion pair*.

Gaseous ion pairs are found at very high temperatures. Ionic bonds are more commonly encountered in *ionic solids,* interlocking structures containing huge numbers of positive and negative ions, not just one of each. In solid NaCl (sodium chloride, table salt) the Na^+ and Cl^- ions are not arranged in pairs; rather, they are ordered in three-dimensions, as shown in Fig. 8-1. On the left is a picture of a space-filling, geometrically accurate model of the NaCl structure. The smaller sodium ions are shown as spheres in contact with the larger chloride ions. On the right the ions have all been "shrunk," and the resulting ball-and-stick model affords a view into the interior of the structure.

In solid NaCl each Na^+ has six Cl^- ions as its nearest neighbors. Similarly, the immediate environment of each Cl^- ion consists of six Na^+ ions. (See Fig. 8-2.) Thus each Na^+ forms not just one but six ionic bonds, as does each Cl^-. (Each of the two kinds of ions forms six bonds; so the formula must be NaCl. Do you see why this is so?)

Lewis structures   As early as 1906 J. J. Thomson expressed his belief that electrons would ultimately be shown to provide the key to chemical bonding. In 1916 the American chemist G. N. Lewis developed a method for keeping track of electrons in atoms and molecules. This method, still useful today, makes use of a diagram, now called a *Lewis structure,* in which the chemical symbol for an atom is surrounded by a number of dots (or, sometimes, little o's or x's, etc.) corresponding to the number of electrons in the valence shell of the atom.[1] A sodium atom has one valence-shell electron; so its Lewis structure is

$$Na \cdot$$

where the dot can be located at the left, right, top, or bottom of the symbol. A chlorine atom, on the other hand, has seven valence-shell electrons; so its Lewis structure is

$$:\overset{\cdot\cdot}{\underset{\cdot\cdot}{Cl}} \cdot$$

It makes no difference on which of the four sides of Cl the single dot is shown.

The symbol in a Lewis structure represents the nucleus plus all inner-shell electrons, this combination often being called the *core,* or *kernel,* of the atom. Thus we have for a magnesium ($Z = 12$) atom

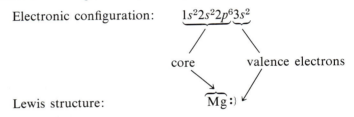

Electronic configuration: $1s^2 2s^2 2p^6 3s^2$

core   valence electrons

Lewis structure: $\overline{Mg:)}$

The dots in a Lewis structure are best grouped to show whether or not the electrons are paired. For an aluminum atom ($Z = 13$) we have

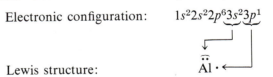

Electronic configuration: $1s^2 2s^2 2p^6 3s^2 3p^1$

Lewis structure: $\overset{\cdot\cdot}{Al} \cdot \leftarrow$

Here the valence electrons are shown as a pair (the two $3s$ electrons) and a single electron (the $3p$). The Lewis structures for the atoms of the third period are shown in Table 8-1, together with their electronic configurations.

Octet rule   As we have mentioned, noble-gas atoms (except for He) have an especially stable valence-shell electronic configuration: $ns^2 np^6$, where n is the principal quantum number of the valence shell. These elements have high ionization energies and low electron affinities, and they show little tendency to undergo chemical reaction. The *octet rule* is a statement of the stability of the $ns^2 np^6$ valence-shell configuration. Atoms which can achieve this configuration by the addition of only a few electrons tend to do so, that is, tend to *complete the octet.* In adding electrons the atom becomes a negative ion. Thus the *chloride ion* (the *-ide*

---

[1] Lewis structures are also sometimes called *electronic,* or *dot, symbols* and *formulas.*

Table 8-1
Lewis structures for
atoms of the third
period

Atom	Electronic configuration	Lewis structure
Na ($Z = 11$)	$1s^2 2s^2 2p^6 3s^1$	Na $\cdot$
Mg ($Z = 12$)	$1s^2 2s^2 2p^6 3s^2$	Mg $:$
Al ($Z = 13$)	$1s^2 2s^2 2p^6 3s^2 3p^1$	$\overset{\cdot}{\text{Al}}:$
Si ($Z = 14$)	$1s^2 2s^2 2p^6 3s^2 3p^2$	$\cdot \overset{\cdot}{\text{Si}}:$
P ($Z = 15$)	$1s^2 2s^2 2p^6 3s^2 3p^3$	$\cdot \overset{\cdot}{\text{P}}:$
S ($Z = 16$)	$1s^2 2s^2 2p^6 3s^2 3p^4$	$\cdot \overset{\cdot\cdot}{\text{S}}:$
Cl ($Z = 17$)	$1s^2 2s^2 2p^6 3s^2 3p^5$	$:\overset{\cdot\cdot}{\text{Cl}}:$
Ar ($Z = 18$)	$1s^2 2s^2 2p^6 3s^2 3p^6$	$:\overset{\cdot\cdot}{\underset{\cdot\cdot}{\text{Ar}}}:$

suffix tells us that it is a *negative* ion) is formed when one electron adds to a chlorine atom:

$$Cl(1s^2 2s^2 2p^6 3s^2 3p^5) + e^- \longrightarrow Cl^-(1s^2 2s^2 2p^6 3s^2 3p^6)$$

or

$$:\overset{\cdot\cdot}{\underset{\cdot\cdot}{Cl}}\cdot + e^- \longrightarrow \left[:\overset{\cdot\cdot}{\underset{\cdot\cdot}{Cl}}:\right]^-$$

Here the negative sign is written to remind us that the resulting particle is an ion, and brackets are used to keep the negative sign from getting lost among the dots.

What about positive ions and the octet rule? When an atom has but a few valence electrons and has an octet in the second shell from the outside, it tends to lose its valence electrons, thereby exposing the octet. In this way the resulting positive ion ends up with an octet in what is now its outer shell. Thus the sodium atom tends to lose its valence electron to form a sodium ion:

$$Na(1s^2 2s^2 2p^6 3s^1) \longrightarrow Na^+(1s^2 2s^2 2p^6) + e^-$$

or

$$Na\cdot \longrightarrow Na^+ + e^-$$

Here the eight electrons in the L shell of Na^+ are not indicated; in the Lewis structure dots are used to count electrons in the valence shell of the *original sodium atom only*.

ADDED COMMENT

Before we go any further it should be made clear that the octet rule is only a rough guideline, useful for making predictions about bonding and stoichiometry in many compounds. It has many exceptions, however. It is common for the transition elements, for example, to "violate" the octet rule. (These atoms use orbitals and electrons in the d subshell of the $n - 1$ shell for bonding.) The octet rule is most commonly obeyed by the atoms of the representative (A-group) elements of periods 2 and 3, but even among some of these violation of the rule is not uncommon. The octet rule is a handy generalization, and it is fortunate that it works as well as it does, but it is not a natural law.

Lewis structures and ionic compounds

To write the Lewis structure for an ionic compound, we write structures for the individual ions. Thus the Lewis structure for NaCl is

$$Na^+ \quad \left[:\ddot{\underset{..}{Cl}}: \right]^-$$

Note that this does not mean that each Na^+ is bonded to only one Cl^-. It means merely that equal numbers of Na^+ and Cl^- ions are present in the compound.

The octet rule helps us to predict stoichiometry, that is, atomic combining ratios, in ionic compounds. In the NaCl example one electron was transferred from one Na atom to one Cl atom:

$$Na \cdot \quad \cdot \ddot{\underset{..}{Cl}}: \Big\} \longrightarrow \Big\{ Na^+ \quad \left[:\ddot{\underset{..}{Cl}}: \right]^- $$

In sodium oxide, however, the situation is different. Oxygen has only *six* valence electrons and thus needs *two* to complete its octet:

$$:\overset{..}{\underset{.}{O}}\cdot \; + \; 2e^- \longrightarrow \left[:\ddot{\underset{..}{O}}: \right]^{2-}$$

Since a sodium atom has only one valence electron to lose (and its *second* ionization energy is very large), *two* sodium atoms are required to furnish *two* electrons to a single oxygen atom:

$$\left. \begin{array}{c} Na\circ \\[2em] \cdot\overset{..}{\underset{.}{O}}: \\[2em] Na\circ \end{array} \right\} \longrightarrow \left\{ \begin{array}{c} Na^+ \\[1em] \left[:\ddot{\underset{..}{O}}: \right]^{2-} \\[1em] Na^+ \end{array} \right.$$

The Lewis structure for sodium oxide can be written as

$$2Na^+ \quad \left[:\ddot{\underset{..}{O}}: \right]^{2-}$$

The empirical formula of sodium oxide is, therefore, Na_2O.

● **EXAMPLE 8-1**

Problem   Write the Lewis structure for calcium chloride.

Solution   Calcium (Ca), in group IIA of the periodic table, has two valence electrons, while chlorine (Cl), in group VIIA, has seven. A calcium atom can, by losing its two valence electrons, convert two Cl atoms to ions.

$$Ca \overset{\displaystyle \cdot \overset{..}{\underset{..}{Cl}}:}{\underset{\displaystyle \cdot \overset{..}{\underset{..}{Cl}}:}{\Big\langle}}$$

Lewis structure: $Ca^{2+} \quad 2\left[:\ddot{\underset{..}{Cl}}: \right]^-$ ●

● **EXAMPLE 8-2**

Problem   Write the Lewis structure for aluminum oxide.

Solution   Aluminum (Al) (group IIIA) has three valence electrons; O (group VIA) has six.   We expect each Al to lose three electrons and O to gain two.   It must be that *two* Al atoms *each* give up three electrons and that this total of six electrons is added to *three* O atoms to form *three* O^{2-} ions:

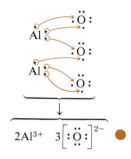

$$2Al^{3+} \quad 3\left[:\ddot{O}:\right]^{2-} ●$$

It is not necessary to work out the Lewis structure for an ionic compound in order to determine the compound's empirical formula *if* the charges of the ions are known.   Note that in a correct empirical formula the sum of all the positive charges due to the cations equals the sum of the negative charges due to the anions.

● **EXAMPLE 8-3**

Problem   Potassium nitride consists of K^+ and N^{3-} ions.   What is its empirical formula?

Solution   The negative charge on one N^{3-} ion can be compensated by the positive charges on three K^+ ions.   Therefore, the empirical formula is K_3N. ●

● **EXAMPLE 8-4**

Problem   Aluminum sulfate consists of Al^{3+} and SO_4^{2-} ions.   What is its empirical formula?

Solution   Two Al^{3+} ions carry a total of six positive charges.   To get six negative charges, three SO_4^{2-} ions are needed.   Thus, we get the formula $Al_2(SO_4)_3$. ●

Energetics and   Using the octet rule we can predict that atoms with only a few valence electrons
ionic bonding   tend to lose them to form positive ions, while atoms with close to eight valence electrons tend to gain in order to complete their octets.   Why should this be true? Part of the answer is found by considering ionization energies and electron affinities.   Atoms with only a few valence electrons (at the left of the periodic table) have low ionization energies and low electron affinities, and so we expect them to have a greater tendency to lose electrons than to gain them.   Similarly, atoms with close to eight electrons (near the right of the table) have high ionization energies and high electron affinities, and so we expect these to gain, rather than lose, electrons.

Let us consider the formation of the ionic bond in an NaCl ion pair.   We start with an isolated, gaseous Na atom and a similar Cl atom.   The bond can be

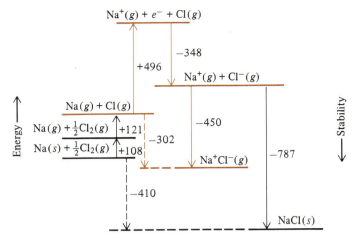

**Figure 8-3
Energy changes
(kJ mol⁻¹) in the forma-
tion of sodium chloride.
Plus signs and upward
arrows indicate absorp-
tion of energy; minus
signs and downward ar-
rows, the release of
energy.**

considered to be formed in three stages, as shown below along with the corresponding energy changes per mole of particles:

Step 1: $\qquad$ $Na\cdot \longrightarrow Na^+ + e^-$ $\qquad$ 496 kJ mol⁻¹ absorbed

Step 2: $\qquad$ $e^- + \cdot\ddot{\underset{\cdot\cdot}{Cl}}: \longrightarrow \left[:\ddot{\underset{\cdot\cdot}{Cl}}:\right]^-$ $\qquad$ 348 kJ mol⁻¹ released

Step 3: $\qquad$ $Na^+ + \left[:\ddot{\underset{\cdot\cdot}{Cl}}:\right]^- \longrightarrow Na^+\left[:\ddot{\underset{\cdot\cdot}{Cl}}:\right]^-$ $\qquad$ 450 kJ mol⁻¹ released

_____

Overall: $\qquad$ $Na\cdot + \cdot\ddot{\underset{\cdot\cdot}{Cl}}: \longrightarrow Na^+\left[:\ddot{\underset{\cdot\cdot}{Cl}}:\right]^-$ $\qquad$ 302 kJ mol⁻¹ released

Step 1 *requires* energy, the *ionization energy* of Na. Step 2 *releases* energy, the
electron affinity of Cl. Step 3, the joining of the oppositely charged Na⁺ and Cl⁻
ions to form a gaseous ion pair, *releases* energy. The energy changes in the
sequence are indicated *in color* in Fig. 8-3. In this drawing energy absorbed is
indicated with a plus sign and an upward-pointing arrow, and energy released,
with a minus sign and a downward-pointing arrow.

In the three-step process described above, the energy required to ionize the Na
atom (step 1) is not quite compensated for by the energy released when the Cl
atom accepts an electron (step 2). The deficit is more than made up in step 3,
when the ion pair is formed. We often associate release of energy with processes
which occur naturally or spontaneously, and so we say that the process

$$Na(g) + Cl(g) \longrightarrow Na^+Cl^-(g) \qquad 302 \text{ kJ mol}^{-1} \text{ released}$$

is an *energetically favorable,* or *favored, process.*[2]

_____

[2]Naturally occurring processes are not always accompanied by a release of energy. For instance, at
temperatures above 0°C ice spontaneously melts and in doing so *absorbs* energy. (In Chap. 18 the
topic of spontaneous change is considered in detail.)

We have shown that a gaseous Na^+Cl^- ion pair is energetically more stable than gaseous Na and Cl atoms. Let us now make a more useful comparison by seeing which is energetically more stable: solid sodium chloride or solid sodium metal plus chlorine gas (Cl_2). These are the states in which these substances are found under ordinary conditions. Consider the following stepwise sequence:

Step A: $Na(s) \longrightarrow$	$Na(g)$	108 kJ mol^{-1} absorbed
Step B: $\frac{1}{2}Cl_2(g) \longrightarrow$	$Cl(g)$	121 kJ mol^{-1} absorbed
Step C: $Na(g) \longrightarrow$	$Na^+(g) + e^-$	496 kJ mol^{-1} absorbed
Step D: $e^- + Cl(g) \longrightarrow$	$Cl^-(g)$	348 kJ mol^{-1} released
Step E: $Na^+(g) + Cl^-(g) \longrightarrow NaCl(s)$		787 kJ mol^{-1} released
Overall: $Na(s) + \frac{1}{2}Cl_2(g) \longrightarrow NaCl(s)$		410 kJ mol^{-1} released

Step A is the *sublimation* (conversion from solid to gas) of one mole of Na atoms. Step B is the *dissociation* of one-half mole of gaseous Cl_2 molecules to form one mole of Cl atoms. Steps C and D are identical to steps 1 and 2 of the previous sequence. Finally, step E is the formation of one mole of *solid* NaCl. Much more energy is released in this step than in the formation of a mole of Na^+Cl^- ion pairs (step 3 of the first sequence), because in the solid each Na^+ forms not one but *six* ionic bonds, as does each Cl^-. (Repulsions among like-charged ions keep the energy released in step E from being even higher.)

The new steps (A, B, and E) are shown *in black* in Fig. 8-3. It can be seen that even though energy is needed to sublime Na and dissociate Cl_2, it is more than made up by the large amount of energy released when solid NaCl is formed from its ions. Solid NaCl is thus energetically more stable than its uncombined elements in their usual states.

8-2 Covalent bonding

We have seen that when one atom has a low ionization energy and the other a high electron affinity, one or more electrons can transfer from the first atom to the second to form an ionic bond. Covalent bonding occurs, on the other hand, when two atoms are more nearly alike in their tendencies to gain and lose electrons. Under these conditions outright transfer of an electron does not occur. Instead, electrons are *shared* between the atoms.

The hydrogen molecule

Consider the hydrogen molecule H_2. This particle is formed from two hydrogen atoms, each of which consists of a nucleus and one electron. Imagine that you are holding two hydrogen atoms in your hands, slowly bringing them toward each other. As they approach each other, you feel an attractive force which tends to draw them still closer together, and their potential energy decreases. As you allow them to approach still more closely, the attractive force first increases and then becomes weaker, finally disappearing altogether. It disappears when the nuclei of the hydrogen atoms are 0.074 nm apart. At this point the potential energy of the pair of atoms has reached a minimum. In order to get the atoms to approach each other still more closely, you must push them together, because when they are closer than 0.074 nm (internuclear distance) apart, a force of repulsion tends to move them out to the 0.074-nm distance. Pushing them together increases their potential energy.

Actually, both the attractive and the repulsive forces exist at *all* distances. When the H atoms are still relatively far apart, however, the force of repulsion is

CHEMICAL BONDING

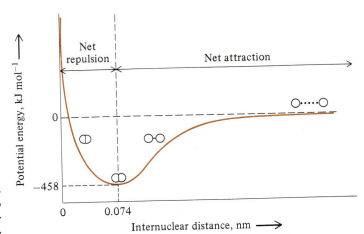

Figure 8-4
Potential energy of two hydrogen atoms at various distances apart.

negligible, and so they experience a *net* attraction. As the atoms move closer, the attractive force at first increases more rapidly than the repulsion, but eventually, at still closer distances, the repulsion starts to increase more rapidly than the attraction. Closer than 0.074 nm the atoms experience a *net* repulsion. At 0.074 nm the attractive force is just balanced by the repulsive force, and the system is at minimum potential energy. The force which pulls the two atoms together is a result of the attraction between *each* nucleus and the electron of the *other* atom. The repulsive force is the electrical repulsion of like charges, i.e., the two positively charged nuclei and the two negatively charged electrons. When the nuclei are 0.074 nm apart, *both* nuclei are attracted equally to *both* electrons. This attraction keeps the hydrogen atoms together and is called the *covalent bond*.

Figure 8-4 shows how the potential energy of two H atoms changes as they are brought together. (Their potential energy when infinitely far apart has been arbitrarily assigned the value zero.) As the H atoms approach each other (moving from *right to left* on the graph) their potential energy at first decreases. But as the atoms approach 0.074 nm, the *net* attraction between them becomes weaker, and so the potential energy curve begins to level off. At 0.074 nm the net attraction

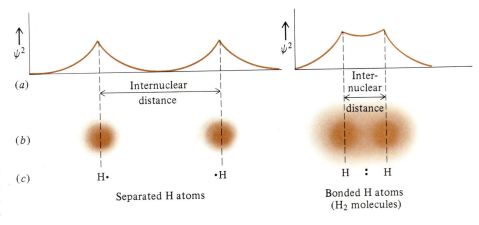

Figure 8-5
The formation of the hydrogen molecule.
(a) Probability-density plots. (b) Electron charge clouds. (c) Lewis structures.

changes to a net repulsion, and so the curve begins to rise. When they are 0.074 nm apart the pair of atoms has a minimum potential energy, -458 kJ mol^{-1}. This means that 458 kJ are necessary to break 1 mol of H—H bonds. Thus 458 kJ mol^{-1} is the *bond dissociation energy* of H_2.

As two hydrogen atoms approach each other each electron begins to "sense" electrostatically the presence of the nucleus of the opposite atom. In terms of quantum mechanics this results in an increase in the probability of finding the first atom's electron near the second atom's nucleus and vice versa. Eventually each electron is equally influenced by the two nuclei, and so the probability of finding *each* electron is the same at *each* nucleus.

Figure 8-5 shows the way electron probability density, Ψ^2, changes as two hydrogen atoms form a covalent bond. This is shown first with probability-density curves (*a*) and then with charge-cloud drawings (*b*). Lewis structures (*c*) are also shown for comparison. Notice that as the two atoms approach each other there is a buildup of electronic charge in the region between the nuclei. (There is also an accompanying decrease in electronic charge at the ends of the molecule.) The increase in charge density tells us that the electron pair in H_2 is simultaneously attracted to both nuclei. The nuclei are therefore indirectly held together.

Because the two electrons in the H_2 molecule are both influenced identically by the two nuclei, they occupy the same region of space. But the Pauli exclusion principle (Sec. 6-5) indicates that this is possible only if the electrons have anti-parallel spins. It can be shown experimentally that an unbonded H atom is paramagnetic (Sec. 6-1), a result of its single unpaired electron. But an H_2 molecule is *not* paramagnetic; its two electrons are *paired,* that is, have antiparallel spins, and are *shared* between the two atoms.

In any covalent bond the distance between the nuclei of the bonded atoms is called the *bond distance* or *bond length*. In the H_2 molecule this distance is 0.074 nm. Actually, this is an *average* distance, because two bonded atoms do not remain motionless with respect to each other. Rather, they vibrate in and out, toward and away from each other. At low temperatures the close-approach distance in H_2 is about 0.06 nm, and at greatest separation the atoms are about 0.09 nm apart. The energy of this vibrational motion is quantized, and at low temperatures the hydrogen molecule is in its ground (lowest energy) vibrational state. The vibration of a diatomic model is similar to the vibration of two balls connected by a spring, as shown in Fig. 8-6, although this model ignores the quantization of vibrational energies.

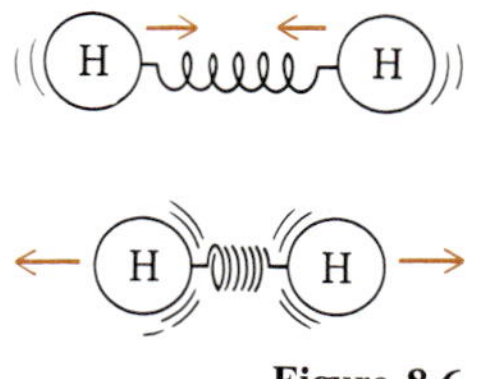

Figure 8-6
Model for the vibration of a hydrogen molecule.

Lewis structures and covalent bonding

A covalent bond consists of a pair of electrons shared between the bonded atoms, and Lewis structures can be written to show this sharing. The formation of the H_2 molecule from two H atoms can be shown as

Isolated H atoms: H·   ·H

H$_2$ molecule: H:H

Here the shared pair of electrons in H_2 is shown as a pair of dots between the symbols.

In the hydrogen fluoride (HF) molecule the formation of the H—F covalent bond can be shown as

Isolated atoms: H·   ·F̈:

HF molecule: H:F̈:

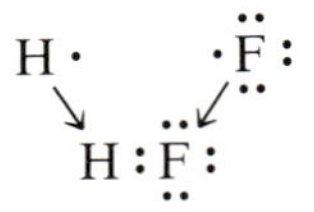

As before, the shared pair of electrons, which constitute the covalent bond, is shown between the two symbols.

Drawing Lewis structures for molecules or polyatomic ions is not difficult. Here are some general rules which are useful:

1 Determine enough of the geometry of the particle so that you know which atoms are bonded to which. (At this point in your chemical education you will often need to be provided with this information.) Then write down the symbol for each atom so that bonded atoms are shown adjacent to each other. *Note:* A single atom in the formula of a simple molecule is often the central atom, surrounded by others to which it is bonded. (Example: CCl_4.)

2 Using the periodic table, determine total number of valence electrons for all the atoms in the molecule. If the particle carries a charge, that is, if it is an *ion*, then add one electron for each negative charge or subtract one electron for each positive charge on the ion.

3 Assign electrons to the symbols written in step 1 *in pairs*, if possible. First, join the bonded atoms with a pair of electrons (dots), and then attempt to place the remaining electrons around the atoms in the molecule (or ion) in such a way that the octet rule is obeyed for each atom (with the exception of hydrogen, which obeys the duet rule). Two electrons shown shared between two atoms are called a *bonding pair*. Unshared pairs of electrons, often called *lone pairs*, are pairs which occupy orbitals of one atom only, and so are represented by dots assigned to one symbol only.

4 For some molecules, following the above rules leads to a situation in which insufficient valence electrons appear to be present to complete each octet. Often this just means that it is necessary to show *two* or *three* pairs of electrons shared between two atoms. These are a *double* and *triple* bond, respectively.

The procedure is best clarified with specific examples:

● **EXAMPLE 8-5**

Problem   Write a Lewis structure for methane, CH_4.

Solution   Since each H can accommodate only two electrons (one pair) in its valence shell, it can form only one bond. Thus each H must be bonded to the C; so we write

H
H C H
H

Each H brings one valence electron to the molecule, and the C brings four. (Carbon is in group IVA of the periodic table.) The total is $4(1) + 4 = 8$. We need a pair for each bond; so the Lewis structure is

$$H$$
$$\overset{..}{H:C:H}$$
$$\overset{..}{H}$$

In this molecule the eight valence electrons are grouped as four bonding, or shared, pairs. ●

● **EXAMPLE 8-6**

Problem   Draw the Lewis structure for ammonia, NH_3.

Solution   Since each H can form only one covalent bond, the arrangement of atoms must be

$$H \quad N \quad H$$
$$H$$

From the periodic table we see that N has five valence electrons. These, plus one electron from each H, give us a total of eight. Bonding the atoms in the molecule requires the use of six valence electrons, as

$$H:N:H$$
$$\overset{..}{H}$$

The remaining two valence electrons are then assigned to N to complete its octet:

$$\overset{..}{H:N:H} \quad ●$$
$$H$$

● **EXAMPLE 8-7**

Problem   Write the Lewis structure for ethane, C_2H_6.

Solution   A little thought will reveal that the two C atoms must be bonded to each other. (Remember: Hydrogen forms only one bond.) Keeping in mind the octet rule, we predict that besides bonding to the other C atom, each C forms three bonds to H atoms:

$$H \quad H$$
$$H \quad C \quad C \quad H$$
$$H \quad H$$

Each C contributes four valence electrons and each H, one, as usual; $2(4) + 6 = 14$ valence electrons. Bonding all atoms in the molecule with pairs of electrons uses all of these. The Lewis structure is thus

$$H \quad H$$
$$\overset{..}{H:C:C:H} \quad ●$$
$$\overset{..}{H} \quad \overset{..}{H}$$

● **EXAMPLE 8-8**

Problem   Write the Lewis structure for ethylene, C_2H_4.

Solution   Here, as in the previous example, the two C atoms must be bonded to each other.

$$H \quad H$$
$$H \quad C \quad C \quad H$$

The total number of valence electrons is $2(4) + 4 = 12$. Bonding all atoms in the molecule uses 10 valence electrons:

$$\begin{array}{cc} \text{H} & \text{H} \\ \text{H}:\overset{..}{\text{C}}:\overset{..}{\text{C}}:\text{H} \end{array}$$

Where shall we put the last two electrons? If we put them in the valence shell of one C, then the octet rule will be violated for the other.

$$\begin{array}{cc} \text{H} & \text{H} \\ \text{H}:\overset{..}{\underset{..}{\text{C}}}:\overset{..}{\text{C}}:\text{H} \end{array}$$

However, placing them between the two C atoms solves the problem.

$$\begin{array}{cc} \text{H} & \text{H} \\ \text{H}:\overset{..}{\text{C}}::\overset{..}{\text{C}}:\text{H} \end{array}$$

The two carbons are thus shown to be bonded via a *double bond,* two shared pairs, or four bonding electrons. ●

Note that in the Lewis structure for ethylene (Example 8-8) the four electrons of the double bond are counted with *each* carbon.

$$\begin{array}{cc} \text{H} \quad \text{H} & \text{H} \quad \text{H} \\ \text{H}:\overset{..}{\text{C}}::\overset{..}{\text{C}}:\text{H} \qquad & \text{H}:\overset{..}{\text{C}}::\overset{..}{\text{C}}:\text{H} \end{array}$$

Thus the octet rule is obeyed for each carbon.

The Lewis structure for ethylene may alternatively be shown as

$$\begin{array}{ccc} \begin{array}{c} \text{H} \ \ \text{H} \\ \overset{..}{\text{C}}::\overset{..}{\text{C}} \\ \text{H} \ \ \text{H} \end{array} \quad \text{or} \quad & \begin{array}{c} \text{H} \ \ \text{H} \\ \overset{..}{\text{C}}::\overset{..}{\text{C}}:\text{H} \\ \text{H} \end{array} \quad \text{or even} \quad & \begin{array}{c} \text{H.} \qquad \text{.H} \\ \overset{.}{\text{C}}::\overset{.}{\text{C}} \\ \text{H.} \qquad \text{.H} \end{array} \end{array}$$

Each of these is equivalent to the version shown in Example 8-8. Lewis structures are not generally expected to represent the geometrical shape of a molecule, one reason being that they must be drawn in a plane, while the molecule itself may not be planar. A Lewis structure is expected to show (a) which atoms in the molecule are bonded together and (b) how the valence electrons are assigned in the molecule, that is, which pairs are bonding pairs and which are lone pairs.

● **EXAMPLE 8-9**

Problem   Write the Lewis structure for the ammonium ion NH_4^+.

Solution   The arrangement of the atoms is

$$\begin{array}{c} \text{H} \\ \text{H N H} \\ \text{H} \end{array}$$

The total number of valence electrons is $(5 + 4) - 1 = 8$. (Here we have subtracted one electron from the total provided by one N and four H atoms, because the ion has a positive charge and therefore has one less electron than a (neutral) molecule.

The Lewis structure is

$$\begin{bmatrix} \quad H \quad \\ \ \ \overset{..}{\ }\ \\ H:N:H \\ \ \ \overset{..}{H} \end{bmatrix}^{+}$$

In the electron counting process one electron of a bonding pair often appears to have come from one of the two bonded atoms, and the other electron from the other atom. Thus we have for the formation of the three covalent bonds in NH_3 (see Example 8-6)

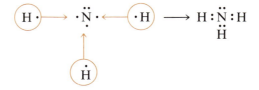

To keep track of electrons' origins small x's and o's are sometimes used in Lewis structures. Writing x's for the electrons from the H atoms and o's for those from N, we get for NH_3

$$H \overset{oo}{\underset{ox}{\overset{x}{N}\overset{x}{}}} H$$
$$H$$

A covalent bond in which each electron of the pair appears to have come from each bonded atom is called a *normal covalent bond*.

In the ammonium ion, again using the x and o symbolism, we can write

$$\begin{bmatrix} \quad H \quad \\ \ \overset{oo}{\ }\ \\ H \overset{x}{\underset{ox}{N}} \overset{x}{} H \\ \ \ H \end{bmatrix}^{+}$$

In this Lewis structure one of the H atoms appears to be bonded to the N by a pair of electrons, both of which originated with the N atom. Such a bond is called a *coordinate covalent bond,* or sometimes, a *dative bond.* But all four N—H bonds in NH_4^+ are identical in all measurable properties. So a *coordinate covalent bond is in no way different from a normal covalent bond.* There appears to be a difference only when we keep track of the electrons' origins.

ADDED COMMENT

Although the x and o symbolism is popular, it should be used with some caution, because it can be taken to mean that electrons are somehow different, depending upon their origin. *This is not true.* Each electron in a molecule belongs to the molecule as a whole. It does not retain a label indicating its origin.

The octet rule and Lewis structures   Wherever possible, Lewis structures should show the octet rule to be obeyed. There are some molecules, however, in which the octet rule is clearly violated.

Consider the molecule of phosphorus pentachloride, PCl_5. In this molecule a phosphorus atom is bonded covalently to *five* chlorine atoms. The total number of valence electrons is 40 (5 from the P plus 35 from the five Cl atoms). Since the P forms five bonds, the Lewis structure is

$$
\begin{array}{c}
:\ddot{\text{Cl}}: \\
:\ddot{\text{Cl}}: \;\; :\ddot{\text{Cl}}: \\
\text{P} \\
:\ddot{\text{Cl}}: \;\; :\ddot{\text{Cl}}:
\end{array}
$$

Here the valence shell of the phosphorus atom is said to have been *expanded* in order to accommodate five electron pairs. The expansion of the valence shell of an atom is possible only if the atom has nd or $(n-1)d$ orbitals which can be added to the ns and three np orbitals normally constituting its valence shell. In the case of PCl_5 the 10 bonding electrons are accommodated in the valence shell of phosphorus which has been expanded by the addition of one of phosphorus' $3d$ orbitals. The valence shells of atoms of periods 1 and 2 cannot be expanded because they contain no $1d$ or $2d$ orbitals. (The $3d$ orbital is unavailable for these atoms because it is of such high energy.)

Sometimes the valence shell of an atom in a molecule contains less than an octet. This is the case with boron trifluoride, BF_3. Its Lewis structure is written as

$$
:\ddot{\text{F}}:\text{B}:\ddot{\text{F}}: \\
:\ddot{\text{F}}:
$$

Here the valence shell of boron holds only three pairs of electrons, and again the octet rule is violated.

The octet rule is a handy generalization, but exceptions to it are numerous. It *must* be violated in molecules having an *odd* number of valence electrons. The Lewis structure for nitric oxide, NO, can be shown as

$$
:\text{N}::\ddot{\text{O}}:
$$

although in this case either resonance theory (below) or MO theory (Sec. 9-4) gives a more satisfactory description of the molecule.

Resonance   A Lewis structure is a schematic summary of the roles played by all valence electrons in a molecule according to valence-bond theory. Sometimes, however, difficulties arise. As an example, consider the case of ozone. The ozone molecule consists of three oxygen atoms having a bent structure as follows

Let us now try to draw its Lewis structure. Each O atom contributes six valence electrons (O is in group VIA), and so the total is $3 \times 6 = 18$ electrons which must be shown. Following the rules, we find that we can write not just one Lewis structure but two:

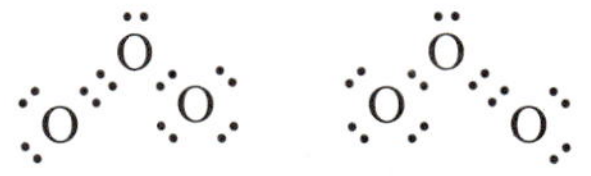

I   II

These two structures are equivalent, but structure I shows the double bond located at the left, while II shows it at the right. In either case one bond (the double bond) should be shorter than the other, but this is not in agreement with the experimental bond length of 0.128 nm for each.

The problem is that simple Lewis structures are really too simple and are not up to the task of showing the bonding in ozone with a single diagram. The structure of the O_3 molecule is actually halfway between I and II. In the language of chemistry the structure of the ozone molecule is said to be a *resonance hybrid* of the two contributing forms I and II. The two forms are usually written side by side with a two-headed arrow between:

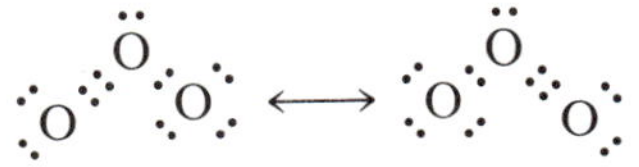

I   II

The word *resonance* and the two-headed arrow are apt to give the wrong impression. *There is no alternation or oscillation between the two contributing forms.* The actual structure is a kind of average or combination of I and II, and so is halfway between.

ADDED COMMENT

Remember that there is no alternation of structures in a resonance hybrid. The analogy between a resonance hybrid and a biological hybrid is often used to illustrate this point. Consider a mule, which is a hybrid, or cross, between a male donkey and a female horse. When you look at a mule, do you see a creature which is rapidly alternating between a horse and a donkey? (No, of course you do not!) The mule has only one "structure," and so too does any resonance hybrid.

Structural formulas   The Lewis structure uses dots to represent electrons. In *structural formulas* bonding electron pairs are represented by short straight lines, and lone electron pairs are sometimes omitted. Compare the Lewis structure with the structural formula of the water molecule below.

Lewis structure 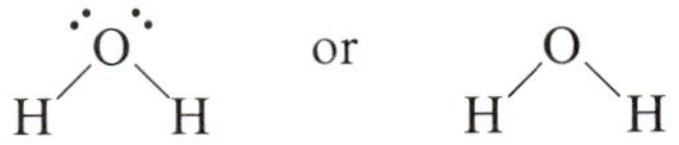   Structural formula

When lone pairs are shown in a structural formula, it is usually done to emphasize their presence and their influence on the properties of the molecule. For example, lone pairs often influence bond angles and overall shape of a molecule. (See Sec. 9-1.)

8-3
Electronegativity

A shared pair of electrons is simultaneously attracted to both bonded atoms, and so the atoms can be considered to be in competition for the electrons. But the electron pair is not shared *equally* unless the two atoms have the same attraction for electrons. This attraction is measured by a quantity known as *electronegativity,* which is defined as the relative tendency shown by a bonded atom to attract electrons to itself.

Bond polarity

Identical atoms have identical electronegativities. In the H_2 molecule

$$H:H$$

the hydrogen atoms attract the electron pair equally. The electronic charge distribution is symmetrical with respect to the two nuclei; that is, it is not pulled closer to one atom than the other. Since one end of the bond is electrostatically just like the other, the bond is said to be *nonpolar*. (This just means that it does not have different poles, or ends.) For the same reason the bond in the fluorine molecule

$$:\ddot{F}:\ddot{F}:$$

is also nonpolar. *Atoms with identical electronegativities form nonpolar covalent bonds.*

Atoms of different elements have different electronegativities. In the hydrogen fluoride molecule

$$H:\ddot{F}:$$

because the F atom has a higher electronegativity than the H atom, the electron pair is shared unequally. The electronic charge cloud of the shared pair is pulled closer to the F atom, as is shown in Fig. 8-7. (An alternate way of describing the situation is to say that the electron spends more time near the F atom than near the H atom.) The resulting bond has negative charge piled up at one end, leaving positive charge at the other. (The hydrogen nucleus is at the other end, and it has been somewhat exposed by the withdrawal of electrons.) A covalent bond in which the electron pair is shared unequally is said to be a *polar covalent bond.*

The polarity of a bond, that is, the degree to which an electron pair is unequally shared, depends on the difference between electronegativities of the two bonded atoms. The greater the electronegativity difference, the more polar is the bond.

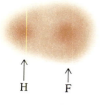

H　　F

$$H:\ddot{F}:$$

Figure 8-7
The polar bond in hydrogen fluoride. (The electronic charge cloud of the bonding electron pair only is shown.)

Partial ionic character

When two atoms of greatly differing electronegativity bond, the result is best classed as an ionic bond. Consider the bond formed between hypothetical atoms A and B. Imagine that B has a high electronegativity. Imagine further that we can alter the electronegativity of A by remote control. Let us start by adjusting the electronegativity of A so that it is exactly the same as that of B. Neglecting all electrons but the shared pair, we can schematically show that the bond is nonpolar by writing two dots exactly in the middle between the symbols A and B.

$$A : B \qquad \text{(nonpolar covalent bond)}$$

Now we turn the dial on the remote controller and gradually decrease the electronegativity of A. We make A *less electronegative,* or *more electropositive.* This decreases A's pull on the pair of electrons, so that their average position moves closer to B, creating partial charges, δ^+ and δ^-, on A and B, respectively:

$$\overset{\delta^+}{A} : \overset{\delta^-}{B} \qquad \text{(polar covalent bond)}$$

The bond has now become polar. It becomes increasingly polar as we further decrease the electronegativity of A, until finally the probability of finding the electron pair on A is very low and on B, very high. The electron pair now largely "belongs" to B. This gives B a net negative charge and leaves A with a positive charge, for A has now transferred an electron to B:

$$A^+ \quad [:B]^- \qquad \text{(ionic bond)}$$

The ionic bond can be seen to be an extremely polar bond, one in which there is essentially no sharing of electrons.

ADDED COMMENT

The idea that we might be able to remotely control the electronegativity of an atom by turning a knob is, of course, absurd. In the previous example we imagined it just for the purpose of illustrating the relationship between bond polarity and electronegativity difference.

Also, it is *not* a common practice to show bond polarity in a Lewis structure by displacing the two dots so that they are closer to one symbol than the other. We did it only to show that a bond of extremely high polarity is essentially an ionic bond.

There is no natural boundary which can be used to distinguish ionic from covalent bonds. The more polar a bond is, the more of the properties of an ionic bond it exhibits. For this reason it is useful to consider that each bond is a mixture of the pure ionic and pure covalent bonds. Looking at it this way, we see that each bond has some ionic and some covalent character. A bond which is called, simply, *covalent* is one which has a high degree (or percent) of covalent character, while one which is called *ionic* has little. Specifying the degree of ionic (or covalent) character of a bond is just another way of indicating something about the bond's polarity. The relationships among electronegativity difference, bond type, and degree of ionic and covalent character are summarized in Table 8-2.

ADDED COMMENT

In a polar bond the accumulation of negative charge at the more electronegative atom is a result of unequal sharing. It does not occur because one atom has more electrons, or more valence electrons, than the other. For instance, the Lewis structure for hydrochloric acid, shows an octet of valence electrons with Cl but only two electrons with H.

$$H : \overset{..}{\underset{..}{Cl}} :$$

The Cl end of the bond is indeed the negative end of this bond, but unequal sharing of the bonding pair cause the effect. Neither the total number nor the number of valence electrons has anything directly to do with it.

In the bromine chloride molecule

$$: \overset{..}{\underset{..}{Br}} : \overset{..}{\underset{..}{Cl}} :$$

Br has a larger total number of electrons than Cl but the same number of valence electrons. Nevertheless, the bond is polar with Cl at the negative end, because Cl is the more electronegative atom and consequently pulls the shared pair closer to itself.

Pauling electronegativities

The concept of *electronegativity* was originally proposed in 1932 by the American chemist Linus Pauling. He pointed out that the distribution of the electronic charge cloud of a bonding pair of electrons should be related to the strength of the bond. A bond which is highly polar (has a high degree of ionic character) should be very strong, as the attraction between the partial negative charge built up on one atom and the partial positive charge left on the other should augment the bond strength. Using measured values of bond energies (bond energy is the energy necessary to break a bond), Pauling devised a set of electronegativity values for most of the elements. These values are shown in Fig. 8-8.

Numerical values of electronegativities are useful for estimating the polarity or degree of ionic character of a bond. The dividing line between predominately ionic and predominately covalent character works out to be an electronegativity difference of about 1.7. This fact is sometimes useful for deciding whether to write a covalent Lewis structure or an ionic one. For HCl, for example, we might write the ionic structure

$$H^+ \left[: \overset{..}{\underset{..}{Cl}} : \right]^-$$

or the covalent structure

$$H : \overset{..}{\underset{..}{Cl}} :$$

**Table 8-2
Electronegativity difference, bond type, and bond character**

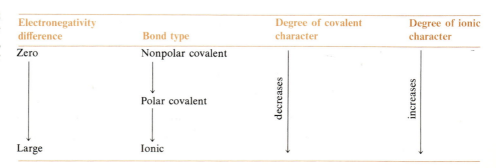

Electronegativity difference	Bond type	Degree of covalent character	Degree of ionic character
Zero	Nonpolar covalent	decreases	increases
	Polar covalent		
Large	Ionic		

1	2	3		4	5	6	7	8	9	10	11	12	13	14	15	16	17	18
1 H 2.1																		2 He —
3 Li 1.0	4 Be 1.5												5 B 2.0	6 C 2.5	7 N 3.0	8 O 3.5	9 F 4.0	10 Ne —
11 Na 0.9	12 Mg 1.2												13 Al 1.5	14 Si 1.8	15 P 2.1	16 S 2.5	17 Cl 3.0	18 Ar —
19 K 0.8	20 Ca 1.0	21 Sc 1.3		22 Ti 1.5	23 V 1.6	24 Cr 1.6	25 Mn 1.5	26 Fe 1.8	27 Co 1.9	28 Ni 1.9	29 Cu 1.9	30 Zn 1.6	31 Ga 1.6	32 Ge 1.8	33 As 2.0	34 Se 2.4	35 Br 2.8	36 Kr —
37 Rb 0.8	38 Sr 1.0	39 Y 1.2		40 Zr 1.4	41 Nb 1.6	42 Mo 1.8	43 Tc 1.9	44 Ru 2.2	45 Rh 2.2	46 Pd 2.2	47 Ag 1.9	48 Cd 1.7	49 In 1.7	50 Sn 1.8	51 Sb 1.9	52 Te 2.1	53 I 2.5	54 Xe —
55 Cs 0.7	56 Ba 0.9	57 La 1.1	58–71 1.1–1.3	72 Hf 1.3	73 Ta 1.5	74 W 1.7	75 Re 1.9	76 Os 2.2	77 Ir 2.2	78 Pt 2.2	79 Au 2.4	80 Hg 1.9	81 Tl 1.8	82 Pb 1.9	83 Bi 1.9	84 Po 2.0	85 At 2.2	86 Rn —
87 Fr 0.7	88 Ra 0.9	89 Ac 1.1	90–103 1.3–1.7	104 Rf —	105 Ha —	106 —												

Figure 8-8

The electronegativities of the elements (Pauling's method).

Since the difference between the electronegativities of Cl and H is $3.0 - 2.1 = 0.9$, the bond is clearly more covalent than ionic, and so the second Lewis structure is appropriate. Numerical values of electronegativities should be used with some caution, however. The values given in Fig. 8-8 are best considered as approximate, because the effective electronegativity of an atom tends to vary somewhat according to the atom's environment (which and how many atoms it is bonded to, etc.).

The variation of electronegativity with atomic number shows the expected periodicity (Fig. 8-9). Note that the effects of nuclear charge, atomic radius, and inner-shell shielding are all apparent. These cause electronegativity to increase across a period (as nuclear charge increases and atomic radius decreases) and decrease down a group (as radius and the number of inner shells both increase). Thus we find that the most electronegative atom is fluorine (F), at the upper right of the periodic table, and the least electronegative (or most electropositive) is francium (Fr), at the lower left.

One of the chemical characteristics of a typical *metal* is a *low electronegativity*. Thus we find the best metals on the left and the best nonmetals on the right in the periodic table. Note that the transition elements all have fairly low electronegativities; they are metals. The change from metallic to nonmetallic properties occurs to the right of these elements in the periodic table. Lastly, note that it is the metals (at the left) which tend to form positive, simple (monatomic) ions, and the nonmetals (at the right) which tend to form negative ones. (Develop the habit of mentally equating *metallic* with *electropositive* and *nonmetallic* with *electronegative*.)

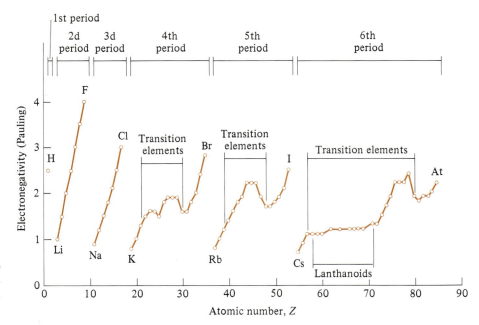

Figure 8-9
Periodicity in electro-
negativity.

8-4
Electronic
bookkeeping

The separation of electrical charges in a complex molecule has a direct bearing on the properties of the molecule. Unfortunately, accurate determination of charge distribution in a molecule is very difficult. Two approximate methods have proven to be quite useful. They are the method of *formal charges* and the method of *oxidation numbers*. These methods are similar; each compares the number of electrons in the valence shell of a bonded atom with the number in the isolated, unbonded atom. The two methods differ in the way the electrons of a bonding pair are divided between the bonded atoms.

Formal charges

The assignment of formal charges starts with a Lewis structure. Each valence electron is assigned to the appropriate atom using the following rules:

1 Assign both electrons of a lone pair to its atom.
2 *Split a shared pair,* assigning one electron to each atom bonded by the pair.

To obtain the formal charge of an atom, subtract the number of valence electrons which the atom appears to have (according to the above assignment) from the number in the isolated atom. The difference is the formal charge; it is the charge which the atom would have if it had exclusive possession of the assigned electrons, instead of sharing them. In other words,

Formal charge =
    (valence electrons in isolated atom) − (valence electrons in bonded atom)

Consider these examples:

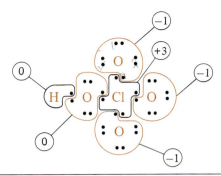

Atom	Number of valence electrons, isolated atom	Number of counted valence electrons, bonded atom	Formal charge
H	1	1	0
O (2 bonds)	6	6	0
O (1 bond)	6	7	−1 (each)
Cl	7	4	+3

H O O O O Cl

Sum of formal charges = 0 0 −1 −1 −1 +3 = 0

Figure 8-10
Formal charges in perchloric acid.

● **EXAMPLE 8-10**

Problem   Assign a formal charge to each atom in the molecule of perchloric acid, $HClO_4$. (The Cl is covalently bonded to the four O atoms.)

Solution   The Lewis structure is shown in Fig. 8-10. To determine formal charges, we assign electrons as is shown schematically in the figure. According to this assignment the Cl atom appears to have four valence electrons, one O atom (the one bonded to the H) appears to have six valence electrons, the other O atoms appear to have seven each, and the H appears to have one.

Since an isolated Cl atom has seven valence electrons (see the periodic table), the formal charge of Cl in $HClO_4$ is 7 − 4, or +3.[3] It is the charge which a Cl would have if it were to lose three electrons *completely*. Similarly, the formal charge of the O in the —OH group is 0; the six valence electrons assigned to it equal the number in the isolated atom. Each of the other three O atoms is assigned one electron more than the isolated atom, and thus each has a formal charge of −1. Lastly, the formal charge of the H atom is 0. Note that *the sum of the formal charges of all the atoms in the neutral molecule is 0.* ●

● **EXAMPLE 8-11**

Problem   Assign a formal charge to each atom in the cyanate ion, OCN^-.

---

[3] In this book we follow the practice of writing both formal charges and oxidation numbers with the plus or minus *before* the number. This contrasts with our practice of showing the actual charge on an ion by writing the number first. (3+ means that the ion carries three positive electrical charges.) It is important to remember that formal charges and oxidation numbers are not actual charges. Each is a charge which the atom *appears* to carry, when electrons are assigned according to some arbitrary rules.

Solution   First draw a Lewis structure for this ion.

$$\left[:\ddot{O}::C::\ddot{N}:\right]^{-}$$

The four electrons in each shared pair are divided evenly between the bonded atoms. Thus the O, C, and N atoms appear to have six, four, and six valence electrons, respectively. Since for O and C this is the same number as in the isolated atom, the formal charge for each of these is 0. But the N atom appears to have one electron more than the isolated atom; so its formal charge is -1. *The sum of the formal charges equals the charge on the ion.* ●

Oxidation numbers: There are two equivalent methods for assigning oxidation numbers.[4] Although the
method 1   first is generally less convenient, it more clearly illustrates the difference between formal charges and oxidation numbers. With this method valence electrons are counted much as they are when formal charges are assigned, except that *both electrons of a bonding pair are assigned to the more electronegative atom.* If the two bonded atoms are identical, the shared pair is split between the two, as with formal charges.

● EXAMPLE 8-12

Problem   Assign an oxidation number to each atom in a molecule of perchloric acid, $HClO_4$.

Solution   Figure 8-11 shows the Lewis structure of $HClO_4$ and the assignment of electrons. Each electron in a shared pair is assigned to the more electronegative atom. (Compare this with Fig. 8-10.)

    Compared with the corresponding isolated, uncombined atoms, each O atom appears to have two extra electrons, the Cl atom to be deficient by seven electrons, and the H atom to be deficient by 1. The oxidation numbers of these atoms are, therefore, -2, $+7$, and $+1$, respectively. In each case the number is the charge the atom would carry, if it had completely gained or lost the specified number of electrons. Note that *the sum of the oxidation numbers of all the atoms in the molecule is 0.* ●

● EXAMPLE 8-13

Problem   Assign an oxidation number to each atom in the cyanate ion, OCN^-.

Solution   The Lewis structure is

$$\left[:\ddot{O}::C::\ddot{N}:\right]^{-}$$

Because both O and N are more electronegative than C, we assign *all four* electrons in the O=C double bond to O, and, similarly, all four in the C=N double bond to N. Then O appears to have eight valence electrons, C to have none, and N to have eight. Since the isolated-atom valence-shell populations for these atoms are six, four, and five, respectively, we can say that O appears to have gained two electrons, C to have lost four electrons, and N to have gained three. Thus we assign oxidation numbers of -2 to oxygen, $+4$ to carbon, and -3 to nitrogen. Note that *the sum of the oxidation numbers is the charge on the ion.* ●

---

[4] Also called *oxidation states.*

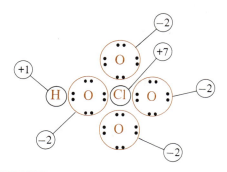

Atom	Number of valence electrons, isolated atom	Number of counted valence electrons, bonded atom	Oxidation number
H	1	0	$+1$
O (2 bonds)	6	8	-2
O (1 bond)	6	8	-2 (each)
Cl	7	0	$+7$

Figure 8-11
Oxidation numbers in perchloric acid.

$$\text{H} \quad \text{O} \quad \text{O} \quad \text{O} \quad \text{O} \quad \text{Cl}$$
Sum of oxidation numbers = $+1 \quad -2 \quad -2 \quad -2 \quad -2 \quad +7 = 0$

Oxidation numbers:
method 2

A set of rules has been developed for the assignment of oxidation numbers. These have the advantage of not requiring the drawing of a Lewis structure, a process which can be tedious in the case of a large molecule. Although it may not seem apparent at first, these rules are essentially equivalent to Method 1, the Lewis-structure method.

RULES FOR ASSIGNING OXIDATION NUMBERS

1 *Fluorine* in its compounds is always assigned an oxidation number of -1.
2 *Oxygen* in its compounds is assigned an oxidation number of -2.
   Exceptions: a *Peroxides and superoxides*. These compounds contain O—O bonds. The oxidation number of each oxygen atom in a peroxide is -1 and in a superoxide is $-\frac{1}{2}$.
   b *Oxygen fluorides*. Rule 1 always takes precedence. Thus in OF_2 and O_2F_2 the oxidation numbers of each oxygen are $+2$ and $+1$, respectively. (See rule 5c.)
3 *Hydrogen* in its compounds is assigned an oxidation number of $+1$.
   Exception: In metallic *hydrides* hydrogen is assigned a value of -1.
4 Combined elements of periodic groups IA (the alkali metals) and IIA (the alkaline-earth metals) are virtually always assigned oxidation numbers of $+1$ and $+2$, respectively. Combined elements of periodic group IIIA are usually assigned the oxidation number $+3$.
5 In the formula of a substance or species (ion, atom, molecule), the sum of the oxidation numbers of all the atoms in the formula equals the electrical charge shown with the formula.

a An atom of any element in the free (uncombined) state has an oxidation number of 0.

b Any simple (monatomic) ion has an oxidation number equal to its charge.

c The sum of the oxidation numbers of all the atoms shown in a formula for an entire compound equals 0. (This applies to both empirical and molecular formulas.)

d The sum of the oxidation numbers of all the atoms shown in the formula for a polyatomic or complex ion equals the electrical charge on the ion.

Here are some examples of correctly assigned oxidation numbers:

Substance	Oxidation number	Rule	Comments
S_8	$S = 0$	5a	Each S is 0
Cu	$Cu = 0$	5a	
HCl	$H = +1$	3	
	$Cl = -1$	5c	By subtraction
CH_4	$H = +1$	3	Each H is $+1$
	$C = -4$	5c	By subtraction
NaH (a hydride)	$Na = +1$	4	
	$H = -1$	3, 5c	
BaO	$Ba = +2$	4	
	$O = -2$	3, 5c	
BaO_2 (a peroxide)	$Ba = +2$	4	
	$O = -1$	2a, 5c	Each O is -1
KNO_3	$K = +1$	4	
	$O = -2$	2	
	$N = +5$	5c	By subtraction
HSO_3^-	$H = +1$	3	
	$O = -2$	2	
	$S = +4$	5d	By subtraction
$Cr_2O_7^{2-}$	$O = -2$	2	
	$Cr = +6$	5d	Each Cr is $+6$
Fe_3O_4	$O = -2$	2	
	$Fe = +\frac{8}{3}$	5c	By subtraction
$C_6H_{12}O_6$	$H = +1$	3	
	$O = -2$	2	
	$C = 0$	5c	By subtraction

ADDED COMMENT

The above rules work for the vast majority of substances. Occasionally you will need to supplement the rules by using the Lewis-structure technique, as discussed previously. (In that event be sure not to confuse oxidation numbers with formal charges.)

Note that oxidation numbers can be fractional. (See the Fe_3O_4 example, above.) This can occur when the compound contains differently bonded atoms of the same element with different oxidation numbers. The rules then give an average oxidation number, which may be a fraction.

Comparisons and uses   Both formal charges and oxidation numbers attempt to say something about the electrical nature of bonded atoms, but each uses a set of rather arbitrary rules; so each must be considered to be approximate. In HCl, for example, the formal charges of H and Cl are each 0. The oxidation number of H is $+1$ and of Cl, -1, however. The formal-charge assignment is based upon the assumption of equal sharing of the bonding electron pair, while the oxidation-number assignment gives the entire pair to chlorine. Neither is correct; the actual charge on the H is somewhere between 0 and $1+$ and on the Cl between 0 and $1-$. In some cases neither approach comes very close to reality. In $HClO_4$, for example, the charge on Cl is probably lower than either the formal charge ($+3$) or the oxidation number ($+7$).

If these quantities are so approximate, what purpose do they serve? An important application of formal charges is the evaluation of alternate Lewis structures for a molecule or ion. Let us reconsider the cyanate ion. Actually, three such structures could be drawn:

$$\left[:\overset{..}{\underset{..}{O}}:C::\overset{}{N}:\right]^{-} \longleftrightarrow \left[:\overset{..}{\underset{..}{O}}::C::\overset{}{N}:\right]^{-} \longleftrightarrow \left[:O:::C:\overset{..}{\underset{..}{N}}:\right]^{-}$$

Formal charges: $-1 \quad 0 \qquad 0 \qquad\qquad 0 \quad 0 \quad -1 \qquad\qquad +1 \quad 0 \quad -2$

The third structure is least likely, because of the placement of a $+1$ formal charge on oxygen, a highly electronegative atom. The actual structure of this ion is best described as a *resonance hybrid* in which each of the above three makes a different contribution. (The contribution made by the last structure is small.) In writing Lewis structures, try to avoid showing formal charges which are inconsistent with known electronegativities, which have like signs on adjacent atoms, or which have unlike signs on greatly separated atoms.

Oxidation numbers are useful for two general purposes, the first of which is in the organization of a great quantity of descriptive chemistry. Knowing that iron exhibits oxidation numbers of $+2$ and $+3$ in most of its compounds, for instance, is of great help in learning and organizing the chemistry of iron and in relating that chemistry to the electronic configuration of the iron atom.

The second use of oxidation numbers is as an important aid in balancing equations for electron-transfer reactions. In such reactions, also called *oxidation-reduction reactions,* the transfer of electrons is accompanied by a change in oxidation numbers. (See Sec. 13-4.)

SUMMARY

Of the various forces of attraction which can exist between atoms, the stronger ones are known as *chemical bonds*. In this chapter we have described two of these, the *ionic bond* and the *covalent bond*. The ionic bond is an electrostatic attraction between oppositely charged ions. Such ions can be formed by the transfer of one or more electrons between atoms, the transfer tending to obey the *octet rule*. This rule states that a valence-shell octet of electrons (ns^2np^6) is a stable configuration and that atoms of many elements, particularly those of the second and third periods, tend to achieve it. *Lewis structures* are useful for indicating valence-shell populations of ions. *Ionic solids* consist of three-dimensional arrays of positive and negative ions alternating in some regular manner in space.

While the formation of an ionic bond can be described in terms of the transfer of electrons, a covalent bond consists of a pair of electrons which is shared between the two bonded atoms. As with ionic compounds, *Lewis structures* are used to keep track of electrons in the aggregate of bonded atoms, and the octet rule is often obeyed. Sometimes an approach known as *resonance* is used to describe a structure which cannot be represented by a single valence-bond structure. In such a case the structure is called a *resonance hybrid* of two or more contributing forms for which Lewis structures can be drawn.

Electronegativity is a measure of the tendency shown by a bonded atom to pull electrons toward itself. *Bond polarity* depends upon the difference between the electronegativities of the bonded atoms. The larger this difference, the more polar and therefore more ionic is the bond.

We concluded this chapter with a description of two methods of keeping track of electrons in molecules: *formal charges* and *oxidation numbers* (*oxidation states*). Each method involves estimating an apparent charge on an atom by counting valence electrons according to a set of somewhat arbitrary rules. The two methods differ in the way bonding pairs are assigned to each covalently bonded atom.

KEY TERMS

Bond distance (Sec. 8-2)
Bonding pair (Sec. 8-2)
Core (Sec. 8-1)
Coordinate covalent bond (Sec. 8-2)
Covalent bond (Sec. 8-2)
Double bond (Sec. 8-2)
Electronegativity (Sec. 8-3)
Electrostatic force (Sec. 8-1)
Expansion of valence shell (Sec. 8-2)
Formal charge (Sec. 8-4)
Ion pair (Sec. 8-1)
Ionic bond (Sec. 8-1)
Kernel (Sec. 8-1)

Lewis structure (Sec. 8-1)
Lone pair (Sec. 8-2)
Nonpolar covalent bond (Sec. 8-3)
Normal covalent bond (Sec. 8-2)
Octet rule (Sec. 8-1)
Oxidation number (Sec. 8-4)
Oxidation state (Sec. 8-4)
Polar covalent bond (Sec. 8-3)
Resonance hybrid (Sec. 8-2)
Shared pair (Sec. 8-2)
Structural formula (Sec. 8-2)
Triple bond (Sec. 8-2)
Valence-bond theory (Sec. 8-2)

QUESTIONS AND PROBLEMS

Ionic bonding

8-1 The phosphate ion is PO_4^{3-}. Use the periodic table to help predict the empirical formula of each of the following ionic compounds: potassium phosphate, aluminum phosphate, magnesium phosphate, cesium phosphate, radium phosphate.

8-2 Use the periodic table to help predict the empirical formula of each of the following ionic compounds: sodium astatide, barium fluoride, potassium sulfide, gallium nitride, rubidium oxide, calcium phosphide.

8-3 Since a sodium ion and a chloride ion are attracted electrostatically, what prevents the two ions from merging to form a single larger atom?

Covalent bonding

8-4 Account for the fact that Cl_2 is a stable molecular species.

8-5 Each of the following contains at least one double bond. Draw the Lewis structure for each: CS_2, C_3H_6, C_4H_6, C_2H_3Cl.

8-6 Each of the following contains at least one triple bond. Draw the Lewis structure for each: CO, C_2H_2, HCN, C_3H_4.

8-7 Show that in ammonium chloride, NH_4Cl, both ionic and covalent bonds are present.

8-8 Show that each of the following contains one coordinate covalent bond: NH_4^+, S_2^{2-}, H_3O^+, H_3PO_4.

8-9 In the hydrogen-molecule ion H_2^+, a *single* electron holds the nuclei together. How would you expect the vibration frequency in H_2^+ to compare with that in H_2?

8-10 Using the periodic table write the molecular formula for the simplest compound (fewest atoms per molecule) formed from chlorine and each of the following elements: sulfur, iodine, silicon, phosphorus, boron.

8-11 The structure of each of the following can be represented as a resonance hybrid. Draw equivalent contributing structures for each: SO_2, SO_3, NO_3^-, NO_2^-.

Electronegativity and bond type

8-12 What are the factors that influence the electronegativity of an atom? Account for the observed variation of electronegativity across a period and down a group in the periodic table.

8-13 Using the periodic table classify each of the following bonds as being either predominately ionic or covalent: O—S, Ca—S, Si—C, H—I, Cl—O, Ga—F, Rb—Br, H—Li, Cs—N.

8-14 Write the electronic configurations for H and Na. Account for the fact that HCl is covalent, while NaCl is ionic.

8-15 Electron affinity and electronegativity each measure the electron-attracting tendency of an atom. Explain clearly how they differ.

8-16 Classify the bonding in each of the following compounds as being either predominately covalent or ionic: CsBr, MgS, NO, SF_4, CaI_2, CS_2, OF_2, KI, Rb_2O.

8-17 How does the overall charge distribution in BrCl differ from that in Cl_2? Draw sketches to illustrate your answer.

8-18 Using the periodic table predict which one of the following bonds should be least polar and which should be most polar: S—Cl, S—Br, Se—Br, Se—Cl.

Lewis structures

8-19 Draw a Lewis structure for each of the following ionic compounds: **(a)** rubidium fluoride **(b)** barium iodide **(c)** magnesium sulfide **(d)** potassium oxide **(e)** cesium nitride **(f)** strontium phosphide.

8-20 Draw a Lewis structure for each of the following molecules: CHI_3, C_2H_5Cl, PCl_3, N_2H_4, HOCl, HOClO, BCl_3, OF_2.

8-21 Consider atoms X, Y, and Z. X is an alkali metal, Y is in periodic group VA, and Z is a halogen. All are in the same period. **(a)** Which of the three atoms is most electronegative? Which is least? **(b)** Of the three bonds X—Y, X—Z, and Y—Z, which would be expected to be ionic? **(c)** Write a Lewis structure for a compound of X and Y, of X and Z, and Y and Z.

8-22 For each of the following ions draw an appropriate Lewis structure: PH_4^+, SO_2^{2-}, $S_2O_3^{2-}$, PO_4^{3-}, HPO_4^{2-}, BF_4^-, CN^-, N_3^-, $CH_3NH_3^+$.

Expanded valence shell

8-23 In each of the following the valence shell of the central atom is expanded to accommodate more than eight electrons. Draw a Lewis structure for each: ICl_4^-, SF_6, SF_4, I_3^-, XeF_4, $XeOF_4$, BrF_3, BrF_5.

8-24 Draw the octet-rule Lewis structure for sulfuric acid, H_2SO_4. **(a)** What is the formal charge on each atom? **(b)** Show how the formal charge on S can be reduced by "allowing" it to expand its valence shell. **(c)** Describe the H_2SO_4 molecule as a resonance hybrid.

8-25 The expansion of a valence shell to include more than eight electrons can occur only if the shell has sufficient orbitals. Keeping this in mind account for the fact that phosphorus forms two chlorides, PCl_3 and PCl_5, while nitrogen forms only one, NCl_3.

Formal charges and oxidation numbers

8-26 Assign a formal charge to each atom in each of the following: BrCl, $BeCl_2$, BF_4^-, IO_3^-, CN^-, $N_2H_5^+$.

8-27 For each of the species shown in Prob. 8-26 assign an oxidation number to each atom.

8-28 Assign an oxidation number to each atom in each of the following: H_3PO_3, $H_2S_2O_7$, $S_2O_3^{2-}$, $S_4O_6^{2-}$, $C_{12}H_{22}O_{11}$, KCl, $KClO_2$, Fe_2O_3, $KHSO_4$, CrF_6^{3-}, HO_2^- (a peroxide).

8-29 Keeping in mind that hydrogen has a -1 oxidation state in hydrides, write the Lewis structure for the hydride ion. What is the empirical formula for sodium hydride? For calcium hydride?

8-30 Assign an oxidation number to each atom in each of the following: C_2Cl_6, $MgSO_4$, $FeSO_4$, $C_3H_6O_2$, $AlCl_3$, $FeCl_3$, $AsCl_5$, ICl, ICl_4^-, I_3^-.

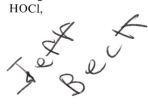

9

THE COVALENT BOND

TO THE STUDENT

The distance between each pair of atoms and the angle between each pair of adjacent covalent bonds are fixed characteristics of a molecule. Stated more simply, each molecule has a characteristic shape. In this chapter we will examine the factors that determine what shape a small molecule adopts. We will also introduce two theories of covalent bonding and see how each of these describes the roles played by electrons in determining the shapes and stabilities of molecules.

9-1 Electron-pair repulsion and molecular geometry

The pairs of dots in the Lewis structure of a molecule (Sec. 8-2) represent pairs of electrons. These electron pairs repel each other electrostatically, and so we can predict their geometrical orientation around the core of the atom by finding the arrangement which is of lowest energy, that is, in which interelectronic repulsions are minimized. This is the basis of the VSEPR method of predicting the geometric shapes of small molecules.

The VSEPR method

In a molecule composed of a central atom bonded covalently to several peripheral atoms the bonding and lone pairs are oriented so that electron-electron repulsions are minimized while electron-nucleus attractions are maximized. The method of determining this orientation is called the *valence-shell electron-pair repulsion* or *VSEPR* method. (The acronym is hard to pronounce accurately; compromise with *vesper*.) The assumptions behind the method are:

1 Electron pairs in the valence shell of an atom tend to orient themselves so that their total energy is minimized. This means that they approach the nucleus as closely as possible, while at the same time staying as far away from each other as possible, thus minimizing interelectronic repulsions.

2 Because lone pairs are spread out more broadly than are bonding pairs, repulsions are greatest between two lone pairs, intermediate between a lone pair and

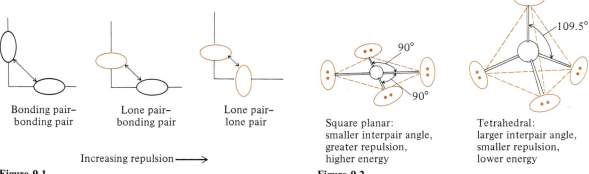

Bonding pair–   Lone pair–   Lone pair–
bonding pair   bonding pair   lone pair

Increasing repulsion ⟶

Figure 9-1
Order of repulsion between electron pairs.

Square planar:
smaller interpair angle,
greater repulsion,
higher energy

Tetrahedral:
larger interpair angle,
smaller repulsion,
lower energy

Figure 9-2
Steric number = 4: two possible orientations.

a bonding pair, and weakest between two bonding pairs. This order of repulsion is shown schematically in Fig. 9-1.

3 Repulsive forces decrease sharply with increasing interpair angle. They are strong at 90°, much weaker at 120°, and very weak at 180°.

Steric number and electron-pair orientation The first step in the VSEPR method for determining the shape of a molecule is to draw its Lewis structure in order to find out how many electron pairs are located around the central atom. Consider arsenic trichloride, $AsCl_3$, and sulfur tetrafluoride, SF_4, as examples. Their Lewis structures are, respectively

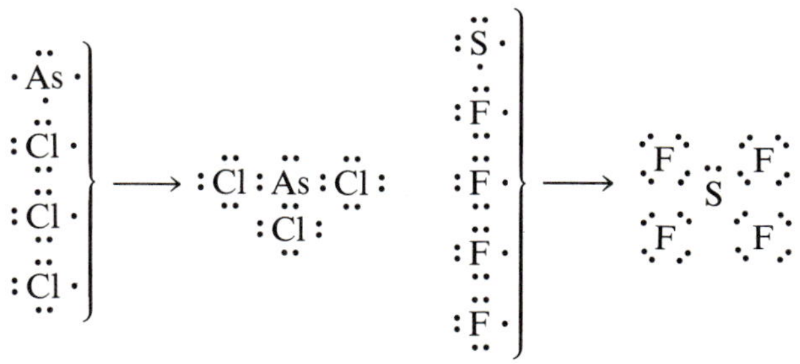

The *steric number* is defined as the total number of electron pairs (lone and bonding) around the central atom. As can be seen from the above Lewis structures, arsenic has a steric number of 4 in $AsCl_3$, while in SF_4 the steric number of sulfur is 5. (The valence shell of sulfur has been expanded to 10 electrons.)

The steric number determines the orientation in space of the valence-shell pairs. Table 9-1 shows the orientations expected for steric numbers of 2, 3, 4, 5, and 6. Each of the orientations is the one which minimizes electron-pair repulsion for that steric number. For example, for a steric number of 4, we might consider a square planar orientation, as shown in Fig. 9-2. But in this orientation the interpair angle is 90°, which produces a greater interpair repulsion than the tetrahedral orientation does. (That is, the pairs are closer together.) Thus, for a steric number of 4, tetrahedral geometry is preferred over square-planar geometry.

In $AsCl_3$ the steric number is 4, and so the orientation of valence-shell electron

Table 9-1
Spatial orientations of electron pairs around a central atom (schematic)

Steric number	Orientation		Angles
2	Linear		180°
3	Trigonal planar		120°
4	Tetrahedral		109.5°
5	Trigonal bipyramidal		90°, 120°
6	Octahedral		90°

pairs around the As atom is predicted to be tetrahedral. In SF_4, with a steric number of 5, the orientation is trigonal bipyramidal, as Table 9-1 shows.

Lone pairs and molecular geometry

The second step is to determine the number and location of *lone* pairs. This is really no problem in the case of $AsCl_3$. The Lewis structure shows that only one pair of electrons is a lone pair. Since all corners of a regular tetrahedron are equivalent, all we need to say is that the lone pair is at a corner. (See Fig. 9-3.) The resulting molecular shape is defined by the location of the four atoms and is called a *trigonal pyramid*.

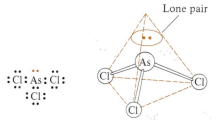

Figure 9-3
The $AsCl_3$ molecule: a trigonal pyramid.

Figure 9-4
Possible structures for SF₄: (a) location of electron pairs (A, axial positions; E, equatorial positions); (b) trigonal pyramid (axial placement of lone pair); (c) see-saw (equatorial placement of lone pair).

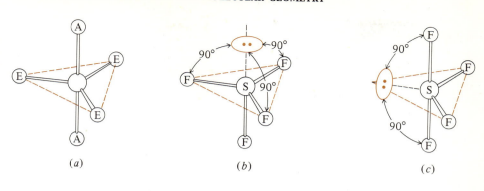

(a)   (b)   (c)

Consider next the second example, sulfur tetrafluoride. Here the Lewis structure shows a steric number of 5 with one lone pair. Thus the lone pair must occupy one of the corners of a *trigonal bipyramid*. (See Table 9-1.) But look at Fig. 9-4a; not all corners of a trigonal bipyramid are equivalent. If the lone pair occupies an *axial* position, as in Fig. 9-4b, there are three lone pair–bonding pair repulsions at 90°. If, on the other hand, the lone pair occupies an *equatorial* position, as in Fig. 9-4c, there are only two 90° lone pair–bonding pair repulsions. With fewer such repulsions, Fig. 9-4c is the predicted (and observed) structure. (We need not consider lone pair–lone pair repulsions as there are none. Also, we can ignore bonding pair–bonding pair repulsions at 90° and *all* repulsions at greater than 90°.)

The structure of SF₄ is sometimes called a *seesaw,* because "resting" on its two equatorial F atoms it resembles a child's teeter-totter. Evidence that the lone pair is actually located in an equatorial position comes from the fact that the axial F—S—F bond angle is a little less than 180°, as shown in Fig. 9-5. This distortion is produced by the repulsion between the two axial pairs and the lone pairs. (Further support comes from the fact that the axial bond lengths are slightly longer than the equatorial bond lengths.)

In summary, the steps in the VSEPR method for determining the geometrical shape of a small molecule are:

1 Draw the Lewis structure for the molecule. Show the octet rule to be obeyed, if it is at all reasonable, but be ready to expand the valence shell of the central atom if necessary to accommodate all the electron pairs.
2 Determine the steric number, the total number of electron pairs in the valence shell of the central atom. Orient these pairs around the central atom in the way which will minimize total interpair repulsion (Table 9-1).
3 Determine the number of lone pairs. Locate these so that the repulsions (a) between lone pairs and (b) between lone pairs and bonding pairs are minimized. Only 90° repulsions need be considered. If there are two or more lone pairs, concentrate on minimizing lone pair–lone pair repulsions. If there is only one lone pair, minimize lone pair–bonding pair repulsions. If there are no lone pairs, the molecular geometry is the same as that of the bonding pairs.

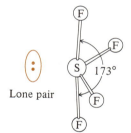

Lone pair

Figure 9-5
The SF₄ molecule.

● **EXAMPLE 9-1**

Problem   Predict the shape of the chlorine trifluoride molecule, ClF_3.

Structure   I   II   III

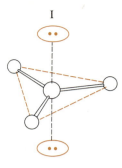   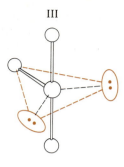

Location of
lone pairs   Both axial   One axial,   Both equatorial
                                                 one equatorial

90° repulsions:
  Lone pair–lone pair   0   1   0
  Lone pair–bonding pair   6   3   4

Figure 9-6
Possible structures of ClF_3.

Solution   We start by drawing the Lewis structure.

$$:\ddot{C}l\cdot$$
$$:\ddot{F}\cdot$$ $$\longrightarrow$$ $$:\ddot{F}:\quad \overset{:\ddot{C}l:}{\underset{:\ddot{F}:}{}} \quad F:$$
$$:\ddot{F}\cdot$$
$$:\ddot{F}\cdot$$

The steric number is 5, so interpair repulsion is least when the five pairs occupy the corners of a trigonal bipyramid. (Table 9-1).

Because the molecule has two lone pairs, these have three possible orientations, as is shown in Fig. 9-6. Structure II in the illustration can be ruled out, because I and III each have fewer lone pair–lone pair repulsions at 90°. Structure III is favored over I, because it has fewer lone pair–bonding pair repulsions at 90°. Therefore, we predict III, a "T-shape," for ClF_3. Experiments show that the ClF_3 molecule does indeed have this shape, but that it is slightly distorted, as shown in Fig. 9-7. The distortion is accounted for by the repulsion between the two lone pairs and the axial bonding pairs. ●

Table 9-2 summarizes the molecular geometries predicted for steric numbers 2 through 6.

Lone pairs

Figure 9-7
The ClF_3 molecule.

ADDED COMMENT

A common mistake is to confuse the geometry of the complete set of electron pairs (bonding plus lone) with the shape of the molecule. In $AsCl_3$, for example, the orientation of electron pairs around the As atom is tetrahedral (see Fig. 9-3). But to call the molecule tetrahedral would be wrong. It is a trigonal pyramid with the As at the apex and a Cl at each corner of the triangular base. The name given to

**Table 9-2
Molecular geometries
according to the
VSEPR method**

Steric number	Number of lone pairs	Molecular geometry	
2	0	Linear	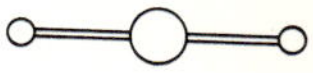
3	0	Trigonal planar	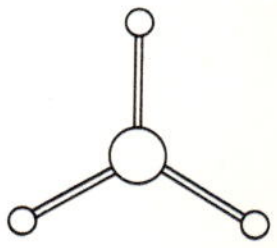
4	0	Tetrahedral	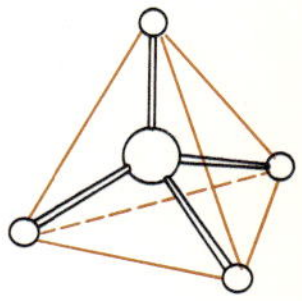
4	1	Trigonal pyramidal	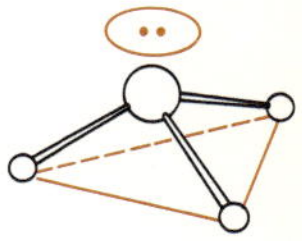
4	2	Bent (angular)	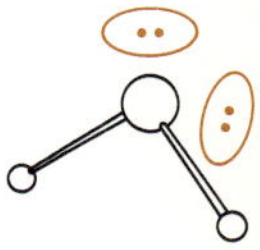
5	0	Trigonal bipyramidal	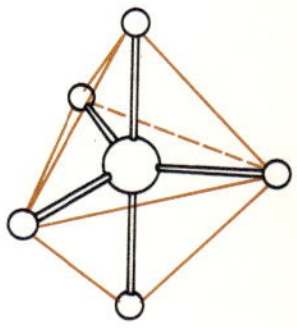
5	1	Seesaw	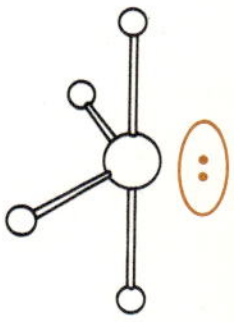

(Continued on next page)

Table 9-2 (Continued)	Steric number	Number of lone pairs	Molecular geometry	
	5	2	T shaped	
	5	3	Linear	
	6	0	Octahedral	
	6	1	Tetragonal pyramidal	
	6	2	Square planar	

describe the shape of a molecule is the name of the geometrical figure obtained by connecting the nuclei of all the "outside" atoms by straight lines. The locations of lone pairs are ignored in this process.

Polar and nonpolar molecules

The concept of polarity is useful for describing entire molecules, as well as bonds (Sec. 8-3). A simple example of a *nonpolar molecule* is the hydrogen molecule, H_2. In this molecule the *center of positive charge coincides with the center of negative charge*. (The term *center of positive charge* refers to the average position of all the positive charges in the molecule, that is, the effective center of the positive charge distribution.) In the H_2 molecule the center of positive charge is

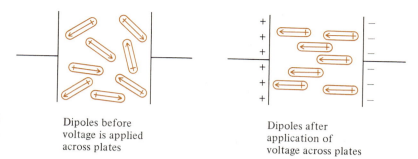

Dipoles before
voltage is applied
across plates

Dipoles after
application of
voltage across plates

a point exactly halfway between the two nuclei, and that point is also the location of the center of negative charge.

In the hydrogen chloride (HCl) molecule, on the other hand, the centers of positive and negative charge do not coincide. They are both located on the bond axis, but the center of negative charge is closer to the Cl atom while the center of positive charge is closer to the H atom. Such a molecule, one in which the *centers of positive and negative charges are in different locations,* is called a *polar molecule,* or *dipole.*

Dipoles can be experimentally distinguished from nonpolar molecules by their behavior in an electric field. When polar molecules are placed between a pair of electrically charged plates, they tend to rotate to line up with the field. The negative end of each dipole tends to point toward the positively charged plate, and the positive end, toward the negatively charged plate. This is schematically shown in Fig. 9-8, where a polar molecule is represented by an arrow with a crossed tail (+——➤). (The arrow points toward the negative end of the molecule.)

The polarity of a molecule is expressed quantitatively by its *dipole moment,* which is the product of the magnitude of the charge at one end of the molecule times the distance between the opposite charges. Dipole moments are determined by measuring the amount of electrical charge which will flow onto a pair of plates (as in Fig. 9-8) when a given voltage is placed across them. The more polar the molecules, the more charge can flow onto the plates as the molecules between them line up with the electric field. Dipole moments are usually expressed in debyes (D).[1] Measured values of some dipole moments are given in Table 9-3.

Once we know the structure of a molecule, it is not difficult to predict whether or not it is polar. In the diatomic case the polarity of the molecule depends only upon the polarity of the bond. If the bond is polar, then the molecule as a whole has positive and negative ends, and so will exhibit a dipole moment. Neither the H_2 nor the Cl_2 molecule has a dipole moment (see Table 9-3), because their bonds are nonpolar. HCl and HBr, on the other hand, are polar molecules. Note that HCl is more polar than HBr, because Cl is more electronegative than Br, and so the separation of the negative and positive charge centers is greater in HCl than in HBr.

With triatomic molecules the shape of the molecule has a bearing on whether or

[1] The SI derived unit for dipole moment is the coulomb-meter (C m): $1 D = 1.602 \times 10^{-29}$ C m. The dipole moment of a negative electronic charge separated from an equivalent positive charge by a distance of 0.100 nm is 4.80 D. (Peter J. W. Debye was a Dutch chemist who did much of the early research in the field of polar molecules.)

Table 9-3
Dipole moments of
some molecules

Molecule	Structure	Dipole moment, D
Hydrogen	H—H	0
Chlorine	Cl—Cl	0
Hydrogen chloride	H—Cl	1.08
Hydrogen bromide	H—Br	0.82
Carbon monoxide	C≡O	0.11
Carbon dioxide	O=C=O	0
Water	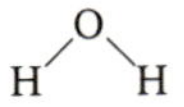 (bent, O with H and H)	1.85
Hydrogen sulfide	 (bent, S with H and H)	0.97
Hydrogen cyanide	H—C≡N	2.98
Ammonia	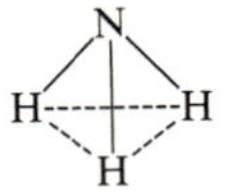 (N with three H)	1.47
Boron trifluoride	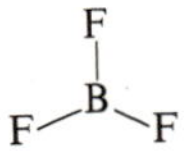 (B with three F)	0

not it is polar. In carbon dioxide, for example, both C—O bonds are polar (O is more electronegative than C), but the molecule is *nonpolar*. This is because it is *linear*. In Fig. 9-9 we see that the shift of electronic charge toward one oxygen atom is exactly compensated by the shift of electronic charge toward the other. The centers of positive and negative charge are in the same location. As far as the whole molecule is concerned, the bond dipoles cancel each other out.

The water molecule is another triatomic molecule, but it is polar, because it has a *bent* structure. Figure 9-10 shows that the individual bond dipoles do not cancel each other out as they do in CO_2. Instead, each contributes to an accumulation of negative charge at the oxygen atom. Thus the O atom is at the negative end of the dipole, and the H atoms are at the positive end, as shown in the illustration. The centers of positive and negative charge are in different locations, but both are on the line bisecting the H—O—H angle. The negative charge center is closer to the oxygen atom, and the positive charge center, farther away.

Another example of how molecular geometry affects molecule polarity is found in a pair of tetratomic molecules: ammonia, NH_3, and boron trifluoride, BF_3. In Fig. 9-11a are shown the Lewis structures of these molecules. In each case there are three covalent bonds from the central atom (N or B) to the other atoms (H or F, respectively). The NH_3 molecule has the shape of a *trigonal pyramid*. BF_3, on the other hand, is a *trigonal planar* molecule. Every bond in each molecule is polar; the shift of electronic charge is toward the N in NH_3 and toward the F

Lewis structure

O ⟵—+ C +—⟶ O

Two polar bonds

Resultant dipole moment = 0

Figure 9-9
The nonpolar CO_2 molecule.

Lewis structure

Two polar bonds

Resultant molecular dipole

Figure 9-10
The polar H_2O molecule.

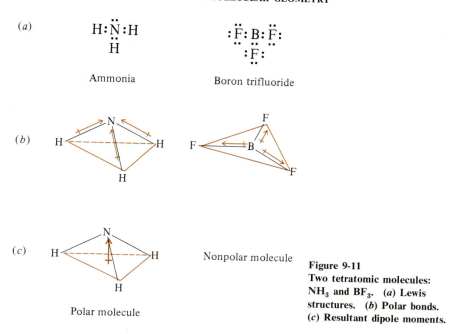

(a)

$$H \!:\! \overset{\cdot\cdot}{N} \!:\! H$$
$$H$$

Ammonia

$$:\overset{\cdot\cdot}{F} \!:\! B \!:\! \overset{\cdot\cdot}{F}:$$
$$:\overset{\cdot\cdot}{F}:$$

Boron trifluoride

(b)

(c)

Polar molecule

Nonpolar molecule

Figure 9-11
**Two tetratomic molecules:
NH_3 and BF_3. (a) Lewis
structures. (b) Polar bonds.
(c) Resultant dipole moments.**

atoms in BF_3, as is shown in Fig. 9-11b. The N end of the NH_3 molecule thus has a partial negative charge, leaving the H end with a partial positive charge. The centers of positive and negative charge do not coincide in NH_3, and the resultant dipole moment is indicated in Fig. 9-11c. In contrast, BF_3 is nonpolar. The planar shape and the symmetry within that plane result in a cancellation of bond dipoles so that the molecule as a whole is nonpolar. (The centers of positive and negative charge coincide, and so this molecule does not exhibit a dipole moment.)

In order to decide whether or not a given molecule is polar, pass an imaginary plane through the center (actually the center of mass, or center of gravity) of the molecule. Do this in as many different ways as you can. If you can find at least one way of passing a plane through the center of the molecule so that the plane unambiguously separates positive from negative charge, then the molecule is polar. Figure 9-12 shows three ways of passing a plane through the center of mass of a water molecule. The first two ways (Fig. 9-12a and b) do not demonstrate a charge separation, but the third (Fig. 9-12c) separates the oxygen (negative) from the hydrogens (positive). This shows that the water molecule must be polar.

Figure 9-12
The water molecule:
geometric test of polar-
ity. (a) Vertical plane,
perpendicular to page:
no polarity shown.
(b) Plane in plane of
page: no polarity
shown. (c) Horizontal
plane, perpendicular to
page: molecule is
shown to be polar.

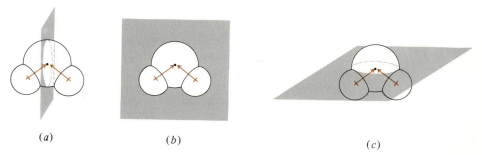

(a)   (b)   (c)

9-2
Valence-bond
theory and orbital
overlap
Two approaches have been used for the purpose of describing the covalent bond and the electronic structures of molecules. At its most sophisticated level each approach employs quantum mechanics, but the basic assumptions of the two methods are quite different. The first approach, called *valence-bond* (VB) *theory,* considers that when a pair of atoms forms a bond, the atomic orbitals of each atom remain essentially unchanged and that a pair of electrons occupies an orbital *in each of the atoms* simultaneously. The second method, *molecular-orbital* (MO) *theory,* assumes that the atomic orbitals of the original unbonded atoms become replaced by a new set of molecular energy levels, called *molecular orbitals,* and that the occupancy of these orbitals determines properties of the resulting molecule.

Although the VB and MO methods appear to be quite different, it turns out that rigorous calculations using each method yield similar results. With the advent of sophisticated electronic computers many such calculations have been successfully completed, and the results support the usefulness of both the VB and MO models for covalent bonding. Throughout this book we will find it valuable to use both descriptions, emphasizing one or the other as the situation warrants. We will first examine the valence-bond model.

The hydrogen molecule

Let us now reconsider the H_2 molecule and once more picture its formation from two isolated, ground-state H atoms. Each H atom has at the start a single electron in a $1s$ atomic orbital. For identification purposes we will call the two H atoms A and B. After the covalent bond has been formed, we find that *each* electron now exists in the $1s$ orbitals of *both* atoms. This can be shown schematically as

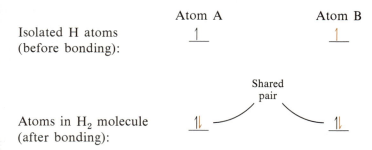

It should be emphasized that we are *not* showing *four* electrons here, but only *two* occupying both orbitals at the same time.

According to valence-bond theory simultaneous occupancy of orbitals of two atoms by a pair of electrons is possible if the orbitals *overlap* each other to an appreciable extent. (Remember, the word *orbital* implies not only an energy level, but also a region of space in which electrons are most probably found.) Figure 9-13 shows the boundary surfaces of the $1s$ orbitals of two bonded hydrogen atoms. The orbital overlap produces a region of enhanced electron probability density located directly between the nuclei. Note that the bond axis (the line connecting the two nuclei) passes through the middle of this region. Furthermore, the overlap region is symmetrical around the bond axis, because each atomic orbital is spherical.

At this point we will borrow a term from MO theory (Sec. 9-4). The bond in H_2 is a *sigma* (σ) *bond,* one in which *the charge-cloud of the shared pair is centered on*

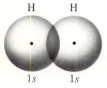

Figure 9-13
Overlap of 1s orbitals in H_2 (σ bond).

and is symmetrical around the bond axis. Such a charge cloud is said to have *axial,* or *cylindrical, symmetry.*

The hydrogen fluoride molecule A sigma bond can also be formed as a result of the overlap of an *s* and a *p* orbital. Consider hydrogen fluoride, HF. Before bonding, a fluorine atom has the following ground-state electronic configuration:

$$\text{F } (Z = 9): \quad \underset{1s}{\underline{\uparrow\downarrow}} \qquad \underset{2s}{\underline{\uparrow\downarrow}} \qquad \underbrace{\underline{\uparrow\downarrow} \;\; \underline{\uparrow\downarrow} \;\; \underline{\uparrow}}_{2p}$$

Two of the three $2p$ orbitals are filled. Assume that the unpaired electron is in the $2p_x$ orbital. This orbital has two lobes (see Fig. 6-23), and if the $1s$ orbital of a hydrogen atom overlaps one of these lobes end-on (Fig. 9-14), then the shared electron pair spends most of its time in a region which is centered on and symmetrical around the bond axis. The bond in HF is therefore a sigma bond.

A σ bond can also be formed as the result of the overlap of two *p* orbitals, but the overlap must be *end-to-end* as in the fluorine molecule, F_2. Here the $2p_x$ orbital of one F atom overlaps the $2p_x$ orbital of the second as is shown in Fig. 9-15.

Pi bonding When *p* orbitals overlap *sideways,* the results are different. If we assume as before that the bond axis is the *x* axis and choose the $2p_z$ orbitals for overlap (Fig. 9-16), the resulting side-to-side overlap produces *enhanced electron probability density in two regions which are on opposite sides of the bond axis.* This is characteristic of a *pi (π) bond,* another term borrowed from MO theory.

Multiple bonds In a double or triple bond one bond is always a σ bond, and the remaining bonds are π bonds. The nitrogen molecule N_2 provides an example of a *triple bond.* The ground-state electronic configuration of a nitrogen atom is

$$\text{N } (Z = 7): \quad \underset{1s}{\underline{\uparrow\downarrow}} \qquad \underset{2s}{\underline{\uparrow\downarrow}} \qquad \underbrace{\underline{\uparrow} \;\; \underline{\uparrow} \;\; \underline{\uparrow}}_{2p}$$

Here the three unpaired electrons are in the $2p_x$, $2p_y$, and $2p_z$ orbitals, respectively. Each of these orbitals overlaps the corresponding orbital of the other atom: the two p_x orbitals overlap end-to-end to form a σ bond, the two $2p_y$ orbitals, side-to-side to form a π bond, and the two $2p_z$ orbitals, side-to-side to form a

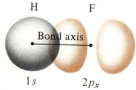

Figure 9-14
Overlap of 1s and $2p_x$ orbitals in HF (σ bond).

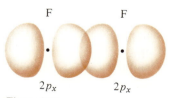

Figure 9-15
Overlap of two $2p_x$ orbitals in F_2 (σ bond).

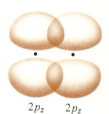

Figure 9-16
Formation of a π bond.

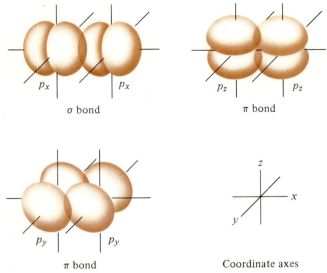

Figure 9-17
The triple bond in N_2.

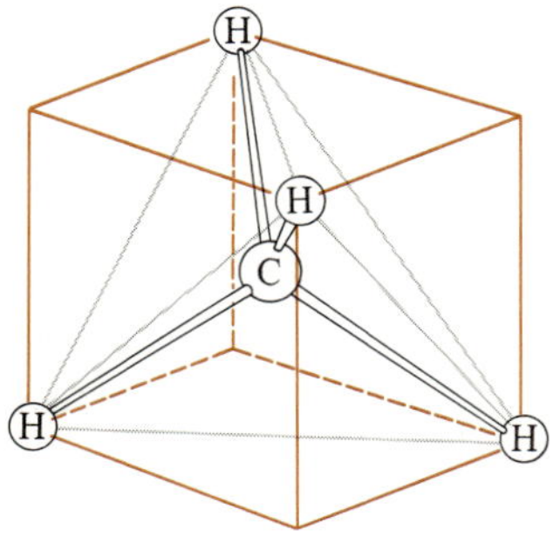

Coordinate axes

Figure 9-18
The tetrahedral methane mole-cule.

second π bond. These three bonds are shown separately as overlapping boundary surfaces in Fig. 9-17. The three overlaps together constitute the triple bond. Compare this with the simple Lewis structure

$$: N : : : N :$$

9-3
Hybrid orbitals

Carbon forms countless compounds in which its atoms bond covalently to four other atoms. The simplest of these is methane, CH_4. How can we describe the four covalent bonds in this molecule in terms of orbital overlap? The ground-state electronic configuration of C is

$$C \ (Z = 6):$$

$\underline{\uparrow\downarrow}$	$\underline{\uparrow\downarrow}$	$\underline{\uparrow} \ \underline{\uparrow} \ \underline{}$
$1s$	$2s$	$2p$

Carbon thus appears to be able to form only two covalent bonds by contributing each of its two unpaired electrons to a shared pair. But the short-lived methylene (CH_2) molecule is much less stable than CH_4.

In the methane molecule (Fig. 9-18) each H atom is located at the corner of a regular tetrahedron, shown inscribed in a cube in the drawing, so that the relationship between these two regular solids can be seen. In CH_4 all bond lengths are the same and the angle between each C—H bond and any of the other three is the *tetrahedral angle,* 109.5°. The observed tetrahedral structure of methane is what we expect after applying VSEPR theory to this molecule.

According to the Lewis structure for methane

$$
\begin{array}{c}
\text{H} \\
\text{\textbf{\textbf{..}}} \\
\text{H} : \overset{..}{\underset{..}{\text{C}}} : \text{H} \\
\text{\textbf{..}} \\
\text{H}
\end{array}
$$

the carbon evidently uses all four of its valence electrons so that four C—H bonds can be formed. It is not too difficult to see how carbon can form four bonds. Suppose that one of the $2s$ electrons is *promoted* to the vacant, but higher energy, $2p$ orbital.

$$
\text{C } (Z = 6): \quad \underline{\underset{1s}{\uparrow\downarrow}} \quad \underline{\underset{2s}{\uparrow}} \quad \underbrace{\underline{\uparrow}\;\underline{\uparrow}\;\underline{\uparrow}}_{2p}
$$

Now the C atom appears to be ready to form four σ bonds by overlap of its $2s$ and $2p$ orbitals with the $1s$ orbitals of four H atoms. The difficulty here is that if the bonding occurred this way, the CH_4 molecule would *not* be tetrahedral. Instead, its shape would be like that shown in Fig. 9-19. In Fig. 9-19a through c are shown the three C—H bonds which would result from overlap of the three $2p$ orbitals of C with the $1s$ orbitals of three H atoms. The fourth bond might go almost anywhere, because an s orbital is spherically symmetrical and good overlap is possible from any direction. If the last H is located as far away as possible from the other H atoms in order to minimize interelectronic repulsion, then it goes in the position indicated in d. The entire proposed CH_4 structure is shown in Fig. 9-19e. *If* these orbitals were used in bonding, methane would evidently have the

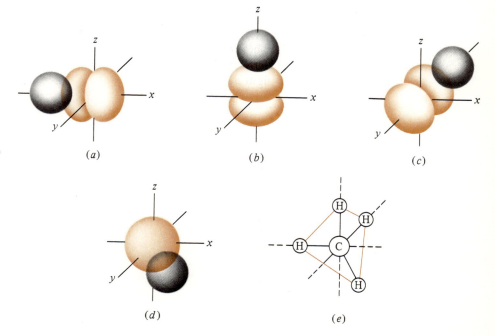

Figure 9-19
Bonding in methane: an *incorrect* model.
(*a*) First bond: $2p_x$ — $1s$ overlap. (*b*) Second bond: $2p_z$ — $1s$ overlap. (*c*) Third bond: $2p_y$ — $1s$ overlap. (*d*) Fourth bond: $2s$ — $1s$ overlap. (*e*) Complete CH_4 molecule: a trigonal pyramid (*incorrect*).

(*a*)   (*b*)   (*c*)

(*d*)   (*e*)

shape of a trigonal pyramid, *but it does not*. In CH_4 all bond angles are equal and all H atoms are equivalent.

   The experimentally determined structure of methane is tetrahedral. How can we account for it using the *s* and *p* orbitals of carbon? The answer is that the ground-state set of *s* and *p* orbitals of carbon is replaced by a new set which is suitable for forming four *equivalent* bonds, each at the tetrahedral angle from each of the others. This may sound like a kind of orbital sleight of hand, and so in order to aid understanding of this replacement, we will pause to consider first two simpler cases, the bonding of beryllium and boron.

sp Hybrid orbitals   Beryllium ($Z = 4$) forms a hydrogen compound which at high temperatures exists as discrete BeH_2 molecules. The ground-state electronic configuration of a Be atom is

$$\text{Be } (Z = 4): \quad \underset{1s}{\underline{\uparrow\downarrow}} \quad \underset{2s}{\underline{\uparrow\downarrow}} \quad \underset{2p}{\underline{\quad}\;\underline{\quad}\;\underline{\quad}}$$

The two bonds in BeH_2 are found to be oriented at 180° from each other; that is, the molecule is linear. How does this come about? When a Be atom forms its two bonds, its 2*s* and one of its 2*p* orbitals are replaced by a pair of new orbitals, and these new orbitals, *hybrid orbitals,* are used for bonding. (Recall from Sec. 6-4 that each orbital corresponds to a solution, a *wave function,* to the Schrödinger wave equation. Because the wave equation is a differential equation, any set of its solutions can be combined mathematically to form a new set of wave functions which are also solutions. These new wave functions are said to be hybrids of the original ones and correspond to a set of hybrid orbitals.)

   Perhaps some pictures will help. At the left of Fig. 9-20 are shown an *s* and a *p* orbital. In the illustration the plus and minus signs *are not charges*. Each is the algebraic sign of the wave function in the designated lobe of the orbital. Now we will combine or *mix* the orbitals, first (upper-right drawing) by *adding* the *p* to the

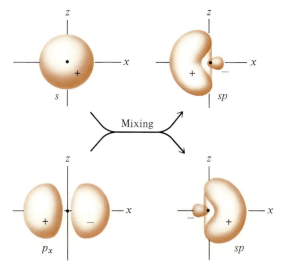

Figure 9-20
Formation of *sp* hybrid orbitals.

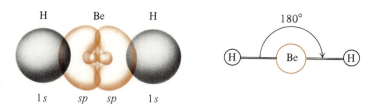

Figure 9-21
The BeH₂ molecule.

s. The result is a hybrid orbital, in which the density of electronic charge has increased where the original wave functions had the same sign and has decreased where they had opposite signs. This hybrid orbital, called an *sp* orbital, is highly directional; overlap is favored in the direction of its large major lobe. *Subtraction of the original s and p orbitals* (lower-right drawing) yields the second hybrid orbital. It is equivalent to the first, but points 180° away. Thus by combining or mixing *two nonequivalent* orbitals (one is an *s* and the other, a *p*) we have obtained *two equivalent sp* hybrid orbitals.

We can keep track of the mixing of the 2*s* and 2*p* orbitals in beryllium as follows:

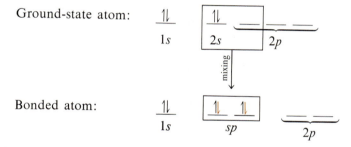

The beryllium *sp* orbitals overlap with hydrogen 1*s* orbitals (the hydrogens' electrons are shown in the above orbital diagram as colored arrows) to produce two σ bonds and a resulting linear molecule, as is shown in Fig. 9-21.

sp² Hybrid orbitals   Other combinations of orbitals may be involved in hybridization. Boron ($Z = 5$) uses its 2*s* orbital and *two* of its 2*p* orbitals to form a set of *three* equivalent *sp²* hybrid orbitals, as shown below:

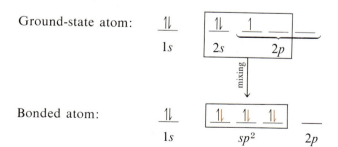

Each of the three *sp²* hybrids has much the same shape as an *sp* hybrid orbital, but the three are oriented at 120° from each other, as is shown in Fig. 9-22.

If each of the *sp²* hybrid orbitals overlaps with a 1*s* orbital of an H atom, the result is a boron hydride, or borane (BH_3), molecule. This molecule, shown in Fig. 9-23, is predicted to have a planar structure in which each H occupies the corner of

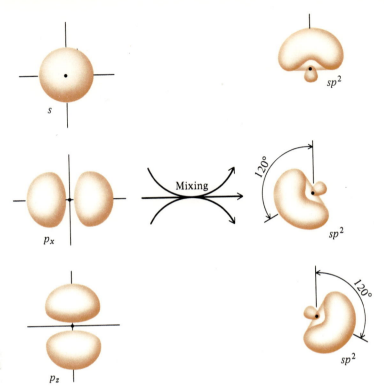

Figure 9-22
Formation of sp^2 hybrid
orbitals.

an equilateral triangle. BH_3 is, however, an almost hypothetical molecule, having been observed only as a short-lived intermediate in certain rapid reactions. We have used it as an example because of the analogy with BeH_2 (just discussed) and with CH_4 (to be discussed next). Boron does form similar trigonal planar molecules of the type BX_3, where X is a halogen atom, F, Cl, Br, or I.

sp^3 Hybrid orbitals   Finally, we return to carbon and its hydrogen compound, methane (CH_4). Here the carbon atom uses its $2s$ and all three of its $2p$ orbitals to form four equivalent sp^3 hybrid orbitals:

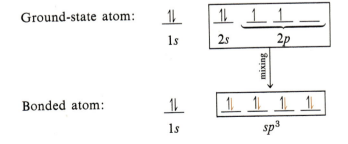

These orbitals look much like the sp and sp^2 hybrids but point toward the corners of a regular tetrahedron (Fig. 9-24). This accounts for the tetrahedral shape of the CH_4 molecule (Fig. 9-18).

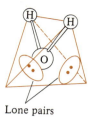

Lone pairs

Figure 9-27
The H₂O molecule.

The water molecule   The bonding in H_2O is similar to that in NH_3. The Lewis structure of water is

$$H:\overset{..}{\underset{..}{O}}:$$
$$H$$

The four orbitals in the valence shell of O $(Z = 8)$ hybridize

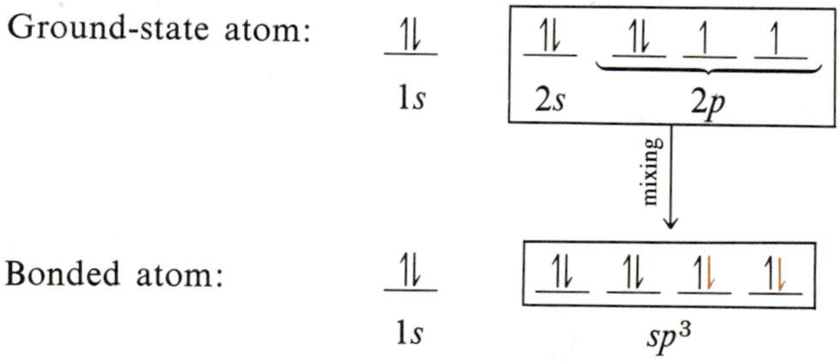

Ground-state atom:

$$\underset{1s}{\underline{\uparrow\downarrow}} \quad \boxed{\underset{2s}{\underline{\uparrow\downarrow}} \quad \underset{}{\underline{\uparrow\downarrow}} \quad \underline{\uparrow} \quad \underline{\uparrow}}$$
$$2p$$

mixing ↓

Bonded atom:

$$\underset{1s}{\underline{\uparrow\downarrow}} \quad \boxed{\underline{\uparrow\downarrow} \quad \underline{\uparrow\downarrow} \quad \underline{\uparrow\downarrow} \quad \underline{\uparrow\downarrow}}$$
$$sp^3$$

and the four sp^3 hybrid orbitals are occupied by two lone pairs and two bonding pairs of electrons. The resulting structure is described as *bent,* or *angular* (Fig. 9-27).

In H_2O there is an even greater shrinking of the bond angle from the tetrahedral angle 109.5° than there is in NH_3, this being a result of the *two* lone pairs in H_2O. The measured bond angle in H_2O is only 104.5°.

Other hybrid   Other possibilities exist for the mixing of pure atomic orbitals to form sets of
orbital sets   hybrid orbitals. The most important of these is the hybridization of one *s*, three *p*, and two *d* orbitals. If the *d* orbitals are from the $n - 1$ shell of the atom, the hybrids are called d^2sp^3. If they are from the valence shell, that is, if they have the same principal quantum number as the *s* and *p* orbitals, then they are called sp^3d^2 orbitals. In either case these orbitals have major lobes pointing out toward the corners of a regular *octahedron,* an eight-sided solid having faces that are identical equilateral triangles. It is evident that with such hybridization the valence shell of the central atom no longer contains just an octet but has been expanded to 12 electrons. Octahedral hybrid orbitals are used to account for the structure of sulfur hexafluoride, SF_6.

Table 9-4 shows a summary of the more important sets of hybrid orbitals, their geometries, and some examples.

Table 9-4
Some important sets of hybrid orbitals

Hybrid orbital set	Geometry*		Examples
sp	Linear		BeF_2 $CdBr_2$ $HgCl_2$
sp^2	Trigonal planar		BF_3 $B(CH_3)_3$ GaI_3
sp^3	Tetrahedral		$TiCl_4$ CCl_4 SiF_4
dsp^3 or sp^3d	Trigonal bipyramidal		PCl_5 $MoCl_5$ $TaCl_5$
d^2sp^3 or sp^3d^2	Octahedral		SF_6 SbF_6^- $CrCl_6^{3-}$

*In these drawings only the major lobes are shown, and these are considerably slimmed down in order to emphasize their directional characteristics.

9-4
The molecular-orbital model

Orbitals in molecules

Molecular-orbital (MO) theory provides an alternative perspective from which to view bonding. According to this approach *all* the valence electrons in a molecule have an influence on the stability of the molecule. (Inner-shell electrons may also make a contribution to the bonding, but for many simple molecules the effect is small.) Furthermore, MO theory considers that valence-shell atomic orbitals (AOs) cease to exist when a molecule is formed. They are replaced by a new set of energy levels with corresponding new charge-cloud (probability-density) distributions. These new energy levels are a property of the molecule as a whole, and are called, consequently, *molecular* orbitals.

Calculating the properties of molecular orbitals is commonly done by assuming that AOs combine to form MOs. The wave functions (Sec. 6-4) of the AOs are combined mathematically to produce wave functions for the resulting MOs.[2] The process is reminiscent of the mixing of pure atomic orbitals to form hybrids, except

[2] The method is known as the *linear combination of atomic orbitals,* or LCAO, method.

that in MO formation atomic orbitals of more than one atom are mixed. Nevertheless, just as in the case of hybridization, the number of new orbitals formed equals the number of original atomic orbitals combined.

As with atomic orbitals, we are interested in two aspects of molecular orbitals: (1) the *shapes* of their probability-density distributions in space and (2) their relative *energies*. First we consider their shapes.

Spatial distributions of MOs

We begin by looking at the MOs which are formed when two atoms bond in a diatomic molecule. Using the simplest approach we consider that *one* AO from one atom combines with *one* AO from a second atom to form *two* MOs. In order for this process to be effective two conditions must be met: (1) the AOs must be of comparable energy, and (2) they must overlap significantly. The quantum-mechanical calculation for combining the original AOs consists of (1) an addition and (2) a subtraction of the AO wave functions. (If the two atoms are different, a factor is included which takes account of the fact that the two AOs will not contribute equally to the formation of the MOs.) The result, then, is two new MO wave functions, one from the addition and one from the subtraction. As always, squaring a wave function for an electron gives us information about the probability of finding that electron. When this is done for the new MOs, the result is probability-density information for electrons in a molecule, and from this information the corresponding boundary surfaces (and also energy levels) can be found.

ADDED COMMENT

Keep in mind that the representations of molecular orbitals are analagous to the AO representations of Chap. 6 and, as before, can be interpreted in two equivalent ways: they show (1) the region(s) in which an electron spends most of its time, that is, the region(s) of high probability of finding the electron, or (2) the region(s) in which the density of electronic charge is high.

In Fig. 9-28 are shown the boundary surfaces of the two molecular orbitals which are formed by combining two 1s atomic orbitals. Shown at the left are the

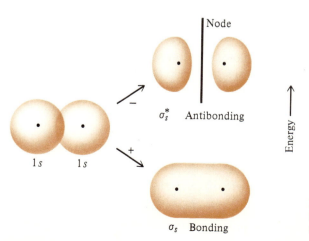

Node

σ_s^* Antibonding

Energy

1s   1s

σ_s Bonding

Figure 9-28
Combination of 1s AOs to form
σ MOs.

two overlapping $1s$ AOs, and at the right, the resulting MOs. The MO formed by subtracting the AO wave functions is labeled σ^* (read "sigma star"), while the one formed by adding them is labeled σ. The contrast between these two MOs is striking. There is an obvious increase in electronic charge density between the nuclei in the σ_s orbital but a decrease in the same region in the σ_s^* orbital. For this reason the σ_s orbital is called a *bonding* orbital and the σ_s^* orbital, an *antibonding* orbital. The former tends to stabilize the bond, while the latter tends to destabilize it. Both of these orbitals are called σ orbitals, because they are both centered on and symmetrical around the bond axis. A cross section of either orbital made perpendicular to the bond axis is circular.

The combination of two p orbitals produces different results depending on which p orbitals are used. If the x axis is the bond axis, then two $2p_x$ orbitals can overlap properly if they approach each other end-to-end, as is shown in Fig. 9-29. The resulting MOs are, as before, a bonding orbital, with electronic charge buildup between the nuclei, and an antibonding MO, with decreased charge between the nuclei. These orbitals are also σ orbitals, because they are symmetrical around the bond axis. They are designated σ_x and σ_x^* to indicate that they have been derived from p_x atomic orbitals.

When $2p_y$ and $2p_z$ orbitals overlap to form MOs, they do so side-to-side, as is shown in Fig. 9-30. In each case the result is a four-lobed antibonding orbital and a two-lobed bonding orbital. These orbitals are not symmetrical around the bond axis. Rather, there are two regions on opposite sides of the bond axis in which the charge-cloud density is high. This is characteristic of a π orbital. Note that as before, the bonding orbital permits a high concentration of electronic charge in the region between the nuclei, while the antibonding orbital shows lowered charge density in this region. (Each antibonding orbital has a nodal plane between the two nuclei.)

Energies of MOs   Whenever two atomic orbitals combine to form two molecular orbitals, the energy of the bonding MO is always lower than that of either AO, while the energy of the

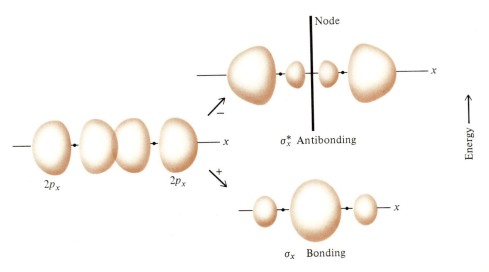

Figure 9-29
Combination of $2p_x$
AOs to form σ MOs.

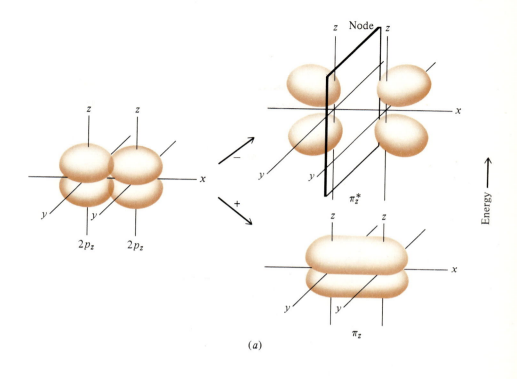

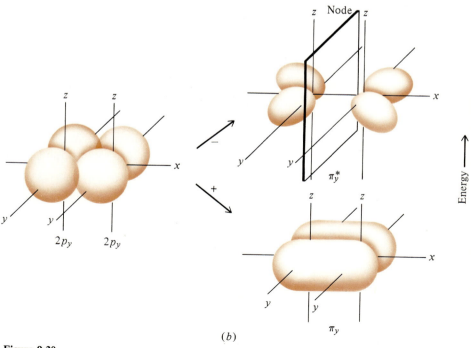

Figure 9-30
Combination of **2p** AOs to form π MOs: (*a*) π_z; (*b*) π_y. Nuclei are at the intersections of the *x*, *y*, and *z* axes.

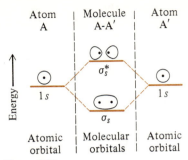

Figure 9-31
Relative energies of σ_s orbitals in homonuclear diatomic molecules.

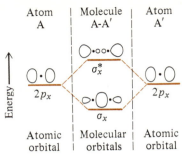

Figure 9-32
Relative energies of σ_x orbitals in homonuclear diatomic molecules.

antibonding MO is higher. Figure 9-31 shows the energy relationships for the $1s$ AO and σ_s MO orbitals for the case of a *homonuclear* diatomic molecule, one in which both atoms are the same. On the left and right are the $1s$ energy levels of two atoms of element A (labeled A and A'). In the center are the σ_s and σ_s^* energy levels of molecule A-A'. The diagonally running broken lines point out that the MOs have been formed from the indicated AOs. Figure 9-31 could be used to show the formation of MOs from a pair of any s orbitals ($2s$, $3s$, $4s$, etc.). In each case an antibonding orbital (of higher energy) and a bonding orbital (of lower energy) are formed.

Consider next the formation of molecular orbitals from a pair of $2p_x$ orbitals, orbitals with lobes pointing along the bonding axis (Fig. 9-32). Again we see the formation of a pair of MOs, one bonding (σ_x) and one antibonding (σ_x^*).

Next, look at the $2p_y$ and $2p_z$ AOs, which overlap side-to-side. The MOs formed from these are shown in Fig. 9-33. The p_y-p_y overlap is exactly like the p_z-p_z overlap (except for orientation), and so the resulting MOs fall into sets of two orbitals of the same energy: the π_y and π_z (bonding) orbitals and the π_y^* and π_z^* (antibonding) orbitals.

The filling of molecular orbitals   In Sec. 6-1 we described the *Aufbau* procedure in which electrons are added one by one to an AO filling diagram in order to build up the electronic configurations of

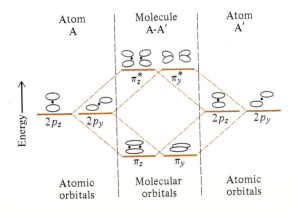

Figure 9-33
Relative energies of π_y and π_z orbitals in homonuclear diatomic molecules.

atoms. We will now use a similar technique, but will add electrons to a filling diagram composed of MO energy levels. We wish to build up the *ground-state* configurations of homonuclear diatomic molecules, and so we will add electrons starting at the bottom of the diagram and work upward toward higher-energy MOs. The MO filling diagram consists of Figs. 9-31 through 9-33. At the start, however, we need consider only the low-energy end of the diagram, that is, those MOs derived from the K shells of the bonded atoms (Fig. 9-31).

H_2. The simplest molecule is hydrogen. Figure 9-34 shows orbital populations for two unbonded ground-state H atoms at the left and right, together with that of the ground-state H_2 molecule, in the middle of the diagram. The two $1s$ electrons end up as a pair (antiparallel spins) in the bonding σ_s orbital of H_2 and constitute a single bond. The electronic configuration of H_2 can be written

$$H_2: \qquad (\sigma_s)^2$$

He_2. Next consider the molecule which might be formed from two atoms of helium, each of which furnishes two electrons to the molecule. This is two more than in H_2, so the MO population would look like that in Fig. 9-35. But the (antibonding) σ_s^* orbital is now filled, and *its destabilizing effect cancels out the stabilizing effect of the two bonding electrons* (in the σ_s orbital). The result is that there is no net attractive force between He atoms, and so He_2 does not exist.

In molecular-orbital theory the *bond order* is defined as

$$\text{Bond order} = \frac{\text{bonding electrons} - \text{antibonding electrons}}{2}$$

Thus the bond order in the H_2 molecule is

$$\frac{2-0}{2} = 1$$

while that in He_2 is

$$\frac{2-2}{2} = 0$$

● **EXAMPLE 9-2**

Problem   Predict the stability of the hydrogen-molecule ion $H_2{}^+$.

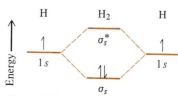

Figure 9-34
MO population diagram for H_2.

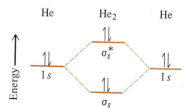

Figure 9-35
MO population diagram for the hypothetical He_2 molecule.

Solution   The H_2^+ ion should have an orbital occupancy like that of H_2 (Fig. 9-34), but with one less electron. Therefore, its electronic configuration is

$$H_2^+: \qquad (\sigma_s)^1$$

The bond order in H_2^+ is $\frac{1}{2}(1-0)$, or $\frac{1}{2}$. This means that the H_2^+ particle should exist, its atoms held together by a *half bond*. The H_2^+ ion does indeed exist; its bond energy is 255 kJ mol^{-1}, a moderately high bond energy. (By comparison the bond energy in H_2 is 433 kJ mol^{-1}.) ●

Li$_2$. Now consider the Li_2 molecule. This molecule has a total of six electrons, but four of these are in the (inner) K shells of the Li atoms, where they contribute little to the bonding. The valence electrons of the two Li atoms are used to populate a new σ_s MO as is shown in Fig. 9-36. The $1s$ atomic orbitals are essentially unperturbed and are not shown in the diagram. The configuration is much like that of H_2, and the bond order, which can be determined from the valence electrons only, is equal to $\frac{1}{2}(2-0)$, or 1. Representing each of the filled $1s$ orbitals by K (for a K shell) the electronic configuration of Li_2 can be shown as

$$Li_2: \qquad KK(\sigma_s)^2$$

With a bond order of 1 the Li_2 molecule is predicted to exist. Neither liquid nor solid Li consists of Li_2 molecules, but diatomic molecules are indeed found in gaseous lithium. The bond energy in Li_2 is 105 kJ mol^{-1}. This is lower than that in H_2 (433 kJ mol^{-1}) because of the shielding of the nucleus by the complete K shell in each atom.

Be$_2$. Moving on to the hypothetical Be_2 molecule, we find a situation like that in He_2. The atomic number of beryllium is 4, and the "seventh" and "eighth" electrons in the Be_2 molecule add to the σ_s^* orbital (See Fig. 9-36). The destabilizing effect of the filled σ_s^* orbital cancels out the stabilizing effect of the filled σ_s orbital, the bond order is zero, and therefore the Be_2 molecule should not be stable. Indeed, Be_2 has not been observed.

Next consider sequence B_2, C_2, N_2, O_2, F_2, and Ne_2, as we work across the rest of the second period constructing homonuclear diatomic molecules. But since we now need some more molecular orbitals, we go to the σ and π orbitals of Figs. 9-32 and 9-33. When we put these two diagrams together, however, we run into a small difficulty. The relative energy of the π_y and π_z orbitals is *less* than that of the σ_x orbital for B_2 through N_2, but *greater* for the remainder of the sequence. Thus the orbital energies for B_2, C_2, and N_2 are as shown in Fig. 9-37a and for O_2, F_2 and Ne_2, as shown in Fig. 9-37b. The main difference between Figs. 9-37a and b is the relative energy of the σ_x compared to the π_y and π_z orbitals.

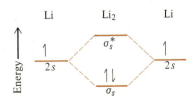

Figure 9-36
MO population diagram for Li$_2$.

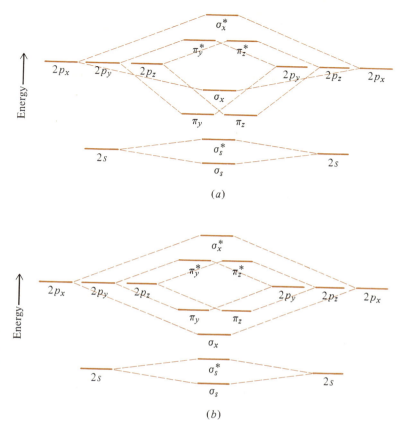

Figure 9-37
MO energies: B₂
through Ne₂: (a) B₂,
C₂, and N₂; (b) O₂, F₂,
and Ne₂.

The change in sequence of MO energies between N_2 and O_2 occurs because the σ_x and σ_x^* MOs actually have some s character, a fact which we had to ignore, when we decided to use the "one AO plus one AO yields two MOs" simplification. (The amount of s character in these orbitals decreases as the nuclear charge increases across the period. Because of this the σ_x energy drops below the π_y-π_z energy at O_2.)

ADDED COMMENT

The important thing to get from this last paragraph is that each MO may actually have the character of more than just two AOs.

B₂. Figure 9-38 shows the MO populations in B_2, C_2, and N_2. In the first molecule, B_2, there is a single electron in each of the π_y and π_z orbitals. Since these are bonding orbitals, and since in all lower energy levels the antibonding electrons have exactly compensated for the bonding electrons, the bond order is 1. (We could call the bond a single bond, but it is perhaps better described as two half bonds.) Note that the π_y and π_z orbitals are equal in energy, and so the two electrons do not pair up in one orbital. By spreading out, the electrons can occupy

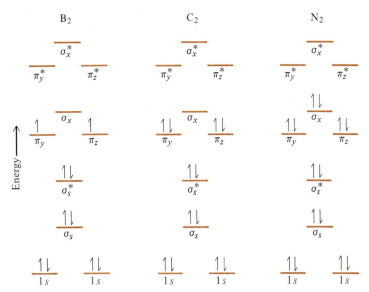

different regions in space, thus avoiding each other and reducing interelectronic repulsion. The electronic configuration of B_2 is written as

$$B_2: \qquad KK(\sigma_s)^2(\sigma_s^*)^2(\pi_y)^1(\pi_z)^1$$

Strong experimental support for this configuration comes from magnetic measurements. B_2 is paramagnetic, and measurements indicate that two unpaired electrons are present per molecule.

Like Li_2, B_2 is not a molecule you can find in a bottle on the shelf of the chemistry stockroom. Elemental boron is usually encountered as a solid in which the arrangement of B atoms is complex. At very high temperatures, however, B_2 molecules can be detected in the gaseous state.

C_2. Adding two more electrons (one from each atom) yields the configuration for C_2 shown in Fig. 9-38. These electrons are added to the π_y and π_z MOs, filling each. All electrons are now paired, and so C_2 is not paramagnetic. The bond order in C_2 can be seen to be 2, because there are four more bonding than antibonding electrons in the molecule. The electronic configuration of C_2 is

$$C_2: \qquad KK(\sigma_s)^2(\sigma_s^*)^2(\pi_y)^2(\pi_z)^2$$

Because of its nonzero bond order, C_2 should, and does, exist, having been detected at high temperatures.

N_2. The last MO population diagram in Fig. 9-38 is that of the nitrogen molecule N_2. It has a *net* excess of six bonding electrons, which corresponds to a bond order of 3. These electrons occupy the π_y, π_z, and σ_x orbitals, as shown, giving N_2 the electronic configuration

$$N_2: \qquad KK(\sigma_s)^2(\sigma_s^*)^2(\pi_y)^2(\pi_z)^2(\sigma_x)^2$$

N₂ is, of course, very stable and is encountered every day in every breath we take. Magnetic measurements confirm that all electrons are paired in N_2; it is not paramagnetic.

The MO model of the N_2 molecule correlates especially well with the VB picture. The six electrons in the π_y, π_z, and σ_x orbitals correspond to the six electrons shown in the VB Lewis structure

$$:N:::N:$$

O₂. The addition of two more electrons to the N_2 configuration gives the orbital population of O_2, as is shown at the left in Fig. 9-39. Note that these two electrons must go into antibonding orbitals resulting in a bond order decrease (from 3 in N_2) to 2. The lower bond order is consistent with the fact that O_2 has a smaller bond energy and a longer bond length than does N_2. The electronic configuration of O_2 is

$$O_2: \quad KK(\sigma_s)^2(\sigma_s^*)^2(\sigma_x)^2(\pi_y)^2(\pi_z)^2(\pi_y^*)^1(\pi_z^*)^1$$

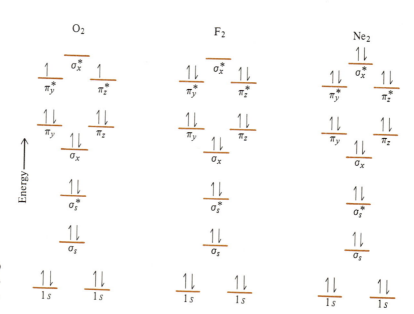

Figure 9-39 MO population diagram: O_2, F_2, and Ne_2.

One of the early triumphs of MO theory was its ability to show that the O_2 molecule ought to be paramagnetic because of its two unpaired electrons. This is in great contrast to simple VB theory which leads to the Lewis structure

$$:\overset{..}{O}::\overset{..}{O}:$$

Of course we could write

$$:\overset{..}{\underset{.}{O}}:\overset{..}{\underset{.}{O}}:$$

which does show the unpaired electrons, but now the bond looks like a single bond. The experimental facts are that the molecule is paramagnetic and has a bond energy which is too high and a bond length too short for a single bond. The MO model readily accounts for the observed magnetic and bond characteristics.

F_2. The addition of two more electrons gives us the population shown in the center of Fig. 9-39. Since the (antibonding) π^* orbitals are both filled, the bond order in F_2 is only 1. This agrees with the experimentally determined bond energy and length, both of which are about what would be expected for a single bond. Also, F_2 is found not to be paramagnetic, which is consistent with the lack of unpaired electrons. The configuration of F_2 is

$$F_2: \quad KK(\sigma_s)^2(\sigma_s^*)^2(\sigma_x)^2(\pi_y)^2(\pi_z)^2(\pi_y^*)^2(\pi_z^*)^2$$

Ne_2. Adding two more electrons fills the σ_x^* orbital, reducing the bond order to zero. Ground-state Ne_2 has never been observed.

Table 9-5 shows a summary of all the homonuclear diatomic molecules, both real and hypothetical, from H_2 to Ne_2. Also shown are the corresponding bond orders, energies, and lengths.

Heteronuclear diatomic molecules　　Diatomic molecules having different atoms are called *heteronuclear*. The difference in electronegativities in these molecules causes the MO energy spacings to be different from those in homonuclear diatomics. In many cases, however, a heteronuclear diatomic has the same electronic configuration as a homonuclear

Table 9-5
Properties of some homonuclear diatomic molecules

Molecule	Electronic configuration	Bond order	Bond energy, kJ mol^{-1}	Bond length, nm
H_2	$(\sigma_s)^2$	1	433	0.074
He_2	$(\sigma_s)^2(\sigma_s^*)^2$	0	—	—
Li_2	$KK(\sigma_s)^2$	1	105	0.267
Be_2	$KK(\sigma_s)^2(\sigma_s^*)^2$	0	—	—
B_2	$KK(\sigma_s)^2(\sigma_s^*)^2(\pi_y)^1(\pi_z)^1$	1	289	0.159
C_2	$KK(\sigma_s)^2(\sigma_s^*)^2(\pi_y)^2(\pi_z)^2$	2	628	0.131
N_2	$KK(\sigma_s)^2(\sigma_s^*)^2(\pi_y)^2(\pi_z)^2(\sigma_x)^2$	3	941	0.110
O_2	$KK(\sigma_s)^2(\sigma_s^*)^2(\sigma_x)^2(\pi_y)^2(\pi_z)^2(\pi_y^*)^1(\pi_z^*)^1$	2	494	0.121
F_2	$KK(\sigma_s)^2(\sigma_s^*)^2(\sigma_x)^2(\pi_y)^2(\pi_z)^2(\pi_y^*)^2(\pi_z^*)^2$	1	151	0.142
Ne_2	$KK(\sigma_s)^2(\sigma_s^*)^2(\sigma_x)^2(\pi_y)^2(\pi_z)^2(\pi_y^*)^2(\pi_z^*)^2(\sigma_x^*)^2$	0	—	—

diatomic which has the same number of electrons. Thus the electronic configuration of CO is the same as that of N_2; CO and N_2 are said to be *isoelectronic*.

ADDED COMMENT

There is much more to be said about molecular orbitals, and many more examples could be given. We have not even mentioned triatomic molecules, for instance. Also, there is an interesting situation which occurs in some molecules: the spreading of an MO over several atoms. Such *delocalized* MOs exist in those molecules which must be described as resonance hybrids in VB theory. (Resonance is not a part of the MO model of bonding.) We describe these and other aspects of MO theory later, however, as it becomes appropriate.

SUMMARY

The geometries of many small molecules can be predicted by means of the *valence-shell electron-pair repulsion* (VSEPR) method. The method is based upon the idea that valence-shell electron pairs tend to orient themselves so as to minimize interelectronic repulsion. Use of the VSEPR approach involves first finding the lowest-energy electron-pair geometry and then deciding which pairs are lone pairs and which are bonding pairs. The locations of the bonding pairs then define the geometry of the molecule. The *polarity* of a molecule is measured by its *dipole moment*. It depends upon the polarity of its bonds and its molecular geometry.

In this chapter we have looked closely at the description of the covalent bond according to both the *valence-bond* (VB) and *molecular-orbital* (MO) theories. According to the VB point of view a covalent bond consists of a pair of electrons shared between the bonded atoms. This means that two orbitals (one from each atom) must *overlap* in such a way that the electron pair simultaneously occupies both orbitals. This concentrates electronic charge in the region between the nuclei and so bonds the atoms together.

A *sigma* (σ) *bond* is one in which the charge distribution of the bonding pair is centered on and symmetrical around the bond axis. Such a bond is formed as a result of *s–s* overlap, *s–p* overlap, or end-to-end *p–p* overlap.

A *pi* (π) *bond* is characterized by two regions of high electron density on opposite sides of the bond axis. A pi bond is commonly formed as a result of side-to-side overlap of *p* orbitals of the bonded atoms.

Single, double, and *triple bonds* are one, two, and three bonding pairs of electrons, respectively. A single bond is a sigma bond, a double bond is a sigma plus a pi bond, and a triple bond consists of a sigma bond plus two pi bonds which are oriented at right angles to each other.

In many molecules the bonding is best described in terms of *hybrid orbitals* which result from the mixing or combining of pure atomic orbitals (*s, p, d,* etc.). In the mixing process the number of orbitals in a hybrid set is equal to the number of pure orbitals combined. In addition, the geometrical relationship among the hybrids in a particular set depends upon the character of that set; for example, two *sp* hybrids have their major lobes pointed in opposite directions, 180° apart, a set of three sp^2 hybrids has major lobes at 120° in a plane, and four sp^3 hybrids show tetrahedral geometry, having lobes at 109.5° from each other.

In *molecular-orbital* (MO) theory atomic orbitals (AOs) from different atoms are considered to combine to form new energy levels (MOs) for the molecule as a whole. As is the case with hybrid orbitals the number of MOs formed equals the number of AOs which have combined. Electrons occupy MOs of molecules in much the same way as they occupy the AOs of atoms; that is, in its ground state a molecule has its lower energy MOs filled. MO electronic configurations can be built up much as are AO configurations. Some MOs are *bonding orbitals,* while others are *antibonding;* when occupied, the former contribute toward stability in the molecule, and the latter to instability.

KEY TERMS

Antibonding orbital (Sec. 9-4)
Bonding orbital (Sec. 9-4)
Delocalized MO (Sec. 9-4)
Dipole (Sec. 9-1)
Hybrid orbital (Sec. 9-3)
Homonuclear (Sec. 9-4)
Half bond (Sec. 9-4)
Molecular orbital (MO) (Sec. 9-4)
Octahedron (Sec. 9-3)

Pi bond (Secs. 9-2; 9-4)
Polar molecule (Sec. 9-1)
Sigma bond (Secs. 9-2; 9-4)
Steric number (Sec. 9-1)
Tetrahedral angle (Sec. 9-3)
Trigonal bipyramid (Sec. 9-1)
Trigonal pyramid (Sec. 9-1)
VSEPR method (Sec. 9-1)

QUESTIONS AND PROBLEMS

VB theory and orbital overlap

9-1 Covalent bonds are said to be directional, while ionic bonds are not. Explain.

9-2 Describe the essential differences between a sigma bond and a pi bond.

9-3 Sketch a contour diagram showing **(a)** a sigma bond and **(b)** a pi bond as each would appear if you were looking down the bond axis.

9-4 Classify the kind of bond which could be formed as a result of each of the following atomic orbital overlaps: **(a)** $s + s$ **(b)** $s + p$ **(c)** $p + p$ (side-to-side) **(d)** $p + p$ (end-to-end).

9-5 Can a bond be formed from the "sideways" overlap of a p orbital with an s orbital? Explain.

9-6 A sigma (bonding) orbital has one major lobe; a pi (bonding) orbital has two. What do you suppose a delta (bonding) orbital looks like? What AOs might be expected to overlap to produce a delta MO?

9-7 The mercurous ion is an unusual homonuclear diatomic ion, Hg_2^{2+}. Comment on the bonding in this ion.

9-8 Which is a better measure of the strength of a bond, bond energy or bond length? Does a short bond length always imply a large bond energy? Explain.

Hybrid orbitals

9-9 What set of hybrid orbitals has a geometrical orientation which is **(a)** trigonal planar **(b)** octahedral **(c)** tetrahedral **(d)** linear?

9-10 What is the percent s character of each orbital in each of the sets in Prob. 9-9?

9-11 Under certain conditions the methylene molecule CH_2 can be detected. Describe the bonding in this short-lived molecule. Why is methane, CH_4, so much more stable than methylene?

9-12 Phosphorus forms two stable fluorides, PF_3 and PF_5. **(a)** Draw a Lewis structure for each. **(b)** Account for the fact that NF_3 exists but NF_5 does not.

9-13 The bond angle in hydrogen sulfide, H_2S, is 92°, indicating that sulfur uses unhybridized orbitals for bonding in this molecule. Contrast this with the bonding in H_2O, and account for the difference.

9-14 The bond angles in ethane, C_2H_6, are 109°, those in ethylene, C_2H_4, are 120°, and those in acetylene, C_2H_2, are 180°. Identify all the σ and π bonds in these molecules. (*Hint:* Consider that the carbon AOs become hybridized before overlap occurs.)

Electron repulsion

9-15 Determine **(a)** the steric number, **(b)** the number of bonding pairs, and **(c)** the number of lone pairs in each of the following: $BeCl_2$, BrF_5, H_2S, BCl_3, AsH_3, XeF_4, I_3^-.

9-16 For each of the following indicate **(a)** the geometrical orientation of the complete set of electron pairs around the central atom and **(b)** the set of hybrid orbitals which provides that orientation: PF_6^-, BF_4^-, $SiCl_4$, SF_2, ClF_3, IF_5, BeH_2.

9-17 Account for the fact that BeF_2 is nonpolar, while OF_2 is polar.

9-18 Predict which of the following will exhibit a dipole moment: I_2, ICl, CCl_4, $CHCl_3$, CH_2Cl_2, PCl_3, $POCl_3$, BF_3, NF_3, H_2S.

9-19 Predict the geometrical shape of each of the following from VSEPR considerations: SiH_4, BrF_5, PCl_5, XeO_4, IF_2^-, PF_6^-, XeO_3, IF_4^-.

9-20 Account for the fact that **(a)** the dipole moment of NF_3 is less than that of NH_3 **(b)** the dipole moment of $SiCl_4$ is less than that of $SeCl_4$.

9-21 Which of the following exhibits a molecular dipole moment? F_2, SeF_4, BCl_3, OF_2, $BeCl_2$, NCl_3.

9-22 Do you predict the ozone molecule, O_3, to be polar or nonpolar? (See Sec. 8-2.) Explain.

Molecular orbitals

9-23 In terms of charge-cloud distributions, explain why antibonding electrons reduce bond strength, while bonding electrons increase it.

9-24 Compare the bond energies in **(a)** O_2, O_2^-, O_2^+ **(b)** N_2, N_2^-, N_2^+.

9-25 Draw a molecular-orbital population diagram for each of the following: Si_2, P_2, Se_2.

9-26 For each of the following molecules determine the bond order and magnetic properties: CN, LiH, ClF, BN, PCl.

9-27 For each of the following ions determine the bond order and magnetic properties: NO^+, NO^-, Cl_2^+, CS^+.

9-28 Arrange the following in order of increasing bond length: O_2, O_2^-, O_2^+, O_2^{2-}.

9-29 Discuss the possibility for the existence of

(a) ground-state He_2 **(b)** excited-state He_2
(c) ground-state He_2^+.

9-30 Sketch contour diagrams for the bonding and antibonding orbitals formed by overlap of each of the following pairs of atomic orbitals:

(a) $1s + 1s$ **(b)** $1s + 2p$ **(c)** $2s + 2s$
(d) $2p_x + 2p_x$ **(e)** $2p_y + 2p_y$ **(f)** $sp^3 + s$

9-31 For each of the MOs formed in Prob. 9-30 determine the number of nodes present in the resultant probability-density distribution. What form or shape does each node have?

9-32 Account for the fact that the charge cloud of *all* electrons in the N_2 molecule is symmetrical around the bond axis.

9-33 What does the MO filling diagram look like for an ion pair, such as LiF?

9-34 Predict the bond order in each of the following: NF, CN^-, BF, PF, NeO^+.

9-35 Predict which of the following has the lowest first ionization energy: N_2, NO, or O_2. Justify your answer.

9-36 The energy of the lowest-lying excited state of C_2 is only slightly above that of the ground state. Account for the fact that this excited state is paramagnetic, while the ground state is not.

10

IDEAL SOLIDS AND LIQUIDS

TO THE STUDENT

Take a moment to recall the structure of a typical gas: tiny moving molecules spaced far apart in a rapidly changing, disordered array (see Chap. 4). Because the typical solid has properties which are quite unlike those of a gas, it is to be expected that the structure of a solid is little like that of a gas. Just a superficial observation of the properties of solids allows us to make some predictions about solid-state structure. For instance, solids are almost incompressible; this tells us that they contain little empty space, that the particles in a solid are very close together. The typical solid's rigidity and hardness tells us that its particles are strongly bound together, not free to move as in a gas. Furthermore, the fact that solids form crystals tells us that their particles are regularly arranged. (How could the beautiful plane faces of a crystal result from a disordered internal arrangement of atoms?) In this chapter we take a close look at solids emphasizing (1) the three-dimensional patterns of their particles, (2) the identities of the particles (atoms, ions, etc.), and (3) the forces between these particles.

In this chapter we also consider the dependence of properties upon structure in the liquid state. Here again the easily observable properties of liquids give us clues to structure. Liquids are almost as incompressible as solids, so we can predict that comparatively little empty space is available between the particles of a liquid. But unlike solids, liquids usually flow readily, and this fact tells us that the forces beween particles in a liquid are not capable of locking the particles into a rigid structure which resists deformation.

**10-1
Solids: some
preliminary
observations**

The traditional definition of a solid goes something like this: *A solid is a substance which maintains a fixed volume and shape.* In other words, a solid's size and shape are not influenced by the size and shape of its container. A more modern definition of a solid is the following: *A solid is a substance which has its constituent particles arranged in a regularly ordered internal array.* The first definition is based on physical properties, the second on internal structure. Are they consistent with

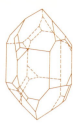

Figure 10-1
A single crystal of
quartz, SiO_2.

each other? No, not always. We will see in Sec. 11-2, that some liquids can be *supercooled*, cooled far below their freezing points without actually freezing. As the temperature of such a liquid drops, it gradually becomes more viscous and unyielding, until finally it seems as hard and rigid as a solid. These highly supercooled liquids maintain a fixed volume and shape, but their atoms are irregularly and randomly arranged. Such substances are said to be *amorphous* (from the Greek, "without form"), while those with an ordered, regular internal atomic arrangement are said to be *crystalline*.

Should we classify amorphous substances as solids because of their hardness and constancy of volume and shape or as liquids because of their irregular internal structures? It is really a matter of choice, but usually we speak of *true,* or *crystalline, solids* versus *amorphous solids,* and simply keep in mind that the latter have structures which are essentially those of liquids. *In this book the word* solid *without a modifier will always refer to a crystalline solid.* Highly supercooled liquids will be called *amorphous solids,* or *glasses.*

Properties of solids Do solids really show fixed volumes and shapes? Actually, no, not exactly. Most solids expand a little with an increase in temperature. (Some contract, however!) Nevertheless, the volume change per degree of temperature rise (the *coefficient of thermal expansion*) is very small in comparison to that of a gas. What about the effect of pressure on the volume of a solid? It is also very small; solids are almost incompressible. For instance, in order to compress a sample of silver metal into one-half its normal volume, a pressure of about 5×10^5 (half a million) atm is needed.

The small dependence of the volume of a solid on temperature and pressure is consistent with a structure which is more compact than that of a gas. The average distance between atoms in a solid is evidently much less than in a gas. In addition, the comparative rigidity of a solid tells us that its atoms must be strongly locked together in this compact arrangement. This is also consistent with the practically unmeasurable rates of flow and diffusion shown by most solids.

Crystals One of the most striking characteristics of solids is their occurrence as crystals. Figure 10-1 shows a drawing of a single perfect crystal of quartz. Crystals of some minerals up to many feet in length have been found in the crust of the earth. They can also be grown in the laboratory. Figure 10-2 shows a photograph of an octahedral crystal of chrome alum, $KCr(SO_4)_2 \cdot 12H_2O$, grown by slow evaporation of a solution of this compound in water. Other methods used to grow crystals include the slow freezing of a liquid.

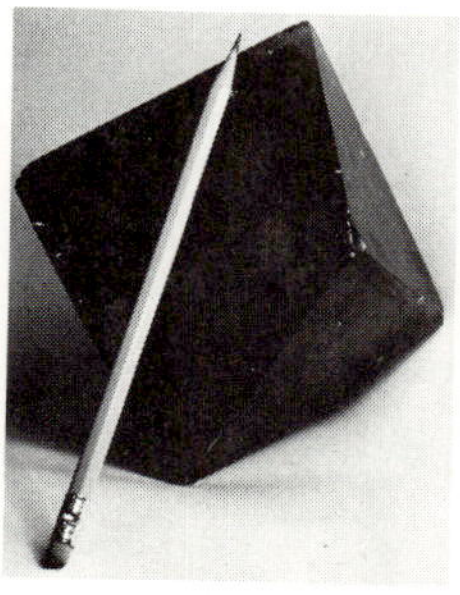

Extremely slow growth is required for large, perfectly formed crystals, and so most large crystals do not have ideal shapes. (Growing very small, nearly perfect crystals is easy, however. It can be done by evaporation of water from a single drop of aqueous solution, and the tiny, growing crystals can be observed under a microscope.) Figure 10-3 shows ideal and distorted forms of chrome alum. Distorted crystals such as these are the result of conditions in the growing crystal's environment which favor faster growth in certain directions than in others. Impurities often affect the form of a crystal. Thus sodium chloride, which usually forms cubic crystals, can be made to form octahedral ones by dissolving urea in the aqueous solution out of which the NaCl crystal is growing. Although irregular

Figure 10-2
A crystal grown in the
laboratory.

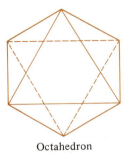
Octahedron

Distorted octahedron

Figure 10-3
Ideal and distorted crystals.

growth alters the geometrical shapes of crystal faces (from triangles to irregular hexagons in the example of Fig. 10-3), the angles between the faces of a given form tend to remain the same. This is summarized in the *law of constancy of interfacial angles,* which states that the angles between adjacent corresponding faces in a given crystal form of a substance are constants characteristic of the substance.

Although the study of crystal forms is interesting, we will leave it to textbooks of mineralogy and crystallography, of which there are many. We are concerned more with the internal structures of crystals, and in this chapter we will consider only *ideal,* or *perfect, crystals,* hypothetical structures which are helpful in describing real crystals. (Most real crystals contain large numbers of imperfections of various kinds which we will consider in Sec. 11-3.)

10-2
X-ray diffraction

Although the external form of a crystal yields valuable clues to the crystal's internal structure, most of our knowledge of the structures of crystalline materials comes from a powerful technique known as x-ray diffraction.

The discovery of
x-ray diffraction

It was mentioned in Sec. 6-1 that any electromagnetic radiation can be diffracted (bent) by a diffraction grating, a series of objects regularly spaced at a distance roughly the same as the wavelength of the radiation. In 1912 the German physicist Max von Laue pointed out that atoms in a crystal are spaced at about the right distance to allow them to serve as the elements of a three-dimensional diffraction grating for x rays. Shortly thereafter two graduate students, Friedrich and Knipping, provided experimental support for von Laue's suggestion by shooting a beam of x rays at a crystal of copper sulfate, $CuSO_4 \cdot 5H_2O$, and verifying that the predicted diffraction actually does take place. This was the birth of x-ray crystallography.

Figure 10-4*a* shows an apparatus which can be used to obtain a diffraction pattern. A beam of monochromatic (single-wavelength) x rays is shot at a single crystal, and the emergent diffracted beams are detected by means of photographic film. The location of the spots on the resulting *Laue pattern* (Fig. 10-4*b*) depends upon the relative locations of the atoms in the crystal.

The mechanism of
diffraction

Diffraction of electromagnetic waves occurs because the elements of a diffraction grating absorb the radiation and then serve as secondary sources, reemitting radiation in all directions. In the case of x-ray diffraction, these secondary sources

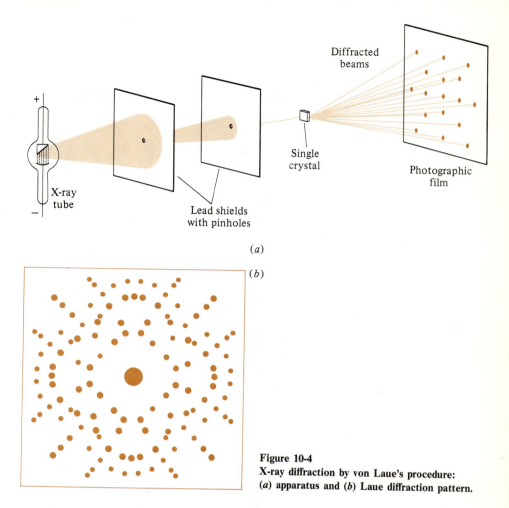

(a)

(b)

Figure 10-4
X-ray diffraction by von Laue's procedure:
(*a*) apparatus and (*b*) Laue diffraction pattern.

are the atoms or, more precisely, their electrons. When a crystal diffracts x rays, the reemitted electromagnetic waves reinforce each other in some directions and cancel each other in others. Consider, the two atoms shown in Fig. 10-5. The incident x radiation is absorbed by the atoms and then reemitted in all directions, only three of which are shown. Some of the radiation passes straight through without being diffracted. At angle *A*, however, the diffracted rays from the two atoms are out of phase with each other, one exactly nullifying or cancelling the other. No x-ray energy is transmitted at this angle. At angle *B* the two waves are in phase and hence reinforce each other. Thus diffraction is said to occur at angle *B*.

The Bragg equation   In 1913 William and Lawrence Bragg, an English father-and-son team, showed that diffraction of x rays can be imagined to occur as if the x rays were reflected by layers of atoms in a crystal, much as light is reflected by a plane mirror. William Bragg showed that there is a very simple relationship between the distance between the layers, the wavelength of the x radiation, and the angle of diffraction. The *Bragg equation* is

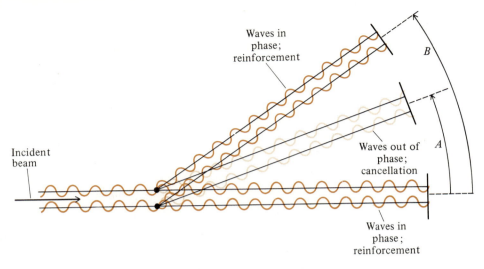

Figure 10-5
Model for diffraction of
x rays by two atoms.

$$n\lambda = 2d \sin \theta$$

where θ = angle between the incident beam (or the "reflected" beam) and the planes of atoms

d = distance between the planes

λ = x-ray wavelength

n = positive integer (often equal to 1)

Although x-ray diffraction is far more complicated than simple reflection by a mirror, this model can be used to derive the Bragg equation. In Fig. 10-6 are shown several layers of atoms at distance d apart. Using the reflection analogy we can say that the angle of reflection θ', equals the angle of incidence θ. Considering rays 1 and 2 shown in the "blowup," we see that in order for these rays to emerge in phase, so that they can reinforce each other, the extra distance traveled by ray 2 must be an integral number of wavelengths. Since this extra distance is $2l$, we can write

$$n\lambda = 2l$$

where n is a positive integer. Using a little trigonometry we see that

$$l = d \sin \theta''$$

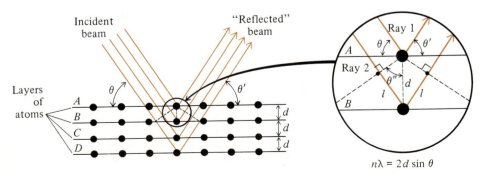

Figure 10-6
Derivation of the Bragg
equation.

From the figure note that $\theta'' = \theta$. Substituting, we have the Bragg equation:

$$n\lambda = 2d \sin \theta$$

Note that the Bragg equation tells us that for any interlayer distance d, diffraction can occur at several angles, each corresponding to a different value of n. The ray which corresponds to $n = 1$ is called the *first-order diffracted ray,* that which corresponds to $n = 2$, the *second-order ray,* and so on. For a given value of d, the first-order diffraction is the one at the smallest angle θ.

● **EXAMPLE 10-1**

Problem   In a diffraction experiment x rays of wavelength 0.154 nm were used on an NaCl crystal. A first-order diffraction occurred at an angle θ of 22.77°. How far apart are the layers of atoms which were responsible for this ray?

Solution   If
$$\theta = 22.77°,$$
then
$$\sin \theta = 0.3870$$
(Use a trigonometry table or pocket calculator.)   Rearranging the Bragg equation, $n\lambda = 2d \sin \theta$, and solving we obtain

$$d = \frac{n\lambda}{2 \sin \theta}$$

$$= \frac{1(0.154 \text{ nm})}{2(0.3870)} = 0.199 \text{ nm}$$ ●

Powder patterns   Several modifications of the von Laue diffraction method are in much more common use than the original technique. In some of these the crystal is rotated or oscillated and the diffracted rays flash out when the Bragg condition is met. One method, the *Debye-Scherrer method,* makes use of a finely ground sample instead of a single crystal. The powdered material is coated on the surface of a fine fiber or packed inside of a fine capillary made of a special glass. Since the sample consists of countless crystal fragments, for each set of layers of atoms, enough fragments are oriented in such a way that the Bragg condition is satisfied, and a diffracted ray emerges, exposing the film. Thus each line on the resulting powder pattern corresponds to a different set of layers. The Debye-Scherrer powder pattern method is often difficult to interpret for the purpose of crystal-structure determination. It is useful for analytical purposes, however, since each substance produces its own unique powder pattern. Powder patterns of many substances have been cataloged and are used for identifying unknown materials.

X-ray diffraction pictures are analyzed in order to determine the orientation of planes of atoms in the substance. From this, the overall three-dimensional crystal structure can be determined. The process of interpreting these pictures was in the past a long, complicated one for most substances. Today, however, the computational part of the determination has been made considerably easier through the use of computers.

10-3   On several occasions we have spoken of the ordered, three-dimensional arrange-
The crystal lattice   ment of atoms in a solid. This arrangement is called a *crystal lattice*. What kinds of crystal lattices are there, and how are they described?

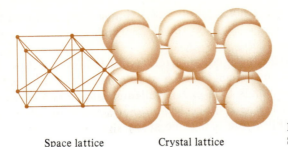

Space lattice   Crystal lattice

Figure 10-7
Space lattice and crystal lattice.

The space lattice   A geometrically regular array of *points* in space is called a *space lattice*. At the left of Fig. 10-7 is shown one kind of space lattice. The points have been enlarged (to make them visible) and connected (to make their locations obvious). Now imagine that iron atoms have been placed in the space lattice in such a way that the nucleus of each atom is centered on a lattice point. The result is a portion of the *crystal lattice* of iron and is shown at the right of Fig. 10-7.

A crystal lattice may be thought of as a space lattice in which the points are occupied by atoms, ions, molecules, or groups of these. The arrangement of particles in a crystal lattice repeats periodically in three dimensions up to the physical boundaries of each single crystal.

The unit cell   In order to simplify the job of describing a crystal lattice, it is useful to specify the *unit cell* of that lattice. A unit cell is a small portion of the lattice which can be used to generate, or construct, the entire lattice by moving the unit cell according to certain rules. But before we study how this is done, we will first look at a two-dimensional analogy: the net.

A *net* is a regular, repeating array of points in a *plane*. Figure 10-8 shows a net. We can choose a unit cell for this net in many different ways. An obvious choice for a unit cell for the net is the square shown in the upper-left-hand corner of Fig. 10-8. The unit cell can be used to generate the entire net by moving it in the plane (1) a distance equal to the length of one of its edges and (2) in a direction

Figure 10-9
Different ways of choosing a unit cell for a net.

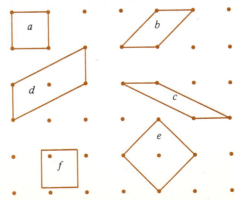

Figure 10-8
A net: Points in a plane arranged in a regular, repeating array. The square unit cell can be used to generate the entire net, as shown.

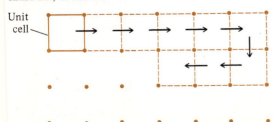

Unit cell

particular net. The same is true in three dimensions. Figure 10-13 shows the (nonprimitive) face-centered cubic unit cell. Also shown is a *rhombohedral* unit cell which is based on the same set of lattice points, but which is primitive. For a given space lattice the unit cell usually chosen is the one which is smallest and most symmetrical. By *most symmetrical* we mean the one with the largest number of perpendicular edges. Thus the face-centered cubic unit cell is preferred over the rhombohedral unit cell.

In 1848 the French scientist A. Bravais showed that there are only 14 fundamentally different ways of regularly arraying points in space. These 14 space lattices, often called *Bravais lattices,* fall into six sets which correspond to the six *crystal systems.* The primitive cell in each of these six sets is shown in Fig. 10-14. Note that these unit cells differ from each other in the angles between their faces and/or the lengths of their edges.

Some representative crystal lattices

The space lattice is composed of points. (The lines on the drawings you have been looking at are not really part of the lattices or unit cells. They were included merely to help guide your eyes.) In a perfect crystal each point is occupied by, or corresponds to, some unit. That unit may be an atom, a molecule, an ion, or a group of several of these particles. Furthermore, in a crystal lattice each point of the space lattice is occupied by the same unit. This is best shown by some specific examples:

ARGON. The simplest unit which can occupy a lattice point is an *atom.* Argon, one of the noble gases, solidifies (at 1 atm pressure) at $-189°C$. As a solid it exhibits a face-centered cubic structure, in which each point in the space lattice is occupied by a single Ar atom. The unit cell is shown in Fig. 10-15a.

ETHYLENE. An example of a crystal lattice in which each point is occupied by a *molecule* is found in solid ethylene, C_2H_4. The crystal lattice in this substance is *body-centered orthorhombic.* The unit cell is shown in Fig. 10-16 and may be considered to be a body-centered cube which has been distorted (the three perpendicular edge lengths are no longer equal) to accommodate the nonspherical C_2H_4 molecule.

Figure 10-15
Solid argon: a face-centered cubic structure.
(a) Space-filling models. (b) Model showing fractions of atoms in unit-cell cube.

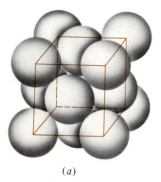

(a)

(b)

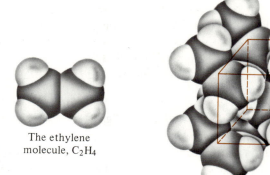

**Figure 10-16
Solid ethylene: a body-centered orthorhombic structure.**

The ethylene molecule, C_2H_4

SODIUM CHLORIDE. In ionic solids the packing of positive and negative ions results in a structure in which groups of ions "occupy" lattice points. (The quotation marks will be justified shortly.) Figure 10-17a shows a ball-and-stick representation of the NaCl unit cell. (See also Fig. 8-1.) The crystal structure of NaCl is face-centered cubic. How can that be? It is clear that Cl^- ions occupy the corners and face centers of a cube, but the Na^+ ions in Fig. 10-17a appear to be at the body and edge centers. The situation becomes clear when we realize that there is one Na^+ per Cl^- in the crystal lattice. In Fig. 10-17b we see the unit cell redrawn with a Cl^- still at each corner and face center but with one Na^+ drawn to the right of each Cl^-, just as it exists in the lattice. When we draw the unit-cell this way we can see that one Cl^- and one Na^+ occupy each face-centered cubic lattice point, even though the Na^+ is actually displaced a distance $e/2$ to the right of the lattice point (where e is the cube edge length). Since there are one Na^+ and one Cl^- per lattice point, the formula is NaCl. Sodium chloride may also be described as two interpenetrating, identical, face-centered cubic lattices, one of Na^+ and the other of Cl^- ions, the lattices being displaced $e/2$ apart.

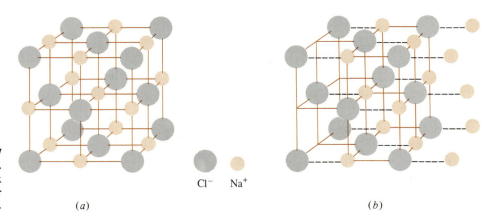

**Figure 10-17
NaCl unit cell.
(a) Ball-and-stick model. (b) Ion-pair model.**

Cl^- Na^+

(a)   (b)

10-4
Close packing

Many elements solidify to form crystal lattices in which single atoms occupy the lattice points. In many such lattices the arrangement of atoms is a *close-packed* one, one in which the *packing efficiency* is the maximum possible. Packing efficiency is a measure of the efficiency with which spheres fill up space when packed in a given structure. It is defined as the percent of a given volume, say one unit cell, which is occupied by spheres.

● **EXAMPLE 10-2**

Problem

Argon crystallizes in the structure shown in Fig. 10-15. Calculate the packing efficiency in solid argon.

Solution

We solve this problem by first finding out the volume $V_{spheres}$ occupied by the spherical atoms in one unit cell. Let us first imagine that countless unit cells are packed together into a large crystal of argon. How many argon atoms are present in a single unit cell? Each unit cell consists of eight corner atoms and six face-centered atoms. But each corner of a unit cell is also the corner of *seven* other unit cells. (Close your eyes and imagine eight small cubical blocks of wood fitted together to form a large cube. There are four small blocks in the bottom layer and four more in the top layer. At the center of the large cube is a point which is at a corner of each of the eight small cubes.) So only one-eighth of each corner atom "belongs" to a given unit cell. Similarly, each of the six face-centered atoms is shared between two adjacent unit cells, so only one-half of each of these belongs to a given unit cell. Figure 10-15b shows the unit cell of argon with the fraction of each kind of atom which is actually *within* the cell.

Since there are eight corner atoms and six face-centered atoms, the total number of argon atoms within each unit cell is

$$8 \times \tfrac{1}{8} = 1 \qquad \text{corner atom}$$
$$\underline{6 \times \tfrac{1}{2} = 3} \qquad \text{face-centered atoms}$$
$$4 \qquad \text{atoms, total}$$

We see that in solid argon, or, for that matter, in any similar face-centered cubic structure, the equivalent of four atoms is present per unit cell.

Since the volume of a sphere is given by the formula $V = \tfrac{4}{3}\pi r^3$, the volume occupied by the equivalent of four spherical argon atoms of radius r is

$$V_{spheres} = 4 \times \tfrac{4}{3}\pi r^3 = \tfrac{16}{3}\pi r^3$$

Our next step is to find the volume of the entire unit cell expressed in terms of r. From Fig. 10-15b it can be seen that the diagonal of one face of the unit cell is $4r$. From this we can find the edge length using the pythagorean theorem. (In a right triangle the square of the hypotenuse is equal to the sum of the squares of the other two sides.)

$$(4r)^2 = e^2 + e^2$$

where e is the edge length of the unit cell. Solving for e we get

$$2e^2 = 16r^2$$
$$e^2 = 8r^2$$
$$e = 2r\sqrt{2}$$

and the volume V_{cell} of the entire unit cell is

$$V_{cell} = e^3$$
$$= (2r\sqrt{2})^3 = 16r^3\sqrt{2}$$

The fraction of the unit cell occupied by the argon atoms is

$$\frac{V_{\text{spheres}}}{V_{\text{cell}}} = \frac{\frac{16}{3}\pi r^3}{16r^3\sqrt{2}} = \frac{\pi}{3\sqrt{2}} = 0.74$$

Packing efficiency $= 0.74 \times 100 = 74\%$ ⬤

Elements which crystallize in close-packed structures include all the noble gases and over 40 metals. Some molecular substances, such as hydrogen (H_2) and methane (CH_4), also show close-packed crystal structures in which the molecules are free to rotate, tumbling over and over, while maintaining their lattice positions.

Close packing of identical spheres

When identical hard spheres are packed together in a single *two-dimensional layer,* several arrangements are possible. One, shown in Fig. 10-18a, is square packing. A second way, hexagonal packing, is shown in Fig. 10-18b. Comparison of the two diagrams shows that the spheres of Fig. 10-18b fit together more compactly than those of Fig. 10-18a. Within a layer hexagonal packing is the most compact packing possible, and so it is therefore called *close,* or *closest, packing.* Notice that each sphere in such a layer is surrounded by six nearest-neighbor spheres.

In *three dimensions* identical spheres can also be packed together in many different ways. A number of kinds of close packing are theoretically possible, but all are combinations of two different basic structures. In each the packing efficiency is the maximum possible, and in each the number of nearest-neighbor spheres surrounding any given sphere is 12.

In order to get a feeling for the nature of close packing in three dimensions start by considering a single close-packed layer of spheres as viewed from directly above in Fig. 10-19a and from an oblique perspective in Fig. 10-19b. Call this layer the *A* layer. Our general procedure will be to add successive layers on top of the *A* layer so that the resulting collection of spheres is close packed in three dimensions.

Notice that the upper surface of the *A* layer shown in Fig. 10-19 is dimpled by a regular array of depressions, each of which is located where three spheres rest against each other. The dimples can be divided into two sets. These are labeled *B* and *C*, respectively, in Fig. 10-19. There is actually no difference between the *B* dimples and the *C* dimples, except that once a sphere has been placed in a *B* dimple, none of the *C* dimples adjacent to it are available for occupancy. This is true because any two *immediately adjacent* dimples are so close together that they cannot be simultaneously occupied. As a result, when a second close-packed layer

(a)

(b)

Figure 10-18
Packing of spheres in two dimensions.
(a) Square packing.
(b) Hexagonal (close) packing.

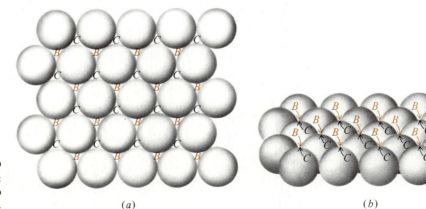

Figure 10-19
Close-packed spheres: one layer (A). (a) Top view. (b) Oblique view.

(a)   (b)

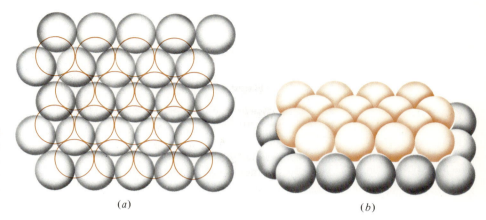

Figure 10-20
Close-packed spheres:
two layers (*AB*).
(*a*) Top view.
(*b*) Oblique view.

(*a*)   (*b*)

is stacked on top of the first, all the spheres must nestle down into one kind of dimple (either *B* or *C*). Putting the spheres of a second layer into *B* dimples, we construct the double-layer structure shown in Fig. 10-20. We will call this second layer the *B* layer.

But where do we place the spheres of the third layer? We have two choices: they can go into either *C* locations or *A* locations, that is, above the *C* dimples of the first layer or directly above the spheres of the first layer themselves. If we place the third-layer spheres in the *A* locations, then we have an *ABA* sequence of layers, since the third layer is positioned directly over the first. This is shown in Fig. 10-21. The fourth layer is then positioned above the second (*ABAB*) and so on. The repeating *ABABAB* · · · layer sequence is called *hexagonal close-packed* (hcp).

If, on the other hand, we position the third-layer spheres in *C* locations as is shown in Fig. 10-22, we have started a different sequence in which the layers repeat: *ABCABCABC* · · ·. This is the *cubic close-packed* (ccp) structure.

What do the unit cells of the hcp and ccp structure look like? Figure 10-23*a* shows a small portion of three layers (*ABA*) of a hexagonal close-packed lattice. The unit cell usually chosen is one-third of a hexagonal prism, shown at the far right. Figure 10-23*b* shows a similar small portion of a cubic close-packed lattice. The space-filling model on the left shows portions of four layers (*ABCA*) in the

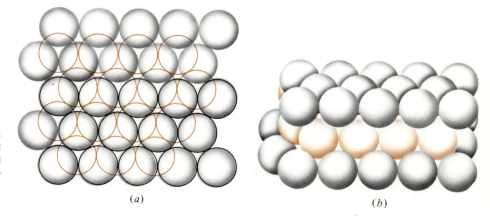

Figure 10-21
Close-packed spheres:
three layers, hexagonal
close-packed (*ABA*) se-
quence. (*a*) Top view.
(*b*) Oblique view.

(*a*)   (*b*)

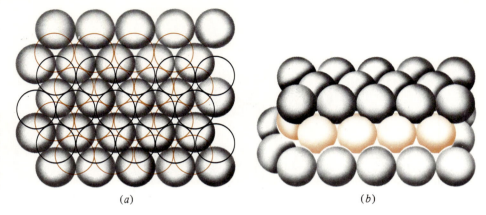

Figure 10-22
Close-packed spheres:
three layers, cubic
close-packed (*ABA*) se-
quence. (*a*) Top view.
(*b*) Oblique view.

(*a*)   (*b*)

structure. The ball-and-stick model at the right clearly shows that this structure is *face-centered cubic*. *Cubic close-packed* (ccp) and *face-centered cubic* (fcc) are different names for the *same* structure.

Crystals with close-packed structures

Which substances crystallize in close-packed structures? In this category we should expect to find atoms and roughly spherical molecules which are attracted to each other by nondirectional forces. The noble gases (group 0) meet these criteria, and so we find that He crystallizes in the hcp structure and the rest (Ne, A, Kr, and Xe) in the ccp form. Many metals also crystallize in close-packed lattices: Cd, Co, Mg, Ti, Zn, and others are hcp; Ag, Al, Ca, Cu, Ni, Pt, and others are ccp. Some metals form close-packed structures which are neither pure hcp nor pure ccp. Neodymium (Nd), for example, crystallizes in a structure in which close-packed layers are arranged in an *ABACABAC* · · · sequence, called *double hexagonal close-packed*. (Some metals, however, form lattices which are not close packed at all; these include Ba, Cr, Fe, K, Na, W, and others.)

Some molecular substances crystallize in close-packed structures. This can occur if the molecule is approximately spherical or if it can rotate at the lattice point, thus becoming effectively spherical. Such substances include CH_4, HCl, and H_2S (all ccp) and H_2 (hcp). The close-packed structure of such a substance is usually stable only above a certain temperature. When cooled below 22 K, for instance, CH_4 changes to a structure which is not close-packed, as the molecules stop rotating and become locked in position. (Even close to absolute zero the H_2 molecule seems to keep rotating, and so solid hydrogen is hcp at all measured temperatures.)

Tetrahedral and octahedral holes

In any close-packed structure there are unoccupied spaces, or voids, among the spheres. The most important of these are the so-called tetrahedral and octahedral holes, examples of which are shown in Fig. 10-24. A *tetrahedral hole* is formed when a second-layer sphere rests in the dimple formed by three spheres in the first layer. The term *tetrahedral* is used because the centers of the four surrounding spheres are at the corners of a regular tetrahedron. In any close-packed structure each sphere nestles into one dimple in the plane of spheres above it and into another in the plane below. There are, consequently, *twice as many tetrahedral holes as there are spheres* in any close-packed structure.

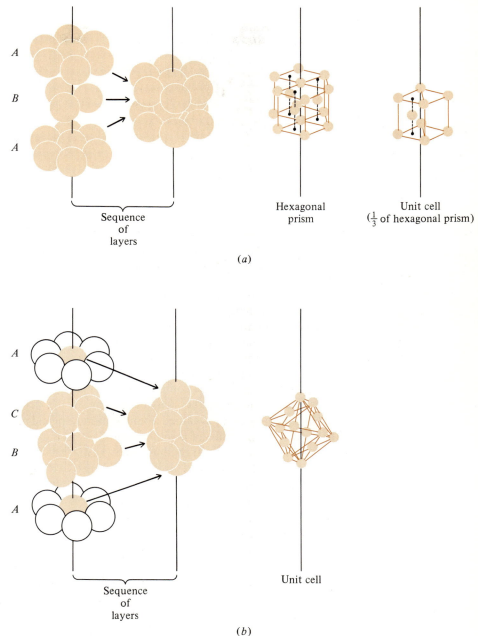

**Figure 10-23
Close-packed unit cells:
(a) hcp, (b) ccp (fcc).**

Octahedral holes are larger than tetrahedral holes. Each may be thought of as a "double dimple," the combination of dimples from two adjacent layers. Figure 10-24 shows that an octahedral hole is surrounded by six spheres, three from one layer and three from the adjacent layer. In any close-packed structure each sphere has six adjacent octahedral holes, and since six spheres are needed to form each such hole, *the number of octahedral holes is equal to the number of spheres.*

Top view

Oblique view

Tetrahedral   Octahedral   **Figure 10-24**
  hole   hole   **Tetrahedral and octahedral holes.**

Crystal structures Many predominately ionic compounds crystallize in structures which can be
based on close described as close-packed arrays of one kind of ion, usually the larger, with the
packing oppositely charged ions occupying holes. For example, the structure shown by
NaCl and many other salts, called the *rock-salt structure,* may be described as a
ccp lattice of the larger ions (Cl⁻ in the case of NaCl) with the smaller ions
occupying the octahedral holes.

Closely related to the rock-salt structure, but less common, is the *nickel arsenide*
structure, named after the compound NiAs. It is based on an hcp lattice of the
larger ions, with the smaller ions occupying all the octahedral holes.

The *fluorite structure* is named after the mineralogical name for calcium fluoride,
CaF_2. It is a ccp lattice of positive ions (Ca^{2+} in the case of CaF_2) with negative
ions occupying all the tetrahedral holes. This structure is shown in Fig. 10-25.

When the locations of the positive and negative ions are reversed from those in
the fluorite structure, the result is the *antifluorite structure.* In the crystal lattice of
lithium oxide, for example, Li^+ ions fill all the tetrahedral holes in a ccp lattice of
O^{2-} ions.

Figure 10-25
The fluorite structure.

Figure 10-26
The zinc blende, or sphalerite, struc-
ture.

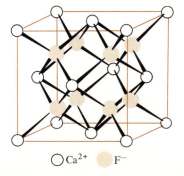

○ Ca^{2+} ● F^-

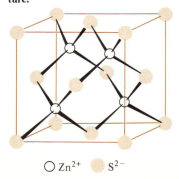

○ Zn^{2+} ● S^{2-}

In some structures not all the holes of a particular type are occupied. In the *zinc blende,* or *sphalerite,* structure positive ions fill only one-half of the tetrahedral holes in a ccp array of negative ions. This structure is shown in Fig. 10-26. Zinc blende and sphalerite are mineralogical names for one form of zinc sulfide, ZnS.

It should be emphasized that imagining the above structures as mere collections of close-packed spherical ions with other ions in some of the holes is an oversimplification. There is often considerable covalent character to the bonding, and often a compound adopts a particular structure because of the directional character of the covalent bond. Still, the concept of *close packing* is very useful as an aid for visualizing many crystal structures.

ADDED COMMENT

One rather common error is to say that the above structures are actually close packed. They are not! Each can be described, however, *in relation to* a true close-packed structure. Imagine, for example, that you have a close-packed (fcc) array of Cl^- ions. Now imagine that you place an Na^+ ion in each of the octahedral holes. In order to do this it is necessary to *separate* the Cl^- ions slightly (so that each hole is large enough to accommodate an Na^+). The original holes in the imaginary Cl^- array were too small to accommodate Na^+ ions. The NaCl crystal is as compact as it can be, but it is not as efficient at filling space as is a pure ccp or hcp lattice.

10-5
Bonding and properties of solids

The properties of solids depend in part on the geometry of the crystal lattice structure. They also depend on the nature of the units (atoms, ions, molecules) at the lattice points and on the forces holding these units together. Examined from this perspective, solids can be classified into four types: ionic, molecular, covalent, and metallic.

Ionic solids

In an ionic solid positive and negative ions are present in the crystal lattice. Since electrostatic forces (ionic bonds) are strong, it is difficult to distort the lattice. Ionic solids are therefore typically hard. It is interesting, however, that they are also brittle. A crystal of NaCl, for example, resists crushing quite strongly, but when it does break, it suddenly splits, rather than gradually distorting or gently crumbling. Figure 10-27 shows enough force being applied to an ionic crystal to start one layer of ions moving across the next. The many strong ionic bonds make this difficult to do—up to the point where ions of opposite charge begin to get close to each other. If enough force is applied, however, interionic repulsions become sufficient to cause the two layers to separate. This point is reached with little warning, and the crystal suddenly splits, or *cleaves,* along this plane. The crystal actually breaks itself apart when attractions are replaced by repulsions.

Ionic solids typically have high melting points. Melting involves a breakdown of the crystal lattice. It occurs when the vibrations of the atoms, ions, or molecules in the solid become so violent that the forces between these particles are no longer strong enough to hold them together. Because ionic bonds are strong, the melting of ionic solids occurs typically at high temperatures. NaCl, for example, melts at $808°C$.

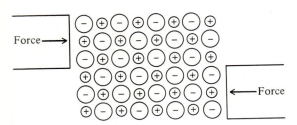

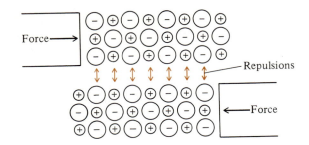

Figure 10-27
Brittleness in ionic solids.

Ionic solids are poor conductors of electricity. Electrical conduction is the passage or movement of charged particles. Ions are certainly charged, but they are not free to move in an ionic solid. (On the other hand, *liquid* NaCl is a good conductor.)

Molecular solids   In molecular solids the units which occupy the lattice points are molecules. *Within* each molecule, covalent bonds, normally strong forces, hold the atoms together. What forces exist *between* the molecules in such a crystal? These forces are comparatively weak and are called *van der Waals forces,* after the nineteenth-century Dutch physicist who first proposed their existence. There are several kinds of van der Waals forces, the most important of which are *dipole-dipole forces* and *London forces.*[1]

Dipole-dipole forces are electrical interactions between polar molecules. Figure 10-28 shows schematically a pair of adjacent polar molecules in a molecular crystal. Two possible orientations are shown, each of which brings oppositely charged ends of the molecules close to each other. Dipole-dipole forces are typically much weaker than ionic (or covalent) bonds.

London forces, also called *dispersion forces,* are usually very weak. They account for the fact that even nonpolar molecules can form a crystal lattice. These forces, first described theoretically by Fritz London in 1930, arise from momentary fluctuations in the electron charge-cloud density in an atom or molecule. For a simple example, see the argon atoms shown in Fig. 10-29. Normally we would depict the electron charge cloud as a symmetrical, spherical blob of negative charge, as shown in Fig. 10-29*a*. London, however, showed that this blob could be interpreted as representing an *average* location whereas the actual charge distribution undergoes rapid (virtually instantaneous) fluctuations in position. At one instant the electronic charge might be distributed as shown for the left atom in Fig. 10-29*b*. At that instant the atom is not symmetrical, but has a momentary dipole moment. (It has more negative charge at the right than at the left.) This momentary charge-cloud distortion tends to repel the electronic charge of the neighboring atom shown. The momentary dipole of the first atom thus *induces* a similar momentary dipole in the second atom, and the result is an attraction. The attraction is very weak, however, and an instant later the charge distribution in

Figure 10-28
Dipole-dipole forces.

[1] Some chemists use the term *van der Waals forces* to mean *London forces* only, and so do not include dipole-dipole forces with this category.

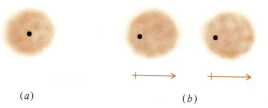

Figure 10-29
London (dispersion) forces between argon atoms. (*a*) One atom: "time-averaged" charge cloud. (*b*) Two atoms: attraction between momentary dipoles.

both molecules has changed. The net result of all of this is what we call the London force.

London forces are weak, short-range forces.[2] They depend, in part, upon the ease of distorting the charge cloud of a molecule, that is, the *polarizability* of the molecule. In general the larger a molecule is and the more electrons it has, the more polarizable it will be, and thus the larger the London forces can be. Molecular shape and other factors are also important, however.

Molecular solids tend to have relatively low melting points, because in these intermolecular forces are weak. They are typically soft, because the molecules can be easily moved around, and are nonconductors of electricity, as no charge carriers are present.

Covalent solids   In a covalent solid, sometimes called an *atomic solid,* the units at the lattice points are covalently bonded atoms. The atoms typically form a three-dimensional network, or "giant molecule," which extends outward to the physical limits of the crystal. One of the simplest examples of a covalent solid is silicon carbide (SiC, or Carborundum). (See Fig. 3-3.) Each silicon is tetrahedrally bonded to four carbons and each carbon similarly to four silicons. The result is a tightly interlocking, rigid structure which accounts for the great hardness of Carborundum and its consequent use as an abrasive. Because the atoms in covalent solids are locked in place by strong bonds, these substances tend to have high melting points. Lastly, no mobile ions or electrons exist in the structure; so these solids are typically electrical nonconductors.

Metallic solids   The units occupying the lattice points in a metallic solid are *positive ions*. In sodium metal, for example, we find Na^+ ions occupying the points of a body-centered cubic (bcc; see Fig. 10-12) lattice. Each Na^+ can be considered to have lost one electron and to have contributed it to an electron cloud which permeates the entire lattice. These electrons are not bound to any one atom or even to a pair, but are *delocalized* over the entire crystal. They are therefore called *free* electrons. They are often also referred to as an *electron gas,* and quantum mechanics treats them as waves which extend over the whole crystal of sodium. (See Fig. 10-30.) In a typical metal, of which sodium is a good example, there is a mutual attraction between the electron gas and the ions. This stabilizes the structure and at the same time allows it to be greatly distorted without falling apart. Thus sodium and some other metals are soft and easily deformable. Some metals are

---

[2] Whereas ordinary electrostatic (coulombic) forces between charged particles vary inversely as the square of the distance between them, London forces vary inversely as the *seventh* power of the distance. This means that doubling the distance reduces the force to $\frac{1}{128}$ of its original value.

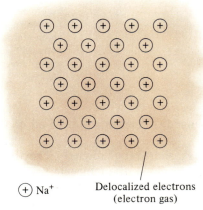

$\oplus$ Na$^+$ Delocalized electrons
(electron gas)

Figure 10-30
Sodium metal.

hard, however. In tungsten and chromium the metallic bonding described above is supplemented by covalent bonds between adjacent atoms. These tend to keep the structure locked in place, preventing easy deformation. The melting points of metals vary greatly because of the variability of bonding. Sodium, for example, has its melting point at 98°C, while tungsten melts at 3410°C.

The delocalized electrons in a metal give rise to electrical and thermal conductivity. Electrons can readily be added to the electron gas at one end of a piece of metal and simultaneously withdrawn from the other end. Passage of electrons from one end to the other constitutes electrical conduction. Metals are also good conductors of heat. Heating one end of a piece of metal increases the average kinetic energy of both the ions, which vibrate more violently, and the electrons, which move more rapidly throughout the metal. The freedom of the electrons to transfer energy rapidly from one end of the metal to the other is responsible for the high thermal conductivity. Lastly, the characteristic appearance of a metal, its *luster,* is also caused by the free electrons. The unbound electrons at the surface of the metal absorb and reradiate light which strikes the surface. This happens in such a way that a smooth metal surface reflects light completely at all angles, giving metals their peculiar luster.

The characteristics of the four principal kinds of solids are summarized in Table 10-1.

ADDED COMMENT

What is a *molecule?* The word usually is taken to mean a small group of atoms covalently bonded together to form a recognizable, larger particle. In how many of the four principal kinds of solids are molecules identifiable? In just one—the molecular solid.

Lattice energy   The strength with which the particles in a solid are bound together may be indicated by specifying the *lattice energy* (or *crystal energy*). This is defined as the amount of energy released when a crystal is formed from its component particles in the gaseous phase; it is generally expressed *per mole* of substance. It may also

Table 10-1
Bonding and properties in solids

	Ionic	Molecular	Covalent	Metallic
Units at lattice points	Positive and negative ions	Molecules	Atoms	Positive ions
Bonding force between units	Ionic bonds	Dipole-dipole forces London forces	Covalent	Attraction between electron gas and positive ions
Hardness	Fairly hard, brittle	Soft	Very hard	Soft to hard
Melting point	Fairly high	Low	Very high	Medium to high
Conductivity	Low	Low	Low	Good to excellent
Examples	NaCl K_2CO_3 $(NH_4)_2SO_4$ Na_3PO_4	CO_2 C_6H_6 H_2O CH_4	SiC SiO_2 (quartz) C (diamond) Al_2O_3	Na Ag Fe W

be thought of as the amount of energy necessary to separate a crystal into a collection of gaseous particles which are the same as those in the solid. (The exception is the conversion of a solid metal to a gas to produce *atoms* of the metal, not ions as are present at the lattice points of the solid.) The lattice energy depends upon the strength of the interactions between the particles in the solid, the number of these interactions per particle, and the geometry of the crystal structure. In Table 10-2 are shown some lattice energies for various kinds of solids. Also shown, for comparison, are the melting points of the solids. (In two cases sublimation points are given; see Sec. 11-5.)

Lattice energies may be determined in two different ways, by theoretical calculation, or by experimental measurement. The theoretical approach involves the calculation of the interaction (attractive and repulsive) energies among all the particles in a crystal. This works well only for crystals which are typically ionic, that is, in which there is little covalent or other contribution to the bonding.

The experimental method for the determination of a lattice energy is by means of a so-called Born-Haber calculation, which is based on measurements of heat

Table 10-2
Lattice energies and melting points

Solid type	Substance	Lattice energy, kJ mol^{-1}	Melting point. °C
Ionic	NaCl	788	801
	CaF_2	2590	1423
	CaO	3520	2614
Molecular	H_2	0.8	−259
	CH_4	9	−182
	CO_2	25	− 78 (sublimes)
Covalent	C	714	3600
	SiC	1235	2700 (sublimes)
	SiO_2	1865	1610
Metallic	Na	108	98
	Ag	285	962
	Cu	340	1083

changes for a set of related chemical processes. But before we look at the details
of such a calculation, let us see how heat changes are indicated for chemical
reactions in general.

When sodium chloride is synthesized by direct combination of sodium metal
with gaseous chlorine at 1 atm pressure, it is found experimentally that 411 kJ of
heat are evolved per mole of NaCl formed. From this information the following
thermochemical equation can be written:

$$Na(s) + \tfrac{1}{2}Cl_2(g) \longrightarrow NaCl(s) + 411 \text{ kJ}$$

A more conventional way of writing this is

$$Na(s) + \tfrac{1}{2}Cl_2(g) \longrightarrow NaCl(s) \qquad \Delta H = -411 \text{ kJ}$$

ΔH specifies the amount of heat *absorbed* during a process taking place at constant
pressure. (It is negative here because heat is *evolved*.) H stands for a quantity
called *enthalpy* (also sometimes called *heat content*), and the Greek capital letter Δ
(delta) always means "change in." ΔH means the change in the enthalpy of a
system in going from reactants to products and is defined as

$$\Delta H = H_{\text{products}} - H_{\text{reactants}}$$

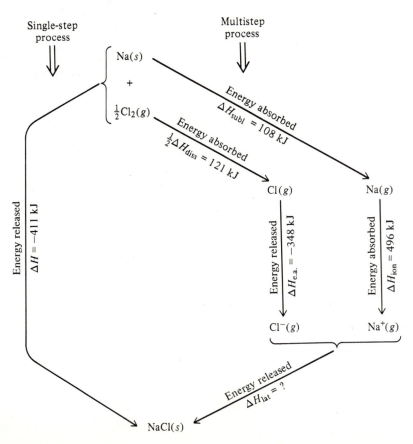

Figure 10-31
Born-Haber calculation.

The enthalpy of 1 mol of NaCl(s) is 411 kJ less than the sum of the enthalpies of 1 mol of Na(s) and 0.5 mol of $Cl_2(g)$, and the heat liberated when 1 mol of NaCl is formed from its elements at constant pressure exactly equals the decrease in enthalpy which occurs during the process.

Just what is enthalpy, H? For the time being we will hedge a bit and just say that it is a quantity which is closely related to energy. (Just how closely will be explained in Chap. 18.) For now it is enough to remember that ΔH gives the heat change for a process taking place at constant pressure. In an *exothermic* reaction heat is *liberated*, so $H_{products}$ must be *less than* $H_{reactants}$, and ΔH is consequently a *negative* number. On the other hand, in an *endothermic* reaction heat is absorbed, $H_{products}$ is greater than $H_{reactants}$, and so ΔH is a positive number.

A *Born-Haber calculation* of the lattice energy of a compound is based upon the principle that the amount of heat released (or absorbed) when the compound is prepared directly from its elements is equal to the net amount of heat released (or absorbed) when the compound is prepared by means of a series of steps. Figure 10-31 shows two methods of preparing NaCl from its elements. The pathway shown on the left of the diagram is the direct one:

$$Na(s) + \tfrac{1}{2}Cl_2(g) \longrightarrow NaCl(s) \qquad \Delta H = -411 \text{ kJ}$$

On the right of Fig. 10-31 is a multistep pathway leading from the same reactants to the same product. The first step is the vaporization, or *sublimation*, of a mole of Na atoms. These atoms are then *ionized* (step 2) to form a mole of Na^+ ions. In step 3 one-half mole of Cl_2 molecules are *dissociated* into Cl atoms, which then (step 4) *accept electrons* to form a mole of Cl^- ions. Finally (step 5) the Na^+ ions from step 2 combine with the Cl^- ions from step 4 to form a mole of crystalline NaCl. These steps and their corresponding enthalpy changes can be summarized as follows:

Step 1:	$Na(s) \longrightarrow Na(g)$	$\Delta H_{subl} = 108$ kJ (sublimation energy)
Step 2:	$Na(g) \longrightarrow Na^+(g) + e^-$	$\Delta H_{ion} = 496$ kJ (ionization energy)
Step 3:	$\tfrac{1}{2}Cl_2(g) \longrightarrow Cl(g)$	$\tfrac{1}{2}\Delta H_{diss} = 121$ kJ ($\tfrac{1}{2}$ dissociation energy)
Step 4:	$e^- + Cl(g) \longrightarrow Cl^-(g)$	$\Delta H_{e.a.} = -348$ kJ (electron affinity)
Step 5:	$Na^+(g) + Cl^-(g) \longrightarrow NaCl(s)$	$\Delta H_{lat} = ?$ (lattice energy)
Overall:	$Na(s) + \tfrac{1}{2}Cl_2(g) \longrightarrow NaCl(s)$	$\Delta H = -411$ kJ

The values of ΔH for steps 1 through 4 are available from experimental measurements, and ΔH for the overall process (the direct reaction) is also known. If we write

$$\Delta H = \Delta H_{subl} + \Delta H_{ion} + \tfrac{1}{2}\Delta H_{diss} + \Delta H_{e.a.} + \Delta H_{lat}$$

then to find the lattice energy, we need only solve for ΔH_{lat}:

$$\Delta H_{lat} = \Delta H - \Delta H_{subl} - \Delta H_{ion} - \tfrac{1}{2}\Delta H_{diss} - \Delta H_{e.a.}$$
$$= -411 - 108 - 496 - 121 - (-348) = -788 \text{ kJ}$$

The sign on ΔH_{lat} is negative, which tells us that in the formation of a mole of NaCl from its ions, energy is *released*, which is certainly what we would expect. Since lattice energies are usually expressed as positive numbers, we can say that the lattice energy of NaCl is 788 kJ mol^{-1}.[3]

10-6
Liquids

A liquid can be formed by either (1) melting a solid or (2) condensing (liquefying) a gas. In the former process energy is absorbed by the solid, and this energy is used to overcome the attractive forces between the particles in the crystal lattice. In the latter process energy is removed as attraction forces are established between the molecules in the liquid. A liquid is much like a solid in that its particles are close together, but it is at the same time much like a gas in that the molecules are not regularly ordered. What are the properties of liquids and how can they be explained in terms of this close-together-but-disordered model?

General properties
of liquids

Liquids flow readily. Because the molecules in a liquid are not locked into a fixed pattern as they are in a solid, they can slip by each other easily. Liquids thus adopt the shape of the container in which they are placed. Intermolecular forces are greater in liquids than in gases, however. This causes the *viscosity* (internal resistance to flow) of a liquid to be considerably larger than that of a gas.

Liquids diffuse more slowly than gases. Gases diffuse comparatively rapidly, because most of their volume is empty space, and the molecules are impeded only by an occasional collision. In a liquid each molecule is hemmed in closely by many others and so cannot make much progress in any direction. An aqueous solution of a colored dye placed in contact with pure water diffuses very slowly into the water. By contrast the molecules of a (vaporized) perfume can be smelled at one side of a room only a few minutes after their source enters the other side. (Solids, of course, diffuse even more slowly than liquids.)

Liquids are much less compressible than gases. Enormous pressures are needed to reduce the volume of a liquid by any significant degree. In a liquid there is little free space, and the repulsions between electron clouds of neighboring molecules strongly resist attempts to push the molecules more closely together.

A liquid maintains a characteristic volume. Although the shape of a liquid depends upon that of its container, a liquid does not expand to fill the entire container as does a gas. Intermolecular attractive forces in a liquid prevent the molecules from flying apart.

Liquids exhibit surface tension. Surface tension is the *tendency shown by a liquid to minimize its surface area.* It arises because molecules at the surface are subject to attractions from within the liquid, but not from the opposite direction. This imbalance of forces gives rise to the surface tension. The fact that a needle or pin can be made to "float" on water depends upon the comparatively high surface tension of water, which, in turn, is a consequence of the strong intermolecular forces in this liquid.

---

[3] In the above calculation the term $\frac{1}{2}\Delta H_{diss}$ represents one-half the energy needed to dissociate a mole of Cl_2 into atoms. It is thus the energy needed to form a mole of Cl atoms from Cl_2 molecules.

Evaporation and
vapor pressure

An important property of a liquid is its tendency to evaporate. In a liquid, as in a solid or gas, the molecules possess a distribution of speeds and hence of kinetic energies. At any instant some molecules are moving very rapidly while others are practically motionless. Figure 10-32 shows the distribution of molecular kinetic energies in a typical liquid. The black curve depicts the distribution at some arbitrary temperature. It shows the fraction of the total number of molecules which have any given kinetic energy at that temperature. Notice that the graph (black line) shows that many more molecules have low kinetic energies than high. This tells us that the temperature is comparatively low. At a higher temperature the distribution shifts to that shown by the colored curve. Here we see that there are fewer slow molecules and more fast ones.

When a liquid evaporates, molecules at the surface of the liquid fly off and leave if they are moving fast enough to overcome the forces which attract them to other nearby molecules. At any instant some fraction of the molecules at the surface have enough kinetic energy to escape from their neighbors. These faster-moving, or "hot," molecules are responsible for evaporation.

A glass of hot water evaporates more rapidly than a glass of cold water. Why? The vertical dashed line in Fig. 10-32 shows the minimum kinetic energy E_{min} which a molecule must possess in order to escape into the gas phase. Note that the fraction of the total number of molecules which have this minimum energy (or greater) is larger at the higher temperature. As the temperature of a liquid is increased, more of its molecules have enough kinetic energy to escape from the attractive forces of their neighbors. Thus the rate of evaporation of a liquid increases as the temperature increases.

The phenomenon of *cooling by evaporation* is one which is familiar to most of us. The sudden chill we feel upon stepping from a shower or swimming pool into a gentle breeze is due, at least in part, to the evaporative cooling of the water on our bodies. Remember that it is the rapidly moving molecules which leave during the evaporation process. The slowly moving molecules are left behind, and so as evaporation occurs, the *average* kinetic energy of the molecules remaining gradually decreases. We sense, or measure, this drop in the average kinetic energy

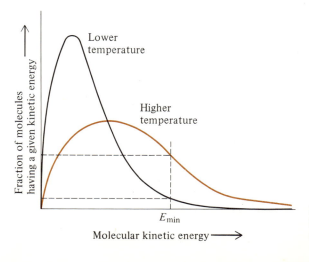

Figure 10-32
Distribution of molecular kinetic energies in a liquid.

accompanying the loss of the higher-kinetic-energy molecules as a decrease in temperature. Evaporative coolers for homes and other buildings, and even automobiles, make use of this effect to cool circulating air; they are more energy efficient than conventional air conditioners.

EQUILIBRIUM VAPOR PRESSURE. In an experiment we place a jar of water under a bell jar fitted with a pressure gauge, as shown in Fig. 10-33. Initially the jar is completely filled with water (no air) and stoppered, and the space under the bell jar is evacuated. Now we remove the stopper by remote control. The water immediately starts to evaporate, and the presence of water molecules in the gas phase causes the gauge to indicate a pressure, as shown in Fig. 10-33a. As time elapses, the pressure in the gas phase rises and approaches a limiting value, as shown in Fig. 10-34. Why does the pressure stop rising? Once there are a few molecules in the gas phase some of these can plunge back into the liquid and be trapped there by intermolecular forces. The rate of this condensation increases as the pressure in the gas phase increases. Eventually the rate at which molecules return to the liquid becomes equal to the rate at which they leave; that is, the rate of condensation equals the rate of evaporation. See Fig. 10-33b. Since there is no *net* increase in number of molecules in the gas phase after this time, the pressure no longer increases.

Figure 10-33b shows a state of *dynamic equilibrium*. This is a condition in which two opposing processes occur at the same rate. In the system shown, once equilibrium has been established, the level of the liquid stops dropping and the pressure of the gas stops rising. One might think that at this point nothing at all is happening inside the bell jar, but this is not so. Both evaporation and condensation are occurring very rapidly. It is only because they occur at the same rate, thus exactly compensating for each other, that no *net* change occurs, and nothing can be observed. We can write an equation representing the liquid water–gaseous water equilibrium:

$$H_2O(l) \rightleftharpoons H_2O(g)$$

Here the double arrow tells us that the evaporation and condensation processes are occurring at the same rate. The equation read from left to right is often said to represent the *forward reaction* and from right to left, the *reverse reaction*. Remember, however, that the terms *forward* and *reverse* have meaning only with respect to the equation as we have written it.

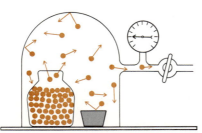

Rate of evaporation >
rate of condensation

(a)

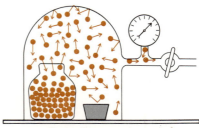

Rate of evaporation =
rate of condensation

(b)

Figure 10-33
Equilibrium vapor pressure. (a) Immediately after removal of stopper. (b) At equilibrium.

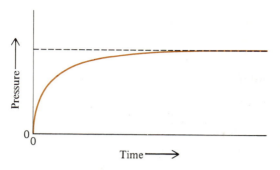

Figure 10-34
Approach to gas-liquid equilibrium.

The pressure exerted by a gas which is in equilibrium with its liquid is called the *equilibrium vapor pressure* of the liquid. Often it is called, more simply, the *vapor pressure* of the liquid, but remember that equilibrium is always implied by this term.

Vapor pressure depends upon the tendency which molecules have to escape from the liquid phase. When intermolecular attractive forces are low, molecules tend to escape readily, and the equilibrium vapor pressure is high. Thus measurement of vapor pressure provides an indication of the size of such forces.

Vapor pressure increases with increasing temperature. Why? As the temperature increases, the fraction of the molecules which can escape from the liquid phase increases, so that at equilibrium the pressure of the gas phase is higher. Figure 10-35 shows plots of the temperature variation of the vapor pressure of three different liquids: ethyl ether, the ether used for anesthesia; ethanol, ethyl alcohol; and water. Notice that the vapor pressure of each liquid increases with temperature and that the three curves have roughly the same shape. At any given temperature the vapor pressure of ethyl ether is highest, that of ethanol inter-

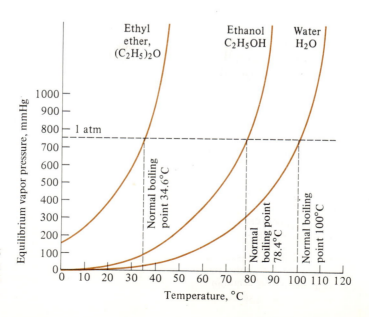

Figure 10-35
Variation of equilibrium vapor pressure with temperature.

mediate, and that of water lowest. This indicates that intermolecular attractive forces in the liquid increase in the order: ethyl ether $<$ ethanol $<$ water.

Boiling

We have seen that the tendency for molecules to escape from the liquid to the gas phase is measured by the vapor pressure. When the temperature of a liquid is raised, the escaping tendency of its molecules may be increased to the point that boiling occurs. Boiling is the formation of bubbles of vapor (gas) within the body of the liquid.[4] These bubbles can form when the vapor pressure becomes equal to the pressure on the liquid, usually atmospheric pressure. Thus the *boiling point* of a liquid is that temperature at which the vapor pressure of the liquid is equal to atmospheric pressure. Figure 10-35 can be used to find the boiling point of each of the three liquids at any pressure. As the (atmospheric) pressure on the liquid increases, it is necessary to heat the liquid to a higher temperature in order to bring its vapor pressure up to atmospheric pressure. Boiling points specified for most liquids are *normal* boiling points. The *normal boiling point* is the *temperature at which the vapor pressure of the liquid equals one atmosphere* (standard pressure). On Fig. 10-35 a broken horizontal line has been drawn at $P = 760$ mmHg (1 atm). The temperature at which the vapor-pressure curve for each liquid crosses this line is the normal boiling point of that liquid.

Clausius-Clapeyron equation

Figure 10-36*a* is a plot of the vapor pressure of water made from data taken at six different temperatures. The graph was made in the usual way: vapor pressure was plotted on the vertical axis and temperature on the horizontal. Now look at Fig. 10-36*b*. Here we have plotted the *logarithm* of the vapor pressure on the vertical axis and the *reciprocal* (or *inverse*) of the *absolute* (kelvin) temperature on the horizontal. We are thus plotting log P against $1/T$. We now get a straight line! What can be deduced from this? A straight line can be described by an equation of the form

$$y = mx + b$$

where y is the vertical coordinate (the ordinate) and x the horizontal (the abscissa). Our straight line evidently represents the equation

$$\log P = m\frac{1}{T} + b$$

By methods which do not concern us yet, it can be shown that the *slope* of the line is a function of the *molar heat of vaporization* of the liquid.

$$m = \frac{-\Delta H_{vap}}{2.303R}$$

Here ΔH_{vap} represents the amount of heat necessary to vaporize one mole of liquid to form an ideal gas at one atmosphere pressure, and R is the ideal gas constant.

---

[4]The bubbles of gas formed during boiling quickly rise to the surface and rapidly grow as they do. They should not be confused with small bubbles of previously dissolved air which may form well below the boiling point of a liquid.

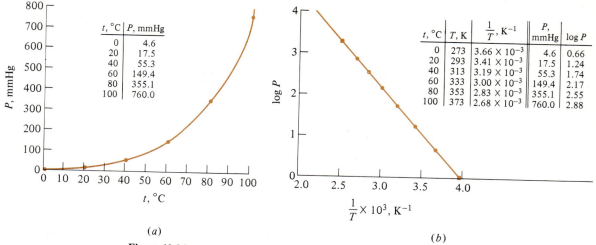

Figure 10-36
Vapor pressure of water: two plots. (a) P versus T. (b) log P versus 1/T.

ΔH is a positive number and represents the enthalpy change for the process

$$H_2O(l) \longrightarrow H_2O(g) \qquad \Delta H_{vap} = H_{gas} - H_{liquid}$$

Now look again at the above relationship between the slope of the straight line of Fig. 10-36b and ΔH_{vap}. If we rearrange it a bit, we get

$$\Delta H_{vap} = -2.303Rm$$

Measuring the slope of the line and then multiplying it by $-2.303R$ should give us ΔH_{vap} for water. The slope is found by selecting two points on the line, generally not experimental points, subtracting their $\log P$ coordinates and dividing this difference by the difference between their $1/T$ coordinates. Using the points marked by squares on the graph we have

$$m = \text{slope} = \frac{(\log P)_1 - (\log P)_2}{(1/T)_1 - (1/T)_2}$$

$$= \frac{3.25 - 0}{2.50 \times 10^{-3}\,\text{K}^{-1} - 3.95 \times 10^{-3}\,\text{K}^{-1}} = -2.24 \times 10^3\,\text{K}$$

Now we multiply this slope by $-2.303R$ in order to compute ΔH_{vap}, but we must be careful to use R in the appropriate units. If we wish ΔH_{vap} to come out in joules per mole, for example, R should be in corresponding units, that is, $R = 8.31\,\text{J K}^{-1}\,\text{mol}^{-1}$. (See App. B.) Thus,

$$\Delta H_{vap} = -2.303Rm$$

$$= -2.303(8.31\,\text{J K}^{-1}\,\text{mol}^{-1})(-2.24 \times 10^3\,\text{K})$$

$$= 4.29 \times 10^4\,\text{J mol}^{-1} = 42.9\,\text{kJ mol}^{-1}$$

The relationship

$$\log P = \frac{-\Delta H_{vap}}{2.303R} \frac{1}{T} + b$$

is one form of the *Clausius-Clapeyron equation* and is useful for finding the heat of vaporization of a liquid from vapor-pressure data, as we have just done. Another form of this equation can be derived if we write the equation twice:

$$\log P_1 = \frac{-\Delta H_{vap}}{2.303R}\left(\frac{1}{T_1}\right) + b$$

$$\log P_2 = \frac{-\Delta H_{vap}}{2.303R}\left(\frac{1}{T_2}\right) + b$$

Here P_1 is the vapor pressure at temperature T_1, and P_2 is the vapor pressure at T_2. Now we subtract the second equation from the first:

$$\log P_1 - \log P_2 = \frac{-\Delta H_{vap}}{2.303R}\left(\frac{1}{T_1} - \frac{1}{T_2}\right)$$

or

$$\log \frac{P_1}{P_2} = \frac{-\Delta H_{vap}}{2.303R}\left(\frac{1}{T_1} - \frac{1}{T_2}\right)$$

This is a handy form of the Clausius-Clapeyron equation. It can be used to

1 Find ΔH_{vap} when the vapor pressure is known at two different temperatures.
2 Find the vapor pressure at a given temperature if ΔH_{vap} is known and if the vapor pressure is known at some other temperature.
3 Find the temperature at which a liquid has a given vapor pressure if ΔH_{vap} and the vapor pressure at one temperature are known.

● **EXAMPLE 10-3**

Problem   The vapor pressure of ethanol is 43.9 mmHg at 20°C and 352.7 mmHg at 60°C. What is the molar heat of vaporization of ethanol?

Solution   Let us call the higher temperature T_2 and the lower one T_1. Then, converting to absolute temperature we have

$$T_1 = 20 + 273 = 293\ K$$

$$T_2 = 60 + 273 = 333\ K$$

$$P_1 = 43.9\ mmHg$$

$$P_2 = 352.7\ mmHg$$

$$R = 8.31\ J\ K^{-1}\ mol^{-1}$$

Before we substitute in the Clausius-Clapeyron equation, and in order to make the algebra a little simpler, we rearrange the equation

$$\log \frac{P_1}{P_2} = \frac{-\Delta H_{vap}}{2.303R}\left(\frac{1}{T_1} - \frac{1}{T_2}\right)$$

$$\frac{-\Delta H_{vap}}{2.303R} = \frac{\log (P_1/P_2)}{(1/T_1 - 1/T_2)}$$

$$\Delta H_{vap} = \frac{2.303R \log (P_1/P_2)}{(1/T_2 - 1/T_1)}$$

$$= \frac{2.303(8.31\ J\ K^{-1}\ mol^{-1}) \log (43.9\ mmHg/352.7\ mmHg)}{(1/333\ K - 1/293\ K)}$$

$$= 4.22 \times 10^4\ J\ mol^{-1} = 42.2\ kJ\ mol^{-1}\ ●$$

● **EXAMPLE 10-4**

Problem   Using the heat of vaporization from Example 10-3, calculate the vapor pressure of ethanol at 0°C.

Solution   We will substitute into the Clausius-Clapeyron equation using the following quantities:

$$T_1 = 20 + 273 = 293 \text{ K}$$
$$T_2 = 0 + 273 = 273 \text{ K}$$
$$\Delta H_{vap} = 4.22 \times 10^4 \text{ J mol}^{-1}$$
$$R = 8.31 \text{ J K}^{-1} \text{ mol}^{-1}$$
$$P_1 = 43.9 \text{ mmHg} \quad \text{(Example 10-3)}$$
$$P_2 = ?$$

$$\log \frac{P_1}{P_2} = \frac{-\Delta H_{vap}}{2.303R}\left(\frac{1}{T_1} - \frac{1}{T_2}\right)$$
$$= \frac{-4.22 \times 10^4 \text{ J mol}^{-1}}{2.303(8.31 \text{ J K}^{-1} \text{ mol}^{-1})}\left(\frac{1}{293 \text{ K}} - \frac{1}{273 \text{ K}}\right) = 0.551$$
$$\frac{P_1}{P_2} = 3.56$$
$$P_2 = \frac{P_1}{3.56} = \frac{43.9 \text{ mmHg}}{3.56} = 12.3 \text{ mmHg} \quad ●$$

● **EXAMPLE 10-5**

Problem   Using data from Example 10-3, determine the normal boiling point of ethanol.

Solution   Since the normal boiling point is the temperature at which the vapor pressure equals 760 mmHg, we can write

$$P_1 = 352.7 \text{ mmHg}$$
$$P_2 = 760 \text{ mmHg}$$
$$T_1 = 60 + 273 = 333 \text{ K}$$
$$T_2 = ?$$
$$\Delta H_{vap} = 4.22 \times 10^4 \text{ J mol}^{-1}$$
$$R = 8.31 \text{ J K}^{-1} \text{ mol}^{-1}$$

$$\log \frac{P_1}{P_2} = \frac{-\Delta H_{vap}}{2.303R}\left(\frac{1}{T_1} - \frac{1}{T_2}\right)$$
$$\frac{1}{T_1} - \frac{1}{T_2} = \frac{\log(P_1/P_2)}{-\Delta H_{vap}/2.303R}$$
$$= \frac{\log(352.7 \text{ mmHg}/760 \text{ mmHg})}{-4.22 \times 10^4 \text{ J mol}^{-1}/2.303(8.31 \text{ J K}^{-1} \text{ mol}^{-1})} = 1.51 \times 10^{-4} \text{ K}^{-1}$$
$$\frac{1}{T_2} = \frac{1}{T_1} - 1.51 \times 10^{-4} \text{ K}^{-1}$$
$$= \frac{1}{333 \text{ K}} - 1.51 \times 10^{-4} \text{ K}^{-1}$$
$$T_2 = 351 \text{ K} \quad \text{or} \quad 351 - 273 = 78°C \quad ●$$

Freezing   When energy is removed from a liquid, the temperature drops as the average kinetic energy of the molecules decreases. Eventually, unless supercooling (Sec. 11-2) occurs, the temperature stops decreasing even though energy continues to be removed from the liquid. Simultaneously, the first few particles of crystalline solid are formed; the liquid has begun to freeze.

The *freezing point* of a liquid is the temperature at which the solid and liquid phases of a substance can be in equilibrium with each other:

$$\text{Liquid} \rightleftharpoons \text{solid}$$

As written above, the forward reaction occurs when heat is removed from the liquid and is the process known as *freezing*. The reverse reaction occurs when heat is added to the solid and is the process known as *melting*. Under true equilibrium conditions the rate of freezing is equal to the rate of melting, and heat is neither absorbed by nor liberated from the entire system. Since the forward and reverse reactions occur at the same temperature, the freezing and melting points of a pure substance are identical.

Just as the boiling point varies with pressure, so also does the freezing point. The change is much less pronounced, however. The freezing point of water, for example, changes less than 0.01°C with a pressure change of 1 atm, a relatively large pressure change. The freezing (or melting) point of a substance at 1 atm pressure is known as the *normal* freezing (or melting) point of the substance.

The amount of heat necessary to melt 1 mol of solid is called the *molar heat* (or *enthalpy*) *of fusion*, ΔH_{fus}. It is the difference between the enthalpy of the liquid and that of the solid. Since it takes 6.02 kJ of heat to melt 1 mol of ice, we write

$$H_2O(s) \longrightarrow H_2O(l) \qquad \Delta H_{fus} = 6.02 \text{ kJ mol}^{-1}$$

For the reverse process, heat must be removed. Thus,

$$H_2O(l) \longrightarrow H_2O(s) \qquad \Delta H_{cryst} = -6.02 \text{ kJ mol}^{-1}$$

where ΔH_{cryst} is the *molar heat of crystallization* (or *solidification*). The minus sign indicates that heat is removed and that the enthalpy of the solid (ice) is less than that of the liquid.

SUMMARY

In this chapter we have talked about condensed phases of matter: solids and liquids. *Solids* are relatively hard, unyielding substances with a tendency to form, or be found as, *crystals*. The properties of solids are best understood in terms of their structures. These are revealed by the powerful technique of *x-ray diffraction*, in which the pattern of x-ray beams diffracted from a crystal is used to determine the spacings between the particles in the *crystal lattice*.

Crystal lattices can exhibit a number of different arrangements or geometries. The *unit cell*, a small portion of a crystal lattice is chosen to show the geo-

metrical structure of the entire lattice. Many lattices are *close-packed lattices*, either cubic close packed (ccp) or hexagonal close packed (hcp), in which a maximum of the space in the crystal is occupied by particles. Other lattices, while not actually close packed, can be described in terms of close-packed lattices in which one kind of particle occupies the lattice sites while other kinds of particles occupy the tetrahedral and/or octahedral holes between the lattice sites.

The properties of solids depend upon the *bonding forces* between the particles at the lattice points of the crystal. These forces can be *ionic bonds, covalent bonds, metallic bonds,* or *van der Waals forces.* The last are usually weak and include *dipole-dipole forces* and *London (dispersion) forces.*

The stability of a particular solid may be indicated by its hardness, its melting point, and, best of all, its *lattice energy.* Lattice energies may be determined by means of *Born-Haber calculations.*

Liquids exhibit greater tendencies to *flow* and *diffuse* than do solids. In a liquid the intermolecular forces are weaker than in a solid. These forces are stronger than in a gas, however, because of the closeness of the molecules in the liquid state. Liquids show a tendency to *evaporate* and to become cooler when they do so.

When a liquid is allowed to evaporate inside a sealed container, a state of *equilibrium* is established in which the rate of evaporation is equal to the rate of condensation. The pressure of the gas in equilibrium with the liquid is called the liquid's *(equilibrium) vapor pressure.* It changes with temperature, and this change can be described by the *Clausius-Clapeyron equation.* The *boiling point* of a liquid is the temperature at which its vapor pressure becomes equal to the external, or atmospheric, pressure. The *freezing point,* or *melting point,* of a liquid is the temperature at which solid and liquid are at equilibrium with each other.

KEY TERMS

Amorphous solid (Sec. 10-1)
Born-Haber calculation (Sec. 10-5)
Bragg relation (Sec. 10-2)
Centered cell (Sec. 10-3)
Clausius-Clapeyron equation (Sec. 10-6)
Close packing (Sec. 10-4)
Covalent solid (Sec. 10-5)
Crystal lattice (Sec. 10-3)
Crystalline solid (Sec. 10-1)
Delocalized electrons (Sec. 10-5)
Dipole-dipole forces (Sec. 10-5)
Dispersion forces (Sec. 10-5)
Electron gas (Sec. 10-5)
Enthalpy (Sec. 10-5)
Equilibrium (Sec. 10-6)
Ionic solid (Sec. 10-5)
Lattice energy (Sec. 10-5)
Laue pattern (Sec. 10-2)
London forces (Sec. 10-5)

Metallic solid (Sec. 10-5)
Molecular solid (Sec. 10-5)
Molar heat of fusion (Sec. 10-6)
Molar heat of vaporization (Sec. 10-6)
Normal boiling point (Sec. 10-6)
Normal freezing point (Sec. 10-6)
Octahedral hole (Sec. 10-4)
Packing efficiency (Sec. 10-4)
Powder pattern (Sec. 10-2)
Primitive cell (Sec. 10-3)
Space lattice (Sec. 10-3)
Surface tension (Sec. 10-6)
Tetrahedral hole (Sec. 10-4)
Thermochemical equation (Sec. 10-5)
Unit cell (Sec. 10-3)
Van der Waals forces (Sec. 10-5)
Vapor pressure (Sec. 10-6)
Viscosity (Sec. 10-6)

QUESTIONS AND PROBLEMS

Solids: general considerations

10-1 Distinguish between a crystalline solid and an amorphous solid. In what way are their properties similar? In what ways are they different?

10-2 Account for each of the following in terms of structure: **(a)** diffusion in solids is much slower than in gases **(b)** solids are much less compressible than gases, but only a little less compressible than liquids **(c)** crystals have plane faces **(d)** liquids flow more rapidly than solids.

10-3 Account for the fact that some solids are very soft, while others are extremely hard.
10-4 How would you design an experiment to measure **(a)** the compressibility and **(b)** the coefficient of thermal expansion of a solid?
10-5 How could you show that a smooth, shiny piece of steel is actually crystalline in nature?
10-6 Draw a diagram of a cube. **(a)** If you connect all *adjacent face centers* by straight lines, what geometric figure have you drawn? **(b)** With a new drawing of a

cube, identify *four* corners, none of which are adjacent to each other. Now connect these corners with straight lines. What geometric figure have you drawn this time?

X-ray diffraction

10-7 In what ways is the diffraction of x rays by a crystal similar to diffraction of light by a diffraction grating? In what ways are they different?

10-8 The first-order diffraction of x rays from some crystal planes occurs at an angle of 11.8° from the planes. If the planes are 0.281 nm apart, what is the wavelength of the x rays?

10-9 X rays of wavelength 0.150 nm are used in a diffraction experiment in which first-order diffraction occurs at 25.0°. How far apart are the diffracting planes?

10-10 In an x-ray diffraction experiment a first-order diffracted ray is observed at $\theta = 48.6°$ when x rays of $\lambda = 0.229$ nm are used. How far apart are the diffracting layers of atoms?

10-11 A set of planes of atoms in a certain crystal gives rise to diffraction at 54.5° when radiation of $\lambda = 0.166$ nm is used. At what angle would similar diffraction occur if the x-ray wavelength were 0.194 nm?

Unit cells

10-12 In a simple cubic lattice of atoms, how many nearest neighbors does each atom have? How many in a body-centered cubic lattice? How many in a face-centered cubic lattice? How many in a hexagonal close-packed lattice?

10-13 Can a cube consisting of Na^+ and Cl^- ions at alternate corners serve as a satisfactory unit cell for the sodium chloride lattice? Explain.

10-14 What is the *net* number of lattice points associated with one unit cell in each of the following lattices: **(a)** simple cubic **(b)** body-centered cubic **(c)** face-centered cubic **(d)** end-centered orthorhombic **(e)** primitive monoclinic.

10-15 Consider a unit cell of NaCl as is drawn in Fig. 10-17a. It has a total of 14 Cl^- ions and 13 Na^+ ions. Should the formula of sodium chloride be $Na_{13}Cl_{14}$? Explain.

10-16 The unit cell of cesium chloride may be chosen as a cube with a chloride ion at each corner and a cesium ion at the body center. This structure is *not* correctly classified as body-centered cubic. Explain. How is the CsCl unit cell correctly classified?

10-17 Copper metal crystallizes in a face-centered cubic lattice. If the radius of a copper atom (actually, ion) in this structure is 0.128 nm, what is the edge length of the unit cell?

10-18 The crystal lattice of sodium metal is body-centered cubic. If the radius of a sodium atom (ion) is 0.185 nm, what is the edge length of the unit cell?

10-19 The unit cell of aluminum metal is cubic with an edge length of 0.405 nm. Determine the type of unit cell (simple, body-centered, or face-centered) if the density of aluminum is 2.70 g cm^{-3}.

10-20 The density of chromium metal is 7.20 g cm^{-3}. If the unit cell is cubic with an edge length of 0.289 nm, determine the type of unit cell (simple, body centered, or face centered).

10-21 Nickel crystallizes in a face-centered cubic lattice. If the radius of a nickel atom (ion) is 0.124 nm, what is the density of nickel?

10-22 The density of fluorite, CaF_2, is 3.180 g cm^{-3}. What is the edge length of its unit cell? (See Fig. 10-25.)

Close packing

10-23 In a three-dimensional, close-packed array of hard spheres, each sphere contacts how many adjacent spheres **(a)** in its own layer and **(b)** in each of the layers above and below?

10-24 What kinds of molecular solids would be expected to crystallize in close-packed structures?

10-25 If the radius of the chloride ion is 0.181 nm, how large a cation will fit in an octahedral hole of a fcc lattice of Cl^- ions?

10-26 In a close-packed array of iodide ions, radius 0.216 nm, how large a cation will fit in each of the tetrahedral holes?

10-27 Ferric oxide crystallizes in a hexagonal close-packed array of oxide (O^{2-}) ions with two out of every three octahedral holes occupied by iron ions. What is the formula for ferric oxide?

10-28 In cadmium iodide every other octahedral hole in an hcp array of iodide (I^-) ions is occupied by a cadmium ion. What is the formula for cadmium iodide?

Bonding in solids

10-29 Covalent solids tend to be hard and to have high melting points, while molecular solids tend to be soft and low-melting. Explain these differences in properties, given that covalent bonding occurs in both kinds of solids.

10-30 Why are ionic solids often quite hard? Why are they typically brittle?

10-31 Account for the fact that although iron and potassium are both metallic, iron is much harder than potassium.

10-32 NaCl and MgS both crystallize in the rock-salt structure. Which of these compounds would you expect to have the higher melting point? Explain.

10-33 The normal melting point of hydrogen is $-259°C$, while that of oxygen is $-219°C$. Account for the difference.

10-34 What physical properties are commonly exhibited by metals? What structural characteristics are believed to account for these properties?

10-35 What kind of intermolecular force must be overcome in order to melt each of the following: CH_4, $CaCl_2$, SiC, CO_2, Ag, Ar, NH_3.

10-36 Account for the fact that the melting point of NaCl is higher than that of NaBr.

Lattice energy

10-37 In ionic compounds how is lattice energy affected by **(a)** the radii of the ions, **(b)** the ionic charges, and **(c)** the crystal structure?

10-38 Although energy is released when an electron adds to an O atom ($\Delta H_{e.a.} = -142$ kJ mol^{-1}), energy is absorbed when the resulting O^- ion adds another electron (for O^- $\Delta H_{e.a.} = +780$ kJ mol^{-1}). In view of the rather large amount of energy needed to form the oxide ion (O^{2-}) how can you account for the fact that NaO is nonexistent, while Na_2O is a stable, well-behaved compound?

10-39 Consider the hypothetical solid compound Na_2Cl (containing the Cl^{2-} ion). Would you expect its lattice energy to be larger than that of NaCl? Explain. Why is it that Na_2Cl is not found?

10-40 Calculate the lattice energy of potassium fluoride, KF, from the following data: ΔH of formation of KF is -563 kJ mol^{-1}; ΔH_{subl} of potassium is 89 kJ mol^{-1}; ΔH_{diss} of F_2 is 158 kJ mol^{-1}; ΔH_{ion} of K is 419 kJ mol^{-1}; and $\Delta H_{e.a.}$ of F is -333 kJ mol^{-1}.

10-41 Account for the fact that none of the following compounds can be found on the shelves of a chemical stockroom: $NaCl_2$, CaCl, Ca_2O, Ca_2Cl.

10-42 What is the relationship between the lattice energy of a molecular solid and its sublimation energy?

10-43 The octet rule leads us to predict that calcium oxide is $Ca^{2+}O^{2-}$, and not Ca^+O^- or $Ca^{3+}O^{3-}$. Why are the second and third formulations unreasonable?

What simple experiment could be employed to prove that they are wrong?

10-44 The same crystal structure is shared by NaCl, MgO, and TiC (titanium carbide). Predict the relative hardness and melting points of these compounds.

Liquids

10-45 In what ways are the properties of liquids similar to those of gases? In what ways are they different?

10-46 Liquids are much less compressible than gases but are in general only a little more compressible than solids. Explain.

10-47 Would you expect the surface tension of a liquid to increase or decrease with increasing temperature? Explain.

10-48 Account for the fact that alcohol produces a greater cooling effect on the skin than does water.

10-49 When alcohol is poured into a beaker at room temperature, the rate of evaporation decreases with time. Explain.

10-50 Predict the ranking of the following compounds in order of increasing normal boiling point: CH_3OH, C_2H_5OH, C_3H_7OH.

10-51 Considering that the carbon tetrachloride molecule is nonpolar, account for the fact that CCl_4 has a higher normal boiling point than chloroform, $CHCl_3$.

10-52 Suppose you had a liquid whose vapor pressure increased with increasing temperature, went through a maximum, and then decreased. What would this tell you about the molar heat of vaporization of the liquid? Is this reasonable? Explain.

10-53 The molar heat of vaporization of water at temperatures near its normal boiling point is 40.7 kJ mol^{-1}. If the vapor pressure of water is 526 mmHg at 90°C, predict its value at 110°C.

10-54 If the normal boiling point of carbon tetrachloride is 77°C and its molar heat of vaporization is 29.8 kJ mol^{-1}, what is the vapor pressure of this compound at 25°C?

10-55 The molar heat of vaporization of benzene is 31.2 kJ mol^{-1}. If its normal boiling point is 80°C, at what temperature will benzene boil if the pressure is 455 mmHg?

10-56 Atmospheric pressure on top of Mt. Everest is sometimes as low as 250 mmHg. What is the boiling point of water at this location? (For H_2O, $\Delta H_{vap} = 40.7$ kJ mol^{-1}.)

10-57 The vapor pressure of chloroform is 10 mmHg at $-30°C$ and 100 mmHg at 10°C. Calculate the molar heat of vaporization of this compound.

10-58 Estimate the molar heat of vaporization of a compound whose vapor pressure doubles when the temperature is raised from 120 to 130°C.

General

10-59 Freezing points are not as dependent upon the pressure as are boiling points. Although we discuss this in Chap. 11, can you suggest a reason at this time for the difference in behavior?

10-60 For a close-packed array of spheres, radius r, calculate the distance between the two planes which pass through the centers of adjacent layers of atoms.

10-61 Tungsten has an atomic radius of 0.137 nm and crystallizes in a body-centered cubic lattice. If the density of tungsten is 19.35 g cm^{-3}, calculate a value for Avogadro's number.

10-62 A certain molecular compound crystallizes in a tetragonal lattice in which the unit cell is primitive and has one molecule of the compound at each corner. If the density of the compound is 4.32 g cm^{-3} and the unit cell dimensions are 0.234 nm $\times$ 0.234 nm $\times$ 0.329 nm, what is the molecular weight of the compound?

10-63 A certain ionic compound AB_2 crystallizes in the fluorite structure. If the edge length of a unit cell is 0.619 nm and the density of AB_2 is 9.17 g cm^{-3}, what is its formula weight?

10-64 When the logarithm of the vapor pressure of a certain liquid is plotted against the reciprocal of the temperature over a wide range, the resulting graph is a curve, rather than a straight line. Explain.

10-65 The vapor pressure of compound A is lower than that of compound B at 25°C, but higher at 50°C. Qualitatively compare the heats of vaporization of A and B and justify your answer.

11
REAL MATTER AND CHANGES OF STATE

TO THE STUDENT

In Chap. 4 we discussed in some detail the properties of an ideal gas. But can an ideal gas really exist? No, not really! Are Boyle's and Charles' laws followed *exactly* by any gas? No, not exactly! In Chap. 10 we described the ideal crystal, a perfectly ordered geometric array of particles. In a crystal of a real substance, such as salt or sugar, are the particles perfectly ordered? Are there no imperfections in the structures of real crystals? The answer to each of these questions is, in fact, No. The study of hypothetical, idealized forms of matter does nevertheless serve as a logical starting point from which we can move into a discussion of real matter, its structures and properties. In this chapter, then, we look at real gases, liquids, and solids, at how some of their properties are best understood in terms of deviations from their ideal or perfect counterparts, and at how transformations from one state to another occur.

11-1
Real gases

We showed in Chap. 4 that ideal behavior is to be expected of a gas if (1) there are no intermolecular forces between its molecules (except at the instant of the perfectly elastic collisions between them) and (2) the volume occupied by the molecules themselves is negligible in comparison to the volume of the container enclosing the gas. In real gases neither of these conditions is entirely fulfilled, and so deviations from ideal-gas behavior result.

Deviations from ideal behavior

For one mole of an ideal gas, $PV = RT$. This can be rewritten as $PV/RT = 1$. One way to compare the behavior of a real gas to that of an ideal one is to make P, V, and T measurements on 1 mol of gas and then plot PV/RT against P. This has been done for nitrogen at three temperatures, and the results are shown in Fig. 11-1. Ideal-gas behavior (horizontal broken line) is shown in the figure for comparison purposes. Notice that the deviation from ideality is most pronounced at high pressures and low temperatures. Similar curves can be obtained for other gases and in each case the ratio PV/RT is essentially equal to 1 at low pressure but

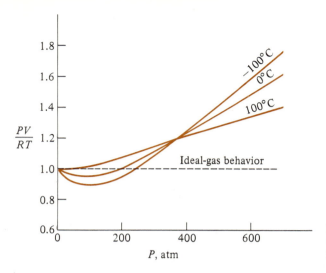

Figure 11-1
One mole of N_2 gas: non-ideal behavior.

deviates significantly at higher pressures. At very high pressures the ratio is always greater than unity.

Why does the ideal-gas law fail at high pressures and low temperatures? At high pressures the molecules of a gas are relatively close together, and since there is less empty space in the gas, molecular volumes are not negligible in comparison with the total gas volume and, furthermore, intermolecular forces are no longer insignificant. Intermolecular forces also become significant at low temperatures. At high temperatures the violence of molecular motion prevents these forces from having an appreciable effect, but as the temperature is lowered the average molecular speed decreases, and so forces of interaction begin to influence molecular motion.

Equations of state
for real gases

For any fixed quantity of gas the three variables P, V, and T are related to each other because specifying two of them automatically fixes the third. For example, 0.1 mol of O_2 at 0.5 atm and 39°C can occupy only a single, definite volume. Any algebraic statement of the relationship between the pressure, volume, and temperature of n moles of a gas is called an *equation of state* for the gas. We have seen that the equation of state for an ideal gas is $PV = nRT$, but no real gas can be described exactly by this equation. Worse yet, no two real gases show exactly the same PVT behavior.

Many equations of state have been worked out for real gases, but each contains at least three constants, as compared to one (R) in the equation of state for an ideal gas. Moreover, at least two of the constants are not universal, but rather have specific values for each gas. The best-known equation of state for real gases is the *van der Waals equation*:

$$\left(P + \frac{n^2a}{V^2}\right)(V - nb) = nRT$$

In this equation P, V, T, R, and n all have their usual meanings, but a and b have values which must be experimentally determined for each gas. In Table 11-1 values of the van der Waals constants a and b are given for a few gases.

Table 11-1
Van der Waals constants

Gas	a, liter^2 atm mol^{-2}	b, liter mol^{-1}
H_2	0.244	0.0266
He	0.034	0.0237
N_2	1.39	0.0391
O_2	1.36	0.0318
CO_2	3.59	0.0427
CH_4	2.25	0.0428
NH_3	4.17	0.0371

● **EXAMPLE 11-1**

Problem   Calculate the pressure exerted by 10.0 g of methane, CH_4, when enclosed in a 1.00-liter container at 25°C by using (a) the ideal-gas law and (b) the van der Waals equation.

Solution   The molecular weight of CH_4 is 16.0; so n, the number of moles of methane, is 10.0 g/16.0 g mol^{-1}, or 0.625 mol.

(a) Considering the gas to be ideal and solving for P, we obtain

$$P = \frac{nRT}{V} = \frac{(0.625 \text{ mol})(0.0821 \text{ liter atm K}^{-1}\text{ mol}^{-1})(298 \text{ K})}{1.00 \text{ liter}} = 15.3 \text{ atm}$$

(b) Treating the gas as a van der Waals gas and solving for P, we have

$$P = \frac{nRT}{V - nb} - \frac{n^2a}{V^2}$$

Using the values of the van der Waals constants a and b from Table 11-1,

$$P = \frac{(0.625 \text{ mol})(0.0821 \text{ liter atm K}^{-1}\text{ mol}^{-1})(298 \text{ K})}{1.00 \text{ liter} - (0.625 \text{ mol})(0.0428 \text{ liter mol}^{-1})}$$

$$- \frac{(0.625 \text{ mol})^2(2.25 \text{ liters}^2\text{ atm mol}^{-2})}{(1.00 \text{ liter})^2} = 14.8 \text{ atm} ●$$

ADDED COMMENT

The van der Waals equation is easy enough to use if you are trying to find a pressure or a temperature, because it can easily be solved for P or T. Using it to find V or n is a bit more time-consuming, however. Usually it is necessary to estimate V or n from the ideal-gas law and then adjust the estimated value upward or downward, substituting the adjusted values successively into the van der Waals equation until the left-hand side equals the right-hand side. This *method of successive approximations* sounds terribly time-consuming, but it really is not, especially if you use a pocket calculator for the arithmetic.

The van der Waals equation does a better job of describing the behavior of real gases than does the ideal-gas law. Figure 11-2 shows the PV behavior of 1 mol of CO_2 gas at 350 K as predicted by the ideal-gas law and the van der Waals equation and as observed experimentally. Other equations of state have been developed for describing the behavior of real gases. Most of these are even more accurate than the van der Waals equation but contain as many as five constants which have values characteristic of a given gas.

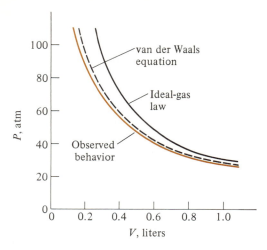

Figure 11-2
Equations of state compared for 1 mol of CO_2 at 350 K.

ADDED COMMENT

All this makes one appreciate the simplicity of the ideal-gas law. Still, it is useful to have available equations such as the van der Waals equation for use when pressures and temperatures are such that the ideal-gas law is an inadequate approximation to real-gas behavior.

Kinetic-molecular theory and real gases

Deviations from ideal-gas behavior exhibited by real gases can be thought of as resulting primarily from two causes: (1) the failure of the gas molecules to occupy a volume which is truly zero and (2) the existence of forces between the molecules of a gas. The van der Waals constants a and b are results of the attempt to take these two molecular characteristics into account.

In our earlier discussions of kinetic molecular theory (Sec. 4-5 and App. E) the assumption was made that the space available to each molecule in a gas is the same as the volume of the container which encloses the gas. This can be true only if the molecules themselves occupy zero volume, so that all parts of the container are accessible to each molecule. In other words, for an ideal gas,

$$V = V_{ideal}$$

where V is the measured container volume and V_{ideal} is the volume accessible to each molecule. In a *real gas*, however, the molecules themselves take up some space. Thus the measured volume occupied by 1 mol of such a gas is *greater than* that which 1 mol of an ideal gas would occupy at the same pressure and temperature. The correction term which takes into account the volume of the molecules themselves is the van der Waals constant b, per mole, and for n moles, nb. Thus the measured volume of a real gas is

$$V = V_{ideal} + nb \qquad \text{or} \qquad V_{ideal} = V - nb$$

We can therefore make a first modification of the ideal-gas law, $P_{ideal}V_{ideal} = nRT$, by writing

$$P_{ideal}(V - nb) = nRT \qquad (11\text{-}1)$$

What about the effect of intermolecular forces? In an ideal gas there are no such forces and the molecular bombardment of the container walls results in the ideal pressure, P_{ideal}. If there are forces of attraction between the molecules, however, the molecules will not strike the walls quite as hard on the average because each molecule will be slightly held back by the other nearby molecules. Thus the measured pressure P is slightly *less than* P_{ideal}. This pressure difference is proportional to the square of the concentration of the molecules, that is, to $(n/V)^2$, and the van der Waals a, a measure of the average force of attraction between molecules, is the constant of proportionality. If we let P represent the measured gas pressure and P_{ideal} its pressure if it behaved ideally, then

$$P = P_{ideal} - a\left(\frac{n}{V}\right)^2 \quad \text{or} \quad P_{ideal} = P + \frac{n^2a}{V^2}$$

Substituting this in Eq (11-1), we have

$$\left(P + \frac{n^2a}{V^2}\right)(V - nb) = nRT$$

which is, of course, the van der Waals equation.

Cooling by expansion   Gases usually become cooler when allowed to undergo free, or unrestrained, adiabatic expansion. An *adiabatic process* is one in which no heat is allowed to enter or leave the system. A gas is able to expand adiabatically if its container is thermally insulated from its surroundings. When any gas undergoes such an expansion *against an opposing force,* such as when it pushes back a piston which fits closely into a cylinder, the gas becomes cooler. This is a result of the fact that in expanding the gas must do work, and the energy expended in order to do that work must come from within the gas itself. Thus the average kinetic energy of the molecules decreases, and we observe this as a drop in temperature. A *free* expansion, on the other hand, is one in which the gas does no external work. This is an expansion against no opposing force, such as expansion into a vacuum. These two kinds of expansion are contrasted in Fig. 11-3.

When most real gases undergo free expansion at room temperature, there is a measurable cooling effect, in spite of the fact that the gas does no work on its surroundings. How can we account for this? When such an expansion occurs, as in Fig. 11-3b, the molecules must move away from each other. In the absence of any intermolecular attractions there would be no cooling, but in a real gas the molecules must pull away from each other against attractive forces; that is, they must do the work of separating themselves. This lowers their average kinetic

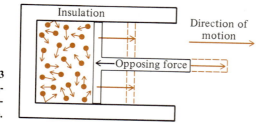
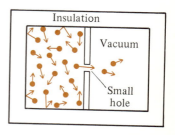

Figure 11-3 Adiabatic gas expansions. (*a*) Against opposing force. (*b*) Free.

(*a*)   (*b*)

energy. The observed cooling indicates that intermolecular attractive forces, such as London and dipole-dipole forces, do indeed exist.

We noted above that *most* gases undergoing free, adiabatic expansions at room temperature show a cooling effect. Hydrogen and helium do not; they become warmer. Warming of a gas during free expansion indicates the presence of forces of *repulsion* between molecules. These repulsions exist because molecules do not have zero volume and because two molecules cannot occupy the same space at the same time. Actually, every gas shows this warming during free expansion, if it is above its so-called *inversion temperature,* which depends upon the pressure and identity of the gas. At ordinary atmospheric pressure and room temperature hydrogen and helium are the only gases which are above their inversion temperatures. All others are below, and so they show cooling upon undergoing free expansion. (Exactly *at* its inversion temperature a gas shows neither warming nor cooling.)

11-2
Real liquids

It is not customary to speak of a hypothetical ideal liquid. Because there are such great variations among the properties of real liquids, we would have to say that most liquids deviate greatly from ideality under most conditions. Liquids have neither the large average intermolecular distances of the ideal gas nor the structural regularity of the perfect crystal. Partly because of this they show several kinds of unexpected or abnormal behavior, which we will consider at this time.

Superheating

In Sec. 10-6 we saw that a liquid boils when its temperature is increased high enough so that its vapor pressure becomes equal to the external, or atmospheric, pressure. Perhaps we should say that theoretically a liquid boils at that temperature, because in practice it may not. In order for boiling actually to start, tiny bubbles, consisting of only a few dozen molecules, must form within the liquid. These tiny bubbles serve as nuclei, growth centers, to which countless other molecules quickly add themselves to form larger, visible bubbles, which rise to the surface of the liquid. In many liquids, however, the probability of forming a bubble nucleus is so small that even at the boiling point one must wait for several seconds, or even, sometimes, minutes, for boiling actually to start. We shall see in Sec. 11-4 that the boiling process serves to keep the temperature of a liquid from rising, even though heat is added. Pure liquids have a fixed, constant boiling point, but when bubble nuclei fail to form, the temperature of the heated liquid rises above the boiling point. This is known as *superheating*.

Some liquids tend to superheat more than others, although most do so rather easily in a glass container having an inside surface which is clean and free from scratches and other flaws. Specks of dirt and tiny imperfections act as sites at which bubble nuclei can form, and so tend to prevent superheating. The way they do this is not entirely understood, although it is thought that often associated with such imperfections are microbubbles of air which serve as bubble nuclei for the boiling process.

Superheating can be a real problem in the laboratory. A liquid does not remain superheated indefinitely. Eventually a bubble nucleus forms, and the resulting bubble then grows with an almost explosive violence. The sudden formation of considerable vapor from liquid abruptly lowers the temperature of the liquid, as

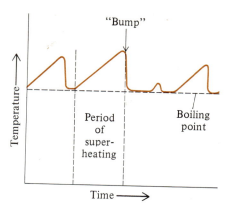

Figure 11-4
Temperature fluctuations in a bumping liquid.

the heat of vaporization, ΔH_{vap}, is used up. After the temperature has dropped to the boiling point, boiling may proceed smoothly, or superheating may again occur, initiating the cycle all over again. Not only does the temperature then fluctuate (see Fig. 11-4), which makes it impossible to use the boiling liquid to provide a constant reference temperature, but the sudden formation of large bubbles in the liquid, a phenomenon known as *bumping,* tends to blow part of the liquid out of its container. This problem is severe when the container is long and narrow, such as a test tube.

Superheating and consequent bumping may be reduced or eliminated by adding one or two *boiling chips,* also known as *boiling stones,* to the liquid. These are usually small pebbles of a ceramic material which has no surface glaze and is hence quite porous. Air trapped in the tiny pores of a boiling chip may serve as bubble nuclei, reducing the tendency for superheating. Even more effective in the case of water and aqueous solutions are small chips of polytetrafluoroethylene, or Teflon. The efficiency of this substance in reducing superheating is apparently somehow related to the fact that water does not readily wet its surface. Teflon boiling chips do not work well in some organic solvents, however.

Supercooling   Liquids exhibit "nonideal" behavior in freezing, too. When energy is removed from a liquid by cooling it, and the temperature drops to the freezing point, the liquid will ideally freeze. In reality the temperature may drop below the freezing point before solidification starts. This is known as *supercooling* or *undercooling.* The cause of supercooling is similar to that of superheating. In order for freezing to start, a few particles of the liquid must by chance fit together to form a tiny growth nucleus, the start of a larger crystal. It is more probable for the liquid molecules to add themselves to the patterned surface of an already-started crystal than it is for them to form that surface in the first place. When a liquid supercools, its temperature drops below the freezing point until a growth nucleus forms somewhere in the liquid. Crystallization follows, sometimes quite rapidly, and is accompanied by the evolution of heat. This heat, the heat of crystallization (ΔH_{cryst}), raises the temperature to the freezing point. Then, as long as heat continues to be withdrawn from the system, the liquid continues to solidify, though more slowly.

Some liquids show such a strong tendency to supercool that getting them to crystallize is quite difficult. In the laboratory supercooled liquids in glass vessels can often be caused to solidify by scratching the inside of the container with a glass stirring rod. An efficient technique is adding a small *seed crystal* of the substance, if one is available; this usually promotes rapid crystallization by providing the required pattern for crystal growth.

Glasses Extreme supercooling of a liquid to temperatures far below the melting point can sometimes be accomplished by cooling very rapidly. Heat is removed at the expense of the average kinetic energy of the molecules, as it is whenever anything is cooled. At temperatures far below the freezing point, however, the average speed of the molecules may be so low that the probability of their coalescing into a crystal lattice in any reasonable length of time is exceedingly low. As the temperature of the liquid drops, its viscosity increases as a result of the increasingly sluggish motion of the molecules. Eventually the viscosity becomes so high that flow is essentially unmeasurable, and the external characteristics of the substance are that of a solid: hard, unyielding, constant in shape, etc. Such highly supercooled liquids are often called *glasses* (ordinary glass is a common example) or *amorphous solids* in order to distinguish them from true, or crystalline, solids (Sec. 10-1).

X-ray diffraction clearly shows that the internal structure of a glass is random, like that of a liquid, rather than regular, like that of a true solid. A Laue pattern (Sec. 10-2) of a liquid shows a series of concentric rings, instead of a pattern of spots obtained from a crystal. The pattern produced by a glass is also composed of a series of rings.

Transformation to a true solid by crystallization can occur in glasses, although the process is often extremely slow because of the very low kinetic energies of the molecules. Bottles and other glass relics from ancient civilizations often show a haze or milkiness, which is caused by countless very tiny crystals grown over a period of many centuries. Crystallization weakens a glass and makes it brittle, because individual crystals are not strongly attached to each other.

Liquid crystals An interesting form of matter is the *liquid crystalline,* or *mesomorphic,* state. This state, although first observed almost 100 years ago, has more recently become the subject of intense investigation. In the liquid crystalline state there is less molecular order than in a solid but more than in an ordinary liquid. Compounds which can form liquid crystals tend to have molecules which are long and fairly rigid. An example of such a molecule is *para*-azoxyanisole, shown below.

$$H-\overset{\overset{\displaystyle H}{|}}{\underset{\underset{\displaystyle H}{|}}{C}}-O-\bigcirc-N=\overset{\overset{\displaystyle O}{|}}{N}-\bigcirc-O-\overset{\overset{\displaystyle H}{|}}{\underset{\underset{\displaystyle H}{|}}{C}}-H$$

(In this structure each circled hexagon represents the cyclic structure which is known as the *benzene ring,* and which is a resonance hybrid of the following contributing forms:

In the benzene, C_6H_6, molecule all carbons are bonded to hydrogens.   See Sec. 23-3.)

Two general classes of liquid crystals have been observed: thermotropic and lyotropic. *Thermotropic* liquid crystals are formed either by heating a solid or by cooling a liquid. *Lyotropic* liquid crystals are not pure substances, but are solutions of a substance in a highly polar liquid such as water. Such solutions show liquid crystalline properties only above a certain concentration. Certain portions of living cells apparently represent matter in the lyotropic liquid crystalline state.

There are three main types of liquid crystals: smectic, nematic, and cholesteric. In the *smectic* (from a Greek word meaning "soap") liquid crystal, the rodlike molecules pack together in layers which stack on top of each other as shown in Fig. 11-5a. This type is thus ordered in two dimensions, although within each layer the rodlike molecules are packed irregularly. Smectic liquid crystals are the most solidlike, being turbid and quite viscous and forming drops with unusual terraced surfaces.

Nematic (from a Greek word meaning "thread") liquid crystals also consist of molecules all of which point in the same direction, as is shown in Fig. 11-5b, but the ordering is strictly one-dimensional, as no layers are present. This type of liquid crystal is typically less viscous than the smectic, but still has a peculiar turbid appearance.

Cholesteric liquid crystals derive their name from the compound cholesterol, because many of them are chemical derivatives of this compound. (Cholesterol itself does not form liquid crystals, however.) These consist of a series of stacked layers within each of which the molecules are packed with their long axes in the plane of the layer, as shown in Fig. 11-5c. The molecular axes in adjacent layers are not parallel, however. Rather, they point in slightly different directions, as is shown in the illustration. The progressive twist of the molecular axes is exaggerated in the drawing; it takes a stack of about 1500 layers in order for the top and bottom layers to be aligned. Cholesteric liquid crystals show vivid, irridescent

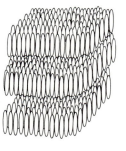  

Figure 11-5
**Types of liquid crystals:
(a) smectic, (b) ne-
matic, (c) cholesteric.**   (a)   (b)   (c)

colors, which change under the influence of temperature, pressure, magnetic and electric fields, and trace additives.

Many interesting practical applications have been found for liquid crystals. Readout displays on electronic devices such as pocket calculators and wristwatches have been made by sandwiching a thin layer of a nematic liquid crystal between two glass plates which have been coated in order to make them electrically conductive. By varying the voltage across the plates, or the frequency, when an alternating current is used, the liquid crystal can be made to change appearance from frosty to invisible, when viewed against a black background. Because of their temperature sensitivity, cholesteric liquid crystals are used in temperature-indicating devices such as household thermometers. A special adhesive-backed plastic tape which has been treated with a cholesteric liquid crystal is used to monitor the skin temperatures of newborn infants. This type of liquid crystal is also used in inexpensive detectors of trace quantities of chemical vapors.

11-3
Real solids

An ideal solid consists of a single perfect crystal with a flawless structure extending outward in three dimensions to the crystal faces. Such crystals are very difficult, perhaps impossible, to grow. A real solid, on the other hand, often consists of a matrix of many small individual crystals, or *grains,* each of which is deformed in appearance because it is packed in with other irregularly shaped grains. The grains in a metal can be seen by microscopically examining its polished and etched surface. But even single crystals which exhibit apparently perfect faces usually have many internal irregularities, called *defects.*

Dislocations

A common kind of crystal defect is the *dislocation,* also known as the *line defect,* of which there are two important types: edge dislocations and screw dislocations. An *edge dislocation* is formed by a layer or plane of particles which stops within the crystal rather than continuing all the way through. Figure 11-6 shows a schematic representation of an edge dislocation. It can be seen that one of the layers stops partway through the crystal, creating the dislocation.

A *screw dislocation* is shown in Fig. 11-7. It consists of an axis running through the crystal around which the layers of atoms are wrapped in the same way that the

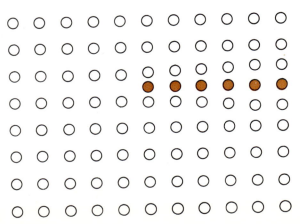

Figure 11-6
Edge dislocation.

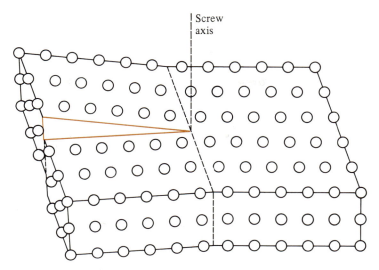

Figure 11-7
Screw dislocation.

threads of a screw are wrapped around its axis. Starting on any thread on a screw you can move to the adjacent thread by going around the screw. Similarly, by starting on any one layer of atoms and traveling around a screw dislocation axis you arrive at an adjacent layer.

Most crystals contain dislocations. The deposition of molecules on a growing crystal is believed to be facilitated by screw dislocations whose axes emerge from the face. This gives rise to beautiful spiral growth patterns which can be seen microscopically. Dislocations are centers of enhanced chemical reactivity; corrosion of metals occurs more rapidly where there is a high surface concentration of dislocations. The softness of many metals is in part a result of the fact that deforming them creates many dislocations, which then move through the crystal and allow layers of atoms to gradually slide over each other. A metal with few dislocations tends to be strong. Whiskers of iron are essentially dislocation-free single crystals and have a tensile strength over four times that of the strongest alloy-steel wire. Excessive cold-working (deforming when cold) of a metal makes it brittle due to the creation of a high concentration of dislocations. (A piece of metal can be broken by bending it many times at the same place.)

Point defects   Another important kind of defect is the *point defect*. These include the *lattice vacancy* and the *lattice interstitial,* illustrated in Fig. 11-8a and b, respectively. A vacancy occurs when an atom, ion, or molecule is missing from the lattice, and an interstitial occurs when a particle is trapped at a location which is not a lattice site.

In ionic solids point defects often occur in two common combinations: the *Schottky defect,* Fig. 11-9a, is a pair of lattice vacancies of opposite charge, and the *Frenkel defect,* Fig. 11-9b, is an ion misplaced in an interstitial position plus the vacancy where it "ought" to be. Note that in each the electrical neutrality of the crystal is preserved, that is, there are still equal numbers of positive and negative charges.

There are several other kinds of point defects known as *color centers.* These also occur in ionic solids and are named because their presence gives colors to other-

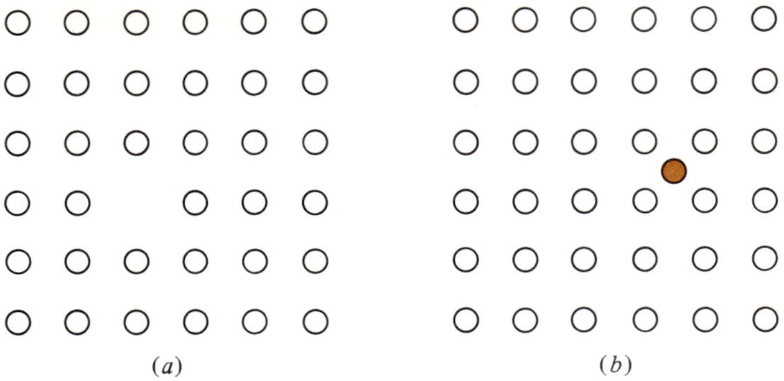

Figure 11-8
Point defects.
(a) Lattice vacancy.
(b) Lattice interstitial.

(a)   (b)

wise colorless crystals. One of these, the F center (from the German, *Farben-zentrum,* "color center"), is an electron located at an anion site.

Also classed as point defects are impurity defects, which consist of foreign particles present at lattice sites. If, for example, molten NaCl to which a few tenths of a percent $CaCl_2$ has been added is crystallized by cooling, the resulting NaCl lattice has an occasional Ca^{2+} where an Na^+ ought to be. In order to preserve electrical neutrality in the crystal, for each Ca^{2+} ion present there will be a cation vacancy (a missing Na^+) somewhere in the crystal. Similarly, ZnS can be prepared with Cl^- ions replacing some of the S^{2-} ions. For every two Cl^- ions a cation vacancy is present. Compounds to which small amounts of impurities have been added are said to have been *doped*. Doping often changes the properties of a substance greatly. The $CaCl_2$-doped sodium chloride mentioned above has, for example, a much higher electrical conductivity than does pure NaCl.

Semiconductors In semiconductors the presence of certain impurity defects has a spectacular effect on electrical conductivity. A *semiconductor* is a substance whose electrical conductivity increases as temperature is increased. (In a metal the opposite is true: increasing the temperature increases the vibrations of the ions in the crystal lattice, and this interferes with the motion of the delocalized electrons, reducing conductivity.) In a semiconductor at low temperature, essentially all electrons are localized, bound to specific atoms. Raising the temperature frees some electrons, which become delocalized, as in a metal. As the temperature goes up, more weakly bound electrons are made available to conduct electric current, that is, to move through the crystal.

Figure 11-9
Schottky and Frenkel
defects in ionic solids.
(a) Schottky defect.
(b) Frenkel defect.

(a)   (b)

Silicon is a good example of a semiconductor. At room temperature the conductivity of silicon is very low, but it can be increased dramatically by doping with any of several other elements. In crystalline Si (periodic group IVA) at room temperature all four valence electrons are used in sp^3 hybrid orbitals to bond to four adjacent Si atoms. By doping the Si crystal with a group V element such as P, As, Sb, or Bi, the structure of the crystal is left unchanged, but every so often in the crystal an atom with five valence electrons is present. This atom uses four of its electrons to bond as Si does, but the fifth becomes delocalized and free to conduct electricity. The group V–doped Si is called an n-type semiconductor, n standing for *negative,* since electrons (negative charges) are responsible for the conductivity.

Doping an Si crystal with a group IIIA element such as B, Al, Ga, or In produces an Si crystal structure in which an occasional atom with only three valence elecrons is present. The electron vacancy is called a *hole.* Under the influence of an electric field, an electron can fall into a hole and annihilate it. But this leaves a hole where the electron was, so that it looks as if the hole has moved in a direction opposite to that in which the electron has actually moved. The hole thus appears to have a positive charge and to move through the silicon lattice as if it were a real particle. The group III–doped Si is called a p-type semiconductor, since *positive* holes appear to be responsible for the conductivity.

ADDED COMMENT

Is it clear how the hole appears to move in one direction while electrons move in the other? An analogy is a string of beads with one bead missing. Imagine that one bead moves over into the hole, leaving a hole where it was. If the bead has moved to the right, the hole has moved to the left. The hole can be made to move along the string to the left by moving a succession of beads each one space to the right.

Electrical conduction in a semiconductor with impurity centers consists of electrons moving in one direction (in an n-type semiconductor) or holes moving in the other (in a p-type semiconductor). This is shown schematically in Fig. 11-10, where the electrons are represented by minus signs and the holes by plus signs. Various combinations of p- and n-type semiconductors are used to make solid-state electronic devices. A solid-state *diode* (Fig. 11-11) is a combination of one of each type of semiconductor. It is used as a rectifier for alternating current, because it will allow current to flow in one direction only. When a voltage is applied in one direction across a diode, as in Fig. 11-11a, electrons migrate through the n-type semiconductor in one direction and holes, through the p type semiconductor in the other. When they reach the junction between the two semiconductors they

Figure 11-10
Conduction in semiconductors: (a) n-type, (b) p-type.

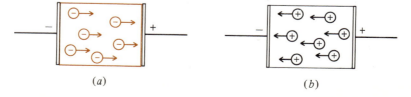

(a)   (b)

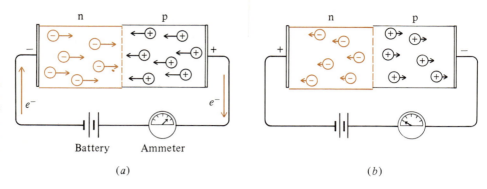

Figure 11-11
A solid-state diode.
(*a*) Current flows
through diode.
(*b*) Polarity reversed;
no current.

annihilate each other. (Actually, the electron just drops into the hole.) Conduction continues because more electrons are fed into the n-type semiconductor from the left (in the illustration), and more holes are created in the p-type semiconductor on the right as electrons are removed. When the polarity of the applied voltage is reversed, however, electrons and holes tend to move away from each other and from the junction, as shown in Fig. 11-11*b*. For most diodes, unless the voltage is high, holes and free electrons cannot be formed at the junction, and so the flow of current through the diode is prevented.

Other applications of semiconductors include the *transistor,* in which a sandwich, either pnp or npn, is used to detect or amplify electrical signals. The *light-emitting diode* (l.e.d.) is a diode in which the energy produced when electrons annihilate holes at the pn junction is released as light. In the *photodiode,* light striking the junction creates equal numbers of holes and electrons on opposite sides of the junction. This causes the diode to act like an electrical cell ("battery") by producing a voltage across its two elements. Similar to the photodiode is the *solar cell,* which is essentially an efficient photodiode, converting radiant energy into electrical energy.

Solid solutions   Impurities may sometimes be incorporated into the host lattice up to very high concentrations. For example, mixtures of potassium chloride and potassium bromide ranging from pure KCl to pure KBr may be recrystallized out of aqueous solution. In each case a single crystalline form with the rock-salt structure is obtained. K^+ ions occupy the cation sites, while Cl^- and Br^- are randomly distributed over the anion sites. The radii of Cl^- (0.181 nm) and Br^- (0.195 nm) are evidently close enough to allow one to substitute for the other. Similarly, I^- ions can be incorporated into the rock-salt structure of AgBr by substituting for Br^- ions, but in this case only a maximum of about 70 percent of the Br^- ions may be replaced by I^- ions. These are examples of *solid solutions* of the *substitutional* type.

In an *interstitial solid solution* foreign atoms, ions, or molecules occupy the nooks and crannies, the *interstices,* in the host lattice. One example of this is *austenite,* an alloy in which carbon atoms ocupy some of the interstices in a face-centered cubic array of iron atoms. Most of the transition elements will form such solid solutions with small atoms such as H, B, C, and N.

Nonstoichiometric compounds

Early in the nineteenth century a controversy raged over whether or not compounds could have variable compositions. The French chemist Claude Berthollet maintained, for example, that zinc oxide, ZnO, could be prepared having various zinc/oxygen ratios, depending on the relative amounts of zinc and oxygen used. Another French chemist, Joseph Proust, who was reported to have about the best analytical techniques of the day, showed that there were no detectable differences in composition among compounds such as ZnO which had been prepared in different ways. Proust's work later became one of the foundations of John Dalton's atomic theory (Sec. 5-1), which relied strongly upon the law of definite composition. Ironically, today we know that Berthollet was correct, that zinc oxide can be prepared with compositions ranging from ZnO to $Zn_{1.0003}O_{1.0000}$.

Nonstoichiometry is certainly an impossibility in gases under ordinary conditions. (A molecule cannot contain fractional atoms.) In solids, however, it is very common. In the zinc oxide example, above, Proust was unable to detect the extra 0.03 atomic percent zinc in the zinc-rich ZnO. Today we sometimes call compounds which follow the law of definite composition *daltonides*, while those which are nonstoichiometric are termed *berthollides*.

The distinction between nonstoichiometric compounds and solid solutions is not always clear-cut. The former term is often taken to refer, however, to compounds in which a limited number of vacancies or interstitial sites can be occupied by "unexpected" atoms. In zinc oxide, for example, extra zinc atoms can be fitted into some of the interstices in the zinc oxide lattice, and this changes the properties of the compound: $Zn_{1.0003}O_{1.0000}$ is orange in color, instead of the white of ZnO, and is an n-type semiconductor instead of an insulator. The semiconduction comes about because some of the interstitial zinc atoms lose one or two electrons, which become delocalized. The previously mentioned compounds containing F centers are also examples of nonstoichiometric compounds.

11-4 Changes of state

Now that we have studied the structures and properties of the states of matter, we can proceed to an examination of changes of state. Our starting point will be a consideration of what happens to a solid when it is uniformly heated.

Heating and cooling curves

Substances do not always get warmer when heat is added to them. At certain temperatures, those at which a change of state occurs, heat added to a substance produces an entirely different result. Consider the slow, steady addition of heat to a solid initially at absolute zero. In Fig. 11-12 we see a graph which shows the way the temperature of the substance changes with time. Before the addition of heat the molecules (or atoms or ions) at the lattice points of the solid weakly vibrate back and forth, possessing only the *zero-point energy*, the residual minimum energy predicted by quantum mechanics for a substance at absolute zero. At time zero we begin the addition of heat, causing the molecules to vibrate more actively about their average positions and raising their average kinetic energy. We observe this as an increase in the temperature of the solid. As we continue to add heat energy, the motion of the molecules becomes progressively more vigorous, and the temperature accordingly rises, until finally the forces between the molecules are no longer strong enough to withstand the violence. At this time (t_1 in Fig. 11-12)

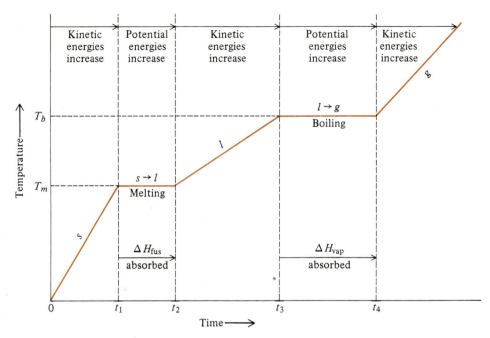

Figure 11-12
Heating curve.

molecules at the surface of the crystal begin to disengage themselves from their neighbors; that is, the solid begins to melt. (We assume here that superheating of the solid, an apparently possible, though rare occurrence, does not occur.)

From t_1 to t_2 the temperature remains constant at T_m, the melting point, while the solid is gradually converted into liquid. The heat absorbed per mole during the melting interval is the *molar heat of fusion,* ΔH_{fus}, and it is a positive number. Since the temperature of the substance does not rise during melting, where does the added heat go? It goes into increased *potential* energy of the molecules.

ADDED COMMENT

There are two ways by which we can arrive at the conclusion that the molecular potential energies must increase during the melting process: First, because the temperature does not rise we know that the average kinetic energy holds constant, so the average molecular potential energy *must* increase. Second, because forces lock the molecules together in the solid, work must be done against these forces in order to overcome them and disorder the molecules. Whenever work is done, potential energy increases. (Doing the work of lifting a book off a desk against the force of gravity increases the potential energy of the book, for example.) Thus *the heat of fusion measures the increase in molecular potential energy which accompanies the melting process.*

At t_2 in Fig. 11-12 all the solid has melted and further addition of heat increases the temperature during the time interval from t_2 to t_3. During this interval kinetic energies are once more on the increase as the molecules become increasingly more active.

At t_3 we find that the temperature again holds constant, this time at the boiling point T_b. (As before, we assume that no superheating occurs.) Now the added energy does the work of separating the molecules against the forces which held them together in the liquid. Virtually free of any forces, the molecules fly apart as a gas. The potential energies of the gas molecules are higher than those of the liquid at the same temperature, but their average kinetic energy is exactly the same. The heat absorbed per mole during the boiling interval is the *molar heat of vaporization*, ΔH_{vap}.

At t_4 the transformation of liquid to gas is complete, and the temperature again rises, as the added heat increases the average kinetic energy of the molecules. The energy changes occurring in the system as it changes from solid to liquid to gas are summarized in Fig. 11-12.

When heat is *withdrawn* from a substance, a temperature versus time plot is called a *cooling curve* (Fig. 11-13). Here we consider that heat is being withdrawn at a uniform rate from a gas. At first the temperature drops as removal of energy causes the molecules of gas to slow down. From t_1 to t_2, condensation (liquefaction) occurs and the molecular potential energies decrease as the heat of condensation, ΔH_{cond}, is removed. ($\Delta H_{cond} = -\Delta H_{vap}$.) Between t_2 and t_3 the average molecular kinetic energy in the liquid decreases, and the temperature falls. At t_3 freezing (crystallization) starts, and as the heat of crystallization, ΔH_{cryst}, is removed, the molecular potential energies decrease. ($\Delta H_{cryst} = -\Delta H_{fus}$.) After freezing has been completed, the temperature once more falls as the molecules again slow down. In the above sequence the condensation point is the same as the

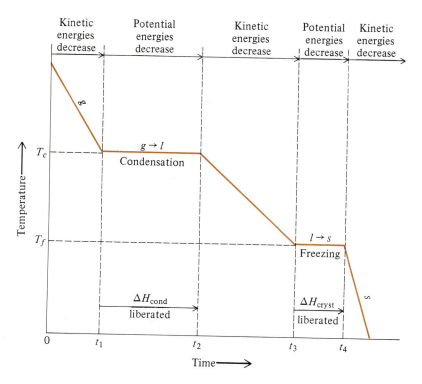

Figure 11-13
Cooling curve.

boiling point, and the freezing point is the same as the melting point. (We have assumed in this discussion that no supercooling occurs.)

Phase equilibria   In Sec. 10-6 we discussed the fact that liquids tend to evaporate and that a liquid-vapor equilibrium can be established in which the rate of evaporation equals the rate of condensation. Such liquid-vapor equilibria are but one type of equilibrium which can exist between two different phases. Solids also tend to transform directly into the gaseous state. This process, analogous to evaporation of a liquid, is called *sublimation.* Sublimation occurs because at any temperature a certain fraction of the molecules at the surface of a crystal are moving energetically enough to fly off. The rate of sublimation of most common solids is low under ordinary temperature and pressure conditions. Snow, for example, can be observed to disappear gradually over a period of several days during which the temperature never rises to "freezing." Mothballs are composed of crystals of napthalene or *para*-dichlorobenzene, which very slowly sublime to release toxic vapors. Just as the rate of evaporation of a liquid can be increased by lowering the external pressure on it, so also can the rate of sublimation of a solid similarly be increased. Fruits, vegetables, meats, and beverages such as coffee can be *freeze-dried* by first lowering the temperature far below 0°C and then subjecting them to a vacuum. The water then rapidly sublimes, leaving a dehydrated product which can be reconstituted later by restoring the water.

When a subliming solid is enclosed in a container, redeposition of gas molecules on the surface is possible and a state of equilibrium is reached:

$$\text{Solid} \rightleftharpoons \text{gas}$$

The pressure of the gas in equilibrium with its solid is called the *equilibrium vapor pressure,* or *sublimation pressure,* of the solid. Just as with a liquid, the vapor pressure of a solid increases with increasing temperature. At the lower left of Fig. 11-14 is shown the variation of the vapor pressure of ice with temperature. The curve rises to the right until it stops at 0.01°C, at which temperature the vapor pressure is still only 4.58 mmHg. It stops at this point, because ice cannot exist above this temperature, called the *triple-point temperature* (for reasons which will become apparent shortly).

The *Clausius-Clapeyron equation* applies to solid-gas equilibria as well as to the liquid-gas equilibria discussed in Sec. 10-6. Thus if we plot the logarithm of the vapor pressure of a solid against the reciprocal of the temperature, we get a straight line of slope

$$\frac{-\Delta H_{sub}}{2.303R}$$

where ΔH_{sub} is the heat of sublimation of the solid.

Figure 11-14 also shows a plot of the vapor pressure of liquid water. Since each of the two curves is an *equilibrium* vapor-pressure curve, they can be viewed from a slightly different viewpoint: the points on each curve represent all the temperature-pressure combinations at which two phases (solid + gas, or liquid + gas, depending on the curve) can be at equilibrium. Liquid and gaseous water, for example, can be at equilibrium *only* at a temperature and pressure represented by a point which is on the liquid-gas equilibrium line.

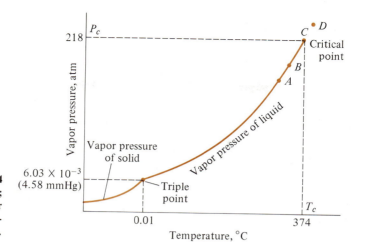

Figure 11-14
Vapor pressure curves;
solid and liquid water
(scale somewhat dis-
torted).

The liquid-gas equilibrium curve starts at the triple point. At the triple-point temperature the vapor pressure of the solid is the same as that of the liquid. Thus at this temperature and pressure *all three phases can coexist stably, in equilibrium.* This is the definition of a *triple point.*

The liquid-gas equilibrium line ends (at the high-temperature end) at a point called the *critical point.* This is *the temperature and pressure above which the distinction between gas and liquid vanishes.* How is this possible? As the temperature and pressure of a liquid-gas pair system are both increased, the increasing temperature tends to make the molecules in the liquid fly apart to form a gas, while the increasing pressure tends to force the molecules of the gas together into a liquid. In other words, the liquid becomes more like the gas and the gas more like the liquid. As the system moves up the liquid-gas equilibrium line toward the critical point, temperature and pressure increasing simultaneously so as to stay on the line, all the properties of the gas and liquid approach each other. At the critical point itself, properties such as density, refractive index, thermal conductivity, viscosity, etc., are barely different in the two phases (and the difference *vanishes* at a slightly higher temperature and/or pressure).

Figure 11-15 shows the appearance of a cell containing a liquid and gas in equilibrium under conditions *A*, *B*, *C*, and *D* close to the critical point. These conditions are indicated in Fig. 11-14. At *A* and *B* the liquid and gas phases can be distinguished, but at the critical point *C* the *interface,* the meniscus at the top of the liquid, is fuzzy and indistinct. (The two phases have almost the same refractive indexes, and so the interface is almost invisible and, indeed, is almost nonexistent.) At *D*, beyond the critical point, only one phase is present.

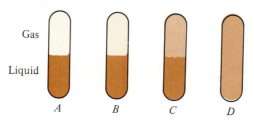

Figure 11-15
Behavior near the critical point.

Table 11-2
Critical constants

Substance	Critical temperature, K	Critical pressure, atm
He	5	2.3
H_2	33	12.8
N_2	126	33.5
O_2	154	49.7
CO_2	304	72.8
HCl	325	81.5
NH_3	406	111.3
Cl_2	417	76.1
H_2O	647	218.3
Hg	1735	1036.0

The temperature at the critical point is called the *critical temperature T_c*. It is *the highest temperature at which a gas can be liquefied by means of compression.* The pressure needed to liquefy a gas at its critical temperature is called the *critical pressure, P_c*. Above the critical temperature, no amount of compression will liquefy a gas, because molecular motion is so violent that no matter how closely together the molecules are crammed, intermolecular forces are not strong enough to hold them together as a liquid. Critical temperatures are high when intermolecular attractive forces are strong. In Table 11-2 are tabulated the critical temperatures and critical pressures of some substances. Helium has the lowest critical temperature of all substances. In helium the intermolecular forces are so weak that above 5 K molecular motion prevents liquefaction, no matter how high the pressure is raised.

In Fig. 11-14 we saw the *P* versus *T* curves for liquid-gas and solid-gas equilibria in water. What about the solid-liquid case? In Fig. 11-16 a third line, the solid-liquid line, has been added to the graph. This line represents all the temperatures and corresponding pressures at which the melting-freezing equilibrium can take place. The line slopes to the left, indicating that the freezing (or melting) point of water *decreases* with *increasing* pressure. Water is very unusual in this respect. It is one of a very few substances whose melting points behave this

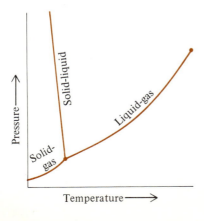

Figure 11-16
Conditions for phase equilibrium in water.

way, and we will see what causes this behavior shortly. Figure 11-16 is an example of a *phase diagram*.

In 1884 the French chemist Henri Le Châtelier (pronounced approximately, *luh-shot'-lee-ay*) suggested that equilibrium systems tend to compensate for the effects of perturbing influences. His principle applies to all kinds of dynamic equilibria and may be stated as follows: *When a system at equilibrium is subjected to a stress, it will tend to adjust in response so as to minimize the effect of the stress.*

Let us consider the liquid-vapor equilibrium system shown in Fig. 11-17a. Here we see a liquid and its vapor enclosed by a cylinder and close-fitting piston. The pressure on the system is held constant by the weight on top of the sliding piston. At equilibrium the rate of evaporation of liquid equals the rate of condensation:

$$\text{Heat} + \text{liquid} \rightleftharpoons \text{gas}$$

Now, what happens if we add heat to the system? We might expect the temperature to rise, but in this case it does not because the system has available a mechanism for adjusting to the stress of added heat. Some of the liquid in the cylinder evaporates and in doing so uses up the heat, so that there is no increase in temperature. After the heat has been added, Fig. 11-17b, more gas and less liquid are present, so we say that the equilibrium *as described by the above equation* has shifted its position, or that it has shifted to the right. Attempting to cool the system causes the reverse change to take place: gas condenses to form more liquid and in so doing heat (of condensation) is liberated, maintaining a constant temperature. Removal of heat shifts the equilibrium (as written above) to the left.

What causes an equilibrium system to shift? In each of the above examples the rate of one process temporarily increases, so that it exceeds the rate of the opposing reaction. Heating the liquid-gas system causes the rate of evaporation to exceed the rate of condensation, using up the added heat. There is thus a *net* transfer of liquid to gas. Afterwards, the system returns to equilibrium, although now more gas and less liquid are present than before.

When we add or remove heat from the liquid-vapor equilibrium system *at constant pressure,* as above, the temperature remains constant. If, on the other hand, we add heat *at constant volume,* the results are a little different. In this case adding heat again causes some liquid to evaporate, but the gas cannot expand, and so the pressure must increase. At the higher pressure the liquid can be at equilibrium with the gas only at a higher temperature (see Fig. 11-14 or 11-16), so that when equilibrium is reestablished, both the temperature and pressure are higher. (The system has moved slightly up the liquid-gas equilibrium line.) The important point here is this: because the system consists of a gas and liquid at equilibrium, *some* of the added heat goes to convert some liquid to gas, instead of raising the temperature of the system. Thus the temperature increase is less than if the system had consisted of liquid or gas only and therefore were not at equilibrium. In other words, when the equilibrium shifts to the right the temperature increase is minimized.

Cooling has the reverse effect. In this case the equilibrium as written above shifts to the left; this means that the system produces heat, and so the temperature does not decrease as much as it would have had it consisted of liquid or gas alone and therefore were not at equilibrium.

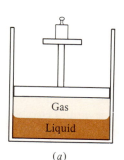

(a)

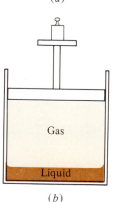

(b)

Figure 11-17
A liquid-vapor equilibrium system at constant pressure. (a) Before addition of heat. (b) After addition of heat.

ADDED COMMENT

Note that shifting "to the right" or "to the left" has meaning only if an equation has been written. We might just as well have described the above equilibrium by writing

$$Gas \rightleftharpoons liquid + heat$$

In this case we should say that adding heat shifts the equilibrium *to the left.* "Right" and "left" really refer to a written equation, not a system.

An increase of *pressure* also causes the liquid-gas equilibrium to shift. Consider, for example, an adiabatic (Sec. 11-1) compression. Figure 11-8a shows the system in an insulated cylinder plus piston at some initial pressure P_1. When we decrease the volume of the system as in Fig. 11-18b, we expect the pressure to increase, and indeed it does, to P_2. But for a given decrease in volume the pressure increase is less than if only gas or liquid were present, and therefore no equilibrium were possible. How does this occur? The density of the liquid is greater than that of the gas. In other words, gram for gram the liquid takes up less room than the gas. Thus the equilibrium

$$Heat + liquid \rightleftharpoons gas$$

shifts to the left as the volume is decreased. This converts gas to liquid and minimizes the pressure increase. But during this process heat is evolved. (See the equation.) This heat cannot escape because of the insulation; so the temperature rises. After the system has returned to equilibrium the pressure is higher, and so is the temperature. Again, the system moved out along the liquid-gas equilibrium line of Fig. 11-16. To summarize, increasing the pressure on a gas-liquid equilibrium system favors the more dense liquid phase, so that after equilibrium has been reestablished at a higher pressure, more liquid and less gas are present, and the temperature is higher.

Solid-gas (sublimation) equilibria behave in much the same way as liquid-gas equilibria; so we need not discuss them separately.

In order to predict solid-liquid (melting-freezing) equilibrium changes, we must

Figure 11-18
Adiabatic compression
of a gas-liquid equilib-
rium system: (a) P_1, T_1,
(b) P_2, T_2.

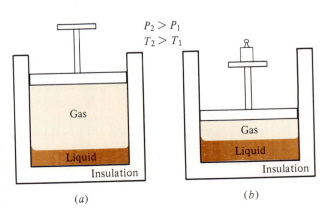

$P_2 > P_1$
$T_2 > T_1$

Gas

Liquid

Insulation

(a)

Gas

Liquid

Insulation

(b)

know the relative densities of the solid and liquid phases. Consider the equilibrium system

$$\text{Heat} + \text{solid} \rightleftharpoons \text{liquid}$$

If we add heat to this system at constant pressure some of the solid will melt, and after equilibrium is reestablished, the temperature will be unchanged. But now consider the addition of heat to the system *at constant volume*. We now expect a pressure change, but it is impossible for us to predict the nature of the change unless we know whether the liquid has a more or less compact structure than the solid. Consider the addition of heat to a system composed of solid and liquid carbon dioxide at equilibrium in a container of constant volume:

$$\text{Heat} + CO_2(s) \rightleftharpoons CO_2(l)$$

By using Le Châtelier's principle we predict a conversion of some solid to liquid, a process which uses up some of the added heat. But what happens to the pressure? Because the liquid is less dense than the solid, the entire system tends to expand, but because the volume is held constant, the pressure increases instead. After we have stopped adding heat and have allowed the system to reestablish equilibrium, we find that the temperature has risen, and so has the pressure. Most substances are like CO_2. Their solid-liquid equilibrium pressures increase as the temperature increases.

The behavior of water contrasts with that of carbon dioxide. The density of water as a liquid is *greater* than its density as a solid. The liquid thus has a *more* compact structure than the solid. (This accounts for the fact that an ice cube floats in liquid water.) If we disturb the equilibrium

$$\text{Heat} + H_2O(s) \rightleftharpoons H_2O(l)$$

by adding heat at constant volume, the temperature goes up as it did with CO_2, but this time the pressure *decreases*. It does so because the liquid is more compact than the solid. Solid and liquid water are at equilibrium at progressively *lower* pressures, as the temperature increases.

To summarize, increasing the pressure on a solid-liquid equilibrium system shifts the equilibrium in the direction of the more compact, more dense phase. Thus the melting, or freezing, point of water decreases with an increase in pressure, while for most other substances it increases.

The decrease in the freezing point of water with an increase in pressure is in part responsible for the fact that ice skating is possible. Micro-irregularities at the surface of ice and skate blade result in high pressures being developed at these points. This lowers the freezing point, and if the temperature of the ice is above this, the ice melts. Even if the temperature is somewhat below the melting point, local heating due to friction between the blade and ice causes melting under the blade, and a thin film of liquid lubricates the blade as it slides along.

11-5
Phase diagrams

The previous section introduced the phase diagram for water. A phase diagram is a graph which shows the conditions under which one phase can be transformed into another. Such diagrams can be constructed for systems of more than one component, but we consider here only one-component systems, and furthermore we examine only pressure-temperature phase diagrams, as these are most commonly encountered.

Water   Each *line* in a phase diagram represents, as we have seen, the conditions of temperature and pressure at which two phases can be at equilibrium. The intersection of three lines at a single *point* (a triple point) represents the *only* condition of pressure and temperature at which three phases can be at equilibrium. An *area* between lines represents all those conditions of pressure and temperature (each point in an area represents a single *PT* condition) at which only one phase can exist stably. In Fig. 11-19 the different areas in the phase diagram for water are indicated by different shading between the lines. The horizontal line drawn at 1 atm pressure allows us to locate the *normal* melting and boiling points of water. The normal melting point T_m is the temperature at which solid and liquid water coexist at equilibrium at 1 atm, and so it is found at the point where the solid-liquid line crosses the 1-atm line. Similarly, the *normal* boiling point is the temperature at which the liquid-gas line crosses the 1-atm line. (The melting and boiling points can, of course, be found for other pressures, too.) Note that the solid-liquid line slopes up to the left in the phase diagram for water. This slope has been exaggerated in Fig. 11-19. (The triple point is only 0.01°C above the normal melting point.)

In a phase diagram the liquid region is clearly separated from the gas region only at temperatures at or below the critical temperature, and at pressures at or below the critical pressure. At the upper right of the diagram the liquid area is continuous with the gas area. In this region the distinction between liquid and gas no longer exists. That it is not always possible to classify a fluid definitely as either a liquid or a gas is a statement of the *principle of continuity of states*.

Phase diagrams can be used to follow temperature and pressure changes in a substance. Look at point *A* in Fig. 11-20. At this temperature and pressure the water is a solid. Imagine that we gradually add heat to the ice *while holding the pressure constant*. As this occurs the system moves horizontally to the right on the phase diagram, as the temperature increases. When point *B* is reached the solid

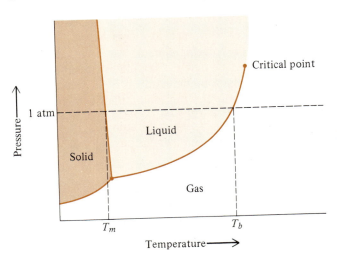

Figure 11-19
Phase diagram of water (not drawn to scale). The liquid-gas equilibrium line stops at the critical point because the distinction between liquid and gas vanishes at higher pressures and temperatures.

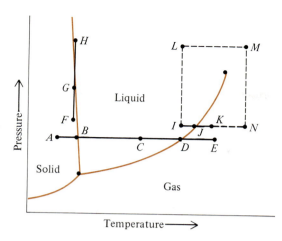

Figure 11-20
Phase changes in water.

starts to melt. (*B* gives us the melting point at this pressure.) The system stays at point *B* until all solid has melted, and only then does it resume its move to the right. (All ice must melt before the temperature can again rise.) Passing through point *C* there is only a liquid phase, which is getting hotter. At point *D* the liquid-gas equilibrium line is reached and the liquid starts to boil. The temperature remains constant until all liquid is gone, and only then can the system move further to the right, that is, to higher temperatures. At point *E* the system is at a much higher temperature, but at the same pressure as at the start.

What happens to ice when its pressure is increased (by decreasing its volume) at constant temperature? Starting at point *F* in Fig. 11-20 the system moves straight upward on the phase diagram until it reaches point *G* on the solid-liquid equilibrium line. At this point the solid starts to melt. (Note that this would not be possible if this line sloped to the right.) At point *G* decreasing the volume no longer results in an increased pressure, as long as any solid is present. When all solid has melted, the pressure can again rise. At point *H* the system is a liquid at much higher pressure, but at the same temperature as at the start.

Now consider transforming the liquid at point *I* (Fig. 11-20) to a gas at the same pressure, but a higher temperature (point *K*). This can be done in a number of ways; one is to go directly to *K* by increasing the temperature at constant pressure. At point *J* vaporization (boiling) occurs, and after it is complete, the temperature moves up to that at *K*. The second way involves making an "end run" around the critical point, following the broken line *ILMNK*. The line *IL* represents increasing the pressure of the gas at constant temperature. *LM* is an increase in pressure at constant temperature. *MN* is a decrease in pressure at constant temperature. Finally, *NK* is a decrease in temperature at constant pressure. We have just changed a liquid to a gas *without* ever boiling it. This process can really be performed and is possible because of the continuity of the gas and liquid states.

Water: high-pressure phase diagram

Many substances can exist in more than one solid phase. (Only one pure substance, helium, exhibits two liquid phases, and because all gases are miscible, no substance can exist in different gaseous phases.) Transformations from one solid phase to another are often very slow, but otherwise they resemble phase changes

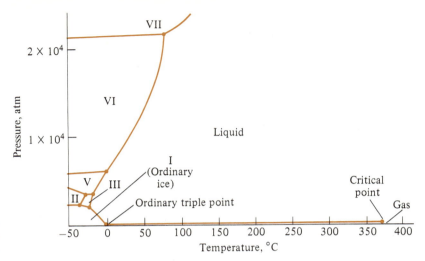

Figure 11-21
High-pressure phase
diagram for water.

involving liquids and gases. Substances which exhibit more than one solid form are said to be *polymorphic*.

Water is polymorphic, exhibiting eight known different forms of ice, each with its own unique crystal structure. Figure 11-21 shows the phase diagram for water extended to very high pressures. In it can be seen the areas which represent conditions under which ices I (ordinary ice), II, III, V, VI, and VII exist. (There was once thought to be an ice IV, but its existence has been disproven.) In order to plot the high-pressure data on this diagram the vertical scale has had to be compressed so much that the liquid-gas equilibrium line appears to be almost lying on the horizontal axis, and the solid-gas line cannot even be seen.

The ordinary (ice I-liquid-gas) triple point and the critical point are indicated on the phase diagram. But notice the other triple points (I-III-liquid, I-II-III, III-V-liquid, II-III-V, V-VI-liquid, and VI-VII-liquid). Each of these represents a temperature and pressure at which three phases can coexist.

Notice that of all the solid-liquid equilibrium lines, only the ice I-liquid line slopes to the left. Ice I is the only form of solid water which is less dense than the liquid. The others are all more dense, and so Le Châtelier's principle predicts that their melting points will *increase* with increasing temperature. An ice cube made from one of these ices will *sink* in liquid water.

Carbon dioxide   The phase diagram for carbon dioxide at lower pressures is typical of that for most substances. It is shown in Fig. 11-22. In it the solid-liquid line slopes to the right, which is the usual case, since most liquids are less dense than their solids.

Note that the triple-point pressure of CO_2 is *greater* than 1 atm. This means that *liquid* CO_2 cannot stably exist at that pressure; so at 1 atm neither melting nor boiling can take place. Carbon dioxide thus has no *normal* melting or boiling point. CO_2 can, of course, sublime, which it does at any pressure below 5.1 atm. Solid CO_2 is called *dry ice*, where the "dry" refers to the fact that it does not melt. At 1 atm the sublimation temperature of dry ice is $-78.2°C$. CO_2 is normally transported as a liquid in pressurized cylinders and tanks.

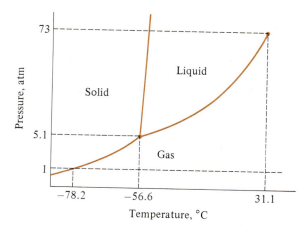

Figure 11-22
Carbon dioxide phase diagram
(not drawn to scale).

SUMMARY

In this chapter we have looked at the properties and behavior of matter as it actually exists, not as it might exist if it "followed all the rules." *Real gases* deviate from the behavior predicted or described by the ideal-gas law because of molecular volumes and intermolecular forces. Many equations of state for real gases have been developed; the most important of these is the *van der Waals equation*. The intermolecular forces in a real gas account for its becoming cooler when undergoing a free, adiabatic expansion under most conditions, although a gas may become warmer in such a change if it is above its *inversion temperature*.

Liquids exhibit a number of kinds of nonideal or abnormal behavior. These include the ability to become *superheated* and *supercooled* in what are unstable, nonequilibrium states. Sometimes a liquid may be supercooled to such an extreme extent that a *glass,* or *amorphous solid,* is formed. Crystallization of a glass to form a true solid may be very slow because of the low molecular kinetic energies. An unusual form of liquid is the *liquid crystal* in which one- or two-dimensional molecular order exists, as compared to three-dimensional order in a solid and little order at all in the usual liquid.

The *solid* phase is almost always characterized by *defects*. These include the structural irregularities known as *dislocations* in which the regularity of the crystal lattice is interrupted along a line. Other defects are *point defects*, which include *lattice vacancies* and *interstitials, color centers,* and *impurity centers. Semiconductors* can be prepared by adding low but controlled concentrations of impurities in certain substances. A solid with a high concentration of impurities may be classed as a *solid solution. Nonstoichiometric compounds* are solids with crystal structures which permit "foreign" particles to substitute for the normal ones, or to be squeezed in among the normal ones, so as to vary the composition of the compound, usually within fixed limits.

Changes of state can be observed by following *heating* or *cooling curves,* on which temperature changes and holds can be correlated with changes in molecular potential and kinetic energies. Solid-liquid, liquid-gas, and solid-gas transitions are all possible, and the pressures and temperatures at which these occur may be summarized on *phase diagrams,* on which solid-solid phase transitions can also be shown.

Le Châtelier's principle describes an important characteristic of phase equilibria and of other equilibria, as well. It states that an equilibrium system tends to absorb a stress placed upon it so as to reduce the effect of that stress. We describe the adjustment made by the system to the imposed stress as a *shift* of the equilibrium which changes the relative amounts of the constituents present.

KEY TERMS

Adiabatic (Sec. 11-1)
Amorphous (Sec. 11-2)

Berthollide (Sec. 11-3)
Boiling chip (Sec. 11-2)

Bumping (Sec. 11-2)
Cholesteric (Sec. 11-2)
Color center (Sec. 11-3)
Continuity of states (Sec. 11-5)
Cooling curve (Sec. 11-4)
Critical point (Sec. 11-4)
Daltonide (Sec. 11-3)
Defect (Sec. 11-3)
Diode (Sec. 11-3)
Dislocation (Sec. 11-3)
Doping (Sec. 11-3)
Edge dislocation (Sec. 11-3)
Equation of state (Sec. 11-1)
F center (Sec. 11-3)
Free expansion (Sec. 11-1)
Frenkel defect (Sec. 11-3)
Glass (Sec. 11-2)
Heating curve (Sec. 11-4)
Hole (Sec. 11-3)
Impurity defect (Sec. 11-3)
Interstitial (Sec. 11-3)
Inversion temperature (Sec. 11-1)
Le Châtelier's principle (Sec. 11-4)
Line defect (Sec. 11-3)
Liquid crystal (Sec. 11-2)

Lyotropic (Sec. 11-2)
Mesomorphic (Sec. 11-2)
n-Type semiconductor (Sec. 11-3)
Nematic (Sec. 11-2)
Nonstoichiometric (Sec. 11-3)
p-Type semiconductor (Sec. 11-3)
Phase diagram (Sec. 11-5)
Point defect (Sec. 11-3)
Polymorphism (Sec. 11-5)
Schottky defect (Sec. 11-3)
Screw dislocation (Sec. 11-3)
Semiconductor (Sec. 11-3)
Shift (equilibrium) (Sec. 11-4)
Smectic (Sec. 11-2)
Solid solution (Sec. 11-3)
Sublimation (Sec. 11-4)
Supercooling (Sec. 11-2)
Superheating (Sec. 11-2)
Temperature hold (Sec. 11-4)
Thermotropic (Sec. 11-2)
Triple point (Secs. 11-4, 11-5)
Vacancy (Sec. 11-3)
Van der Waals equation (Sec. 11-1)
Zero-point energy (Sec. 11-4)

QUESTIONS AND PROBLEMS

Gases

11-1 What two factors contribute to nonideality in a gas? How does each of these affect the observed pressure of a real gas enclosed in a container of a certain volume at a certain temperature?

11-2 Explain in terms of kinetic-molecular theory why it is that most gases become cooler upon free expansion.

11-3 One mole of N_2 gas is to be expanded from a volume of 15 liters to one of 150 liters. Which method, free expansion or expansion against a piston, will cool the gas more? Explain.

11-4 What is the physical significance of each of the van der Waals constants a and b?

11-5 Which of each of the following pairs would you expect to have the larger van der Waals a constant? **(a)** CH_4 or C_2H_6 **(b)** NH_3 or CH_4 **(c)** Ar or Ne.

11-6 Which of each of the following pairs would you expect to have the larger van der Waals b? **(a)** CH_4 or C_2H_6 **(b)** NH_3 or CH_4 **(c)** Ar or Ne.

11-7 Calculate the pressure exerted by 10.0 g of N_2 enclosed in a container having a volume of 0.0200 liter at 25°C using **(a)** the ideal gas law and **(b)** van der Waals equation.

11-8 Repeat the calculations of Prob. 11-7 but for a container volume of 20.0 liters.

11-9 Using **(a)** the ideal-gas law and **(b)** van der Waals equation, calculate the temperature to which 10.0 g of methane, CH_4, in a 1.00-liter container must be heated in order for its pressure to be 25.0 atm.

11-10 When the volume of a certain gas not near its condensation point is halved, its pressure is observed to rise to only 1.6 times its original value if the temperature is held constant. Suggest a reason for this very large deviation from Boyle's law.

11-11 Consider a gas composed of molecules which occupy no volume but between which attractive forces exist. How would the behavior of this gas differ from that of an ideal gas? Deduce an equation of state for this gas.

11-12 Consider a gas composed of molecules which occupy a finite volume but between which no forces operate. How would the behavior of this gas differ from that of an ideal gas? Deduce an equation of state for this gas.

11-13 Calculate the volume occupied by 1.00 mol of carbon dioxide at 4.00×10^2 K and 5.00×10^2 atm using the van der Waals equation.

Liquids

11-14 Water is often seen to boil more smoothly in an old beaker than in a new one. Explain.

11-15 Account for the fact that a boiling stone prevents bumping in a boiling liquid.

11-16 Small crystals of silicon carbide, a very hard substance, are often effective as boiling stones, in spite of the fact that they are not porous. Explain.

11-17 What is a glass? How can it be shown that a glass has the internal structure of a liquid?

11-18 What is a liquid crystal? What kinds of molecules form liquid crystals?

Solids

11-19 What is a dislocation? Distinguish between an edge dislocation and a screw dislocation.

11-20 Draw some sketches and use them to show how an edge dislocation can move to the surface of a crystal when the crystal is placed under mechanical stress.

11-21 Suggest a reason for the fact that many crystalline minerals are observed to have slightly curved crystal faces.

11-22 What is a semiconductor? Describe the two main types of semiconductors and contrast their conduction mechanisms.

11-23 Since an arsenic atom has five valence electrons as compared to silicon with four, why is it that an arsenic-doped silicon crystal does not have an overall negative charge?

11-24 How does the presence of a great many **(a)** Schottky defects and **(b)** Frenkel defects affect the density of a crystal?

11-25 Nonstoichiometric cuprous oxide, Cu_2O, may be prepared in which the copper/oxygen atomic ratio is less than 2:1. Account for the fact that this substance is a p-type semiconductor.

11-26 Classify each of the following as either a p-type or an n-type semiconductor: **(a)** Ge doped with In **(b)** B doped with Si **(c)** NaCl doped with Na **(d)** $NiO_{1.0001}$ **(e)** $Zn_{1.0001}O$.

11-27 Copper and nickel form a continuous series of solid solutions from pure Cu to pure Ni. Would you expect these solutions to be interstitial or substitutional? Justify your answer in terms of the probable packing in these metals.

Changes of state

11-28 Draw a typical gas-liquid-solid cooling curve. Indicate what happens to **(a)** the kinetic energy and **(b)** the potential energy of the molecules during each time interval on the curve.

11-29 Draw a cooling curve for each of the following: **(a)** a liquid which freezes without supercooling **(b)** a liquid which freezes after a small amount of supercooling **(c)** a liquid which freezes after extreme supercooling **(d)** a liquid which cools to form a glass.

11-30 Account for the fact that the temperature of a supercooled liquid rises when solidification occurs.

11-31 Account for the fact that when the temperature of a freezing supercooled liquid rises, it never goes above the melting point.

11-32 When a liquid crystallizes its molecules become bound at specific sites in the crystal lattice. How can you reconcile this with the fact that there is no decrease in average molecular kinetic energy associated with freezing?

11-33 Account for the fact that skin burns from steam are often more severe than burns from liquid water, even when both are at the same temperature.

11-34 Account for the fact that when liquid water is placed in a flask which is then rapidly evacuated, the water freezes before it has all evaporated.

11-35 What is the highest possible melting point of ice I? What is the lowest possible freezing point of carbon dioxide?

11-36 Naphthalene has a sublimation pressure of 1.70×10^{-2} mmHg at 10°C and 5.37×10^{-2} mmHg at 21°C. Calculate the molar heat of sublimation of naphthalene.

11-37 The compound 2,4,6-trinitrotoluene (TNT) has a sublimation pressure of 1.61×10^{-4} mmHg at 50°C and a molar heat of sublimation of 118 kJ mol^{-1}. Calculate the sublimation pressure of TNT at 140°C.

11-38 What are the critical temperature and the critical pressure? How are they related to the principle of continuity of states and to the strength of intermolecular forces?

11-39 Which substance will exhibit a greater evaporative cooling effect at room temperature: one with a high or a low critical temperature? Explain.

11-40 What approximate relationship would you expect to exist between the heats of sublimation, vaporization, and fusion for a substance?

Le Châtelier's principle

11-41 Explain in terms of Le Châtelier's principle why

it is that the boiling point of *any* liquid increases with increasing pressure.

11-42 Explain in terms of Le Châtelier's principle why it is that the melting points of some solids increase with increasing pressure, while those of others decrease.

11-43 Can a solid-gas equilibrium temperature decrease with increasing pressure? Explain in terms of Le Châtelier's principle.

11-44 An insulated container of fixed volume contains liquid and gaseous water at equilibrium at 90°C. Tell what will happen to the pressure and temperature of the system if **(a)** more liquid water at 90°C is added **(b)** some water vapor is removed **(c)** an inert gas at 90°C is added.

11-45 An insulated container of constant volume contains solid and liquid water at 0°C. What will happen to the pressure and temperature of the system if **(a)** some additional solid is added **(b)** additional liquid is added **(c)** an inert solid substance is added.

Phase diagrams

11-46 When the pressure of water at its normal melting point is increased at constant temperature, what phase results?

11-47 When the pressure of carbon dioxide at its triple point is increased at constant temperature, what phase results?

11-48 If ice is heated at a constant rate, how many temperature holds will the heating curve show if the pressure is held constant **(a)** at 1 mmHg? **(b)** at 400 mmHg? **(c)** at 400 atm?

11-49 A certain solid can exist in two different crystalline forms at 1 atm depending upon the temperature. **(a)** Draw a phase diagram for this substance. **(b)** Draw a heating curve for this substance, assuming that the critical pressure is greater than 1 atm. **(c)** How would your curve differ if the critical pressure were below 1 atm?

11-50 In order for a solid-gas critical point to exist for a substance, what would its phase diagram have to look like?

11-51 Draw a phase diagram for a substance which has the following properties: the density of the liquid is less than that of the solid, the normal melting point is 50°C, the normal boiling point is 200°C, and the triple point pressure is 0.1 atm. Estimate the freezing and boiling points at 0.5 atm.

11-52 Water vapor at 1.0×10^{-3} atm and $-0.10°C$ is slowly compressed at constant temperature until the final pressure is 100 atm. Describe all the changes which would occur.

11-53 Draw a phase diagram for the following compound. The substance exists as two solid forms, I and II, both of which are more dense than the liquid. I is the stable phase at lower pressures. Solid I melts at 20°C under its own sublimation pressure of 10 mmHg. Phases I, II, and liquid are at equilibrium at 50°C and 1000 atm. The I–II transition temperature decreases with increasing pressure. The normal boiling point is 120°C, and the critical point is 280°C and 30 atm.

12

SOLUTIONS

TO THE STUDENT

The properties of a *heterogeneous mixture* (Sec. 2-2) are a composite of the properties of its individual components. A mixture of sand and salt, for instance, exhibits both the properties of sand and those of salt. On the other hand, the properties of a *homogeneous mixture*, a *solution*, often appear to be quite unrelated to those of the individual components. A sample of saltwater, for example, freezes at a temperature which is below the freezing point of either pure water or pure salt. (Actually, it freezes over a range of temperatures. As pure ice forms, the concentration of salt in the remaining solution increases, and, as we will see shortly, this further decreases the freezing point.) In this chapter our attention will be focused on the nature of solutions and on how their properties and compositions are related.

12-1 Mixtures

A mixture may be classified as either heterogeneous or homogeneous, but because there is no sharp distinction between these two classes, categorizing a system as one or the other is sometimes difficult.

Differentiating heterogeneous mixtures from homogeneous mixtures

Visual inspection is often adequate for deciding whether a mixture is heterogeneous or homogeneous. It is easy to see that a piece of granite rock is heterogeneous, because the quartz, feldspar, and mica phases can be distinguished, usually by the naked eye. What about a mixture of sand and salt? Here again there is likely to be little problem in recognizing the existence of two phases, because grains of sand usually look different from grains of salt. If necessary, a magnifying glass could be used to see that two kinds of particles are present. If a heterogeneous mixture is composed of extremely small particles, however, even a high-powered light microscope might be inadequate for detecting the presence of more than one phase. (An electron microscope would probably do the trick, however.) In cases in which visual techniques are inadequate for characterizing a mixture as homogeneous or heterogeneous, measurements of physical properties are often useful.

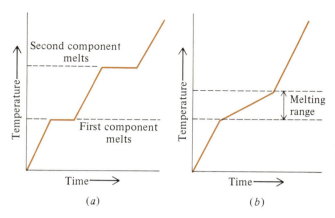

Figure 12-1
Melting-point heating curves. (*a*) **Two-component heterogeneous mixture.** (*b*) **Two-component solution.**

For example, the phase-change temperature of one of two components in a heterogeneous mixture is unaffected by the presence of the other. A melting-point heating curve, for example, shows two temperature holds, each characteristic of one component.

The phase-change temperature of a solution is different from that of each component and, in addition, changes as the phase change progresses. The melting-point heating curve of a solid solution shows no temperature hold, but only a *break* (change of slope) in the curve when melting starts and another break when it ends. The contrast between the melting-point heating curve of a heterogeneous mixture of solids and that of a solid solution is shown in Fig. 12-1. When a substance melts over a range, as is shown in Fig. 12-1*b*, it indicates that the substance is not pure.

Colloids   When comparatively large particles, such as grains of sand, are shaken up with water, the resulting system is clearly heterogeneous: the water and sand phases can be seen individually, and they are rapidly separated by gravity, the more dense sand ending up on the bottom. When much smaller particles, such as sugar molecules, are dispersed in water, they form a homogeneous mixture, a solution, in which the dispersed particles are invisible and are prevented from settling out by thermal molecular motion.

In between the heterogeneous and homogeneous categories is a gray area, the *colloidal dispersion,* or, simply, *colloid.* Here the dispersed particles are too small to be individually seen, often even under the microscope, yet they are larger than what we usually think of as a molecule. The dispersed particles of a colloid do not settle out, nor can they be separated by ordinary filtration. The typical colloidal particle has at least one of its dimensions smaller than 10^{-5} cm. Colloidal particles may consist of up to hundreds of thousands of atoms or molecules. (A single protein molecule may be considered to be a colloidal particle. Some proteins have molecular weights in the millions.) Different types of colloids include the *sol* (a dispersion of solid particles in a liquid), the *emulsion* (a dispersion of liquid droplets in another liquid), the *smoke* (solid particles dispersed in a gas), and the *fog* (liquid droplets in a gas). Smokes and fogs are also called *aerosols.*

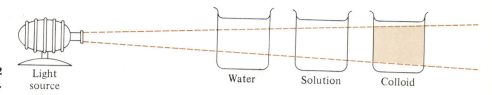

Figure 12-2
Tyndall effect.   Light   Water   Solution   Colloid
                            source

An additional type of colloid is the *gel* in which both the dispersed and dispersing phases extend continuously throughout the system. In a gel the dispersed phase forms fine filaments or thin mats which trap the dispersing phase in a semirigid structure. Jellies are examples of gels. The three-dimensional structure of the dispersed phase in some gels may be temporarily broken down by applying stress or agitation. Such a gel then reverts to a sol which is no longer viscous or semirigid but can flow readily. Left undisturbed, the system reforms a gel. This phenomenon is known as *thixotropy*. Some paints are thixotropic gels; they are thick and viscous in the can, "liquefy" when a brush is dipped in, thicken up on the brush so as to minimize dripping, liquefy when brushed on a wall or ceiling, so as to go on smoothly, and become viscous once more on the painted surface, where they dry without running or dripping.

Many colloids can not be identified as such by appearance alone. A colloidal dispersion of gold particles in water, for example, is a beautiful red (or sometimes blue or purple, depending upon particle size) liquid with no trace of turbidity (cloudiness). Such a system can readily be shown to be a colloid by making use of the *Tyndall effect*, the sideways scattering of a beam of light as it passes through the mixture (Fig. 12-2). (The particles in a solution are too small to scatter light.)

When two colloidal particles collide, they may adhere to form a larger particle, and this process may continue until the particles are heavy enough to settle out. Such coagulation of colloids may be prevented in a number of ways. Colloidal particles can adsorb ions from the surrounding solution. (In *ad*sorption the sorbed substance sticks to the surface of the sorbing solid and is not drawn into the interior as in *ab*sorption.) Thus colloidal particles of ferric oxide, Fe_2O_3, can adsorb H^+ ions to acquire a positive charge, and the resulting electrostatic repulsion then keeps the particles apart. Some colloidal particles adsorb molecules of the dispersing medium on their surfaces, effectively insulating them from each other. In water certain substances such as gelatin can be used to coat colloidal particles in a thin layer which then adsorbs water molecules. The gelatin, which also forms gels by itself, is called a *protective colloid* in this case.

12-2
Types of solutions

Considering *binary* (two component) solutions only, there appear to be nine possible types of solutions. An example of each of these is shown in Table 12-1.

It should be pointed out, however, that in any solution the dispersed particles are no longer part of their original solid, liquid, or gas. The condition of alcohol molecules dissolved in water is not much different from that of sugar molecules in water, for example. In each case the substance which is dispersed has disappeared. Reasoning from this standpoint it is perhaps better to consider that there are only *three* types of solution: solid, liquid, and gaseous.

Table 12-1
Types of solutions

Solution type	Example
Gaseous solutions	
Gas dissolved in gas	Oxygen dissolved in nitrogen
Liquid dissolved in gas	Chloroform dissolved in (evaporated into) nitrogen
Solid dissolved in gas	Dry ice dissolved in (sublimed into) nitrogen
Liquid solutions	
Gas dissolved in liquid	Carbon dioxide dissolved in water
Liquid dissolved in liquid	Alcohol dissolved in water
Solid dissolved in liquid	Sugar dissolved in water
Solid solutions	
Gas dissolved in solid	Hydrogen dissolved in palladium
Liquid dissolved in solid	Mercury dissolved in gold
Solid dissolved in solid	Copper dissolved in nickel

ADDED COMMENT

More succinctly, after a molecule of alcohol dissolves in water, it does not matter whether it came from the gaseous, liquid, or solid state. It is merely part of the new phase, the solution.

Gaseous solutions   It is not possible to have a heterogeneous mixture of two gases, because gases mix with each other in all proportions. (We say they are completely *miscible*.) A gaseous solution has the typical gas structure, but the molecules are not all alike. Air, the gaseous solution with which we have the greatest contact (literally), is composed primarily of oxygen and nitrogen, with smaller quantities of argon, carbon dioxide, neon, helium, and at least several dozen other substances in small, sometimes varying, concentrations. Also, there are many more substances, considered pollutants, at appreciable concentrations in the air today than there were 100 years ago.

Liquid solutions   Liquid solutions have the molecular arrangement which is typical of a pure liquid, but not all the particles are the same. Much of this chapter is devoted to the properties of liquid solutions. In the special case in which the dispersing substance is water, a solution is called an *aqueous* solution.

Solid solutions   As we have already seen (Sec. 11-3), a solid solution has an ordered crystal lattice in which the structure has been made more random either by the existence of different particles occupying the lattice points (a substitutional solid solution) or by the presence of particles occupying some of the spaces between lattice points (an interstitial solid solution).

   If it is desired to prepare a solid solution of two components, both of which are solids when pure, simply mixing the two is usually not practical. The two components may tend to dissolve in each other spontaneously, but because solid-state diffusion (of each component into the crystal lattice of the other) is so slow, a more practical procedure is to melt the two solids together and then freeze the mixture. (This assumes, of course, that the two components are miscible in *both* the liquid and solid states.)

**12-3
Concentration
and solubility**

The properties of a solution depend upon its composition and upon the nature of each of its components. When one component is present in considerable excess over the other, it is usually referred to as the *solvent*. Other components, present in smaller proportions, are called *solutes*. If we dissolve 1 g each of sodium chloride and sugar in 100 g of water, we would refer to the water as the solvent and to the NaCl and sugar as solutes. The solvent may be thought of as the component throughout which the particles of solute are randomly dispersed.

Two more terms which are in common chemical use are *concentrated* and *dilute*. These are relative terms and are usually used to give a qualitative indication of the concentration of a solute in a solution. A "concentrated solution of NaCl" thus has a higher proportion of NaCl in it than does a "dilute solution of NaCl." The word *concentrated* is also used to specify certain common, commercially available solutions. Thus "concentrated sulfuric acid" usually refers to the solution which is readily available from laboratory chemical manufacturers and which is about 95 percent sulfuric acid and 5 percent water by mass.

ADDED COMMENT

The terms *concentrated* and *dilute* refer to a high concentration and a low concentration, respectively, of solute. Terms often used by the nonchemist to mean the same thing are *strong* and *weak*. "This coffee is really strong!" and "I'd like a cup of weak tea." are perfectly understandable statements, for example. We will avoid using strong and weak, however, since these terms have different, though related, chemical meanings. (See Sec. 12-5.)

Concentration units

The composition of a solution can be described quantitatively by specifying its concentration. For our purposes the most important of many *concentration units* are mole fraction, mole percent, molarity, molality, percent by mass (commonly called *percent by weight*), and normality. Definitions and explanations of these terms follow, except for those of normality, which is discussed in Sec. 13-5.

Mole fraction, represented by the symbol X, is the ratio of the number of moles of one component to the total number of moles. Letting n represent number of moles and designating the various components as $A, B, C, \ldots$, we can write

$$X_A = \frac{n_A}{n_A + n_B + n_C + \cdots}$$

and

$$X_B = \frac{n_B}{n_A + n_B + n_C + \cdots}$$

and so on. Note that $X_A + X_B + X_C + \cdots = 1$.

● **EXAMPLE 12-1**

Problem   Five grams of NaCl is dissolved in 25.0 g of H_2O. What is the mole fraction of NaCl in the solution?

Solution   The formula weight of NaCl is 58.44, so 5.00 g of NaCl is

$$5.00 \text{ g} \left(\frac{1 \text{ mol}}{58.44 \text{ g}}\right) = 8.56 \times 10^{-2} \text{ mol NaCl}$$

The molecular weight of water is 18.02; so 25.0 g of H_2O is

$$\frac{25.0 \text{ g}}{18.02 \text{ g mol}^{-1}} = 1.39 \text{ mol } H_2O$$

$$X_{NaCl} = \frac{n_{NaCl}}{n_{NaCl} + n_{H_2O}}$$

$$= \frac{8.56 \times 10^{-2} \text{ mol}}{8.56 \times 10^{-2} \text{ mol} + 1.39 \text{ mol}} = 5.80 \times 10^{-2}$$

In this solution, then, the mole fraction of NaCl is 0.0580. (The mole fraction of water is 1.0000 − 0.0580, or 0.9420.) ●

Mole percent (mol %) is the percent of the total number of moles that is of one component. It is simply mole fraction times 100, or

$$\text{mol \% } A = X_A \times 100$$

● **EXAMPLE 12-2**

Problem   What is the mole percent NaCl in the solution of Example 12-1?

Solution $\text{mol \% NaCl} = X_{NaCl} \times 100 = 5.80 \times 10^{-2} \times 100 = 5.80\%$

The solution is 5.80 mol % NaCl and 94.20 mol % H_2O. ●

Molarity (M) is the number of moles of solute dissolved per liter of solution. It is calculated by taking the ratio of the number of moles of solute to the volume of solution in liters. In other words,

$$M_A = \frac{n_A}{V_{\text{soln, liters}}}$$

A solution that contains 2 mol of solute per liter of solution is said to be a 2-*molar* (2 M) solution.

● **EXAMPLE 12-3**

Problem   Ten grams of ascorbic acid (vitamin C), $C_6H_8O_6$, is dissolved in enough water to make 125 ml of solution. What is the molarity of the ascorbic acid?

Solution   The molecular weight of ascorbic acid is 6(12.0) + 8(1.0) + 6(16.0), or 176.0. Thus 10.0 g of ascorbic acid is

$$10.0 \text{ g} \frac{1 \text{ mol}}{176.0 \text{ g}} = 5.68 \times 10^{-2} \text{ mol}$$

The volume of the solution in liters is

$$125 \text{ ml} \frac{1 \text{ liter}}{10^3 \text{ ml}} = 0.125 \text{ liter}$$

Thus the molarity is

$$M_{C_6H_8O_6} = \frac{n_{C_6H_8O_6}}{V_{\text{soln, liters}}}$$

$$= \frac{5.68 \times 10^{-2} \text{ mol}}{0.125 \text{ liter}} = 0.454 \text{ mol liter}^{-1} \quad \text{or} \quad 0.454 \text{ } M$$

The solution is 0.454-molar. ●

● **EXAMPLE 12-4**

Problem Calculate the molarity of NaCl in the solution of Example 12-1. The density of the solution is 1.12 g ml^{-1}.

Solution We know the number of moles of NaCl from Example 12-1, but we need to find the volume of the solution in liters. The mass of the solution is 5.00 g (NaCl) + 25.0 g (H_2O) = 30.0 g (total). From the density we see that each milliliter weighs 1.12 g; so

$$\text{Volume of solution} = 30.0 \text{ g} \left(\frac{1 \text{ ml}}{1.12 \text{ g}} \right) = 26.8 \text{ ml}$$

In liters this is

$$26.8 \text{ ml} \frac{1 \text{ liter}}{10^3 \text{ ml}} = 2.68 \times 10^{-2} \text{ liter}$$

$$\text{Molarity} = \frac{n_{NaCl}}{V_{\text{soln, liters}}}$$

$$= \frac{8.56 \times 10^{-2} \text{ mol}}{2.68 \times 10^{-2} \text{ liter}} = 3.19 \text{ mol liter}^{-1} \quad \text{or} \quad 3.19 \text{ } M$$

The solution is 3.19-molar. ●

Molality (*m*) is the number of moles of solute dissolved per kilogram of solvent. Molality is calculated by taking the ratio of the number of moles of solute to the number of kilograms of solvent.

$$m_A = \frac{n_A}{\text{mass}_{\text{solvent, kg}}}$$

A solution containing 3 mol of solute per kilogram of solvent is said to be 3-molal (3 *m*).

● **EXAMPLE 12-5**

Problem What is the molality of NaCl in the solution of Example 12-1?

Solution We know the number of moles of NaCl (from Example 12-1), but what is the mass of the water in kilograms? Since 1 g is 10^{-3} kg, 25.0 g of H_2O is 0.0250, or 2.50×10^{-2} kg.

$$\text{Molality} = \frac{n_{NaCl}}{\text{mass}_{H_2O, \text{ kg}}}$$

$$= \frac{8.56 \times 10^{-2} \text{ mol}}{2.50 \times 10^{-2} \text{ kg}} = 3.42 \text{ mol kg}^{-1} \quad \text{or} \quad 3.42 \text{ } m$$

The solution is 3.42-molal. ●

Percent by mass (*mass %*) is the percent of the total mass of a solution that is of one component. It is 100 times the mass fraction, which is the ratio of the mass of one component to the total mass. (The masses can be in any units as long as they are the same.)

$$\text{Mass \% } A = \frac{\text{mass}_A}{\text{mass}_A + \text{mass}_B + \text{mass}_C + \cdots} \times 100$$

● **EXAMPLE 12-6**

Problem   What is the mass percent of NaCl in the solution of Example 12-1?

Solution

$$\text{Mass \% NaCl} = \frac{\text{mass}_{\text{NaCl}}}{\text{mass}_{\text{NaCl}} + \text{mass}_{\text{H}_2\text{O}}} \times 100$$

$$= \frac{5.00 \text{ g}}{5.00 \text{ g} + 25.0 \text{ g}} \times 100 = 16.7\%$$

The solution is 16.7 percent NaCl and 83.3 percent H_2O by mass. ●

Percent by mass is commonly called *percent by weight (or weight percent,* wt. %).

The mechanism   Just how do solutions form?   We will answer this question by concentrating our
of dissolving   attention on structural changes which take place during the dissolving process and
on the various forces acting among solute and solvent particles.   Consider what
happens when a solid is added to a liquid to form a (liquid) solution.   When the
solute is first added, the process of destruction of the solute's solid-state structure
begins.   Little by little, solvent particles chip away at the surface of the crystal
lattice, prying out solute particles, surrounding them, and finally dispersing them.
The result is the disintegration of the structure of the solute and the alteration, at
least, of the structure of the solvent.   (There are now some solute particles where
there was once only solvent.)   The ease with which all this takes place depends on
the relative strengths of the forces between adjacent solute particles (*solute-solute
interactions*) and between solvent particles (*solvent-solvent interactions*) *before* the
dissolving process, and the forces between solute and solvent particles (*solute-
solvent interactions*) *after* dissolving.   As dissolving takes place solute-solute and
solvent-solvent forces are replaced by solute-solvent forces.   We can conclude,
then, that high solubility is generally favored by low solute-solute (lattice-energy)
and solvent-solvent forces and also by high solute-solvent forces.   There is an old
generalization which goes: "Like dissolves like."   This means that a solvent will
dissolve a solute if they have similar structures.   More specifically, polar solvents
tend to dissolve polar solutes, and nonpolar solvents to dissolve nonpolar solutes.
Although not perfect, this rule is nonetheless very useful.   Let us see how it can be
justified in terms of the structural changes which take place during the dissolving
process.

First consider carbon tetrachloride (CCl_4), a typical *nonpolar solvent.*   (CCl_4 is
tetrahedral, and so has no dipole moment.)   Solutes which are found to have
relatively high solubilities in CCl_4 are *nonpolar,* iodine (I_2) and sulfur (S_8), for
example.   In nonpolar molecular substances the only intermolecular attractions
are the comparatively weak London forces, so that when solutions of these
components are formed, solute-solute and solvent-solvent interactions are easily
replaced by solute-solvent interactions.   When nonpolar solute molecules are
inserted among nonpolar solvent molecules, there is little change in environment
for either, and so the solution is readily formed.   On the other hand, CCl_4 and
other nonpolar liquids are *poor* solvents for *polar* and *ionic* compounds.   Lattice
energies in such solutes tend to be higher than in nonpolar solutes and because of
the lack of polarity of the solvent molecules, solute-solvent interactions are weak,
and so solute-solute forces are not readily overcome.

Now consider the solvent properties of a *polar solvent*, water. (Recall the high electronegativity of oxygen and the bent shape of the H_2O molecule.) Water is a good solvent for *polar solutes* such as sucrose (cane sugar, $C_{12}H_{22}O_{11}$) and also for *ionic solutes* such as sodium chloride (NaCl). In such cases the establishment of strong solute-solvent interactions provides the energy needed to overcome the strong forces of interaction in the solute. Water is an especially good solvent for many ionic compounds, because the ion-dipole forces established in solution are strong, and many H_2O molecules surround each ion. However, water is a poor solvent for *nonpolar solutes*. The solvent-solute forces in water (and in other highly polar solvents) are strong, and with nonpolar solutes the weak solute-solvent interactions provide insufficient energy to compensate for that required to separate the solvent dipoles.

The like-dissolves-like rule of thumb is a handy generalization, but it, like most generalizations, tends to oversimplify the situation. Not all ionic substances, for example, are highly soluble in water; NaCl is, but $CaSO_4$ is not. Evidently we must look closely at the specific interactions in solute, solvent, and solution. (In $CaSO_4$ the ionic charges are $2+$ and $2-$, which would tend to make the lattice energy higher than in NaCl.) Another case in point: the ethanol molecule (C_2H_5OH) is less polar (dipole moment, 1.70 D) than ethyl chloride (C_2H_5Cl; 2.05 D), yet ethanol is *completely* miscible with water, while the solubility of ethyl chloride in water is very low. Here again considering only polarity of solute and solvent is an error; specific strong interactions between ethanol and water (described below) account for their mutual solubility.

Water as a solvent   Alchemists dreamed of a universal solvent, a liquid which would dissolve anything. (It is probably fortunate that none exists. How would it be contained?) In spite of the fact that water is the commonest substance on the earth's surface, this liquid has some unusual properties. One of the most important of these is its ability to dissolve many kinds of substances. Though not the once-sought-for universal solvent, water dissolves many ionic compounds, many polar organic and inorganic substances, and even some substances of low polarity with which it can form specific interactions.

One reason water dissolves ionic substances is its ability to stabilize ions in solution, to keep them separated from each other. This is largely due to the high dielectric constant of water. The *dielectric constant* of a substance is a measure of the *polarizability* of its molecules, i.e., their ability to distort and/or orient themselves so as to "neutralize" a charge located near them. The force between two charged particles is reduced if a medium of high dielectric constant is between them. Figure 12-3 schematically shows a pair of oppositely charged ions in (a) a vacuum and (b) a medium of high dielectric constant. The medium allows positive ($+$) charges to build up around the negative ion and negative ($-$) charges around the positive ion, partially neutralizing the ionic charges. The electrostatic force of attraction between the ions in Fig. 12-3b is lower than in a. Without worrying about units, we can write the force (F) between two charges, q_1 and q_2, at a distance r apart as

$$F = \frac{1}{\epsilon} \frac{q_1 q_2}{r^2}$$

Figure 12-3
Electric charges.
(*a*) In a vacuum.
(*b*) In a medium of high
dielectric constant
(schematic).

(*a*)

(*b*)

where ϵ (epsilon) is the dielectric constant of the medium between the charges. The dielectric constant of a vacuum is 1, and that of air is 1.0005, but for H_2O the dielectric constant is about 80 at room temperature. This means that the attractive force between two oppositely charged ions in water is one-eightieth of what it would be in a vacuum. Ions can be individually stable in water, then, because the water greatly reduces the forces between them.

The high dielectric constant of water is due in part to the high polarity of its molecules and in part to the fact that in water intermolecular forces are abnormally high because of a phenomenon known as *hydrogen bonding*. Hydrogen bonding is an unexpectedly strong intermolecular attraction in which H atoms form "bridge bonds" between atoms of high electronegativity in adjacent molecules. It occurs in compounds in which H atoms are bonded to electronegative atoms such as N, O, and F (as in NH_3, H_2O, and HF, respectively). The hydrogen bond is a weaker force than the covalent or ionic bond, but usually stronger than the London force and the usual dipole-dipole attraction. How does the hydrogen bond form? A useful though somewhat oversimplified explanation is that the electron pair which bonds an H atom to a highly electronegative atom is effectively withdrawn from the H atom, greatly reducing the electron charge-cloud density around the hydrogen nucleus, which is a mere proton. Thus the unshielded proton attracts electrons surrounding an electronegative atom *in an adjacent molecule*. In this way the H atom serves to bond two other atoms together indirectly.

In liquid water (and in ice) both H atoms of a water molecule can form hydrogen bonds to adjacent molecules. These H atoms serve to "bridge bond" the oxygen of the first H_2O to the oxygens of the two adjacent H_2O molecules. At the same time two other H_2O molecules can similarly hydrogen bond to the first H_2O. This leads to the structure schematically shown in Fig. 12-4, in which each O atom is surrounded by four H atoms, two with which it is covalently bonded and two with which it is hydrogen bonded. In the illustration the hydrogen bonds are shown as dashed lines.

In *ice* each water molecule has its O atom at the center of a regular tetrahedron with the oxygen atoms of four adjacent water molecules at the corners. In between each pair of oxygens is a hydrogen atom, closer to the central oxygen in two cases, and farther away in the other two. The shorter distance is that of the

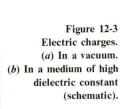

Figure 12-4
Hydrogen bonding in
water (dashed lines rep-
resent hydrogen bonds).

covalent O—H bond, and the longer, of the hydrogen bond. In *liquid water* the structure is less regular and is constantly changing, as the molecules move around, and bonds are broken and formed.

Water is like most hydrogen-bonded systems in that the hydrogen is closer to one O atom than the other

$$\text{-----H—O-----H—O-----H—O-----H—}$$
$$\qquad\quad|\qquad\quad|\qquad\quad|$$
$$\qquad\quad\text{H}\qquad\quad\text{H}\qquad\quad\text{H}$$

but each H atom can jump to the O atom of the adjacent molecule

$$\text{—H-----O—H-----O—H-----O—H-----}$$
$$\qquad\quad|\qquad\quad|\qquad\quad|$$
$$\qquad\quad\text{H}\qquad\quad\text{H}\qquad\quad\text{H}$$

This is called *unsymmetrical* hydrogen bonding, and the H atoms are believed to jump rapidly back and forth, as covalent and hydrogen bonds interchange positions. In some other systems, however, the hydrogen is *symmetrically* located halfway between the two electronegative atoms.

Hydrogen bonding is not completely understood. It is believed that a major component of the interaction is the electrostatic attraction between the exposed hydrogen nucleus and a lone pair on the adjacent electronegative atom. There is evidence, however, that the bond contains some covalent character, as well. The electron charge cloud of the lone pair may be smeared out to partially shield the exposed proton.

In hydrogen-bonded liquids intermolecular forces are abnormally high, and so these liquids are said to be highly *associated*. Perhaps the most striking evidence for hydrogen bonding in a pure liquid comes from an anomalously high boiling point. Figure 12-5 shows the normal boiling points of hydrogen compounds of the periodic-group IV, V, VI, and VII representative elements. It can be seen that NH_3, H_2O, and HF do not conform to the trend in their respective series. In each

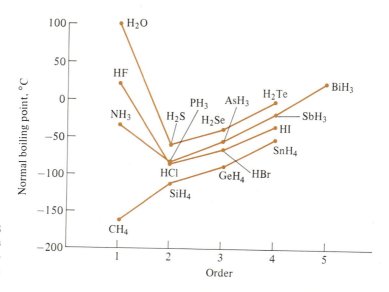

**Figure 12-5
Normal boiling-points
of some binary hydro-
gen compounds.**

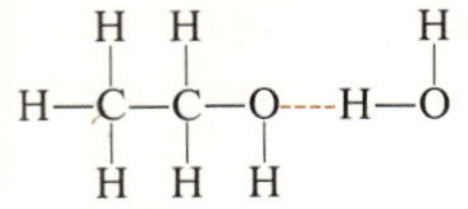

H—C—C—O---H—O (with H atoms labeled above and below)

Figure 12-6
Hydrogen bonding in water-ethanol.

case the boiling point appears to be inconsistently high. The boiling point of CH_4, however, is *not* out of line with the others in its series. In NH_3, H_2O, and HF, the presence of one or more lone pairs on a highly electronegative atom results in hydrogen bonding, which raises the boiling point. Carbon is not very electronegative and has no lone pairs in CH_4 so in this compound the effect is absent.

Hydrogen bonding accounts for the fact that many compounds with hydroxyl (—OH) groups are soluble in water. As previously mentioned, ethanol (C_2H_5OH) mixes with water in all proportions, while ethyl chloride (C_2H_5Cl), which has about the same dipole moment, is essentially insoluble. The explanation for the apparent anomaly is that ethanol molecules can hydrogen bond with water molecules, as shown in Fig. 12-6. The high solubility of sucrose and of other sugars is also a result of solute-solvent hydrogen bonding. Sucrose has eight —OH groups per molecule, each of which can hydrogen bond to water.

Saturation and solubility

Let us do a thought experiment: Imagine a beaker containing 1.000 kg of water at 25°C. (The volume occupied by the water is 1.002 liters.) Now imagine that we start adding sodium chloride to the water a little at a time, stirring the mixture after each addition. The first pinch of salt rapidly dissolves

$$NaCl(s) \longrightarrow Na^+(aq) + Cl^-(aq)$$

In the above equation, (*aq*) serves to emphasize that the ions are dissolved in aqueous solution. Most of the time in this book, however, we omit this for simplicity's sake:

$$NaCl(s) \longrightarrow Na^+ + Cl^-$$

Additional NaCl dissolves rapidly, too, but gradually the time necessary for a given-sized pinch to dissolve increases until finally a last "micropinch" of salt fails to dissolve at all. At this point all change appears to have ceased in the beaker. But our eyes do not tell us the whole story. At the surfaces of the crystals of NaCl two reactions are occurring: one is the continued dissolving of Na^+ and Cl^- ions,

$$NaCl(s) \longrightarrow Na^+ + Cl^-$$

and the other is the redeposition of Na^+ and Cl^- ions from the solution on the surfaces of the solid.

$$Na^+ + Cl^- \longrightarrow NaCl(s)$$

These reactions occur simultaneously and at the same rate, so that they compensate for each other. The system is, in other words, at *equilibrium:*

$$NaCl(s) \rightleftharpoons Na^+ + Cl^-$$

Support for the fact that there really are two reactions occurring in this system comes from a number of different experiments which can be performed. Perhaps the most direct evidence comes from placing the system (solution plus excess crystals of NaCl) in a sealed container and then just waiting for a long time, 30 to 40 years. At the end of this time, as long as temperature fluctuations have not been too large, the tiny crystals will have been replaced by a single large symmetrical crystal of NaCl weighing *exactly* what the small crystals weighed. This change cannot be explained except in terms of the two opposing reactions. (The single crystal of salt has a very slightly lower energy than the collection of small crystals; so its formation is favored.) A second and more rapid way of showing the existence of equilibrium in the above solution is to add to the excess undissolved sodium chloride a few crystals of NaCl in which some of the chloride ions are the radioactive isotope $^{36}_{17}Cl$ (a β-particle emitter). A short time later radioactivity can be detected from Cl^- ions *in the solution,* showing the existence of the equilibrium

$$NaCl(s) \rightleftharpoons Na^+ + Cl^-$$

When an equilibrium exists between the solution and excess solute, the solution is said to be *saturated.* But because we like to be able to refer to any solution of this solute at this concentration as being saturated, we define a saturated solution as follows: a *saturated solution* is one which is in equilibrium with excess solute, or one which would be, if excess solute were present. (In this way we may still refer to the solution as saturated, even if we filter off the excess solute.)

An *unsaturated solution* is one which has a concentration of solute *less than* that of a saturated solution. A *supersaturated* solution is one in which the solute concenration is *greater than* that of a saturated solution.

The *solubility* of a solute in a given solvent is defined as the *concentration* of that solute *in its saturated solution.* Some solutes are infinitely soluble in a given solvent, while others have solubilities so low that they are not measurable by direct methods. Although there is probably no such thing as complete insolubility, the term *insoluble* is often applied to a substance whose solubility is very low. *Sparingly soluble* and *slightly soluble* are terms which are also used. Solubility depends upon the properties of solute and solvent, as we have already discussed, and it also depends upon temperature and pressure.

Solubility and temperature

Since solubility is an equilibrium concentration, we can apply Le Châtelier's principle (Sec. 11-4) in order to find out what happens when the temperature of a saturated solution is changed. It is important to know whether the dissolving process is *exothermic*

$$Solute + solvent \longrightarrow solution + heat$$

or *endothermic,*

$$Solute + solvent + heat \longrightarrow solution$$

The *heat of solution* is defined as ΔH for the dissolving process, and so is equal to $H_{solution} - (H_{solute} + H_{solvent})$. Therefore, for the *exothermic* case, ΔH_{soln} is *negative,* and for the *endothermic, positive.*[1]

Consider now a saturated aqueous solution of potassium iodide with excess $KI(s)$ present. For KI, $\Delta H_{soln} = 21$ kJ mol^{-1}; so we know that KI dissolves in water with the absorption of heat and can write the saturation equilibrium equation as

$$21 \text{ kJ} + KI(s) \rightleftharpoons K^+ + I^-$$

---

[1] Sometimes heat of solution is defined as the heat *liberated* when the solute dissolves. This definition gives the heat of solution the opposite sign from our definition. One obviously must be careful when encountering this term in strange places.

If we raise the temperature of the saturated KI solution, we predict according to Le Châtelier's principle that the above equilibrium will *shift to the right* (1) using up some of the added heat (and some of the excess solid KI) and (2) increasing the concentration of K^+ and I^- ions in solution. After equilibrium has been reestablished at a higher temperature, the concentration of dissolved KI has become higher, that is, the solubility of KI increases with increasing temperature.

An example of an exothermic process is the dissolving of lithium iodide (LiI) in water, for which $\Delta H_{soln} = -71$ kJ mol^{-1}. We can write the saturation equilibrium equation as

$$LiI(s) \rightleftharpoons Li^+ + I^- + 71 \text{ kJ}$$

If we raise the temperature of a saturated solution of LiI, the equilibrium shifts to the left (1) using up some of the added heat (and Li^+ and I^- ions in solution) and (2) forming more solid LiI. (We observe the precipitation of some LiI out of solution.) After equilibrium has been reestablished at a higher temperature, the concentration of dissolved LiI is lower; so we can say that the solubility of lithium iodide decreases with an increase in temperature.

Why should ΔH_{soln} vary from substance to substance? The problem is best tackled by imagining that the dissolving process occurs in two steps: first, the breakup of the structure of the solute followed by the insertion of solute particles into pockets formed by separation of solvent molecules. The latter process is known generally as *solvation;* when water is the solvent, it is called *hydration.* The relative energy changes in the two steps determine whether the overall process is

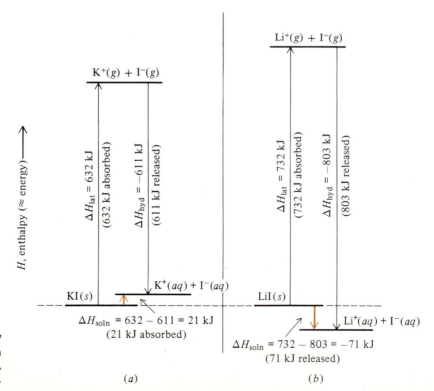

Figure 12-7
Molar heats of solution compared: (*a*) **KI,** (*b*) **LiI.**

endothermic or exothermic. Consider, for example, the contrasting cases of KI and LiI. In each case the energy necessary to break up the crystal lattice is the lattice energy ΔH_{lat}. The hydration of the ions is accompanied by a release of energy. ΔH for this process is called the *hydration energy* ΔH_{hyd}. The relative sizes of the ΔH_{lat}, ΔH_{hyd}, and ΔH_{soln} for KI and LiI are shown in Fig. 12-7. It can be seen that in each case the ΔH_{soln} is a relatively small difference between two larger numbers. It also can be seen that when the magnitude of the lattice energy is larger than that of the hydration energy, the dissolving process is endothermic. When the reverse is true, the process is exothermic.

ΔH_{soln} for dissolving solids or liquids in liquids can be either positive or negative. For aqueous solutions it is more frequently positive; so the majority of substances have solubilities which increase with temperature. (This should not be used as a rule of thumb, however; there are too many exceptions.) In Fig. 12-8 are plotted the solubilities of several solutes in water as a function of temperature.

When *gases* dissolve in liquids, ΔH_{soln} is usually negative; that is, heat is liberated. (The solvation energy usually exceeds the energy necessary to separate the molecules in the liquid.) Thus in the majority of cases the solubility of gas decreases with temperature. This nearly always is true in water. Boiled water, for example, tastes "flat," in part because dissolved air (and chlorine!) is less soluble at the boiling point and is removed from the water.

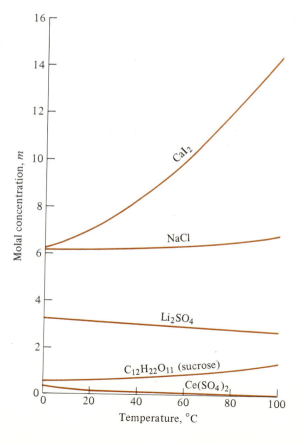

Figure 12-8
Change of solubility with temperature.

Because of the change of solubility with temperature it is possible to prepare a *supersaturated solution,* one in which the solute concentration is *higher* than that in a saturated solution. Such a solution is made by first preparing a saturated solution, then removing excess solute by filtration or other means, and finally changing conditions so that the solubility is lowered. Potassium nitrate, KNO_3, for example, has a solubility in water which *increases* with temperature. Expressed in terms of molality its solubility at 20°C is 3.1 m, while at 40°C it is 6.3 m. If we prepare a saturated solution of KNO_3 at 40°C, filter off the excess solid, and cool it to 20°C, the concentration of the solution *may* remain at 6.3 m, even though the solubility at the lower temperature is only 3.1 m. The word *may* has been emphasized because the supersaturated state is unstable and tends to revert to the saturated state by precipitating some solute out of solution. The situation is reminiscent of that of a supercooled liquid (Sec. 11-2). The probability that crystallization (precipitation) starts may be low enough so that the supersaturated solution exists for some time. Eventually, however, it will start to precipitate, and once the first tiny crystal has formed, precipitation may occur rapidly until equilibrium has been established. As in the case of a supercooled liquid, crystallization may be initiated by agitation, by scratching the inside of the container, or most effectively of all, by adding a small seed crystal of solute.

In the case of a solute whose solubility *decreases* with *increasing* temperature, preparing a saturated solution at one temperature, filtering off excess solute, and then *heating* it may produce a supersaturated solution.

A supersaturated solution may form as a result of a chemical reaction which occurs in solution. If the product of a reaction is formed rapidly, and its concentration rises above saturation, a supersaturated solution results.

Solubility and pressure

The solubility of solids and liquids in liquid solvents is practically independent of pressure. According to Le Châtelier's principle, an increase in pressure should favor the dissolving process if the solution volume is less than that of the unmixed solute and solvent. In this case, then, the solubility would increase with increasing pressure. The volume change which accompanies the dissolving process can be determined from density information, but it is always small, so the effect of pressure on solid and liquid solubilities is almost always negligible.

Figure 12-9
Increase in solubility of a gas in a liquid as pressure increases. (*a*) Low pressure. (*b*) Pressure is increased; piston moves down. (*c*) High pressure.

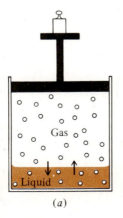
(*a*)

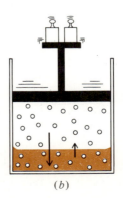
(*b*)

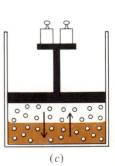
(*c*)

Table 12-2
Values of the Henry's
law constant in water

Gas	$K \times 10^5$ atm^{-1}			
	0°C	20°C	40°C	60°C
H_2	1.72	1.46	1.31	1.21
N_2	1.86	1.32	1.00	0.874
O_2	3.98	2.58	1.84	1.57

Gases always dissolve in liquids with a considerable decrease in total volume. This means that in the equilibrium

$$\text{Solute}(g) + \text{solvent}(l) \rightleftharpoons \text{solution}(l)$$

increasing the pressure favors the solution. If a gas and its saturated solution are enclosed in a cylinder with a tight-fitting piston as in Fig. 12-9a, pushing in the piston (b) causes more gas to dissolve, so that when the system returns to equilibrium (c) the pressure is not as high as it would have been if the gas had not been able to dissolve in the liquid, and hence the system were not at equilibrium. At the higher pressure the solubility of the gas is greater than it was at the start. A decrease in pressure accordingly causes a decrease in the solubility of a gas. When a bottle of soda pop is opened, for example, the partial pressure of the carbon dioxide at the top of the bottle is decreased, and because this decreases the solubility of CO_2 in the soda pop, the CO_2 starts to come out of the solution which is now supersaturated. The result is foam.

The solubility of a gas dissolved in a liquid is proportional to the partial pressure of the gas above the liquid. This is a statement of Henry's law, which can be written

$$X = KP$$

where X is the equilibrium mole fraction of the gas in solution (its solubility), P is its partial pressure in the gas phase, and K is the constant of proportionality, usually called the *Henry's law constant.*

Henry's law applies only when the concentration of the solute and its partial pressure are both low, that is, when the gas and its solution are essentially ideal, and when the solute does not interact strongly in any way with the solvent.

Values of the Henry's law constant for several gases at several temperatures are given in Table 12-2. Notice that the units of the constant are in atm^{-1}. (Note also that the values have been multiplied by 10^5. This means that the value of the Henry's law constant for H_2 at 0°C is 1.72×10^{-5} atm^{-1}.)

**12-4
Colligative
properties**
Properties of a solution which depend upon the concentration of solute particles and not upon their identity are known as *colligative properties.*[2] Each of these properties depends upon the lowering of the *escaping tendency* of solvent molecules by the addition of solute particles. The colligative properties include vapor-pressure lowering, boiling-point elevation, freezing-point depression, and osmotic pressure.

---

[2] *Colligative* comes from a Latin word meaning "collected" or "bound together."

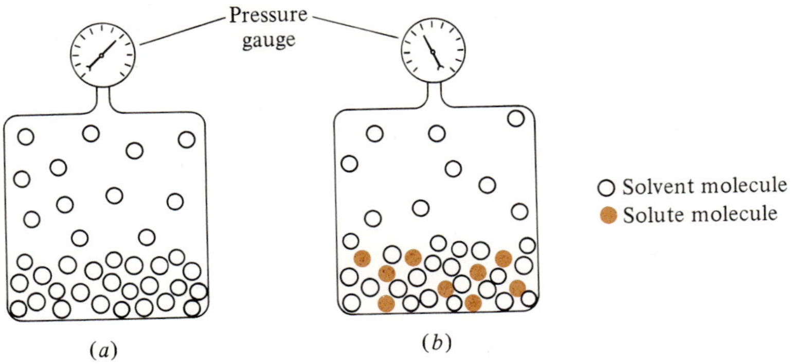

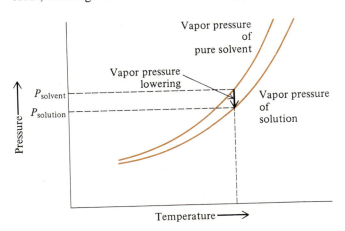

Pressure gauge

○ Solvent molecule
● Solute molecule

Figure 12-10
Mechanism of vapor-
pressure lowering.
(a) Pure solvent.
(b) Solution.

(a)   (b)

Vapor-pressure
lowering

The escaping tendency of a solvent is measured by its vapor pressure. In Sec. 10-6 we saw that at any given temperature the vapor pressure of a pure liquid depends upon the fraction of the molecules at the surface which have sufficient kinetic energy to escape from the attractions of their neighbors. Figure 12-10a shows a pure liquid solvent in equilibrium with its vapor. The vapor pressure measures the concentration of solvent molecules in the gas phase. Figure 12-10b shows a solution at the same temperature in equilibrium with its vapor. The solute is assumed to be nonvolatile; so just as in Fig. 12-10a, the only molecules in the gas phase are those of the solvent. But in Fig. 12-10b there are fewer molecules in the gas phase than in Fig. 12-10a. Why? In the solution not all molecules at the surface are solvent molecules. Since a smaller proportion of the molecules at the surface can leave the surface, the rate of evaporation from the solution must be less than that from the pure solvent. Furthermore, since the system is at equilibrium, the rate of condensation must be lower in Fig. 12-10b than in Fig. 12-10a. The only way this can occur is for the concentration of solvent molecules in the gas phase above the solution to be less than that above the pure solvent. The vapor pressure of the solution is hence lower.

Figure 12-11 shows two vapor pressure curves, one for a pure solvent and another for its solution. The vertical distance between the two curves shows the magnitude of the lowering at each temperature at a constant solution composition. The vapor pressure of the solution can be seen to be lower at all temperatures, although the size of the lowering is smaller at lower temperatures.

Vapor pressure
of
pure solvent

Vapor pressure
lowering

$P_{solvent}$
$P_{solution}$

Vapor pressure
of
solution

Pressure →

Temperature →

Figure 12-11
Vapor-pressure lowering.

Raoult's law The quantitative relationship between vapor-pressure lowering and concentration *in an ideal solution* is stated in Raoult's law: the partial vapor pressure of a component in solution is equal to the mole fraction of that component times its vapor pressure when pure. For a solution of a nonvolatile solute (component 2) in a volatile solvent (component 1), we can write

$$P_1 = X_1 P_1^\circ$$

where P_1 and P_1° are the vapor pressures of the solution and the pure solvent, respectively, and X_1 is the mole fraction of solvent. Since $X_1 = 1 - X_2$,

$$P_1 = (1 - X_2)P_1^\circ$$

or, solving for X_2, $$X_2 = \frac{P_1^\circ - P_1}{P_1^\circ}$$

This says that the *fractional vapor-pressure lowering* is equal to the mole fraction of the solute.

● **EXAMPLE 12-7**

Problem The nonvolatile compound sulfanilamide ($C_6H_8O_2N_2S$) dissolves readily in acetone (C_3H_6O). What is the vapor pressure at 39.5°C of a solution containing 1.00 g of sulfanilamide dissolved in 10.0 g of acetone, if the vapor pressure of pure acetone at this temperature is 4.00×10^2 mmHg?

Solution The molecular weight of sulfanilamide is $6(12.0) + 8(1.0) + 2(16.0) + 2(14.0) + 32.1$, or 172.1, and that of acetone is $3(12.0) + 6(1.0) + 16.0$, or 58.0. One gram of sulfanilamide is

$$1.00 \text{ g} \frac{1 \text{ mol}}{172.1 \text{ g}} = 5.81 \times 10^{-3} \text{ mol}$$

And 10.0 g of acetone is

$$10.0 \text{ g} \frac{1 \text{ mol}}{58.0 \text{ g}} = 0.172 \text{ mol}$$

The mole fraction of acetone, X_1, in the solution is

$$X_1 = \frac{0.172 \text{ mol}}{0.172 \text{ mol} + 5.81 \times 10^{-3} \text{ mol}} = 0.967$$

By Raoult's law we have

$$P_1 = X_1 P_1^\circ$$
$$= 0.967(4.00 \times 10^2 \text{ mmHg}) = 3.87 \times 10^2 \text{ mmHg} ●$$

Raoult's law provides us with a way of determining molecular weights of nonvolatile solutes. The method is valid only if the solution is dilute enough to be essentially ideal, and if the solute does not react with the solvent.

● **EXAMPLE 12-8**

Problem Five grams of the nonvolatile compound formamide was dissolved in 1.00×10^2 g of water at 30°C. The vapor pressure of the solution was found to be 31.20 mmHg. If the vapor pressure of pure water is 31.82 mmHg at this temperature, what is the molecular weight of formamide?

Solution   From the definition of mole fraction we have

$$X_2 = \frac{n_2}{n_1 + n_2}$$

When this is solved for n_2, we get

$$n_2 = \frac{X_2 n_1}{1 - X_2}$$

n_1, the number of moles of H_2O (molecular weight = 18.0), is

$$n_1 = \frac{1.00 \times 10^2 \text{ g}}{18.0 \text{ g mol}^{-1}} = 5.56 \text{ mol}$$

X_2, the mole fraction of formamide, is found from Raoult's law to be

$$X_2 = \frac{P_1^\circ - P_1}{P_1^\circ}$$

$$= \frac{31.82 \text{ mmHg} - 31.20 \text{ mmHg}}{31.82 \text{ mmHg}} = 1.9 \times 10^{-2}$$

Substituting these values in the above relationship we get

$$n_2 = \frac{X_2 n_1}{1 - X_2} = \frac{(1.9 \times 10^{-2})(5.56 \text{ mol})}{1 - 1.9 \times 10^{-2}} = 0.11 \text{ mol}$$

Evidently we have 0.11 mol of formamide in solution. But we know that it weighs 5.00 g; so

$$\text{Molecular weight} = \frac{5.00 \text{ g}}{0.11 \text{ mol}} = 45 \text{ g mol}^{-1} \quad \bullet$$

This method of determining molecular weights is limited by the fact that in dilute solutions the magnitude of the vapor-pressure lowering is small and may therefore be difficult to measure, while more concentrated solutions may deviate considerably from ideal behavior.

When *both* components of an ideal binary solution are volatile liquids, Raoult's law applies to each. Figure 12-12*a* shows how the vapor pressure of component B in a binary solution is lowered by the presence of a component A, as expressed by the Raoult's law relationship

$$P_B = X_B P_B^\circ$$

Figure 12-12*b* similarly shows the lowering of the vapor pressure of A by B.

$$P_A = X_A P_A^\circ$$

Finally, Fig. 12-12*c* shows the total vapor pressure (solid line) of the solution. At each composition the total solution vapor pressure P_T equals the sum of the partial vapor pressures, in accordance with Dalton's law of partial pressures.

$$P_T = P_A + P_B$$

An *ideal solution* is one which obeys Raoult's law. Although essentially ideal solutions do exist, most solutions show deviations from Raoult's law behavior. Figure 12-13 shows the partial and total vapor pressure plots for (*a*) a system which

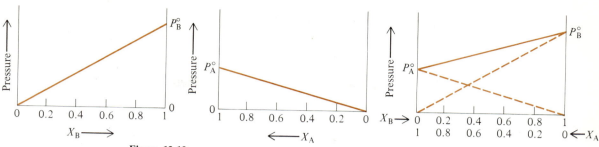

Figure 12-12
Vapor pressure of a solution of two volatile components (constant temperature). (*a*) **Partial vapor pressure of component B.** (*b*) **Partial vapor pressure of component A.** (*c*) **Combined graph showing total vapor pressure of solution.**

shows *positive deviation* and (*b*) one which shows *negative deviation* from Raoult's law. (The broken lines in the diagrams show Raoult's law behavior.) A positive deviation from Raoult's law is one in which the partial vapor pressure of each component, and hence the total vapor pressure, is higher than that in an ideal solution. This means that the escaping tendency of each component is abnormally high. In such a solution the intermolecular attractions must be weaker than those in the pure components. The process of mixing to form the solution must therefore be endothermic, as weaker forces replace stronger ones. (ΔH_{soln} is positive.)

A negative deviation from Raoult's law shows that the escaping tendency of each component in solution is abnormally low. Thus intermolecular forces in solution must be stronger than in the pure components. Mixing the components to form the solution in this case is exothermic, as stronger forces replace weaker ones. (ΔH_{soln} is negative.) The relationship between deviation from Raoult's law and ΔH_{soln} is summarized in Table 12-3.

For a solution of two volatile liquids how does the composition of the vapor phase compare with that of the solution? You might guess that the vapor would be richer in the more volatile component, and you would be right, but how can we

Figure 12-13
Deviation from Raoult's law (constant temperature). (*a*) **Positive deviation.** (*b*) **Negative deviation.**

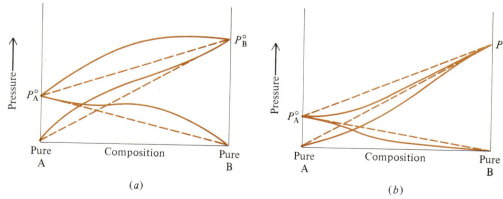

(*a*)   (*b*)

Table 12-3
Deviation from Raoult's
law and ΔH of solution

Deviation	Relative strengths of intermolecular forces		ΔH_{soln}	Example
	Pure components	Solution		
Positive	Stronger	Weaker	Positive	Acetone + carbon disulfide
Negative	Weaker	Stronger	Negative	Acetone + chloroform
None (ideal)			Zero	Benzene + toluene

prove it? Suppose we have two miscible components A and B with vapor pressures P_A° and P_B°, respectively, when pure. After they are mixed, their respective partial vapor pressures are

$$P_A = X_A P_A^\circ \quad \text{and} \quad P_B = X_B P_B^\circ$$

But in any mixture of gases the *partial pressure* of each gas equals the product of its *mole fraction times the total pressure*. (Can you prove this by assuming ideal-gas behavior?) If we let Y represent mole fraction *in the gas phase* and P_T the total pressure, we can write

$$P_A = Y_A P_T = X_A P_A^\circ \quad \text{and} \quad P_B = Y_B P_T = X_B P_B^\circ$$

Dividing the first equation by the second

$$\frac{Y_A}{Y_B} = \frac{X_A}{X_B} \frac{P_A^\circ}{P_B^\circ}$$

Now look closely at this relationship: If component A is more volatile than component B (P_A° is greater than P_B°), then the ratio of the mole fractions (A to B) in the gas phase must be *larger* than the corresponding ratio of mole fractions in the liquid. In other words if A is more volatile than B, the vapor will be relatively richer in A than the liquid.

● **EXAMPLE 12-9**

Problem   Toluene and benzene form ideal solutions. At 60°C the vapor pressure of toluene is 139 mmHg and of benzene is 392 mmHg. In a liquid solution in which the mole fraction of toluene is 0.600, what is (a) the partial vapor pressure of each component, (b) the total vapor pressure, and (c) the mole fraction of toluene in the vapor phase.

Solution   (a) Letting A represent toluene and B, benzene, we have

$$P_A = X_A P_A^\circ$$
$$= 0.600(139 \text{ mmHg})$$
$$= 83.4 \text{ mmHg} \quad \text{(partial vapor pressure of toluene)}$$

$$P_B = X_B P_B^\circ$$
$$= (1.000 - 0.600)(392 \text{ mmHg})$$
$$= 157 \text{ mmHg} \quad \text{(partial vapor pressure of benzene)}$$

(b) The total vapor pressure of the solution is

$$P_T = P_A + P_B$$
$$= 83.4 \text{ mmHg} + 157 \text{ mmHg} = 240 \text{ mmHg}$$

(c) Since $P_A = Y_A P_T$ and $P_B = Y_B P_T$, we have

$$Y_A = \frac{P_A}{P_T}$$

$$= \frac{83.4 \text{ mmHg}}{240 \text{ mmHg}} = 0.348 \qquad \text{(vapor-phase mole fraction of toluene)}$$

$$Y_B = \frac{P_B}{P_T}$$

$$= \frac{157 \text{ mmHg}}{240 \text{ mmHg}} = 0.654 \qquad \text{(vapor-phase mole fraction of benzene)} \quad \bullet$$

By using the method of Example 12-9 for an ideal solution of two components it is possible to determine the composition of the vapor phase in equilibrium with the liquid phase of any known composition. The upper line in Fig. 12-14 represents the total vapor pressure of an A-B mixture in which A is the more volatile component. It also gives the composition of the liquid which corresponds to each total vapor pressure. The lower (curved) line represents the composition of the vapor at the same total vapor pressure. A horizontal line, called a *tie line*, can be drawn from each liquid composition extending to the left to the composition of the vapor in equilibrium with that liquid. Thus if the composition of the liquid is X', the total vapor pressure is P', and using the tie-line CD, we see that the vapor composition is Y'. The vapor is thus poorer in B and richer in A (the more volatile component) than is the liquid.

The preceding curves have all been *pressure versus composition* curves for systems at *constant temperature*. It is also useful to plot *temperature versus composition* curves for a system at *constant pressure,* because the temperature given by the curve is the *boiling point* of the mixture. This has been done in Fig. 12-15 for the same substances A and B as in Fig. 12-14. (Note that A, the more volatile component, has the lower boiling point.)

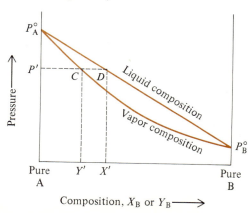

Figure 12-14
Liquid and vapor composition curves (ideal solution, constant temperature).

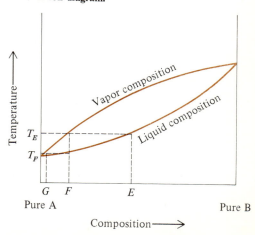

Figure 12-15
Distillation diagram.

The lower curve gives the boiling point of any liquid mixture between pure A and pure B, while the upper curve gives the composition of the vapor in equilibrium with that liquid. When, for instance, liquid of composition E is boiled, the boiling point is T_E and the vapor first produced has composition F.

Figure 12-15 is often called a *distillation diagram,* because it shows how a solution of two volatile liquids can be separated by means of a series of repeated distillations known as *fractional distillation.* If we start, for example, with liquid of composition E and raise its temperature, it will start to boil at T_E and the first vapor formed will be richer in A, having composition F. If this vapor is now cooled and redistilled (at temperature T_F) the first vapor formed has composition G, richer still in A. By successive distillations of the first fractions the composition of the final *distillate* (condensed vapor) can be made to approach pure A, while the final residue approaches the composition of pure B. Fractional distillation is a useful laboratory separation technique. In industry large quantities of complex liquid mixtures, such as petroleum, can be separated using systems of fractionating columns which are often many stories high.

Boiling-point elevation

A liquid boils at the temperature at which its vapor pressure equals atmospheric pressure. At each temperature the vapor pressure of a solution is lowered as a result of the presence of a solute, and so it is necessary to heat the solution to a higher temperature, in order for it to reach its boiling point. This is shown in Fig. 12-16, in which the horizontal broken line represents atmospheric pressure. It can be seen that the boiling point T_b of the solution is higher than that of the pure solvent.

The vapor-pressure lowering in dilute solutions is a function of concentration of solute particles and does not depend upon their identity; so we should expect boiling-point elevation to be similar. In fact, it can be shown that in dilute solution the *boiling-point elevation is proportional to the molality of the solute particles.* (We assume at this point that the solute is nonvolatile.) In other words, if ΔT_b represents the *boiling-point elevation* and equals $(T_b)_{solution} - (T_b)_{solvent}$, then

$$\Delta T_b = K_b m$$

where m is the molality and K_b is the proportionality constant, known as the *molal boiling-point elevation constant.* The value of K_b depends only upon the solvent and represents the increase in the boiling point caused by the addition of one mole

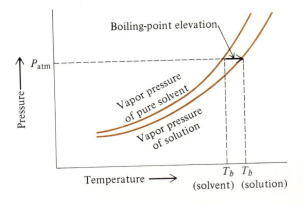

Figure 12-16
Boiling-point elevation.

Table 12-4
Boiling-point elevation and freezing-point depression constants

Solvent	Normal boiling point, °C	K_b, °C m^{-1}	Normal freezing point, °C	K_f, °C m^{-1}
Water	100.000	0.512	0.000	1.855
Benzene	80.2	2.53	5.5	5.12
Acetic acid	118.5	3.07	16.6	3.90
Naphthalene	218.0	5.65	80.2	6.9
Camphor	208.3	5.95	178.4	40.

of solute particles to one kilogram of solvent, if the solution behaves ideally. Table 12-4 lists values for a few solvents.

● **EXAMPLE 12-10**

Problem   A quantity of urea (molecular weight 60.1) weighing 0.300 g is dissolved in 10.0 g of water. Assuming ideal-solution behavior calculate the normal boiling point of the solution.

Solution

$$0.300 \text{ g of urea} = \frac{0.300 \text{ g}}{60.1 \text{ g mol}^{-1}} = 4.99 \times 10^{-3} \text{ mol}$$

$$\text{Molality of the solution} = \frac{4.99 \times 10^{-3} \text{ mol}}{1.00 \times 10^{-2} \text{ kg}} = 0.499 \ m$$

$$K_b \text{ for H}_2\text{O} = 0.512°\text{C } m^{-1}$$

$$\Delta T_b = K_b m$$
$$= (0.512°\text{C } m^{-1})(0.499 \ m) = 0.255°\text{C}$$

$$\text{Normal boiling point} = 100.00°\text{C} + 0.255°\text{C} = 100.25°\text{C} \ ●$$

The proportionality between molality and boiling-point elevation provides a useful method of determining molecular weights of nonvolatile solutes. In practice, determinations are often made at higher than ideal-solution concentrations, but even so, the approximate values obtained are useful.

● **EXAMPLE 12-11**

Problem   Suppose 7.39 g of a nonvolatile solute is added to 85.0 g of benzene. The resulting solution boils at 82.6°C at standard pressure. Calculate the approximate molecular weight of the solute.

Solution   Since 85.0 g = 0.0850 kg, the number of grams of solute which would be dissolved *per kilogram* of benzene is

$$\frac{7.39 \text{ g solute}}{0.0850 \text{ kg benzene}} = 86.9 \ \frac{\text{g solute}}{\text{kg benzene}}$$

The molality of the solution can be found from the boiling-point elevation, which is 82.6°C − 80.2°C = 2.4°C, where the normal boiling point of pure benzene is taken from Table 12-4.

$$\Delta T_b = K_b m$$

$$m = \frac{\Delta T_b}{K_b}$$

$$= \frac{2.4°\text{C}}{2.53°\text{C } m^{-1}} = 0.95 \ m$$

Now we know that in this solution the amount of solute present per kilogram of benzene may be expressed as 86.9 g or as 0.95 mol. In other words 0.95 mol weighs 86.9 g. Therefore, 1.00 mol of solute must weigh

$$\frac{86.9 \text{ g}}{0.95 \text{ mol}} = 91 \text{ g mol}^{-1}$$

The (approximate) molecular weight is 91. ●

Freezing-point depression   One way to explain the phenomenon of boiling-point elevation is to say that because solute particles lower the escaping tendency of the solvent, we need to compensate for this by raising the temperature higher in order to get it to boil. But escaping tendency means tendency to escape to *any* other phase; so we can use a similar argument to justify the fact that a solute *lowers* the freezing point of a solvent; that is, in order to freeze the solvent we must cool it to a lower temperature in order to compensate for its lowered escaping tendency. The presence of a solute always lowers the freezing point if the solute is insoluble in the solid phase. The lowering, or depression, of the freezing point causes the solid-liquid equilibrium line to be moved to the left in the phase diagram. Figure 12-17 shows two phase diagrams: the first is for pure H_2O, and the second for H_2O plus a solute which is soluble in the liquid phase only. Note that the solid-gas (sublimation) equilibrium line is unaffected. The diagram shows that the temperature at which ice and liquid water can coexist under a given atmospheric pressure, P_{atm}, is lowered. (The elevation of the boiling point is indicated also.)

The relationship between the freezing-point depression and the molality is a direct proportionality in dilute solutions, as it is in the case of the boiling-point elevation. Mathematically it is written as

$$\Delta T_f = -K_f m$$

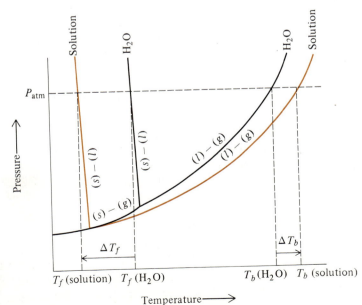

Figure 12-17
Alteration of the phase diagram for water by a liquid-soluble solute.

where m = molality
K_f = molal freezing-point depression constant
ΔT_f = freezing point depression = $(T_f)_{\text{solution}} - (T_f)_{\text{solvent}}$

(The minus sign thus indicates that the solute *lowers* the freezing point.)   The value of K_f depends only on the solvent and represents the decrease in the freezing point brought about by the addition of one mole of solute particles to one kilogram of solvent.   Table 12-4 lists values of K_f for a few solvents.

As is the case with boiling-point elevation, freezing-point depression measurements can be used for determining approximate molecular weights of solutes.

● EXAMPLE 12-12

Problem   Suppose 1.42 g of a solute is added to 25.0 g of benzene, and the solution is found to freeze at a temperature 1.96°C below that of pure benzene. Calculate the apparent molecular weight of the solute.

Solution   From Table 12-4 we see that K_f for benzene equals 5.12°C m⁻¹. This allows us to find the molality of the solution

$$\Delta T_f = -K_f m$$

$$m = -\frac{\Delta T_f}{K_f}$$

$$= -\frac{-1.96°C}{5.12°C\ m^{-1}} = 0.383\ m$$

This tells us that there is 0.383 mol of solute per kilogram of solvent. Now we need to find out independently how many grams of solute are present in this amount of solvent. Since there are 1.42 g per 25.0 g, which is 0.0250 kg, of solvent, the number of grams of solute per kilogram of solvent is

$$\frac{1.42\ \text{g solute}}{0.0250\ \text{kg solvent}} = 56.8\ \frac{\text{g solute}}{\text{kg solvent}}$$

Now we know that per kilogram of benzene there is 0.383 mol of solute and that this quantity must weigh 56.8 g. Therefore, 1.00 mol of solute weighs

$$\frac{56.8\ \text{g}}{0.383\ \text{mol}} = 148\ \text{g mol}^{-1}$$

The apparent molecular weight of the solute is 148. ●

Osmotic pressure   We have seen that in a solution the escaping tendency of the solvent is less than that in the pure solvent, and that is why the vapor pressure of the solution is less than that of the pure solvent. The difference in vapor pressures can lead to a transfer of solvent molecules from pure solvent to solution in an experiment illustrated in Fig. 12-18. One beaker contains pure solvent, and the other, a solution of nonvolatile solute. Under the conditions shown, the system is not at equilibrium because of the difference in escaping tendencies. Both liquids are in contact with the same vapor at some pressure, so the rate of condensation into the two beakers is the same. But the rate of evaporation from the pure solvent is greater than that from the solution, as is shown schematically by the arrows in the diagram. This means that over a period of time solvent molecules will be

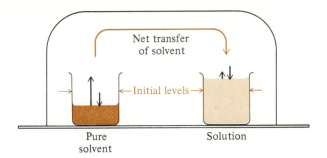

Figure 12-18
Transfer of solvent due to vapor pressure difference.

transferred from solvent to solution. This system can reach equilibrium only after all the pure solvent has evaporated.

In Fig. 12-19 we see another way of observing the difference between escaping tendencies in solvent and solution. The two liquids are shown separated by a semipermeable membrane. A *semipermeable membrane* is a thin barrier which allows certain atomic, ionic, or molecular species to pass but which stops others. In this case it allows solvent molecules to pass in both directions but is impermeable to solute particles. (Depending upon the system, parchment, gelatin, plastic, or other membrane can be used.) Since the escaping tendency at the left is greater than that at the right, however, the rate of transfer of solvent molecules from left to right is greater than that from right to left. Thus after a while there will have been a net transfer of solvent molecules from pure solvent to solution. This process of transfer of solvent molecules across a semipermeable membrane is known as *osmosis*. In order for osmosis to occur, the concentrations of solute particles must be different in the two liquids.

Figure 12-20 shows how the process of osmosis can be stopped by pushing in on a piston with a pressure, π, just large enough to compensate for the difference in escaping tendencies, that is, to stop osmosis from occurring. This pressure is called the *osmotic pressure*.

Osmotic pressure is another colligative property. In dilute solution it is directly proportional to the *molarity* of the solution and to the absolute temperature.

Figure 12-19
Osmosis

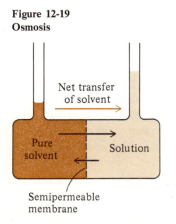

Figure 12-20
Osmotic pressure.

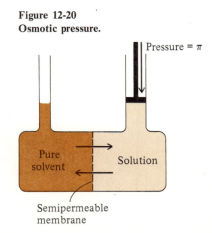

Surprisingly enough, the proportionality constant is the ideal-gas constant R, as can be shown theoretically. The complete relationship is

$$\pi = MRT$$

Since the molarity M equals the number of moles of solute, n_2, per liter volume of solution, V, that is,

$$M = \frac{n_2}{V}$$

therefore
$$\pi = \frac{n_2}{V} RT \quad \text{or} \quad \pi V = n_2 RT$$

The obvious similarity between this equation, sometimes called the *van't Hoff equation,* and the ideal-gas law suggests that the mechanism for the exertion of pressure by a gas may be closely related to the mechanism for osmotic pressure. This has been shown *not* to be the case, however.

The magnitude of osmotic pressures produced by comparatively dilute solutions is very high. For example, a 1.00 M solution at 0°C would exhibit an osmotic pressure of

$$\pi = MRT = (1.00 \text{ mol liter}^{-1})(0.0821 \text{ liter atm K}^{-1} \text{ mol}^{-1})(273 \text{ K})$$
$$= 22.4 \text{ atm}$$

This is a large pressure, enough to support a column of mercury about 17 m (or a column of water about 230 m) high! Even a 1.00×10^{-4} M solution exhibits an osmotic pressure of 2.24×10^{-3} atm, which is 1.70 mmHg. If a water-filled manometer were used to measure this pressure, the height of the water column would still be 23.0 mm, a height which can easily be measured. It is because of the usefulness of this technique at low molar concentrations that it is employed for the determination of very high molecular weights.

● **EXAMPLE 12-13**

Problem　A sample of the protein hemoglobin weighing 0.500 g was dissolved in enough water to make 100.0 ml of solution. The osmotic pressure of the solution was measured at 25°C and found to be 1.35 mmHg. What was the molecular weight of the hemoglobin?

Solution　We will start by using the osmotic-pressure relationship in order to find the molarity of the solution. The osmotic pressure in atmospheres is

$$\pi = 1.35 \text{ mmHg} \frac{1 \text{ atm}}{760 \text{ mmHg}} = 1.78 \times 10^{-3} \text{ atm}$$

Since $\pi = MRT$,

$$M = \frac{\pi}{RT}$$

$$= \frac{1.78 \times 10^{-3} \text{ atm}}{(0.0821 \text{ liter atm K}^{-1} \text{ mol}^{-1})(298 \text{ K})} = 7.28 \times 10^{-5} \ M$$

In 1 liter of solution there would be 7.28×10^{-5} mol of hemoglobin. How many *grams* would there be in this volume? There are 0.500 g of hemoglobin dissolved per 100.0 ml, or 0.1000 liter, of solution. The number of grams dissolved *per liter* is

$$\frac{0.500 \text{ g hemoglobin}}{0.1000 \text{ liter solution}} = 5.00 \frac{\text{g hemoglobin}}{\text{liter solution}}$$

Now we know that the amount of hemoglobin dissolved per liter is 7.28×10^{-5} mol and that it weighs 5.00 g. Therefore 1 mol weighs

$$\frac{5.00 \text{ g}}{7.28 \times 10^{-5} \text{ mol}} = 6.87 \times 10^4 \text{ g mol}^{-1}$$

The molecular weight of hemoglobin is about 68,700. ●

Reverse osmosis   If the pressure applied to a solution in an osmotic-pressure apparatus (Fig. 12-20) is increased above the osmotic pressure, the net transfer of solvent is reversed. Using a pressure *higher* than the osmotic pressure to remove solvent from a solution is called *reverse osmosis*. Figure 12-21 shows a schematic of a cell designed for reverse osmosis. Such cells are now in use for reclaiming saline water in certain arid areas of the world, where the expense is justified by the need for fresh water. Saline water is forced into the cell at a pressure greater than the osmotic pressure. Ions and other dissolved particles are essentially filtered out by the semipermeable membrane and fresh water emerges on the opposite side. Provision is also made for bleeding off waste water from the cell. This water has been made very concentrated in dissolved substances and is thrown away, into the ocean, for example. As population increases and more water supplies are contaminated, reverse osmosis may prove to be an increasingly important process.

12-5
Electrolytes   We have seen that some substances dissolve in solution as ions. These solutes are called *electrolytes*, and their solutions conduct electricity better than the pure solvent does. A substance which is not an electrolyte, a *nonelectrolyte*, does not release ions to the solution as it dissolves, and so does not add to the conductivity

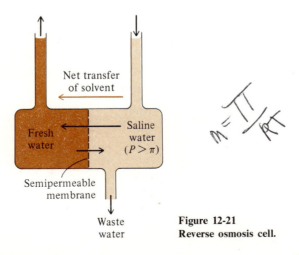

Figure 12-21
Reverse osmosis cell.

Figure 12-22
Dissolving and dissocia-
tion of an ionic solid
(schematic).

of the solvent.   (Electrical conductivity involves the passage of electrically charged particles, such as ions, through a medium.)

Dissociation   The process by which ions are released into solution when a solute dissolves is called *dissociation*.[3]   Although electrolytic dissociation occurs in solvents besides water, this solvent is so important that in the remainder of this chapter we will consider dissociation in aqueous solutions only.   Both ionic and molecular solutes can dissociate in water.   An example of an *ionic electrolyte* is sodium chloride. When NaCl dissolves, the Na^+ and Cl^- ions are pried out of the crystal lattice by the water dipoles.   The resulting solution contains ions, each surrounded by a cluster of hydrating (Sec. 12-3) H_2O molecules; it contains no NaCl molecules, which is really not unexpected, since no molecules exist even in the solid.   Figure 12-22 shows schematically the dissolving and dissociation of an ionic solid such as NaCl.   The process is represented by the equation

$$NaCl(s) \longrightarrow Na^+ + Cl^-$$

ADDED COMMENT

It is true that H_2O does not show up anywhere in the above equation, but that does not mean that H_2O molecules are not involved.   They pull apart the NaCl lattice and hydrate the ions.   Water molecules can be shown in the above equation, but it makes it unnecessarily cluttered to do so.   (More about this shortly.)

[3]It is also called *ionization*.   Dissociation implies a process of separation, while ionization emphasizes that ions are the result.   Both terms are appropriate, but we will use the former almost exclusively, because ionization has another meaning: the formation of an ion from an atom or molecule by the loss of an electron.

An example of a *molecular electrolyte* is hydrogen chloride, HCl. It is a gas at room temperature, but is quite soluble in water. The HCl molecule is quite polar, and as a result H_2O dipoles are so strongly attracted to each end that it splits apart, that is, dissociates, into an H^+ ion and a Cl^- ion. This process can be written as

$$HCl(g) \longrightarrow H^+ + Cl^-$$

and is shown schematically in Fig. 12-23. The ions released to the solution in any dissociation are hydrated, surrounded by a cluster of water dipoles. The hydrogen ion, being only a proton, is especially strongly hydrated because it can form hydrogen bonds with water molecules. This is often emphasized by writing the hydrated hydrogen ion as H_3O^+, called the *hydronium,* or *oxonium, ion.* This formulation has both advantages and disadvantages, as we will see shortly.

When solutions of NaCl or HCl are tested they show a high electrical conductivity. On the other hand HF shows a conductivity which, although higher than that of pure water, is much lower than that of the NaCl or HCl solutions. This and other evidence indicates that for the same number of moles of solute dissolved per liter of solution the HF solution has fewer ions than do the other two. The conclusion to be drawn is that in solution not all the HF molecules are dissociated into ions. Thus we have two categories of electrolytes: (1) *strong electrolytes,* which exist only as ions in solution, and (2) *weak electrolytes,* which exist as a *mixture of ions and undissociated molecules* in solution.

In a solution of a weak electrolyte ions are in equilibrium with undissociated molecules. In order to see how this comes about, look closely at what happens when HF dissolves in water. At first the HF simply dissolves because of the high HF–H_2O dipole-dipole attractions:

$$HF(g) \longrightarrow HF(aq)$$

Then the water molecules begin splitting dissolved HF molecules into their ions:

$$HF(aq) \longrightarrow H^+(aq) + F^-(aq)$$

But the H—F bond is not as easy to break as the H—Cl bond and the H^+—F^- attraction in solution is stronger than the H^+—Cl^- attraction. As a result H^+ and F^- ions in solution can recombine to form HF molecules:

$$H^+(aq) + F^-(aq) \longrightarrow HF(aq)$$

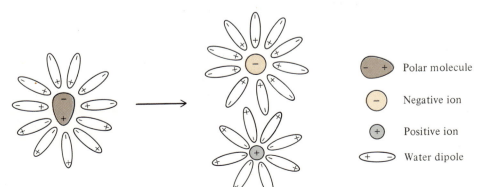

Figure 12-23
Dissociation of a polar molecule (schematic).

As the concentrations of H⁺ and F⁻ build up, the rate of recombination increases, while the rate of dissociation of HF molecules decreases as fewer of them are left. Eventually, although all this takes only a second or less, the rising rate of recombination of H⁺ and F⁻ ions reaches the falling rate of dissociation of HF molecules. From this time on no further *net* change occurs; the system is at equilibrium:

$$HF(aq) \rightleftharpoons H^+(aq) + F^-(aq)$$

Degree of dissociation   In a 1.0 *M* solution of HF (one prepared by adding 1.0 mol of HF molecules to enough water to bring the total volume to 1.0 liter), on the average, out of every 100 HF molecules originally added 97 are still present as molecules, and the other 3 as hydrated H⁺ and F⁻ ions. (This explains the relatively low conductivity of an HF solution.) The HF in a 1.0 *M* solution is thus only about 3 percent dissociated.

The degree to which a given electrolyte is dissociated in solution is not constant, but increases as the solution becomes more dilute. Figure 12-24 shows the variation of the percent dissociation of HF with concentration. At 0.1 *M* the percent dissociation has risen to 8 percent, and at 0.01 *M* it is 23 percent. The increase in percent dissociation with dilution is characteristic of all electrolytes. It is, in fact, predicted by Le Châtelier's principle: Upon diluting any solution, the *concentration* of solute particles decreases. Dissociation, on the other hand, increases the *number* of solute particles. (In the case of HF *two* ions replace *one* molecule.) When a solution of a weak electrolyte is diluted, the position of equilibrium shifts favoring increased dissociation. In the case of HF,

$$HF \rightleftharpoons H^+ + F^-$$

the equilibrium shifts to the right, creating a larger total number of solute particles, and partially compensating for the dilution effect. In doing so the percent of the HF in ionic form increases. Percent dissociation of a weak electrolyte at a given concentration varies from one electrolyte to another and from one solvent to another. (HCl is a strong electrolyte in water, but a weak one in benzene, for example.) It also depends upon temperature.

It is important to distinguish between a strong and a weak electrolyte. But the degree of dissociation of a weak electrolyte increases with dilution and approaches 100 percent dissociation at very low concentrations. What is usually done is to

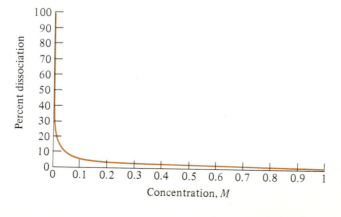

Figure 12-24
Concentration dependence of percent dissociation of HF.

specify an arbitrary concentration, usually 1 M, and specify that a strong electrolyte is one which is essentially 100 percent dissociated at a 1 M concentration, while a weak electrolyte is one which is less than 100 percent dissociated at this concentration. This distinction works out pretty well in practice. Most electrolytes are either clearly weak or strong. In addition, most weak electrolytes remain weak at all but quite low concentrations, while most strong electrolytes remain strong at all but very high concentrations.

Table 12-5 summarizes the differences among the three classes of solutes: nonelectrolytes, weak electrolytes, and strong electrolytes. Examples of each class are also given.

● **EXAMPLE 12-14**

Problem A $5.00 \times 10^{-2}\,m$ solution of HF in water shows a normal freezing point of $-0.103\,°C$. What is the percent dissociation of HF at this concentration?

Solution The molal freezing-point depression constant for water is $1.86\,°C\,m^{-1}$ (Table 12-4). Therefore the solute-particle molality (the *total* number of moles of solute-derived particles *of all kinds* per kilogram of water) is

$$m = -\frac{\Delta T_f}{K_f}$$

$$= -\frac{(-0.103\,°C)}{1.86\,°C\,m^{-1}} = 5.54 \times 10^{-2}\,m$$

Per kilogram of H_2O 5.00×10^{-2} mol of HF was added, but after dissolving, 5.54×10^{-2} mol of particles are present. The number of moles of particles has apparently increased during the dissolving process by $5.54 \times 10^{-2} - 5.00 \times 10^{-2}$, or 0.54×10^{-2}. Now, every time *one* HF molecule dissociates, *two* ions are formed. Therefore, the number of particles goes up by *one* for each molecule which dissociates. If the number of moles of particles has increased by 0.54×10^{-2} this means 0.54×10^{-2} mol of HF molecules must have dissociated. In other words,

	HF	⇌	H⁺	+	F⁻	
Before dissociation:	5.00×10^{-2}		0		0	mol kg⁻¹ H₂O
Dissociated:	-0.54×10^{-2}					mol kg⁻¹ H₂O
After dissociation:	4.46×10^{-2}		0.54×10^{-2}		0.54×10^{-2}	mol kg⁻¹ H₂O

(Check: Total number of moles $= 4.46 \times 10^{-2} + 0.54 \times 10^{-2} + 0.54 \times 10^{-2}$

$$= 5.54 \times 10^{-2}\,mol)$$

Table 12-5
Classification of solutes

	Nonelectrolytes	Weak electrolytes	Strong electrolytes
Solute particles	Molecules	Molecules + ions	Ions
Equilibrium among particles?	No	Yes	No, or essentially no
Percent dissociation	0	Between 0 and 100 (usually small)	~100
Examples (for aqueous solutions)	Ethanol (C_2H_5OH) Sucrose ($C_{12}H_{22}O_{11}$) Acetone [$(CH_3)_2CO$] Oxygen (O_2)	Hydrogen fluoride (HF) Acetic acid ($HC_2H_3O_2$) Thallous hydroxide (TlOH) Mercuric chloride ($HgCl_2$)	Hydrogen chloride (HCl) Sodium chloride (NaCl) Sodium hydroxide (NaOH) Potassium fluoride (KF)

If 0.54×10^{-2} mol of HF has dissociated, the percent dissociation is

$$\% \text{ dissoc} = \frac{\text{moles dissociated}}{\text{total moles}} \times 100$$

$$= \frac{0.54 \times 10^{-2}}{5.00 \times 10^{-2}} \times 100 = 11 \text{ percent} \quad \bullet$$

**12-6
Ions in aqueous
solution**

A solution of a binary compound which is a strong electrolyte may be considered to be a solution not of just one substance but of two. Saltwater, for example, has one set of properties which are a result of the presence of Na^+ ions in solution and another independent set, a consequence of the Cl^- ions' presence. The solvent, water, because of its high dielectric constant, is quite unusual in its ability to keep oppositely charged ions apart. Solvents of lower dielectric constant are less efficient at separating ions; in such solvents there is evidence of the existence of *ion pairs* (essentially, molecules). (Ion pairs are thought to exist even in water if the solute concentration is so high that insufficient water molecules are available to keep the ions apart.)

*Acids, bases,
and salts*

The word *acid* is an old one coming from a Latin word meaning "sour." (Solutions of acids have a sour taste.) The meaning of the word has gone through several stages of evolution over the years until today it has several useful and related meanings. At this point we offer a *provisional* definition of an acid, one which will be considerably broadened in the next chapter. This definition was first proposed by a Swedish chemist, Svante Arrhenius, who in 1887 first showed that strong electrolytes were completely dissociated. The *Arrhenius definition* states that *an acid is a compound of hydrogen which dissociates in water to give hydrogen ions* H^+. Thus hydrogen chloride is an Arrhenius acid

$$HCl(g) \longrightarrow H^+ + Cl^-$$

It is often named as *hydrochloric acid,* a name also used to describe its aqueous solution.

Bases have also been known for centuries. (An older name, *alkali,* is still in use.) Years ago it was recognized that bases have the ability to cancel out, or *neutralize,* the properties of an acid. The *Arrhenius definition* of a base states that *a base is a hydroxide compound which dissociates in water to give hydroxide ions.* An example of an Arrhenius base is sodium hydroxide, NaOH, which is a strong electrolyte and dissociates as follows:

$$NaOH(s) \longrightarrow Na^+ + OH^-$$

In a *neutralization reaction* hydrogen ions (responsible for the characteristic properties of an acid) combine with hydroxide ions (responsible for the characteristic properties of a base) to form water:

$$H^+ + OH^- \longrightarrow H_2O$$

Thus when a solution of sodium hydroxide (containing Na^+ and OH^- ions) is mixed with a solution of hydrochloric acid (containing H^+ and Cl^- ions) the following occurs:

HCl solution: (H$^+$)   and Cl^-

NaOH solution: Na^+ and   (OH$^-$) $\longrightarrow$ H$_2$O

ADDED COMMENT

What happens to the Na^+ and Cl^- ions in the above neutralization reaction? Not a thing! Na^+ and Cl^- might be expected to combine to form NaCl, but this could happen only if NaCl were a weak electrolyte, which it is not, or if it were insoluble in water, which it also is not. After mixing the solutions of HCl and NaOH, we could cause NaCl to be formed by evaporating the water from solution. This would eventually produce a solution saturated with respect to NaCl, and further removal of water would force NaCl to crystallize:

$$Na^+ + Cl^- \longrightarrow NaCl(s)$$

But this is a second reaction and does not occur unless the water is evaporated.

A third category of electrolyte was also proposed by Arrhenius: the *salt.* A *salt,* according to Arrhenius, is a compound whose ions are left after an acid is neutralized by a base. NaCl is thus a salt, for its ions Na^+ and Cl^- are left in solution after the solutions of HCl and NaOH react. Usually a salt is an ionic solid in which neither H^+ nor OH^- ions are present. In a neutralization reaction, then, water molecules are formed and the ions of a salt are left in solution. In the special case in which the solubility of the salt is low, its ions may combine and the salt itself precipitate out of solution.

Table 12-6 lists the names of some acids, bases, and salts and the reaction which each undergoes when dissolved in water.

Hydration of ions   Both positive and negative ions are hydrated in aqueous solution. This means that they are surrounded by an approximately spherical shell of water dipoles, each with its oxygen end pointing in toward a positive ion or its hydrogen end pointing in toward a negative ion. The extent of hydration of an ion can be measured by the hydration energy of the ion, ΔH_{hyd}. This is the energy released

---

Table 12-6
Some Arrhenius acids, bases, and salts

	Name	Dissolving process
Acids	Hydrochloric acid (hydrogen chloride)	$HCl(g) \longrightarrow H^+ + Cl^-$
	Hydrofluoric acid (hydrogen fluoride)	$HF(g) \longrightarrow HF \rightleftharpoons H^+ + F^-$
	Nitric acid (hydrogen nitrate)	$HNO_3(l) \longrightarrow H^+ + NO_3^-$
Bases	Sodium hydroxide	$NaOH(s) \longrightarrow Na^+ + OH^-$
	Calcium hydroxide	$Ca(OH)_2(s) \longrightarrow Ca^{2+} + 2OH^-$
	Lanthanum hydroxide	$La(OH)_3(s) \longrightarrow La^{3+} + 3OH^-$
Salts	Sodium fluoride	$NaF(s) \longrightarrow Na^+ + F^-$
	Calcium chloride	$CaCl_2(s) \longrightarrow Ca^{2+} + 2Cl^-$
	Lanthanum nitrate	$La(NO_3)_3(s) \longrightarrow La^{3+} + 3NO_3^-$

when the ion leaves the gas phase and enters water to become hydrated. Hydration energies are always negative and depend upon *ionic size* and *charge*. Table 12-7 shows that hydration energies are high for an ion which has (1) a small size and/or (2) a high charge. In the sequence $Li^+–Na^+–K^+$, for example, the hydration energy decreases because the ionic radius increases. Also note that $2+$ ions in general have higher hydration energies than do $1+$ ions. High charge and small size favor high ion-dipole interaction.

Ions in dilute solution are probably surrounded by two or three concentric, approximately spherical shells of water molecules, but the H_2O molecules in the innermost sphere, the *first hydration sphere*, are most tightly held. The number of molecules in this sphere is known as the *hydration number* of the ion. Experimental evidence indicates that for an ion which has a high hydration energy the hydration number is essentially a constant. For other ions the hydration number apparently changes with concentration and temperature, since water molecules are more loosely bound and can move into and out of the hydration sphere easily.

Because of its hydration sphere an ion in water has an *effective* radius which is much larger than that of a gaseous ion or an ion in a crystal. In general the larger the hydration energy of an ion, the larger will be the effective radius of that ion. Thus the hydrated K^+ ion (radius about 0.16 nm) is actually *smaller* than the hydrated Na^+ ion (about 0.24 nm), which partially accounts for the fact that K^+ ions pass more readily through living cell membranes.

A hydrogen ion is the smallest of all ions, being the mere nucleus of a hydrogen atom. Although it carries only a $1+$ charge, this charge is concentrated in such a tiny volume, that the degree of interaction between a hydrogen ion and water molecules is extremely high. The proton can be thought of as attracting a lone pair of electrons on a nearby oxygen (of a water molecule) and forming a *hydronium ion* (also called *oxonium ion*):

$$H^+ + \overset{\displaystyle ..}{\underset{\displaystyle H}{:\!O\!:}}H \longrightarrow \left[H \overset{\displaystyle ..}{\underset{\displaystyle H}{:\!O\!:}} H \right]^+$$

The hydronium ion is a proton hydrated by one water molecule. The process continues and complex ions containing varying numbers of water molecules are

Ion	Ionic radius, nm	ΔH_{hyd}, kJ mol^{-1}
Li^+	0.068	-506
Na^+	0.097	-397
K^+	0.133	-314
Mg^{2+}	0.066	-1910
Ca^{2+}	0.099	-1580
Sr^{2+}	0.112	-1430
Al^{3+}	0.051	-4640
F^-	0.133	-506
Cl^-	0.181	-377
I^-	0.220	-297

Table 12-7
Hydration energies of some ions

formed. There is evidence that $H_9O_4^+$ is a common ionic species, for instance. This is a proton combined with four water molecules by hydrogen bonding.

Because of hydrogen bonding water is a highly associated liquid. In water hydrogen bonds are continually being formed and broken in an ever-changing structure. For this reason the number of H_2O molecules which hydrate a proton is not constant, and so it is impossible to write an accurate formula for the hydrated hydrogen ion. Writing H^+ is misleading because it appears that the hydrogen ion is not hydrated by any water molecules at all. H_3O^+, the hydronium ion formulation, seems to show that exactly one H_2O molecule hydrates a proton. $H_9O_4^+$ is similarly misleading and is unnecessarily complicated, as well. What shall we do about the formula for the hydrated hydrogen ion? Throughout the majority of this book we will not include waters of hydration in formulas for *any* ions in aqueous solution. Thus most of the time we will write, H^+, Cl^-, Na^+, S^{2-}, Al^{3+}, etc., to represent such ions. It is important, however, to keep in mind that *all ions in aqueous solution are hydrated;* the hydrogen ion most of all. There are times, however, when it is desirable to emphasize the role of hydrating water molecules. Then we write $Li(H_2O)_4^+$, $Al(H_2O)_6^{3+}$, or possibly $Na(H_2O)_x^+$. For the hydrogen ion in such situations we adopt the H_3O^+ formulation, as it keeps us reminded of the waters of hydration but is much less cumbersome than $H_9O_4^+$ or $H(H_2O)_x^+$.

Colligative properties and Debye-Hückel effects

We have said that a dilute solution, one in which interactions between solute particles are at a minimum, is an ideal solution. The relations developed in Sec. 12-4 for colligative properties are limited to such solutions. In more concentrated solutions interactions between solute particles produce deviations from ideal behavior. When the solute is an electrolyte, especially a strong electrolyte, interactions between charged ions of the solute produce nonideal behavior even at fairly low concentrations.

The nonideality of solutions of electrolytes can be readily shown by examining the colligative properties of such solutions. In a $0.100\ m$ solution of NaCl, for example, there are 0.100 mol of Na^+ ions and 0.100 mol of Cl^- ions per kilogram of water

$$NaCl(s) \longrightarrow \quad Na^+ \quad + \quad Cl^-$$

$$0.100\ \text{mol} \longrightarrow 0.100\ \text{mol} + 0.100\ \text{mol}$$

The total particle molality is thus $0.100 + 0.100$, or $0.200, m$. Since the molal freezing-point depression constant is $1.86°C\ m^{-1}$, we would expect the freezing point lowering to be

$$\Delta T_f = -K_f m$$

$$= -(1.86°C\ m^{-1})(0.200\ m) = 0.372°C$$

and the normal freezing point to be $-0.372°C$. In fact, the measured freezing point of a $0.100\ m$ solution of NaCl is $-0.347°C$. For years it was thought that strong electrolytes are not quite 100 percent dissociated and that a $0.100\ m$ NaCl solution actually contains less than 0.200 mol of particles per kilogram of water. Old chemistry texts and, unfortunately, some newer books in other fields discuss the "degree of dissociation" of a strong electrolyte as determined from colligative properties, for example. But strong electrolytes are indeed completely dissociated, except, in some cases, in more concentrated solutions. How can we explain the fact that a solution which really contains 0.200 mol of ions appears from its freezing point to contain less than this number?

In 1923 Peter Debye and Ernst Hückel proposed what is now called the *Debye-Hückel theory of strong electrolytes*. This theory assumes that strong electrolytes are completely dissociated in solution, but that the hydrated ions have enough residual attraction for each other that they are not truly independent of each other. The theory assumes that each ion has surrounding it *on the average* more ions of opposite charge, *counterions,* than of the same charge, and that the number of counterions near a given ion increases with increasing concentration. The presence of this diffuse *ionic atmosphere* relatively rich in counterions restricts the movement of each ion in the solution.

The existence of the ionic atmosphere can be shown by electrical conductivity measurements. The conductivity of a solution of a strong electrolyte such as NaCl increases as the concentration of NaCl increases, but is not quite proportional. Thus doubling the concentration of NaCl less than doubles the conductivity. This can be explained by assuming that as the counterions become more abundant at higher concentrations, the "drag" produced on an ion by its ionic atmosphere becomes stronger. The Debye-Hückel theory permits a quantitative prediction of how the conductivity should change with concentration, a prediction which agrees with experimental results.

The Debye-Hückel theory also permits a calculation of the ionic-atmosphere effect on colligative properties. The presence of the ionic atmosphere can be shown to restrict the effectiveness of each ion in lowering the escaping tendency of the solvent, and hence in depressing the freezing point, elevating the boiling point, etc. And as with the conductivity there is quantitative agreement between theoretical prediction and experimental result, at least in dilute solutions.

It is clear that as its concentration is raised, a solution of an electrolyte begins to deviate from ideal-solution behavior at lower concentrations than does a solution of a nonelectrolyte. In concentrated solutions electrolytes behave very nonideally, and even the Debye-Hückel theory fails; it has, however, provided added evidence that at least over a range of comparatively dilute solutions strong electrolytes are indeed 100 percent dissociated.

SUMMARY

We started this chapter with a brief excursion into the field of *colloids*. These are systems which fall between the conventional heterogeneous and homogeneous classifications. In a colloid particles of larger-than-molecular dimensions are dispersed throughout a second medium.

Most of the chapter has been about *solutions,* defined as *homogeneous mixtures*. Solutions may be *solid, liquid,* or *gaseous*. When a solution is very rich in one component, that component is usually called the *solvent,* while the others are called *solutes*. The composition of a solution may be expressed quantitatively by specifying the concentration of one or more components; several *concentration units* are important. These include *mole fraction, mole percent, molarity, molality,* and *percent by mass*.

A dissolved solute may be at a high enough concentration to be in a state of equilibrium with excess undissolved solute. The solution under these circumstances is said to be *saturated*. Any solution with a lower solute concentration than this is said to be *unsaturated*. The *solubility* of a solute is its concentration in a saturated solution. The solubility of a given solute in a given solvent depends upon the relative sizes of the forces which operate between particles of the solute, between particles of the solvent, and between solute and solvent particles. Polar and ionic substances tend to be soluble in each other, as do nonpolar substances in each other, but often specific interactions must be considered.

The solubility of a substance depends upon temperature and pressure. *Le Châtelier's* principle is useful for predicting changes in solubility with temperature and pressure changes. The change of solubility with temperature depends upon the *heat of solution* at saturation. The heat of solution of a solid in water depends, in turn, on the *lattice energy* of the solid and the *hydration energy* of its particles.

Water is an unusually good solvent for many ionic and polar solutes. This is true partly because of its ability to form *hydrogen bonds* with itself and with other molecules. In such a bond a hydrogen atom acts as a bridge between two highly electronegative atoms.

Colligative properties are properties of solutions which depend only on solute particle concentration, and not on the identity of the particles. These include *vapor-pressure lowering, freezing-point depression, boiling-point elevation,* and *osmotic pressure.* Each of these

properties depends on the ability of solute particles to lower the escaping tendency of the solvent.

An important type of solute is the *electrolyte,* which dissociates to release ions into solution. Electrolytes can be classified as *strong* or *weak,* depending upon the degree to which they are dissociated. In a solution of a weak electrolyte, undissociated solute molecules are in equilibrium with ions produced by the dissociation. According to the *Arrhenius* viewpoint, electrolytes can also be classed as *acids, bases,* or *salts.* In solution the positive ion of an acid is the hydrogen ion, the negative ion of a base is the hydroxide ion, and the ions of a salt are neither of these. In water all ions are *hydrated,* that is, surrounded by a shell of water molecules. The hydrogen ion is very strongly hydrated because of its small size; it is often represented in solution as H_3O^+, called the *hydronium ion.*

Although strong electrolytes are completely dissociated in aqueous solution, their ions attract each other enough to produce anomalous colligative and other properties. The *Debye-Hückel theory* both qualitatively and quantitatively provides an explanation for these effects.

KEY TERMS

Acid (Arrhenius) (Sec. 12-6)
Aerosol (Sec. 12-1)
Associated liquid (Sec. 12-3)
Base (Arrhenius) (Sec. 12-6)
Boiling-point elevation (Sec. 12-4)
Colligative property (Sec. 12-4)
Colloid (Sec. 12-1)
Concentrated solution (Sec. 12-3)
Counterion (Sec. 12-6)
Debye-Hückel theory (Sec. 12-6)
Deviation from Raoult's law (Sec. 12-4)
Dielectric constant (Sec. 12-3)
Dilute solution (Sec. 12-3)
Dissociation (Sec. 12-5)
Distillation diagram (Sec. 12-4)
Electrolyte (Sec. 12-5)
Electrolyte strength (Sec. 12-5)
Emulsion (Sec. 12-1)
Escaping tendency (Sec. 12-4)
Fog (Sec. 12-1)
Fractional distillation (Sec. 12-4)
Freezing-point depression (Sec. 12-4)
Gel (Sec. 12-1)
Heat of solution (Sec. 12-3)
Henry's law (Sec. 12-3)

Heterogeneous mixture (Sec. 12-1)
Homogeneous mixture (Sec. 12-1)
Hydration (Secs. 12-3, 12-5, 12-6)
Hydration energy (Secs. 12-3, 12-6)
Hydrogen bond (Sec. 12-3)
Hydronium ion (Secs. 12-5, 12-6)
Ideal solution (Sec. 12-4)
Ionic atmosphere (Sec. 12-6)
Ionization (Sec. 12-5)
Molality (Sec. 12-3)
Molarity (Sec. 12-3)
Mole fraction (Sec. 12-3)
Mole percent (Sec. 12-3)
Nonelectrolyte (Sec. 12-5)
Neutralization reaction (Sec. 12-6)
Osmosis (Sec. 12-4)
Osmotic pressure (Sec. 12-4)
Oxonium ion (Sec. 12-5)
Percent by mass (Sec. 12-3)
Polarizability (Sec. 12-3)
Raoult's law (Sec. 12-4)
Reverse osmosis (Sec. 12-4)
Salt (Sec. 12-6)
Saturated solution (Sec. 12-3)
Semipermeable membrane (Sec. 12-4)

Smoke (Sec. 12-1)
Sol (Sec. 12-1)
Solubility (Sec. 12-3)
Solute (Sec. 12-3)
Solution (Sec. 12-1)
Solvation (Sec. 12-3)
Solvent (Sec. 12-3)
Strong electrolyte (Sec. 12-5)

Supersaturated solution (Sec. 12-3)
Tyndall effect (Sec. 12-1)
Unsaturated solution (Sec. 12-3)
Vapor-pressure lowering (Sec. 12-4)
Vapor-pressure curve (Sec. 12-4)
Weak electrolyte (Sec. 12-5)

QUESTIONS AND PROBLEMS

Colloids

12-1 What is a colloid? How can a colloid be distinguished from a solution?

12-2 Considering the "in-between" sizes of colloidal particles, can you think of two general methods of preparing colloids?

12-3 Describe two ways in which a colloid can be stablized, that is, prevented from settling out.

12-4 Why do the particles of a sol frequently settle out faster than do solute particles in a solution?

12-5 Sols of arsenious sulfide, As_2S_3, are stabilized by the adsorption of OH^- ions. What might you add to such a sol in order to coagulate it?

12-6 The osmotic pressure of a colloidal dispersion of polystyrene in toluene was found to be 0.51 mmHg at 25°C. If the dispersion contained 2.0×10^{-3} g of polystyrene per milliliter, what was the molecular weight of the polystyrene?

12-7 A colloid consists of roughly spherical molecules having an average density of 0.94 g ml^{-1}. If the radius of the particles is 120 nm, calculate the molecular weight.

Solutions—general considerations

12-8 Given a homogeneous liquid, how could you find out if it is a pure substance or a solution?

12-9 Gold will suck up small amounts of mercury almost as a sponge sucks up water. Where does the mercury go?

12-10 What could you do to find out whether a given solution of calcium acetate is unsaturated, saturated, or supersaturated?

12-11 Discuss the expression, Like dissolves like.

12-12 In what ways are an ideal gas and an ideal solution similar?

12-13 Define or explain each of the following terms: hydration, dissociation, solubility, supersaturation, hydrogen bond.

12-14 Discuss the mechanism of hydration. Can particles other than ions be hydrated? How might some of the properties of a solution of sodium chloride in water be changed, if the ions were not hydrated?

12-15 How do lattice energy and hydration energy determine the heat of solution of a substance in water.

Concentration units

12-16 Calculate the molarity of each of the following solutions: **(a)** 1.00 g of NaCl dissolved in 1.00 liter of solution **(b)** 1.00 g of H_2SO_4 dissolved in 1.00 liter of solution **(c)** 4.00 g of NaOH dissolved in 55.0 ml of solution.

12-17 Calculate the molality of each of the following solutions: **(a)** 6.00 g of methanol, CH_3OH, dissolved in 1.00 kg of water **(b)** 6.00 g of CH_3OH dissolved in 1.00 kg of carbon tetrachloride, CCl_4 **(c)** 14.0 g of benzene, C_6H_6, dissolved in 25.0 g of CCl_4.

12-18 Calculate the mole fraction of benzene, C_6H_6, in each of the following solutions: **(a)** 1.00 g of C_6H_6 + 1.00 g CCl_4 **(b)** 4.00 g of C_6H_6 + 4.00 g CCl_4 + 4.00 g CS_2.

12-19 Suppose 52.0 g of sucrose, $C_{12}H_{22}O_{11}$, is added to 48.0 g of water to form a solution having a density of 1.24 g ml^{-1}. Calculate **(a)** the percent by mass **(b)** the mole percent **(c)** the molarity **(d)** the molality of the sucrose in the solution.

12-20 Suppose 144 g of ethanol, C_2H_5OH, and 96.0 g of water are mixed to form a solution having a density of 0.891 g ml^{-1}. Calculate **(a)** the mass percent **(b)** the mole fraction **(c)** the molality **(d)** the molarity of ethanol in the solution.

12-21 Suppose 30.0 g of citric acid, $C_6H_8O_7$, is dissolved in 70.0 g of water to form a solution having a density of 1.13 g ml^{-1}. Calculate the molality of **(a)** citric acid and **(b)** water in the mixture.

12-22 Calculate the molarity of **(a)** citric acid and **(b)** water in the solution described in Prob. 12-21.

12-23 Calculate **(a)** the mass percent and **(b)** mole fraction of calcium chloride in a $0.100\ m\ CaCl_2$ solution in water.

12-24 Calculate **(a)** the mole fraction and **(b)** the molarity of a $1.59\ m$ aqueous solution of calcium chloride, $CaCl_2$. The density of the solution is $1.13\ g\ ml^{-1}$.

12-25 Calculate the molar concentration of chloride ions in **(a)** $0.15\ M\ NaCl$ **(b)** $0.15\ M\ CaCl_2$ **(c)** $0.28\ M\ AlCl_3$.

12-26 Suppose 0.10 mol of $NaCl$, 0.20 mol of $MgCl_2$, and 0.30 mol of $FeCl_3$ are added to enough water to make 0.500 liter of solution. What is the molarity of Cl^- ions in the solution?

12-27 Suppose 15.0 ml of a $0.240\ M\ NaCl$ solution is mixed with 35.0 ml of water. If the final volume is 50.0 ml, what is the final molarity?

12-28 Calculate the final molarity of sucrose in a solution prepared by mixing 25.0 ml of $0.150\ M$ sucrose with **(a)** 25.0 ml of water **(b)** 38.0 ml of water **(c)** 25.0 ml of $0.100\ M$ sucrose **(d)** 91.0 ml of $0.109\ M$ sucrose. Assume in each case that the volumes are additive.

Solubility and heat of solution

12-29 If ΔH_{soln} for a substance is a positive number, will its solubility increase or decrease with an increase in temperature?

12-30 Using data from Table 12-2 calculate the solubility (mole fraction) of oxygen in water at 5.00 atm and $20°C$.

12-31 From your answer to Prob. 12-30, calculate how many milliliters of O_2 at 5.00 atm and $20°C$ will dissolve in 1.00 kg of water at the same temperature.

12-32 Rank the following substances in order of their increasing solubility in water: CH_3OH, CH_3F, NaF, CH_4.

12-33 Which of each of the following pairs of ions should have the higher hydration energy (most negative ΔH_{hyd})? **(a)** Na^+ or Li^+ **(b)** B^{3+} or Al^{3+} **(c)** Ca^{2+} or Ga^{3+} **(d)** S^{2-} or Se^{2-} **(e)** S^{2-} or Cl^-.

12-34 The lattice energy of $AgCl$ is $904\ kJ\ mol^{-1}$, the hydration energy of Ag^+ is $-469\ kJ\ mol^{-1}$, and the hydration energy of Cl^- is $-377\ kJ\ mol^{-1}$. Calculate the heat of solution of $AgCl$. Should the solubility of $AgCl$ increase or decrease with temperature?

12-35 Describe how you would attempt to prepare a saturated solution of a solute whose solubility **(a)** increases **(b)** decreases with temperature.

12-36 If ΔH_{soln} for a substance is negative, what will happen to the temperature of a supersaturated solution of the substance if a seed crystal is added?

12-37 The solubility of some electrolytes in water can be greatly decreased if some ethanol is added to the water. Explain.

12-38 The mole-fraction solubility of ethylene gas in water at $0°C$ is 2.05×10^{-4} at 1.00 atm. What is its solubility at 5.00 atm at the same temperature?

12-39 How much pressure would you need to apply to 0.100 mol of oxygen gas in order to get it all to dissolve in 1.00 kg of water at $20°C$?

12-40 How much pressure would you need to apply to 1.00 liter of N_2 gas at $20°C$ and originally at 1.00 atm in order to get it to all dissolve in 1.00 kg of water at $20°C$?

12-41 The solubility of ammonium chloride, NH_4Cl, in water is $8.56\ m$ at $40°C$ and $6.95\ m$ at $20°C$. If 25.0 g of water is saturated with NH_4Cl at $40°C$ and then cooled to $20°C$, how much NH_4Cl will crystallize out?

Colligative properties

12-42 Calculate **(a)** the normal freezing point and **(b)** the normal boiling point of a $0.262\ m$ solution of sucrose in water.

12-43 Calculate the **(a)** vapor pressure lowering and **(b)** osmotic pressure at $25°C$ of a 10.0 mass % sucrose solution in water. The density of the solution is $1.04\ g\ ml^{-1}$, and the vapor pressure of pure water is $23.76\ mmHg$ at $25°C$.

12-44 Calculate the molecular weight of a nonvolatile compound 1.15 g of which raises the boiling point of 75.0 g of benzene, C_6H_6, $0.275°C$.

12-45 Suppose 2.14 g of a compound is added to 55.0 g of acetic acid and the freezing point is found to be $0.429°C$ below that of pure acetic acid. What is the molecular weight of the compound?

12-46 Suppose 7.65 g of a nonvolatile compound is dissolved in 90.0 g of ethanol, C_2H_5OH. This lowers the vapor pressure of the ethanol from 242.61 to $241.19\ mmHg$ at $30°C$. Calculate the molecular weight of the compound.

12-47 What is the freezing point of an aqueous solution which boils (at 1 atm) at $101.214°C$?

12-48 What is the molecular weight of substance if a solution containing 3.50 g liter^{-1} has an osmotic pressure of $0.337\ mmHg$ at $30°C$?

12-49 Benzene, C_6H_6, and toluene, C_7H_8, form essentially ideal solutions. For a solution prepared by mixing 1.00 mol of benzene with 2.00 mol of toluene, calculate **(a)** the partial vapor pressure of each component and **(b)** the total vapor pressure of the solution at 20°C. At this temperature the vapor pressure of pure benzene is 74.7 mmHg and that of toluene is 22.2 mmHg.

12-50 Bromobenzene, C_6H_5Br, and chlorobenzene, C_6H_5Cl, form essentially ideal solutions. At 100°C the vapor pressure of bromobenzene is 137 mmHg and that of chlorobenzene is 285 mmHg. Calculate the vapor pressure of a 50.0 mass % solution of bromobenzene in chlorobenzene at this temperature.

Electrolytes

12-51 In terms of Le Châtelier's principle, why does the degree of dissociation of a weak electrolyte increase with dilution?

12-52 Account for the fact that some *molecular* substances are electrolytes in water.

12-53 Calculate the molarity of Na^+, Cl^-, K^+, and SO_4^{2-} present in a solution prepared by adding 0.100 mol of NaCl, 0.200 mol of Na_2SO_4, and 0.300 mol of K_2SO_4 together to enough water to make 0.725 liter of solution. All of the salts are strong electrolytes and none react with each other.

12-54 Suppose 10.0 g of each of the following strong electrolytes is added to enough water to make 424 ml of solution: HCl, $MgCl_2$, NaCl, HBr. None of the compounds reacts. Calculate the molar concentration of each solute-derived ion in the solution.

12-55 Acetic acid, $HC_2H_3O_2$, is a weak acid which dissociates to form H^+ and $C_2H_3O_2^-$ (acetate) ions. In a 0.150 m solution of acetic acid in water the H^+ ion concentration is $1.6 \times 10^{-3}\,m$. What is the freezing point of this solution?

12-56 A 0.25 m aqueous solution of a hypothetical weak acid HA freezes at −0.651°C. Calculate the apparent percent dissociation of the acid.

General

12-57 The vapor pressure of a solution is found to be greater than that of the pure solvent from which it was prepared. Suggest two possible explanations.

12-58 Solutions which are injected into the bloodstream for medical purposes can do physiological damage if they are either too dilute or too concentrated. Explain.

12-59 Would you expect a weak electrolyte to be more or less dissociated in a solvent of low dielectric constant than it is in water? Explain.

12-60 At 120°C the vapor pressure of chlorobenzene is 560 mmHg and that of bromobenzene is 275 mmHg. Assuming that Raoult's law is obeyed, **(a)** calculate the total vapor pressure of a solution of these compounds which is 40.0 mol % chlorobenzene, **(b)** find the composition (mol %) of the vapor in equilibrium with the solution in part **a**, and **(c)** determine the composition of the liquid which would be in equilibrium with a gaseous mixture of chlorobenzene and bromobenzene having equal partial pressures at 120°C.

12-61 Which solution would you expect to have the highest normal boiling point: 0.1 M $CaCl_2$, 0.3 M sucrose, or 0.15 M NaCl? Explain.

12-62 Stories abound of automobile engines which have been ruined by a vandal's putting sugar (sucrose) in the gas tank. Given that gasoline is composed primarily of hydrocarbons of low polarity, what damage can the sugar do?

12-63 Suppose 1.00 g of glycerol, a nonvolatile nonelectrolyte, is added to 25.0 g of water, and the resulting solution is found to freeze at −0.806°C. The composition of glycerol is 39.1 percent C, 8.76 percent H, and 52.1 percent O by mass. What is the molecular formula of glycerol?

12-64 Suppose 1.50 mol of HCl and 1.00 mol of NaOH are both added to enough water to make 725 ml of solution. Calculate the molar concentrations of H^+, Cl^-, and Na^+ in the resulting solution.

12-65 One hundred-proof "vodka" can be made by mixing equal volumes of pure ethanol, C_2H_5OH, density 0.789 g ml^{-1}, and water, 0.998 g ml^{-1} (at 20°C). What is the molarity of the ethanol in 100-proof "vodka," if its density is 0.926 g ml^{-1}? (*Caution:* It is unwise to make "vodka" this way. Commercially available "pure" ethanol generally contains traces of benzene, which is toxic and carcinogenic.)

13 AQUEOUS-SOLUTION REACTIONS

TO THE STUDENT

We live in a watery world. Three-quarters of the earth's surface is covered by liquid water, and in addition, gaseous and solid water occur naturally in great abundance. Many chemical processes in the natural environment take place in aqueous solution. In fact, most living organisms, both plant and animal, are composed predominately of water, and their biochemistry is largely water-based. Because of its participation in ion-dipole, dipole-dipole, and hydrogen-bond interactions, water is an excellent solvent for a wide variety of compounds and for this reason has been extensively used as a reaction medium for investigating the nature of chemical change. We now turn to a consideration of reactions which occur in this important solvent.

**13-1
Acid-base reactions**

The early French chemist Antoine Lavoisier (Sec. 1-2) believed that all acids contain oxygen. This belief was shared by many others until it was finally shown that hydrochloric acid, HCl, contains no oxygen. Later, the suggestion that combined hydrogen was responsible for acidic behavior became popular, but the existence of many hydrogen-containing compounds which were not acids could not be satisfactorily explained. It remained for Svante Arrhenius to first construct a consistent picture of electrolytic solutions which would do a satisfactory job of explaining acid-base behavior.

*The Arrhenius
definition*

In 1884 Arrhenius defined an *acid* as a hydrogen-containing substance which produces hydrogen ions in solution and a *base* as a hydroxide-containing substance which produces hydroxide ions in solution (Sec. 12-6). *Neutralization* was described by Arrhenius as the combination of these ions to form water:

$$H^+ + OH^- \longrightarrow H_2O$$

This is consistent with the observation that when a dilute solution of HCl, HBr, HI, HNO_3, or $HClO_4$ is mixed with a dilute solution of NaOH, KOH, RbOH,

$Ba(OH)_2$, or $La(OH)_3$, the molar *heat of neutralization*, ΔH_{neut}, is always the same: -55.90 kJ per mole of water formed (i.e., 55.90 kJ of heat are liberated per mole). The above acids are all strong (completely dissociated) acids, and the bases similarly are all strong bases, so that no matter which acid-base pair is chosen, the reaction is the same: the combination of a hydrogen ion with a hydroxide ion to form a water molecule.

However, some molar heats of neutralization are much less than the above value. When HCN and NaOH solutions are mixed, for example, $\Delta H = -10.3$ kJ mol^{-1}. The fact that so much less heat is given off is consistent with the observation that HCN is a weak acid, and so most of the HCN present is dissolved as HCN molecules, not as H^+ and CN^- ions. Energy is consequently required to break the H—CN bonds in the undissociated molecules, and so less energy is left over for release to the surroundings. This time the reaction is represented as

$$HCN + OH^- \longrightarrow H_2O + CN^-$$

The Arrhenius picture of acids and bases, strong and weak, is a useful one even today. It has at least three important shortcomings, however, one being that it is limited to behavior in aqueous solutions. Another is that it ignores many substances and dissolved species other than OH^- ions which will combine with H^+ ions, and many besides H^+ ions which will combine with OH^- ions. Lastly, many substances which are not hydrogen or hydroxide compounds increase the H^+ or OH^- concentration when added to water. This suggests the need for broader definitions of acid and base.

The solvent-system definition

The ability of water to stabilize ions by hydrating them is great enough that water can dissociate into hydrogen and hydroxide ions. The *autodissociation*, or *self-dissociation*, equation is written according to the Arrhenius view as

$$H_2O \rightleftharpoons H^+ + OH^-$$

Water is a very weak electrolyte, being dissociated only about 2×10^{-7} percent at 25°C. As a result, pure water conducts electricity slightly. Other solvents undergo similar autodissociation. For example, ammonia, a liquid below -33°C at 1 atm, dissociates to form a hydrogen ion and an amide ion, NH_2^-:

$$NH_3 \rightleftharpoons H^+ + NH_2^-$$

The solvation (*ammoniation* in this case) of the hydrogen ion is almost always indicated by writing it as NH_4^+, the ammonium ion, analogous to H_3O^+ written instead of H^+ in water. In other words, the dissociation equilibrium is usually shown as

$$2NH_3 \rightleftharpoons NH_4^+ + NH_2^-$$

Liquid sulfur dioxide, a nonprotonic solvent, undergoes similar autodissociation to form the thionyl, SO^{2+}, and sulfite, SO_3^{2-}, ions:

$$2SO_2 \rightleftharpoons SO^{2+} + SO_3^{2-}$$

The *solvent-system definition* of acids and bases is intended to extend the Arrhenius concept to other self-dissociating solvents. According to this definition an *acid* is a substance which increases the concentration of cations related to the solvent, and a *base* is a substance which increases the concentration of anions related to the solvent. (Remember, a cation is positively charged, while an anion is negatively charged.) In water, then, an acid is something which produces H^+ (H_3O^+), and a base is something which produces OH^-. In NH_3 an acid produces NH_4^+, and a base, NH_2^-; in SO_2 an acid produces SO^{2+}, and a base, SO_3^{2-}.

The Brønsted-Lowry definition   In 1923 Brønsted in Denmark and Lowry in England independently suggested an acid-base definition which has proved very useful. The *Brønsted-Lowry definition is a protonic definition.* According to it, an *acid* is a species which tends to donate a proton, and a *base* is a species which tends to accept a proton. In addition, an *acid-base reaction* is a *proton-transfer reaction.* The Brønsted-Lowry definition is a very general one in many respects. HCl, for instance, is found to be an acid in aqueous solution, just as it is according to the Arrhenius definition. But in addition, HCl is a Brønsted-Lowry acid *in any other solvent,* and even when no solvent at all is present. It is an acid simply because it can donate a proton.

According to the Brønsted-Lowry picture, an acid-base reaction involves the competition for a proton between two bases. When hydrogen chloride dissolves in water, for instance, an HCl molecule (an acid) donates a proton to H_2O (a base) to form H_3O^+ (an acid) and Cl^- (a base). In other words

$$HCl + H_2O \longrightarrow H_3O^+ + Cl^-$$

$$Acid_1 + base_2 \longrightarrow acid_2 + base_1$$

Protons (H^+) do not show up explicitly. Instead, the equation shows an acid transferring a proton to a base to form their *conjugate* base and acid, respectively. In the above equation HCl and Cl^- constitute a *conjugate acid-base pair,* and H_2O and H_3O^+ are another such pair.

Ammonia is a base. In aqueous solution it accepts a proton

$$NH_3 + H_2O \longrightarrow NH_4^+ + OH^-$$

$$Base_1 + acid_2 \longrightarrow acid_1 + base_2$$

NH_4^+ and NH_3 are one conjugate acid-base pair and H_2O and OH^- are the other in this reaction.

The reaction between HCl and NH_3 is another example of a Brønsted-Lowry acid-base reaction:

$$HCl + NH_3 \longrightarrow NH_4^+ + Cl^-$$

$$Acid_1 + base_2 \longrightarrow acid_2 + base_1$$

This reaction occurs in the absence of any solvent at all. (Gaseous HCl and NH_3 form solid NH_4Cl, ammonium chloride, its crystal lattice being composed of NH_4^+ and Cl^- ions.)

Notice that according to the Brønsted-Lowry picture, water can react either as an acid (to form its conjugate base OH^-) or as a base (to form its conjugate acid H_3O^+). Both processes occur in the autodissociation of water:

$$H_2O + H_2O \longrightarrow H_3O^+ + OH^-$$

$$Acid_1 + base_2 \longrightarrow acid_2 + base_1$$

According to the Brønsted-Lowry picture, the *strength* of an acid is its tendency to donate a proton, while the strength of a base is its tendency to accept a proton. Consider the reaction

$$HCl + H_2O \longrightarrow H_3O^+ + Cl^-$$

This reaction goes essentially to completion, which means that no appreciable HCl is left. Thus HCl is a stronger acid than H_3O^+, and H_2O is a stronger base than Cl^-. In general the stronger an acid is, the weaker is its conjugate base. For example, HCl has a great tendency to lose a proton; so Cl^- has but a slight tendency to gain one. The stronger a base is, the weaker is its conjugate acid. For example, the O^{2-} ion greatly tends to gain a proton; so OH^- tends to lose one only slightly.

The relationship between conjugate acids and bases allows us to list them (Table 13-1) with acid strength decreasing from top to bottom while conjugate base strength correspondingly increases. Acids are shown on the left and bases on the right, and the farther above a base an acid is, the more the pair will tend to react. (In brief, an acid above reacts with a base below.) Even if an acid is below a base, however, the pair will react slightly. Thus H_2O reacts slightly with SO_3^{2-} even though their relative positions in the table are not favorable.

Look in the table at the positions of the four acids $HClO_4$, H_2SO_4, HCl, and HNO_3 (on the left) with respect to that of H_2O (on the right). Each of these acids has a great tendency to donate a proton to water. In Arrhenius language we refer to these, simply, as *strong acids*. The acids (on the left) which are below H_2O (on the right) show much smaller tendencies to transfer a proton to water. These are, then, *weak acids* from the Arrhenius viewpoint.

The fact that the so-called strong acids are all above H_2O in Table 13-1 is an illustration of the *leveling effect*. Water is a strong enough base to remove the protons from the acids above it almost completely. In other words, because of the basicity of H_2O, each of the following reactions reaches a position of equilibrium far to the right:

$$HClO_4 + H_2O \rightleftharpoons H_3O^+ + ClO_4^-$$

$$H_2SO_4 + H_2O \rightleftharpoons H_3O^+ + HSO_4^-$$

$$HCl + H_2O \rightleftharpoons H_3O^+ + Cl^-$$

$$HNO_3 + H_2O \rightleftharpoons H_3O^+ + NO_3^-$$

In a less basic solvent the tendency for each acid to transfer its proton is smaller. In acetic acid, $HC_2H_3O_2$, for example, the corresponding equilibrium positions of the following reactions are all different:

$$HClO_4 + HC_2H_3O_2 \rightleftharpoons H_2C_2H_3O_2^+ + ClO_4^-$$

$$H_2SO_4 + HC_2H_3O_2 \rightleftharpoons H_2C_2H_3O_2^+ + HSO_4^-$$

$$HCl + HC_2H_3O_2 \rightleftharpoons H_2C_2H_3O_2^+ + Cl^-$$

$$HNO_3 + HC_2H_3O_2 \rightleftharpoons H_2C_3H_3O_2^+ + NO_3^-$$

Measurements made of the electrical conductivities of solutions of equal concentration of these acids dissolved in acetic acid show a gradation in conductivity from

Table 13-1
Relative strengths
of Brønsted-Lowry
acid-base pairs

	Conjugate acid	Conjugate base	
Stronger			Weaker
	$HClO_4$	ClO_4^-	
	H_2SO_4	HSO_4^-	
	HCl	Cl^-	
	HNO_3	NO_3^-	
	H_3O^+	H_2O	
	H_2SO_3	HSO_3^-	
	HSO_4^-	SO_4^{2-}	
	HF	F^-	
	$HC_2H_3O_2$	$C_2H_3O_2^-$	
	H_2S	HS^-	
	HSO_3^-	SO_3^{2-}	
	H_2O	OH^-	
	HS^-	S^{2-}	
	OH^-	O^{2-}	
Weaker			Stronger

$HClO_4$, the highest, to HNO_3, the lowest. Higher conductivity means more ions in solution; so $HClO_4$ is a stronger acid than HNO_3. Because acetic acid can differentiate among the strengths of these acids, it is called a *differentiating solvent*.

The Lewis definition   A still broader acid-base definition was suggested by the American chemist G. N. Lewis in 1923, the same year that Brønsted and Lowry made their proposals. According to Lewis, an acid is a species with a vacant orbital capable of accepting an electron pair, while a base is a species which can donate an electron pair to form a coordinate covalent bond (Sec. 8-2). More briefly, an *acid* is an electron-pair acceptor, and a *base* is an electron-pair donor.

Each Lewis acid-base reaction consists of the formation of a coordinate covalent bond. The reaction between hydrogen and hydroxide ions can be viewed from this perspective:

$$H^+ + \left[:\overset{..}{\underset{H}{O}}: \right]^- \longrightarrow H:\overset{..}{\underset{H}{O}}:$$

Acid   Base

Here the electron pair (circled) on the oxygen forms a covalent bond with the proton. The OH^- ion is a base because it has donated an electron pair to the bond, while the H^+ is an acid because it has accepted (a share of) the electron pair.

The following are also Lewis acid-base reactions. In each case the important electron pair is circled.

$$H^+ + \left[:\overset{..}{\underset{..}{F}}: \right]^- \longrightarrow H:\overset{..}{\underset{..}{F}}:$$

Acid   Base

$$H^+ + :\overset{H}{\underset{H}{N}}:H \longrightarrow \left[H:\overset{H}{\underset{H}{N}}:H \right]^+$$

Acid   Base

$$Ag^+ + 2:\overset{H}{\underset{H}{N}}:H \longrightarrow \left[H:\overset{H}{\underset{H}{N}}:Ag:\overset{H}{\underset{H}{N}}:H \right]^+$$

Acid   Base

The last reaction is an example of the formation of a complex ion and would not be classed as an acid-base reaction according to the Brønsted-Lowry (or Arrhenius) definition.

The Lewis definition has many important applications because of its great generality. Many substances which do not fit the Arrhenius or Brønsted-Lowry criteria quite logically become classified as Lewis acids or bases. For aqueous systems, however, the terms *acid* and *base* are usually used in either the Arrhenius or Brønsted-Lowry context. The terms *Lewis acid* and *Lewis base* are usually used when the broader viewpoint is adopted.

**13-2
Precipitation and complexation reactions**

Although the hydrating shells of water molecules tend to keep oppositely charged ions apart in aqueous solution, in some cases such ions can combine with each other. Two important types of such reactions are precipitation and complexation reactions. In a *precipitation* reaction the product is only slightly soluble in water; in a *complexation* reaction the product is a soluble complex ion.

Solubility and precipitation

When the concentration of a product begins to exceed the solubility of that substance, then any further product formed precipitates out of solution, as long as the solution does not supersaturate. Consider the formation of barium sulfate, $BaSO_4$, for example. $BaSO_4$ has a low solubility in water, about $4 \times 10^{-5} M$. This means that only 4×10^{-5} mol of $BaSO_4$ dissolves in water to make a liter of saturated solution. $BaSO_4$ is a *strong electrolyte,* and so when it dissolves, it dissociates completely:

$$BaSO_4(s) \longrightarrow Ba^{2+} + SO_4^{2-}$$
$$4 \times 10^{-5} \text{ mol} \longrightarrow 4 \times 10^{-5} \text{ mol} + 4 \times 10^{-5} \text{ mol} \qquad \text{(per liter)}$$

so that the resulting solution is really one which is $4 \times 10^{-5} M$ in Ba^{2+} and $4 \times 10^{-5} M$ in SO_4^{2-}. (Remember: This is a *saturated* solution.)

Now imagine that we mix 0.5 liter of 1 M barium chloride, $BaCl_2$, with 0.5 liter of 1 M sodium sulfate, Na_2SO_4. The first of these solutions contains Ba^{2+} at a 1 M concentration, because

$$BaCl_2(s) \longrightarrow \boxed{Ba^{2+}} + 2Cl^-$$
$$1 \text{ mol} \longrightarrow \boxed{1 \text{ mol}} + 2 \text{ mol} \qquad \text{(per liter)}$$

and the second contains SO_4^{2-} also at a 1 M concentration:

$$Na_2SO_4(s) \longrightarrow 2Na^+ + \boxed{SO_4^{2-}}$$
$$1 \text{ mol} \longrightarrow 2 \text{ mol} + \boxed{1 \text{ mol}} \qquad \text{(per liter)}$$

In 0.5 liter of the 1 M Ba^{2+} solution there is 0.5 mol of Ba^{2+}:

$$\frac{1 \text{ mol } Ba^{2+}}{1 \text{ liter solution}} 0.5 \text{ liter solution} = 0.5 \text{ mol } Ba^{2+}$$

Similarly the 0.5 liter of 1 M SO_4^{2-} solution contains 0.5 mol of SO_4^{2-}. Assuming that the final volume is 1 liter, the Ba^{2+} concentration after mixing appears to be

$$\frac{0.5 \text{ mol Ba}^{2+}}{1 \text{ liter solution}} = 0.5 \frac{\text{mol Ba}^{2+}}{\text{liter solution}} = 0.5 \, M$$

Likewise, the SO_4^{2-} concentration after mixing appears to be 0.5 M. But is this possible? In a saturated solution the concentrations of Ba^{2+} and SO_4^{2-} are each only $4 \times 10^{-5} \, M$. The solution prepared by mixing the $BaCl_2$ and Na_2SO_4 solutions is evidently supersaturated and will soon begin to precipitate solid $BaSO_4$. (In all likelihood the precipitation will begin slightly after the two original solutions contact each other.) $BaSO_4$ will continue precipitating until the concentration of Ba^{2+} and of SO_4^{2-} have been reduced to $4 \times 10^{-5} \, M$, at which point the solution is just saturated. *Whenever the ions of an insoluble electrolyte are introduced separately into a solution in such a way that the final concentration of the electrolyte exceeds its solubility, then some of it will precipitate out of solution.*

ADDED COMMENT

Remember that there is no such thing as complete insolubility. *Insoluble* really means "slightly soluble."

In addition, in the above example, the Ba^{2+} and SO_4^{2-} concentrations were equal. If they were not, the same reaction would still take place, unless the concentration of one ion were very low. (We look into this situation carefully in Sec. 17-2.)

● **EXAMPLE 13-1**

Problem Sodium fluoride (NaF), sodium chloride (NaCl), calcium chloride ($CaCl_2$), and calcium fluoride (CaF_2) are all strong electrolytes, but only the last, CaF_2, is insoluble in water. Write the equation for the reaction which occurs when moderately concentrated aqueous solutions of NaF and $CaCl_2$ are mixed.

Solution Present in the NaF solution are Na^+ and F^- ions, because when the solution was originally prepared, the solute NaF dissociated as it dissolved:

$$NaF(s) \longrightarrow Na^+ + F^-$$

In the $CaCl_2$ solution there are Ca^{2+} and Cl^- ions from the dissociation:

$$CaCl_2(s) \longrightarrow Ca^{2+} + 2Cl^-$$

So immediately after mixing, Na^+, F^-, Ca^{2+}, and Cl^- ions are present together in one solution. The fact that Ca^{2+} and F^- ions are both present in moderate concentration means that the solution, if it stayed like this, would be supersaturated with respect to CaF_2. Thus we know that CaF_2 will precipitate out until the supersaturation has been relieved:

$$Ca^{2+} + 2F^- \longrightarrow CaF_2(s)$$

Since none of the other combinations of ions, in particular, Na^+ and Cl^-, are ions of an insoluble substance, there is no other reaction. ●

Each of the preceding equations is an example of a *net ionic equation*. A *net equation* is one which shows on the left only those substances or species in solution which actually react (are used up) and on the right only those which are formed. An *ionic* equation is, of course, one which shows ions. Note in particular that in the last equation the Na^+ and Cl^- ions and H_2O molecules are not shown. This does not mean that they are not present, only that they are not consumed or formed in the reaction. (We will have more to say about net equations in Sec. 13-3.)

ADDED COMMENT

How did we know that the $CaCl_2$ solution (Example 13-1) contained Ca^{2+} and Cl^- ions? Why not Ca^{2+} and either Cl_2^- or Cl_2^{2-} ions? In the first place, if one Ca^{2+} and one Cl_2^- were combined, the result would be $CaCl_2^+$, not $CaCl_2$. (Look at those charges.) Secondly, Cl_2^{2-} ions balance the charge satisfactorily, but since the compound is called calcium *chloride*, it must contain *chloride* ions, and these should be old friends by now: Cl^-. (Besides, what kind of bonding, and Lewis structure, could Cl_2^{2-} have? What is the bond order in Cl_2^{2-}?)

Complexation   Silver chloride (AgCl) has a very low solubility in water, so that when Ag^+ ions in a solution of silver nitrate ($AgNO_3$) are mixed with Cl^- ions in a solution of sodium chloride (NaCl), white silver chloride precipitates:

$$Ag^+ + Cl^- \longrightarrow AgCl(s)$$

If NH_3 is now added to the AgCl, the AgCl redissolves, forming a colorless solution. That solution contains the ionic species $Ag(NH_3)_2^+$, and it is formed in the reaction

$$AgCl(s) + 2NH_3 \longrightarrow Ag(NH_3)_2^+ + Cl^-$$

This ion is an example of a *complex ion*, commonly called the *silver-ammonia complex ion*. In this process the silver ion has been *complexed* by the two NH_3 molecules, the process itself is called *complexation*, and NH_3 is called a *complexing agent*.

The term *complex ion* is difficult to define. It was originally defined as an ion composed of two or more atoms, but today this definition is usually considered too broad, because it includes as complex ions, species which almost never dissociate into their fragments, such as nitrate (NO_3^-), sulfate (SO_4^{2-}), phosphate (PO_4^{3-}), and cyanide (CN^-). Each of these maintains its integrity and reacts as a single, unchanging unit in many of its reactions. Today *complex ion* usually is taken to mean an aggregate formed when a metal ion (or, sometimes, an atom) bonds to several other ions or molecules which cluster around it. These molecules of the complexing agent are called *ligands* in the complex and can usually be removed or replaced by other ligands, at least more easily than the oxygen atoms in NO_3^-, for instance. Some examples of complex ions are shown in Table 13-2 along with two similar complexes which are neutral (uncharged).

**Table 13-2
Some common
complexes**

$Co(NH_3)_6{}^{3+}$	$Cr(H_2O)_6{}^{3+}$
$Fe(NCS)(H_2O)_5{}^{2+}$	$Ag(NH_3)_2{}^+$
$Cu(NH_3)_4{}^{2+}$	$Cu(CN)_2{}^-$
$Fe(CN)_6{}^{4-}$	$Zn(OH)_4{}^{2-}$
$CoCl_3(NH_3)_3$	$PtCl_2(NH_3)_2$

The hydration of an ion may be considered a kind of complexation, and the resulting complex ion is called an *aquo complex*. The chromic ion forms such a complex in water:

$$Cr^{3+} + 6H_2O \longrightarrow Cr(H_2O)_6{}^{3+}$$

In addition to the six water molecules in the first hydration sphere of this ion, other water molecules are more weakly bound to the complex.

*Complexation as
Lewis acid-base
behavior*

The formation of a complex may be viewed as an illustration of Lewis acid-base behavior. Thus when the copper ion is complexed by four ammonia ligands,

$$Cu^{2+} + 4NH_3 \longrightarrow Cu(NH_3)_4{}^{2+}$$

each of the ammonia molecules donates an electron pair to form a coordinate covalent bond with the copper ion, as is shown in Fig. 13-1. In this reaction Cu^{2+} acts as a Lewis acid and NH_3 as a Lewis base.

Because bonding in complexes can be described in terms of coordinate covalent bond formation, the number of electron pairs accepted by the central ion (or atom) is called the *coordination number* of the ion (or atom). We will discuss the bonding and structures of complexes in detail in Chap. 22.

**13-3
Net equations for
aqueous-solution
reactions**

A net equation is a useful kind of equation because it is usually simple and free from extraneous or misleading information. In order to write a net equation, one must either *know* the identities of all reactants and product species, or one must know some of them and be able to figure out the rest. "Figuring out the rest" is really what this section is about.

*Classifying and
predicting reactions*

The following is a useful classification scheme for reactions which take place in aqueous solution:

AQUEOUS SOLUTION REACTIONS

I Reactions with no electron transfer
   A Production of a new phase
      1 Solid (precipitation reactions)
      2 Liquid (not common)
      3 Gas
   B Production of a soluble weak electrolyte
      1 Molecules
      2 Complex ions
II Reactions with electron transfer (redox reactions)

The above classification scheme has a major weakness in that many reactions

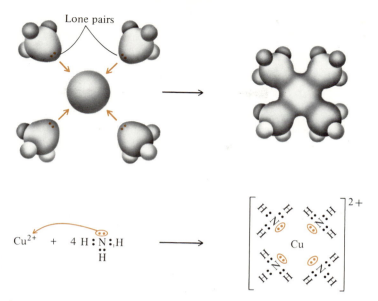

**Figure 13-1
The formation of a
complex ion: a Lewis
acid-base reaction.**

fall into several categories. Still, for simple reactions it can be used to help predict whether or not a specific reaction might take place and what the products are likely to be. In order to use the scheme, however, you need to know something about (1) solubilities, (2) strengths of electrolytes, and (3) strengths of oxidizing and reducing agents. We will consider redox reactions in Sec. 13-4 and Chap. 19, but for other types of reactions the following two sets of guidelines should prove useful:

SOLUBLE AND INSOLUBLE COMPOUNDS

1 All common *inorganic acids* are soluble.
2 All common compounds of the *alkali metals* are soluble.
  Exceptions: Some compounds of Li are insoluble.
3 All common *nitrates* (NO_3^-) are soluble.
4 Most common *acetates* ($C_2H_3O_2^-$) are soluble.
  Exceptions: $AgC_2H_3O_2$ and $Hg_2(C_2H_3O_2)_2$ are insoluble.
5 Most common *sulfates* (SO_4^{2-}) are soluble.
  Exceptions: $CaSO_4$, $SrSO_4$, $BaSO_4$, $PbSO_4$, Ag_2SO_4, and Hg_2SO_4 are insoluble.
6 Most common *halides* (salts containing a negative halogen ion) are soluble.
  Exceptions: (a) Chlorides, bromides, and iodides of Ag(I), Pb(II), and Hg(I) are insoluble;[1] mercuric iodide (HgI_2) is also insoluble; (b) fluorides of Mg, Ca, Sr, Ba, and Pb(II) are insoluble.
7 Most *carbonates* (CO_3^{2-}), *chromates* (CrO_4^{2-}), *oxalates* ($C_2O_4^{2-}$), *phosphates* (PO_4^{3-}), and *sulfites* (SO_3^{2-}) are insoluble.
  Exceptions: See items 1 and 2 above.
8 Most *hydroxides* (OH^-) are insoluble.
  Exceptions: (a) See item 2, above; (b) the hydroxides of Sr and Ba are fairly soluble.

---

[1] A Roman numeral in parentheses indicates oxidation state. (See Sec. 20-1.)

Note: Not listed above as soluble are certain compounds which undergo considerable reaction when they dissolve. Also, the above generalizations are incomplete and oversimplified and reflect some arbitrary decisions necessitated by having only two categories: "soluble" and "insoluble."

STRENGTHS OF ELECTROLYTES

A Acids

  1 The *common strong acids* are hydrochloric (HCl), nitric (HNO_3), and, for the loss of the first proton, sulfuric (H_2SO_4). Other strong acids include $HClO_3$, $HClO_4$, HBr, and HI.

  2 *Most acids are weak.* Unless evidence to the contrary is available, an unfamiliar acid should be assumed to be weak.

  3 An anion which is formed when a polyprotic acid (Sec. 16-1) loses *some* of its protons is a weak acid. Examples include HSO_4^-, HCO_3^-, and $H_2PO_4^{2-}$.

B Bases

  1 The *common strong bases* are the hydroxides of the alkali metals and the alkaline-earth metals, except for $Be(OH)_2$, which is weak.

  2 It is difficult to generalize about other hydroxides. The approach usually used is to consider them as strong. Those which are weak are often so insoluble that their saturated solution is very dilute, and under these conditions their degree of dissociation is high.

  3 A commonly encountered weak base is ammonia, NH_3 (Brønsted-Lowry), also sometimes known as ammonium hydroxide, NH_4OH (Arrhenius).

C Salts

  1 *Almost all salts are strong electrolytes.* Unless evidence to the contrary is available, an unfamiliar salt should be assumed to be strong.

  2 Some weak salts are $HgCl_2$, $CdSO_4$, and $Pb(C_2H_3O_2)_2$.

D Water is a weak electrolyte.

E Complex ions are weak electrolytes. Examples include $Ag(NH_3)_2^+$, $CuCl_4^{2-}$, $CrCl_2(NH_3)_4^+$, and countless other complexes of the type described in Sec. 13-2.

Using the guidelines One can use the above solubility guidelines (1 through 8) and electrolyte-strength guidelines (A through E) for predicting the product(s) formed from a given set of reactants or for deciding what reactants will produce a given product or set of products. Consider the following example.

● **EXAMPLE 13-2**

Problem Solutions of silver nitrate ($AgNO_3$) and potassium bromide (KBr) are mixed producing a light cream-colored precipitate. Write the net equation for the reaction.

Solution The first step is to find out what species are present in the two solutions. $AgNO_3$ and KBr are salts, so we will assume that they are strong electrolytes (guideline C). In the first solution, then, we have silver ions (Ag^+), nitrate ions (NO_3^-) and, of course, water molecules. In the second we have potassium ions (K^+), bromide ions (Br^-), and water molecules.

Because a precipitate is formed we look for a compound which can be formed by precipitation from some of the species which we know are present. Guideline 6a tells us that silver bromide (AgBr) is insoluble. Thus the reaction must be:

$$Ag^+ + Br^- \longrightarrow AgBr(s)$$

Do the NO_3^- and K^+ ions react? No, because potassium nitrate, KNO_3, is a completely dissociated (guideline C), soluble (guidelines 2 and 3) salt.

The above can be shown schematically as:

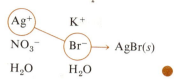

Although the net equation is the most useful type of equation for our purposes, other types of equations are sometimes used. One of these is the so-called molecular equation, in which products and reactants are shown as if they all consisted of molecules. For the reaction of Example 13-2, that is, the mixing of aqueous solutions of $AgNO_3$ and KBr, the "molecular equation" is

$$AgNO_3(aq) + KBr(aq) \longrightarrow AgBr(s) + KNO_3(aq)$$

Instead of specifying reacting species, as the net equation does, the "molecular equation" indicates the nature of the solutes from which the reacting solutions were prepared. It also indicates that the solution remaining after AgBr has precipitated is essentially one of potassium nitrate, KNO_3.

Since $AgNO_3$, KBr, and KNO_3 are all completely dissociated in solution, the above reaction can be represented by a *total ionic equation*

$$Ag^+ + NO_3^- + K^+ + Br^- \longrightarrow AgBr(s) + K^+ + NO_3^-$$

in which *all* ions are shown, whether or not they actually react. In this equation NO_3^- and K^+ are often referred to as *spectator ions*. These are ions which appear on both sides of the equation and hence are neither true reactants nor true products.

If the spectator ions are removed from both sides of a total ionic equation, the result is the *net* (or *net ionic*) *equation*, which for this reaction is, as we have seen,

$$Ag^+ + Br^- \longrightarrow AgBr(s)$$

We will make little use of "molecular equations" for ionic reactions, because they tend to conceal the true nature of the reaction. The "molecular equation" for the precipitation of AgBr makes it look as if KNO_3 were formed, whereas in reality its ions are merely left in solution. Although spectator ions must be present in order to maintain the electrical neutrality of the solution, they are not themselves involved in the reaction and so are best deleted from the equation.

● EXAMPLE 13-3

Problem Will a reaction occur if solutions of barium chloride, ($BaCl_2$) and aluminum sulfate [$Al_2(SO_4)_3$] are mixed? If so, write the net equation.

Solution By guideline 5 we can predict that insoluble $BaSO_4$ will be formed. Al^{3+} and Cl^- will not react (guidelines 6 and C). The net equation is

$$Ba^{2+} + SO_4^{2-} \longrightarrow BaSO_4(s)$$

Schematically,

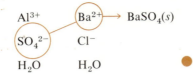

EXAMPLE 13-4

Problem   Write an equation for the reaction, if any, which occurs when dilute solutions of $SrCl_2$ and HNO_3 are mixed.

Solution   The species present are Sr^{2+}, Cl^-, H^+, NO_3^-, and H_2O. $Sr(NO_3)_2$ will not form, because it is a strong electrolyte (guideline C) and is soluble (guideline 3). HCl will not form because it is also a strong electrolyte (guideline A) and is soluble (guideline 1). No reaction takes place.

EXAMPLE 13-5

Problem   Solid calcium fluoride, CaF_2, although insoluble in water, dissolves in a dilute solution of hydrochloric acid, HCl. Write a net equation for the reaction.

Solution   The possible reactants are

$$CaF_2(s)$$
$$H^+ \qquad Cl^- \qquad H_2O$$

In order for the CaF_2 to dissolve, something in the HCl solution must have a strong affinity for either Ca^{2-} or F^-. We can eliminate Cl^- as a reactant, because $CaCl_2$ is a soluble (guideline 6), strong (guideline C) electrolyte. How about H^+? Guideline A allows us to predict that hydrofluoric acid, HF, is a weak acid. Thus there must be a strong affinity between H^+ and F^- to form HF. But our reactant is not F^- (in solution); it is solid CaF_2; so we must show it as such in the equation:

$$CaF_2(s) + 2H^+ \longrightarrow Ca^{2+} + 2HF$$

The Ca^{2+} remains in solution, as does the weak acid HF. (In this reaction if enough HF is formed, it may bubble off as a gas.)

ADDED COMMENT

It is crucial that you understand why the knowledge that HF is a weak acid helps you to account for the dissolving of CaF_2 in HCl. When you say, "HF is a *weak* acid," it means that HF has only a small tendency to dissociate. If this is true, then the bond between H and F must be strong (hard for water to break). This means *there must be a strong tendency for H^+ and F^- to combine to form HF*, strong enough to pull F^- ions out of the CaF_2 crystal lattice and make it dissolve.

Here are a few more examples given without explanation: (See if you can figure out *why* the reaction takes place.)

1 Solutions of Na_2CO_3 and $Ba(NO_3)_2$ are mixed:

$$CO_3^{2-} + Ba^{2+} \longrightarrow BaCO_3(s)$$

2 Solid $BaCO_3$ is dissolved in a minimum of dilute HCl:

$$BaCO_3(s) + H^+ \longrightarrow Ba^{2+} + HCO_3^-$$

3 More dilute HCl is added to the solution resulting from item 2:

$$HCO_3^- + H^+ \longrightarrow H_2CO_3$$

4 The solution from item 3 is warmed and a gas is evolved:

$$H_2CO_3 \longrightarrow H_2O + CO_2(g)$$

5 Solutions of HCl and $Ba(OH)_2$ are mixed:

$$H^+ + OH^- \longrightarrow H_2O$$

6 Solutions of HCN and $Ba(OH)_2$ are mixed:

$$HCN + OH^- \longrightarrow H_2O + CN^-$$

Note: Why does this equation look different from the one for item 5? The difference results from the fact that HCN is a weak acid (HCl is strong). In the HCN solution most of the solute is present as undissociated molecules; there are relatively few H^+ and CN^- ions. *When faced with making a choice among several reasonable reactant species, choose the one which is present in greatest abundance,* in this case HCN, instead of H^+. Actually, even if the OH^- reacts with the few H^+ ions present,

$$H^+ + OH^- \longrightarrow H_2O$$

the HCN will dissociate further to replace most of the H^+ removed (Le Châtelier's principle),

$$HCN \longrightarrow H^+ + CN^-$$

so that the net (overall) change is

$$HCN + OH^- \longrightarrow H_2O + CN^-$$

7 Solutions of $AgNO_3$ and HCl are mixed:

$$Ag^+ + Cl^- \longrightarrow AgCl(s)$$

8 The mixture from item 7 is treated with a solution of NH_3:

$$AgCl(s) + 2NH_3 \longrightarrow Ag(NH_3)_2^+ + Cl^-$$

9 The mixture from item 8 is treated with a solution of HNO_3:

$$Ag(NH_3)_2^+ + 2H^+ + Cl^- \longrightarrow AgCl(s) + 2NH_4^+$$

10 Solid $BaCO_3$ is treated with an excess of dilute HCl and warmed:

$$BaCO_3(s) + 2H^+ \longrightarrow Ba^{2+} + H_2O + CO_2(g)$$

The net equation has a limitation which should be mentioned. By placing emphasis on reactants and products, the steps by which products are formed from reactants are ignored. For example, reaction 10 goes through the stages of reactions 1, 2, 3, and 4. (Actually, each of these is a multistep process in itself.)

**13-4
Electron-transfer
reactions**

An important reaction type is the *electron-transfer reaction,* also known as the *oxidation-reduction,* or *redox, reaction.* In such a reaction one or more electrons appear to be transferred from one atom to another. The word *appear* is used because assigning electrons to individual atoms often involves the somewhat arbitrary bookkeeping technique of using oxidation numbers (Sec. 8-4).

*Oxidation and
reduction*

Oxidation is a word which originally meant combination with oxygen gas, but so many other reactions were seen to resemble reactions with oxygen that the term was eventually broadened to refer to *any reaction in which a substance or species loses electrons.* The following are all oxidations:

$$\text{Na} \cdot \longrightarrow \text{Na}^+ + e^-$$

$$2\left[:\ddot{\text{Cl}}:\right]^- \longrightarrow :\ddot{\text{Cl}}:\ddot{\text{Cl}}: + 2e^-$$

$$2\left[:\ddot{\text{O}}:\text{H}\right]^- + \left[:\ddot{\text{Cl}}:\right]^- \longrightarrow \left[:\ddot{\text{Cl}}:\ddot{\text{O}}:\right]^- + \text{H}:\ddot{\text{O}}: + 2e^-$$
$$\text{H}$$

In the first example it is clear that an Na atom loses an electron, and so we say that the sodium becomes *oxidized.* In the second example, each Cl appears to lose one electron, if you consider that in the Cl_2 molecule each Cl atom has a one-half share in the bonding pair. The oxidation number of Cl in this change goes from -1 to 0, corresponding to a one-electron loss. In the third example, the Cl^- ion *appears* to lose two electrons to an O atom, as its oxidation number changes from -1 to $+1$. (Remember that in the assignment of oxidation numbers a pair of electrons shared between Cl and O are both *counted* with the O, because of its higher electronegativity.) In summary, *oxidation is a loss of electrons.*

Reduction is a gain of electrons. (The term seems to have its origins in metallurgical terminology: the *reduction* of an ore to its metal.) Reduction is just the opposite of oxidation, and so if each of the examples given in the preceding paragraph were reversed, it would be a reduction. Some additional examples are

$$\text{Ca}^{2+} + 2e^- \longrightarrow \text{Ca}:$$

$$\begin{array}{c}\text{H}\\:\ddot{\text{O}}:\ddot{\text{O}}:\\\text{H}\end{array} + 2e^- \longrightarrow 2\left[:\ddot{\text{O}}:\text{H}\right]^-$$

$$3\text{H}:\ddot{\text{O}}: + \left[:\ddot{\text{O}}:\overset{\displaystyle :\ddot{\text{O}}:}{\text{S}}:\ddot{\text{O}}:\right]^{2-} + 6e^- \longrightarrow \left[:\ddot{\text{S}}:\right]^{2-} + 6\left[:\ddot{\text{O}}:\text{H}\right]^-$$
$$\text{H}$$

Fortunately, it is not necessary to write Lewis-structure equations in order to be aware of the loss or gain of electrons in a reaction; oxidation numbers make the job much easier. Just remember that when a substance is *oxidized,* the *oxidation number* of at least one of its atoms *increases* (becomes more positive), as electrons are lost. Similarly, when a substance is *reduced,* the *oxidation number* of at least one of its atoms decreases (becomes more negative), as electrons are gained. Some examples of oxidations are given below. In each case notice that the oxidation number of an atom (shown below that atom) increases:

$$\underset{+2}{Sn^{2+}} \longrightarrow \underset{+4}{Sn^{4+}} + 2e^-$$

$$2\underset{-2}{H_2O} \longrightarrow \underset{0}{O_2} + 4H^+ + 4e^-$$

$$4OH^- + \underset{+2}{Mn^{2+}} \longrightarrow \underset{+4}{MnO_2} + 2H_2O + 2e^-$$

The following are examples of reductions, shown similarly:

$$\underset{+3}{Fe^{3+}} + e^- \longrightarrow \underset{+2}{Fe^{2+}}$$

$$2\underset{+1}{H^+} + 2e^- \longrightarrow \underset{0}{H_2}$$

$$8H^+ + \underset{+7}{MnO_4^-} + 5e^- \longrightarrow \underset{+2}{Mn^{2+}} + 4H_2O$$

ADDED COMMENT

If you find yourself rusty on the rules for the assignment of oxidation numbers, refer to Sec. 8-4. Make sure you understand these rules before you go on.

An oxidation cannot take place without its having a reduction coupled with it; that is, electrons cannot be lost unless something else gains them. In any redox reaction the substance which oxidizes another substance, that is, which takes electrons from it, is called the *oxidizing agent,* or *oxidant.* The substance which donates electrons to another substance, and hence reduces it, is called the *reducing agent,* or *reductant.* An example of the use of these terms is found below:

$$\underset{0}{Fe} + 2\underset{+1}{H^+} \longrightarrow \underset{+2}{Fe^{2+}} + \underset{0}{H_2}$$

Fe: Increases in oxidation number (from 0 to +2).
      Loses electrons (two per atom).
      Is oxidized (by H^+).
      Is the reducing agent (reduces H^+ to H_2).
H^+:   Decreases in oxidation number (from +1 to 0).
      Gains electrons (one per ion).
      Is reduced (by Fe).
      Is the oxidizing agent (oxidizes Fe to Fe^{2+}).

Balancing redox equations   A balanced equation for a redox reaction must show both the oxidation and the reduction. Balancing can sometimes be accomplished by inspection, but it is usually easier if a systematic procedure is followed. We will consider the balancing of equations for two kinds of redox reactions: (1) those which occur without a solvent and (2) those which occur in aqueous solution.

BALANCING EQUATIONS FOR REDOX REACTIONS—NO SOLVENT PRESENT.   In general when you are faced with balancing an equation for a nonaqueous redox reaction, you will be provided with formulas for all reactants and products.   The method used, the *oxidation-number method,* is outlined below:

BALANCING REDOX EQUATIONS: OXIDATION-NUMBER METHOD

Step 1 Assign oxidation numbers to all atoms.

Step 2 Note which atoms appear to lose and which appear to gain electrons, and determine how many electrons are lost and gained.

Step 3 If there is more than one atom losing or gaining electrons in a formula unit, determine the total loss or gain of electrons per formula unit.

Step 4 Make the gain of electrons by the oxidizing agent equal to the loss by the reducing agent by inserting the appropriate coefficient before the formula of each on the left-hand side of the equation.   (See the following examples.)

Step 5 Complete the balancing of the equation by inspection.   First balance atoms which have gained or lost electrons; second, atoms other than O or H; third, O atoms; and last, H atoms.

● **EXAMPLE 13-6**

Problem   Balance the following equation:

$$MnO_2 + KClO_3 + KOH \longrightarrow K_2MnO_4 + KCl + H_2O$$

Solution   Step 1: $\underset{+4\ -2}{MnO_2} + \underset{+1+5-2}{K\ ClO_3} + \underset{+1-2+1}{K\ O\ H} \longrightarrow \underset{+1\ +6\ -2}{K_2MnO_4} + \underset{+1-1}{K\ Cl} + \underset{+1-2}{H_2O}$

Step 2: Manganese, Mn, changes oxidation number from $+4$ to $+6$ and so appears to lose two electrons.   Cl changes from $+5$ to -1 and so appears to gain six electrons.

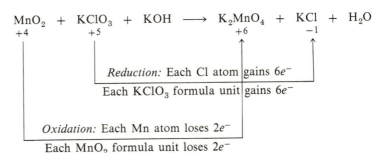

Step 3: As is shown above, since there is only one Mn atom in each MnO_2 formula unit, the electron loss per formula unit is two.   Also, since there is but one Cl atom in each $KClO_3$ formula unit, the electron gain per formula unit is six.

Step 4: $3MnO_2 + KClO_3 + KOH \longrightarrow K_2MnO_4 + KCl + H_2O$

Just look at the left-hand side for now.   By placing a 3 before MnO_2, we are saying that if one MnO_2 formula unit loses two electrons, then three will lose 3×2, or six, electrons. This will balance the electrons lost (six) against the electrons gained (six) by one $KClO_3$ formula unit.   Now that this $3:1$ ratio ($MnO_2/KClO_3$) has been established, *it must not be changed.*

Step 5: Add coefficients to the right-hand side to balance the Mn and Cl atoms, because they were involved in the electron transfer.

$$3MnO_2 + KClO_3 + KOH \longrightarrow 3K_2MnO_4 + KCl + H_2O$$

Then, balance the K atoms (note: the coefficient before $KClO_3$ must not be changed)

$$3MnO_2 + KClO_3 + 6KOH \longrightarrow 3K_2MnO_4 + KCl + H_2O$$

then, O atoms

$$3MnO_2 + KClO_3 + 6KOH \longrightarrow 3K_2MnO_4 + KCl + 3H_2O$$

and last, H atoms. But note that they are already balanced. ●

● **EXAMPLE 13-7**

Problem   Balance the following equation:

$$H_2C_2O_4 + KMnO_4 \longrightarrow CO_2 + MnO + K_2O + H_2O$$

Solution   Step 1:

$$\underset{+1 +3 -2}{H_2C_2O_4} + \underset{+1 +7 -2}{K\,MnO_4} \longrightarrow \underset{+4 -2}{CO_2} + \underset{+2 -2}{Mn\,O} + \underset{+1 -2}{K_2O} + \underset{+1 -2}{H_2O}$$

Step 2:

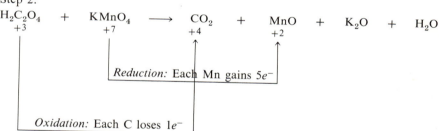

Step 3:

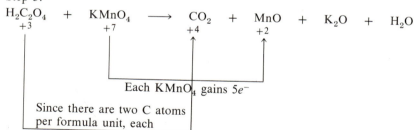

Step 4: $5H_2C_2O_4 + 2KMnO_4 \longrightarrow CO_2 + MnO + K_2O + H_2O$

Step 5: $5H_2C_2O_4 + 2KMnO_4 \longrightarrow 10CO_2 + 2MnO + K_2O + 5H_2O$ ●

ADDED COMMENT

Step 3 is the one which you are most likely to overlook. Make sure you understand its purpose.

BALANCING EQUATIONS FOR REDOX REACTIONS—AQUEOUS SOLUTIONS.   For reactions which occur in aqueous solution you may be given only a skeletal equation, one in which only the principal reactants and products are given.   You must complete the equation, as well as balance it.   There are two methods for doing this.   These methods, the *oxidation-number method* and the *half-reaction method,* are different and should not be confused with each other.   Each has its advantages and disadvantages.

BALANCING REDOX EQUATIONS (AQUEOUS SOLUTIONS):
OXIDATION-NUMBER METHOD

Step 1 Assign oxidation numbers to all atoms.
Step 2 Note which atoms appear to lose and which appear to gain electrons, and determine how many electrons are lost and gained.
Step 3 If more than one atom in a formula unit loses or gains electrons, determine the total loss or gain per formula unit.
Step 4 Make the gain of electrons by the oxidizing agent equal to the loss by the reducing agent by inserting an appropriate coefficient before the formula of each on the left-hand side of the equation.   (See the following examples.)
Step 5 Balance the atoms which have gained or lost electrons by adding appropriate coefficients on the right.
Step 6 Balance all other atoms except for O and H.
Step 7 Balance the charge (the sum of all the ionic charges) so that it is the same on both sides, by adding either H^+ or OH^-.
    (a) If the reaction takes place in *acidic* solution, add H^+ ions to the side deficient in positive charges.
    (b) If the reaction takes place in *basic* solution, add OH^- ions to the side deficient in negative charges.
Step 8 Balance O atoms by adding H_2O to the appropriate side.   Check to see that the H atoms are now balanced.   (They will be, if you have made no mistakes.)

● **EXAMPLE 13-8**

Problem   Complete and balance the following equation for a reaction which takes place in acidic solution:

$$Cr_2O_7^{2-} + Fe^{2+} \longrightarrow Cr^{3+} + Fe^{3+} \quad \text{(acidic)}$$

Solution   Step 1:
$$\underset{+6 \; -2}{Cr_2O_7^{2-}} + \underset{+2}{Fe^{2+}} \longrightarrow \underset{+3}{Cr^{3+}} + \underset{+3}{Fe^{3+}}$$

Step 2:
$$\underset{+6}{Cr_2O_7^{2-}} + \underset{+2}{Fe^{2+}} \longrightarrow \underset{+3}{Cr^{3+}} + \underset{+3}{Fe^{3+}}$$

Oxidation: Loss of $1e^-$ per Fe

Reduction: Gain of $3e^-$ per Cr

Step 3: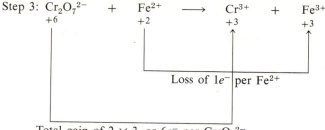

Total gain of 2 × 3, or $6e^-$ per $Cr_2O_7^{2-}$

Step 4: $$Cr_2O_7^{2-} + 6Fe^{2+} \longrightarrow Cr^{3+} + Fe^{3+}$$

Step 5: $$Cr_2O_7^{2-} + 6Fe^{2+} \longrightarrow 2Cr^{3+} + 6Fe^{3+}$$

Step 6: Done!

Step 7:
  Total charge on left $= -2 + 6(+2) = +10$.
  Total charge on right $= 2(+3) + 6(+3) = +24$.
  Added positive charge needed *on left* $= +14$.

$$14H^+ + Cr_2O_7^{2-} + 6Fe^{2+} \longrightarrow 2Cr^{3+} + 6Fe^{3+}$$

Step 8: $$14H^+ + Cr_2O_7^{2-} + 6Fe^{2+} \longrightarrow 2Cr^{3+} + 6Fe^{3+} + 7H_2O$$ ●

● **EXAMPLE 13-9**

Problem   Complete and balance the following equation for a reaction in *basic* solution:

$$CrO_4^{2-} + Fe(OH)_2 \longrightarrow CrO_2^- + Fe(OH)_3 \quad \text{(basic)}$$

Solution   Step 1: $$\underset{+6\ -2}{Cr\ O_4^{2-}} + \underset{+2\ -2+1}{Fe(O\ H)_2} \longrightarrow \underset{+3\ -2}{Cr\ O_2^-} + \underset{+3\ -2+1}{Fe(O\ H)_3}$$

Steps 2 and 3:

$$\underset{+6}{CrO_4^{2-}} + \underset{+2}{Fe(OH)_2} \longrightarrow \underset{+3}{CrO_2^-} + \underset{+3}{Fe(OH)_3}$$

Oxidation: Loss of $1e^-$ per Fe
Loss of $1e^-$ per $Fe(OH)_2$

Reduction: Gain of $3e^-$ per Cr
Gain of $3e^-$ per CrO_4^{2-}

Step 4: $$CrO_4^{2-} + 3Fe(OH)_2 \longrightarrow CrO_2^- + Fe(OH)_3$$

Step 5: $$CrO_4^{2-} + 3Fe(OH)_2 \longrightarrow CrO_2^- + 3Fe(OH)_3$$

Step 6: Done!

Step 7:
  Total charge on left $= -2 + 3(0) = -2$.
  Total charge on right $= -1 + 3(0) = -1$.
  Added negative charge needed *on right* $= -1$.

$$CrO_4^{2-} + 3Fe(OH)_2 \longrightarrow CrO_2^- + 3Fe(OH)_3 + OH^-$$

Step 8: $2H_2O + CrO_4^{2-} + 3Fe(OH)_2 \longrightarrow CrO_2^- + 3Fe(OH)_3 + OH^-$ ●

The second method for balancing redox equations for reactions which occur in aqueous solution is the *half-reaction method,* also known as the *ion-electron method.* Note that in it you do *not* assign any oxidation numbers.

BALANCING REDOX EQUATIONS (AQUEOUS SOLUTIONS):
HALF-REACTION METHOD

Step 1 Separate the skeletal equation into two half-reactions, one an oxidation and the other a reduction.

Step 2 Balance each half-reaction separately according to this sequence:
    a Balance all atoms other than H and O by inspection.
    b Balance O atoms by adding H_2O to the appropriate side.
    c Balance the H atoms. The way this is done depends on whether the solution is *acidic* or *basic.*
      i In *acidic* solution, add the appropriate number of H^+ to the side deficient in H.
      ii In *basic* solution, instead of adding H^+ ions, *add one H_2O molecule to the side deficient in H plus one OH^- ion to the opposite side, for each needed H atom.* You may wish to cancel out H_2O molecules duplicated on each side at this point.
    d Balance the charge by adding electrons (e^-) to the side deficient in *negative* charge.

Step 3 Multiply each balanced half-reaction by an appropriate number in order to balance the electron loss against the electron gain. Then add the two half-reactions.

Step 4 Subtract out (cancel) anything which appears on both sides. (The electrons should all disappear in this step.)

● **EXAMPLE 13-10**

Problem Using the half-reaction method complete and balance the following equation for a reaction occurring in *acidic* solution.

$$Cr_2O_7^{2-} + Fe^{2+} \longrightarrow Cr^{3+} + Fe^{3+} \quad \text{(acidic)}$$

Solution Step 1: $Cr_2O_7^{2-} \longrightarrow Cr^{3+}$ | $Fe^{2+} \longrightarrow Fe^{3+}$

Step 2: a $Cr_2O_7^{2-} \longrightarrow 2Cr^{3+}$ (done!) | (done!)
      b $Cr_2O_7^{2-} \longrightarrow 2Cr^{3+} + 7H_2O$ (done!)
      c $14H^+ + Cr_2O_7^{2-} \longrightarrow 2Cr^{3+} + 7H_2O$ (done!)
      d $6e^- + 14H^+ + Cr_2O_7^{2-} \longrightarrow 2Cr^{3+} + 7H_2O$ | $Fe^{2+} \longrightarrow Fe^{3+} + e^-$

Step 3: $\underbrace{6e^- + 14H^+ + Cr_2O_7^{2-} \longrightarrow 2Cr^{3+} + 7H_2O}$ | $\underbrace{[Fe^{2+} \longrightarrow Fe^{3+} + e^-] \times 6}$

$$\longrightarrow 6e^- + 14H^+ + Cr_2O_7^{2-} \longrightarrow 2Cr^{3+} + 7H_2O$$
$$6Fe^{2+} \longrightarrow 6Fe^{3+} + 6e^- \longleftarrow$$

$$6e^- + 14H^+ + Cr_2O_7^{2-} + 6Fe^{2+} \longrightarrow 2Cr^{3+} + 6Fe^{3+} + 7H_2O + 6e^-$$

Step 4: $14H^+ + Cr_2O_7^{2-} + 6Fe^{2+} \longrightarrow 2Cr^{3+} + 6Fe^{3+} + 7H_2O$ ●

● **EXAMPLE 13-11**

Problem   Complete and balance the following equation for a reaction taking place in *basic* solution using the half-reaction method:

$$CrO_4^{2-} + Fe(OH)_2 \longrightarrow CrO_2^- + Fe(OH)_3 \quad \text{(basic)}$$

Solution   Step 1: $CrO_4^{2-} \longrightarrow CrO_2^-$ $Fe(OH)_2 \longrightarrow Fe(OH)_3$

Step 2: a (done!)   (done!)

       b $CrO_4^{2-} \longrightarrow CrO_2^- + 2H_2O$ $H_2O + Fe(OH)_2 \longrightarrow Fe(OH)_3$

       c $4H_2O + CrO_4^{2-} \longrightarrow$ $OH^- + H_2O + Fe(OH)_2 \longrightarrow$

           $CrO_2^- + 2H_2O + 4OH^-$ $Fe(OH)_3 + H_2O$

          $2H_2O + CrO_4^{2-} \longrightarrow$ $OH^- + Fe(OH)_2 \longrightarrow Fe(OH)_3$

           $CrO_2^- + 4OH^-$

       d $3e^- + 2H_2O + CrO_4^{2-} \longrightarrow$ $OH^- + Fe(OH)_2 \longrightarrow Fe(OH)_3 + e^-$

           $CrO_2^- + 4OH^-$

Step 3: $3e^- + 2H_2O + CrO_4^{2-} \longrightarrow$ $[OH^- + Fe(OH)_2 \longrightarrow$

           $CrO_2^- + 4OH^-$ $Fe(OH)_3 + e^-] \times 3$

$$\longrightarrow 3e^- + 2H_2O + CrO_4^{2-} \longrightarrow CrO_2^- + 4OH^-$$
$$3OH^- + 3Fe(OH)_2 \longrightarrow 3Fe(OH)_3 + 3e^-$$

$$3e^- + 2H_2O + 3OH^- + CrO_4^{2-} + 3Fe(OH)_2 \longrightarrow$$
$$CrO_2^- + 3Fe(OH)_3 + 4OH^- + 3e^-$$

Step 4: $2H_2O + CrO_4^{2-} + 3Fe(OH)_2 \longrightarrow CrO_2^- + 3Fe(OH)_3 + OH^-$ ●

At this point a comparison of the oxidation-number and half-reaction methods is in order. For *nonsolution reactions* the oxidation-number method is usually superior to the half-reaction method. For *reactions in aqueous solution* the half-reaction method is the slower of the two, but for most people the surer (you are less likely to make mistakes and more likely to find those you do make). A major advantage of the half-reaction method is the lack of necessity of assigning oxidation numbers. Individual half-reactions are used in describing electrode reactions in electrochemical cells (Chap. 19). The biggest advantage of the oxidation-number method is its potential speed. With practice you will find that you can do much of it in your head.

13-5
Solution
stoichiometry

Acid-base
stoichiometry

In an Arrhenius acid-base reaction water molecules are products. Since it takes *one* H⁺ plus *one* OH⁻ to form *one* H₂O molecule, the stoichiometry of these reactions is based on the simple 1:1:1 ratio.

We will introduce at this time a common convention, that of using *square brackets* to indicate the concentration, expressed in molarity, of a molecular or ionic species in solution. Thus [Cl⁻] means "the molar concentration of chloride ions."

● **EXAMPLE 13-12**

Problem   How many milliliters of a 0.100 *M* solution of NaOH are required to neutralize 25.0 ml of 0.300 *M* HCl?

Solution   A 0.300 *M* solution of HCl contains 0.300 mol of H⁺ and 0.300 mol of Cl⁻ per liter because

HCl is a strong acid. In other words, in the solution $[H^+] = 0.300\ M$ and $[Cl^-] = 0.300\ M$. Therefore, the number of moles of H^+ in 25.0 ml, or 0.0250 liter, of solution is

$$0.0250 \text{ liter } (0.300 \text{ mol liter}^{-1}) = 7.50 \times 10^{-3} \text{ mol}$$

The equation for the reaction is

$$H^+ + OH^- \longrightarrow H_2O$$

Because OH^- and H^+ react in a 1:1 ratio, the number of moles of OH^- needed must also be 7.50×10^{-3} mol. How many liters of the NaOH solution will supply this number of moles? Since NaOH is a strong base, there is 0.100 mol of OH^- per liter, and so the number of liters is

$$\frac{7.50 \times 10^{-3} \text{ mol}}{0.100 \text{ mol liter}^{-1}} = 7.50 \times 10^{-2} \text{ liter}$$

This is 0.0750 liter, or 75.0 ml, of the NaOH solution. ●

● **EXAMPLE 13-13**

Problem   How many milliliters of 0.120 M NaOH are needed for complete neutralization of 50.0 ml of 0.100 M HF?

Solution   HF is a weak acid, so that before the addition of OH^- ions, it is present mainly as undissociated HF molecules. As neutralization proceeds, however, the HF dissociates further until essentially all the protons have combined with OH^- ions. The net equation is, therefore,

$$HF + OH^- \longrightarrow H_2O + F^-$$

(See also the note after item 6, Sec. 13-3.)

Number of moles of HF = 0.050 liter $(0.100 \text{ mol liter}^{-1}) = 5.00 \times 10^{-3}$ mol

Total number of moles of available protons = 5.00×10^{-3} mol

Number of moles of OH^- needed = 5.00×10^{-3} mol

The NaOH solution is 0.120 M; so in it $[OH^-] = 0.120\ M$.

Number of liters of NaOH solution needed = $\dfrac{5.00 \times 10^{-3} \text{ mol}}{0.120 \text{ mol liter}^{-1}} = 0.0417$ liter

Therefore, 41.7 ml of the solution of base are needed. ●

● **EXAMPLE 13-14**

Problem   How many milliliters of 0.210 M NaOH are needed for complete neutralization of 10.0 ml of 0.100 M H_3PO_4 (phosphoric acid)?

Solution H_3PO_4 is a *triprotic acid;* that is, there are *three* protons (H^+) available from one H_3PO_4 molecule. (This can usually, though not always, be inferred from the formula.) The equation for the reaction is

$$H_3PO_4 + 3OH^- \longrightarrow 3H_2O + PO_4{}^{3-}$$

Number of moles of H_3PO_4 = 0.0100 liter $(0.100 \text{ mol liter}^{-1}) = 1.00 \times 10^{-3}$ mol

Number of moles of available H^+ = $3 \times 1.00 \times 10^{-3}$ mol = 3.00×10^{-3} mol

Number of moles of OH^- needed = 3.00×10^{-3} mol

$$\text{Number of liters of NaOH solution needed} = \frac{3.00 \times 10^{-3} \text{ mol}}{0.210 \text{ mol liter}^{-1}} = 0.0143 \text{ liter}$$

which is 14.3 ml. ●

Equivalents of acids and bases   A convenient way of handling acid-base calculations makes use of equivalents. *One equivalent of an acid is the quantity of that acid which will supply or donate one mole of H^+ ions (protons). One equivalent of a base is the quantity which furnishes one mole of OH^- ions. Finally, one equivalent of acid reacts with (neutralizes) one equivalent of base.*

● **EXAMPLE 13-15**

Problem   How many equivalents for complete neutralization are there in (a) 4.00 mol of H_2SO_4? (b) 0.176 mol of $Ca(OH)_2$? (c) 1.60 mol of H_3PO_4?

Solution   (a) The formula H_2SO_4 tells us that there are two H^+ available from each molecule. Thus 1 mol of H_2SO_4 can furnish 2 mol of H^+. This means that 1 mol of H_2SO_4 is really the same quantity as 2 equiv of H_2SO_4. Therefore, 4.00 mol of H_2SO_4 are

$$4.00 \text{ mol} \frac{2 \text{ equiv}}{1 \text{ mol}} = 8.00 \text{ equiv of } H_2SO_4$$

(b) $Ca(OH)_2$ can supply 2 mol of OH^- per mole of $Ca(OH)_2$. Therefore there are 2 equiv per mole of $Ca(OH)_2$. Thus 0.176 mol of $Ca(OH)_2$ is

$$1.76 \text{ mol} \frac{2 \text{ equiv}}{1 \text{ mol}} = 0.352 \text{ equiv of } Ca(OH)_2$$

(c) Since there are 3 available H^+ per mole of H_3PO_4, 1.60 mol of H_3PO_4 is

$$1.60 \frac{3 \text{ equiv}}{1 \text{ mol}} = 4.80 \text{ equiv of } H_3PO_4 \; ●$$

Equivalent weight is defined as the mass in grams of one equivalent.

● **EXAMPLE 13-16**

Problem   What is the equivalent weight of H_3PO_4?

Solution   Since 1 mol of H_3PO_4 is 3 equiv, 1 equiv is $\frac{1}{3}$ mol and so has a mass one-third that of 1 mol. One mole of H_3PO_4 weighs 98.0 g; so its equivalent weight is

$$98.0 \text{ g mol}^{-1} \frac{1 \text{ mol}}{3 \text{ equiv}} = 32.7 \text{ g equiv}^{-1} \; ●$$

A solution which contains one equivalent per liter of solution is said to be a one-normal solution. In other words, *normality (N)* is a concentration unit which is analogous to molarity, except that it is defined in terms of equivalents, rather than moles.

$$\text{Normality} = \frac{\text{no. of equivalents of solute}}{\text{no. of liters of solution}}$$

● **EXAMPLE 13-17**

Problem   What is the normality of a 1.4 M H_2SO_4 solution?

Solution   Since each mole of H_2SO_4 is 2 equiv, the number of *equivalents* per liter must be *twice* the number of *moles* per liter, or

$$1.4 \text{ mol liter}^{-1} \frac{2 \text{ equiv}}{1 \text{ mol}} = 2.8 \text{ equiv liter}^{-1} \quad \text{or} \quad 2.8 \ N \ \bullet$$

Acid-base stoichiometry problems can be solved in terms of equivalents of the acid and base. The next problem is the same as Example 13-14, and a comparison of the two approaches should prove valuable.

● **EXAMPLE 13-18**

Problem   How many milliliters of 0.210 M NaOH are needed for complete neutralization of 10.0 ml of 0.100 M H_3PO_4?

Solution   Since 0.100 M H_3PO_4 is 0.300 N H_3PO_4, then 10.0 ml (0.100 liter) of 0.300 N H_3PO_4 contains

$$0.0100 \text{ liter } (0.300 \text{ equiv liter}^{-1}) = 3.00 \times 10^{-3} \text{ equiv}$$

This requires 3.00×10^{-3} equiv of NaOH. Since a 0.210 M NaOH solution is 0.210 N, the volume of this solution needed is

$$\frac{3.00 \times 10^{-3} \text{ equiv}}{0.210 \text{ equiv liter}^{-1}} = 0.0143 \text{ liter} \quad \text{or} \quad 14.3 \text{ ml} \ \bullet$$

Redox stoichiometry   One method of working redox stoichiometry calculations relies on a balanced equation for the reaction.

● **EXAMPLE 13-19**

Problem   How many milliliters of 0.0500 M $KMnO_4$ are needed to oxidize 25.0 ml of 0.0400 M H_2SO_3 in acidic solution? (MnO_4^- oxidizes H_2SO_3 and forms Mn^{2+} and HSO_4^-.)

Solution   The balanced equation is

$$H^+ + 2MnO_4^- + 5H_2SO_3 \longrightarrow 2Mn^{2+} + 5HSO_4^- + 3H_2O$$

In 25.0 ml of 0.0400 M H_2SO_3 there is

$$0.0250 \text{ liter } (0.0400 \text{ mol liter}^{-1}) = 1.00 \times 10^{-3} \text{ mol } H_2SO_3$$

From the balanced equation we see that 1.00×10^{-3} mol of H_2SO_3 requires

$$1.00 \times 10^{-3} \text{ mol } H_2SO_3 \frac{2 \text{ mol } MnO_4^-}{5 \text{ mol } H_2SO_3} = 4.00 \times 10^{-4} \text{ mol } MnO_4^-$$

In the 0.0500 M $KMnO_4$ solution, $[MnO_4^-] = 0.0500 \ M$, and so the volume of this solution needed is

$$\frac{4.00 \times 10^{-4} \text{ mol}}{0.0500 \text{ mol liter}^{-1}} = 8.00 \times 10^{-3} \text{ liter} \quad \text{or} \quad 8.00 \text{ ml} \ \bullet$$

Equivalents of   Equivalents are also useful for redox stoichiometry. *One equivalent of an oxidiz-*
oxidizing and   *ing agent is that quantity which accepts one mole of electrons. One equivalent of a*
reducing agents   *reducing agent is that quantity which furnishes one mole of electrons. One equiva-*
lent of an oxidizing agents reacts with one equivalent of a reducing agent.

● **EXAMPLE 13-20**

Problem How many equivalents are there in (a) 1.00 mol of MnO_4^- and (b) 1.00 mol of H_2SO_3 in a reaction in which these form Mn^{2+} and HSO_4^-, respectively?

Solution (a) When MnO_4^- forms Mn^{2+}, the Mn atom appears to gain $5e^-$ as Mn(VII) changes to Mn(II). One mole of MnO_4^- gains 5.00 mol of elecrons and is, consequently, 5.00 equiv of MnO_4^-. (b) When H_2SO_3 forms HSO_4^-, the S atom appears to lose $2e^-$ as S(IV) changes to S(VI). One mole of H_2SO_3 is therefore 2.00 equiv of H_2SO_3. ●

● **EXAMPLE 13-21** (Compare with Example 13-19.)

Problem How many milliliters of $0.0500\ M$ $KMnO_4$ are needed to oxidize 25.0 ml of $0.0400\ M$ H_2SO_4 in acidic solution? (MnO_4^- oxidizes H_2SO_3 and forms Mn^{2+} and HSO_4^-.)

Solution $0.0400\ M$ H_2SO_3 is $0.0800\ N$ H_2SO_3. (See Example 13-20.) The number of equivalents of H_2SO_3 present is

$$0.0250\text{ liter }(0.0800\text{ equiv liter}^{-1}) = 2.00 \times 10^{-3}\text{ equiv}$$

The number of equivalents of MnO_4^- needed is 2.00×10^{-3} equiv. The $0.0500\ M$ $KMnO_4$ solution is 0.250 N. (See Example 13-20.) Therefore, the volume of the $KMnO_4$ solution needed is

$$\frac{2.00 \times 10^{-3}\text{ equiv}}{0.250\text{ equiv liter}^{-1}} = 8.00 \times 10^{-3}\text{ liter} \quad\text{or}\quad 8.00\text{ ml} ●$$

Other solution stoichiometry Stoichiometric calculations for solution reactions other than neutralization and redox reactions are usually carried out by making use of a balanced equation.

● **EXAMPLE 13-22**

Problem Suppose 25.0 ml each of $0.200\ M$ $Ca(NO_3)_2$ and $0.100\ M$ Na_3PO_4 are mixed. (a) Assuming complete precipitation, how many grams of $Ca_3(PO_4)_2$ are formed? (b) Calculate the concentrations of Ca^{2+}, NO_3^-, Na^+, and PO_4^{3-} left in solution. (Assume a final volume of 50.0 ml.)

Solution (a) In $0.200\ M$ $Ca(NO_3)_2$ $[Ca^{2+}] = 0.200\ M$ and $[NO_3^-] = 0.400\ M$. In $0.100\ M$ Na_3PO_4 $[Na^+] = 0.300\ M$ and $[PO_4^{3-}] = 0.100\ M$. The number of moles of Ca^{2+} is

$$0.0250\text{ liter }(0.200\text{ mol liter}^{-1}) = 5.00 \times 10^{-3}\text{ mol}$$

The number of moles of PO_4^{3-} is

$$0.0250\text{ liter }(0.100\text{ mol liter}^{-1}) = 2.50 \times 10^{-3}\text{ mol}$$

The balanced equation for the precipitation is

$$3Ca^{2+} + 2PO_4^{3-} = Ca_3(PO_4)_2(s)$$

The number of moles of Ca^{2+} required to react with 2.50×10^{-3} mol of PO_4^{3-} is

$$2.50 \times 10^{-3}\text{ mol PO}_4^{3-}\,\frac{3\text{ mol Ca}^{2+}}{2\text{ mol PO}_4^{3-}} = 3.75 \times 10^{-3}\text{ mol Ca}^{2+}$$

Since more than this (5.00×10^{-3} mol) is available, all PO_4^{3-} is used up, and $(5.00 \times 10^{-3}\text{ mol}) - (3.75 \times 10^{-3}\text{ mol}) = 1.25 \times 10^{-3}$ mol of Ca^{2+} is left over in excess.

The number of moles of $Ca_3(PO_4)_2$ formed is

$$2.50 \times 10^{-3} \text{ mol } PO_4^{3-} \frac{1 \text{ mol } Ca_3(PO_4)_2}{2 \text{ mol } PO_4^{3-}} = 1.25 \times 10^{-3} \text{ mol } Ca_3(PO_4)_2$$

This weighs $(1.25 \times 10^{-3} \text{ mol})(310 \text{ g mol}^{-1}) = 0.388$ g.

(b) The number of moles of NO_3^- and Na^+ are unaffected by the reaction. They are

Number of moles of $NO_3^- = 0.0250$ liter $(0.400 \text{ mol liter}^{-1}) = 1.00 \times 10^{-2}$ mol

Number of moles of $Na^+ = 0.0250$ liter $(0.300 \text{ mol liter}^{-1}) = 7.50 \times 10^{-3}$ mol

Number of moles of $PO_4^{3-} = 0$

Number of moles of Ca^{2+} (left over) $= 1.25 \times 10^{-3}$ mol

The final concentrations are

$$[NO_3^-] = \frac{1.00 \times 10^{-2} \text{ mol}}{0.050 \text{ liter}} = 0.200 \ M$$

$$[Na^+] = \frac{7.50 \times 10^{-3} \text{ mol}}{0.0500 \text{ liter}} = 0.150 \ M$$

$$[PO_4^{3-}] = 0$$

$$[Ca^{2+}] = \frac{1.25 \times 10^{-3} \text{ mol}}{0.050 \text{ liter}} = 0.0250 \ M$$

Note: This calculation ignores the low solubility of $Ca_3(PO_4)_2$ in water. This effect is small, however, as we will see in Chap. 17.

SUMMARY

Acids and *bases* have been defined in a number of useful ways. The most important of these are the *Arrhenius, solvent-system, Brønsted-Lowry,* and *Lewis* definitions. Each definition has strengths and weaknesses. In aqueous solutions the Arrhenius and Brønsted-Lowry definitions are often the most useful.

Aqueous-solution reactions include *neutralization, precipitation, complexation* and *redox reactions. Equations for redox* (oxidation-reduction, or electron-transfer) *reactions* can be balanced by using either the *oxidation-number* or the *half-reaction* method.

Writing *net equations* for many aqueous-solution reactions is aided by a knowledge of what kinds of reactions can occur and also of solubilities and strengths of electrolytes.

The principles of *reaction stoichiometry* can be applied to *aqueous-solution reactions.* Calculations can be performed in many cases by using either *moles and molarities* or *equivalents and normalities.* Equivalents (and normalities) are defined for acids and bases and also for oxidizing and reducing agents.

KEY TERMS

Acid (Sec. 13-1)
Acid strength (Sec. 13-1)
Aquo complex (Sec. 13-2)
Arrhenius definition (Sec. 13-1)
Autodissociation (Sec. 13-1)
Base (Sec. 13-1)
Brønsted-Lowry definition (Sec. 13-1)
Conjugate acid-base pair (Sec. 13-1)

Complex (Sec. 13-2)
Complexation (Sec. 13-2)
Coordination number (Sec. 13-2)
Differentiating solvent (Sec. 13-1)
Equivalent: acid-base (Sec. 13-5)
Equivalent: redox (Sec. 13-5)
Equivalent weight (Sec. 13-5)
Half-reaction (Sec. 13-4)

Heat of neutralization (Sec. 13-1)
"Insoluble" (Sec. 13-2)
Leveling effect (Sec. 13-1)
Lewis definition (Sec. 13-1)
Ligand (Sec. 13-2)
Net ionic equations (Secs. 13-2, 13-3)
Neutralization (Sec. 13-1)
Normality (Sec. 13-5)

Oxidation (Sec. 13-4)
Oxidizing agent (Sec. 13-4)
Precipitation (Sec. 13-2)
Reducing agent (Sec. 13-4)
Reduction (Sec. 13-4)
Self-dissociation (Sec. 13-1)
Solvent-system definition (Sec. 13-1)

QUESTIONS AND PROBLEMS

Acids and bases

13-1 Define *acid* and *base* in terms of each of the four perspectives discussed in this chapter.

13-2 Each of the following is a Brønsted-Lowry acid. For each give the conjugate base: H_3PO_4, NH_4^+, H_2O, HCO_3^-, $Al(H_2O)_6^{3+}$.

13-3 How is the term *dissociation* (of an acid) defined in Arrhenius terms? In Brønsted-Lowry terms?

13-4 Classify each of the following as either a Brønsted-Lowry acid and/or base: NH_3, $H_2PO_4^-$, SO_4^{2-}, OCl^-, NH_2^-. Write equations supporting your classifications.

13-5 Using examples explain what is meant by the leveling effect.

13-6 Classify each of the following as a Lewis acid or Lewis base: H_2O, NH_3, Al^{3+}, CO, $AlCl_3$. Write Lewis structures supporting your classifications.

13-7 Write an equation for the autodissociation of each of the following solvents: H_2O, NH_3, $HC_2H_3O_2$, H_2SO_4.

13-8 How can $HClO_4$ be classified as a stronger acid than HNO_3 when they are both essentially 100 percent dissociated in aqueous solution?

13-9 Rank the following reactions in order of increasing tendency to take place:

$$HSO_4^- + SO_3^{2-} \longrightarrow HSO_3^- + SO_4^{2-}$$

$$HF + H_2O \longrightarrow H_3O^+ + F^-$$

$$H_2S + C_2H_3O_2^- \longrightarrow HC_2H_3O_2 + HS^-$$

13-10 Describe how heats of neutralization can be used to estimate the relative strengths of acids and bases.

13-11 Write **(a)** Arrhenius and **(b)** Brønsted-Lowry equations showing the stepwise dissociation of H_3PO_4.

Precipitation and complexation

13-12 Suggest three solutions, each of which could be used to precipitate insoluble $PbCl_2$ from a solution of $Pb(NO_3)_2$.

13-13 Suggest three solutions, each of which could be used to precipitate insoluble $BaCO_3$ from a solution of Na_2CO_3.

13-14 What is the formula of the complex ion formed when *six* of each of the following ligands complex *one* Co^{3+} ion? **(a)** NH_3 **(b)** Cl^-.

13-15 Write the formula of the complex formed when the following species combine:
(a) $Cr^{3+} + 3NH_3 + 3Cl^-$
(b) $Cr^{3+} + 2NH_3 + 4Cl^-$
(c) $Cr^{3+} + 5NH_3 + Cl^-$.

13-16 Write the formula of the complex formed when the following species combine: **(a)** $Ag^+ + 2Cl^-$
(b) $Cu^{2+} + 4NH_3$ **(c)** $Ni + 4CO$ **(d)** $Fe^{3+} + 6CN^-$
(e) $Co^{3+} + 3C_2O_4^{2-}$.

Net ionic equations

13-17 List all of the solute species present in each of the following $1 M$ solutions: $MgBr_2$, HNO_3, HNO_2, $Ba(OH)_2$, H_2SO_3.

13-18 List all of the solute species present in each of the following $1 M$ solutions: BaI_2, H_2SO_4, $La(OH)_3$, $HOCl$, $HClO_3$.

13-19 Write a balanced net equation for the reaction which occurs when each of the following pairs of solutions is mixed: **(a)** HCl and $AgNO_3$ **(b)** $SrCl_2$ and Na_2CO_3 **(c)** $FeCl_3$ and NaOH **(d)** HNO_3 and NaCN **(e)** HCl and RbOH **(f)** Na_2SO_4 and $Pb(NO_3)_2$.

13-20 Write a balanced net equation for the reaction which occurs when each of the following pairs of solutions is mixed: **(a)** $CuSO_4$ and NaOH **(b)** HF and $CaCl_2$ **(c)** $Pb(NO_3)_2$ and Na_2CrO_4 **(d)** HCl and NaOBr **(e)** H_2SO_3 and $BaCl_2$ **(f)** $HClO_4$ and LiOH.

13-21 Each of the following solids is insoluble in water but dissolves in a solution of excess strong acid. Write a net equation for each such dissolving reaction: **(a)** BaF_2 **(b)** $AgC_2H_3O_2$ **(c)** CaC_2O_4 **(d)** $SrCO_3$ **(e)** $Cu_3(PO_4)_2$.

13-22 A solution of ammonia in water will neutralize acids. Write an equation for the neutralization of a dilute HCl solution representing the dissolved ammonia as **(a)** NH_3 **(b)** NH_4OH.

13-23 Write a balanced net equation for each of the following: **(a)** Solid calcium carbonate dissolves in dilute HCl with the evolution of CO_2 gas. **(b)** Solid $BaSO_3$ dissolves in dilute HCl with the evolution of SO_2 gas. **(c)** Solid CdS dissolves in dilute HCl with the evolution of H_2S gas. **(d)** A solution of NH_4Cl is mixed with one of NaOH to form NH_3 gas.

13-24 Comment on the statement, "In aqueous solutions only weak acids tend to be formed."

Oxidation-reduction

13-25 Balance each of the following equations for nonsolution reactions:
(a) $KNO_3 + Cr_2(SO_4)_3 + Na_2CO_3 \longrightarrow$
$\quad\quad\quad\quad KNO_2 + Na_2CrO_4 + Na_2SO_4 + CO_2$
(b) $KClO_3 + H_2C_2O_4 \longrightarrow ClO_2 + KHCO_3$
(c) $Fe_2O_3 + CO \longrightarrow FeO + CO_2$
(d) $Fe_2O_3 + CO \longrightarrow Fe_3O_4 + CO_2$
(e) $CuO + NH_3 \longrightarrow N_2 + H_2O + Cu$
(f) $MnO_2 + Na_2CO_3 + O_2 \longrightarrow CO_2 + Na_2MnO_4$
(g) $KClO_3 \longrightarrow KClO_4 + KCl$

13-26 In each of the following reactions identify the oxidizing agent and the reducing agent:
(a) $N_2 + O_2 \longrightarrow 2NO$
(b) $H_2 + Cl_2 \longrightarrow 2HCl$
(c) $2Fe^{3+} + 2I^- \longrightarrow 2Fe^{2+} + I_2$
(d) $4NH_3 + 7O_2 \longrightarrow 2N_2O_4 + 6H_2O$
(e) $2KClO_3 \longrightarrow 2KCl + 3O_2$

13-27 In each of the following reactions identify the substance oxidized and the substance reduced:
(a) $H_2S + H_2SO_4 \longrightarrow SO_2 + S + 2H_2O$
(b) $N_2H_4 + H_2O_2 \longrightarrow N_2 + H_2O$
(c) $Cl_2 + HgO \longrightarrow HgCl_2 + Cl_2O$
(d) $2V_2O_5 + 6Cl_2 \longrightarrow 4VOCl_3 + 3O_2$
(e) $Pb(NO_3)_2 \longrightarrow PbO + NO_2 + O_2$

13-28 Using the oxidation-number method balance each of the following equations for reactions occurring in *acidic* solution:
(a) $NO_2 + HOCl \longrightarrow NO_3^- + Cl^-$
(b) $Cr_2O_7^{2-} + H_2SO_3 \longrightarrow Cr^{3+} + HSO_4^-$
(c) $MnO_4^- + H_2C_2O_4 \longrightarrow Mn^{2+} + CO_2$
(d) $Mn^{3+} \longrightarrow Mn^{2+} + MnO_2$
(e) $MnO_2 + PbO_2 \longrightarrow Pb^{2+} + MnO_4^-$
(f) $Cr_2O_7^{2-} + C_3H_7OH \longrightarrow C_2H_5COOH + Cr^{3+}$

13-29 Balance each equation listed in question 13-28 by the half-reaction method.

13-30 Using the oxidation-number method balance each of the following equations for reactions taking place in *basic* solution:
(a) $ClO^- + I^- \longrightarrow Cl^- + I_2$
(b) $Sn(OH)_4^{2-} + CrO_4^{2-} \longrightarrow Sn(OH)_6^{2-} + CrO_2^-$
(c) $SeO_3^{2-} + Cl_2 \longrightarrow SeO_4^{2-} + Cl^-$
(d) $S^{2-} + SO_3^{2-} \longrightarrow S_8$
(e) $SbO_3^{3-} + ClO_2 \longrightarrow ClO_2^- + Sb(OH)_6^-$
(f) $Fe_3O_4 + MnO_4^- \longrightarrow Fe_2O_3 + MnO_2$

13-31 Balance each of the equations listed in question 13-30 by the half-reaction method.

13-32 In hydrogen peroxide, H_2O_2, the oxidation number of each oxygen atom is -1. H_2O_2 can act as either an oxidizing agent or a reducing agent. **(a)** When it acts as an oxidizing agent, what product is formed? **(b)** What product is formed when it acts as a reducing agent? **(c)** Write a balanced equation showing how H_2O_2 can oxidize and reduce itself.

Solution stoichiometry: acid-base

13-33 You are given 225 ml of 0.150 M HCl. How many **(a)** moles and **(b)** equivalents of HCl are present?

13-34 How many **(a)** moles and **(b)** equivalents are present in 44.0 ml of 1.28 M H_2SO_4?

13-35 What is the equivalent weight (for complete neutralization) of each of the following acids and bases: **(a)** HBr **(b)** H_2SO_3 **(c)** H_3PO_4 **(d)** $H_4P_2O_7$ **(e)** KOH **(f)** $Ba(OH)_2$.

13-36 How many moles of HCl are needed to neutralize 75.0 ml of 1.46 M NaOH?

13-37 How many milliliters of 0.140 M HCl are needed to neutralize 0.100 mol of NaOH?

13-38 How many milliliters of 0.125 M HCl are needed to neutralize 45.0 ml of 0.340 M NaOH?

13-39 What is the normality of each of the following: (for complete neutralization) **(a)** 0.10 M HCl **(b)** 0.10 M H_2SO_4 **(c)** 0.10 M H_3PO_4.

13-40 Calculate the number of milliliters of 0.200 N H_2SO_4 needed to neutralize completely **(a)** 125 ml of 0.150 M NaOH **(b)** 40.0 ml of 0.100 M $Ba(OH)_2$ **(c)** 50.0 ml of 0.100 N $Ba(OH)_2$.

13-41 Suppose 37.9 ml of an unknown base is used to neutralize 50.0 ml of 0.200 N H_2SO_4. What is the normality of the base?

13-42 Suppose 44.8 ml of 0.128 M H_2SO_4 is needed to neutralize a 25.0 ml portion of an unknown base. What is the normality of the base?

13-43 Suppose 25.0 ml of a sulfuric acid solution of unknown concentration is neutralized using 29.2 ml of 0.108 M KOH. What is **(a)** the normality and **(b)** the molarity of the H_2SO_4?

Solution stoichiometry: oxidation-reduction

13-44 How many equivalents are present in 0.100 mol of $KMnO_4$ for a reaction in which MnO_4^- is reduced to **(a)** MnO_4^{2-} **(b)** MnO_2 **(c)** Mn^{3+} **(d)** Mn^{2+}?

13-45 What is the molarity of a 0.500 N $K_2Cr_2O_7$ solution if the $Cr_2O_7^{2-}$ is reduced to **(a)** CrO_2 **(b)** Cr^{3+} **(c)** Cr^{2+}?

13-46 How many milliliters of 0.100 N $KMnO_4$ are needed to oxidize 38.0 ml of 0.149 N H_2SO_3?

13-47 How many milliliters of 0.100 M $KMnO_4$ are needed to oxidize 125 ml of 0.115 M H_2SO_3 if the products include Mn^{2+} and HSO_4^-?

13-48 How many milliliters of 0.250 M $K_2Cr_2O_7$ are needed to oxidize 1.00×10^{-3} equiv of a reducing agent if the $Cr_2O_7^{2-}$ is reduced to **(a)** CrO_2^- **(b)** Cr^{2+}?

13-49 Suppose 25.0 ml of an unknown oxidizing agent is reduced with 46.2 ml of 0.104 N KI. What is the normality of the oxidizing agent?

13-50 Suppose 1.60 g of an unknown reducing agent requires 43.6 ml of 0.129 N $KMnO_4$ for complete oxidation. What is the equivalent weight of the reducing agent?

Solution stoichiometry: general

13-51 Suppose 124 ml of 0.129 M HCl is added to 101 ml of 0.151 M NaOH. **(a)** Which ion is in excess: H^+ or OH^-? **(b)** What is the final molarity of this ion? **(c)** What are the molarities of Na^+ and Cl^- in the final solution? (Assume the final volume to be 225 ml.)

13-52 You are given 125 ml of 0.200 M $CaCl_2$. **(a)** What is $[Cl^-]$ in this solution? Assuming that the volumes are additive, calculate $[Cl^-]$ after adding *successively* **(b)** 75.0 ml of 0.100 M $AlCl_3$ **(c)** 115 ml of water **(d)** 150 ml of 0.100 M $AgNO_3$.

13-53 Suppose 125 ml of 2.0 M $HClO_4$, 135 ml of 0.100 M $KMnO_4$, and 270 ml of 0.350 M $FeSO_4$ are mixed. In the reaction which takes place, MnO_4^- is reduced to Mn^{2+} and Fe^{2+} is oxidized to Fe^{3+}. Assuming that no other reaction occurs and the final volume is 530 ml, calculate the final concentration of H^+, ClO_4^-, K^+, MnO_4^-, Fe^{2+}, SO_4^{2-}, Mn^{2+}, and Fe^{3+}.

13-54 Suppose 0.500 g of sodium vanadate ($NaVO_3$) is reduced by 24.6 ml of a 0.500 N solution of a reducing agent to form the vanadous ion. If the vanadous ion is a simple (monatomic) ion, what is its charge?

14

CHEMICAL KINETICS

TO THE STUDENT

We see evidence of chemical change all around us. Some of these changes occur too fast for our liking. The cracking and peeling of paint exposed to sun and rain, the rusting of an automobile body, the spoiling of food, the burning of a candle: who has not wished for one or more of these changes to occur more slowly? On the other hand, it would be convenient if some reactions went faster: the growth of plants and animals used for food, the healing of a wound, the boiling of a potato, the hardening of concrete, and the disintegration of litter. Studying the factors which influence chemical change has obvious practical applications. In addition, the study of reaction rates yields valuable insights into *how* chemical reactions actually take place.

    Chemical kinetics is the study of the rates and mechanisms of chemical reactions. The *rate* of a reaction is a measure of how fast products are formed and reactants consumed. The *mechanism* of a reaction is the detailed sequence of single, elementary steps which lead from reactants to products. A net equation shows none of these steps, but rather shows the net change which is the overall result of all the steps in the mechanism. Much of our knowledge of reaction mechanisms stems from studies of reaction rates and how they are influenced by different variables. In general, the rate of a given reaction is determined by (1) the properties of the reactants, (2) the reactant concentrations, and (3) the temperature. It may also be influenced by (4) concentrations of species other than reactants, and by (5) areas of surfaces in contact with the reactants.

14-1
Reaction rates and mechanisms

Reaction rate, or *reaction velocity,* is a measure of how fast a reactant is used up or a product formed. In order to see how reaction rates can be described quantitatively, consider the hypothetical, *homogeneous* (single-phase) reaction:

$$A + B \longrightarrow C + D$$

The rate of a reaction

Assume that A and B are mixed at time $t = 0$, and that the initial concentration of

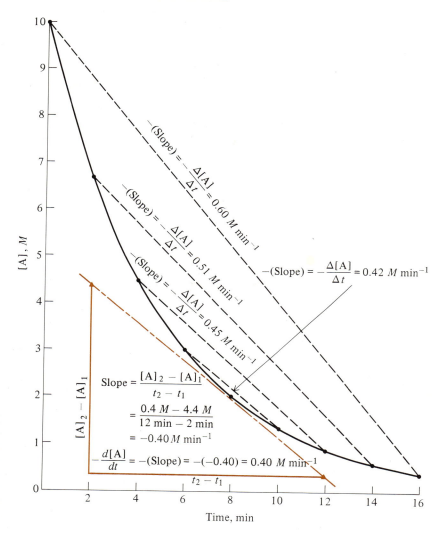

Figure 14-1 Average and instantaneous rates. Black broken lines indicate average rates and the colored broken line indicates instantaneous rate.

A is 10 M. As the reaction proceeds, [A] decreases, as is shown by the curve in Fig. 14-1. Expressing the rate of the reaction by stating how rapidly [A] decreases is not easy because the rate keeps changing as reactants are used up. But we can express the *average* rate of disappearance of A over any given time interval. The following table shows [A] at 2-min intervals:

t (time), min	[A], M
0	10.00
2	6.69
4	4.48
6	3.00
8	2.00
10	1.34
12	0.90
14	0.60
16	0.40

The average rate of disappearance of A equals the change in [A] divided by the corresponding time interval, or

$$\text{Average rate} = -\frac{\Delta[A]}{\Delta t} = -\frac{[A]_2 - [A]_1}{t_2 - t_1}$$

Here the negative sign indicates that the concentration of A is *decreasing* with time. Over the interval from $t = 0$ to $t = 16$, [A] decreases from 10 to 0.40 M, and so over these 16 min the average rate is

$$\text{Average rate} = -\frac{\Delta[A]}{\Delta t} = -\frac{0.40\ M - 10\ M}{16\ \text{min} - 0\ \text{min}} = 0.60\ M\ \text{min}^{-1}$$

Since the 8-min point is at the midpoint of this time interval, we could say that the average rate of the reaction at the 8-min mark is $-0.60\ M\ \text{min}^{-1}$. The average rate over the time interval from t_1 to t_2,

$$-\frac{\Delta[A]}{\Delta t} = -\frac{[A]_2 - [A]_1}{t_2 - t_1}$$

is the *negative* of the *slope* of the straight line connecting the point $([A]_2, t_2)$ with the point $([A]_1, t_1)$ on the graph. (See App. D.) Four of these straight lines have been drawn in Fig. 14-1, each from a different pair of data points on the curve. The slope of each (with its sign changed) represents the average rate over the corresponding time interval. Note that each straight line has a different slope.

How do we find a rate *at one specific instant* rather than an average rate over an interval? What we need is the slope of the straight line which is *tangent* to the concentration-time curve at the desired point. Such a tangent line has been drawn at the 8-min point in Fig. 14-1. The negative of the slope of this is called the *instantaneous rate* and may be evaluated by using the coordinates ([A], t) of any two points on the line, such as (2, 4.4) and (12, 0.4).

$$-\text{Slope} = -\frac{[A]_2 - [A]_1}{t_2 - t_1}$$

$$= -\frac{0.4\ M - 4.4\ M}{12\ \text{min} - 2\ \text{min}} = 0.40\ M\ \text{min}^{-1}$$

Since we have been using $\Delta[A]/\Delta t$ to represent an average change, we need a different symbol for the instantaneous rate. The one which is nearly always used for this purpose is[1]

$$\frac{d[A]}{dt}$$

In this symbol d/dt means *rate of increase of;* so $d[A]/dt$ stands for the rate of increase of A (with time). In order to express a rate of *decrease,* we put a minus sign in front:

$$-\frac{d[A]}{dt} \quad \text{(a decrease)}$$

---

[1] Those of you who have studied calculus will recognize this as a derivative.

This stands for the rate of *decrease* in [A]. This symbolism comes from calculus, and although we will not be using any calculus in this book, it is important that you properly interpret the d/dt symbol. (Note particularly that the letter d is not an algebraic symbol at all, and so dt does not mean d times t.) Thus the instantaneous rate at the 8-min mark is given by

$$\text{Instantaneous rate} = -\frac{d[A]}{dt} = -\text{slope} = 0.40 \ M \ \text{min}^{-1}$$

Now consider the reaction

$$A + 2B \longrightarrow 3C + 4D$$

The expression, *rate of reaction,* is vague unless we indicate the rate of a *specific change.* Thus the rate can be indicated by any one of the following:

$$-\frac{d[A]}{dt} \qquad \text{(rate of decrease of [A])}$$

$$-\frac{d[B]}{dt} \qquad \text{(rate of decrease of [B])}$$

$$\frac{d[C]}{dt} \qquad \text{(rate of increase of [C])}$$

$$\frac{d[D]}{dt} \qquad \text{(rate of increase of [D])}$$

But a glance at the equation for the reaction shows that the rates of these four changes are not all the same. For instance, B is used up twice as fast as A. (*One mole of A consumes two moles of B.*) The different rates are related to each other by

$$-\frac{d[A]}{dt} = -\frac{1}{2}\frac{d[B]}{dt} = \frac{1}{3}\frac{d[C]}{dt} = \frac{1}{4}\frac{d[D]}{dt}$$

For any reaction

$$aA + bB \longrightarrow cC + dD$$

the rates are related by

$$-\frac{1}{a}\frac{d[A]}{dt} = -\frac{1}{b}\frac{d[B]}{dt} = \frac{1}{c}\frac{d[C]}{dt} = \frac{1}{d}\frac{d[D]}{dt}$$

The mechanism of a reaction

Most reactions take place not in a single step described by the net equation but in a number of steps. Sometimes these steps are ordered in a simple sequence, while for other reactions they are interrelated in a complex way. The steps which lead from reactants to products, and the relationship among these steps comprise the *mechanism* of a reaction.

One reaction which proceeds by a simple two-step mechanism is the reaction of iodine monochloride with hydrogen, a homogeneous, gas-phase reaction:

$$2ICl(g) + H_2(g) \longrightarrow 2HCl(g) + I_2(g) \qquad \text{(net equation)}$$

The mechanism of this reaction has been found by experiment to be

Step 1: $ICl + H_2 \longrightarrow HI + HCl$

Step 2: $ICl + HI \longrightarrow I_2 + HCl$

The first step in this mechanism is the collision of an ICl molecule with an H_2 molecule. They react to form an HI molecule and an HCl molecule. The HI molecule then collides and reacts with a second ICl molecule forming an I_2 molecule and a second HCl molecule. The overall change is described by the stoichiometry of the net equation, which can be obtained by adding the equations for the two steps.

ADDED COMMENT

A not-so-obvious point here is that even though we say that step 1 in the mechanism precedes step 2, they actually both occur simultaneously in the reaction mixture. This apparent contradiction is resolved when one realizes that in a large collection of ICl and H_2 molecules it would be unreasonable to expect all step-2 collisions to wait until all step-1 collisions had taken place. It is true that a step-2 collision cannot take place until an HI molecule has been formed by step 1, but after the initial mixing of ICl and H_2 molecules, many step-2 collisions occur before all H_2 molecules are used up in step 1.

The above mechanism illustrates the formation of an *intermediate,* a species which is formed in one step, only to be used up in a subsequent step. HI is such a species in this reaction mechanism and does not show up as a final product.

We will consider reaction mechanisms in more detail in Sec. 14-5.

**14-2
The rate law**

Concentrations of reacting species influence the rate of a reaction, and the algebraic statement of the relationship between concentration and rate is known as the *rate law* for the reaction. It should be understood from the start that *a rate law cannot be determined from the net equation for a reaction,* but must be determined from experimental measurements of reaction rates.

*The measurement of
reaction rates*

Reaction-rate experiments are often difficult to design and carry out, because accurate measurement of a *changing* concentration is often difficult. In principle one should be able to measure the rate of a reaction by first mixing the reactants and then periodically withdrawing samples from the reaction mixture for analysis. In this way one can measure the change in the concentration of a reactant or product during the preceding time interval. This procedure may be satisfactory if the reaction does not take place too rapidly. But if the reaction is so fast that appreciable changes in concentration occur during the sampling-and-analysis period, then experimental results will be quite imprecise. One way to combat this *sampling-and-analysis-time problem* is to slow the reaction suddenly (to *quench* it) either by suddenly cooling the sample or by rapidly removing a reactant. In the latter case, addition of a substance which rapidly combines with one of the reactants effectively stops the reaction. Another critical problem in the case of fast reactions is the *mixing-time* problem. It is not possible to mix reactants instantaneously. In the case of a slow reaction the time necessary to accomplish mixing may be negligible in comparison with the time interval from mixing to sampling. With a fast reaction, however, special methods for rapid mixing must be devised.

Rather than making conventional chemical analyses of reaction mixtures, *instrumental methods* are often employed to measure physical properties which change during the course of a reaction. Thus, for example, the rate of the gas-phase decomposition of ethylene oxide into methane and carbon monoxide,

$$C_2H_4O(g) \longrightarrow CH_4(g) + CO(g)$$

can be followed by observing how the pressure of the mixture increases with time. Note that the progress of such a gas-phase reaction can be followed this way only if the stoichiometry of the reaction is such that the number of moles of products is different from the number of moles of reactants.

In principle, any changing physical property can be used to follow the progress of a reaction. The absorption of light by iodine can be used to follow its gas-phase reaction with hydrogen,

$$H_2(g) + I_2(g) \longrightarrow 2HI(g)$$

which could not be followed by watching pressure changes. Usually a *spectrophotometer* is employed in the experiment. This is a device which measures the absorption of light at various wavelengths. By choosing a wavelength at which one component strongly absorbs energy, its change in concentration can be monitored as the reaction proceeds. Other techniques which can be used to follow a reaction's progress include measurement of electrical conductivity, density, refractive index, and viscosity.

Concentration dependence of reaction rates

The rate of a reaction usually depends in some way upon the concentration of one or more reactants, but it may also depend upon the concentrations of products or even of substances which do not show up in the net equation at all. Consider, for example, the homogeneous reaction,

$$A + B + 2C \longrightarrow D + E$$

Suppose that it is found that doubling [A] doubles the measured rate of the reaction, and that tripling [A] triples the rate. Mathematically this describes a direct proportionality between the rate and [A]. If we use the expression $-d[A]/dt$ to represent the rate of disappearance of A, then the proportionality is

$$-\frac{d[A]}{dt} \propto [A]$$

Suppose that it is also found experimentally that the rate is proportional to [B] but independent of all other concentrations, including [C]. Then we can write

$$-\frac{d[A]}{dt} \propto [B]$$

Combining these two relationships we have

$$-\frac{d[A]}{dt} \propto [A][B]$$

Replacing the proportionality sign by an equals sign gives us

$$-\frac{d[A]}{dt} = k[A][B]$$

where k, the constant of proportionality, is called the *specific rate constant* for the reaction. This is an example of a *rate law*. It expresses the relationship between the rate of a reaction and the concentrations of species which influence the rate. The specific rate constant k has a fixed value for all concentrations of A and B at a given temperature, but varies with temperature. It is very important to note that *there is no necessary relationship between the stoichiometric net equation for a reaction and the rate law*. This means that it is impossible merely to look at a net equation for a reaction and from it alone write the rate law for the reaction. Rate laws must be determined from experimental data.

The *order of a reaction* is the sum of the exponents on the concentration terms in the rate law. The *order with respect to one species* is the exponent on the concentration term for that species alone. Thus, for example, a reaction for which the rate law is

$$-\frac{d[A]}{dt} = k[A][B]^2$$

is said to be *first order in A, second order in B,* and, therefore, *third order overall*.

A rate law describes the concentration-rate relationship at one temperature. Since reaction rates do depend additionally on temperature, it follows that k, the specific rate constant, is a function of temperature, usually increasing with increasing temperature. (We will talk about temperature dependence of reaction rates in Sec. 14-3.)

First-order reactions

Several methods are useful for determining rate laws and, hence, reaction orders. One, the *initial rate method,* involves running a series of separate experiments at a given temperature, determining the rate at the start of each, and then mathematically analyzing the relationship between initial concentration and initial rate. Consider, for example, the hypothetical reaction

$$A(g) \longrightarrow \text{products}$$

Suppose that the reaction is very slow at room temperature, but fast at some elevated temperature, say, 500°C. Suppose, further, that we run an experiment by injecting A into a container at 500°C and following the decrease in [A] as time elapses.

Time, min	[A], M
0.0	1.30
5.0	1.08
10.0	0.90
15.0	0.75
20.0	0.62
25.0	0.52
30.0	0.43

If the resulting data are plotted as in Fig. 14-2, the instantaneous rate of the reaction at any time can be found from the slope of the tangent to the curve at that time. In particular, the *initial rate* of the reaction is found by drawing the tangent to the curve at $t = 0$. In Fig. 14-2 this tangent has been drawn, and its slope found.

$$\text{Slope} = \frac{[A]_2 - [A]_1}{t_2 - t_1}$$

$$= \frac{0\ M - 1.30\ M}{27.2\ \text{min} - 0\ \text{min}} = -4.78 \times 10^{-2}\ M\,\text{min}^{-1}$$

This is the instantaneous rate of change of [A] at $t = 0$. In other words, at $t = 0$ (when $[A] = 1.30\ M$)

$$-\frac{d[A]}{dt} = 4.78 \times 10^{-2}\ M\,\text{min}^{-1}$$

If additional experiments were run at the same temperature, the initial rate of the reaction could be obtained for other initial concentrations, as summarized below:

Experiment no.	Initial [A], M	Initial rate, $-\dfrac{d[A]}{dt}$, $M\,\text{min}^{-1}$
1	1.30	4.78×10^{-2}
2	2.60	9.56×10^{-2}
3	3.90	1.43×10^{-1}
4	0.891	3.28×10^{-2}

If we compare the results of experiments 1 and 2, we see that a *doubling* of the initial concentration of A produces a *doubling* of the initial rate. Similarly, a

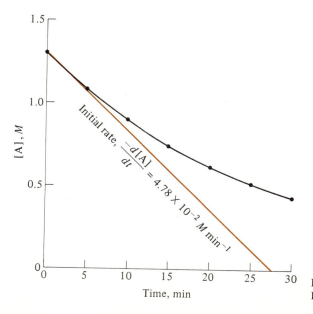

Figure 14-2
Determination of initial rate.

comparison of experiments 1 and 3 shows that a *tripling* of the initial concentration of A triples the rate. This indicates a direct proportionality between rate and concentration. In other words,

$$-\frac{d[A]}{dt} \propto [A]$$

or, written as a rate law,

$$-\frac{d[A]}{dt} = k_1[A]$$

(The subscript on k is not essential. It serves to remind us that in this case the specific rate constant is for a *first*-order reaction.)

In experiments 1, 2, and 3 (above) the initial concentrations of A can be seen by simple inspection to be in the ratio $1:2:3$. What if the ratio of two similar concentrations were not an integer? We see such a situation when we compare experiments 1 and 4. Here the initial concentration is decreased by a factor of $0.891/1.30$, or 0.685, and the initial rate becomes decreased by a factor of $(3.28 \times 10^{-2})/(4.78 \times 10^{-2})$, or 0.686. Allowing for the small rounding-off deviation, the ratios are the same: an indication of a first-order reaction.

A more general method of finding reaction order from the above data is as follows. First write the rate law as

$$-\frac{d[A]}{dt} = k[A]^x$$

where x is a number to be determined. Substituting data from experiment 1 we write

$$4.78 \times 10^{-2} = k(1.30)^x$$

and from experiment 4,

$$3.28 \times 10^{-2} = k(0.891)^x$$

Now we divide one of these equalities by the other,

$$\frac{4.78 \times 10^{-2}}{3.28 \times 10^{-2}} = \frac{k(1.30)^x}{k(0.891)^x}$$

$$1.46 = 1.46^x$$

We can see that x must equal 1. If this were not obvious, we could take logarithms of both sides, and then easily solve for x. (We illustrate this technique when we discuss second-order reactions.) Again we find that the exponent on [A] equals 1. The reaction is first-order.

Having determined the rate law, we can now use data from *any experiment* to find the numerical value of the specific rate constant k_1.

$$-\frac{d[A]}{dt} = k_1[A]$$

$$k_1 = \frac{-d[A]/dt}{[A]}$$

Substituting from experiment 1, we have

$$k_1 = \frac{4.78 \times 10^{-2}\ M\,\mathrm{min}^{-1}}{1.30\ M} = 3.68 \times 10^{-2}\ \mathrm{min}^{-1}$$

First-order rate constants always have dimensions of reciprocal time: s^{-1}, min^{-1}, h^{-1}, etc.

● **EXAMPLE 14-1**

Problem   Azomethane, $C_2H_6N_2$, decomposes according to the equation

$$C_2H_6N_2(g) \longrightarrow C_2H_6(g) + N_2(g)$$

Determine the order of the reaction and evaluate the specific rate constant from the following data:

Experiment no.	Initial $[C_2H_6N_2]$, M	Initial $-\dfrac{d[C_2H_6N_2]}{dt}$, $M\,\mathrm{min}^{-1}$
1	1.96×10^{-2}	3.14×10^{-4}
2	2.57×10^{-2}	4.11×10^{-4}

Solution   If the initial concentration of azomethane is increased by a factor of

$$\frac{2.57 \times 10^{-2}}{1.96 \times 10^{-2}} = 1.31$$

the initial rate becomes faster by a factor of

$$\frac{4.11 \times 10^{-4}}{3.14 \times 10^{-4}} = 1.31$$

The rate and concentration of azomethane are clearly proportional; so the reaction is first-order, and the rate law is

$$-\frac{d[C_2H_6N_2]}{dt} = k_1[C_2H_6N_2]$$

$$k_1 = \frac{-d[C_2H_6N_2]/dt}{[C_2H_6N_2]}$$

$$= \frac{4.11 \times 10^{-4}\ M\,\mathrm{min}^{-1}}{2.57 \times 10^{-2}\ M} = 1.60 \times 10^{-2}\ \mathrm{min}^{-1}\ ●$$

A second method, the *graphical method,* is often used to show that a reaction is first order.   By using calculus it is possible to show that for a first-order reaction such as

$$A \longrightarrow \text{products}$$

the following relationship is true:

$$\log[A] = \left(-\frac{k_1}{2.303}\right)t + \log[A]_0$$

where $[A]$ is the concentration of A at any time t, and $[A]_0$ is the *initial* concentration of A (at t equals 0).   (It is not necessary for you to be able to derive this relationship, but it is a useful one to know.)   This is the basis for a *graphical method* of showing that a reaction is first-order and finding its specific rate constant.

The above relationship has the form of an equation for a straight line:

$$y = mx + b$$

in which $\log[A] = y$ and $t = x$. Thus, plotting $\log[A]$ against t for a first-order reaction should yield a straight line of slope

$$-\frac{k_1}{2.303}$$

Using data introduced earlier in this section for the hypothetical reaction

$$A \longrightarrow products$$

we will calculate $\log[A]$ at 5-min intervals through 30 min.

Time, min	[A], M	log [A]
0.0	1.30	0.11
5.0	1.08	0.03
10.0	0.90	−0.05
15.0	0.75	−0.13
20.0	0.62	−0.21
25.0	0.52	−0.28
30.0	0.43	−0.37

Now we plot $\log[A]$ (on the y axis) against time (on the x axis) as shown in Fig. 14-3. The result is a straight line. This kind of plot will result in a straight line *only for a first-order reaction;* so we have demonstrated that the reaction is indeed first-order. The specific rate constant can be found from the slope, which is found to be $-1.6 \times 10^{-2} \, \text{min}^{-1}$. Since this is equal to $-k_1/2.303$,

$$k_1 = -2.303(\text{slope})$$
$$= -2.303(-1.6 \times 10^{-2} \, \text{min}^{-1}) = 3.7 \times 10^{-2} \, \text{min}^{-1}$$

essentially the same result as was found from the initial-rate method.

The graphical method has two advantages over the initial-rate method. First, only one experimental run is needed to supply the data necessary to determine the reaction order and evaluate the rate constant. Second, the difficulty of determining initial rates by estimating tangents to curves at $t = 0$ is avoided.

A third method, the *half-life method,* is sometimes useful, expecially for first-order reactions. The *half-life* of any reaction is defined as the period of time necessary for one-half of a reactant to be consumed. If we use the relationship presented above,

$$\log[A] = -\frac{k_1}{2.303}t + \log[A]_0$$

and let $[A]_{1/2}$ be equal to the concentration of A remaining at the end of the half-life period, that is, at time $t_{1/2}$, then

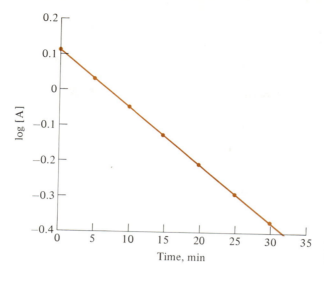

Figure 14-3
First-order plot.

$$\log [A]_{1/2} = -\left(\frac{k_1}{2.303}\right)t_{1/2} + \log [A]_0$$

$$\frac{k_1}{2.303}t_{1/2} = \log [A]_0 - \log [A]_{1/2}$$

$$= \log \frac{[A]_0}{[A]_{1/2}}$$

But $[A]_{1/2} = \frac{1}{2}[A]_0$; so

$$\frac{k_1}{2.303}t_{1/2} = \log \frac{[A]_0}{\frac{1}{2}[A]_0} = \log 2$$

$$t_{1/2} = \frac{2.303 \log 2}{k_1} = \frac{0.693}{k_1}$$

This says that *the half-life of a first-order reaction is a constant*, equal to $(2.303 \log 2)/k_1$. (All the terms on the right are constants.) In other words, if it can be shown that the half-life of a reaction is a constant, the reaction must be first-order.

ADDED COMMENT

Note that for any first-order reaction, A $\longrightarrow$ products, the time required for [A] to be reduced to one-half its initial value is the same as that required for it to be again reduced by half, that is, down to one-quarter of its initial value. (And so on!)

Second-order reactions   The *initial-rate method* can be used to verify that a reaction is second-order. For the hypothetical reaction

$$C + 2D \longrightarrow products$$

consider the following data:

Experiment no.	Initial [C], M	Initial [D], M	Initial $-\dfrac{d[A]}{dt}$, $M\,min^{-1}$
1	0.346	0.369	0.123
2	0.692	0.369	0.492
3	0.346	0.738	0.123

In this case doubling [C] while holding [D] constant (experiments 1 and 2) has caused the rate to increase by a factor not of 2, but of 2^2, or 4. $(0.123 \times 4 = 0.492.)$ The rate is thus proportional to $[C]^2$. But changing the concentration of D has no influence on the reaction rate. The rate is thus independent of [D] (the reaction is *zero* order in D), and so the rate law is

$$-\frac{d[C]}{dt} = k[C]^2$$

Now, consider another example,

$$2E + F \longrightarrow products$$

for which the following data were obtained:

Experiment no.	Initial [E], M	Initial [F], M	Initial $-\dfrac{d[E]}{dt}$, $M\,s^{-1}$
1	0.0167	0.234	3.61×10^{-2}
2	0.0569	0.234	4.20×10^{-1}
3	0.0569	0.361	4.20×10^{-1}

Here we see from a comparison of experiments 2 and 3 that the rate is independent of [F]. It is probably proportional to [E] raised to some power; that is,

$$-\frac{d[E]}{dt} = k[E]^x$$

We now need to evaluate x, that is, to find the order of the reaction. If we substitute the data from experiments 1 and 2 separately in this relationship and then divide one by the other, we obtain

$$\frac{3.61 \times 10^{-2}}{4.20 \times 10^{-1}} = \frac{k(0.0167)^x}{k(0.0569)^x}$$

$$8.60 \times 10^{-2} = \left(\frac{0.0167}{0.0569}\right)^x$$

$$8.60 \times 10^{-2} = (2.93 \times 10^{-1})^x$$

Using logarithms to solve this, we get

$$x \log (2.93 \times 10^{-1}) = \log (8.60 \times 10^{-2})$$

$$x = \frac{\log (8.60 \times 10^{-2})}{\log (2.93 \times 10^{-1})} = 2.00$$

Thus the reaction is *second order* in E, and the rate law is

$$-\frac{d[E]}{dt} = k_2[E]^2$$

A *graphical method* can also be used to demonstrate that a reaction is second-order. Using calculus it is possible to show that for a second-order reaction for which the rate law is

$$\text{Rate} = k_2[A]^2$$

the following relationship exists:

$$\frac{1}{[A]} = k_2t + \frac{1}{[A]_0}$$

where [A] is the concentration of A at time t, and $[A]_0$ is its concentration at $t = 0$. Comparing this with the $y = mx + b$ equation, we see that for a second-order reaction a plot of $1/[A]$ (on the y axis) against t (on the x axis) will be a straight line of slope k_2.

● **EXAMPLE 14-2**

Problem   At 383°C measurements of the decomposition of NO_2 (to form NO and O_2) provided the following data:

Time, s	0	5	10	15	20
$[NO_2]$, M	0.10	0.017	0.0090	0.0062	0.0047

Show that the reaction is second-order and evaluate the specific rate constant.

Solution   In order to make a second-order plot, we need to find the reciprocal of each concentration.

Time, s	0	5	10	15	20
$1/[NO_2]$, M^{-1}	10	59	111	161	213

Now we plot $1/[NO_2]$ against time, as shown in Fig. 14-4. Since the points fall on a straight line, the reaction must be second-order. The rate law is, therefore,

$$-\frac{d[NO_2]}{dt} = k_2[NO_2]^2$$

and $k_2 = \text{slope} = 10.1\ M^{-1}\,s^{-1}$. ●

The half-life $t_{1/2}$ of a second-order reaction is *not* independent of initial concentration. For the reaction A $\longrightarrow$ products for which $-(d[A]/dt) = k_2[A]^2$, the half-life can be shown to be

$$t_{1/2} = \frac{1}{k_2[A]_0}$$

where $[A]_0$ is the initial concentration of A. *The half-life is independent of initial concentration only for a first-order reaction.*

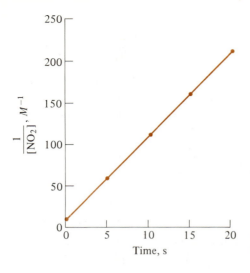

Figure 14-4
Decomposition of nitrogen dioxide.

Reactions of Many reactions show orders other than one or two.   As an example, the reaction of
other orders nitric oxide with oxygen to form nitrogen dioxide,

$$2NO(g) + O_2(g) \longrightarrow 2NO_2(g)$$

obeys a third-order rate law,

$$-\frac{d[NO]}{dt} = k_3[NO]^2[O_2]$$

This reaction can occur in the atmosphere; the brown color of photochemical smog
is at least in part due to NO_2.

A few reactions show a *zero* (overall) *order*.   One of these is the heterogeneous
decomposition of ammonia on the surface of tungsten metal,

$$2NH_3(g) \longrightarrow N_2(g) + 3H_2(g)$$

The rate law for this reaction is

$$-\frac{d[NH_3]}{dt} = k_0$$

This means that the decomposition of ammonia occurs at a rate which is *independent of any concentration.*

Some reactions show *fractional orders*.   Thus, the thermal decomposition of
acetaldehyde, CH_3CHO, to form methane and carbon monoxide follows the
rate law

$$-\frac{d[CH_3CHO]}{dt} = k[CH_3CHO]^{3/2}$$

The order with respect to a species may even be *negative*.   An example is the
decomposition of ozone,

$$2O_3(g) \longrightarrow 3O_2(g)$$

for which the rate law is

$$-\frac{d[O_3]}{dt} = k\frac{[O_3]^2}{[O_2]} = k[O_3]^2[O_2]^{-1}$$

Note that this reaction is first-order (overall).

Finally, it should be noted that some rate laws are sufficiently complex that no assignment of overall reaction order is possible. For example, the reaction between hydrogen and bromine gases,

$$H_2(g) + Br_2(g) \longrightarrow 2HBr(g)$$

is described by the rate law

$$-\frac{d[H_2]}{dt} = \frac{k[H_2][Br_2]^{1/2}}{1 + k'[HBr]/[Br_2]}$$

(See Sec. 14-5.) In this case the reaction can be said to be first order in H_2, but the dependence on $[Br_2]$ and $[HBr]$ is complex.

14-3 Collision theory

During the early part of the twentieth century the first successful theory of reaction rates was developed from the kinetic-molecular theory of gases. It assumes that for gas molecules to react they must collide, and so the theory is known as *collision theory*.

Elementary processes and molecularity

We have said that most reactions consist of a number of steps. Each individual step is called an *elementary process* (or an *elementary reaction*). An elementary process is described by an equation which specifies the reactant and the product particles for that step. Thus, for example, the equation for the elementary process

$$2ClO(g) \longrightarrow Cl_2(g) + O_2(g)$$

shows two ClO molecules colliding with each other and reacting to form one Cl_2 molecule and one O_2 molecule. Each elementary process can be characterized by specifying the *molecularity* of the process. When a single particle (molecule, atom, ion) is the only reactant in such a process, the process is said to be *unimolecular* (molecularity = 1). When two particles collide and react, the process is said to be *bimolecular* (molecularity = 2). *Termolecular* (also called *trimolecular*) processes are believed to be rare. (The simultaneous collision of three particles, a so-called three-body collision, is an unlikely event.)

The terms *unimolecular, bimolecular,* and *termolecular* indicate the number of reacting particles in a single elementary process. Properly speaking, these terms should not be applied to an overall reaction. However, in common usage they sometimes are. For example, if a reaction has one slow, or rate-determining, step (Sec. 14-5) which is bimolecular, then the entire reaction might be called a bimolecular reaction, even though this usage is not strictly correct.

Note that molecularities, unlike reaction orders, must be integers. Note, also, that the molecularity of a single step bears no necessary relation to the order of the overall reaction.

Bimolecular
gas-phase processes

Consider the gas-phase elementary process,

$$A_2(g) + B_2(g) \longrightarrow 2AB(g)$$

In order to react, an A_2 molecule must collide with a B_2 molecule; so we know that the rate of the reaction depends upon the *frequency of collision Z* (number of collisions per second) between A_2 and B_2 molecules. Since the rate should double if A_2 and B_2 collide with each other twice as frequently, we can say that the reaction rate is directly proportional to Z, or

$$Rate \propto Z$$

Further, the collision frequency depends in turn on the *concentrations* of A_2 and of B_2. If, for example, we were to *double* the concentration of A_2 molecules, this would double the probability of an A_2-B_2 collision, and hence of the collision frequency. Since the same logic applies to doubling the concentration of B_2 molecules, we see that the collision frequency Z is proportional to $[A_2]$ and to $[B_2]$. In other words,

$$Z \propto [A_2] \quad \text{and} \quad Z \propto [B_2]$$

Therefore, $$Z \propto [A_2][B_2]$$

Calling the proportionality constant Z_0, we can write

$$Z = Z_0[A_2][B_2]$$

Z_0 represents the collision frequency when $[A_2] = [B_2] = 1$. Now we can rewrite our rate expression as

$$Rate \propto Z_0[A_2][B_2]$$

What else besides collision frequency does the rate depend on? Imagine the collision of a single A_2 molecule with a single B_2 molecule. If the collision is to be effective in producing the product AB molecules, the molecules must hit each other hard enough to break the A—A and B—B bonds, so that A—B bonds can then take their places. Thus the colliding molecules must have a certain minimum energy, called the *activation energy E_a*, in order for the collisions to be effective in producing reaction.

During the latter half of the nineteenth century the Scottish mathematician James Clerk Maxwell and the Austrian physicist Ludwig Boltzmann independently developed relationships describing the distribution of molecular speeds and energies in a gas. Using the Maxwell-Boltzmann equations, it can be shown that in a large collection of reactant molecules, the fraction of the molecules which have energies at least equal to the molar activation energy E_a is

$$e^{-E_a/RT}$$

where e = base of natural logarithms (App. D),
                 a constant equal to 2.71828 ...
            R = ideal-gas constant
            T = absolute temperature

According to collision theory, the rate of *successful* molecular collision, collision

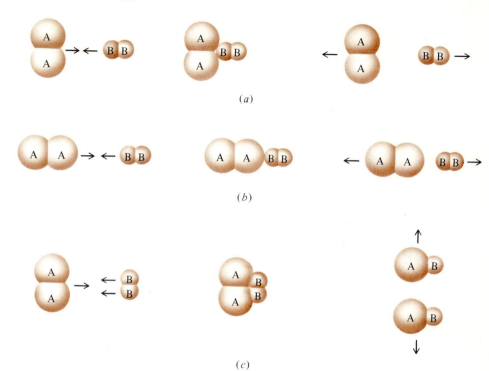

Figure 14-5
Steric effect in colli-
sions. (a) No reac-
tion. (b) No reaction.
(c) Reaction.

leading to the formation of reaction products, is proportional to this fraction. In other words,

$$\text{Rate} \propto e^{-E_a/RT}$$

Although not included in the original version of collision theory, one more factor must be considered. It is not enough that an A_2 molecule collides with a B_2 molecule; it is not even enough that they have the activation energy before collision. No reaction will occur unless the *relative orientation* of the molecules at the instant of collision is favorable for rupturing the A—A and B—B bonds and forming A—B bonds. Figure 14-5 shows a number of different orientations for A_2-B_2 collisions. Only one of those pictured, Fig. 14-5c, produces reaction. In Fig. 14-5a and b, the molecules merely bounce off each other. In Fig. 14-5c, however, electrons rearrange themselves so as to favor the new AB molecules. This effect, the *steric* (or *orientation*) *effect,* limits successful collisions to the ones in which the molecules are favorably oriented. The *steric factor p,* also called *probability factor,* is the fraction of the collisions in which the molecules have an orientation favorable for reaction. Values of steric factors for a few reactions are given in Table 14-1. A steric factor of, say, 0.1 means that only one collision out of 10 has the right orientation for reaction. When a very specific orientation is required for reaction, p is very small.

According to collision theory, the rate of a reaction is equal to the product of three terms:

Table 14-1
Steric factors for some bimolecular collisions

Reaction	p
$CH_4 + H \longrightarrow CH_3 + H_2$	0.5
$H_2 + I_2 \longrightarrow 2HI$	0.2
$2NO_2 \longrightarrow 2NO + O_2$	0.06
$CO + O_2 \longrightarrow CO_2 + O$	0.004
$2ClO \longrightarrow Cl_2 + O_2$	0.002

1 The steric factor p

2 The fraction of the collisions which are successful in producing reaction, $e^{-E_a/RT}$

3 The collision frequency Z

In other words,

$$\text{Rate} = pe^{-E_a/RT}Z = pe^{-E_a/RT}Z_0[A_2][B_2]$$

Note that for a given bimolecular process at a given temperature, all the terms preceding $[A_2][B_2]$ in the last expression are constant; and so lumping them together and calling them k, we have

$$\text{Rate} = k[A_2][B_2]$$

which is the rate law for the bimolecular process

$$A_2(g) + B_2(g) \longrightarrow 2AB(g)$$

and where

$$k = pe^{-E_a/RT}Z_0$$

For a bimolecular process of the type

$$2C \longrightarrow \text{products}$$

the probability of a C-C collision is quadrupled when [C] is *doubled,* because *both* colliding particles are C molecules. Collision theory thus predicts that the rate law for such a process is

$$\text{Rate} = k[C]^2$$

Note that for each of the above cases, $A_2 + B_2 \longrightarrow 2AB$ and $2C \longrightarrow$ products, the exponents on the concentration terms correspond to the coefficients written before the colliding species in the equation for the process.

Unimolecular and termolecular processes

A *unimolecular* process,

$$A \longrightarrow \text{products}$$

appears to involve the spontaneous breakup of an A molecule without any external cause. This is actually not so, because in order for the bonds in an A molecule to break, energy must be supplied to the molecule. This can happen when A collides with another A molecule or with a molecule of any inert substance which may be present. The energized A molecule may then *either* lose energy to some other molecule by colliding with it *or* fly apart into products. According to collision theory, a unimolecular process is one in which the probability of the molecule's splitting is low in comparison with the probability of de-energization occurring. Thus most energized A molecules merely become de-energized through collision, and only relatively infrequently does one form products. Using collision theory it is possible to show that for such a reaction the rate law is

$$\text{Rate} = k[A]$$

(Note again, the correspondence between the exponent in the rate law and the coefficient in the equation for the process.)

Because of the low probability of a simultaneous three-body collision *termolecular processes* are uncommon. A few processes of the type

$$2A + B \longrightarrow \text{products}$$

are known. Collision theory predicts the rate law for these to be

$$\text{Rate} = k[A]^2[B]$$

For each of the above elementary processes the exponents in the rate law correspond to the coefficients in the equation for the process. In general, *for any elementary process* the rate law can be written from its stoichiometry. In other words, if the equation for such a process is

$$xA + yB \longrightarrow \text{products} \qquad \text{(elementary process)}$$

then the rate law is

$$\text{Rate} = k[A]^x[B]^y$$

ADDED COMMENT

Note the restriction on the above generalization. A common error is to apply the rule to the net equation for an overall reaction. *Do not!* It is valid *only* when applied to an elementary process, that is, a single step.

Activation energy and temperature dependence

In most cases the observed rate of a chemical reaction increases with increasing temperature, but the extent of the increase varies greatly among reactions. (According to an old rule of thumb, the rate of a reaction approximately doubles with each 10°C temperature rise. Unfortunately the rule is so approximate as to be of limited use.)

In terms of the rate law the reason for the variation of reaction rate with temperature is that the specific rate constant k varies as temperature changes. The relationship between the two was first found in 1887 by van't Hoff and in 1889, independently by Arrhenius. Arrhenius made an extensive study of its application to many reactions. The relationship, known as the *Arrhenius equation,* is

$$k = Ae^{-E_a/RT}$$

where A is called the *frequency factor* and E_a, R, and T all have their previously specified meanings. A comparison of the Arrhenius equation with the collision-theory equation derived above for the rate of a bimolecular process shows that

$$k = pe^{-E_a/RT}Z_0$$

According to the Arrhenius equation, the value of the specific rate constant k

increases as temperature increases. (Look at the equation: as *T increases, E_a/RT decreases,* which causes $-E_a/RT$ to *increase,* and this makes the whole right-hand side of the equation *increase.*) This means that an increase in temperature must produce an increase in the rate of a reaction, as is usually observed. Why should this be so? The answer is found in the fact that at any temperature there is a distribution, the *Maxwell-Boltzmann distribution,* of molecular kinetic energies in a substance, and at higher temperatures this distribution is shifted so that there are more fast molecules and fewer slow ones. Figure 14-6 shows the distribution of molecular kinetic energies at two different temperatures. The activation energy E_a is marked on the diagram, and it can be seen that the total fraction of the molecules which have E_a or higher (the shaded area under each curve) increases as the temperature rises.

The Arrhenius equation is useful because it expresses a quantitative relationship between temperature, activation energy, and rate constant. Perhaps its most valuable use is in determining the activation energy of a reaction from rate experiments at different temperatures. This is best done graphically: If we take the *natural* logarithms (log to the base *e*; see App. *D*) of both sides of the Arrhenius equation

$$k = Ae^{-E_a/RT}$$

we get

$$\ln k = \ln A - \frac{E_a}{RT}$$

where ln means "natural logarithm of." Since a natural logarithm can be converted into a *common* logarithm (log to the base 10) by the relationship

$$\ln x = 2.303 \log x$$

we can write

$$2.303 \log k = 2.303 \log A - \frac{E_a}{RT}$$

or

$$\log k = -\frac{E_a}{2.303R}\frac{1}{T} + \log A$$

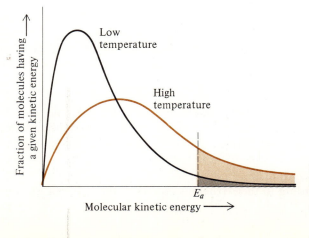

Figure 14-6
Effect of temperature on fraction of molecules having the activation energy.

This is in the form of an equation for a straight line, $y = mx + b$, where $y = \log k$, $m = E_a/2.303R$, $x = 1/T$, and $b = \log A$. Therefore, if we have values of the specific rate constant k at different temperatures, a plot of $\log k$ against $1/T$ should give us a straight line of slope $-E_a/2.303R$.

● **EXAMPLE 14-3**

Problem   The rate of decomposition of N_2O_5 was studied at a series of different temperatures. The following values of the specific rate constant were found:

t, °C	k, s^{-1}
0	7.86×10^{-7}
25	3.46×10^{-5}
35	1.35×10^{-4}
45	4.98×10^{-4}
55	1.50×10^{-3}
65	4.87×10^{-3}

Calculate the value of the activation energy for this reaction.

Solution   We must first find the reciprocal of the absolute temperature and the logarithm of each value of k. The results of these calculations are

t, °C	T, K	$\frac{1}{T}$, K^{-1}	k, s^{-1}	$\log k$
0	273	3.66×10^{-3}	7.86×10^{-7}	-6.10
25	298	3.36×10^{-3}	3.46×10^{-5}	-4.46
35	308	3.24×10^{-3}	1.35×10^{-4}	-3.87
45	318	3.14×10^{-3}	4.98×10^{-4}	-3.30
55	328	3.05×10^{-3}	1.50×10^{-3}	-2.82
65	338	2.96×10^{-3}	4.87×10^{-3}	-2.31

The result of plotting $\log k$ against $1/T$ is the graph shown in Fig. 14-7, usually known as an *Arrhenius plot*. The slope of the straight line is determined to be -5.39×10^3 K^{-1}. Since

$$\text{Slope} = -\frac{E_a}{2.303R}$$

$$E_a = (-2.303R)(\text{slope})$$
$$= -2.303(8.314 \text{ J K}^{-1} \text{ mol}^{-1})(-5.39 \times 10^3 \text{ K}^{-1})$$
$$= 1.03 \times 10^5 \text{ J mol}^{-1} \quad \text{or} \quad 103 \text{ kJ mol}^{-1} \ ●$$

Solving problems like the preceding one by using a graphical approach is a wise procedure, because experimental scatter of data is often "averaged out" in the process of drawing the best straight line through the data points, assuming that more than just two data points can be plotted. But when values of k are available at only two different temperatures, the graphical method is no better than the following analytical method.

Suppose that we know the rate constants k_1 and k_2 at temperatures T_1 and T_2, respectively. Then

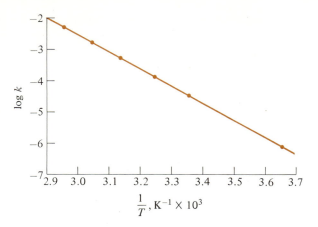

Figure 14-7
Arrhenius plot.

$$\log k_1 = -\frac{E_a}{2.303R}\frac{1}{T_1} + \log A$$

and

$$\log k_2 = -\frac{E_a}{2.303R}\frac{1}{T_2} + \log A$$

Subtracting the first equation from the second, we get

$$\log k_2 - \log k_1 = -\frac{E_a}{2.303R}\left(\frac{1}{T_2} - \frac{1}{T_1}\right)$$

or

$$\log\frac{k_2}{k_1} = \frac{E_a}{2.303R}\left(\frac{1}{T_1} - \frac{1}{T_2}\right)$$

Using this equation we can find any one of the five variables, k_1, k_2, T_1, T_2, and E_a if we know the other four.

● EXAMPLE 14-4

Problem   The specific rate constant for the combination of H_2 with I_2 to form HI is 0.0234 $M^{-1}s^{-1}$ at 400°C and 0.750 $M^{-1}s^{-1}$ at 500°C.   Calculate the activation energy for this reaction.

Solution   Let $k_2 = 0.750\ M^{-1}s^{-1}$ at 500 + 273, or 773, K; and let $k_1 = 0.0234\ M^{-1}s^{-1}$ at 400 + 273, or 673, K.   Then

$$\log\frac{k_2}{k_1} = \frac{E_a}{2.303R}\left(\frac{1}{T_1} - \frac{1}{T_2}\right)$$

$$E_a = \frac{2.303R\log(k_2/k_1)}{(1/T_1 - 1/T_2)}$$

$$= \frac{2.303(8.314\ \text{J K}^{-1}\ \text{mol}^{-1})\log[(0.750\ M^{-1}\ s^{-1})/(0.0234\ M^{-1}\ s^{-1})]}{(1/673\ \text{K} - 1/773\ \text{K})}$$

$$= 1.50 \times 10^5\ \text{J mol}^{-1}\qquad\text{or}\qquad 150\ \text{kJ mol}^{-1}\qquad\qquad ●$$

Reactions in   In some respects the kinetics of liquid-solution reactions are much like those of
liquid solution   gaseous reactions.   In nonpolar solvents, especially, reaction rates, frequency
factors, and activation energies are often similar to those for gas-phase reactions.

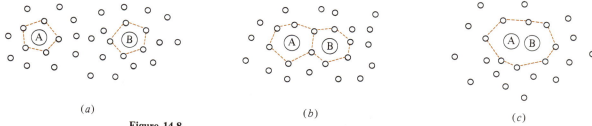

Figure 14-8
Cage effect. (*a*) A and B in separate cages. (*b*) Cages merge. (*c*) A and B in the same cage.

In many cases, however, the solvent plays a major role in the mechanism of a reaction, and so affects the rate. In liquid solution the rate may depend on the number of *encounters,* instead of collisions, which one reactant species has with another. Consider the sequence of events shown in Fig. 14-8. A solute species such as A or B in Fig. 14-8*a* may be considered to be surrounded by a *cage* of solvent molecules (dashed lines in the illustration). Hemmed in by the closeness of these molecules, the solute particle slowly diffuses through the solution with much jostling and bumping as it goes. This diffusion can be described as a movement of the solute particle from one cage to another. Eventually two reactant species may end up in the same cage (Fig. 14-8*c*). This is called an *encounter* and lasts much longer than a gas-phase collision. Once in the same cage the reactant particles may collide up to several hundred times before either reacting with or escaping from each other. Because of this effect, the *cage* effect, collisions between reactants are bunched up (in time) into encounters consisting of many collisions, but with long intervals between successive encounters. Elementary reactions with low steric (*p*) factors and large activation energies will have rates which depend on the number of collisions per second as in the gas phase. Reactions with high steric factors or low activation energies are different, however. In these cases one encounter provides so many opportunities for reaction that virtually every encounter leads to formation of products. These reactions occur as fast as reactant species can diffuse through the solvent to each other; they are thus called *diffusion-controlled reactions.*

14-4 The activated complex

When the reactant species in a bimolecular process collide with a favorable orientation and with energy at least equal to the activation energy, a highly unstable and hence short-lived composite particle is formed. This is referred to as the *activated complex,* or the *transition state,* of the reaction. The detailed study of the formation and decomposition of the activated complex is the focus of a theory called by several names: *transition-state theory, absolute-reaction-rate theory,* and *activated-complex theory.*

Transition-state theory

Let us again consider the hypothetical elementary process (Sec. 14-3),

$$A_2(g) + B_2(g) \longrightarrow 2AB(g)$$

The reactant molecules will form the activated complex $[A_2B_2]^{\ddagger}$ if they have sufficient energy and if they collide in a way which is geometrically favorable for

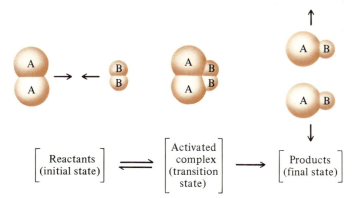

Figure 14-9
The activated complex.

$$\begin{bmatrix} \text{Reactants} \\ \text{(initial state)} \end{bmatrix} \rightleftharpoons \begin{bmatrix} \text{Activated} \\ \text{complex} \\ \text{(transition} \\ \text{state)} \end{bmatrix} \longrightarrow \begin{bmatrix} \text{Products} \\ \text{(final state)} \end{bmatrix}$$

forming the complex. Once the complex is formed, it may decompose to form the original reactants, or it may decompose to form the product AB molecules, instead. The entire process can be represented as

$$A_2(g) + B_2(g) \rightleftharpoons [A_2B_2]^{\ddagger} \longrightarrow 2AB(g)$$

Figure 14-9 represents this process. It should be kept in mind that the activated complex is not a stable molecule, and so it exists for only an instant before flying apart one way or the other. It flies apart because its potential energy is higher than that of either the reactant or the product molecules. (Whenever two particles collide, the potential energy of the pair is higher at the instant of collision than either before or after. Some or all of the kinetic energy of the moving particles is converted to potential energy during the short interval of collision and then back to kinetic energy as the particles fly apart.)

Figure 14-10a shows how the potential energy of the reacting system changes during the course of the reaction. (Position on the horizontal axis measures degree of progress of the reaction.) The rapidly moving A_2 and B_2 molecules collide, and the potential energy of the combination rises up to the top of the potential-energy "hill." E_a represents the *increase in potential energy* associated with the formation of $[A_2B_2]^{\ddagger}$, the activated complex, and so it must also equal the *kinetic energy*

Figure 14-10
Potential-energy changes during the course of a reaction. (a) Exothermic reaction. (b) Endothermic reaction.

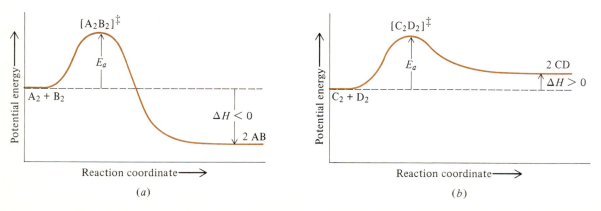

(a)   (b)

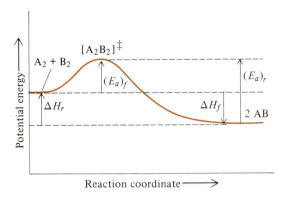

Figure 14-11
Activation energies for forward and reverse reactions.

which A_2 and B_2 must have in order to get to the "summit." After this, it is
"downhill all the way," either back to the reactant A_2 and B_2 molecules or, if the
A—A and B—B bonds break, on to the product AB molecules. In Fig. 14-10a the
potential energy of the products is less than that of the reactants, as shown; so
there is a net release of energy, that is, ΔH for the reaction is negative. In the
endothermic case, Fig. 14-10b, the potential energy of the products is greater than
that of the reactants; so $\Delta H_{\text{reaction}}$ is positive.

If A_2 and B_2 can react to form 2AB, then two AB molecules can react to form A_2
and B_2. The relationships among the activation energies and heats of reaction for
forward (f) and reverse (r) reactions are shown in Fig. 14-11.

Potential-energy diagrams like those in Fig. 14-10 are helpful in visualizing why
it is that highly exothermic reactions can be very slow. There is clearly no direct
correlation between E_a and $\Delta H_{\text{reaction}}$. If E_a is large, then colliding molecules
need to have considerable energy in order to get over the potential-energy hill, and
only a small fraction may actually have this energy (depending on the temperature). Such a reaction may be strongly exothermic; if the descent from the
potential-energy maximum is greater than the ascent, the energy released is equal
to the difference between E_a and the energy released when $[A_2B_2]^{\ddagger}$ forms the
reaction products.

14-5 Reaction mechanisms

As more kinetic knowledge is accumulated, it appears that the majority of reactions are complex; that is, they have multistep mechanisms. Figuring out the
correct mechanism from experimental data is often not an easy task. It may not
be difficult to arrive at a mechanism which is consistent with both the rate law and
the stoichiometry of the balanced net equation for the reaction. The problem is
that it is often possible to figure out several such mechanisms, and designing
experiments which will distinguish among reasonable alternatives is often a
challenge.

Stepwise reactions

For a complex reaction which has a mechanism consisting of an uncomplicated
sequence of steps, it is often found that one of these steps is considerably slower
than any of the others. In such a case the observed rate of the (overall) reaction is
determined by the rate of the slow step.[2] The slow step in a mechanism is called
the *rate-determining step* (or *rate-limiting step*). The following is an example.

[2]This is often called the *bottleneck principle*.

Step 1: $\qquad\qquad$ A + B $\xrightarrow{k_1}$ C + I $\quad$ (slow)

Step 2: $\qquad\qquad$ A + I $\xrightarrow{k_2}$ D $\qquad$ (fast)

where k_1 and k_2 are the respective rate constants. I is an intermediate, a species which is formed in step 1 but is used up in step 2. The net equation is found by adding the two steps

$$2A + B \longrightarrow C + D$$

The rate of formation of products is simply the rate of the first step, which is rate-determining:

$$\text{Rate} = \frac{d[D]}{dt} = k_1[A][B]$$

What is the effect on the rate law if the sequence of steps is the same, but the second step is rate-determining?

Step 1: $\qquad\qquad$ A + B $\xrightarrow{k_1}$ C + I $\quad$ (fast)

Step 2: $\qquad\qquad$ A + I $\xrightarrow{k_2}$ D $\qquad$ (slow)

In this case the rate of the overall reaction is the rate of the second step:

$$\text{Rate} = k_2[A][I]$$

But rate laws for overall reactions are *not* written in terms of concentrations of intermediates. What we need to do, therefore, is to use a combination of chemistry and algebra, so that we can eliminate [I] from the rate law. This can be done if we recognize that the first step in the mechanism, being fast, will begin to reverse itself soon after the start of the reaction, before step 2 has had a chance to form much of its product. Thus we should rewrite the mechanism as

Step 1a: $\qquad\qquad$ A + B $\xrightarrow{k_{1a}}$ C + I $\quad$ (fast)

Step 1b: $\qquad\qquad$ C + I $\xrightarrow{k_{1b}}$ A + B $\quad$ (fast)

Step 2: $\qquad\qquad$ A + I $\xrightarrow{k_2}$ D $\qquad$ (slow)

where steps 1a and 1b are the forward and reverse reactions of (old) step 1. For these three steps the respective rate laws are

$$\text{Rate}_{1a} = k_{1a}[A][B]$$

$$\text{Rate}_{1b} = k_{1b}[C][I]$$

$$\text{Rate}_2 = k_2[A][I]$$

If we now assume that steps 1a and 1b are equal in rate (they are at equilibrium), then we can write

$$k_{1a}[A][B] = k_{1b}[C][I]$$

and solving for [I], we have

$$[I] = \frac{k_{1a}[A][B]}{k_{1b}[C]}$$

Substituting this in the rate law for step 2, we get

$$Rate_2 = k_2[A][I]$$

$$= k_2[A]\left(\frac{k_{1a}[A][B]}{k_{1b}[C]}\right) = \left(\frac{k_{1a}k_2}{k_{1b}}\right)\frac{[A]^2[B]}{[C]} = k_{expt}\frac{[A]^2[B]}{[C]}$$

Here k_{expt} is the rate constant which would be obtained from an experimental study of the rate-concentration dependence. It can be seen to be a composite of the k's for the individual steps. The above example shows how a relatively complex rate law may result from a comparatively simple mechanism.

Figuring out reaction mechanisms often involves a process of trial and error. What is usually done is to make an educated guess and then see if the proposed mechanism is consistent with (1) the net equation for the overall reaction and (2) the experimentally determined rate law. Unfortunately, just because a mechanism meets both of the criteria, it does not necessarily mean that it is the correct one, as there may be other mechanisms which "work" as well. In kinetics it may be possible to prove a mechanism incorrect, but it is virtually impossible to prove one correct. In fact, as experimental knowledge grows and theoretical awareness increases, accepted simple mechanisms seem to be replaced by increasingly complex ones. There is an old adage quoted by chemical kineticists: "The only simple reactions are the ones which need more study!"

Chain reactions   Many mechanisms do not consist of an ordered sequence of steps, one of which is rate-determining, but rather they are more complex. One of the best examples of this is the *chain reaction*, a classic example of which is the reaction

$$H_2(g) + Br_2(g) \longrightarrow 2HBr(g)$$

The steps in the mechanism for this reaction have been found to be

Step 1:	$Br_2 \xrightarrow{k_1} 2Br$	(chain initiation)
Step 2:	$Br + H_2 \xrightarrow{k_2} HBr + H$	(chain propagation)
Step 3:	$H + Br_2 \xrightarrow{k_3} HBr + Br$	
Step 4:	$H + HBr \xrightarrow{k_4} H_2 + Br$	(chain inhibition)
Step 5:	$2Br \xrightarrow{k_5} Br_2$	(chain termination)

In step 1, called *chain initiation*, a Br_2 molecule dissociates to form two Br atoms, which are highly reactive intermediates. A Br atom then attacks an H_2 molecule (step 2) to form a product HBr molecule and an H atom, another reactive intermediate. The H atom then attacks a Br_2 molecule (step 3) to form another HBr molecule and another Br atom. Notice that one product of step 3 (Br) is a reactant for step 2, and one product of step 2 (H) is a reactant for step 3. Thus steps 2 and 3 can repeat again and again in the sequence 2, 3, 2, 3, $\cdots$. These steps are called the *chain-propagation* steps.

The above mechanism is called a *chain reaction* because a single pair of Br atoms from step 1 can initiate a chain of events (steps 2 and 3) leading to the production of many HBr molecules. In this mechanism, H and Br atoms are called *chain carriers*. Step 4 removes a product HBr molecule, and so it is called a *chain-inhibition* step. Step 5 (the reverse of step 1) removes Br chain carriers from the reaction mixture, and so it is called a *chain-termination* step. The rate law for this reaction is complicated. It has been found to be

$$\frac{d[\text{HBr}]}{dt} = \frac{2k_2(k_1/k_5)^{1/2}[\text{H}_2][\text{Br}_2]^{1/2}}{1 + (k_4/k_5)[\text{HBr}]/[\text{Br}_2]}$$

In some chain reactions a chain-propagation step produces more chain carriers than it consumes. This is known as *chain branching*, and the resulting increase in the concentration of chain carriers in the reacting mixture sends the rate sky-high. It is believed that some explosions are branching-chain reactions.[3] In hydrogen-oxygen explosions, for example, chain-branching steps are believed to include

$$\text{H} + \text{O}_2 \longrightarrow \text{OH} + \text{O}$$

and

$$\text{O} + \text{H}_2 \longrightarrow \text{OH} + \text{H}$$

Even though much work has been done on the kinetics of explosions, they are difficult to investigate and remain imperfectly understood.

14-6
Catalysis

Catalysts, rates, and mechanisms

The rates of many reactions are increased by the presence of a *catalyst,* a substance which increases the rate of a reaction without being consumed by it. This is possible because it is used up in one step in the reaction mechanism and regenerated in a later step. A catalyst acts by making available a new reaction mechanism with a lower activation energy. Figure 14-12 shows the uncatalyzed path of a reaction contrasted with a catalyzed path. (Each potential-energy maximum corresponds to the formation of an activated complex.) Note that ΔH for the reaction is independent of reaction mechanism, and depends only upon the reactants and products. However, the activation energy for the catalyzed path is lower than that for the uncatalyzed path. Thus at any given temperature, more reactant molecules possess the activation energy for the catalyzed reaction than for the uncatalyzed one. A catalyst does not eliminate a reaction mechanism; it offers a new, faster one. More molecules, often essentially all, will follow the new mechanism instead of the old.

In *homogeneous catalysis* the catalyst is in solution along with the reactants. Consider the elementary process

$$\text{A} + \text{B} \longrightarrow \text{products} \qquad \text{(slow)}$$

Assume that this process has a high activation energy. If we now add catalyst C to the reaction mixture, a new, two-step mechanism is possible:

Step 1: $\text{A} + \text{C} \longrightarrow \text{AC}$ (fast)
Step 2: $\text{AC} + \text{B} \longrightarrow \text{products} + \text{C}$ (faster)

---

[3] Another type of explosion is the thermal explosion, in which the heat liberated by a highly exothermic reaction cannot be dissipated fast enough to prevent a sudden temperature rise and consequent increase in reaction rate.

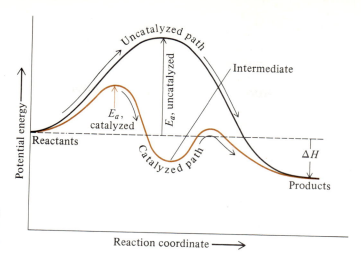

Figure 14-12
Lowering of activation
energy by a catalyst.

is which both activation energies are low, and each reaction is faster than the original, uncatalyzed reaction. Notice that the overall net equation is unchanged and that while C is used in step 1, it is regenerated in step 2. The rate law for the uncatalyzed reaction is

$$\text{Rate} = k[A][B]$$

and for the catalyzed reaction it is

$$\text{Rate} = k'[A][C]$$

An example of homogeneous catalysis is found in the oxidation of sulfur dioxide to sulfur trioxide by oxygen, nitric oxide being the catalyst. The net equation for the reaction is

$$2SO_2(g) + O_2(g) \longrightarrow 2SO_3(g)$$

The uncatalyzed reaction is very slow either because it is termolecular (unlikely) or because one step in its reaction mechanism has a very high activation energy.

Addition of nitric oxide, NO, to the mixture greatly speeds the reaction by providing the mechanism

$$O_2(g) + 2NO(g) \longrightarrow 2NO_2(g)$$

$$[NO_2(g) + SO_2(g) \longrightarrow NO(g) + SO_3(g)] \times 2$$

The sum of these gives the original net equation, and each step is a reasonably fast process.

Heterogeneous
catalysis

A *heterogeneous catalyst* is one which provides a *surface* on which molecules can combine more readily. Heterogeneous catalysis begins with the *ad*sorption of a molecule on the surface of a catalyst. There are two general types of adsorption: the relatively weak *physical,* or van der Waals, adsorption and the stronger *chemisorption.* Evidence that the chemisorbed molecule is relatively strongly bonded at the surface comes from the fact that considerably more heat is evolved during chemisorption than during physical adsorption.

Chemisorption is involved in surface catalysis and apparently takes place preferentially at certain sites on the surface, called *active sites,* or *active centers.* The nature of these is not entirely understood; they may be related to surface defects and emergences of dislocations (Sec. 11-3). In any event, the chemisorbed molecule is somehow changed at the active site so that it can more easily react with another molecule. There is evidence that some molecules become dissociated into highly reactive fragments. On some metal surfaces hydrogen, for example, may be dissociated into atoms and so can react more rapidly. The reaction of ethylene with hydrogen is thought to be surface-catalyzed by nickel metal in this way:

$$H_2(g) + C_2H_4(g) \longrightarrow C_2H_6(g)$$

Enzymes   Enzymes are very large, complex protein molecules which act as catalysts for reactions in biological systems. Like heterogeneous catalysts, enzyme molecules are believed to have *active sites* on their surfaces. These sites are capable of bonding only to certain very specific molecules, called *substrates.* An active site is often a depression or cavity on the surface of the enzyme molecule into which a substrate molecule, or part of it, can fit. For years the *lock-and-key* (enzyme-and-substrate) theory was used to explain the specificity of catalysis by enzymes. According to this theory, the active site is a cavity of fixed conformation (shape), and only a specific substrate molecule will fit into it properly to undergo reaction. A more recent proposal, the *induced-fit* theory, suggests that in some cases, at least, the cavity changes conformation as it accepts the substrate molecule. (The substrate may also change conformation.) Figure 14-13 schematically depicts how an active site on an enzyme accepts a substrate molecule (*a*) to form the enzyme-substrate complex (*b*), which produces the reaction products (*c*). (*Note:* The drawings in Fig. 14-13 are *very* schematic.)

Enzymes are amazingly efficient catalysts. A measure of this efficiency is the *turnover number,* the number of substrate molecules which one enzyme molecule converts into products per unit time. The turnover number of the enzyme carbonic anhydrase, found in all animal organisms, is about 6×10^5 s^{-1}. This means that one molecule of carbonic anhydrase can produce 6×10^5 molecules of H_2CO_3 (from CO_2 and H_2O) per second.

Figure 14-13
Schematic representation of an enzyme-catalyzed reaction. (*a*) **Substrate molecule approaches enzyme.** (*b*) **Enzyme-substrate complex is formed.** (*c*) **Reaction products leave enzyme.**

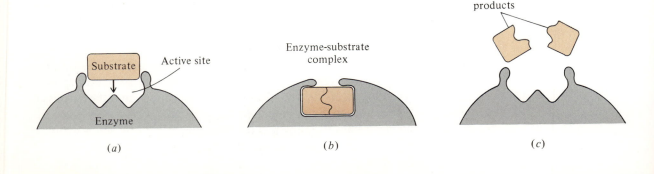

Inhibitors   *Inhibitors,* also sometimes called *negative catalysts,* are substances which, when added to a reaction mixture, slow down the reaction. Inhibitors can act in a number of ways. One kind of inhibition occurs when the added substance combines with a potential catalyst, rendering it inactive and thus slowing the rate. For example, inhibition of a surface-catalyzed reaction can occur when foreign molecules bond at the active sites, blocking them for substrate molecules. Such inhibition is frequently called *poisoning* and the inhibitor, a *poison.* Another inhibition mechanism occurs in the case of chain reactions. Here the inhibitor combines with a chain carrier, breaks the chain, and thus slows the rate.

SUMMARY

The focus of this chapter has been on the *rates* and *mechanisms* of chemical reactions. Rates of reactions are often difficult to determine, because they require the measurement of a changing quantity: concentration. *Instrumental methods* are often useful; they generally measure a change in concentration by a change in some physical property, such as pressure, light absorption, conductivity, etc.

Reaction rates are usually found to depend on the concentrations of various species in the reaction mixture. These usually include one or more reactants, but they may also include products or other substances. The *rate law* expresses the nature of the proportionality between the rate and these concentrations. The *order of a reaction* is the sum of the exponents on the concentration terms in the rate law. *A rate law for an overall reaction cannot be inferred from the net equation, only from experimental rate data.*

The mechanism of a reaction often consists of a series of individual steps called *elementary processes.* These can be treated theoretically by means of collision theory, which postulates that the rate of an elementary process depends upon (1) the number of collisions of reactant molecules per second, (2) the fraction of those which have the *activation energy,* the energy required for "successful" rearrangement of bonds, and (3) the fraction of those collisions which are geometrically oriented so as to produce products (the *steric factor*). The *molecularity* of an elementary process is the number of reactant molecules which collide in that step; molecularities can be one, two, or occasionally, three. An *intermediate* is a species which is formed in one step

and used up in a subsequent step of the reaction mechanism.

The activation energy of a reaction determines its temperature dependence. It is the energy which colliding reactant molecules must have in order to form the *activated complex,* an unstable collection of atoms which are weakly bonded together and which can decompose into either reactant molecules or products. The magnitude of the activation energy can be determined from an *Arrhenius plot.*

Transition-state theory focuses on the nature of the activated complex, the way it is formed, and the way it decomposes. The path of a reaction can be followed by plotting the potential energy of the system as it passes from reactants to products.

A *reaction mechanism* must be consistent both with the stoichiometry of the balanced net equation and with the experimentally determined rate law. Many reactions occur by a sequential series of steps, one of which is slower than the rest. This, the *rate-determining step,* acts as a bottleneck and limits the rate of the overall reaction. Some mechanisms are more complex and have no single rate-determining step; *chain reactions* are good examples of these.

Catalysts increase the rates of reactions by providing alternative mechanisms with lower activation energies. A catalyst is not consumed in a reaction, although it is used up in an early step and is regenerated in a later one. *Heterogeneous catalysts* are surfaces with *active sites* which bind reactant molecules in a more reactive state. Similar to these are enzymes, biochemical catalysts, which are large protein molecules.

KEY TERMS

Activated complex (Sec. 14-4)
Activation energy (Secs. 14-3, 14-4)
Active site (Sec. 14-6)

Arrhenius equation (Sec. 14-3)
Arrhenius plot (Sec. 14-3)
Average rate (Sec. 14-1)

Bimolecular process (Sec. 14-3)
Cage effect (Sec. 14-3)
Catalyst (Sec. 14-6)
Chain branching (Sec. 14-5)
Chain reaction (Sec. 14-5)
Chemisorption (Sec. 14-6)
Collision frequency (Sec. 14-3)
Collision theory (Sec. 14-3)
Diffusion-controlled reaction (Sec. 14-3)
Elementary process (Sec. 14-3)
Encounter (Sec. 14-3)
Enzyme (Sec. 14-6)
First-order reaction (Sec. 14-2)
Frequency factor (Sec. 14-3)
Half-life (Sec. 14-2)
Heterogeneous catalysis (Sec. 14-6)
Homogeneous catalysis (Sec. 14-6)
Initial-rate method (Sec. 14-2)
Inhibitor (Sec. 14-6)

Instantaneous rate (Sec. 14-1)
Mechanism (Sec. 14-1, 14-5)
Molecularity (Sec. 14-3)
Physical adsorption (Sec. 14-6)
Rate-determining step (Sec. 14-5)
Rate law (Sec. 14-2)
Reaction order (Sec. 14-2)
Reaction rate (Sec. 14-1)
Second-order reaction (Sec. 14-2)
Specific rate constant (Sec. 14-2)
Steric factor (Sec. 14-3)
Substrate (Sec. 14-6)
Surface catalysis (Sec. 14-6)
Third-order reaction (Sec. 14-2)
Transition-state theory (Sec. 14-4)
Termolecular process (Sec. 14-3)
Unimolecular process (Sec. 14-3)
Zero-order reaction (Sec. 14-2)

QUESTIONS AND PROBLEMS

Reaction rates

14-1 Sketch a curve showing how the concentration of a reactant might decrease with time. Using your curve show the distinction between average rate and instantaneous rate.

14-2 List the factors which determine the rate of a reaction.

14-3 Write an expression which symbolizes each of the following instantaneous rates: **(a)** the rate of decrease of H_2 **(b)** the rate of increase of ClO^-.

14-4 What is the fastest chemical reaction you can think of? What is the slowest? How would you go about designing an experiment to measure the rate of each?

14-5 If $-d[N_2]/dt$ for the gas-phase reaction $N_2 + 3H_2 \longrightarrow 2NH_3$ is $2.60 \times 10^{-3}\ M\,s^{-1}$, what is $-d[H_2]/dt$?

14-6 What is $d[NH_3]/dt$ for the reaction described in Prob. 14-5, above?

14-7 In one experiment the partial pressure of N_2O_5 decreased by 34.0 mmHg in 1 min as a result of the gas-phase reaction $2N_2O_5 \longrightarrow 4NO_2 + O_2$. What was the change in the total pressure during this interval?

14-8 The rate of decrease in [A] in a reaction was measured as follows:

Time, min	0.0	20.0	40.0	60.0	80.0	100.0
[A], M	1.00	0.819	0.670	0.549	0.449	0.368

Calculate the average rate of the reaction $(-\Delta[A]/\Delta t)$ between **(a)** 40.0 and 60.0 min **(b)** 20.0 and 80.0 min **(c)** 0.0 and 100.0 min.

14-19 From the data given in Prob. 14-8, find the instantaneous rate of the reaction $-d[A]/dt$ at **(a)** 50.0 min **(b)** 0.0 min.

14-10 Under what conditions are average rates essentially the same as instantaneous rates?

Rate laws

14-11 Distinguish clearly between reaction rate, rate law, and specific rate constant.

14-12 A certain complex reaction is first order in A, second order in B, and third order in C. What will be the effect on the rate of the reaction of doubling **(a)** A **(b)** B **(c)** C?

14-13 A certain reaction is one-half order in D, three-halves order in E, and zero order in F. What will be the effect on the rate of the reaction of doubling **(a)** D **(b)** E **(c)** F?

14-14 For the reaction described by each of the following rate laws indicate the order with respect to each species and the overall reaction order: **(a)** Rate = $k[A][B]^2$ **(b)** Rate = $k[A]^2$ **(c)** Rate = $k[A][B]/[C]$.

14-15 The following initial-rate data were obtained for the reaction $2A + B \longrightarrow C + 3D$:

Initial [A], M	Initial [B], M	Initial $-\dfrac{d[A]}{dt}$, $M\,s^{-1}$
0.127	0.346	1.64×10^{-6}
0.254	0.346	3.28×10^{-6}
0.254	0.692	1.31×10^{-5}

(a) Write the rate law for the reaction. **(b)** Calculate the value of the rate constant. **(c)** Calculate the rate of disappearance of A if $[A] = 0.100\ M$ and $[B] = 0.200\ M$. **(d)** Calculate the rate of formation of D under the conditions of **(c)**.

14-16 The following initial-rate data were obtained for the reaction $A + 2B \longrightarrow 3C + 4D$:

Initial [A], M	Initial [B], M	Initial [X], M	Initial $-\dfrac{d[A]}{dt}$, $M\,s^{-1}$
0.671	0.238	0.127	1.41×10^{-3}
0.839	0.238	0.127	1.41×10^{-3}
0.421	0.476	0.127	5.64×10^{-3}
0.911	0.238	0.254	2.82×10^{-3}

(a) Write the rate law for the reaction. **(b)** Evaluate the specific rate constant. **(c)** Calculate $-d[A]/dt$ if $[A] = [B] = [X] = 0.500\ M$. **(d)** Calculate $d[C]/dt$ if $[A] = [B] = [X] = 0.200\ M$. **(e)** Calculate $d[X]/dt$ if $[A] = [B] = [X] = 0.100\ M$.

14-17 The following initial-rate data were obtained for the reaction $A + B \longrightarrow C$:

Initial [A], M	Initial [B], M	Initial $-\dfrac{d[A]}{dt}$, $M\,s^{-1}$
0.245	0.128	1.46×10^{-4}
0.490	0.128	2.92×10^{-4}
0.735	0.256	8.76×10^{-4}

Write the rate law for the reaction.

14-18 Write the rate law and calculate the rate constant for reaction $A + B \longrightarrow C$ using the following initial-rate data:

Initial [A], M	Initial [B], M	Initial $-\dfrac{d[A]}{dt}$, $M\,s^{-1}$
0.395	0.284	1.67×10^{-5}
0.482	0.284	2.04×10^{-5}
0.482	0.482	5.88×10^{-5}

14-19 Write the rate law and calculate the rate constant for the reaction $A + B \longrightarrow C$ from the following initial-rate data:

Initial [A], M	Initial [B], M	Initial [C], M	Initial $-\dfrac{d[A]}{dt}$, $M\,s^{-1}$
0.918	0.216	0.712	1.46×10^{-4}
0.621	0.216	0.712	9.88×10^{-5}
0.420	0.719	0.712	6.68×10^{-5}
0.514	0.319	0.448	1.30×10^{-4}

14-20 The following data were obtained at 320°C for the reaction $SO_2Cl_2 \longrightarrow SO_2 + Cl_2$:

Time, h	0.00	1.00	2.00	3.00	4.00
$[SO_2Cl_2]$, M	1.200	1.109	1.024	0.946	0.874

Using a graphical method determine the order of the reaction and the specific rate constant at 320°C.

14-21 The thermal decomposition of methyl ether proceeds according to the equation

$$(CH_3)_2O(g) \longrightarrow CH_4(g) + H_2(g) + CO(g)$$

A sample of methyl ether was placed in a container and quickly heated to 504°C. During the reaction the following changes in the total pressure took place:

Time, min	0.00	6.50	12.95	19.92	52.58
Total pressure, mmHG	312	408	488	562	779

Calculate the order of the reaction and the specific rate constant at 504°C.

14-22 At 310°C the thermal decomposition of arsine occurs according to the equation

$$2AsH_3(g) \longrightarrow 2As(s) + 3H_2(g)$$

At this temperature the following data were obtained:

Time, h	0.0	3.0	4.0	5.0
$[AsH_3]$, M	0.0216	0.0164	0.0151	0.0137
Time, h	6.0	7.0	8.0	
$[AsH_3]$, M	0.0126	0.0115	0.0105	

(a) What is the order of the reaction? **(b)** What is the value of the specific rate constant?

14-23 The decomposition of N_2O_5 in CCl_4 solution to form NO_2 and O_2 was studied at 30°C. From the following results:

Time, min	0	80	160	240	320
$[N_2O_5]$, M	0.170	0.114	0.078	0.053	0.036

find the order of the reaction and evaluate the specific rate constant.

14-24 The reaction $2HI \longrightarrow H_2 + I_2$ was studied at 600 K. The following data were obtained:

Time, h	0.0	1.0	2.0	3.0	4.0	5.0
$[HI]$, M	3.95	3.74	3.55	3.37	3.22	3.08

Write the rate law as the rate of disappearance of HI and calculate the value of the rate constant at 600 K.

14-25 What is the half-life of a first-order reaction for which $k = 1.4 \times 10^{-2}$ min^{-1}?

14-26 The half-life of a first-order reaction is independent of initial concentration. How does the half-life of a second-order reaction of the type $2A \longrightarrow$ products depend on initial concentration? How about a zero-order reaction?

14-27 The gas-phase decomposition of N_2O_3 to form NO_2 and NO is first-order with $k_1 = 3.2 \times 10^{-4}$ s^{-1}. In a reaction in which the initial $[N_2O_3]$ is 1.00 M, how long would it take this concentration to be reduced to 0.125 M?

14-28 The gas-phase decomposition of acetaldehyde, $CH_3CHO \longrightarrow CH_4 + CO$, is under certain conditions a second-order reaction with $k_2 = 0.25$ M^{-1} s^{-1}. How long would it take $[CH_3CHO]$ to decrease from 0.0300 to 0.0100 M?

14-29 The first-order decomposition of hydrogen peroxide, $2H_2O_2 \longrightarrow 2H_2O + O_2(g)$, has a rate constant equal to 2.25×10^{-6} s^{-1} at a certain temperature. In a given solution $[H_2O_2] = 0.800$ M. **(a)** What will the H_2O_2 concentration be after 1.00 day? **(b)** How long will it take the H_2O_2 concentration to decrease to 0.750 M?

14-30 The half-life of the gas-phase reaction $A \longrightarrow 2B$ is 35 min. Enough A is placed in a container so that its pressure is 725 mmHg. After 140 min, what is **(a)** the partial pressure of A **(b)** the total pressure?

Collision theory

14-31 How does molecularity differ from reaction order? What would you need to know in order to predict reaction order from molecularity?

14-32 What is a unimolecular process? Can a molecule ever undergo reaction without colliding with another? How?

14-33 For a bimolecular process what three factors determine the rate of formation of products?

14-34 Rank the following bimolecular processes in order of decreasing steric factor, p:
(a) $O_3 + NO \longrightarrow NO_2 + O_2$
(b) $CH_3 + CH_3 \longrightarrow C_2H_6$
(c) $I^+ + I^- \longrightarrow I_2$
(d) $CH_3CH_2CH_2COOH + HOCH_3 \longrightarrow$
$\qquad\qquad\qquad CH_3CH_2CH_2COOCH_3 + H_2O$

14-35 Account for the fact that unimolecular reactions follow first-order rate laws at higher pressure but second-order rate laws at lower ones.

14-36 Why are termolecular reations rare?

Temperature dependence

14-37 Describe activation energy **(a)** from a theoretical standpoint **(b)** from an experimental standpoint.

14-38 If the rate of a reaction is approximately doubled by a 10°C increase in temperature, what must the approximate activation energy be if the lower temperature is **(a)** 25°C **(b)** 500°C?

14-39 A certain reaction has $E_a = 146$ kJ mol^{-1}. If the specific rate constant is 4.25×10^{-4} s^{-1} at 25.0°C, what is its value at 100.0°C?

14-40 If E_a for a reaction is 198 kJ mol^{-1} and $k = 5.00 \times 10^{-6}$ s^{-1} at 25°C, at what temperature will k be equal to 5.00×10^{-5} s^{-1}?

14-41 What kinds of elementary processes would you expect to have very low or zero activation energies?

14-42 Can an elementary process have a negative activation energy? Explain.

14-43 A few reactions decrease in rate with increasing temperature. What does this mean about the experimentally determined activation energy? How can this behavior be accounted for in terms of the mechanisms of such reactions?

14-44 A certain reaction shows an experimental activation energy of 50 kJ mol^{-1} at 100°C and 200 kJ mol^{-1} at 25°C. How can the difference in activation energies be accounted for?

14-45 The value of the specific rate constant for the decomposition of NO_2 is 0.755 M^{-1} s^{-1} at 330°C and 4.02 M^{-1} s^{-1} at 378°C. What is the activation energy for this reaction?

14-46 The decomposition of N_2O_5 occurs with a rate constant of 4.87×10^{-3} s^{-1} at 65°C and 3.38×10^{-5} s^{-1} at 25°C. Calculate the activation energy for the reaction.

Transition-state theory

14-47 What is an activated complex? What are the two general ways in which an activated complex can decompose?

14-48 What relationship exists among the activation energies for forward and reverse reactions and ΔH for an elementary process?

14-49 In a reaction which occurs by a stepwise mechanism, what relation is there between E_a for the overall reaction and E_a for the rate-determining step?

14-50 Draw a potential energy versus reaction coordinate curve for a two-step mechanism in which the activation energy of the second step is greater than that for the first step. Indicate on your drawing the activation energy for the overall reaction.

14-51 Repeat Prob. 14-50 for the case in which the second step has the lower activation energy.

Reaction mechanisms

14-52 Write the rate law for each of the following *elementary* processes:
(a) $X + Y \longrightarrow$ products
(b) $2X \longrightarrow$ products
(c) $2X + Y \longrightarrow$ products
(d) $X \longrightarrow$ products

14-53 For each of the following mechanisms write the net equation and the rate law for the overall reaction:

(a) (1) $X + Y \xrightarrow{k_1} Z + I$ (slow)

    (2) $I + Y \xrightarrow{k_2} A$ (fast)

(b) (1) $A \xrightarrow{k_1} B + C$ (slow)

    (2) $C + D \xrightarrow{k_2} E$ (fast)

(c) (1) $A \underset{k_{1b}}{\overset{k_{1a}}{\rightleftharpoons}} B + C$ (fast equilibrium)

    (2) $B + D \xrightarrow{k_2} E$ (slow)

(d) (1) $A + B \underset{k_{1b}}{\overset{k_{1a}}{\rightleftharpoons}} B + C$ (fast equilibrium)

    (2) $C + B \xrightarrow{k_2} D$ (slow)

    (3) $D + B \xrightarrow{k_3} A + E$ (fast)

14-54 In a sequential (stepwise) mechanism, how do the steps which follow the rate-determining step affect the overall rate?

14-55 What is a chain reaction? Describe the role of chain carriers in a chain reaction. What is a branching-chain reaction?

14-56 Write the rate law for the $H_2 + Br_2$ reaction assuming that the reactants have just been mixed, and very little HBr has been formed. What is the overall order of this reaction?

14-57 Explain why it is that the catalyst shows up in the rate law for a catalyzed reaction.

14-58 If a catalyst lowers the activation energy of a reaction, does a negative catalyst (inhibitor) raise it? Explain.

General

14-59 The rate law for a certain reaction consists of two separate terms which are *added* together. What does this tell you about the mechanism of the reaction?

14-60 The thermal decomposition of potassium chlorate, $KClO_3$, to form KCl and O_2 is catalyzed by solid manganese dioxide. How would you go about trying to find out if this catalysis is homogeneous or heterogeneous?

14-61 In the fourth step ($H + HBr \longrightarrow H_2 + Br$) of the H_2–Br_2 mechanism, one chain carrier produces another; so the chain is not terminated. Why is the term *inhibition* used here?

14-62 Comment on the old adage: "The only simple reactions are the ones which need more study!"

14-63 What general feature would the rate law for an *autocatalytic*, or *self-catalytic, reaction* have?

14-64 Why must a candle be lit in order to make it burn? After it is lit, why does it keep on burning?

14-65 If a reaction is zero-order in one of its reactants, this means that the rate of the reaction is independent of the concentration of that reactant. How can this ever be the case, considering that if that reactant is removed from the reaction mixture altogether, the reaction must stop?

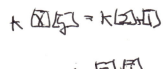

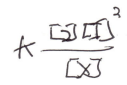

15

CHEMICAL EQUILIBRIUM

TO THE STUDENT

In earlier chapters we considered several kinds of systems at equilibrium, for example, a liquid in equilibrium with its vapor (Chap. 10), a saturated solution in equilibrium with excess solute (Chap. 12), and a weak electrolyte in equilibrium with its ions in solution (Chap. 12). Each of these is a typical dynamic equilibrium, one in which two opposing processes occur at exactly the same rate, one process exactly counteracting the other.

All chemical reactions tend toward equilibrium, although this is not always apparent. When a reaction reaches equilibrium, the quantity of at least one remaining reactant is sometimes so small that it is almost unmeasurable. In this case we usually say that the substance has been used up, because for practical purposes it has. (The reaction has gone essentially *to completion.*) But strictly speaking there is no such thing as a reaction which uses up *all* reactants, that is, which goes 100 percent to completion. All reacting systems end up in a state of equilibrium in which at least small amounts of reactants are left. Even when a 2:1 molar mixture of H_2 and O_2 gases is exploded, the reaction

$$2H_2(g) + O_2(g) \longrightarrow 2H_2O(g)$$

proceeds (rapidly!) to a state of equilibrium

$$2H_2(g) + O_2(g) \rightleftharpoons 2H_2O(g)$$

in which small quantities of H_2 and O_2 remain. So each time you hear the expression "goes to completion," remember to qualify it in your mind.

15-1
Homogeneous chemical equilibrium

Before studying the quantitative aspects of equilibrium, you should try to gain a qualitative feeling for how the equilibrium state can be approached and how it responds to disturbances.

The equilibrium
state

Consider the hypothetical reaction

$$A(g) + B(g) \longrightarrow C(g) + D(g)$$

Suppose that some A and some B are in a container and that we have available a device which allows us to follow the course of the reaction. What we observe is summarized in Fig. 15-1, which shows how the concentrations of all species change with the passage of time. The reaction starts at t_0. By time t_1, the concentrations of A and B have decreased and those of C and D have increased. (At all times [C] equals [D] because they are formed in a 1:1 molar ratio, as the equation shows.) By time t_2, [A] and [B] have decreased and [C] and [D] have increased still further, but their rate of change is now less than it was at the start of the reaction. After time t_3, there is essentially no further change in any concentration.

At time t_0, only the forward reaction can take place:

$$A(g) + B(g) \longrightarrow C(g) + D(g)$$

By time t_1, however, some C and D have been formed, and so the reverse reaction can get under way:

$$A(g) + B(g) \longleftarrow C(g) + D(g)$$

As the forward reaction proceeds, its rate decreases, because [A] and [B] are decreasing. At the same time the rate of the reverse reaction increases, as more and more [C] and [D] are formed by the forward reaction. Finally, at t_3, the rate of the forward reaction has decreased and that of the reverse reaction has increased to the point that they are equal. From this time on *no further change occurs in any concentration,* since reactants and products are formed and consumed at equal rates:

$$A(g) + B(g) \rightleftharpoons C(g) + D(g)$$

ADDED COMMENT

Remember that although no change is apparent in a system at equilibrium, considerable activity may be taking place. Both forward and reverse reactions may be very rapid, but their equal rates mask this fact. Equilibrium is a *dynamic* state.

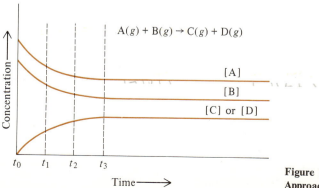

Figure 15-1
Approach to equilibrium.

The approach to equilibrium

Equilibrium can usually be established in a number of different ways. Consider the following reaction:

$$CO_2(g) + H_2(g) \rightleftharpoons CO(g) + H_2O(g)$$

One way to establish equilibrium in this case is to add equal quantities, say, 1 mol each, of CO_2 and H_2 to a container and wait for all concentrations to reach constant values. The changes in the concentrations of reactants and products are shown in Fig. 15-2a. A second way to attain equilibrium is to add equal quantities, say, 1 mol each, of CO and H_2O to the container. The concentration changes for this approach are shown in Fig. 15-2b. In this case the initial reaction is the reverse reaction, but the end result is the same: the final equilibrium concentrations in Fig. 15-2b are the same as in a. Figure 15-2c shows what happens when equilibrium is established by adding *unequal* numbers of moles of H_2 and CO_2 to the container. In this case $[H_2]$ and $[CO_2]$ both decrease and $[CO]$ and $[H_2O]$ increase, but the final equilibrium concentrations in Fig. 15-2c are different from those in a. A fourth variation is shown in Fig. 15-2d. In this case equal numbers of moles of CO_2 and H_2 are added along with some CO. Again, a new set of final equilibrium concentrations results. Equilibrium can be established, at least in principle, by starting with any combination of the reactants and products in any concentrations, as long as all of the reactants *or* all of the products are present in the initial mixture. If this condition is not met, neither forward nor reverse reaction can proceed.

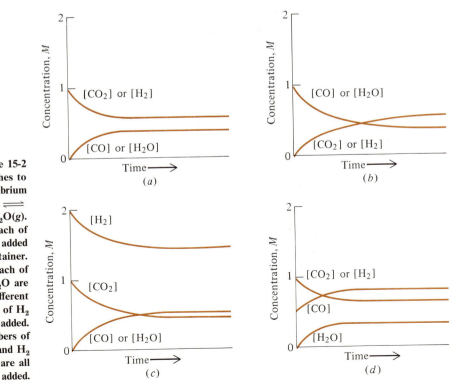

Figure 15-2 Alternate approaches to the equilibrium $CO_2(g) + H_2(g) \rightleftharpoons CO(g) + H_2O(g)$. (a) One mole each of CO_2 and H_2 are added to an empty container. (b) One mole each of CO and H_2O are added. (c) Different numbers of moles of H_2 and CO_2 are added. (d) Equal numbers of moles of CO_2 and H_2 plus some CO are all added.

ADDED COMMENT

Remember: Reactants, products, forward reaction, and *reverse reaction* have meaning only with respect to a given written equation.

Chemical equilibrium and Le Châtelier's principle

When equilibrium systems are disturbed, they adjust to minimize the effect of the disturbance. This is a statement of *Le Châtelier's principle,* which we introduced in Sec. 11-4. As the system adjusts, the position of equilibrium *shifts* to favor either more products or more reactants. A shift to favor more products is called a *shift to the right,* since products are on the right in a written equation. A shift to favor more reactants is called a *shift to the left.*

Consider as an example the equilibrium

$$N_2(g) + 3H_2(g) \rightleftharpoons 2NH_3(g)$$

Suppose that we have added N_2, H_2, and NH_3 to a container maintained at constant temperature and have waited until the system has reached equilibrium. Suppose, further, that the concentrations of these three substances are as shown at time t_0 in Fig. 15-3. Now in a sequence of six separate steps we will disturb the equilibrium first by adding more of each of the three substances in sequence and then by removing some of each from the reaction container. Figure 15-3 summarizes the results.

Time t_1: *More N_2 is added.* The effect of this change is to increase $[N_2]$. This causes the equilibrium to shift to the *right,* using up *some* of the added N_2,

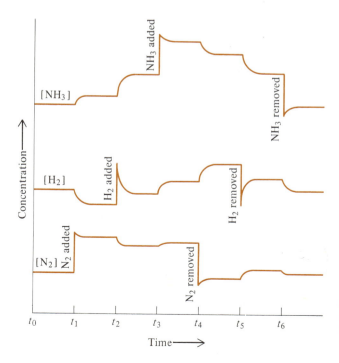

Figure 15-3
Shifting the equilibrium
$N_2(g) + 3H_2(g) \rightleftharpoons 2NH_3(g)$.

using up some H_2, and forming more NH_3. After equilibrium has been reestablished, $[N_2]$ is higher, $[H_2]$ is lower, and $[NH_3]$ is higher than they were before the N_2 was added.

Time t_2: *More H_2 is added.* This again shifts the equilibrium to the *right,* using up *some* of the added H_2. After equilibrium has been reestablished, $[N_2]$ is lower, $[H_2]$ is higher, and $[NH_3]$ is higher than before the H_2 was added.

Time t_3: *More NH_3 is added.* This causes $[NH_3]$ to increase, so that the equilibrium shifts to the *left,* using up *some* of the added NH_3. Once the system has reestablished equilibrium, $[N_2]$ is found to be higher, $[H_2]$ to be higher, and $[NH_3]$ also to be higher than before the NH_3 was added.

Time t_4: *Some N_2 is removed.* This reduces $[N_2]$ which causes the equilibrium to shift to the *left,* producing more N_2. After the system has returned to equilibrium, $[N_2]$ is found to be lower (the equilibrium shift did not produce as much N_2 as was removed), $[H_2]$ to be higher, and $[NH_3]$ to be lower than before the removal of N_2.

Time t_5: *Some H_2 is removed.* This causes the equilibrium to shift again to the *left,* partially compensating for the removed H_2. After the return to equilibrium, $[N_2]$ is higher, $[H_2]$ is lower, and $[NH_3]$ is lower than before the removal of H_2.

Time t_6: *Some NH_3 is removed.* This shifts the equilibrium to the *right.* After return to equilibrium, $[N_2]$, $[H_2]$, and $[NH_3]$ are all found to be lower than before the NH_3 was removed.

Note that in this sequence the amounts of each component added and later removed were the same. (This can be seen by examining Fig. 15-3 closely.) As a result the final equilibrium concentrations equal the initial ones.

The response made by an equilibrium system to the addition or removal of a component can be shown to be a response to a changed *concentration,* rather than a changed quantity. In order to see that this is true, consider again the equilibrium

$$N_2(g) + 3H_2(g) \rightleftharpoons 2NH_3(g)$$

If we suddenly *decrease the container volume,* the *amounts* of N_2, H_2, and NH_3 are not immediately affected, but their *concentrations* are all increased. In this case the equilibrium does indeed respond. More NH_3 is produced and less N_2 and H_2 are present after return to equilibrium. The response of the system must be concentration-related.

How does Le Châtelier's principle account for the formation of more NH_3 in the above equilibrium? The equilibrium shifts to the *right* because this reduces the total number of molecules in the container. (Look at the equation—in the forward reaction *two* molecules take the place of *four.*) Decreasing the volume of a mixture of gases tends to increase the total pressure (Boyle's law). But in this case the pressure increase is minimized by the decrease in the number of gas molecules. Note, however, that after equilibrium has been reestablished, although *more* NH_3 and *less* N_2 and H_2 are present, the *concentrations* of all have been *increased,* a consequence of the smaller volume.

It can easily be seen that a container-volume *increase* produces results all of which are opposite from the above.

A change in container volume does not always produce an equilibrium shift. In a reaction such as

$$2HI(g) \rightleftharpoons H_2(g) + I_2(g)$$

in which the number of gas molecules is the same on both sides of the equation, the equilibrium system does not respond at all to a volume decrease. In this case no mechanism is available for minimizing the pressure increase.

What happens to an equilibrium if *the total pressure is increased by adding an inert gas?* In this case we might expect the equilibrium to shift in the direction of fewer molecules, but it does not. The reason is that the concentrations of N_2, H_2, and NH_3 are unaffected by the added inert gas, as long as the container volume is held constant.

Finally, we consider the effort of temperature changes on an equilibrium system. Le Châtelier's principle predicts that an increase in temperature favors the reaction which is endothermic. Let us see how this comes about. The formation of ammonia from its elements is an exothermic reaction.

$$N_2(g) + 3H_2(g) \rightleftharpoons 2NH_3(g) \qquad \Delta H = -92.2 \text{ kJ}$$

This could be rewritten as

$$N_2(g) + 3H_2(g) \rightleftharpoons 2NH_3(g) + 92.2 \text{ kJ}$$

Thus the forward reaction is exothermic, and the reverse one, endothermic. Adding heat to this equilibrium system causes a shift to the *left*. The endothermic (reverse) reaction uses some of the added heat to produce more N_2 and H_2 from NH_3, and so the temperature does not increase as much as it otherwise would. At a higher temperature, then, the equilibrium values of $[H_2]$ and $[N_2]$ are higher and that of $[NH_3]$ is lower. Lowering the temperature reverses all of the above effects as the exothermic reaction is favored. This produces heat which partially compensates for the heat withdrawn from the system.

15-2
The law of
chemical
equilibrium

We have described *how* an equilibrium shifts in response to a stress or disturbance, but not *how much*. Fortunately, the quantitative treatment of equilibrium depends largely upon a single simple relationship known as the *law of chemical equilibrium*. But first we must define a few quantities.

The mass-action
expression

Consider a hypothetical gas-phase reaction

$$A(g) + B(g) \longrightarrow C(g) + D(g)$$

For this reaction a quantity, Q, known as the *mass-action expression* is defined as

$$Q = \frac{[C][D]}{[A][B]}$$

For a second reaction $E(g) + F(g) \longrightarrow 2G(g)$

$$Q = \frac{[G]^2}{[E][F]}$$

In general the mass-action expression Q is a quotient, a fraction with the product of the concentrations of the products in the numerator and the product of the concentrations of the reactants in the denominator. Each concentration is raised to an exponential power equal to its coefficient in the balanced equation. For the general reaction

$$p\text{H} + q\text{I} + \cdots \longrightarrow r\text{J} + s\text{K} + \cdots$$

the mass-action expression is

$$Q = \frac{[\text{J}]^r[\text{K}]^s \cdots}{[\text{H}]^p[\text{I}]^q \cdots}$$

The mass-action expression can have any value, because it depends on the extent of the reaction. For example, if 1 mol each of N_2 and H_2 gases are mixed in a 1-liter container held at $350°C$, the mass-action expression for the reaction

$$N_2(g) + 3H_2(g) \longrightarrow 2NH_3(g)$$

has the value

$$\frac{[NH_3]^2}{[N_2][H_2]^3} = \frac{(0)^2}{1(1)^3} = 0$$

before the reaction has proceeded appreciably (at time t_0). As the reaction progresses, however, its mass-action expression increases, as shown in the following table:

Time	$[N_2]$	$[H_2]$	$[NH_3]$	$Q = \dfrac{[NH_3]^2}{[N_2][H_2]^3}$
t_0	1.000	1.000	0	0
t_1	0.874	0.634	0.252	0.285
t_2	0.814	0.442	0.372	1.97
t_3	0.786	0.358	0.428	5.08
t_4	0.781	0.343	0.438	6.09
t_5	0.781	0.343	0.438	6.09

As can be seen, the value of Q increases as $[NH_3]$ increases and $[N_2]$ and $[H_2]$ decrease, until the system reaches equilibrium (by time t_4), after which no further change in Q occurs.

Table 15-1 shows the mass-action expressions for a number of different reactions.

The equilibrium constant   If the mass-action expression for a reaction can have various values, of what use is it? Consider the following three experiments. In each experiment the equilibrium

$$N_2(g) + 3H_2(g) \rightleftharpoons 2NH_3(g)$$

is established by adding at least two of the above gases to a 1.00-liter container maintained at a constant $350°C$.

Experiment 1: 1.000 mol of N_2 and 3.000 mol of H_2 are added to the container.

Table 15-1
Some mass-action
expressions

Reaction	Mass-action expression, Q
$2HI(g) \longrightarrow H_2(g) + I_2(g)$	$\dfrac{[H_2][I_2]}{[HI]^2}$
$PCl_5(g) \longrightarrow PCl_3(g) + Cl_2(g)$	$\dfrac{[PCl_3][Cl_2]}{[PCl_5]}$
$2NO(g) + O_2(g) \longrightarrow 2NO_2(g)$	$\dfrac{[NO_2]^2}{[NO]^2[O_2]}$
$N_2O_4(g) \longrightarrow 2NO_2(g)$	$\dfrac{[NO_2]^2}{[N_2O_4]}$
$CS_2(g) + 4H_2(g) \longrightarrow CH_4(g) + 2H_2S(g)$	$\dfrac{[CH_4][H_2S]^2}{[CS_2][H_2]^4}$

After equilibrium has been established, the contents are analyzed. It is found that $[N_2] = 0.325\ M$, $[H_2] = 0.975\ M$, and $[NH_3] = 1.350\ M$.

Experiment 2: This differs from experiment 1, in that initially 1.000 mol *each* of N_2 and H_2 are added to the container. At equilibrium the concentrations are found to be $[N_2] = 0.781\ M$, $[H_2] = 0.343\ M$, and $[NH_3] = 0.438\ M$.

Experiment 3: In this experiment 1.000 mol *each* of N_2, H_2, and NH_3 are added to the container. The equilibrium concentrations this time are found to be $[N_2] = 0.885\ M$, $[H_2] = 0.655\ M$, and $[NH_3] = 1.230\ M$.

Table 15-2 summarizes the results of the three experiments. At first glance the equilibrium concentrations of N_2, H_2, and NH_3 may appear to be unrelated. But look at the value of Q, the mass-action expression, after equilibrium has been established. In each case the equilibrium value of Q is the same. (The slight differences in the third significant figure result from roundings-off in the computations.) Further experiments would confirm this conclusion. No matter how the equilibrium is established, the value of Q is a constant at equilibrium at 350°C.

Similar results could be obtained for any equilibrium at any given temperature, and so we can make the following generalization, called the *law of chemical equilibrium: At any given temperature the value of the mass-action expression for a given reaction at equilibrium is a constant.*[1] This constant value is known as the *equilibrium constant K* for the reaction at that temperature. In other words, at equilibrium

---

[1] The law of chemical equilibrium was first derived by a somewhat unrigorous procedure by the Norwegian chemists Cato Maximilian Guldberg and Peter Waage in 1864. They spoke of the "active masses" of reacting substances and from this we get our term *mass-action expression*. Goldberg and Waage's active masses are approximately our concentrations.

Table 15-2   Equilibrium study at 350°C: $N_2(g) + 3H_2(g) \rightleftharpoons 2NH_3(g)$

Experiment	Initial concentration, mol liter^{-1}			Equilibrium concentration, mol liter^{-1}			Q at equilibrium, $\dfrac{[NH_3]^2}{[N_2][H_2]^3}$
	N_2	H_2	NH_3	N_2	H_2	NH_3	
1	1.000	3.000	0	0.325	0.975	1.350	6.05
2	1.000	1.000	0	0.781	0.343	0.438	6.09
3	1.000	1.000	1.000	0.885	0.655	1.230	6.08

$$Q = K$$

This equality describes a condition which is met by a system at equilibrium, and so it is referred to as the *equilibrium condition.*

The actual numerical value of the equilibrium constant for a reaction depends upon the units used in the mass-action expression. Using concentrations in moles per liter we can write the equilibrium condition for

$$N_2(g) + 3H_2(g) \rightleftharpoons 2NH_3(g)$$

as
$$\frac{[NH_3]^2}{[N_2][H_2]^3} = K_c$$

where the subscript c is used to indicate that concentrations (molarities) are being used in the mass-action expression.

For *gases* the mass-action expression is often written as a function of *partial pressures,* usually in *atmospheres.* Thus the equilibrium condition for the above reaction can alternatively be written as

$$\frac{P^2_{NH_3}}{P_{N_2}P^3_{H_2}} = K_P$$

where the various P's are the respective partial pressures of the three substances at equilibrium.

Although both K_c and K_P are constants at any temperature, they are not necessarily equal. The relationship between the two is easily found. For the general reaction

$$kA(g) + lB(g) \rightleftharpoons mC(g) + nD(g)$$

$$K_P = \frac{P^m_C P^n_D}{P^k_A P^l_B}$$

From the ideal-gas law, $PV = nRT$, we have

$$P = \frac{nRT}{V}$$

$$K_P = \frac{(n_C RT/V)^m (n_D RT/V)^n}{(n_A RT/V)^k (n_B RT/V)^l} = \frac{(n_C/V)^m (m_D/V)^n}{(n_A/V)^k (n_B/V)^l} (RT)^{[(m+n)-(k+l)]}$$

But n/V is simply molar concentration, and $(m + n) - (k + l)$ is the change in the number of moles of gas, Δn, shown in the balanced equation. Therefore

$$K_P = \frac{[C]^m [D]^n}{[A]^k [B]^l} (RT)^{\Delta n}$$

or
$$K_P = K_c (RT)^{\Delta n}$$

● **EXAMPLE 15-1**

Problem   For the equilibrium

$$2NOCl(g) \rightleftharpoons 2NO + Cl_2$$

the value of the equilibrium constant K_c is 3.75×10^{-6} at 796°C. Calculate K_P for this reaction at this temperature.

Solution   For this reaction $\Delta n = (2 + 1) - 2 = 1$

$$K_P = K_c(RT)^{\Delta n}$$
$$= (3.75 \times 10^{-6})[0.0821(796 + 273)]^1 = 3.29 \times 10^{-4} \quad \bullet$$

A special comment is in order here. Equilibrium constants look as if they ought to have units (dimensions). Actually, some books do give units with K_c and K_P. We will not do so, however, because whether appearing to be expressed in atmospheres or in molar concentrations, the mass-action expression, and hence the equilibrium constant, is actually *dimensionless*. Completely resolving this seeming paradox is inappropriate for this book, but we will point out one fact: in the mass-action expression for K_P a term P_A is really a *ratio* of the partial pressure of A to its *standard-state* pressure, which is 1 atm. In other words, if the partial pressure of A is 2 atm, then in the mass-action expression P_A really means

$$\frac{2 \text{ atm}}{1 \text{ atm}} \quad \text{or} \quad 2$$

which has no units. K_P therefore has no units. Similarly, the standard-state *concentration* is $1\,M$; so if A is present at a $3\,M$ concentration, then in the mass-action expression [A] really means

$$\frac{3\,M}{1\,M} \quad \text{or} \quad 3$$

which has no units; so K_c is dimensionless also.

(*Note:* In this book we will be using K_c's more frequently than K_P's. For this reason we will often drop the subscript c from K_c, unless it is needed for emphasis.)

The magnitude of an equilibrium constant gives an indication of a reaction's position of equilibrium. For example, for the equilibrium

$$A(g) + B(g) \rightleftharpoons C(g) + D(g)$$

if K is large, say 100 or greater, it means that in the mass-action expression

$$\frac{[C][D]}{[A][B]}$$

the numerator is at least 100 times greater than the denominator at equilibrium. Thus at equilibrium the product concentrations tend to be large, and so we say that the reaction tends to go to completion. On the other hand, if K is very small, this tells us that in moving toward equilibrium the system forms only relatively small amounts of products.

Only if a system is at equilibrium does the mass-action expression equal the equilibrium constant. Thus for the reaction

$$A + B \rightleftharpoons C + D$$

the equilibrium condition is that

$$\frac{[C][D]}{[A][B]} = K$$

If
$$\frac{[C][D]}{[A][B]} < K$$

this tells us that the reaction will proceed *to the right,* increasing [C] and [D] and decreasing [A] and [B], until $Q = K$. On the other hand, if

$$\frac{[C][D]}{[A][B]} > K$$

then the reaction will proceed to the left, increasing [A] and [B] and decreasing [C] and [D], until equilibrium is established.

● **EXAMPLE 15-2**

Problem   For the equilibrium

$$2SO_3(g) \rightleftharpoons 2SO_2(g) + O_2(g)$$

the value of the equilibrium constant K is 4.8×10^{-3} at 700°C. If the concentrations of the above three substances in a container are

$$[SO_3] = 0.60 \ M \qquad [SO_2] = 0.15 \ M \qquad [O_2] = 0.025 \ M$$

how will these concentrations change as the system approaches equilibrium if the temperature is maintained at 700°C?

Solution   First we find the value of the mass-action expression.

$$Q = \frac{[SO_2]^2[O_2]}{[SO_3]^2}$$

$$= \frac{(0.15)^2(0.025)}{(0.60)^2} = 1.6 \times 10^{-3}$$

Since this is less than the value of the equilibrium constant, 4.8×10^{-3}, the reaction proceeds to the right, so that [SO_2] increases, [O_2] increases, and [SO_3] decreases until $Q = K$. ●

Kinetics and equilibrium   An explanation for the behavior of systems at or approaching equilibrium comes from a kinetic analysis of such systems. At equilibrium the rates of the forward and reverse reactions are equal. As a result, the rate constants for forward and reverse reactions and the equilibrium constant are all related. This relationship is easiest to show in the case of an *elementary process.* Consider the reaction at equilibrium

$$A + B \underset{k_{-1}}{\overset{k_1}{\rightleftharpoons}} C + D$$

where k_1 and k_{-1} are the rate constants for the forward and reverse reactions, respectively. The forward and reverse rates are

$$\text{Forward rate} = k_1[A][B]$$

$$\text{Reverse rate} = k_{-1}[C][D]$$

At equilibrium, these rates are equal, and so

$$k_1[A][B] = k_{-1}[C][D]$$

or

$$\frac{[C][D]}{[A][B]} = \frac{k_1}{k_{-1}} = K$$

The expression on the left is the mass-action expression for the reaction, and so this constitutes the kinetic proof that this expression is a constant for an elementary reaction at equilibrium. A similar proof can be made for any *complex reaction*. Suppose, for example, that the reaction

$$A + B \rightleftharpoons C + D$$

proceeds by the two-step mechanism

Step 1:

$$2A \underset{k_{-1}}{\overset{k_1}{\rightleftharpoons}} A_2$$

Step 2:

$$A_2 + B \underset{k_{-2}}{\overset{k_2}{\rightleftharpoons}} C + D + A$$

The *principle of detailed balancing* states that when a complex reaction is at equilibrium, each of its steps must be at equilibrium. This means that from step 1 we can say

$$k_1[A]^2 = k_{-1}[A_2]$$

and from step 2,

$$k_2[A_2][B] = k_{-2}[C][D][A]$$

From the first of these relationships we get

$$\frac{[A_2]}{[A]^2} = \frac{k_1}{k_{-1}}$$

and from the second,

$$\frac{[C][D][A]}{[A_2][B]} = \frac{k_2}{k_{-2}}$$

Multiplying these together we have

$$\frac{[A_2]}{[A]^2} \times \frac{[C][D][A]}{[A_2][B]} = \frac{k_1 k_2}{k_{-1} k_{-2}}$$

or

$$\frac{[C][D]}{[A][B]} = K$$

where the equilibrium constant K is the product of the forward rate constants divided by the product of the reverse rate constants. Thus for this two-step reaction we have shown that the mass-action expression is a constant at equilibrium. By using the principle of detailed balancing, it can be similarly shown that for any reaction, no matter how complex, $Q = K$ at equilibrium.

We have seen that for an elementary process

$$K = \frac{k_1}{k_{-1}}$$

that is, the equilibrium constant is equal to the ratio of the specific rate constants of the forward and reverse reactions. Although we will not prove it here, the relationship applies to complex reactions; that is,

$$K = \frac{k_f}{k_r}$$

where k_f and k_r are the experimental rate constants for forward and reverse reactions, provided only that the mechanisms of the forward and reverse reactions are the same at equilibrium as they are when k_f and k_r are evaluated.

Heterogeneous equilibria   All of the equilibria which we have discussed have been *homogeneous*. A *heterogeneous equilibrium* is one which involves two or more phases. An example is the high-temperature equilibrium

$$C(s) + S_2(g) \rightleftharpoons CS_2(g)$$

for which we may write the equilibrium condition as

$$\frac{[CS_2]}{[C][S_2]} = K'$$

(The reason for the prime will be evident in a moment.)

Stating the equilibrium condition this way is not incorrect, but the above statement can be (and usually is) simplified. Each of the brackets in a mass-action expression is the numerical value of the concentration of a component *in a certain phase*. In the mass-action expression

$$\frac{[CS_2]}{[C][S_2]}$$

$[CS_2]$ and $[S_2]$ refer to the respective concentrations of CS_2 and S_2 in the gas phase. But $[C]$ means the concentration of solid carbon in *the phase which is pure solid carbon*. It is the number of moles of carbon atoms in 1 liter of carbon. But the concentration of carbon in pure carbon cannot be significantly changed under any normal conditions. It is a constant. Therefore, if we rewrite the equilibrium condition as

$$\frac{[CS_2]}{[S_2]} = [C]K'$$

the right-hand side is the product of two constant numbers. If we combine them and represent the product by K, the equilibrium condition is

$$\frac{[CS_2]}{[S_2]} = K$$

This follows the general convention of "absorbing" the concentrations of any pure solid or liquid phases into the value of K, as we have done above. You may assume that this has been done when given a numerical value for an equilibrium constant, and you should write the mass-action expression accordingly.

Table 15-3 lists some heterogeneous reactions and their equilibrium conditions.

Equilibrium constants for forward and reverse reactions

It may have become apparent to you that if for the equilibrium

$$A(g) + B(g) \rightleftharpoons C(g) + D(g)$$

it is true that

$$\frac{[C][D]}{[A][D]} = K$$

then it must be equally true that

$$\frac{[A][B]}{[C][D]} = K'$$

Furthermore,

$$K' = \frac{1}{K}$$

By general agreement the *numerator* of the mass-action expression contains the *products* (on the right of the equation) and the *denominator,* the *reactants* (on the left). Much confusion is avoided by always following this convention. Note, however, that K' is the equilibrium constant for the above reaction written in reverse:

$$C(g) + D(g) \rightleftharpoons A(g) + B(g)$$

Thus it is essential to know how an equilibrium equation has been written when specifying or using an equilibrium constant or mass-action expression.

Variation of K with temperature

We saw in Sec. 10-6 that the variation of vapor pressure with temperature is described by the Clausius-Clapeyron equation

$$\log \frac{P_1}{P_2} = \frac{-\Delta H_{vap}}{2.303R}\left(\frac{1}{T_1} - \frac{1}{T_2}\right)$$

where P_1, P_2, T_1, and T_2 are the initial and final pressures and temperatures,

Table 15-3 Some heterogeneous equilibria

Reaction	Equilibrium condition
$2SO_3(g) + S(s) \rightleftharpoons 3SO_2(g)$	$\dfrac{[SO_2]^3}{[SO_3]^2} = K$
$2HgO(s) \rightleftharpoons 2Hg(g) + O_2(g)$	$[Hg]^2[O_2] = K$
$Fe_3O_4(s) + H_2(g) \rightleftharpoons 3FeO(s) + H_2O(g)$	$\dfrac{[H_2O]}{[H_2]} = K$
$N_2(g) + 2H_2O(g) \rightleftharpoons NH_4NO_2(s)$	$\dfrac{1}{[N_2][H_2O]^2} = K$
$4CuO(s) \rightleftharpoons 2Cu_2O(s) + O_2(g)$	$[O_2] = K$

respectively, and ΔH_{vap} is the molar heat of vaporization. For a vaporization equilibrium

$$\text{Liquid} \rightleftharpoons \text{gas}$$

the equilibrium condition can be written as

$$\frac{[\text{Gas}]}{[\text{Liquid}]} = K'$$

But since the denominator is a constant,

$$[\text{Gas}] = K'[\text{liquid}] = K_c$$

In the above equilibrium, Δn, the change in the number of moles of gas, is $1 - 0$, or 1; so

$$K_P = K_c(RT)^{\Delta n}$$

$$= [\text{gas}](RT)^1 = \frac{n}{V}RT = P$$

We see, then, that the equilibrium constant K_P is just the partial pressure of the gas, which is, of course, the equilibrium vapor pressure of the liquid and is a constant at any given temperature.

The Clausius-Clapeyron equation can be thought of as a specialized form of a more general relationship which applies to any system at equilibrium, not just a liquid and its vapor. For a gas-phase equilibrium this can be written as

$$\log \frac{(K_P)_1}{(K_P)_2} = \frac{-\Delta H°}{2.303R} \left(\frac{1}{T_1} - \frac{1}{T_2} \right)$$

This is one form of the *van't Hoff* equation.[2] In it $\Delta H°$ is the *standard heat of reaction,* that is, the heat of reaction when reactants and products are in their standard states (each gas at 1 atm pressure, assuming ideal behavior).

Not only can the van't Hoff equation be used to find values of equilibrium constants at various temperatures, but it also offers a means of finding standard heats of reaction when for some reason it is difficult or impossible to measure them directly.

● **EXAMPLE 15-3**

Problem   For the equilibrium

$$2H_2S(g) \rightleftharpoons 2H_2(g) + S_2(g)$$

$K_P = 1.18 \times 10^{-2}$ at 1065°C and 5.09×10^{-2} at 1200°C. Calculate $\Delta H°$ for the reaction.

Solution   Rearranging the van't Hoff equation to solve for $\Delta H°$, we get

$$\Delta H° = \frac{2.303R \log [(K_P)_2/(K_P)_1]}{(1/T_1) - (1/T_2)}$$

[2] This equation should not be confused with the van't Hoff osmotic-pressure equation (Sec. 12-4).

Now letting $T_1 = 1065 + 273 = 1338$ K, $T_2 = 1200 + 273 = 1473$ K, $(K_P)_1 = 1.18 \times 10^{-2}$, and $(K_P)_2 = 5.09 \times 10^{-2}$, and substituting:

$$\Delta H° = \frac{2.303(8.314 \text{ J K}^{-1} \text{ mol}^{-1}) \log (5.09 \times 10^{-2}/1.18 \times 10^{-2})}{1/1338 \text{ K} - 1/1473 \text{ K}}$$

$$= 1.77 \times 10^5 \text{ J mol}^{-1} \quad \text{or} \quad 177 \text{ kJ mol}^{-1}$$

(Note that we have used R in *joules* per kelvin-mole, so that $\Delta H°$ would come out in *joules* per mole.) ●

Look closely at the van't Hoff equation, above, and observe the correlation between it and the temperature-induced equilibrium shift predicted by Le Châtelier's principle. If the temperature is *increased* for an *exothermic* reaction ($\Delta H°$ is *negative*), then

$$T_2 > T_1$$

and so

$$\frac{1}{T_2} < \frac{1}{T_1}$$

and therefore,

$$\frac{1}{T_1} - \frac{1}{T_2} > 0$$

Since $\Delta H°$ is negative, $-\Delta H°$ is a positive quantity. Thus the entire right-hand side of the equation is positive. This means that

$$\log \frac{(K_P)_1}{(K_P)_2} > 0$$

The only way this can be so is for

$$(K_P)_1 > (K_P)_2$$

This means that for an exothermic reaction K_P *decreases* as the temperature increases. In other words, the equilibrium shifts to the left, which is exactly what Le Châtelier's principle predicts.

15-3 Equilibrium calculations

● **EXAMPLE 15-4**

Problem Hydrogen iodide is injected into a container at 458°C. The HI dissociates to form H_2 and I_2. After equilibrium has been established at this temperature, samples are taken and analyzed. [HI] is found to be 0.421 M, while [H_2] and [I_2] are both 6.04×10^{-2} M. Calculate the value of the equilibrium constant for the dissociation of HI at 458°C.

Solution The equilibrium equation is

$$2\text{HI}(g) \rightleftharpoons \text{H}_2(g) + \text{I}_2(g)$$

The mass-action expression for this is

$$\frac{[\text{H}_2][\text{I}_2]}{[\text{HI}]^2}$$

In order to find the value of the equilibrium constant, we substitute the equilibrium concentrations in the mass-action expression

$$\frac{(6.04 \times 10^{-2})(6.04 \times 10^{-2})}{(0.421)^2} = 2.06 \times 10^{-2}$$

$$K = 2.06 \times 10^{-2}$$

● EXAMPLE 15-5

Problem   HI, H_2, and I_2 are all placed in a container at 458°C. At equilibrium $[HI] = 0.360\ M$ and $[I_2] = 0.150\ M$. What is the equilibrium concentration of H_2 at this temperature?

Solution   Since we know the value of K from Example 15-4, we can write

$$2HI(g) \rightleftharpoons H_2(g) + I_2(g)$$

$$\frac{[H_2][I_2]}{[HI]^2} = K$$

$$[H_2] = \frac{K[HI]^2}{[I_2]}$$

$$= \frac{(2.06 \times 10^{-2})(0.360)^2}{0.150} = 1.78 \times 10^{-2}\ M$$

● EXAMPLE 15-6

Problem   In another experiment 1.00 mol of HI is placed in a 5.00-liter container at 458°C. What are the concentrations of HI, H_2, and I_2 after equilibrium has been established at this temperature?

Solution   Again the equation is

$$2HI(g) \rightleftharpoons H_2(g) + I_2(g)$$

Immediately after injecting the HI into the container, its concentration is

$$\frac{1.00\ \text{mol}}{5.00\ \text{liters}} = 0.200\ \text{mol liter}^{-1} \quad \text{or} \quad 0.200\ M$$

At the same time the initial concentrations of H_2 and I_2 are both 0. The reaction proceeds forward, increasing $[H_2]$ and $[I_2]$ and decreasing $[HI]$ until the equilibrium condition is met. Let x equal the increase in the concentration of H_2 necessary to reach equilibrium. Then x must also equal the increase in the concentration of I_2. The *decrease* in the concentration of HI is $2x$, because 2 mol of HI are used to make 1 mol each of H_2 and I_2. Thus at equilibrium the concentration of HI is $0.200 - 2x$, that of H_2 is x, and that of I_2 is x.

	Initial concentration, M	Concentration change due to reaction, M	Equilibrium concentration, M
$[H_2]$	0	x	x
$[I_2]$	0	x	x
$[HI]$	0.200	$-2x$	$0.200 - 2x$

The equilibrium condition is

$$\frac{[H_2][I_2]}{[HI]^2} = K$$

$$\frac{x(x)}{(0.200 - 2x)^2} = 2{:}06 \times 10^{-2}$$

Taking the square root of both sides, we obtain

$$\frac{x}{0.200 - 2x} = 1.44 \times 10^{-1}$$

Solving for x, we get

$$x = 2.24 \times 10^{-2}$$

At equilibrium, therefore,

$$[H_2] = x = 2.24 \times 10^{-2} \, M$$

$$[I_2] = x = 2.24 \times 10^{-2} \, M$$

$$[HI] = 0.200 - 2x = 0.155 \, M \quad \bullet$$

● **EXAMPLE 15-7**

Problem   Suppose 3.00 mol of HI, 2.00 mol of H_2, and 1.00 mol of I_2 are placed together in a 1.00-liter container at 458°C. After equilibrium has been established, what are the concentrations of all species?

Solution   Since the container volume this time is 1.00 liter, the initial concentrations are numerically the same as the number of moles added:

$$[H_2] = 2.00 \, M \qquad [I_2] = 1.00 \, M \qquad [HI] = 3.00 \, M$$

Again the equilibrium equation is

$$2HI(g) \rightleftharpoons H_2(g) + I_2(g)$$

Let x = the *increase* in $[H_2]$ in going from the initial, just-mixed state to the final equilibrium state. Then x = the *increase* in $[I_2]$, and $2x$ = the *decrease* in $[HI]$.

Note: We have assumed that the net reaction will take place *to the right.* If we are wrong x will turn out to be a negative number, meaning that $[H_2]$ and $[I_2]$ both actually *decrease,* while $[HI]$ *increases,* and we will get the right answer anyway. (There is an easy way to tell ahead of time which of the two possibilities actually occurs: just compare the size of the mass-action expression to that of K.)

At equilibrium

$$[H_2] = 2.00 + x \qquad [I_2] = 1.00 + x \qquad [HI] = 3.00 - 2x$$

The equilibrium condition is

$$\frac{[H_2][I_2]}{[HI]^2} = K$$

$$\frac{(2.00 + x)(1.00 + x)}{(3.00 - 2x)^2} = 2.06 \times 10^{-2}$$

This takes a little time to solve. By multiplying out and collecting terms, we get

$$0.918x^2 + 3.25x + 1.81 = 0$$

This can be solved using the quadratic formula for finding the roots of an equation in the form $ax^2 + bx + c = 0$ (see App. D):

$$x = \frac{-b \pm \sqrt{b^2 - 4ac}}{2a}$$

In this case

$$x = \frac{-3.25 \pm \sqrt{(3.25)^2 - 4(0.918)(1.81)}}{2(0.918)}$$

from which we get the roots

$$x = -0.690 \quad \text{and} \quad x = -2.84$$

Since the second root leads to negative concentrations of $[H_2]$ and $[I_2]$, it is extraneous and we will reject it. The other root, $x = -0.690$, is physically meaningful. Since it is negative, it means that our original assumption was wrong, and that $[H_2]$ and $[I_2]$ actually decrease and $[HI]$ increases. Thus the final equilibrium concentrations are

$$[H_2] = 2.00 + x = 2.00 - 0.690 = 1.31 \ M$$

$$[I_2] = 1.00 + x = 1.00 - 0.690 = 0.31 \ M$$

$$[HI] = 3.00 - 2x = 3.00 - 2(-0.690) = 4.38 \ M \quad \bullet$$

● **EXAMPLE 15-8**

Problem   One mole of NOCl gas is added to a 4.0-liter container at 25°C. The NOCl undergoes slight decomposition to form NO and Cl_2 gases. If the equilibrium constant for

$$2NOCl(g) \rightleftharpoons 2NO(g) + Cl_2(g)$$

is 2.0×10^{-10} at 25°C, what are the concentrations of all species at equilibrium at this temperature?

Solution   The initial concentrations are

$$[NOCl] = \frac{1.0 \ \text{mol}}{4.0 \ \text{liters}} = 0.25 \ M$$

$$[NO] = 0 \qquad [Cl_2] = 0$$

Let $x =$ the *increase* in $[Cl_2]$. Then $2x =$ the *increase* in $[NO]$, and $2x =$ the *decrease* in $[NOCl]$. At equilibrium,

$$[NO] = 2x \qquad [Cl_2] = x \qquad [NOCl] = 0.25 - 2x$$

$$\frac{[NO]^2[Cl_2]}{[NOCl]^2} = K$$

$$\frac{(2x)^2 x}{(0.25 - 2x)^2} = 2.0 \times 10^{-10}$$

This is a cubic equation, but the application of a little chemical common sense makes it easy to solve. The value of K, 2.0×10^{-10}, is very small, meaning that at equilibrium the numerator of the mass-action expression is much smaller than the denominator. (The reaction proceeds forward to only a very small extent.) Thus x ought to be a small number, so small that in the denominator $2x$ is negligible in comparison with 0.25. We can therefore simplify the above equation to

$$\frac{(2x)^2(x)}{(0.25)^2} \approx 2.0 \times 10^{-10}$$

where $\approx$ means "is approximately equal to." This equation is easy to solve:

$$4x^3 \approx 1.25 \times 10^{-11}$$

$$x \approx 1.5 \times 10^{-4}$$

Now our assumption that $2x$ is negligible in comparison with 0.25 can be checked:

$$0.25 - 2x = 0.25 - 2(1.5 \times 10^{-4}) = 0.25 - 0.00030 = 0.25$$

Clearly, 0.00030 is negligible in comparison with 0.25.
At equilibrium, then,

$$[NO] = 2x = 2(1.5 \times 10^{-4}) = 3.0 \times 10^{-4}\ M$$

$$[Cl_2] = x = 1.5 \times 10^{-4}\ M$$

$$[NOCl] = 0.25 - 2x = 0.25 - 2(1.5 \times 10^{-4}) = 0.25\ M \quad \bullet$$

Note: If we had solved the above cubic equation without making any approximations, we would have arrived at the same result.

SUMMARY

We study equilibrium because all reactions proceed in a direction which approaches the equilibrium state. In this chapter we considered equilibrium both qualitatively and quantitatively. *Le Châtelier's principle,* a qualitative description, can be applied to an equilibrium system in order to predict how the system will respond to a stress, such as the addition or removal of reacting components, changes in pressure, and changes in temperature.

The law of chemical equilibrium states that although the *mass-action expression* for a reaction may have an unlimited number of values at each temperature, when the reaction is at equilibrium, it has only one value. This value is called the *equilibrium constant K* for the reaction. This relationship provides a way of describing equilibria quantitatively. The value of K is equal to the ratio of the specific rate constants for the forward and reverse reactions.

For *heterogeneous* reactions the mass-action expression is nearly always simplified so that concentrations of pure liquid and solid phases are not shown. At equilibrium these constant concentrations are "absorbed" into" the value of the equilibrium constant.

Equilibrium constants do change with temperature. This is described quantitatively by the *van't Hoff equation,* which describes this change as a function of $\Delta H°$ of the reaction.

The technique of solving equilibrium problems was demonstrated in the last part of this chapter. This technique is applicable to all equilibrium systems, including the aqueous systems which we will be studying in Chaps. 16 and 17.

KEY TERMS

Concentration equilibrium constant, K_c (Sec. 15-2)
Equilibrium condition (Sec. 15-2)
Equilibrium constant (Sec. 15-2)
Heterogeneous equilibrium (Sec. 15-2)
Homogeneous equilibrium (Sec. 15-1)

Law of chemical equilibrium (Sec. 15-2)
Le Châtelier's principle (Sec. 15-1)
Mass-action expression (Sec. 15-2)
Pressure equilibrium constant, K_P (Sec. 15-2)
Van't Hoff equation (Sec. 15-2)

QUESTIONS AND PROBLEMS

Approach to equilibrium

15-1 It is desired to establish the following equilibrium:

$$FeO(s) + CO(g) \rightleftharpoons Fe(l) + CO_2(g)$$

Which of the following combinations, when added to a container, would *not* work? **(a)** FeO and CO

(b) FeO, CO, and Fe **(c)** FeO, Fe, and CO **(d)** CO and Fe **(e)** FeO and CO_2. (Assume no other reactions are possible.)

15-2 Substances A, B, and C are mixed together in a container, which is then sealed. After several weeks the container is opened and the contents analyzed. It is found that the quantities of A, B, and C have not changed. What are three possible explanations?

15-3 Consider the reaction

$$A(g) + 2B(g) \longrightarrow C(g) + 2D(g)$$

Draw a set of curves similar to those in Fig. 15-1 showing how concentrations change with time, assuming that initially (a) $[A] = [B]$ (b) $[A] = \frac{1}{2}[B]$.

15-4 For the equilibrium

$$A(g) \rightleftharpoons B(g) + C(g), \quad K_c = 2.$$

Tell what will happen in each independent case to the concentration of A in a 1-liter box if initially (a) 1 mol each of A, B, and C are added to the box (b) 2 mol each of A, B, and C are added (c) 3 mol each of A, B, and C are added.

15-5 All reactions *tend to* approach equilibrium. What are some of the reasons why they sometimes do not get there?

Equilibrium changes

15-6 Consider the reaction

$$2Cl_2(g) + 2H_2O(g) \rightleftharpoons 4HCl(g) + O_2(g)$$
$$\Delta H° = 113 \text{ kJ}$$

Assume that the system has come to equilibrium. Tell what will happen to the number of moles of H_2O in the container if (a) some O_2 is added (b) some Cl_2 is added (c) some HCl is removed (d) the volume of the container is decreased (e) the temperature is lowered (f) some helium is added.

15-7 The following equilibrium is established:

$$2C(s) + O_2(g) \rightleftharpoons 2CO(g) \quad \Delta H° = -221 \text{ kJ}$$

Tell what will be the effect on the equilibrium concentration of O_2 if (a) CO is added (b) O_2 is added (c) C is added (d) the container volume is increased (e) the temperature is raised.

15-8 When the volume of a system at equilibrium is decreased, the number of moles of each component remains unchanged. What do you know about the reaction?

15-9 When the temperature of a certain system at equilibrium is increased, the number of moles of each component remains essentially unchanged. What do you know about the reaction?

15-10 N_2, H_2, and NH_3 are all placed in a container and allowed to come to equilibrium. Then one additional mole each of NH_3 and N_2 are simultaneously added. What happens to $[H_2]$? Explain.

15-11 Discuss Henry's law (Sec. 12-3) as an illustration of Le Châtelier's principle.

15-12 Consider the following system at equilibrium:

$$2NOBr(g) \rightleftharpoons 2NO(g) + Br_2(g)$$

Suggest a way of increasing the pressure of the system that will (a) decrease the number of moles of Br_2 at equilibrium (b) increase the number of moles of Br_2 at equilibrium (c) leave the number of moles of Br_2 at equilibrium unchanged.

Equilibrium condition

15-13 Distinguish clearly among the following: *mass-action expression, equilibrium constant,* and *equilibrium condition.*

15-14 Write the equilibrium condition for each of the following:
(a) $2H_2O(g) \rightleftharpoons 2H_2(g) + O_2(g)$
(b) $2NO(g) + O_2(g) \rightleftharpoons 2NO_2(g)$
(c) $O_2(g) + 2SO_2(g) \rightleftharpoons 2SO_3(g)$
(d) $4HCl(g) + O_2(g) \rightleftharpoons 2H_2O(g) + 2Cl_2(g)$
(e) $NOCl(g) \rightleftharpoons NO(g) + \frac{1}{2}Cl_2(g)$

15-15 Write the equilibrium condition for each of the following:
(a) $NH_4NO_2(s) \rightleftharpoons N_2(g) + 2H_2O(g)$
(b) $Fe_3O_4(s) + H_2(g) \rightleftharpoons 3FeO(s) + H_2O(g)$
(c) $CaCO_3(s) \rightleftharpoons CaO(s) + CO_2(g)$
(d) $H_2O(l) \rightleftharpoons H_2O(g)$
(e) $CO_2(g) + 2NH_3(g) \rightleftharpoons NH_4CO_2NH_2(s)$

15-16 Write the equilibrium condition for each of the following:
(a) $2SO_2(g) \rightleftharpoons S_2(g) + 2O_2(g)$
(b) $SO_2(g) \rightleftharpoons \frac{1}{2}S_2(g) + O_2(g)$
(c) $S_2(g) + 2O_2(g) \rightleftharpoons 2SO_2(g)$
(d) $\frac{1}{2}S_2(g) + O_2(g) \rightleftharpoons SO_2(g)$

15-17 The equilibrium constant K_c for

$$2SO_2(g) + O_2(g) \rightleftharpoons 2SO_3(g)$$

is found to be 249 at a certain temperature. An analysis of the contents of a container holding these three components at that temperature gave the following results: $[SO_3] = 2.62 \ M$, $[SO_2] = 0.149 \ M$, and $[O_2] = 0.449 \ M$. Is the system at equilibrium?

15-18 K_c for $PCl_5(g) \rightleftharpoons PCl_3(g) + Cl_2(g)$ at 250°C is 1.77. A 4.50-liter container holds 2.57 mol of PCl_5, 6.39 mol of PCl_3, and 3.20 mol of Cl_2 at 250°C. Is the system at equilibrium?

15-19 K_c for $2FeBr_3(s) \rightleftharpoons 2FeBr_2(g) + Br_2(g)$ is 0.983 at a certain temperature. A container of volume 6.00 liters holds 4.12 mol of $FeBr_3$, 7.26 mol of $FeBr_2$, and 4.03 mol of Br_2 at this temperature. Is the system at equilibrium?

15-20 At 796°C K_P for $2NOCl(g) \rightleftharpoons 2NO(g) + Cl_2(g)$ is 3.29×10^{-4}. These components in a container at this temperature have the following partial pressures: $P_{NOCl} = 3.46$ atm, $P_{NO} = 0.110$ atm, and $P_{Cl_2} = 0.430$ atm. Is the system at equilibrium?

K_c and K_P

15-21 At 1205°C K_c for $2CO(g) + O_2(g) \rightleftharpoons 2CO_2(g)$ is 7.09×10^{12}. Calculate K_P at this temperature.

15-22 The equilibrium $CO_2(g) + H_2(g) \rightleftharpoons CO(g) + H_2O(g)$ has $K_P = 4.40$ at 2000 K. Calculate K_c at this temperature.

15-23 K_P for the equilibrium

$$NH_4HS(s) \rightleftharpoons NH_3(g) + H_2S(g)$$

is 0.11 at 25°C. Calculate K_c at this temperature.

15-24 K_c for the equilibrium

$$SO_2(g) + \tfrac{1}{2}O_2(g) \rightleftharpoons SO_3(g) \text{ at } 727°C$$

is 16.7. Calculate K_P at this temperature.

Temperature and K

15-25 K_P for the equilibrium

$$I_2(g) + Cl_2(g) \rightleftharpoons 2ICl(g)$$

is 2.0×10^5 at 298 K. If $\Delta H°$ for the reaction is -26.9 kJ, what is K_P at 673 K?

15-26 For the equilibrium

$$CaCO_3(s) \rightleftharpoons CaO(s) + CO_2(g)$$

$K_P = 1.5 \times 10^{-23}$ at 298 K. Calculate K_P at 1273 K, assuming that $\Delta H°$ for the reaction stays constant at 178 kJ.

15-27 At 298 K K_P for the equilibrium

$$2NaHSO_4(s) \rightleftharpoons Na_2S_2O_7(s) + H_2O(g)$$

is 2.5×10^{-7}. If $\Delta H°$ for the reaction is 82.8 kJ, what is the equilibrium pressure of water vapor in this system at 773 K?

15-28 K_P for $H_2(g) + Cl_2(g) \rightleftharpoons 2HCl(g)$ is 1.08 at 298 K and 1.15×10^{-12} at 473 K. Calculate $\Delta H°$ for the reaction.

15-29 For the equilibrium

$$NO_2(g) + CO(g) \rightleftharpoons NO(g) + CO_2(g)$$

$K_P = 6.4 \times 10^{28}$ at 127°C and 9.4×10^{18} at 327°C. Calculate $\Delta H°$ for the reaction.

Equilibrium calculations

15-30 The value of K_c for the equilibrium:

$$CO(g) + H_2O(g) \rightleftharpoons CO_2(g) + H_2(g)$$

at 600 K is 302. A 1.00-liter container contains 0.100 mol of CO, 0.200 mol of H_2O, and 0.300 mol of CO_2 at 600 K at equilibrium. Calculate $[H_2]$.

15-31 At 600 K the value of K_c for

$$CO(g) + H_2O(g) \rightleftharpoons CO_2(g) + H_2(g)$$

is 302. Equal numbers of moles of CO and H_2O are added to a container at 600 K. After equilibrium has been established $[CO_2]$ is found to be 4.60 M. What is $[CO]$ at equilibrium?

15-32 Calculate the original number of moles of CO added in Prob. 15-31 if the container volume was 5.00 liters.

15-33 At 600 K the value of K_c for

$$CO(g) + H_2O(g) \rightleftharpoons CO_2(g) + H_2(g)$$

is 302. Suppose 2.00 mol each of CO and H_2O are added to a 1.00-liter container at 600 K. What is the equilibrium concentration of **(a)** CO_2 **(b)** H_2 **(c)** CO **(d)** H_2O?

15-34 At 600 K the value of K_c for

$$CO(g) + H_2O(g) \rightleftharpoons CO_2(g) + H_2(g)$$

is 302. Suppose 1.67 mol each of CO_2 and H_2 are added to a 4.00-liter container at 600 K. What is the equilibrium concentration of **(a)** CO_2 **(b)** H_2 **(c)** CO **(d)** H_2O?

15-35 At 600 K the value of K_c for

$$CO(g) + H_2O(g) \rightleftharpoons CO_2(g) + H_2(g)$$

is 302. Suppose 0.400 mol each of CO_2, H_2O, CO, and H_2 are added to a 2.00-liter container at 600 K. What are the equilibrium concentrations of all species?

15-36 At 600 K the value of K_c for

$$CO(g) + H_2O \rightleftharpoons CO_2(g) + H_2(g)$$

is 302. Suppose 0.600 mol of CO_2 and 0.400 mol of H_2 are added to a 1.00-liter container at 600 K. What are the equilibrium concentrations of all species?

15-37 At 1000 K the value of K_c for

$$2CO_2(g) \rightleftharpoons 2CO(g) + O_2(g)$$

is 4.5×10^{-23}. The equilibrium concentrations of CO_2 and O_2 are each 4.3×10^{-1} M at 1000 K. Calculate $[CO]$ at this temperature.

15-38 At 1000 K the value of K_c for

$$2CO_2(g) \rightleftharpoons 2CO(g) + O_2(g)$$

is 4.5×10^{-23}. Suppose 1.00 mol of CO_2 is added to a 1.00-liter container at 1000 K. At equilibrium what are the concentrations of all species present?

15-39 At 25°C the value of K_c for

$$N_2(g) + O_2(g) \rightleftharpoons 2NO(g)$$

is 4.5×10^{-31}. Suppose 0.100 mol each of N_2 and O_2 are placed in a 1.00-liter container at 25°C. What is the equilibrium concentration of NO?

15-40 At 25°C the value of K_c for

$$N_2(g) + O_2(g) \rightleftharpoons 2NO(g)$$

is 4.5×10^{-31}. In a 1.00-liter container at 25°C are placed 0.050 mol of N_2 and 0.090 mol of O_2. What is the equilibrium concentration of NO?

15-41 At 150°C the value of K_P for

$$PCl_5(g) \rightleftharpoons PCl_3(g) + Cl_2(g)$$

is 8.2×10^{-3}. Suppose 1.0 mol of PCl_5 is placed in a 10.0-liter container at 150°C. Calculate the equilibrium partial pressure of Cl_2.

15-42 At 1400 K the value of K_c for

$$2HBr(g) \rightleftharpoons H_2(g) + Br_2(g)$$

is 1.5×10^{-5}. Calculate the equilibrium concentration of H_2 in a 0.500-liter container in which 0.118 mol of HBr has been placed at 1400 K.

15-43 At 1400 K the value of K_c for

$$2HBr \rightleftharpoons H_2(g) + Br_2(g)$$

is 1.5×10^{-5}. Calculate the concentration of H_2 in a 0.37-liter container in which 0.10 mol of HBr and 0.15 mol of Br_2 have been placed and allowed to reach equilibrium at 1400°C.

15-44 At 800°C K_P for $NOCl(g) \rightleftharpoons NO(g) + \frac{1}{2}Cl_2(g)$ is 1.8×10^{-2}. A certain container was filled with NOCl and the system allowed to reach equilibrium. If P_{NOCl} ended up at 0.657 atm, what was the partial pressure of NO?

Equilibrium—general

15-45 Describe two different kinds of situations in which $K_c = K_P$.

15-46 Nitric oxide, NO, is produced in the atmosphere both as a result of human activity and from "natural" causes, such as lightning. The minimum possible concentration of NO can be determined from the equilibrium: $N_2(g) + O_2(g) \rightleftharpoons 2NO(g)$ for which $\Delta H° = 180$ kJ and $K_c = 4.5 \times 10^{-31}$ at 25°C. Calculate the "background" partial pressure of NO in "pure" air at **(a)** 25°C and **(b)** 1000°C. Assume that air is 21 percent O_2 and 78 percent N_2 *by volume*.

15-47 Suppose 1.0 mol each of SO_2, O_2, and SO_3 are added to a 1.0-liter container at 600 K and allowed to reach equilibrium: $2SO_2(g) + O_2(g) \rightleftharpoons 2SO_3(g)$. If K_c is 8.3×10^2 at this temperature, what are the equilibrium concentrations of all species?

15-48 Dimers ("double molecules") of trifluoroacetic acid dissociate according to the equilibrium

$$(CF_3COOH)_2(g) \rightleftharpoons 2CF_3COOH(g)$$

If the density of trifluoroacetic acid is 4.30 g liter^{-1} at 0.908 atm and 118°C, what is K_c for the above dissociation?

15-49 Solid NH_4HS is placed in a container, where it decomposes to form NH_3 and H_2S:

$$NH_4HS(s) \rightleftharpoons NH_3(g) + H_2S(g)$$

The total gas pressure is measured as 0.659 atm. If NH_3 is added to bring the total equilibrium pressure to 1.250 atm what is the new partial pressure of H_2S? (Assume that the temperature remains constant.)

15-50 At 1000 K the equilibrium constant K_P for the decomposition of solid $CaCO_3$ to form solid CaO and gaseous CO_2 is 4.0×10^{-2}. Some $CaCO_3$ is added to a 5.00-liter container at 1000 K. After equilibrium has been established, how many grams of CaO are present?

15-51 Solid ammonium carbamate, $NH_4CO_2NH_2$, decomposes to form gaseous NH_3 and CO_2. At 37°C K_c for this decomposition is 4.1. A 0.40-liter container is maintained at 37°C and ammonium carbamate is slowly introduced. How many grams must be added before any solid will remain at equilibrium?

16

AQUEOUS SOLUTIONS: ACID-BASE EQUILIBRIA

TO THE STUDENT

In this chapter and the next we consider the special case of equilibrium systems in the solvent water. All the properties of equilibrium systems discussed in Chap. 15 are exhibited in these systems, too. Le Châtelier's principle, for example, is of great value in predicting equilibrium changes in aqueous solutions. And as is the case with gaseous systems, all reactions in aqueous solution tend to go toward equilibrium.

We first consider those equilibria which are classed as acid-base equilibria, according to the Arrhenius or Brønsted-Lowry approaches. These equilibria are all homogeneous: all reacting species are in solution. In Chap. 17 we consider solubility (heterogeneous) and complex-ion (homogeneous) equilibria and also some more complicated systems consisting of several interrelated simultaneous equilibria.

16-1
The dissociation of weak acids

Dissociation constants for weak acids

We have seen (Secs. 12.5, 13-1) that a weak acid is one which is not completely dissociated. The *Arrhenius* equation for the dissociation of the weak acid HA is

$$HA \rightleftharpoons H^+ + A^-$$

for which the equilibrium condition is

$$\frac{[H^+][A^-]}{[HA]} = K$$

Because this equilibrium is called a *dissociation equilibrium,* the equilibrium constant is called a *dissociation constant,* and is often designated as K_{diss}, K_d, or K_a (*a* stands for acid). It is also called an *ionization constant,* K_i.

The *Brønsted-Lowry* equation for the dissociation of HA emphasizes the fact that a proton is transferred to water,

$$HA + H_2O \rightleftharpoons H_3O^+ + A^-$$

and so the equilibrium condition is written as

$$\frac{[H_3O^+][A^-]}{[HA][H_2O]} = K'$$

In dilute solutions, however, the concentration of H_2O molecules is essentially the same as that in pure water, about 55 M. This is a large concentration and because *comparatively* few H_2O molecules are needed for hydration, reaction, etc., the concentration of H_2O in dilute solutions remains essentially constant. (For example, the concentration of hydrated protons in 0.10 M acetic acid is only about 1×10^{-3} M; so the number of water molecules involved in this hydration is small in comparison with the total number present.) Considering H_2O to be a constant, then, we can write

$$\frac{[H_3O^+][A^-]}{[HA]} = [H_2O]K' = K$$

where K is a constant by virtue of the fact that it is the product of two constant quantities.

The two expressions

$$\frac{[H^+][A^-]}{[HA]} = K \qquad \text{(Arrhenius)}$$

and
$$\frac{[H_3O^+][A^-]}{[HA]} = K \qquad \text{(Brønsted-Lowry)}$$

say exactly the same thing: that at equilibrium the product of the concentrations of (hydrated) hydrogen ions and A^- ions divided by the concentration of undissociated HA molecules is a constant, the *dissociation constant* of HA.

It should be emphasized at this point that we are talking about dilute solutions. We will see in Chap. 18 that, strictly speaking, the mass-action expression is a constant at equilibrium only if it is expressed in terms of *chemical activities*. The chemical activity of a species in solution is approximately equal to its concentration, this approximation being better the more dilute the solution is. When solute

Table 16-1 Dissociation constants of weak acids (25°C)	Acid	Dissociation reaction (a) Arrhenius (b) Brønsted-Lowry	K_a
	Acetic	(a) $HC_2H_3O_2 \rightleftharpoons H^+ + C_2H_3O_2^-$ (b) $HC_2H_3O_2 + H_2O \rightleftharpoons H_3O^+ + C_2H_3O_2^-$	1.8×10^{-5}
	Chlorous	(a) $HClO_2 \rightleftharpoons H^+ + ClO_2^-$ (b) $HClO_2 + H_2O \rightleftharpoons H_3O^+ + ClO_2^-$	1.1×10^{-2}
	Hydrocyanic	(a) $HCN \rightleftharpoons H^+ + CN^-$ (b) $HCN + H_2O \rightleftharpoons H_3O^+ + CN^-$	4.0×10^{-10}
	Hydrofluoric	(a) $HF \rightleftharpoons H^+ + F^-$ (b) $HF + H_2O \rightleftharpoons H_3O^+ + F^-$	6.7×10^{-4}
	Hypochlorous	(a) $HOCl \rightleftharpoons H^+ + OCl^-$ (b) $HOCl + H_2O \rightleftharpoons H_3O^+ + OCl^-$	3.2×10^{-8}

particles are ions, a solution must be quite dilute in order for activity and concentration to be essentially the same.

The *strength* of an acid, the extent to which it dissociates in solution, is indicated by the size of its dissociation constant. The weaker an acid is, the smaller is its dissociation constant. Table 16-1 shows values of K_a for a few weak acids. In each case the dissociation is shown according to (a) the Arrhenius description and (b) the Brønsted-Lowry description. Additional dissociation constants for weak acids are shown in App. H.

Polyprotic acids   Many acids have more than one available proton. These acids are called *diprotic* if there are two available protons per molecule, *triprotic* if there are three available protons, etc. Thus sulfurous acid, H_2SO_3, is a diprotic acid and dissociates in two stages, or steps, each with its own dissociation constant. These steps can be written as Arrhenius dissociations:

$$H_2SO_3 \rightleftharpoons H^+ + HSO_3^- \qquad \frac{[H^+][HSO_3^-]}{[H_2SO_3]} = K_1 = 1.3 \times 10^{-2}$$

$$HSO_3^- \rightleftharpoons H^+ + SO_3^{2-} \qquad \frac{[H^+][SO_3^{2-}]}{[HSO_3^-]} = K_2 = 6.3 \times 10^{-8}$$

or as Brønsted-Lowry proton transfers:

$$H_2SO_3 + H_2O \rightleftharpoons H_3O^+ + HSO_3^- \qquad \frac{[H_3O^+][HSO_3^-]}{[H_2SO_3]} = K_1 = 1.3 \times 10^{-2}$$

$$HSO_3^- + H_2O \rightleftharpoons H_3O^+ + SO_3^{2-} \qquad \frac{[H_3O^+][SO_3^{2-}]}{[HSO_3^-]} = K_2 = 6.3 \times 10^{-8}$$

Table 16-2 gives dissociation constants for a few polyprotic acids. (More are given in App. H.) Note that for any given acid, K_2 is smaller than K_1. This is true because the tendency for a proton to leave a negatively charged ion is less than for one to leave the corresponding neutral molecule.

Weak-acid   The methods introduced in Sec. 15-3 are easily applied to problems involving the
calculations   dissociation of weak acids.

Table 16-2
Dissociation constants
of polyprotic acids
(25°C)

Acid	Formula	K_a	
Ascorbic	$H_2C_6H_6O_6$	K_1	5.0×10^{-5}
		K_2	1.5×10^{-12}
Carbonic	H_2CO_3	$\{K_1$	4.2×10^{-7}
	$(H_2O + CO_2)\}$	K_2	5.6×10^{-11}
Phosphoric	H_3PO_4	K_1	7.6×10^{-3}
		K_2	6.3×10^{-8}
		K_3	4.4×10^{-13}
Sulfuric	H_2SO_4	K_1	(Large; strong acid)
		K_2	1.2×10^{-2}
Sulfurous	H_2SO_3	$\{K_1$	1.3×10^{-2}
	$(H_2O + SO_2)\}$	K_2	6.3×10^{-8}

● **EXAMPLE 16-1**

Problem What is the concentration of each solute-derived species in a 0.50 M solution of acetic acid, $HC_2H_3O_2$?

Solution The dissociation equilibrium is

$$HC_2H_3O_2 \rightleftharpoons H^+ + C_2H_3O_2^-$$

for which $K_a = 1.8 \times 10^{-5}$ (Table 14-1). Let x equal the number of moles of $HC_2H_3O_2$ in 1 liter which must dissociate in order to establish equilibrium. Then the number of moles of $HC_2H_3O_2$ per liter at equilibrium is $0.50 - x$. Since *one* mole of $HC_2H_3O_2$ dissociates to give *one* mole of H^+ plus *one* mole of $C_2H_3O_2^-$, the number of moles of H^+ and the number of moles of $C_2H_3O_2^-$ per liter at equilibrium must *each* be equal to x. The initial concentrations (before any dissociation has occurred) and the final equilibrium concentrations are tabulated below:

Solute species	Initial concentration, M	Equilibrium concentration, M
$HC_2H_3O_2$	0.50	$0.50 - x$
H^+	0	x
$C_2H_3O_2^-$	0	x

The equilibrium condition is

$$\frac{[H^+][C_2H_3O_2^-]}{[HC_2H_3O_2]} = K_a$$

$$\frac{x(x)}{0.50 - x} = 1.8 \times 10^{-5}$$

$$x^2 = 9.0 \times 10^{-6} - (1.8 \times 10^{-5})x$$

$$x^2 + (1.8 \times 10^{-5})x - 9.0 \times 10^{-6} = 0$$

Substituting in the quadratic formula (App. D)

$$x = \frac{-(1.8 \times 10^{-5}) \pm \sqrt{(1.8 \times 10^{-5})^2 - 4(1)(-9.0 \times 10^{-6})}}{2(1)}$$

and solving for x, we get

$$x = 3.0 \times 10^{-3} \quad \text{or} \quad x = -3.0 \times 10^{-3}$$

We reject the negative root because it corresponds to a negative concentration, and so

$$[HC_2H_3O_2] = 0.50 - x = 0.50 - 3.0 \times 10^{-3} = 0.50 \; M$$

$$[H^+] = x = 3.0 \times 10^{-3} \; M$$

$$[C_2H_3O_2^-] = x = 3.0 \times 10^{-3} \; M$$

Using the quadratic formula in this way gives the answer, but we could have made the job easier by applying the same kind of chemical common sense that we applied in Example 15-8. Because acetic acid is quite weak (K is small), we assume that the number of moles of $HC_2H_3O_2$ dissociated per liter is small in comparison with the total number of moles present. This means that in the expression

$$\frac{x(x)}{0.50 - x} = 1.8 \times 10^{-5}$$

if $x \ll$ (is much less than) 0.50, then $0.50 - x \approx 0.50$ and

$$\frac{x(x)}{0.50} \approx 1.8 \times 10^{-5}$$

This is easier to solve:

$$x^2 \approx 9.0 \times 10^{-6}$$

$$x \approx 3.0 \times 10^{-3}$$

Now we can check our assumption that $x \ll 0.50$. It is indeed, because $0.50 - x = 0.50 - 3.0 \times 10^{-3} = 0.50$. In other words, x is negligible in comparison with 0.50. ●

● **EXAMPLE 16-2**

Problem   In 0.50 M $HC_2H_3O_2$, what percent of the $HC_2H_3O_2$ is dissociated?

Solution   In our solution to the problem of Example 16-1 we let x equal the number of moles of $HC_2H_3O_2$ that dissociate per liter. Percent dissociation is the number of moles dissociated per liter divided by the total number of moles per liter, all times 100, or

$$\% \text{ diss} = \frac{3.0 \times 10^{-3}}{0.50} \times 100 = 0.60\% \quad ●$$

A point that is sometimes missed at this stage is that the equality expressed by the equilibrium condition

$$\frac{[H^+][A^-]}{[HA]} = K_a$$

is true no matter how the equilibrium is established and what the individual concentrations are. More specifically, it is true whether or not $[H^+] = [A^-]$.

● **EXAMPLE 16-3**

Problem   A solution is prepared by adding 0.40 mol of sodium acetate, $NaC_2H_3O_2$, and 0.50 mol of acetic acid to enough water to make 1.0 liter. Calculate the concentrations of all solute species and the percent dissociation of acetic acid in this solution.

Solution   Sodium acetate, a salt, dissociates completely as it dissolves,

$$NaC_2H_3O_2(s) \longrightarrow Na^+ + C_2H_3O_2^-$$

making the solution 0.40 M in Na^+ and 0.40 M in $C_2H_3O_2^-$. The acetic acid dissolves but dissociates incompletely, coming to equilibrium.

$$HC_2H_3O_2 \rightleftharpoons H^+ + C_2H_3O_2^-$$

Now, as before, let x equal the number of moles of $HC_2H_3O_2$ dissociated per liter. This then means $[H^+] = x$. But $[C_2H_3O_2^-]$ is different this time, because there are *two* sources for $C_2H_3O_2^-$ ions: the completely dissociated salt $NaC_2H_3O_2$ and the slightly dissociated acid $HC_2H_3O_2$. The total $[C_2H_3O_2^-]$ is the sum of the contributions of the two: 0.40 M (from the salt) plus x molar from the acid. As before, $[HC_2H_3O_2]$ at equilibrium equals $0.50 - x$. In summary,

Solute species	Initial concentration, M	Equilibrium concentration, M
$HC_2H_3O_2$	0.50	$0.50 - x$
H^+	0	x
$C_2H_3O_2^-$	0.40	$0.40 + x$

The equilibrium condition is

$$\frac{[H^+][C_2H_3O_2^-]}{[HC_2H_3O_2]} = K_a$$

$$\frac{x(0.40 + x)}{0.50 - x} = 1.8 \times 10^{-5}$$

Rather than face the rigors of the quadratic formula again, we simply observe the low value of K_a and predict that the number of moles of $HC_2H_3O_2$ dissociated is likely to be much less than the total number of moles of $HC_2H_3O_2$, so that if

$$x \ll 0.50$$

then, probably,

$$x \ll 0.40$$

and so

$$\frac{x(0.40)}{0.50} \approx 1.8 \times 10^{-5}$$

Solving for x, we get

$$x \approx 2.3 \times 10^{-5}$$

(Were our approximations justified? Yes, 2.3×10^{-5} is indeed so much smaller than 0.50 and than 0.40 that it is negligible in comparison with each of these numbers.) Thus,

$$[HC_2H_3O_2] = 0.50 - x = 0.50 - (2.3 \times 10^{-5}) = 0.50 \ M$$

$$[H^+] = x = 2.3 \times 10^{-5} \ M$$

$$[C_2H_3O_2^-] = 0.40 + x = 0.40 + (2.3 \times 10^{-5}) = 0.40 \ M$$

$$[Na^+] = 0.40 \ M \quad \text{(Do not forget this one!)}$$

$$\% \text{ diss} = \frac{\text{moles dissociated}}{\text{total moles}} \times 100 = \frac{2.3 \times 10^{-5}}{0.50} \times 100 = 4.6 \times 10^{-3} \ \%$$

Compare the percent dissociation found in Example 16-3 with that found in Example 16-2.

Example	Solution	% of $HC_2H_3O_2$ dissociated
16-2	$0.50 \ M \ HC_2H_3O_2$	0.60
16-3	$\begin{cases} 0.50 \ M \ HC_2H_3O_2 \\ + \\ 0.40 \ M \ C_2H_3O_2^- \end{cases}$	0.0046

Again we see an illustration of Le Châtelier's principle. The extra $C_2H_3O_2^-$ ions (Example 16-3) have kept the equilibrium

$$HC_2H_3O_2 \rightleftharpoons H^+ + C_2H_3O_2^-$$

shifted well to the left, repressing the dissociation of $HC_2H_3O_2$. This is an example of the so-called *common-ion effect* in which the presence of extra ions in solution represses a dissociation. The dissociation of $HC_2H_3O_2$ can also be repressed by the presence of added H^+, as from the strong acid HCl.

Le Châtelier's principle also predicts that the degree of dissociation of a weak electrolyte is greater the more dilute its solution is.

● **EXAMPLE 16-4**

Problem   Calculate the percent dissociation in a $0.10\ M\ HC_2H_3O_2$ solution.

Solution
$$HC_2H_3O_2 \rightleftharpoons H^+ + C_2H_3O_2^-$$

Let x equal the number of moles of $HC_2H_3O_2$ dissociated per liter. Then at equilibrium:

$$[H^+] = x \qquad [C_2H_3O_2^-] = x$$

and
$$[HC_2H_3O_2] = 0.10 - x$$

The equilibrium condition is

$$\frac{[H^+][C_2H_3O_2^-]}{[HC_2H_3O_2]} = K_a$$

$$\frac{x(x)}{0.10 - x} = 1.8 \times 10^{-5}$$

Now we will make the assumption that $x \ll 0.10$, so that $0.10 - x \approx 0.10$. Then

$$\frac{x(x)}{0.10} \approx 1.8 \times 10^{-5}$$

which gives

$$x \approx 1.3 \times 10^{-3}$$

(Was our assumption justified? Yes; $1.3 \times 10^{-3} \ll 0.10$.)

$$\% \text{ diss} = \frac{1.3 \times 10^{-3}}{0.10} \times 100 = 1.3\% \ ●$$

Now compare the results of Examples 16-2 and 16-4:

Example	$HC_2H_3O_2$	% diss
16-2	0.50 M	0.60
16-4	0.10 M	1.3

As Le Châtelier principle predicts, percent dissociation increases as a solution is diluted. The total number of solute particles increases and partially compensates for the dilution effect.

● **EXAMPLE 16-5**

Problem   Calculate the concentrations of all solute species in a $0.10\ M\ H_2SO_3$ solution.

Solution   Sulfurous acid, H_2SO_3, is diprotic; so we have two dissociations to take into account:

$$H_2SO_3 \rightleftharpoons H^+ + HSO_3^- \qquad K_1 = 1.3 \times 10^{-2}$$
$$HSO_3^- \rightleftharpoons H^+ + SO_3^{2-} \qquad K_2 = 6.3 \times 10^{-8}$$

Observe that K_2 is much smaller than K_1. This means that although both dissociations produce H^+, the contribution made by the second dissociation may be neglected in comparison with that made by the first. In addition, we can neglect the HSO_3^- used up in the second dissociation in comparison with that formed by the first. Now let x equal the number of moles of H_2SO_3 dissociated per liter. Then at equilibrium

$$[H^+] = x \qquad [HSO_3^-] = x$$

$$[H_2SO_3] = 0.10 - x$$

(Notice that in setting $[H^+] = x$ we are ignoring the contribution made by the second dissociation.) The equilibrium condition is

$$\frac{[H^+][HSO_3^-]}{[H_2SO_3]} = K_1$$

$$\frac{x(x)}{0.10 - x} = 1.3 \times 10^{-2}$$

But in this case K_1 is large enough to tell us that x cannot be neglected in comparison with 0.10 in the denominator. (Try neglecting it and solving the resulting approximated equation. x is then equal to 0.036, which is not negligible compared to 0.10.) Therefore,

$$x^2 = (0.10 - x)(1.3 \times 10^{-2})$$

By rearranging, solving with the use of the quadratic formula, and selecting the positive root, we get

$$x = 3.0 \times 10^{-2}$$

and so
$$[H^+] = x = 3.0 \times 10^{-2}\ M$$

$$[HSO_3^-] = x = 3.0 \times 10^{-2}\ M$$

$$[H_2SO_3] = 0.10 - (3.0 \times 10^{-2}) = 0.07\ M$$

We are not finished, however, because a small amount of SO_3^{2-} (sulfite ion) is present. What is its concentration? Looking at the second dissociation, we see that its equilibrium condition is

$$\frac{[H^+][SO_3^{2-}]}{[HSO_3^-]} = K_2$$

Now $[H^+]$ and $[HSO_3^-]$ refer to *total* concentrations in solution, *not* just the proportions produced by one or the other step. Moreover, we found from our first calculation that $[H^+] = [HSO_3^-] = 3.0 \times 10^{-2}\ M$, and so

$$\frac{(3.0 \times 10^{-2})[SO_3^{2-}]}{3.0 \times 10^{-2}} = 6.3 \times 10^{-8}$$

$$[SO_3^{2-}] = 6.3 \times 10^{-8}\ M$$

Added note: Look at the second dissociation equation. 6.3×10^{-8} must also be the number of moles of H^+ per liter *produced by the second dissociation only*. It is indeed negligible compared to that produced by the first, $3.0 \times 10^{-2}\ M$. Also, 6.3×10^{-8} represents the number of moles of HSO_3^- per liter used up in the second dissociation. This is certainly small compared with the 3.0×10^{-2} mol liter^{-1} formed in the first, thus fully justifying our neglecting it in our first calculation. ●

ADDED COMMENT

You may not be familiar with the technique of solving an approximate equation and, in fact, may be suspicious of it. It is perfectly legitimate, however, and is foolproof if used correctly. For example, when you assume that you can "neglect x in comparison with 0.50" in the term $0.50 - x$, if the assumption is not justified, you will soon find out, because you can compare x with 0.50 after you have evaluated it. If x should come out to be 0.04 for example, it is clearly *not* negligible in comparison with 0.50 because $0.50 - 0.04 = 0.46$. This means that you must use the quadratic formula. It turns out that "negligible" usually means less than 5 percent of the original concentration.

16-2 The dissociation of weak bases

The dissociation of a weak base in aqueous solution is similar to that of a weak acid except that attention is focused on the production of OH^- ions.

Dissociation constants for bases

If we consider the hydroxide BOH to be an *Arrhenius* weak base, then its dissociation is shown as

$$BOH \rightleftharpoons B^+ + OH^-$$

for which the equilibrium condition is

$$\frac{[B^+][OH^-]}{[BOH]} = K$$

K, the dissociation constant, is often designated K_{diss}, K_d, K_b (*b* for base), or K_i.

A *Brønsted-Lowry* base is a proton acceptor. If C is such a base, then the dissociation may be represented as

$$C + H_2O \rightleftharpoons CH^+ + OH^-$$

for which the equilibrium condition is

$$\frac{[CH^+][OH^-]}{[C]} = K_b$$

where, as before, we have absorbed the essentially constant concentration of water into K_b.

Although the Brønsted-Lowry viewpoint is generally the more useful one for weak bases, the two approaches are equivalent. Consider, for example, the substance ammonia, NH_3. Ammonia is a gas at ordinary temperatures and pressures and is very soluble in water. Its aqueous solution is basic. In the past this led to the hypothesis that ammonia reacts with water to form a weak Arrhenius base, formulated NH_4OH, and called *ammonium hydroxide*,

$$NH_4OH \rightleftharpoons NH_4^+ + OH^-$$

for which the equilibrium condition is

$$\frac{[NH_4^+][OH^-]}{[NH_4OH]} = K_b$$

Table 16-3
Dissociation constants
of weak bases (25°C)

Base	Dissociation reaction (Brønsted-Lowry)	K_b
Ammonia	$NH_3 + H_2O \rightleftharpoons NH_4^+ + OH^-$	1.8×10^{-5}
Hydroxylamine	$NH_2OH + H_2O \rightleftharpoons NH_3OH^+ + OH^-$	9.1×10^{-9}
Methylamine	$CH_3NH_2 + H_2O \rightleftharpoons CH_3NH_3^+ + OH^-$	4.4×10^{-4}
Nicotine	$C_{10}H_{14}N_2 + H_2O \rightleftharpoons C_{10}H_{14}N_2H^+ + OH^-$	7.4×10^{-7}
	$C_{10}H_{14}N_2H^+ + H_2O \rightleftharpoons C_{10}H_{14}N_2H_2^{2+} + OH^-$	1.4×10^{-11}
Phosphine	$PH_3 + H_2O \rightleftharpoons PH_4^+ + OH^-$	1×10^{-14}

Not only can NH_4OH *not* be isolated as a pure substance, however, but it can also be shown that NH_4OH molecules do not exist even in solution.

One way around the problem is to treat ammonia as a Brønsted-Lowry base:

$$NH_3 + H_2O \rightleftharpoons NH_4^+ + OH^-$$

for which the equilibrium condition is

$$\frac{[NH_4^+][OH^-]}{[NH_3]} = K_b$$

It is clear that whether we consider the base to be NH_4OH or NH_3, the result is essentially the same. At equilibrium

$$\frac{[NH_4^+][OH^-]}{[\text{Undissociated species}]} = K_b$$

and the actual nature of the undissociated species is not important. (It is actually more complex than either NH_3 or NH_4OH.) This is why the dissociation constant K_b is sometimes listed under NH_3 and in other references under NH_4OH. Its numerical value is 1.8×10^{-5} and is included among the dissociation constants of several bases listed in Table 16-3 and in App. H. Commercially available solutions of ammonia in water supplied for use in the laboratory are invariably labeled "ammonium hydroxide." Nevertheless, the Brønsted-Lowry perspective is usually more convenient for describing weak bases in water, and so the dissociation equations have been written accordingly in Table 16-3.

Weak-base
calculations   ● **EXAMPLE 16-6**

Problem   Calculate the concentration of each solute species in a 0.40 M NH_3 solution. What is the percent dissociation in this solution?

Solution   The dissociation equilibrium is

$$NH_3 + H_2O \rightleftharpoons NH_4^+ + OH^-$$

If x is the number of moles of NH_3 per liter which have dissociated (". . . have accepted protons" might be better language to use), then at equilibrium

$$[NH_4^+] = x \qquad [OH^-] = x \qquad [NH_3] = 0.40 - x$$

The equilibrium condition is $\dfrac{[NH_4^+][OH^-]}{[NH_3]} = K_b$

$$\frac{x(x)}{0.40 - x} = 1.8 \times 10^{-5}$$

K_b is a small number, and so we will assume that $x \ll 0.40$, so that $0.40 - x \approx 0.40$. Then

$$\frac{x(x)}{0.40} \approx 1.8 \times 10^{-5}$$

which gives us

$$x \approx 2.7 \times 10^{-3}$$

(A check shows that our assumption was valid; $2.7 \times 10^{-3} \ll 0.40$.) Therefore

$$[NH_4^+] = x = 2.7 \times 10^{-3}\,M$$

$$[OH^-] = x = 2.7 \times 10^{-3}\,M$$

$$[NH_3] = 0.40 - x = 0.40 - (2.7 \times 10^{-3}) = 0.40\,M$$

$$\% \text{ diss} = \frac{\text{moles per liter dissociated}}{\text{total moles per liter}} \times 100$$

$$= \frac{2.7 \times 10^{-3}}{0.40} \times 100 = 0.68\%$$

ADDED COMMENT

Do you see that it would have made no difference at all if we had written the dissociation as

$$NH_4OH \rightleftharpoons NH_4^+ + OH^-$$

We would have written

$$[NH_4^+] = x \qquad [OH^-] = x \qquad [NH_4OH] = 0.40 - x$$

which, when substituted into the equilibrium condition

$$\frac{[NH_4^+][OH^-]}{[NH_4OH]} = K$$

gives us

$$\frac{x(x)}{0.40 - x} = 1.8 \times 10^{-5}$$

so that

$$[NH_4^+] = x = 2.7 \times 10^{-3}\,M$$

$$[OH^-] = x = 2.7 \times 10^{-3}\,M$$

$$[NH_4OH] = 0.40 - x = 0.40 - (2.7 \times 10^{-3}) = 0.40\,M$$

The only difference is whether we wish to call the undissociated species NH_3 or NH_4OH.

EXAMPLE 16-7

Problem   Suppose 0.10 mol of NH_3 is added to 1.0 liter of 0.10 M NaOH. What is the NH_4^+ ion concentration, if the volume of the solution remains unchanged?

Solution   The dissociation is

$$NH_3 + H_2O \rightleftharpoons NH_4^+ + OH^-$$

but this time most of the OH^- ions come from the strong base NaOH.

$$NaOH(s) \longrightarrow Na^+ + OH^-$$

Since this is a 0.10 M NaOH solution, $[OH^-]$ before adding the NH_3 is 0.10 M. Now let x equal the number of moles of NH_3 per liter which dissociate. Then

$$[NH_4^+] = x \qquad [OH^-] = 0.10 + x \qquad [NH_3] = 0.10 - x$$

The equilibrium condition is

$$\frac{[NH_4^+][OH^-]}{[NH_3]} = K_b$$

Substituting, we get

$$\frac{x(0.10 + x)}{0.10 - x} = 1.8 \times 10^{-5}$$

If we assume that $x \ll 0.10$, then $0.10 + x \approx 0.10$, and $0.10 - x \approx 0.10$, and so

$$\frac{x(0.10)}{0.10} \approx 1.8 \times 10^{-5}$$

$$x \approx 1.8 \times 10^{-5}$$

(The assumption was justified; $1.8 \times 10^{-5} \ll 0.10$.) Thus

$$[NH_4^+] = x = 1.8 \times 10^{-5} \ M \quad ●$$

16-3
The dissociation of water

Pure water shows a definite, though low, electrical conductivity, a consequence of its ability to undergo self-dissociation (autodissociation).

The ion product for water

The dissociation of water may be written as

$$H_2O \rightleftharpoons H^+ + OH^-$$

or as

$$H_2O + H_2O \rightleftharpoons H_3O^+ + OH^-$$

For this dissociation the equilibrium condition may be written as

$$\frac{[H^+][OH^-]}{[H_2O]} = K'$$

or as

$$\frac{[H_3O^+][OH^-]}{[H_2O]^2} = K''$$

In either case since the concentration of H_2O molecules is essentially constant,

$$[H^+][OH^-] = K'[H_2O] = K_w$$

or

$$[H_3O^+][OH^-] = K''[H_2O]^2 = K_w$$

The constant K_w, the *dissociation constant for water*, has the value of 1.0×10^{-14} at 25°C. (K_w is also called the *ion product for water*.) In water, then, whether it is the purest of pure distilled water, or the dirtiest from the Limpopo River, the product of the concentrations of the (hydrated) hydrogen ion and the hydroxide ion is a constant: 1.0×10^{-14} at 25°C.

Acidic, basic, and neutral solutions   An *acidic solution* is defined as one in which the hydrogen- (hydronium-) ion concentration is greater than the hydroxide-ion concentration. A *basic solution* is one in which the reverse is true ($[OH^-]$ exceeds $[H^+]$). Finally, a *neutral solution* is one in which these two concentrations are equal ($[OH^-]$ equals $[H^+]$).

Because $[H^+][OH^-]$ equals a constant, these two concentrations can be considered to be "balanced" against each other: when either one rises, the other must fall. They are not independent, but are linked by the relationship $[H^+][OH^-] = K_w$, which allows us to calculate one concentration from the other.

● **EXAMPLE 16-8**

Problem   What is the hydroxide-ion concentration in a 0.020 M HCl solution?

Solution   Because HCl is a strong acid, it is essentially completely dissociated into H^+ and Cl^-. Furthermore, essentially all the H^+ in the solution comes from the HCl, since H_2O is such a weak electrolyte. Thus, $[H^+] = 0.020$ M.

$$[H^+][OH^-] = K_w$$

$$[OH^-] = \frac{K_w}{[H^+]} = \frac{1.0 \times 10^{-14}}{0.020} = 5.0 \times 10^{-13} \ M \ ●$$

● **EXAMPLE 16-9**

Problem   What is the hydrogen-ion concentration in a neutral solution?

Solution   In a neutral solution,

$$[H^+] = [OH^-]$$

Therefore

$$[H^+][OH^-] = [H^+]^2 = K_w = 1.0 \times 10^{-14}$$

$$[H^+] = \sqrt{1.0 \times 10^{-14}} = 1.0 \times 10^{-7} \ M \ ●$$

pH   The hydrogen-ion concentration in a solution can range from more than 10 M to less than 1×10^{-15} M. The pH scale was devised to express such a wide range of acidities in a more convenient way. *pH* is defined as the *negative logarithm of the hydrogen-ion* (or *hydronium-ion*) *concentration*.[1]

$$pH = -\log[H^+] \quad \text{or} \quad -\log[H_3O^+]$$

Thus for a solution in which $[H^+] = 1 \times 10^{-3}$ M, pH equals 3.0. A neutral solution, in which $[H^+] = 1 \times 10^{-7}$ M, has a pH of 7.0. (Logarithms are reviewed in App. D.)

● **EXAMPLE 16-10**

Problem   What is the pH of a 4.6×10^{-3} M solution of HCl?

Solution   Since HCl is a strong acid and the contribution of H^+ from H_2O is so small, $[H^+] = 4.6 \times 10^{-3}$ M

$$pH = -\log(4.6 \times 10^{-3}) = -\log 4.6 - \log 10^{-3}$$

$$= -0.66 + 3 = 2.34 \qquad ●$$

[1] Again, we are assuming that the chemical activity of a species can be replaced by its concentration. Strictly speaking, pH $= -\log a_{H^+}$, where a represents activity.

● EXAMPLE 16-11

Problem   The pH of a solution is 11.68.   What is the hydrogen-ion concentration?

Solution   This time we have to go backward.

$$pH = -\log [H^+] = 11.68 = 12 - 0.32$$

$$\log [H^+] = 0.32 - 12 = 0.32 + (-12)$$

Now from a table of logarithms or a pocket calculator we can find that 0.32 is the logarithm of 2.1, and since -12 is the logarithm of 10^{-12},

$$[H^+] = 2.1 \times 10^{-12} \, M \quad ●$$

Since at 25°C $[H^+]$ in a neutral solution is $1.0 \times 10^{-7} \, M$, the pH of such a solution is 7.00.   Since an acidic solution has a *higher* $[H^+]$, it must have a pH *less than* 7.00, and since a basic solution has a *lower* $[H^+]$, it must have a pH *greater than* 7.00.

The p__ symbolism has been extended to cover other quantities.   For example,

$$pOH = -\log [OH^-] \qquad \text{and} \qquad pCl = -\log [Cl^-]$$

pOH is useful because of the following relationship:

$$[H^+][OH^-] = K_w$$

Taking the logarithms of both sides, we get

$$\log [H^+] + \log [OH^-] = \log K_w$$

and

$$-\log [H^+] - \log [OH^-] = -\log K_w$$

and so

$$pH + pOH = pK_w$$

At 25°C K_w is 1.0×10^{-14}; and so $pK_w = 14.00$.   Therefore,

$$pH + pOH = 14.00 \qquad \text{(at 25°C)}$$

● EXAMPLE 16-12

Problem   What are the pH and pOH of a 0.016 M NaOH solution?

Solution   NaOH is a strong base.   Ignoring the negligible contribution made to $[OH^-]$ by the dissociation of water, we can say that $[OH^-] = 0.016 \, M = 1.6 \times 10^{-2} \, M$.   Then

$$pOH = -\log (1.6 \times 10^{-2}) = 1.80$$

To find pH we have a choice of methods:

Method I: $[H^+][OH^-] = K_w$

$$[H^+] = \frac{K_w}{[OH^-]}$$

$$= \frac{1.0 \times 10^{-14}}{1.6 \times 10^{-2}} = 6.3 \times 10^{-13} \, M$$

$$pH = -\log (6.3 \times 10^{-13}) = 12.20$$

Method II: $pH + pOH = 14.0$

$$pH = 14.0 - pOH = 14.0 - 1.80 = 12.20 \quad ●$$

Table 16-4
pH, pOH, and acidity
(25°C)

	[H$^+$], M	pH	[OH$^-$], M	pOH	
More acidic	10	-1	10^{-15}	15	More acidic
	1	0	10^{-14}	14	
	10^{-1}	1	10^{-13}	13	
	10^{-2}	2	10^{-12}	12	
	10^{-3}	3	10^{-11}	11	
	10^{-4}	4	10^{-10}	10	
	10^{-5}	5	10^{-9}	9	
	10^{-6}	6	10^{-8}	8	
Neutral	10^{-7}	7	10^{-7}	7	Neutral
More basic	10^{-8}	8	10^{-6}	6	More basic
	10^{-9}	9	10^{-5}	5	
	10^{-10}	10	10^{-4}	4	
	10^{-11}	11	10^{-3}	3	
	10^{-12}	12	10^{-2}	2	
	10^{-13}	13	10^{-1}	1	
	10^{-14}	14	1	0	
	10^{-15}	15	10	-1	

The relationship among [H$^+$], [OH$^-$], pH, pOH, acidity, and basicity are summarized in Table 16-4. Note that each line in the table represents an *increase* in [H$^+$] and a *decrease* in [OH$^-$] *by a factor of 10* compared to the line below. One tends to forget that a change of only one pH unit corresponds to such a large change in [OH$^-$] and [H$^+$].

In Fig. 16-1 are listed the pH values of a number of common solutions.

16-4 Hydrolysis

A useful term which arises out of the Arrhenius acid-base definition is *hydrolysis*. (The word means "water-splitting.") A hydrolysis reaction is a reaction between an ion and water.

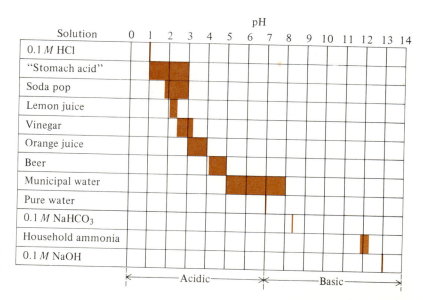

Figure 16-1
pH values of some
common substances.

Anion hydrolysis   The hydrolysis of an anion can be represented as

$$A^- + H_2O \rightleftharpoons HA + OH^-$$

The reaction is thus a removal of protons from water molecules to form HA molecules and hydroxide ions. The latter make the solution basic. Why should this reaction take place? From the Arrhenius perspective it occurs because HA is a weak acid. In other words, stating that HA is a weak acid is equivalent to saying that the bond in the molecule is strong enough to prevent HA from dissociating completely. This means that when A^- is in solution, it has a strong affinity for a proton, strong enough to remove one from a water molecule.

From the Brønsted-Lowry perspective anion hydrolysis is simply a proton transfer from H_2O to an anion. Anion hydrolysis occurs when A^- is a strong enough base to remove a proton from water and establish the above equilibrium.

How can the extent of a hydrolysis reaction be predicted qualitatively? A weak acid is the product of an anion hydrolysis, and the weaker this acid is, the greater the extent of hydrolysis of the anion. For example, the cyanide ion CN^- can be predicted to hydrolyze more than the fluoride ion F^- because HCN is a weaker acid ($K_a = 1.0 \times 10^{-10}$) than HF ($K_a = 6.7 \times 10^{-4}$).

The weaker an acid is, the more strongly its proton is bonded in the molecule, and therefore the more strongly its anion tends to hydrolyze to form the acid. More succinctly, *the weaker an acid is, the stronger is its conjugate base.*

● EXAMPLE 16-13

Problem   Does the chloride ion Cl^- hydrolyze in aqueous solution?

Solution   *If* the chloride ion hydrolyzed, it would do so to form the acid HCl:

$$Cl^- + H_2O \xrightarrow{\ ?\ } HCl + OH^-$$

But this reaction does not take place to any appreciable extent. How do we know this? Hydrochloric acid, HCl, is a strong acid. Because it dissociates completely, there is negligible attraction between H^+ and Cl^-; so the hydrolysis of Cl^- does not occur. ●

ADDED COMMENT

Do not think that "HCl forms and then dissociates completely because it is a strong acid." Instead, think that "because HCl is a strong acid, there is no tendency for it to form in solution in the first place."

When the anion A^- hydrolyzes the equilibrium established is

$$A^- + H_2O \rightleftharpoons HA + OH^-$$

for which the equilibrium condition is

$$\frac{[HA][OH^-]}{[A^-]} = K_h$$

where K_h is called the *hydrolysis constant* (or *hydrolytic constant*) and in which $[H_2O]$ has been absorbed into K_h. Hydrolysis constants are not often tabulated because it is easy to calculate their values from other data, as follows: by multiplying the numerator and denominator in the above equilibrium condition by $[H^+]$, we get

$$\frac{[HA][OH^-]}{[A^-]} \times \frac{[H^+]}{[H^+]} = K_h$$

Rearranging this a little makes it look like

$$\frac{[HA]}{[H^+][A^-]} \times [H^+][OH^-] = K_h$$

or

$$\frac{1}{K_a} K_w = K_h$$

Most simply,

$$K_h = \frac{K_w}{K_a}$$

Notice that K_a is the *dissociation constant of the weak acid formed in the hydrolysis*. Since such dissociation constants are readily available, and since K_w is known, it is easy to calculate K_h for use in a hydrolysis calculation.

From the Brønsted-Lowry point of view, since the above hydrolysis is just a reaction of the base A^- with water to form its conjugate acid HA, we can write

$$K_b = \frac{K_w}{K_a}$$

This general relationship shows how the dissociation constant of a base is related to that of its conjugate acid.

● **EXAMPLE 16-14**

Problem   Calculate the pH and percent hydrolysis in a 1.0 M NaCN solution.

Solution   Sodium cyanide, NaCN, is a salt and is thus completely dissociated in solution to form Na^+ (1.0 M) and CN^- (1.0 M). The cyanide ion CN^- hydrolyzes because it is the anion of a weak acid, HCN.

$$CN^- + H_2O \rightleftharpoons HCN + OH^-$$

for which the equilibrium condition is

$$\frac{[HCN][OH^-]}{[CN^-]} = K_h$$

First, let us find the value of K_h.

$$K_h = \frac{K_w}{K_a}$$

$$= \frac{1.0 \times 10^{-14}}{4.0 \times 10^{-10}} = 2.5 \times 10^{-5}$$

where K_a for HCN has been taken from Table 16-1. Now that we know K_h we can proceed. Let x equal the number of moles of CN^- which hydrolyze per liter. Then at equilibrium

$$[HCN] = x \qquad [OH^-] = x \qquad [CN^-] = 1.0 - x$$

Now we substitute into the equilibrium condition.

$$\frac{x(x)}{1.0 - x} = 2.5 \times 10^{-5}$$

We will assume, as usual, that $x \ll 1.00$, so that $1.0 - x \approx 1.0$, and

$$\frac{x(x)}{1.0} \approx 2.5 \times 10^{-5}$$

$$x \approx \sqrt{2.5 \times 10^{-5}} = 5.0 \times 10^{-3}$$

(The assumption was justified: $5.0 \times 10^{-3} \ll 1.0$.)

$$[OH^-] = x = 5.0 \times 10^{-3} \, M$$

$$[CN^-] = 1.0 - x = 1.0 - (5.0 \times 10^{-3}) = 1.0 \, M$$

From $[OH^-]$ we can find $[H^+]$, and from this, the pH:

$$[H^+] = \frac{K_w}{[OH^-]} = \frac{1.0 \times 10^{-14}}{5.0 \times 10^{-3}} = 2.0 \times 10^{-12}$$

$$pH = -\log(2.0 \times 10^{-12}) = 11.70$$

$$\% \text{ hydrolysis} = \frac{\text{moles of } CN^- \text{ hydrolyzed}}{\text{total moles of } CN^-} \times 100$$

$$= \frac{5.0 \times 10^{-3}}{1.0} \times 100 = 0.50\%$$

Cation hydrolysis   Cations also undergo hydrolysis. In this case, however, a weak *base* is formed along with H^+ ions. The hydrolysis of a cation, M^+, can generally be represented by the equation:

$$M^+ + H_2O \rightleftharpoons MOH + H^+$$

if the cation is a simple metal ion. In the case of hydrolysis of the ammonium ion, however, writing the hydrolysis in this way leads to showing the formation of the questionable NH_4OH species (Sec. 16-2).

$$NH_4^+ + H_2O \rightleftharpoons NH_4OH + H^+$$

for which the equilibrium condition is

$$\frac{[NH_4OH][H^+]}{[NH_4^+]} = K_h$$

The Brønsted-Lowry equation for this reaction is

$$NH_4^+ + H_2O \rightleftharpoons NH_3 + H_3O^+$$

for which the equilibrium condition is written as

$$\frac{[NH_3][H_3O^+]}{[NH_4^+]} = K_a$$

The two ways of writing the equations and their equilibrium conditions are entirely equivalent.

As before, a hydrolysis constant can be calculated from K_w and the dissociation constant of the weak electrolyte formed.

$$K_h = \frac{K_w}{K_b}$$

In Brønsted-Lowry language this translates into

$$K_a = \frac{K_w}{K_b}$$

In general, $$K_a K_b = K_w$$

⬤ **EXAMPLE 16-15**

Problem   What is the pH of a 0.20 M solution of ammonium chloride, NH_4Cl?

Solution   Ammonium chloride is a salt, and so it is completely dissociated into NH_4^+ (0.20 M) and Cl^- (0.20 M). The NH_4^+ ions hydrolyze.

$$NH_4^+ + H_2O \rightleftharpoons NH_3 + H_3O^+$$

The equilibrium condition is

$$\frac{[NH_3][H_3O^+]}{[NH_4^+]} = K_h$$

First we evaluate K_h.

$$K_h = \frac{K_w}{K_b} = \frac{1.0 \times 10^{-14}}{1.8 \times 10^{-5}} = 5.6 \times 10^{-10}$$

where the value K_b has been taken from Table 16-3.

Now if we let x equal the number of moles of NH_4^+ per liter that hydrolyze, then

$$[NH_3] = x \qquad [H_3O^+] = x \qquad [NH_4^+] = 0.20 - x$$

Substituting these in the equilibrium condition, we get

$$\frac{x(x)}{0.20 - x} = 5.6 \times 10^{-10}$$

If $x \ll 0.20$, then $0.20 - x \approx 0.20$, and

$$\frac{x(x)}{0.20} \approx 5.6 \times 10^{-10}$$

$$x^2 \approx 1.1 \times 10^{-10}$$

$$x \approx 1.1 \times 10^{-5}$$

($1.1 \times 10^{-5} \ll 0.20$; so our approximation was justified.)

$$[NH_3] = x = 1.1 \times 10^{-5} \, M$$

$$[H_3O^+] = x = 1.1 \times 10^{-5} \, M$$

$$[NH_4^+] = 0.20 - x = 0.20 - 1.1 \times 10^{-5} = 0.20 \, M$$

and the pH of the solution is

$$pH = -\log [H_3O^+]$$
$$= -\log (1.1 \times 10^{-5}) = 4.96 \quad ⬤$$

The mechanism of cation hydrolysis

Cations which have a high charge-to-size ratio undergo hydrolysis. In the case of a solution of chromic nitrate, $Cr(NO_3)_3$, the chromic ion Cr^{3+} hydrolyzes in a reaction written most simply as

$$Cr^{3+} + H_2O \rightleftharpoons CrOH^{2+} + H^+$$

But this is really an oversimplification. In aqueous solution the chromic ion is hydrated by six water molecules,

$$Cr(H_2O)_6{}^{3+}$$

and this aquo complex undergoes dissociation, donating a proton to water:

$$Cr(H_2O)_6{}^{3+} + H_2O \rightleftharpoons Cr(OH)(H_2O)_5{}^{2+} + H_3O^+$$

The six water molecules which hydrate the chromic ion are clustered around it at the corners of a regular octahedron, as is shown in Fig. 16-2. The high positive charge on the chromic ion tends to pull electrons in from the water molecules, weakening the O—H bonds, so that one (or more) proton(s) can be transferred to solvent H_2O molecules.

The sulfate ion $SO_4{}^{2-}$ may be imagined as being the product of the loss of eight protons from a hypothetical hydrated S^{6+} cation:

$$S^{6+} + 4H_2O \longrightarrow \left[S(H_2O)_4\right]^{6+}$$

$$\left[S(H_2O)_4\right]^{6+} \xrightarrow{\text{in 8 steps}} SO_4{}^{2-} + 8H^+$$

The charge on the sulfur is not quite enough to let the last proton go completely; so the following equilibrium is established:

$$HSO_4{}^- \rightleftharpoons SO_4{}^{2-} + H^+ \qquad K = 1.2 \times 10^{-2}$$

where the above K is usually called K_2 for H_2SO_4.

● **EXAMPLE 16-16**

Problem   Calculate the pH of a 0.10 M $Cr(NO_3)_3$ solution. Consider only the first step in the hydrolysis, for which $K_h = 1.3 \times 10^{-4}$.

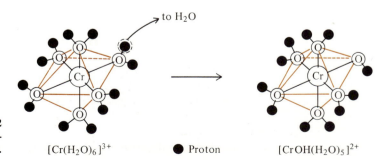

Figure 16-2
Hydrolysis of the hydrated chromium ion.

$[Cr(H_2O)_6]^{3+}$ ● Proton $[CrOH(H_2O)_5]^{2+}$

Solution   The simplest way to write the hydrolysis equation is

$$Cr^{3+} + H_2O \rightleftharpoons CrOH^{2+} + H^+$$

Let x equal the number of moles of Cr^{3+} per liter which hydrolyze. Then

$$x = [CrOH^{2+}] \qquad x = [H^+] \qquad 0.10 - x = [Cr^{3+}]$$

The equilibrium condition is

$$\frac{[CrOH^{2+}][H^+]}{[Cr^{3+}]} = K_h$$

$$\frac{x(x)}{0.10 - x} = 1.3 \times 10^{-4}$$

Making the usual assumption that $x \ll 0.10$, we get

$$\frac{x(x)}{0.10} \approx 1.3 \times 10^{-4}$$

which yields

$$x \approx 3.6 \times 10^{-3}$$

(Our assumption was justified; $3.6 \times 10^{-3} \ll 0.10$.)

$$[H^+] = 3.6 \times 10^{-3} \, M$$

$$pH = -\log(3.6 \times 10^{-3}) = 2.44 \quad \bullet$$

Hydrolysis and pH   Hydrolysis may be considered to be a disturbance of the water self-dissociation equilibrium. Thus the hydrolysis of an anion, A^-, of a weak acid may be considered to be composed of two steps: first, the combination of the anion with H^+ ions from the dissociation of water:

$$A^- + H^+ \rightleftharpoons HA$$

and second, the shifting of the water equilibrium so as to replace some of the lost H^+ ions and to produce more OH^- at the same time:

$$H_2O \rightleftharpoons H^+ + OH^-$$

These two changes occur simultaneously, so that the overall result is shown by the sum of the two equations:

$$A^- + H_2O \rightleftharpoons HA + OH^-$$

When the water equilibrium is disturbed, the $[H^+] = [OH^-]$ equality of pure water is destroyed, so that the solution is no longer neutral. Hydrolysis of a *cation* tends to *lower* the pH of a solution, and hydrolysis of an *anion* tends to *raise* the pH.

The pH of solutions   When a salt dissolves in water, the resulting solution may be acidic, basic, or
of salts   neutral, depending on the nature of the salt. If it is a salt of a *strong acid* and a *strong base,* its solution is neutral. For example, NaCl, the salt of the strong acid HCl and the strong base NaOH, dissolves to form a neutral solution. Neither the Na^+ ion nor the Cl^- ion hydrolyze.

When the salt of a *weak acid* and a *strong base* dissolves, its solution is *basic.* An example is NaF, the salt of the weak acid HF and the strong base NaOH. The Na^+ ion does not hydrolyze, but the F^- ion does:

$$F^- + H_2O \rightleftharpoons HF + OH^-$$

and so a solution of NaF is basic.

When the salt of a *strong acid* and a *weak base* dissolves, its solution is *acidic*. NH_4Cl, the salt of the strong acid HCl and the weak base NH_3, is such a salt. The Cl^- ion does not hydrolyze, but the NH_4^+ does:

$$NH_4^+ + H_2O \rightleftharpoons NH_3 + H_3O^+$$

or $$NH_4^+ + H_2O \rightleftharpoons NH_4OH + H^+$$

and so the solution is acidic.

There is one more combination: the salt of a *weak acid* and a *weak base*. Here it is impossible to give a single generalization. If the acid is a stronger electrolyte than is the base, the solution of the salt will be acidic. If the base is a stronger electrolyte than the acid, the solution will be basic. Only if the electrolyte strength of the acid is the same as that of the base will the solution be neutral. Consider the hydrolysis of a solution of ammonium fluoride, NH_4F. In this solution the cation hydrolyzes

$$NH_4^+ + H_2O \rightleftharpoons NH_3 + H_3O^+$$

for which $$K_h = \frac{K_w}{K_b} = \frac{1.0 \times 10^{-14}}{1.8 \times 10^{-5}} = 5.6 \times 10^{-10}$$

The anion also hydrolyzes

$$F^- + H_2O \rightleftharpoons HF + OH^-$$

for which $$K_h = \frac{K_w}{K_a} = \frac{1.0 \times 10^{-14}}{6.7 \times 10^{-4}} = 1.5 \times 10^{-11}$$

Since K_h for the cation hydrolysis (which tends to make the solution acidic) is a little larger than K_h for the anion hydrolysis (which tends to make the solution basic), the solution ends up with a small excess of H^+ (H_3O^+) ions and is hence acidic.

In the case of the hydrolysis of NH_4CN the production of OH^- ions from the hydrolysis of CN^-

$$CN^- + H_2O \rightleftharpoons HCN + OH^-$$

for which $$K_h = \frac{K_w}{K_a} = \frac{1.0 \times 10^{-14}}{4.0 \times 10^{-10}} = 2.5 \times 10^{-5}$$

is greater than the production of H^+ ions from the hydrolysis of NH_4^+ (K_h was found above to be 5.6×10^{-10}). This time the solution is basic.

The hydrolysis of ammonium acetate is an interesting case. $NH_4C_2H_3O_2$ is the salt of an acid ($HC_2H_3O_2$) and a base (NH_3) which are just about equally weak. Their dissociation constants are almost exactly equal, 1.8×10^{-5}, to two significant figures. Thus the hydrolysis constants of the anion and cation are essentially the same, 5.6×10^{-10}. The production of H^+ ions by the cation hydrolysis is essentially matched by the production of OH^- ions by the anion hydrolysis, and so a solution of $NH_4C_2H_3O_2$ is almost exactly neutral at any concentration.

16-5
Acid-base
indicators
and titration

The basis for the analysis of solutions of acids and bases has been described in Sec. 13-5. Since one equivalent of an acid exactly neutralizes one equivalent of a base, all you need to do to analyze a solution of a base (or acid) is to add an acid (or base, respectively) to it until equivalent amounts have reacted. Then from the volumes of the two solutions and the normality of one, the normality of the other can be computed. This much was described in Sec. 13-5. What was not mentioned was *how you know* when you have mixed equal numbers of equivalents of acid and base. In this section we will describe a process known as *titration,* which provides the needed information.

Indicators

An indicator is a Brønsted-Lowry conjugate acid-base pair of which the acid has one color and the base, another. At least one of the colors is intense enough to be visible in dilute solutions. Most indicators are organic molecules with fairly complex structures; so we will just use the abbreviation HIn to represent the *acid form* and In⁻ the conjugate *base form* of an indicator. Thus in aqueous solution,

$$\underset{\text{Acid form}}{HIn} + H_2O \rightleftharpoons \underset{\text{Base form}}{In^-} + H_3O^+$$

As can be seen, in this equilibrium the indicator exists in the acid form in more acidic solutions and in the base form in solutions which are less acidic, or more basic. For the indicator phenolphthalein, for example, HIn is colorless, while In⁻ is pink. For bromthymol blue HIn is yellow, while In⁻ is blue.

The equilibrium condition for the dissociation of HIn is

$$\frac{[In^-][H_3O^+]}{[HIn]} = K_{In}$$

Rearranging this, we have

$$\frac{[In^-]}{[HIn]} = \frac{K_{In}}{[H_3O^+]}$$

Now suppose that we observe the color of the indicator chlorophenol red in solutions of various pH values. The acid form of chlorophenol red, HIn, is yellow, while the basic form, In⁻, is red. K_{In} is 1×10^{-6}. Thus, we have

$$\frac{[In^-]}{[HIn]} = \frac{1 \times 10^{-6}}{[H_3O^+]}$$

The color we observe in a solution of this indicator depends on the ratio of $[In^-]$ to $[HIn]$. In a solution of pH 4.0, $[H_3O^+]$ is 1×10^{-4}; so

$$\frac{[In^-]}{[HIn]} = \frac{1 \times 10^{-6}}{1 \times 10^{-4}} = 1 \times 10^{-2} = \tfrac{1}{100}$$

This means that the concentration of HIn is 100 times that of In⁻, and so the

Table 16-5
Color change in the
indicator chlorophenol
red*

pH	$\dfrac{[In^-]}{[HIn]}$	Observed color
4	$\dfrac{1}{100}$	Yellow
5	$\dfrac{1}{10}$	Yellow
6	$\dfrac{1}{1}$	Orange
7	$\dfrac{10}{1}$	Red
8	$\dfrac{100}{1}$	Red

*$K_{In} = 1 \times 10^{-6}$; HIn, yellow; In⁻, red

solution appears yellow. Let us now make the solution more basic. At a pH of 5.0 the ratio is

$$\frac{[In^-]}{[HIn]} = \frac{1 \times 10^{-6}}{1 \times 10^{-5}} = 1 \times 10^{-1} = \tfrac{1}{10}$$

and the concentration of the red In⁻ form is beginning to increase. The solution still appears yellow, but with perhaps a tinge of orange. At a pH of 6.0,

$$\frac{[In^-]}{[HIn]} = \frac{1 \times 10^{-6}}{1 \times 10^{-6}} = \tfrac{1}{1}$$

Equal concentrations of In⁻ and HIn give the solution an orange color. As the pH rises farther the color gets redder, as the In⁻/HIn ratio increases. These changes are summarized in Table 16-5.

An indicator such as chlorophenol red can be used to determine the approximate pH of a solution. Since different indicators have different K_{In} values, the pH range over which the color changes varies from one indicator to another. This means that by testing a solution with a number of different indicators, its pH can be determined to one pH unit or less. The pH range over which the eye can see an indicator color change varies from one indicator to another and from one person to another. (The ability to discriminate between colors varies somewhat even among persons with so-called perfect vision.) Thus the pH range over which chlorophenol red changes from yellow to red is *about* 5.2 to 6.8, although some people might not be able to detect the start of the yellow-orange transition until a pH of 5.4 has been reached, for example. Table 16-6 lists a number of common indicators, their pK_{In} values, color changes, and the approximate pH range over which the color change can be seen. The pK_{In} of an indicator is best thought of as the pH at which $[HIn] = [In^-]$.

Acid-base titrations

The acid-base analysis of a solution of unknown concentration is usually done by a procedure known as *titration*. In the titration of a solution of an acid of unknown concentration a measured volume of the acid is added to a flask, and a *titrant,* a solution of a base of known concentration, is added until the *equivalence*

**Table 16-6
Indicators and their
color changes**

Indicator	pK_{In}	Approximate pH range for color change	Corresponding color change
meta-Cresol purple	1.5	1.2– 2.8	Red to yellow
Methyl orange	3.4	3.1– 4.4	Red to orange
Bromophenol blue	3.8	3.0– 4.6	Yellow to blue
Methyl red	4.9	4.4– 6.2	Red to yellow
Chlorophenol red	6.0	5.2– 6.8	Yellow to red
Bromothymol blue	7.1	6.2– 7.6	Yellow to blue
meta-Cresol purple	8.3	7.6– 9.2	Yellow to purple
Phenolphthalein	9.4	8.0–10.0	Colorless to red
Thymolpthalein	10.0	9.4–10.6	Colorless to blue
Alizarin yellow R	11.2	10.0–12.0	Yellow to violet

point is reached. This is the point at which equal numbers of equivalents of acid and base have been mixed. In the simplest procedure the equivalence point is signaled by the color change of an indicator added before the start of the titration. The pH at the equivalence point normally changes rapidly as very small amounts of titrant are added; so a sharply defined color change provides a clear indication of the equivalence point.

The equivalence point in an acid-base titration is not necessarily at a pH of 7. This means that the proper indicator must be chosen before the start of the titration. Usually the approximate pH at the equivalence point can be predicted ahead of time; so the problem reduces to choosing an indicator from a list such as that in Table 16-6.

⬤ **EXAMPLE 16-17**

Problem   What is the pH at the equivalence point in a titration of 25 ml of 0.10 M HCl with 0.10 M NaOH?

Solution   The strong acid–strong base neutralization reaction is

$$H^+ + OH^- \longrightarrow H_2O$$

At the equivalence point equal numbers of moles of H^+ and OH^- have been added, and the solution contains the Na^+ left over from the NaOH solution and the Cl^- left over from the HCl. Neither of these hydrolyze, and so the water self-dissociation equilibrium is not disturbed. The pH of the solution is consequently that of pure water, 7.00. ⬤

⬤ **EXAMPLE 16-18**

Problem   What is the pH at the equivalence point of a titration of

$$\text{25 ml of 0.10 } M \text{ HC}_2\text{H}_3\text{O}_2 \text{ with 0.10 } M \text{ NaOH?}$$

Solution   The weak acid–strong base neutralization reaction is

$$HC_2H_3O_2 + OH^- \longrightarrow H_2O + C_2H_3O_2^-$$

At the equivalence point acetate ions, $C_2H_3O_2^-$, are thus left in solution, along with Na^+ ions from the NaOH. But since the acetate ion is the anion of a weak acid, it hydrolyzes according to the reaction

$$C_2H_3O_2^- + H_2O \rightleftharpoons HC_2H_3O_2 + OH^-$$

and so the solution is *basic* at the equivalence point. Finding the pH of this solution is simply finding the pH of a sodium acetate solution. First we find the concentration of acetate ions. At the start the number of moles of acetic acid is

$$0.025 \text{ liter } (0.10 \text{ mol liter}^{-1}) = 2.5 \times 10^{-3} \text{ mol}$$

The neutralization equation tells us that 2.5×10^{-3} mol of $C_2H_3O_2^-$ is formed. But now the volume is 50 ml, because we have added 25 ml of 0.10 M NaOH to reach the equivalence point. (We assume here that the volumes are additive.) So neglecting any hydrolysis, the final concentration of acetate ions is

$$\frac{2.5 \times 10^{-3} \text{ mol}}{0.050 \text{ liter}} = 0.050 \; M$$

Now we calculate the pH of a solution in which $[C_2H_3O_2^-] = 0.050 \; M$. The hydrolysis equation is

$$C_2H_3O_2^- + H_2O \rightleftharpoons HC_2H_3O_2 + OH^-$$

If we let x equal the number of moles of $C_2H_3O_2^-$ per liter which hydrolyze,

$$[HC_2H_3O_2] = x$$

$$[OH^-] = x$$

$$[C_2H_3O_2^-] = 0.050 - x$$

The equilibrium condition is

$$\frac{[HC_2H_3O_2][OH^-]}{[C_2H_3O_2^-]} = K_h$$

We can find K_h from the value of K_a for $HC_2H_3O_2$:

$$K_h = \frac{K_w}{K_a} = \frac{1.0 \times 10^{-14}}{1.8 \times 10^{-5}} = 5.6 \times 10^{-10}$$

Therefore

$$\frac{x(x)}{0.050 - x} = 5.6 \times 10^{-10}$$

Assuming $x \ll 0.050$, we have $\dfrac{x(x)}{0.050} \approx 5.6 \times 10^{-10}$

$$x \approx 5.3 \times 10^{-6}$$

Having checked to see that our assumption was justified, we conclude that

$$[OH^-] = x = 5.3 \times 10^{-6} \; M$$

$$pOH = -\log (5.3 \times 10^{-6}) = 5.28$$

$$pH = 14.00 - 5.28 = 8.72 \quad \text{(basic solution)}$$

A similar calculation shows that the titration of a weak base with a strong acid yields a solution which is acidic because of the hydrolysis of the cation. Furthermore, different weak acids and bases yield different final pH values. Therefore, for each titration the right indicator must be chosen, one which will signal a color change at the appropriate pH.

Titration curves   It is possible to calculate the pH of a solution being titrated at any stage in the titration, given the concentration of both acid and base. The following example illustrates the technique.

● **EXAMPLE 16-19**

Problem In a titration of 25 ml of 0.10 M $HC_2H_3O_2$ with 0.10 M NaOH, what is the pH of the solution after the addition of 10 ml of the base?

Solution This starts out as a problem in stoichiometry. In 25 ml of 0.10 M $HC_2H_3O_2$ the number of moles of $HC_2H_3O_2$ is

$$0.025 \text{ liter } (0.10 \text{ mol liter}^{-1}) = 2.5 \times 10^{-3} \text{ mol}$$

In 10 ml of 0.10 M NaOH the number of moles of OH^- is

$$0.010 \text{ liter } (0.10 \text{ mol liter}^{-1}) = 1.0 \times 10^{-3} \text{ mol}$$

From the equation for the neutralization reaction we can find the numbers of moles of $HC_2H_3O_2$ and of $C_2H_3O_2^-$ after mixing.

$$HC_2H_3O_2 + OH^- \longrightarrow C_2H_3O_2^- + H_2O$$

Moles at start: 2.5×10^{-3} 1.0×10^{-3} ~ 0

Moles at end: 1.5×10^{-3} ~ 0 1.0×10^{-3}

Thus the concentrations after mixing are

$$[HC_2H_3O_2] = \frac{1.5 \times 10^{-3} \text{ mol}}{0.035 \text{ liter}} = 4.3 \times 10^{-2} \ M$$

$$[C_2H_3O_2^-] = \frac{1.0 \times 10^{-3} \text{ mol}}{0.035 \text{ liter}} = 2.9 \times 10^{-2} \ M$$

where the final volume is assumed to be 25 ml + 10 ml, or 35 ml. But the remaining $HC_2H_3O_2$ undergoes dissociation.

$$HC_2H_3O_2 \rightleftharpoons H^+ + C_2H_3O_2^-$$

If we let x equal the number of moles of $HC_2H_3O_2$ which dissociate per liter, then

$$[H^+] = x$$

$$[C_2H_3O_2^-] = 2.9 \times 10^{-2} + x$$

$$[HC_2H_3O_2] = 4.3 \times 10^{-2} - x$$

The equilibrium condition is

$$\frac{[H^+][C_2H_3O_2^-]}{[HC_2H_3O_2]} = K_a$$

$$\frac{x[(2.9 \times 10^{-2}) + x]}{(4.3 \times 10^{-2}) - x} = 1.8 \times 10^{-5}$$

If we assume that x is negligible in comparison with 2.9×10^{-2} and with 4.3×10^{-2}, we have

$$\frac{x(2.9 \times 10^{-2})}{4.3 \times 10^{-2}} \approx 1.8 \times 10^{-5}$$

$$x \approx 2.7 \times 10^{-5}$$

Having checked that our assumption was justified, we can conclude that

$$[H^+] = x = 2.7 \times 10^{-5} \ M$$

$$pH = -\log [H^+] = -\log (2.7 \times 10^{-5}) = 4.57 \ ●$$

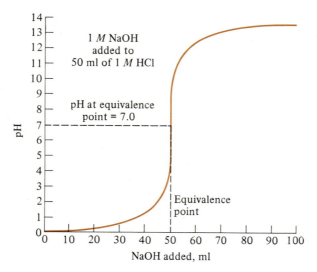

Figure 16-3
Titration curve: strong acid–strong base.

The pH of a solution being titrated can be plotted in the form of a *titration curve* which shows the pH of the solution as a function of the volume of titrant added. Titration curves clearly show the need for choosing the right indicator for a specific titration. Figure 16-3 shows the titration curve for the titration of 25 ml of 1 M HCl with 1 M NaOH. All strong acid–strong base curves look like this because the reaction for each is the same:

$$H^+ + OH^- \longrightarrow H_2O$$

For this titration a number of indicators would satisfactorily serve to signal the equivalence point, because the pH rises extremely sharply as the equivalence point is passed. Certainly any indicator which shows a color change in the range from 4 to 10 would work.

Figure 16-4 shows the titration curve for a *weak acid–strong base* titration. In

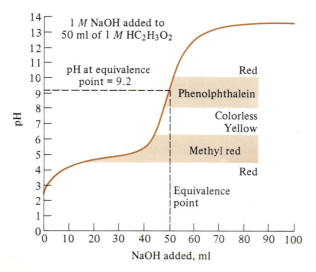

Figure 16-4
Titration curve: weak acid–strong base.

this case 50 ml of 1 M $HC_2H_3O_2$ are being titrated with 1 M NaOH. There are a number of differences between this curve and that in Fig. 16-3. Note especially that the pH at the equivalence point is greater than 7 and that the change of pH at the equivalence point is not as sharp as in the strong acid–strong base titration. For the sharpest indicator color change we should choose one which has the middle of its color-change span about at the equivalence point of the titration. Phenolphthalein is often used for titrating $HC_2H_3O_2$ with NaOH. If an inappropriate indicator were chosen, methyl red, for instance, the color would change very gradually over a considerable volume of added NaOH solution. In addition, the color change would be completed before the equivalence point was even reached. Thus the *end point* signaled by the indicator would be both "slushy" (in common chemical jargon) and incorrect, not corresponding to the equivalence point.

A *strong acid–weak base* titration curve is shown in Fig. 16-5. This represents the titration of 50 ml of 1 M ammonia with 1 M HCl. The equivalence point is at pH = 4.8, and so an indicator should be chosen with this pH within its color-change range. Methyl red would serve this time, as its end-point signal would be very close to the equivalence point. Phenolphthalein, on the other hand, would be totally unsatisfactory, as Fig. 16-5 shows.

A *polyprotic acid* has more than one equivalence point. Sulfurous acid, H_2SO_3, has one for the reaction

$$H_2SO_3 + OH^- \longrightarrow HSO_3^- + H_2O$$

and another for

$$HSO_3^- + OH^- \longrightarrow SO_3^{2-} + H_2O$$

The titration curve for the titration of 50 ml of 1 M H_2SO_3 with 1 M NaOH is shown in Fig. 16-6. At each equivalence point there is a more-or-less sharp rise in the pH. As is shown, chlorophenol red can be used to signal the first equivalence point, and thymolphthalein would probably be the choice for the second, but the end point signaled by the indicator this time will be rather slushy, as the pH does not rise very sharply at the second equivalence point.

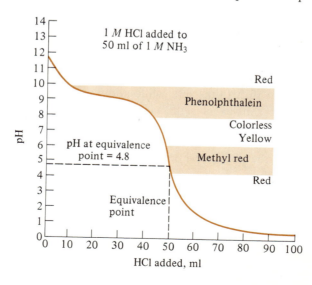

Figure 16-5
Titration curve: strong acid–weak base.

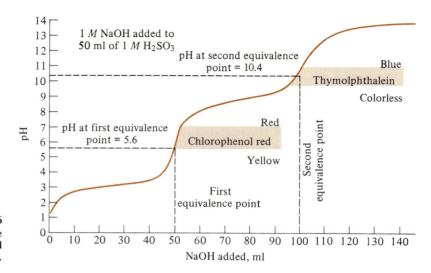

Figure 16-6
Titration curve for the
weak diprotic acid
H$_2$SO$_3$.

16-6
Buffers

One of the characteristics of a weak acid–strong base titration curve is an initial sharp pH rise followed by an interval during which the pH stays relatively constant, in spite of the fact that base is being added. (See Fig. 16-4.) A similar situation occurs in a strong acid–weak base titration curve: an initial drop in pH is followed by an interval during which the pH stays relatively constant. (See Fig. 16-5.) In each of these cases the sluggish response of the pH to added acid or base is an illustration of the *buffering* action of the solution.

The buffer effect

A *buffer*, or *buffer solution*, is a solution which undergoes only a slight change in pH when H$^+$ or OH$^-$ ions are added to it. It is a solution which contains *an acid plus its conjugate base*, present at approximately equal concentrations. An example of a buffer is a solution which has present both acetic acid and acetate ions at about equal concentrations. How does the HC$_2$H$_3$O$_2$–C$_2$H$_3$O$_2^-$ combination serve to buffer the solution? Consider the equilibrium

$$HC_2H_3O_2 \rightleftharpoons C_2H_3O_2^- + H^+$$

If HC$_2$H$_3$O$_2$ and C$_2$H$_3$O$_2^-$ are both present at reasonable concentrations, then the above equilibrium can readily shift in either direction. Adding H$^+$ will cause the equilibrium to shift to the left, while adding OH$^-$ will remove H$^+$, causing the equilibrium to shift to the right. In each case much of the added H$^+$ or OH$^-$ is used up without greatly altering the pH of the solution.

To describe buffer action quantitatively we start by writing the equilibrium condition for the above reaction.

$$\frac{[H^+][C_2H_3O_2^-]}{[HC_2H_3O_2]} = K_a$$

Rearranging this a little gives us

$$[H^+] = K_a \frac{[HC_2H_3O_2]}{[C_2H_3O_2^-]}$$

Now if we take the negative logarithm of each side, we have

$$-\log [\text{H}^+] = -\log K_a - \log \frac{[\text{HC}_2\text{H}_3\text{O}_2]}{[\text{C}_2\text{H}_3\text{O}_2^-]}$$

which can be written as

$$\text{pH} = \text{p}K_a - \log \frac{[\text{HC}_2\text{H}_3\text{O}_2]}{[\text{C}_2\text{H}_3\text{O}_2^-]}$$

This is known as the *Henderson-Hasselbalch equation.*
For a solution in which $[\text{HC}_2\text{H}_3\text{O}_2] = [\text{C}_2\text{H}_3\text{O}_2^-]$,

$$\log \frac{[\text{HC}_2\text{H}_3\text{O}_2]}{[\text{C}_2\text{H}_3\text{O}_2^-]} = 0$$

so $$\text{pH} = \text{p}K_a = -\log (1.8 \times 10^{-5}) = 4.74$$

If small amounts of H^+ or OH^- are added to this solution, the result will be to convert some $\text{C}_2\text{H}_3\text{O}_2^-$ to $\text{HC}_2\text{H}_3\text{O}_2$, or vice versa. But the ratio of concentrations of acetic acid to acetate will not change greatly. For instance, if at the start $[\text{HC}_2\text{H}_3\text{O}_2] = [\text{C}_2\text{H}_3\text{O}_2^-] = 1.00\ M$, then addition of 0.10 mol of OH^- per liter changes the ratio to

$$\frac{[\text{HC}_2\text{H}_3\text{O}_2]}{[\text{C}_2\text{H}_3\text{O}_2^-]} = \frac{1.00 - 0.10}{1.00 + 0.10} = 0.82$$

And since $\log 0.82 = -0.09$ this means that the new pH is

$$\text{pH} = 4.74 - (-0.09) = 4.83$$

Thus 0.10 mol of base has raised the pH by only 0.09 units. As long as $[\text{HC}_2\text{H}_3\text{O}_2]$ is of the same order of magnitude as $[\text{C}_2\text{H}_3\text{O}_2^-]$, the ratio of the two will stay fairly close to unity; so the pH will change little upon addition of acid or base. Of course, the solution is *best* buffered when $[\text{HC}_2\text{H}_3\text{O}_2] = [\text{C}_2\text{H}_3\text{O}_2^-]$.

● **EXAMPLE 16-20**

Problem   Compare the effect of adding 0.10 mol of H^+ to 1.0 liter of (a) a formic acid–formate buffer in which the concentrations of formic acid (HCHO_2) and formate ion (CHO_2^-) are each 1.00 M; (b) pure water. (K_a for formic acid is 1.8×10^{-4}.)

Solution   (a) For formic acid,

$$\text{p}K_a = -\log (1.8 \times 10^{-4}) = 3.74$$

Using the Henderson-Hasselbalch equation,

$$\text{pH} = \text{p}K_a - \log \frac{[\text{HCHO}_2]}{[\text{CHO}_2^-]}$$

$$= 3.74 - \log \frac{1.00}{1.00} = 3.74 - 0 = 3.74$$

The initial pH of the buffer is thus 3.74. Addition of 0.10 mol of H^+ will convert 0.10 mol of CHO_2^- to HCHO_2:

$$H^+ + CHO_2^- \longrightarrow HCHO_2$$

Moles at start:	0.10	1.00	1.00
Moles at end:	0	0.90	1.10

The new pH is

$$pH = 3.74 - \log \frac{1.10}{0.90} = 3.74 - \log 1.22$$

$$= 3.74 - 0.09 = 3.65$$

(b) Pure water is almost completely unbuffered, because in the equilibrium

$$H_2O + H_2O \rightleftharpoons H_3O^+ + OH^-$$

no conjugate acid-base pair (either H_2O and H_3O^+ or H_2O and OH^-) is present with *both* members of the pair at roughly equal concentrations. Pure water is therefore extremely responsive in pH to added acid or base. When 0.10 mol of H^+ is added to 1.0 liter of H_2O (initial pH = 7.00), the resulting $[H^+]$ becomes 0.10 M, because no base is present to remove an appreciable quantity of H^+. Thus the pH becomes 1.00. This means that the completely unbuffered water has changed pH by 6 units, as compared with only 0.09 units by the buffer in part (a). ●

Buffers in biological systems
Biochemical reactions in both plants and animals are often very sensitive to pH changes, either because critical equilibria are affected or, more often, because reaction rates are greatly altered by a change in the pH of the reaction medium. These pH changes ordinarily do not occur in a healthy organism, however, because its internal fluids are usually well buffered. Wide variations in what you eat and drink and the way you live, though producing considerable internal change in your body, effect the pH of your blood very little. Even most illnesses produce only very slight changes.

Human blood is buffered by a number of buffer systems including

$$H_2PO_4^- \rightleftharpoons HPO_4^{2-} + H^+ \qquad pK_a = 7.20$$

and

$$H_2CO_3 \rightleftharpoons HCO_3^- + H^+ \qquad pK_a = 6.38$$

The H_2CO_3-HCO_3^- system in the blood is especially interesting. Gaseous carbon dioxide dissolves in water, where a small portion of it, about 1 percent, combines with water to form carbonic acid, H_2CO_3:

$$CO_2 + H_2O \rightleftharpoons H_2CO_3$$

H_2CO_3 cannot be isolated pure, but in solution acts as a weak diprotic acid. The loss of its first proton yields the hydrogen carbonate ion HCO_3^-, usually called by its traditional name, the *bicarbonate ion:*

$$H_2CO_3 \rightleftharpoons H^+ + HCO_3^- \qquad K_1 = 4.2 \times 10^{-7}$$

But the value of K_1 given here (and usually given) reflects a mass-action expression in which the total concentration of dissolved CO_2, including that combined as H_2CO_3, is in the denominator

$$\frac{[H^+][HCO_3^-]}{[CO_2 + H_2CO_3]} = K_1$$

Since most of the CO_2 is not combined as H_2CO_3, this is usually written

$$\frac{[H^+][HCO_3^-]}{[CO_2]} = K_1$$

and the equilibrium as

$$CO_2 + H_2O \rightleftharpoons H^+ + HCO_3^-$$

Addition of CO_2 to water thus produces an acidic solution, and dissolved CO_2 plus HCO_3^- constitute a buffer system. In the human body cells produce CO_2 which dissolves in the venous blood returning to the heart and lungs. In the lungs some of the CO_2 is lost through exhalation; so the pH rises a little. Actually, if it were not for the presence of other buffer systems in the blood, the pH change would be excessive. As it is, the change is normally slight. Excessive loss of CO_2 from the blood can be produced by hyperventilation, rapid deep breathing. Since breathing is stimulated by the presence of dissolved CO_2 in the blood, this makes it possible to hold one's breath for long periods, even until the onset of unconsciousness due to lack of oxygen. (Apparently the CO_2 buildup is too slow to force one to take a breath.) Hyperventilation can raise the pH of blood by as much as 0.05 units, enough to produce light-headedness at best and severe heart attack–mimicking chest pains at worst. The reverse effect, a lowering of the pH of the blood due to excessive CO_2 buildup, sometimes occurs in forms of pneumonia in which the lungs begin to fail. This condition, called *acidosis*, causes severe disruption in the functioning of various body tissues.

● **EXAMPLE 16-21**

Problem   Human blood has a pH which is invariably close to 7.4. Calculate the $[CO_2]/[HCO_3^-]$ ratio in blood of this pH.

Solution   Applying the Henderson-Hasselbalch equation, we have

$$pH = pK_a - \log \frac{[CO_2]}{[HCO_3^-]}$$

or

$$\log \frac{[CO_2]}{[HCO_3^-]} = pK_a - pH$$

$$= 6.38 - 7.4 = -1.0$$

Taking antilogs, we have

$$\frac{[CO_2]}{[HCO_3^-]} = 0.1$$

Since this ratio is not very close to unity, the CO_2–HCO_3^- buffer system by itself would not be very effective. Actually, hemoglobin is thought to be the major buffer in the blood. ●

**16-7
Simultaneous
acid-base
equilibria**

Several equilibria often occur simultaneously in aqueous solution. Fortunately, either because of a large equilibrium constant or high concentrations of certain species, one equilibrium usually predominates. Thus in a solution of acetic acid the following two equilibria are established:

$$HC_2H_3O_2 \rightleftharpoons H^+ + C_2H_3O^- \qquad K_a = 1.8 \times 10^{-5}$$

$$H_2O \rightleftharpoons H^+ + OH^- \qquad K_w = 1.0 \times 10^{-14}$$

The conditions for both equilibria are met simultaneously, but in the calculation of the hydrogen-ion concentration usually only the first equilibrium needs to be considered, since the second contributes only a very small amount to $[H^+]$.

Mixtures of weak acids   In a solution containing two weak acids the stronger of the two may be considered to be the primary source of hydrogen ions, unless the two dissociation constants are numerically close together.

● **EXAMPLE 16-22**

Problem   Calculate $[H^+]$ in a solution prepared by adding 0.10 mol each of $HC_2H_3O_2$ and HCN to enough water to make 1.0 liter of solution.

Solution   In this solution these equilibria occur simultaneously:

$$HC_2H_3O_2 \rightleftharpoons H^+ + C_2H_3O_2^- \qquad K_a = 1.8 \times 10^{-5}$$

$$HCN \rightleftharpoons H^+ + CN^- \qquad K_a = 4.0 \times 10^{-10}$$

$$H_2O \rightleftharpoons H^+ + OH^- \qquad K_w = 1.0 \times 10^{-14}$$

The equilibrium conditions for all three must be met simultaneously, but since acetic acid is the strongest of these acids, we can ignore the contributions made by the HCN and H_2O to the hydrogen-ion concentration. If we let x equal the number of moles of $HC_2H_3O_2$ dissociated per liter, then at equilibrium

$$[H^+] = x \qquad [C_2H_3O_2^-] = x \qquad [HC_2H_3O_2] = 0.10 - x$$

Substituting in the equilibrium condition

$$\frac{[H^+][C_2H_3O_2^-]}{[HC_2H_3O_2^-]} = K_a$$

we get

$$\frac{x(x)}{0.10 - x} = 1.8 \times 10^{-5}$$

$$x = 1.3 \times 10^{-3}$$

$$[H^+] = x = 1.3 \times 10^{-3} \ M \ ●$$

● **EXAMPLE 16-23**

Problem   In the solution described in Example 16-22, what percent of the hydrogen ions present have been produced by (a) the dissociation of HCN; (b) the dissociation of H_2O?

Solution   Although in the previous example we assumed that the H^+ produced by the HCN and H_2O were negligible compared with those produced by the $HC_2H_3O_2$, this does not mean that we assumed they were zero.
    (a) Using the HCN dissociation

$$HCN \rightleftharpoons H^+ + CN^-$$

let y equal the number of moles of HCN dissociated per liter. Then

$$y = [H^+] \qquad \textit{from the HCN dissociation only}$$

But the *total* $[H^+]$ is the sum of the contributions of the *three* equilibria. Ignoring the contribution from water, which must be very small, we can write that at equilibrium:

$$[H^+] = (1.3 \times 10^{-3}) + y$$

$$[CN^-] = y \qquad [HCN] = 0.10 - y$$

Since

$$\frac{[H^+][CN^-]}{[HCN]} = K_a$$

$$\frac{(1.3 \times 10^{-3} + y)(y)}{0.10 - y} = 4.0 \times 10^{-10}$$

We can assume that $y \ll 0.10$ and can even assume that $y \ll 1.3 \times 10^{-3}$, because HCN is such a weak acid, and because its dissociation has been further repressed by the H^+ from $HC_2H_3O_2$. So we have

$$\frac{(1.3 \times 10^{-3})(y)}{0.10} \approx 4.0 \times 10^{-10}$$

which gives us

$$y \approx 3.1 \times 10^{-8}$$

and we see that our assumptions are justified. Therefore

$$[H^+] \text{ from HCN only} = 3.0 \times 10^{-8} \ M$$

The fraction of the $[H^+]$ contributed by HCN is thus

$$\frac{[H^+] \text{ from HCN}}{\text{Total } [H^+]} \approx \frac{[H^+] \text{ from HCN}}{[H^+] \text{ from } HC_2H_3O_2} = \frac{3.1 \times 10^{-8} \ M}{1.3 \times 10^{-3} \ M}$$

and this times 100 gives us the percent of the $[H^+]$ which comes from the dissociation of HCN:

$$\frac{3.1 \times 10^{-8} \ M}{1.3 \times 10^{-3} \ M} \times 100 = 2.4 \times 10^{-3} \%$$

(b) We can find the contribution to the $[H^+]$ made by the dissociation of water similarly. Let z equal the number of moles of H_2O which dissociate per liter. Then, since

$$H_2O \rightleftharpoons H^+ + OH^-$$

$$z = [H^+] \qquad \textit{from } H_2O \textit{ only}$$

so that the total

$$[H^+] = 1.3 \times 10^{-3} + 3.1 \times 10^{-8} + z$$

$$= 1.3 \times 10^{-3} + z$$

Now, since

$$[OH^-] = z$$

and

$$[H^+][OH] = K_w$$

$$(1.3 \times 10^{-3} + z)(z) = 1.0 \times 10^{-14}$$

$$z = 7.7 \times 10^{-12}$$

$$[H^+] \text{ from } H_2O \text{ only} = 7.7 \times 10^{-12} \ M$$

$$\text{Percent of } [H^+] \text{ from } H_2O \text{ only} = \frac{[H^+] \text{ from } H_2O}{\text{total } [H^+]} \times 100$$

$$= \frac{7.7 \times 10^{-12}}{1.3 \times 10^{-3} + 3.1 \times 10^{-8} + 7.7 \times 10^{-12}} \times 100$$

$$= 5.9 \times 10^{-7} \%$$

Polyprotic acids   Polyprotic acids in solution provide another situation in which several equilibrium conditions are simultaneously satisfied. But, as in the case of a mixture of weak acids of considerably different strengths, usually one dissociation predominates. This can be judged by comparing successive K_a values. Usually K_1 will be 10^4 or more times larger than K_2, and under these conditions the calculation can be handled as was shown in Example 16-5.

Acid salts   The term *acid salt* is applied to a salt which is formed from the removal of only some of the protons of a polyprotic acid. $NaHCO_3$, $NaHSO_3$, NaH_2PO_4, and Na_2HPO_4 are examples. An interesting situation occurs when an acid salt is dissolved in water: the anion can both dissociate and hydrolyze. In the case of the bicarbonate ion HCO_3^-, the reactions are

Dissociation: $HCO_3^- \rightleftharpoons H^+ + CO_3^{2-}$

Hydrolysis: $HCO_3^- + H_2O \rightleftharpoons H_2CO_3 + OH^-$

(Here H_2CO_3 represents total dissolved CO_2, as before.)

● **EXAMPLE 16-24**

Problem   What is $[H^+]$ in a 0.10 M $NaHCO_3$ solution?

Solution   Solving this problem exactly (no assumptions) turns out to be extremely tedious, but by applying a little chemical common sense, we can make the process much less complicated. Look at the two equilibria, above, and observe that H^+ produced in the first equilibrium combine with the OH^- produced in the second to form water. This means that both equilibria proceed to the right to about the same extent. (If either $[H^+]$ or $[OH^-]$ were to build up to a significant excess, the appropriate equilibrium would shift to lower the concentration.)

Because the dissociation and hydrolysis equilibria take place to about the same degree, we can add the two equations to get a combined equilibrium equation:

$$2HCO_3^- \rightleftharpoons H_2CO_3 + CO_3^{2-}$$

for which the equilibrium condition is

$$\frac{[H_2CO_3][CO_3^{2-}]}{[HCO_3^-]^2} = K$$

How can we evaluate K? All we need to do is multiply both numerator and denominator of the mass-action expression by H^+.

$$K = \frac{[H_2CO_3][CO_3^{2-}]}{[HCO_3^-]^2} \times \frac{[H^+]}{[H^+]} = \frac{[H_2CO_3]}{[H^+][HCO_3^-]} \times \frac{[H^+][CO_3^{2-}]}{[HCO_3^-]}$$

This is just the product of the reciprocal of the first dissociation constant of H_2CO_3 times its second dissociation constant, or

$$K = \frac{K_2}{K_1}$$

Thus (taking data from Table 16-2) we have

$$K = \frac{5.6 \times 10^{-11}}{4.2 \times 10^{-7}} = 1.3 \times 10^{-4}$$

If we let x equal the number of moles of H_2CO_3 formed per liter, then the number of moles of HCO_3^- used up is $2x$; so at equilibrium,

$$[H_2CO_3] = x \qquad [CO_3^{2-}] = x \qquad [HCO_3^-] = 0.10 - 2x$$

Substituting these in the equilibrium condition, we have

$$\frac{[H_2CO_3][CO_3^{2-}]}{[HCO_3^-]^2} = \frac{x(x)}{(0.10 - 2x)^2} = 1.3 \times 10^{-4}$$

Assuming that $2x \ll 0.10$, we get

$$\frac{x^2}{(0.10)^2} \approx 1.3 \times 10^{-4}$$

$$x \approx 1.1 \times 10^{-3}$$

Our assumption is justified; so

$$[H_2CO_3] = x = 1.1 \times 10^{-3} \, M$$

Now to get $[H^+]$ we need merely to use K_1 or K_2. From the first dissociation,

$$H_2CO_3 \rightleftharpoons H^+ + HCO_3^-$$

$$\frac{[H^+][HCO_3^-]}{[H_2CO_3]} = K_1$$

$$[H^+] = \frac{[H_2CO_3]}{[HCO_3^-]} K_1$$

$$= \frac{1.1 \times 10^{-3}}{0.10} (4.2 \times 10^{-7}) = 4.6 \times 10^{-9} \, M \quad \bullet$$

SUMMARY

This has been the first of two chapters about ionic equilibria in aqueous solutions. We started by considering the *dissociation of weak acids*. These dissociations can be described in either Arrhenius or Brønsted-Lowry terms. In either case the equilibrium condition is that the mass-action expression is equal to the *dissociation constant* K_a. The weaker an acid is, the smaller is its dissociation constant. *Polyprotic acids* have more than one available proton per molecule of acid and so undergo stepwise dissociation, each step having its own dissociation constant.

The equilibrium condition for a weak-acid dissociation can be used to calculate the concentrations of solute species present in solutions of these acids. By means of such calculations it can be shown that the presence in solution of an appreciable concentration of the anion of a weak acid represses the dissociation of the acid. This is a special illustration of Le Châtelier's principle called the *common-ion effect*. Weak-acid calculations also show that the degree of dissociation of a weak acid increases with decreasing concentration— another effect predicted by Le Châtelier's principle.

Weak bases dissociate in water much as weak acids do, except that they raise the hydroxide-ion concentration, instead of the hydrogen- (hydronium-) ion concentration. An important and common weak base is ammonia. In its aqueous solutions the undissociated molecular species have been formulated as NH_3 and as NH_4OH, although each of these is now believed to be an oversimplification. The formula NH_3 is usually preferable, however.

Weak-base calculations are based upon the equality of the mass-action expression and the base dissociation constant K_b at equilibrium. Such calculations are similar to weak-acid calculations.

Water undergoes *self-dissociation* to release a low concentration of hydrated hydrogen and hydroxide ions. The mass-action expression for this reaction is simply $[H^+][OH^-]$, or $[H_3O^+][OH^-]$, and at equilibrium (as is usually the case) is equal to the *dissociation constant for water* K_w, also called the *ion product for water*. It has the value 1.0×10^{-14} at 25°C. The addition of an acid disturbs this equilibrium so that $[H^+] > [OH^-]$ (the solution is *acidic*), and the addition of a base makes

$[OH^-] > [H^+]$ (the solution is *basic*). In a *neutral* solution $[H^+] = [OH^-]$.

The *pH* of a solution is $-\log[H^+]$. The pH scale provides a convenient way of expressing the acidity (or basicity) of a solution, which can vary over a range of more than 10^{16}.

Hydrolysis is described in Arrhenius language as the reaction of an ion with water to form a weak acid or base and OH^- or H^+, respectively. The weaker the acid or base is, the greater is the tendency for the ion to undergo hydrolysis. An anion hydrolyzes by accepting a proton from water. The resulting OH^- ion makes the solution basic. A cation hydrolyzes by donating a proton, sometimes from its shell of hydrating water molecules. This proton is accepted by the solvent water, and so the solution becomes acidic. Hydrolysis equilibria can be treated quantitatively by using hydrolysis constants, K_h. The value of K_h for a given hydrolysis is calculated by dividing K_w by the value of the dissociation constant for the weak electrolyte formed in the hydrolysis. In Brønsted-Lowry termi-

nology this relationship becomes $K_a K_b = K_w$, where K_a and K_b are the dissociation constants of the respective acid and conjugate base.

Acids and bases can be analyzed in solution by means of *titration*. The equivalence point in an acid-base titration is commonly signaled by the color change of an *indicator* added to the solution. A plot of the variation of pH as an acid is gradually added to a base (or vice versa) is known as a *titration curve*. The pH and the slope of the titration curve at the equivalence point are dependent upon the strength of the acid and of the base.

A *buffer* is a solution which contains an acid and its conjugate base in approximately equal concentrations. Such a solution changes pH very sluggishly in response to added H^+ or OH^-, and so buffers are used when it is desired to maintain pH within narrow limits.

The final topic in this chapter was that of *simultaneous equilibria*. Although all equilibrium conditions in a complex system must be simultaneously satisfied, it is often possible to make a few judicious approximations which greatly simplify the calculations.

KEY TERMS

Acid salt (Sec. 16-7)
Acidic solution (Sec. 16-3)
Basic solution (Sec. 16-3)
Buffer (Sec. 16-6)
Common-ion effect (Sec. 16-1)
Dissociation constant (Sec. 16-1)
Equivalence point (Sec. 16-5)
Henderson-Hasselbalch equation (Sec. 16-6)
Hydrolysis (Sec. 16-4)
Hydrolysis constant (Sec. 16-4)
Indicator (Sec. 16-5)
Ion product for water (Sec. 16-3)
K_a (Sec. 16-1)
K_b (Sec. 16-2)

K_{diss} (Secs. 16-1, 16-2)
K_h (Sec. 16-4)
K_w (Sec. 16-3)
K_1 (Secs. 16-1, 16-7)
K_2 (Secs. 16-1, 16-7)
Neutral solution (Sec. 16-3)
pH (Sec. 16-3)
pK (Sec. 16-3)
pOH (Sec. 16-3)
Polyprotic acid (Secs. 16-1, 16-7)
Titrant (Sec. 16-5)
Titration (Sec. 16-5)
Titration curve (Sec. 16-5)

QUESTIONS AND PROBLEMS[2]

pH and the dissociation of water

16-1 Calculate the pH of a solution in which the hydrogen-ion concentration is **(a)** $1.0\ M$ **(b)** $1.6 \times 10^{-4}\ M$ **(c)** $9.0 \times 10^{-7}\ M$ **(d)** $6.4 \times 10^{-11}\ M$.

16-2 Calculate the hydrogen-ion concentration in a solution which has a pH of **(a)** 4.62 **(b)** 7.97 **(c)** 9.41 **(d)** 13.58.

16-3 Calculate the hydroxide-ion concentration in a

solution which has a pH of **(a)** 3.19 **(b)** 9.87 **(c)** 1.00 **(d)** 11.41.

16-4 Calculate the pOH of a solution which has a hydrogen-ion concentration of **(a)** $1.1 \times 10^{-2}\ M$ **(b)** $3.6 \times 10^{-4}\ M$ **(c)** $5.0 \times 10^{-10}\ M$ **(d)** $1.9 \times 10^{-14}\ M$.

16-5 Calculate the pH of each of the following solutions: **(a)** $0.26\ M$ HCl **(b)** $1.4 \times 10^{-3}\ M$ NaOH **(c)** $0.098\ M$ HNO$_3$ **(d)** $0.019\ M$ Sr(OH)$_2$.

[2] Assume $t = 25°C$, unless otherwise specified.

16-6 How much 6.0 M HCl should be added to 1.000 liter of water in order to obtain a solution which has a pH of 1.50?

16-7 How much 1.0 M NaOH should be added to 475 ml of water in order to obtain a solution with a pH of 10.90?

16-8 At 50°C the ion product of water K_w is 5.5×10^{-14}. What is the pH of a neutral solution at 50°C?

16-9 The self-dissociation constant for liquid ammonia is 1.0×10^{-33} at -33°C. What is the pH of a neutral solution in liquid ammonia at this temperature?

Weak acids

16-10 A solution prepared by dissolving acetic acid in water has a pH of 4.31. What is the acetate-ion concentration in the solution?

16-11 A solution of acetic acid in water has an acetate-ion concentration of 3.64×10^{-3} M. What is the pH of the solution?

16-12 In a certain solution the equilibrium concentration of $HC_2H_3O_2$ is 0.30 M and $[C_2H_3O_2^-]$ is 0.50 M. What is the pH of the solution?

16-13 Calculate the concentration of acetate ions in a solution of pH 4.40 in which $[HC_2H_3O_2] = 0.10$ M.

16-14 Calculate the pH of each of the following solutions: **(a)** 4.5×10^{-1} M $HC_2H_3O_2$ **(b)** 0.90 M HOCl **(c)** 0.10 M $HClO_2$.

16-15 Calculate the pH of each of the following solutions: **(a)** 6.3×10^{-1} M $HC_2H_3O_2$ **(b)** 6.3×10^{-4} M $HC_2H_3O_2$.

16-16 Calculate the percent dissociation in **(a)** 0.35 M $HC_2H_3O_2$ **(b)** 0.035 M $HC_2H_3O_2$ **(c)** 0.00035 M $HC_2H_3O_2$.

16-17 Calculate the concentrations of all dissolved molecular and ionic species in each of the following: **(a)** 1.2 M HCN **(b)** 1.2 M H_2CO_3.

16-18 Suppose 0.23 mol of an unknown monoprotic acid is dissolved in enough water to make 2.55 liters of solution. If the pH of the solution is 3.62, what is the dissociation constant of the acid?

16-19 Equal numbers of moles of the weak acid HA and its salt NaA are dissolved in a glass of water. If the pH of the solution is 3.29, what is K_a?

Weak bases

16-20 A solution is prepared by dissolving some NH_3 in water. If the pH of the solution is 10.90, what is the NH_4^+ ion concentration?

16-21 A solution is prepared by dissolving some NH_3 in water. If the pH of the solution is 11.27, how many moles of NH_3 were dissolved per liter?

16-22 Calculate $[NH_4^+]$ in a solution in which $[NH_3] = 8.9 \times 10^{-2}$ M and the pH is 9.00.

16-23 Calculate the pH of each of the following solutions: **(a)** 0.62 M NH_3 **(b)** 0.92 M CH_3NH_2.

16-24 How many moles of ammonium chloride would need to be added to 25 ml of 0.10 M NH_3 in order to bring the pH down to 8.50?

Hydrolysis

16-25 Classify each of the following 1 M solutions as being acidic, basic, or neutral. Write one or more equations justifying each answer: **(a)** NH_4Cl **(b)** KCN **(c)** Na_2SO_3 **(d)** NH_4CN **(e)** KBr **(f)** $KHSO_4$.

16-26 For each of the following pairs tell which 1 M solution would be more basic, and explain your answer: **(a)** $NaC_2H_3O_2$ or NaCN **(b)** Na_2SO_4 or Na_2SO_3 **(c)** Na_2SO_4 or $NaHSO_4$ **(d)** H_2SO_3 or $NaHSO_3$ **(e)** NH_4CN or NaCN.

16-27 Calculate the pH of each of the following: **(a)** 0.25 M $NaC_2H_3O_2$ **(b)** 0.25 M NaCN **(c)** 0.25 M NH_4Cl.

16-28 Calculate the pH of each of the following: **(a)** 0.10 M Na_2SO_3 **(b)** 0.10 M Na_3PO_4.

16-29 Calculate the percent hydrolysis in each of the following: **(a)** 0.25 M $NaC_2H_3O_2$ **(b)** 0.25 M NaCN.

16-30 The pH of a 1.0 M solution of sodium nitrite, $NaNO_2$, is 8.65. Calculate K_a for nitrous acid, HNO_2.

16-31 The pH of a 1.0 M solution of sodium selenide, Na_2Se, is 12.51. Calculate K_2 for hydroselenic acid, H_2Se.

16-32 Calculate the concentration of all molecular and ionic species in a 0.71 M NH_4Cl solution.

Titrations and indicators

16-33 What will be the pH at the equivalence point of a titration in which 35 ml of 0.25 M HNO_3 are titrated with 0.25 M KOH?

16-34 Suppose 25.0 ml of 0.24 M $HC_2H_3O_2$ are titrated with 0.24 M NaOH. What is the pH at the equivalence point?

16-35 Suppose 35.0 ml of 0.10 M NH_3 are titrated with 0.10 M HNO_3. What is the pH at the equivalence point?

16-36 Suppose 25.0 ml of 0.17 M NH_3 are titrated with 0.14 M HCl. What is the pH at the equivalence point?

16-37 Suppose 25.0 ml of 0.28 M H_2SO_3 are titrated with 0.50 M NaOH. What is the pH at each equivalence point?

16-38 Suppose 25.0 ml of 0.1 M HCl are titrated with 0.1 M NaOH. If the indicator used is phenolphthalein, how many milliliters past the equivalence point will the indicator end point occur? Assume the pH at the indicator end point to be equal to pK_{In}.

16-39 Suppose 10.0 ml of 0.10 M NaOH are titrated with 0.10 M HCl. Calculate the pH of the solution after adding   **(a)** 1.0 ml   **(b)** 5.0 ml   **(c)** 9.0 ml   **(d)** 9.9 ml   **(e)** 10.0 ml of the HCl.

16-40 Suppose 10.0 ml of a 0.10 M NH$_3$ solution are titrated with 0.10 M HCl. Calculate the pH of the solution after adding   **(a)** 1.0 ml   **(b)** 5.0 ml   **(c)** 9.0 ml   **(d)** 9.9 ml   **(e)** 10.0 ml of the HCl.

16-41 Plot a titration curve for the titration of 10.0 ml of 0.10 M HC$_2$H$_3$O$_2$ with 0.10 M NaOH. For your points on the curve calculate the pH of the solution after adding 1.0 ml at a time from the beginning until a total of 20.0 ml of base has been added.

16-42 A 0.25 M solution of a weak acid is titrated with some 0.25 M NaOH. At the half-equivalence point, that is, when half the base needed to reach the equivalence point has been added, the pH of the solution is 4.41. What is the dissociation constant of the acid?

16-43 Choose an appropriate indicator for the titration of a 0.1 M solution of each of the following acids with 0.1 M NaOH: **(a)** aspartic $(K_1 = 1.4 \times 10^{-4})$ **(b)** barbituric $(K_a = 9.1 \times 10^{-5})$ **(c)** phenol $(K_a = 1.0 \times 10^{-10})$ **(d)** oxalic $(K_1 = 5.4 \times 10^{-2})$ **(e)** folic $(K_a = 5.5 \times 10^{-9})$.

Buffers

16-44 What is the pH of each of the following buffers?   **(a)** 0.4 M HC$_2$H$_3$O$_2$ + 0.4 M NaC$_2$H$_3$O$_2$ **(b)** 0.7 M NH$_3$ + 0.7 M NH$_4$NO$_3$ **(c)** 0.1 M CO$_2$ + 0.1 M NaHCO$_3$ **(d)** 0.1 M NaHCO$_3$ + 0.1 M Na$_2$CO$_3$.

16-45 Buffer capacity is the ability of a buffer to resist pH change. For a given system buffered at a certain pH what determines the buffer capacity?

16-46 How many moles of NaC$_2$H$_3$O$_2$ should be added to 375 ml of 0.30 M HC$_2$H$_3$O$_2$ in order to prepare a buffer with pH = 4.50? Assume no volume change.

16-47 How many moles of H$^+$ may be added to 100 ml of a buffer which is 0.5 M in both HC$_2$H$_3$O$_2$ and NaC$_2$H$_3$O$_2$ before the pH changes by 1 unit?

16-48 Calculate the pH before and after adding 0.010 mol of HCl to 0.100 liter of each of the following: **(a)** pure H$_2$O   **(b)** 0.10 M NaOH   **(c)** 0.10 M HCl   **(d)** 0.2 M HC$_2$H$_3$O$_2$ + 0.2 M NaC$_2$H$_3$O$_2$ **(e)** 1.0 M HC$_2$H$_3$O$_2$ + 1.0 M NaC$_2$H$_3$O$_2$.

Acid-base equilibrium—general

16-49 Suppose 215 ml of a solution contains 0.10 mol of HC$_2$H$_3$O$_2$.   **(a)** What is the pH of the solution?   **(b)** What is the pH after 0.12 mol of NaC$_2$H$_3$O$_2$ is added? (Assume no volume change.)   **(c)** What is the pH if 0.050 mol of NaOH is added to the original solution? (Assume no volume change.)

16-50 **(a)** What is the pH of a 0.20 M NH$_3$ solution?   **(b)** What will be the final pH if 0.020 mol of NH$_4$Cl is added to 333 ml of 0.20 M NH$_3$? (Assume no volume change.)   **(c)** What will be the final pH if 0.020 mol of HCl is added to 333 ml of 0.20 NH$_3$? (Assume no volume change.)

16-51 Calculate the pH of a solution prepared by mixing equal volumes of   **(a)** water and 0.020 M HCl   **(b)** 0.020 M HCl and 0.020 M NaOH   **(c)** 0.020 M HCl and 0.040 M NaOH   **(d)** 0.020 M HCN and 0.020 M NaOH.

16-52 Calculate the pH of a solution prepared by mixing 211 ml of 0.50 M HCN with 211 ml of   **(a)** water   **(b)** 0.50 M NaCN   **(c)** 0.25 M NaOH   **(d)** 0.50 M NaOH   **(e)** 0.50 M HCl.

16-53 Arrange the following in order of decreasing concentration in 1 M H$_2$SO$_3$: H$^+$, HSO$_3^-$, H$_2$SO$_3$, SO$_3^{2-}$, OH$^-$.

16-54 Arrange the following in order of decreasing concentration in 1 M Na$_2$SO$_3$: H$^+$, HSO$_3^-$, H$_2$SO$_3$, SO$_3^{2-}$, OH$^-$, Na$^+$.

16-55 Arrange the following in order of decreasing concentration in 1 M NaHSO$_3$: H$^+$, HSO$_3^-$, SO$_3^{2-}$, OH$^-$, Na$^+$, H$_2$SO$_3$.

16-56 In 0.1 M HC$_2$H$_3$O$_2$ what percent of the hydrogen ions present come from the dissociation of H$_2$O?

16-57 Calculate [H$^+$], [H$_2$PO$_4^-$], [HPO$_4^{2-}$], and [PO$_4^{3-}$] in a 1.0 M solution of H$_3$PO$_4$.

16-58 The equilibrium constant for the dissociation of AlOH^{2+} into Al^{3+} and OH$^-$ is 7.1 $\times$ 10^{-10}. Calculate the pH of a 0.22 M solution of AlCl$_3$.

16-59 It is desired to prepare 100.0 ml of a buffer having a pH of 5.20. Assume 0.10 M solutions of HC$_2$H$_3$O$_2$ and NaC$_2$H$_3$O$_2$ are available. What volumes should be mixed?

16-60 You are given 1.0 liter of 0.10 M CO$_2$.   **(a)** What is [CO$_3^{2-}$] in the solution?   **(b)** Suppose 1.0 liter of 0.20 M NaOH is added to the original solution. What is the new [CO$_3^{2-}$]?   **(c)** Suppose 1.0 liter of 0.10 M HCl is added to the solution from part **(b)**. What is the [CO$_3^{2-}$] now?

16-61 Calculate the pH of a 1.00 $\times$ 10^{-7} M solution of HCl.

17

AQUEOUS SOLUTIONS: SOLUBILITY AND COMPLEX-ION EQUILIBRIA

TO THE STUDENT

In this chapter we continue our study of ionic equilibria in aqueous solutions. We first study *solubility equilibria*. These are examples of *heterogeneous equilibria*, because each involves the equilibrium of a solid with its ions in solution, a two-phase system. Then we consider the (homogeneous) equilibrium between a complex ion and its dissociation products. Finally, we consider some systems in which solubility and other equilibria interact so that several equilibrium conditions are met simultaneously.

17-1
The solubility of ionic solids

When a solid *nonelectrolyte* dissolves in water, the resulting solution contains but one kind of solute species. Thus a saturated solution of sucrose, $C_{12}H_{22}O_{11}$, contains only sucrose molecules in equilibrium with excess undissolved solute.

$$C_{12}H_{22}O_{11}(s) \rightleftharpoons C_{12}H_{22}O_{11}(aq)$$

When a solid *electrolyte* dissolves, however, at least two kinds of particles (ions) are released to the solution, and so at saturation the equilibrium is more complex. Thus in a saturated solution of NaCl, both sodium ions and chloride ions in solution are in equilibrium with excess solid NaCl.

$$NaCl(s) \rightleftharpoons Na^+(aq) + Cl^-(aq)$$

The solubility product

Consider a slightly soluble ionic solid, MA, composed of M^+ and A^- ions in a crystal lattice. Suppose that enough MA is dissolved in water to produce a saturated solution with some solid MA left over. This establishes a *solubility equilibrium*, which can be written

$$MA(s) \rightleftharpoons M^+ + A^-$$

for which the equilibrium condition is

$$\frac{[M^+][A^-]}{[MA]} = K'$$

But this expression can be simplified.   As we showed in Sec. 15-2, the concentration of a substance in its pure solid phase is a constant.   In this case, $[MA]$ is not affected by anything which goes on in solution, and so the above equilibrium condition can be rewritten as

$$[M^+][A^-] = K'[MA]$$

or

$$[M^+][A^-] = K_{sp}$$

where K_{sp} represents the product of the two constant terms K and $[MA]$.   The mass-action expression on the left, $[M^+][A^-]$, is called the *ion product,* and K_{sp} is called the *solubility product,* or *solubility product constant,* of the substance MA. *At equilibrium the ion product equals the solubility product.*

As with all reactions the form of the mass-action expression depends on the stoichiometry of the reaction.   Thus for

$$CaF_2(s) \rightleftharpoons Ca^{2+} + 2F^-$$

the equilibrium condition is

$$[Ca^{2+}][F^-]^2 = K_{sp}$$

and for

$$Ca_3(PO_4)_2(s) \rightleftharpoons 3Ca^{2+} + 2PO_4^{3-}$$

the equilibrium condition is

$$[Ca^{2+}]^3[PO_4^{3-}]^2 = K_{sp}$$

Numerical values of solubility products can be calculated from measurements of solubilities, although for substances which have very low solubilities, indirect methods are usually used.

● **EXAMPLE 17-1**

Problem   The solubility of calcium sulfate in water is $4.9 \times 10^{-3}\ M$ at 25°C.   Calculate K_{sp} for $CaSO_4$ at this temperature.

Solution   The solubility equilibrium is

$$CaSO_4(s) \rightleftharpoons Ca^{2+} + SO_4^{2-}$$

and so the equilibrium condition is

$$[Ca^{2+}][SO_4^{2-}] = K_{sp}$$

From the equation we see that when 4.9×10^{-3} mol of $CaSO_4$ dissolves per liter, the resulting ionic concentrations are

$$[Ca^{2+}] = 4.9 \times 10^{-3}\ M$$

and

$$[SO_4^{2-}] = 4.9 \times 10^{-3}\ M$$

Therefore, $[Ca^{2+}][SO_4^{2-}] = (4.9 \times 10^{-3})(4.9 \times 10^{-3}) = 2.4 \times 10^{-5}$

K_{sp} for $CaSO_4$ is 2.4×10^{-5} at 25°C. ●

● **EXAMPLE 17-2**

Problem   The solubility of lead chloride is $1.6 \times 10^{-2}\ M$ at 25°C.   What is K_{sp} for $PbCl_2$ at this temperature?

Solution   The solubility equilibrium is

$$PbCl_2(s) \rightleftharpoons Pb^{2+} + 2Cl$$

When 1.6×10^{-2} mol of $PbCl_2$ dissolves per liter, the resulting ionic concentrations are

$$[Pb^{2+}] = 1.6 \times 10^{-2} M$$

and
$$[Cl^-] = 2(1.6 \times 10^{-2}) = 3.2 \times 10^{-2} M$$

Therefore,
$$K_{sp} = [Pb^{2+}][Cl^-]^2$$
$$= (1.6 \times 10^{-2})(3.2 \times 10^{-2})^2 = 1.6 \times 10^{-5}$$

Values of some solubility products at 25°C are shown in Table 17-1. (A more extensive list is given in App. H.) These are useful for calculating molar solubilities.

● **EXAMPLE 17-3**

Problem   The solubility product for silver iodide, AgI, is 8.5×10^{-17} at 25°C. What is the solubility of AgI in water at this temperature?

Solution   Silver iodide dissolves according to the equation

$$AgI(s) \rightleftharpoons Ag^+ + I^-$$

In this example the Ag^+ and I^- ions are present in a 1:1 ratio, since the only source of each is the AgI which has dissolved. If x equals the number of moles of AgI dissolved per liter (the solubility of AgI), then

$$[Ag^+] = x \quad \text{and} \quad [I^-] = x$$

At equilibrium,

$$[Ag^+][I^-] = K_{sp}$$
$$x(x) = 8.5 \times 10^{-17}$$
$$x = \sqrt{8.5 \times 10^{-17}} = 9.2 \times 10^{-9}$$

The solubility of AgI in water is $9.2 \times 10^{-9} M$ at 25°C. ●

Table 17-1
Solubility products (25°C)

Compound	K_{sp}
AgCl	1.7×10^{-10}
Ag_2CrO_4	1.9×10^{-12}
Ag_2S	5.5×10^{-51}
Al(OH)_3	5×10^{-33}
BaF_2	1.7×10^{-6}
BaSO_4	1.5×10^{-9}
CaF_2	1.7×10^{-10}
Ca(OH)_2	1.3×10^{-6}
CaSO_4	2.4×10^{-5}
Cu(OH)_2	1.6×10^{-19}
CuS	8×10^{-37}
Fe(OH)_2	2×10^{-15}
Mg(OH)_2	8.9×10^{-12}
PbCl_2	1.6×10^{-5}
ZnS	1.2×10^{-23}

● **EXAMPLE 17-4**

Problem K_{sp} for strontium fluoride is 2.5×10^{-9} at 25°C. Calculate the solubility of SrF_2 in water at this temperature.

Solution   The solubility equilibrium is

$$SrF_2(s) \rightleftharpoons Sr^{2+} + 2F^-$$

This shows us that in a saturated solution of SrF_2 $[F^-]$ is twice $[Sr^{2+}]$. If x equals the number of moles of SrF_2 dissolved per liter, then

$$[Sr^{2+}] = x \quad \text{and} \quad [F^-] = 2x$$

At equilibrium,

$$[Sr^{2+}][F^-]^2 = K_{sp}$$
$$x(2x)^2 = 2.5 \times 10^{-9}$$
$$4x^3 = 2.5 \times 10^{-9}$$
$$x = 8.5 \times 10^{-4}$$

The solubility of SrF_2 in water is 8.5×10^{-4} M at 25°C. ●

The common-ion effect   In Sec. 16-1 we showed that the dissociation of a weak acid, HA,

$$HA \rightleftharpoons H^+ + A^-$$

is repressed by the presence of additional A^- in the solution, an effect called the *common-ion effect*. This effect also accounts for the reduction in solubility of an ionic solid brought about by the presence of additional cations or anions in common with those of the solid.

In Example 17-3 we calculated that at 25°C the solubility of silver iodide, AgI, in water is 9.2×10^{-9} M. Le Châtelier's principle predicts that a high $[I^-]$ will shift the equilibrium

$$AgI(s) \rightleftharpoons Ag^+ + I^-$$

to the left, lowering $[Ag^+]$. Thus if we dissolve AgI in a solution which *already contains some* I^-, equilibrium will be established at a lower $[Ag^+]$. The solubility of AgI is thus predicted to be less in a solution of NaI than in pure water.

● **EXAMPLE 17-5**

Problem   Calculate the solubility of AgI in 0.10 M NaI at 25°C. K_{sp} for AgI is 8.5×10^{-17} at 25°C.

Solution   In this example the ratio of $[Ag^+]$ to $[I^-]$ is *not* 1:1, because of the additional I^- ions in solution (from the NaI). Before any AgI has dissolved $[I^-] = 0.10$ M. (Remember: NaI is a strong electrolyte.) Let x equal the number of moles of AgI which dissolve per liter. Then at equilibrium

$$[Ag^+] = x \quad \text{and} \quad [I^-] = 0.10 + x$$

The ionic concentrations before and after the AgI dissolves can be summarized as

$$AgI(s) \rightleftharpoons Ag^+ + \quad I^-$$

	Ag^+	I^-
Concentration *before* adding AgI (M):	0	0.10
Increase (M):	x	x
Equilibrium concentration (M):	x	$0.10 + x$

Substituting these in the equilibrium condition

$$[Ag^+][I^-] = K_{sp}$$

we get
$$x(0.10 + x) = 8.5 \times 10^{-17}$$

If we now assume that $x \ll 0.10$, then $0.10 + x \approx 0.10$, and so

$$x(0.10) \approx 8.5 \times 10^{-17}$$

$$x \approx 8.5 \times 10^{-16}$$

This number is clearly negligible in comparison with 0.10, and so our assumption is justified. Therefore, the solubility of AgI in 0.10 M NaI is 8.5×10^{-16} M at 25°C. ●

If we compare the results of Examples 17-3 and 17-5, we see quantitative evidence of reduction in solubility according to the common-ion effect.

Example	Solvent	Solubility of AgI, M
17-3	Water	9.2×10^{-9}
17-5	0.10 M NaI	8.5×10^{-16}

● **EXAMPLE 17-6**

Problem   Calculate the solubility of magnesium hydroxide, $Mg(OH)_2$, at 25°C in (a) pure water and (b) a solution having a pH equal to 12.00. K_{sp} for $Mg(OH)_2$ is 8.9×10^{-12} at this temperature.

Solution   (a) The solubility equilibrium in this case is

$$Mg(OH)_2(s) \rightleftharpoons Mg^{2+} + 2OH^-$$

If x equals the number of moles of $Mg(OH)_2$ which disssolve per liter, then

$$[Mg^{2+}] = x \quad \text{and} \quad [OH^-] = 2x$$

But what about the OH^- ions present due to the dissociation of water? Water is such a very weak electrolyte that we will assume that $[OH^-]$ due to its dissociation is negligible in comparison with $[OH^-]$ from the $Mg(OH)_2$. At equilibrium,

$$[Mg^{2+}][OH^-]^2 = K_{sp}$$

Substituting,
$$x(2x)^2 = 8.9 \times 10^{-12}$$

$$4x^3 = 8.9 \times 10^{-12}$$

and so
$$x = 1.3 \times 10^{-4}$$

The hydroxide-ion concentration is

$$[OH^-] = 2x = 2(1.3 \times 10^{-4}) = 2.6 \times 10^{-4} \ M$$

In pure water $[OH^-]$ from the self-dissociation

$$H_2O \rightleftharpoons H^+ + OH^-$$

is only 1.0×10^{-7} M, and in this solution $[OH^-]$ *from this self-dissociation only* is even lower, because the equilibrium has been shifted to the left by the OH^- ions from the $Mg(OH)_2$. We were therefore safe in assuming that $[OH^-]$ from the water is negligible in this case. Finally, the molar solubility of $Mg(OH)_2$ in water, x, equals 1.3×10^{-4} M.

  (b) In this case $[OH^-]$ is high at the start. Since pH = 12.00,

$$[H^+] = 1.0 \times 10^{-12} \ M$$

and so $\qquad$ $[OH^-] = \dfrac{K_w}{[H^+]} = \dfrac{1.0 \times 10^{-14}}{1.0 \times 10^{-12}} = 0.010\ M$

If x equals the number of moles of $Mg(OH)_2$ which dissolve per liter, then

$$Mg(OH)_2(s) \rightleftharpoons Mg^{2+} + \quad 2OH^-$$

	Mg^{2+}	$2OH^-$
Concentration at start (M):	0	0.010
Change (M):	x	$2x$
Equilibrium concentration (M):	x	$0.010 + 2x$

At equilibrium $\qquad$ $[Mg^{2+}][OH^-]^2 = K_{sp}$

Substituting, $\qquad$ $x(0.010 + 2x)^2 = 8.9 \times 10^{-12}$

Assuming that $2x \ll 0.010$, we may simplify this equation.

$$x(0.010)^2 \approx 8.9 \times 10^{-12}$$

$$x \approx 8.9 \times 10^{-8}$$

We can see that our assumption was valid. The solubility of $Mg(OH)_2$ in a solution of pH 12.00 is $8.9 \times 10^{-8}\ M$. ●

Example 17-6 is another clear demonstration of the common-ion effect.

Example	Solvent	Solubility of $Mg(OH)_2$, M
17-6(a)	Water	1.3×10^{-4}
17-6(b)	Solution, pH = 12.00	8.9×10^{-8}

The presence of additional OH^- ions in solution greatly reduces the solubility of $Mg(OH)_2$.

17-2 Precipitation reactions

Consider again the solubility equilibrium for silver iodide

$$AgI(s) \rightleftharpoons Ag^+ + I^-$$

$$[Ag^+][I^-] = K_{sp}$$

This condition is met at equilibrium, which means in a *saturated solution*. If the solution is *unsaturated*, the ion product, $[Ag^+][I^-]$, is *less than* K_{sp}. In the unstable condition of supersaturation $[Ag^+][I^-]$ is *greater than* K_{sp}.

Suppose now that we have a solution of 0.50 M NaI to which we very slowly add solid $AgNO_3$, one "speck" at a time. Silver nitrate, a salt and strong electrolyte, releases Ag^+ to the solution. As we add the $AgNO_3$, $[Ag^+]$ gradually rises, and so the numerical value of the ion product also increases. Nothing visible happens until the value of the ion product is finally brought up to that of K_{sp}. At this point the next speck of added $AgNO_3$ causes onset of precipitation of AgI. Further addition of Ag^+ causes further precipitation, which lowers $[I^-]$ so that $[Ag^+][I^-]$ remains equal to K_{sp}. The results of this imaginary experiment are summarized in Table 17-2. The $[Ag^+]$ is at first seen to rise irregularly, as specks of $AgNO_3$ of irregular size are added. When $[Ag^+]$ reaches 1.7×10^{-16}, the ion product equals

	$[Ag^+]$, M	$[I^-]$, M	Ion product $[Ag^+][I^-]$		K_{sp} for AgI
Unsaturation	0	0.50	0	<	8.5×10^{-17}
	1.0×10^{-50}	0.50	5.0×10^{-51}	<	8.5×10^{-17}
	3.6×10^{-38}	0.50	1.8×10^{-38}	<	8.5×10^{-17}
	6.4×10^{-29}	0.50	3.2×10^{-29}	<	8.5×10^{-17}
	3.2×10^{-19}	0.50	1.6×10^{-19}	<	8.5×10^{-17}
Saturation	1.7×10^{-16}	0.50	8.5×10^{-17}	=	8.5×10^{-17}
Saturation and precipitation	1.8×10^{-16}	0.47	8.5×10^{-17}	=	8.5×10^{-17}
	2.0×10^{-16}	0.43	8.5×10^{-17}	=	8.5×10^{-17}
	7.1×10^{-16}	0.12	8.5×10^{-17}	=	8.5×10^{-17}
	4.3×10^{-15}	0.02	8.5×10^{-17}	=	8.5×10^{-17}
	6.5×10^{-12}	1.3×10^{-5}	8.5×10^{-17}	=	8.5×10^{-17}

K_{sp}, and so additional Ag^+ initiates precipitation, since the ion product can not exceed K_{sp}. (We are assuming that supersaturation does not occur.) Further increase in $[Ag^+]$ lowers $[I^-]$ through precipitation, so that the equilibrium condition

$$[Ag^+][I^-] = K_{sp}$$

is maintained.

Predicting the occurrence of precipitation

When a solution of $AgNO_3$ is mixed with one of NaI, no precipitate forms as long as the ion product remains below K_{sp}. But if the ion product (calculated as if no precipitation occurred) can be shown to exceed K_{sp}, then precipitation will take place until the concentrations of both ions are reduced to the point at which the ion product equals K_{sp}. In order to determine whether or not a precipitate can form, one must therefore merely calculate the value which the ion product *would have* if no precipitation occurred and compare it with the solubility product. Of course, unless a supersaturated solution is formed, the ion product never really gets a chance to exceed K_{sp}. Nevertheless, we say that *if the calculated ion product exceeds the solubility product,* precipitation will occur.

● **EXAMPLE 17-7**

Problem   Will a precipitate form if 25.0 ml of 1.4×10^{-9} M NaI and 35.0 ml of 7.9×10^{-7} M $AgNO_3$ are mixed? (K_{sp} for AgI is 8.5×10^{-17}.)

Solution   In the original NaI solution $[I^-] = 1.4 \times 10^{-9}$ M, and in the $AgNO_3$ solution $[Ag^+] = 7.9 \times 10^{-7}$ M. Now we need to calculate these concentrations after mixing, assuming that no precipitation occurs. If the volumes are additive, always a good assumption when solutions are dilute, then the volume of the final mixture is 25.0 ml + 35.0 ml, or 60.0 ml. Each concentration becomes reduced by dilution when the two solutions are mixed. Since the volume of the iodide-containing solution increases from 25.0 to 60.0 ml, the final iodide concentration is

$$[I^-] = (\text{initial } [I^-]) \times (\text{ratio of volumes})$$

$$= 1.4 \times 10^{-9} \, M \frac{25.0 \text{ ml}}{60.0 \text{ ml}} = 5.8 \times 10^{-10} \, M$$

The volume of the silver-containing solution increases from 35.0 to 60.0 ml, and so the final silver ion concentration is

$$[Ag^+] = (\text{initial } [Ag^+]) \times (\text{ratio of volumes})$$

$$= 7.9 \times 10^{-7} \, M \, \frac{35.0 \text{ ml}}{60.0 \text{ ml}} = 4.6 \times 10^{-7} \, M$$

The ion product is therefore

$$[Ag^+][I^-] = (4.6 \times 10^{-7})(5.8 \times 10^{-10}) = 2.7 \times 10^{-16}$$

This is *greater* than K_{sp} for AgI; so *precipitation will occur.* ●

● **EXAMPLE 17-8**

Problem

The solubility product for CaF_2 is 1.7×10^{-10}, and that for $CaCO_3$ is 4.7×10^{-9}. A solution contains F^- and CO_3^{2-}, both at $5.0 \times 10^{-5} \, M$ concentration. Solid $CaCl_2$ is slowly added. Which solid precipitates first: CaF_2 or $CaCO_3$?

Solution

We need to determine which solubility product is exceeded first. Just looking at the K_{sp} values we might jump to the conclusion that since 1.7×10^{-10} is less than 4.7×10^{-9}, the former K_{sp} would be the first to be exceeded as $[Ca^{2+}]$ rises. But that is not the case. In order to exceed K_{sp} for CaF_2, what must $[Ca^{2+}]$ be?

$$[Ca^{2+}][F]^2 > 1.7 \times 10^{-10}$$

and so

$$[Ca^{2+}] > \frac{1.7 \times 10^{-10}}{[F^-]^2} = \frac{1.7 \times 10^{-10}}{(5.0 \times 10^{-5})^2}$$

$$[Ca^{2+}] > 6.8 \times 10^{-2} \, M$$

In order to exceed K_{sp} for $CaCO_3$,

$$[Ca^{2+}][CO_3^{2-}] > 4.7 \times 10^{-9}$$

and so

$$[Ca^{2+}] > \frac{4.7 \times 10^{-9}}{[CO_3^{2-}]} = \frac{4.7 \times 10^{-9}}{5.0 \times 10^{-5}}$$

$$[Ca^{2+}] > 9.4 \times 10^{-5} \, M$$

When $[Ca^{2+}]$ reaches $9.4 \times 10^{-5} \, M$, $CaCO_3$ begins to precipitate. At this point K_{sp} for CaF_2 has not been exceeded; so this substance does not precipitate. Not until enough $CaCl_2$ has been added to increase $[Ca^{2+}]$ to $6.8 \times 10^{-2} \, M$ will CaF_2 start to precipitate. ●

17-3
**Complex-ion
equilibria**

As was stated in Sec. 13-2 the term *complex ion* usually means a charged particle which is composed of a central ion surrounded by a number of ions or molecules called *ligands*. The number of bonds formed by ligands to the central ion is called the *coordination number* of the ion. In Chap. 22 we study geometrical structure and bonding in these complexes. We now consider the equilibrium processes by which such complexes lose their ligands, that is, dissociate, in solution.

*The dissociation of
complex ions*

When cupric sulfate, $CuSO_4$, is dissolved in water, the cupric ion Cu^{2+} becomes hydrated. Four water molecules are most strongly bound to the Cu^{2+}, and so we can write the process of hydration as

$$Cu^{2+} + 4H_2O \longrightarrow Cu(H_2O)_4^{2+}$$

The hydrated cupric ion is a complex ion and is a medium-blue color. If we now add ammonia, the solution turns a striking deep blue. The new color indicates the presence of a new complex ion, in which NH_3 molecules have replaced the water molecules as ligands. The net reaction can be written

$$Cu(H_2O)_4^{2+} + 4NH_3 \longrightarrow Cu(NH_3)_4^{2+} + 4H_2O$$

A complex ion such as $Cu(NH_3)_4^{2+}$ has a tendency to exchange ligands with the solvent in stepwise equilibrium processes. Thus when $Cu(NH_3)_4^{2+}$ is in solution, the following sequential equilibria are established:

$$Cu(NH_3)_4^{2+} + H_2O \rightleftharpoons Cu(H_2O)(NH_3)_3^{2+} + NH_3$$

$$K_1 = \frac{[Cu(H_2O)(NH_3)_3^{2+}][NH_3]}{[Cu(NH_3)_4^{2+}]}$$

$$Cu(H_2O)(NH_3)_3^{2+} + H_2O \rightleftharpoons Cu(H_2O)_2(NH_3)_2^{2+} + NH_3$$

$$K_2 = \frac{[Cu(H_2O)_2(NH_3)_2^{2+}][NH_3]}{[Cu(H_2O)(NH_3)_3^{2+}]}$$

$$Cu(H_2O)_2(NH_3)_2^{2+} + H_2O \rightleftharpoons Cu(H_2O)_3(NH_3)^{2+} + NH_3$$

$$K_3 = \frac{[Cu(H_2O)_3(NH_3)^{2+}][NH_3]}{[Cu(H_2O)_2(NH_3)_2^{2+}]}$$

$$Cu(H_2O)_3(NH_3)^{2+} + H_2O \rightleftharpoons Cu(H_2O)_4^{2+} + NH_3$$

$$K_4 = \frac{[Cu(H_2O)_4^{2+}][NH_3]}{[Cu(H_2O)_3(NH_3)^{2+}]}$$

Each of these reactions is an *exchange reaction,* because NH_3 and H_2O molecules exchange places. In practice each is usually called a *dissociation,* because if the H_2O molecules are omitted, each equation shows the loss of one NH_3 molecule from a complex.

$$Cu(NH_3)_4^{2+} \rightleftharpoons Cu(NH_3)_3^{2+} + NH_3$$

$$K_1 = \frac{[Cu(NH_3)_3^{2+}][NH_3]}{[Cu(NH_3)_4^{2+}]}$$

$$Cu(NH_3)_3^{2+} \rightleftharpoons Cu(NH_3)_2^{2+} + NH_3$$

$$K_2 = \frac{[Cu(NH_3)_2^{2+}][NH_3]}{[Cu(NH_3)_3^{2+}]}$$

$$Cu(NH_3)_2^{2+} \rightleftharpoons Cu(NH_3)^{2+} + NH_3$$

$$K_3 = \frac{[Cu(NH_3)^{2+}][NH_3]}{[Cu(NH_3)_2^{2+}]}$$

$$Cu(NH_3)^{2+} \rightleftharpoons Cu^{2+} + NH_3$$

$$K_4 = \frac{[Cu^{2+}][NH_3]}{[Cu(NH_3)^{2+}]}$$

By multiplying together the equilibrium conditions for the stepwise dissociations, we obtain

$$K_1K_2K_3K_4 = \frac{[Cu^{2+}][NH_3]^4}{[Cu(NH_3)_4{}^{2+}]}$$

The product of the individual dissociation constants is often called the *cumulative dissociation constant* K_{diss} of the complex ion. We therefore have the equilibrium condition

$$K_{diss} = \frac{[Cu^{2+}][NH_3]^4}{[Cu(NH_3)_4{}^{2+}]}$$

K_{diss} is called a cumulative constant because it *appears* to be the constant for the cumulative dissociation reaction

$$Cu(NH_3)_4{}^{2+} \rightleftharpoons Cu^{2+} + 4NH_3$$

Remember, however, that the four NH_3 molecules leave the complex *separately*.

Cumulative dissociation constants for some complex ions are given in Table 17-3. Values of such constants can be used to estimate the stability of complexes.[1]

Complex-ion dissociation calculations

It is clear that because of multiple stepwise dissociations the exact calculation of the concentrations of all species in solutions of complexes such as $Cu(NH_3)_4{}^{2+}$ is a difficult task. In some situations, however, the concentrations of the most important dissolved species can be calculated by making some reasonable assumptions.

● **EXAMPLE 17-9**

Problem   Suppose 0.10 mol of cupric sulfate, $CuSO_4$, is added to 1.0 liter of 2.0 M NH_3. Calculate the concentration of Cu^{2+} in the resulting solution.

Solution   Table 17-3 shows that the cumulative dissociation constant for $Cu(NH_3)_4{}^{2+}$ is 1.0×10^{-12}, and since this is a small number, we conclude that this complex will be formed.

$$Cu^2 + 4NH_3 \longrightarrow Cu(NH_3)_4{}^{2+}$$

But we know that this reaction does not go to completion, and so at equilibrium a small concentration of Cu^{2+} will be left. In order to calculate that concentration we perform a thought experiment. We first assume that the reaction actually goes to completion, using up all the Cu^{2+}.

$$Cu^{2+} + 4NH_3 \longrightarrow Cu(NH_3)_4{}^{2+}$$

Moles at start:	0.10	2.0	0
Moles after reaction:	0	1.6	0.10

(*Note:* Two moles of NH_3 were present at the start because 1.0 liter of a 2.0 M solution was present.)

---

[1] Complex-ion dissociation constants are often referred to as *instability constants,* because the larger the constant, the more unstable is the complex. Sometimes the reciprocals of these constants are tabulated. These correspond to the reactions written in reverse, that is, as *associations*. When tabulated this way, the constants are known as *stability constants* or *formation constants*.

Table 17-3
Cumulative dissociation
(instability) constants
for some complex ions
(25°C)

Ion	Cumulative reaction	K_{diss}
$Ag(NH_3)_2{}^+$	$Ag(NH_3)_2{}^+ \rightleftharpoons Ag^+ + 2NH_3$	5.9×10^{-8}
$Ag(S_2O_3)_2{}^{3-}$	$Ag(S_2O_3)_2{}^{3-} \rightleftharpoons Ag^+ + 2S_2O_3{}^{2-}$	6×10^{-14}
$Co(NH_3)_6{}^{3+}$	$Co(NH_3)_6{}^{3+} \rightleftharpoons Co^{3+} + 6NH_3$	6.3×10^{-36}
$Cu(NH_3)_4{}^{2+}$	$Cu(NH_3)_4{}^{2+} \rightleftharpoons Cu^{2+} + 4NH_3$	1×10^{-12}
$Cu(CN)_4{}^{2-}$	$Cu(CN)_4{}^{2-} \rightleftharpoons Cu^{2+} + 4CN^-$	1×10^{-25}
$Fe(CN)_6{}^{3-}$	$Fe(CN)_6{}^{3-} \rightleftharpoons Fe^{3+} + 6CN^-$	1×10^{-42}

Now we imagine that the $Cu(NH_3)_4{}^{2+}$ dissociates slightly to provide a little Cu^{2+} to the solution in order to establish the final equilibrium. Let x equal the number of moles of the complex which dissociate. Then x also equals the number of moles of Cu^{2+} ions released to the solution. Since K_{diss} is small, we know that the complex dissociates very little, and so the number of moles of Cu^{2+} and of NH_3 will be essentially those before dissociation. Because the volume of the solution is 1.0 liter, at equilibrium,

$$[Cu(NH_3)_4{}^{2+}] = 0.10 \ M$$

$$[NH_3] = 1.6 \ M \quad \text{and} \quad [Cu^{2+}] = x$$

The equilibrium condition for the dissociation is

$$\frac{[Cu^{2+}][NH_3]^4}{[Cu(NH_3)_4{}^{2+}]} = K_{diss}$$

Substituting, we have

$$\frac{x(1.6)^4}{0.10} = 1.0 \times 10^{-12}$$

$$x = 1.5 \times 10^{-14}$$

Thus the presence of the ammonia has reduced the concentration of Cu^{2+}, actually $Cu(H_2O)_4{}^{2+}$, to

$$[Cu^{2+}] = 1.5 \times 10^{-14} \ M$$

The
amphoterism of
metal hydroxides

When a solution of NaOH is added to one of $ZnCl_2$, a precipitate of zinc hydroxide forms. The net equation can be written as

$$Zn^{2+} + 2OH^- \longrightarrow Zn(OH)_2(s)$$

The product, zinc hydroxide, will dissolve after the addition of either excess base,

$$Zn(OH)_2(s) + OH^- \longrightarrow Zn(OH)_3{}^-$$

or acid,

$$Zn(OH)_2(s) + H^+ \longrightarrow Zn(OH)^+ + H_2O$$

Zinc hydroxide is thus capable of acting either as an Arrhenius acid (by reacting with OH^-) or as an Arrhenius base (by reacting with H^+). Such substances are said to be *amphoteric*.

How can the behavior of amphoteric hydroxides be explained? In all of the above zinc-containing species the zinc is believed to have a coordination number of 4. Zn^{2+} is really $Zn(H_2O)_4{}^{2+}$, and $Zn(OH)_3{}^-$ is really $Zn(OH)_3(H_2O)^-$, for example. The sparingly soluble zinc hydroxide dissolves in acidic solution because it reacts with H^+.

$$Zn(OH)_2(H_2O)_2(s) + H^+ \longrightarrow Zn(OH)(H_2O)_3{}^+$$

Zinc hydroxide dissolves in basic solution because OH^- pulls a proton off one of the water molecules:

$$Zn(OH)_2(H_2O)_2 + OH^- \longrightarrow Zn(OH)_3(H_2O)^- + H_2O$$

The experimental criterion for amphoterism in such a hydroxide is that it is soluble in both strong acids and strong bases.

Hydroxides can be either acidic, basic, or amphoteric. The difference depends on the relative strength of the bond from the central atom to the oxygen and that of the bond from the oxygen to the hydrogen. If we schematically indicate a hydroxide as

then acidic behavior can be described as the loss of a proton by a water ligand.

Basic behavior is described as the gain of a proton by a hydroxide ligand.

In Brønsted-Lowry terminology a species which can either donate or accept protons is said to be *amphiprotic*.

In order for a hydroxide to act as an acid, the O—H bond must be weak. For it to act as a base, the O—H bond must be strong. Whether one or the other or both of these occurs depends in turn upon the ability of the central atom in the complex to draw electrons toward itself. If it tends to attract electrons strongly, the O—H bond is weakened, and a proton is lost, so that acidic behavior results. If, on the other hand, the electron-attracting ability of the central atom is weak, the oxygen atoms can bond to additional protons, and so basic behavior is the consequence.

The electron-attracting tendency of an atom is measured by its electronegativity (Sec. 8-3). The hydroxide of a highly electropositive metal is therefore basic, while that of a good nonmetal is acidic. Thus NaOH is basic,

$$NaOH \longrightarrow Na^+ + OH^-$$

but ClOH is acidic,

$$ClOH \rightleftharpoons ClO^- + H^+$$

Zinc hydroxide represents the intermediate case, amphoterism.

17-4
Simultaneous
equilibria

Acid-base, solubility, and complex-ion equilibria may simultaneously compete for one or more species in solution. Although making exact calculations of concentrations in such systems may involve solving four, five, six, or more equations in as many unknowns, it is often possible to make simplifying approximations, so that a computer is not required.

● **EXAMPLE 17-10**

Problem

A solution contains $0.10\ M\ Cl^-$ and $1.0 \times 10^{-8}\ M\ CrO_4{}^{2-}$. Solid $AgNO_3$ is slowly added. Assuming that the solution volume remains constant, calculate (a) $[Ag^+]$ when AgCl first starts to precipitate; (b) $[Ag^+]$ when Ag_2CrO_4 first starts to precipitate; (c) $[Cl^-]$ when Ag_2CrO_4 first starts to precipitate. (K_{sp} values are given in Table 17-1.)

Solution

(a) The AgCl will first start to precipitate when the ion product becomes equal to K_{sp}.

$$[Ag^+][Cl^-] = K_{sp}$$

$$[Ag^+] = \frac{K_{sp}}{[Cl^-]} = \frac{1.7 \times 10^{-10}}{0.10} = 1.7 \times 10^{-9}\ M$$

(b) As Ag^+ continues to be added, $[Ag^+]$ increases and $[Cl^-]$ decreases, as AgCl is precipitated out of the solution. Finally, $[Ag^+]$ becomes high enough so that the solubility product for Ag_2CrO_4 is exceeded by its ion product.

$$Ag_2CrO_4(s) \rightleftharpoons 2Ag^+ + CrO_4{}^{2-}$$

$$[Ag^+]^2[CrO_4{}^{2-}] = K_{sp}$$

$$[Ag^+] = \sqrt{\frac{K_{sp}}{[CrO_4{}^{2-}]}}$$

$$= \sqrt{\frac{1.9 \times 10^{-12}}{1.0 \times 10^{-8}}} = 1.4 \times 10^{-2}\ M$$

(c) By the time $[Ag^+]$ has increased to $1.4 \times 10^{-2}\ M$, $[Cl^-]$ has decreased to

$$[Cl^-] = \frac{K_{sp}}{[Ag^+]} = \frac{1.7 \times 10^{-10}}{1.4 \times 10^{-2}} = 1.2 \times 10^{-8}\ M\ ●$$

● **EXAMPLE 17-11**

Problem

If 0.050 mole of NH_3 is added to 1.0 liter of $0.020\ M\ MgCl_2$, how many moles of NH_4Cl must be first added in order to prevent the precipitation of $Mg(OH)_2$?

Solution

$Mg(OH)_2$ will precipitate if $[OH^-]$ becomes high enough to exceed K_{sp}. OH^- ions are produced by the dissociation of NH_3, but this dissociation can be repressed by adding $NH_4{}^+$ to the solution. We start by finding out at what point K_{sp} for $Mg(OH)_2$ is exceeded.

$$Mg(OH)_2(s) \rightleftharpoons Mg^{2+} + 2OH^-$$

$$[Mg^{2+}][OH^-]^2 = K_{sp}$$

$$[OH^-] = \sqrt{\frac{K_{sp}}{[Mg^{2+}]}}$$

$$= \sqrt{\frac{8.9 \times 10^{-12}}{0.020}} = 2.1 \times 10^{-5} \, M$$

We must add enough NH_4^+ to repress the dissociation of NH_3 so that $[OH^-]$ does not rise above $2.1 \times 10^{-5} \, M$

$$NH_3 + H_2O \rightleftharpoons NH_4^+ + OH^-$$

$$\frac{[NH_4^+][OH^-]}{[NH_3]} = K_b$$

If $[OH^-]$ is to be kept below $2.1 \times 10^{-5} \, M$, then $[NH_4^+]$ must be kept above

$$[NH_4^+] = \frac{[NH_3]K_b}{[OH^-]}$$

$$= \frac{0.050(1.8 \times 10^{-5})}{2.1 \times 10^{-5}} = 0.043 \, M$$

Thus adding at least 0.043 mol of NH_4Cl will prevent the precipitation of $Mg(OH)_2$. ●

One of the most interesting illustrations of simultaneous equilibria is found in the selective precipitation of insoluble sulfides in qualitative inorganic analysis. Cupric sulfide (CuS; $K_{sp} = 8 \times 10^{-37}$) and zinc sulfide (ZnS; $K_{sp} = 1.2 \times 10^{-23}$) can each be precipitated from a solution containing both Cu^{2+} and Zn^{2+} by providing enough S^{2-} in solution so that the two K_{sp} values are both exceeded. By controlling the sulfide-ion concentration so that K_{sp} for CuS is exceeded, but that for ZnS is not, the Cu^{2+} ions may be effectively removed from the solution, leaving Zn^{2+} ions behind.

Consider that we have a solution which is 0.020 M in both Zn^{2+} and Cu^{2+}. In order to exceed K_{sp} for CuS,

$$CuS(s) \rightleftharpoons Cu^{2+} + S^{2-}$$

$$[Cu^{2+}][S^{2-}] = K_{sp}$$

$$[S^{2-}] = \frac{K_{sp}}{[Cu^{2+}]}$$

$$= \frac{8 \times 10^{-37}}{0.020} = 4 \times 10^{-35} \, M$$

In order to exceed K_{sp} for ZnS,

$$ZnS(s) \rightleftharpoons Zn^{2+} + S^{2-}$$

$$[Zn^{2+}][S^{2-}] = K_{sp}$$

$$[S^{2-}] = \frac{K_{sp}}{[Zn^{2+}]}$$

$$= \frac{1.2 \times 10^{-23}}{0.020} = 6.0 \times 10^{-22} \, M$$

If we regulate the sulfide-ion concentration so that it is greater than 4×10^{-35} M, but less than 6.0×10^{-22} M, then only CuS will precipitate.

In theory we could perform this separation by adding just enough sulfide ions, from Na_2S, for example, to get the desired $[S^{2-}]$ in solution, but think of the practical impossibility of trying to measure out 4×10^{-35} mole of sodium sulfide.[2] Fortunately there is an easier way. We can regulate the sulfide-ion concentration in solution indirectly by making use of the dissociation of the weak acid H_2S.

Hydrogen sulfide, or hydrosulfuric acid, is a weak diprotic acid.

$$H_2S \rightleftharpoons H^+ + HS^- \qquad K_1 = 1.1 \times 10^{-7}$$

$$HS^- \rightleftharpoons H^+ + S^{2-} \qquad K_2 = 1.0 \times 10^{-14}$$

It is evident from these equations that increasing $[H^+]$ will shift both equilibria to the left, reducing $[S^{2-}]$. Lowering $[H^+]$ will raise $[S^{2-}]$. Thus we can regulate $[S^{2-}]$ by merely adjusting the pH of the solution, which is much easier than adjusting $[S^{2-}]$ directly.

● **EXAMPLE 17-12**

Problem A solution contains Cu^{2+} and Zn^{2+}, each at 0.020 M. It is desired to separate the two by adjusting the pH and then saturating the solution with H_2S so that CuS precipitates, but ZnS does not. Calculate (a) the *lowest* pH which could be used to precipitate CuS and (b) the *highest* pH which could be used without precipitating any ZnS. (*Note:* A saturated solution of H_2S is 0.10 M.)

Solution (a) At equilibrium,

$$\frac{[H^+][HS^-]}{[H_2S]} = K_1 \qquad \text{and} \qquad \frac{[H^+][S^{2-}]}{[HS^-]} = K_2$$

Multiplying these equalities together, we have

$$\frac{[H^+][HS^-]}{[H_2S]} \times \frac{[H^+][S^{2-}]}{[HS^-]} = K_1 K_2$$

or

$$\frac{[H^+]^2[S^{2-}]}{[H_2S]} = K_1 K_2$$

Since $[H_2S]$ in a saturated solution is 0.10 M,

$$[H^+]^2[S^{2-}] = [H_2S]K_1K_2$$
$$= 0.10(1.1 \times 10^{-7})(1.0 \times 10^{-14}) = 1.1 \times 10^{-22}$$

We have shown that in order to precipitate CuS from a 0.020 M Cu^{2+} solution, $[S^{2-}]$ must exceed 4×10^{-35} M. This means that $[H^+]$ must not be higher than

$$[H^+] = \sqrt{\frac{1.1 \times 10^{-22}}{4 \times 10^{-35}}} = 2 \times 10^6 \ M$$

In other words the pH must not be lower than

$$pH = -\log(2 \times 10^6) = -6.3$$

---

[2] How many sulfide ions are present in 4×10^{-35} mol?

A solution of this high acidity can not be prepared, and so we conclude that CuS is so insoluble that no matter how low the pH is, the sulfide-ion concentration will be high enough to precipitate CuS.

(b) In order not to precipitate ZnS, the sulfide-ion concentration must be kept lower than 6.0×10^{-22} M, as we have previously calculated. In order to keep it below this value, the hydrogen-ion concentration must be above

$$[H^+] = \sqrt{\frac{1.1 \times 10^{-22}}{6.0 \times 10^{-22}}} = 0.43 \ M$$

This corresponds to a pH of $-\log 0.43 = 0.37$.

In summary, any pH less than 0.37 will allow CuS to precipitate but will prevent ZnS from precipitating out of a saturated H_2S solution. ●

ADDED COMMENT

Look at the relationship

$$\frac{[H^+]^2[S^{2-}]}{[H_2S]} = K_1 K_2$$

This relationship is correct, but it looks as if it might be the equilibrium condition for

$$H_2S \rightleftharpoons 2H^+ + S^{2-}$$

This equation does not represent an equilibrium. What is wrong with it? It appears from the above equation that *both* protons are lost from the H_2S simultaneously. This is not true! If it were, $[H^+]$ in a solution of H_2S would be twice $[S^{2-}]$. In fact, $[H^+]$ is many times greater than this. (Use K_1 and K_2 *separately* to calculate $[H^+]$ and $[S^{2-}]$ in a 0.10 M H_2S solution, and you will see.) Anytime you need it, use the relationship

$$\frac{[H^+]^2[S^{2-}]}{[H_2S]} = K_1 K_2$$

but do not use the incorrect stoichiometry implied by the pseudo-equilibrium equation above.

SUMMARY

In this chapter we studied equilibria between sparingly soluble electrolytes and their ions in solution. Such *solubility equilibria* are heterogeneous, and so the concentration of excess solid does not show up in the mass-action expression. The equilibrium constant for such an equilibrium is called a *solubility product*. One of its uses is in predicting whether or not precipitation will occur in a mixture of solutions. If the solubility product of a compound is exceeded by its ion product in solution, precipitation will take place.

The second type of equilibrium described in this

chapter is *complex-ion equilibrium*. Calculations based on such equilibria are often complicated to handle mathematically, but the presence of a large excess of one component often simplifies matters considerably.

A brief discussion of the *amphoteric behavior* of hydroxides was presented. The acid-base behavior of such compounds can be correlated with the electronegativity of the central atom in the complex.

The chapter concluded with several examples of *simultaneous equilibria*, equilibria which simultaneously compete for one or more solute species.

KEY TERMS

Amphoterism (Sec. 17-3)
Amphiprotism (Sec. 17-3)
Common-ion effect (Sec. 17-1)
Complex ion (Sec. 17-3)
Coordination number (Sec. 17-3)
Cumulative dissociation constant (Sec. 17-3)
Formation constant (Sec. 17-3)
Instability constant (Sec. 17-3)

Ion product (Sec. 17-1)
K_{sp} (Sec. 17-1)
Ligand (Sec. 17-3)
Simultaneous equilibria (Sec. 17-4)
Solubility equilibria (Sec. 17-1)
Solubility product (Sec. 17-1)
Stability constant (Sec. 17-3)

QUESTIONS AND PROBLEMS

Solubility equilibria

17-1 In a saturated solution of $BaCrO_4$, $[Ba^{2+}] = 9.2 \times 10^{-6}$. Calculate the solubility product of $BaCrO_4$.

17-2 Solid silver bromide, AgBr, is added to a 0.10 M NaBr solution until the solution is saturated. $[Ag^+]$ at this point is 5.0×10^{-12}. Calculate K_{sp} for AgBr.

17-3 In a saturated solution of magnesium fluoride, MgF_2, in water, $[Mg^{2+}] = 2.7 \times 10^{-3}\ M$. What is K_{sp} for MgF_2?

17-4 The solubility of barium fluoride, BaF_2, in 0.10 M NaF is $1.7 \times 10^{-4}\ M$. Calculate K_{sp} for BaF_2.

17-5 The solubility of silver cyanide, AgCN, in water is $1.3 \times 10^{-7}\ M$. What is K_{sp} for AgCN?

17-6 The solubility of lead iodate, $Pb(IO_3)_2$, in water is $3.1 \times 10^{-5}\ M$. What is K_{sp} for this compound?

17-7 The solubility of manganous hydroxide, $Mn(OH)_2$, in a solution of pH 12.50 is $2.0 \times 10^{-10}\ M$. Calculate K_{sp} for $Mn(OH)_2$.

17-8 Calculate the solubility of zinc sulfide, ZnS, in pure water. (Ignore hydrolyses.)

17-9 Calculate the solubility of ZnS in 0.25 M $ZnCl_2$. (Ignore hydrolyses.)

17-10 What is the solubility of cupric hydroxide, $Cu(OH)_2$, in pure water?

17-11 What is the solubility of $Cu(OH)_2$ in a solution of pH 10.80?

17-12 What is the concentration of sulfate ions in a saturated $BaSO_4$ solution?

17-13 What is the concentration of silver ions in a saturated solution of Ag_2CrO_4?

17-14 The mercurous ion is Hg_2^{2+}. Calculate $[Hg_2^{2+}]$ in a saturated solution of mercurous chloride, Hg_2Cl_2. K_{sp} for Hg_2Cl_2 is 1.1×10^{-18}.

17-15 What is the minimum concentration of sulfate ions necessary to begin the precipitation of calcium sulfate, $CaSO_4$, from a 0.50 M $CaCl_2$ solution?

17-16 K_{sp} for lead bromide, $PbBr_2$, is 4.0×10^{-5}. What is the minimum concentration of bromide ions necessary to precipitate $PbBr_2$ from a 0.080 M $Pb(NO_3)_2$ solution?

17-17 HCl and $AgNO_3$ solutions, each $2.0 \times 10^{-5}\ M$, are mixed in equal volumes. Will AgCl precipitate?

17-18 Suppose 25.0 ml of $1.8 \times 10^{-2}\ M$ $Ba(NO_3)_2$ is mixed with 35.0 ml of $3.0 \times 10^{-2}\ M$ NaF. Will BaF_2 precipitate?

17-19 Suppose 0.10 M NaF is slowly added to a solution in which the Ba^{2+} and Ca^{2+} concentrations are each $1.0 \times 10^{-4}\ M$. What substance precipitates first?

17-20 A solution contains SO_4^{2-} at $1.0 \times 10^{-5}\ M$ and some F^- ions. When some solid $BaCl_2$ is added, BaF_2 and $BaSO_4$ form simultaneously. What was the fluoride-ion concentration before precipitation?

17-21 Is it possible to precipitate $CaSO_4$ from a 1.0×10^{-4} solution of Na_2SO_4 by adding 0.020 M $CaCl_2$? Explain.

Complex-ion equilibria

17-22 Suppose 0.10 mol of $AgNO_3$ and 1.0 mol of NH_3 are dissolved in enough water to make 1.0 liter of solution. Calculate $[Ag^+]$ in the solution.

17-23 Suppose 0.10 mol of potassium ferricyanide, $K_3Fe(CN)_6$, and 0.10 mol of potassium cyanide are both dissolved in enough water to make 200 ml of solution. Calculate the concentration of Fe^{3+} in the resulting solution. (Ignore all hydrolyses.)

17-24 To what value should the ammonia concentration be adjusted in order to lower $[Ag^+]$ in a 0.10 M solution of $AgNO_3$ to $5.0 \times 10^{-10}\ M$? Assume no volume change upon addition of NH_3.

17-25 In a certain solution prepared by adding $CuSO_4$ and NH_3 to enough water to make 1.0 liter, $[NH_3]$ is 2.0 M and $[Cu^{2+}]$ is $5.0 \times 10^{-15}\ M$. How many moles of $CuSO_4$ were added to the solution?

Simultaneous equilibria

17-26 Will cadmium sulfide (CdS; $K_{sp} = 1.0 \times 10^{-28}$) precipitate if 1.5×10^{-2} mol of $CdCl_2$ is added to 500 ml of saturated (0.10 M) H_2S? (Ignore the hydrolysis of Cd^{2+}.)

17-27 A 5.0×10^{-3} M solution of $SnCl_2$ buffered at pH 2.0 is saturated with H_2S (0.10 M). Will SnS precipitate? (K_{sp} for SnS is 1×10^{-26}.)

17-28 What is the lowest pH at which 0.020 M Co^{2+} can be precipitated as CoS ($K_{sp} = 5.0 \times 10^{-22}$) from a saturated H_2S solution (0.10 M)?

17-29 Solid Na_2CO_3 is slowly added to a 1.0×10^{-3} M solution of $MgCl_2$. Which precipitates first: $MgCO_3$ or $Mg(OH)_2$? (K_{sp} for $MgCO_3$ is 2.1×10^{-5} and for $Mg(OH)_2$ is 8.9×10^{-12}.)

17-30 Will 0.10 mol of AgCl dissolve in 0.10 liter of 4.0 M sodium thiosulfate, $Na_2S_2O_3$?

17-31 A given solution is prepared by mixing 120 ml of 0.40 M NaF and 360 ml of 0.40 M Na_2SO_4. Then some solid $BaCl_2$ is slowly added to the solution. **(a)** Calculate $[Ba^{2+}]$ when $BaSO_4$ starts to precipitate. **(b)** Calculate $[Ba^{2+}]$ when BaF_2 starts to precipitate. **(c)** Calculate $[SO_4^{2-}]$ when BaF_2 starts to precipitate.

17-32 K_{sp} for $Pb(OH)_2$ is 4.0×10^{-15}. NH_3 is slowly added to a 1.0×10^{-3} M $Pb(NO_3)_2$ solution without change in volume. Calculate $[NH_3]$ when the first precipitate appears.

17-33 K_{sp} for $Pb(OH)_2$ is 4.0×10^{-15}. In a solution with $[Pb^{2+}] = 0.10$ M and $[NH_3] = 1.0 \times 10^{-2}$ M, what is the minimum $[NH_4^+]$ necessary to prevent precipitation of $Pb(OH)_2$?

18 CHEMICAL THERMODYNAMICS

In everyday life we observe many changes which, once initiated, proceed naturally, that is, without continued outside assistance. For example, once given a nudge, a book falls naturally from the edge of a table to the floor. On the other hand, the opposite process, that of the book rising unaided from floor to table, is an unnatural one; experience tells us that it cannot occur. Natural changes, changes which can occur, are called *spontaneous changes.*

$$\text{Book}_{(table)} \longrightarrow \text{book}_{(floor)} \qquad \text{(spontaneous)}$$

The opposite, changes which cannot occur, are called *nonspontaneous changes.*

$$\text{Book}_{(floor)} \longrightarrow \text{book}_{(table)} \qquad \text{(nonspontaneous)}$$

The study of how spontaneous and nonspontaneous changes differ is part of the discipline known as *thermodynamics,* defined as *the study of the changes or transformations of energy which accompany physical and chemical changes in matter.* Applied to chemistry, thermodynamics provides us with ways of predicting whether or not a chemical change can possibly occur under a given set of conditions, that is, whether or not it is spontaneous under those conditions. For example, using thermodynamics it is possible to predict that at 25°C and 1 atm sodium metal can react with chlorine gas to form sodium chloride:

$$2Na(s) + Cl_2(g) \longrightarrow 2NaCl(s) \qquad \text{(spontaneous)}$$

Indeed, merely exposing a clean surface of sodium metal to chlorine gas initiates this reaction. At the same time we predict that under the same conditions the reverse reaction cannot possibly occur:

$$2NaCl(s) \longrightarrow 2Na(s) + Cl_2(g) \qquad \text{(nonspontaneous)}$$

But thermodynamics tells us that we can *force* the second reaction to occur (much as we can force a book to rise from floor to table by lifting it) by using energy appropriately. In fact, sodium metal and chlorine gas are formed if NaCl is melted and an electric current is passed through the liquid.

Thermodynamics is a powerful discipline with applications in all of the sciences. In this chapter we will present a brief introduction to chemical thermodynamics.

18-1
The first law

Thermodynamics describes the behavior of macroscopic systems, rather than individual molecules. It is a logical system based on a few generalizations known as the *laws of thermodynamics*. The *first law of thermodynamics* should be familiar; it is simply the *law of conservation of energy*.

Heat, work, and energy changes

The *energy E* of a system may increase or decrease in several ways, but at this point we will consider only two. One is by having heat added to it (or removed from it). If we let q represent *added* heat; E_1, the *initial* energy of the system; and E_2, its *final* energy, then

$$E_2 - E_1 = \Delta E = q$$

The second way is by doing work on its surroundings (or having work done on it by its surroundings). If a system gains or loses no heat ($q = 0$), but does work w on its surroundings, then it *loses* energy in the amount w. In other words,

$$E_2 - E_1 = \Delta E = -w$$

If a system absorbs heat q from its surroundings and simultaneously does work w on its surroundings, then the energy change, $E_2 - E_1$, is

$$\Delta E = q - w$$

This is an algebraic statement of the *first law*. In summary, when a system gains heat, its energy increases; when a system does work, its energy decreases.

ADDED COMMENT

Be sure to make note of the algebraic sign conventions we have used. q represents the heat *gained by* a system (from its surroundings) and w, the work *done by* a system (on its surroundings). Also, ΔE, the change in the energy of the system, is *positive* for an energy *increase* ($E_2 > E_1$) and *negative* for an energy decrease ($E_2 < E_1$). In summary:

Quantity	Algebraic sign	Meaning
q	+	Heat is *absorbed* by the system from its surroundings
	−	Heat is *released* by the system to its surroundings
w	+	Work is done *by* the system on its surroundings
	−	Work is done *on* the system by its surroundings
ΔE ($= E_2 - E_1$)	+	The energy of the system *increases*
	−	The energy of the system *decreases*

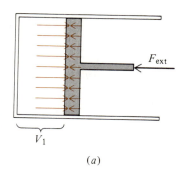
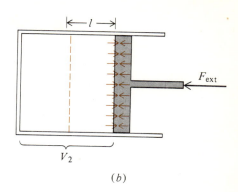

F_{ext}

F_{ext}

V_1

V_2

Figure 18-1
Work of expansion.
(a) Initial state.
(b) Final state.

(a)

(b)

How can a system do work on its surroundings? One way is by moving something against an opposing force. Consider, for example, the expansion of a gas against a close-fitting piston in a cylinder, as is shown in Fig. 18-1. (Actually, what we are about to say is not restricted to gaseous systems, but these are especially easy to imagine.) At the start of the expansion, Fig. 18-1a, the volume occupied by the gas is V_1. Now let us assume that the external force F_{ext}, against which the gas expands, remains constant during the expansion and is less than the force exerted against the piston by the gas. This causes the piston to move out. As the gas expands its pressure decreases, so that the force exerted against the piston also becomes less. After it has dropped to the value F_{ext}, the piston stops moving, because now the two forces are equal. This state is shown in Fig. 18-1b.

Let l equal the distance the piston is moved by the expanding gas. Mechanical work is defined as the product of the distance moved times the force opposing the motion, and so in the expansion shown in Fig. 18-1, the work done by the gas can be written

$$w = l \times F_{ext}$$

If the surface area of the piston is A, then the pressure P_{ext} exerted by the piston on the gas is

$$P_{ext} = \frac{F_{ext}}{A}$$

because pressure is defined as force per unit area. Therefore,

$$F_{ext} = P_{ext}A$$

and so

$$w = l \times P_{ext}A = P_{ext} \times Al$$

but Al is the volume of a cylindrical space with cross-sectional area A and length l. From Fig. 18-1 we see that this is just the increase in the volume of the gas, $V_2 - V_1$, or ΔV. Therefore,

$$w = P_{ext} \Delta V$$

● EXAMPLE 18-1

Problem   Calculate the work done by a substance when it expands from a volume of 14.00 liters to one of 18.00 liters against a constant external pressure of 1.00 atm. Express your answer in (a) liter-atmospheres and (b) joules. (1 liter atm = 101.3 J.)

Solution   (a)
$$w = P_{ext}\,\Delta V$$
$$= 1.00\ \text{atm}(18.00 - 14.00\ \text{liters}) = 4.00\ \text{liter atm}$$

(b) Although the liter-atmosphere is a perfectly good unit of work or energy, it is a little unconventional; 4.00 liter atm is

$$4.00\ \text{liter atm}\ \frac{101.3\ \text{J}}{1\ \text{liter atm}} = 405\ \text{J} \quad●$$

Consider, now, a system which can absorb energy only in the form of heat, q, and which can do work, w, on its surroundings only by expanding against them. For such a system, then,

$$\Delta E = q - w = q - P_{ext}\,\Delta V$$

This relationship tells us that the energy change of such a system depends on the amount of heat absorbed and the amount of expansion work done.

● EXAMPLE 18-2

Problem   A system having a volume of 25.00 liters absorbs exactly 1000 J of heat. Calculate ΔE for the system if (a) the heat is absorbed at constant volume; (b) as the heat is absorbed, the system expands against a constant 1.00 atm pressure to a volume of 28.95 liters; (c) as the heat is absorbed, the system expands against a constant 0.560 atm pressure to a volume of 42.63 liters.

Solution   (a) If the volume of the system remains constant, no work is done. ($P_{ext}\,\Delta V = 0$.) Therefore,

$$\Delta E = q - w = q = +1000\ \text{J}$$

(b) This time the system does work as it expands from $V_1 = 25.00$ liters to $V_2 = 28.95$ liters against an external pressure $P_{ext} = 1.00$ atm.

$$w = P_{ext}\,\Delta V$$
$$= 1.00\ \text{atm}(28.95\ \text{liters} - 25.00\ \text{liters}) = 3.95\ \text{liter atm}$$

which is
$$3.95\ \text{liter atm}\ \frac{101.3\ \text{J}}{1\ \text{liter atm}} = 400\ \text{J}$$

$$\Delta E = q - w$$
$$= 1000\ \text{J} - 400\ \text{J} = 600\ \text{J}$$

(c)
$$w = P_{ext}\,\Delta V$$
$$= 0.560\ \text{atm}(42.63\ \text{liters} - 25.00\ \text{liters}) = 9.87\ \text{liter atm}$$

which is
$$9.87\ \text{liter atm}\ \frac{101.3\ \text{J}}{1\ \text{liter atm}} = 1000\ \text{J}$$

$$\Delta E = q - w$$
$$= 1000\ \text{J} - 1000\ \text{J} = 0$$

It is clear from a comparison of the results of parts (a), (b), and (c) that the energy change which a system experiences when it absorbs heat depends upon how much work the system does. ●

**18-2
Enthalpy and
heat capacity**

Back in Sec. 10-5 we introduced enthalpy, H, and stated that ΔH for a system represents the heat absorbed by a system undergoing a change *at constant pressure*. We will now see how enthalpy and energy are related.

Enthalpy

Consider a system maintained at a constant pressure P. For such a system the external pressure P_{ext} exerted on the system must be equal to the pressure of the system itself. This system can be made to expand at constant pressure (for instance, by heating it), and when it does, the work done on the surroundings is

$$w = P_{ext}\,\Delta V = P\,\Delta V$$

Using a subscript P to indicate a process which occurs under constant pressure conditions, we can write

$$\Delta E = q_P - w_P = q_P - P\,\Delta V$$

or

$$q_P = \Delta E + P\,\Delta V \qquad (18\text{-}1)$$

This says that the heat absorbed by a system during a constant-pressure process goes to (a) increase the energy of the system and (b) do work of expansion. (We are assuming here that no other kind of work, such as electrical work, can be done.)

Enthalpy, H, is defined as

$$H = E + PV$$

The enthalpy of a system thus equals the sum of its energy and the product of its pressure and volume. Enthalpy is defined this way for the following reason: For a change of a system from state 1 to state 2,

$$H_1 = E_1 + P_1 V_1$$

and

$$H_2 = E_2 + P_2 V_2$$

and so

$$H_2 - H_1 = (E_2 - E_1) + (P_2 V_2 - P_1 V_1)$$

or

$$\Delta H = \Delta E + \Delta(PV)$$

At constant pressure $P_2 = P_1$, which we will just call P, so we have

$$H_2 - H_1 = (E_2 - E_1) + P(V_2 - V_1)$$

or

$$\Delta H = \Delta E + P\,\Delta V \qquad (18\text{-}2)$$

Comparing Eqs. (18-1) and (18-2), we see that

$$q_P = \Delta H_P$$

Thus the increase in enthalpy of a system undergoing a transformation at constant pressure is equal to the heat absorbed during the process. (Note that if heat is *liberated* during such a process, ΔH and q_P are *negative numbers*.) For many processes ΔH and ΔE are roughly the same in magnitude, as the following example illustrates.

● **EXAMPLE 18-3**

Problem   One mole of liquid water at 100°C and 1.00 atm has a density of 0.958 g ml⁻¹. If ΔH_{vap} of water is 40.66 kJ mol⁻¹, calculate ΔE_{vap} under these conditions. (Assume that water vapor behaves as an ideal gas.)

Solution   First we must find ΔV for the process

$$H_2O(l) \longrightarrow H_2O(g)$$

Since the molecular weight of water is 18.0, the molar volume of the liquid is

$$V_{liquid} = \frac{18.0 \text{ g}}{1 \text{ mol}} \frac{1 \text{ ml}}{0.958 \text{ g}} = 18.8 \text{ ml mol}^{-1} \quad \text{or} \quad 0.0188 \text{ liter mol}^{-1}$$

We can find the molar volume of the gaseous water by using the ideal-gas law.

$$PV_{gas} = nRT$$

$$\frac{V_{gas}}{n} = \frac{RT}{P}$$

$$= \frac{(0.0821 \text{ liter atm K}^{-1} \text{ mol}^{-1})(373 \text{ K})}{1.00 \text{ atm}} = 30.6 \text{ liter mol}^{-1}$$

$$\Delta V = V_{gas} - V_{liquid}$$

$$= 30.6 \text{ liter mol}^{-1} - 0.0188 \text{ liter mol}^{-1} = 30.6 \text{ liter mol}^{-1}$$

The work done by the water as it vaporizes against the constant 1.00 atm pressure is

$$w = P\,\Delta V$$

$$= 1.00 \text{ atm}(30.6 \text{ liter mol}^{-1})(101.3 \text{ J liter}^{-1} \text{ atm}^{-1}) = 3.10 \times 10^3 \text{ J mol}^{-1}$$

For any change, $\Delta H = \Delta E + \Delta(PV)$

At constant pressure, $\Delta H = \Delta E + P\,\Delta V$

and so $\Delta E = \Delta H - P\,\Delta V$

$$= 40.66 \times 10^3 \text{ J mol}^{-1} - 3.10 \times 10^3 \text{ J mol}^{-1} = 37.56 \times 10^3 \text{ J mol}^{-1}$$

This is 37.56 kJ mol⁻¹. We see that of the 40.66 kJ of heat absorbed by the water, 37.56 kJ went to convert the liquid to gas, and 3.10 kJ to do the work of pushing back the atmosphere as the water expanded. ●

Heat capacity   When a substance not undergoing a phase change or chemical reaction is heated, its temperature rises. The heat necessary to increase the temperature of a substance by one kelvin (or 1°C) is called the *heat capacity* of the substance. If *one mole* of substance is involved, this is called the *molar heat capacity, C*.[1] Heat capacities are often expressed in J K⁻¹ mol⁻¹.

If q joules of heat are added to n moles of a substance, and the resulting temperature increase is $T_2 - T_1$, or ΔT,

$$q = nC\,\Delta T$$

If the volume is held constant during this process, then from the first law,

$$q = \Delta E + w = \Delta E + P_{ext}\,\Delta V = \Delta E$$

---

[1] Heat capacity per *gram* of substance is often called *specific heat*.

Table 18-1
Molar heat capacities
at 25°C

Substance	C_P, J K^{-1} mol^{-1}
Al(s)	24.3
Ag(s)	25.4
AgCl(s)	50.8
Au(s)	25.4
Cu(s)	24.4
Fe(s)	25.1
H$_2$(g)	28.8
H$_2$O(l)	75.3
Mn(s)	26.3
MnCl$_2$(s)	72.9
MnO$_2$(s)	54.1
N$_2$(g)	29.1
NH$_3$(g)	35.1
SiO$_2$(s)	44.4

Using a subscript V to represent this constant-volume condition, we can write

$$q_V = \Delta E = nC_V\,\Delta T$$

while at constant pressure

$$q_P = \Delta H = nC_P\,\Delta T$$

where C_V and C_P are called the *molar heat capacities* at constant volume and pressure, respectively. Usually, values of C_P are more useful than those of C_V, because many processes take place at pressures which are either constant or nearly constant, at atmospheric pressure, for instance. Values of C_P for a few substances are given in Table 18-1.

● **EXAMPLE 18-4**

Problem   Suppose 15.0 g of gold (Au) is heated from 16.1 to 49.3°C under atmospheric pressure. Calculate ΔH for the process, assuming that C_P for gold remains constant over this range.

Solution   Since the atomic weight of gold is 197, 15.0 g of gold is

$$15.0\text{ g}\,\frac{1\text{ mol}}{197\text{ g}} = 7.61 \times 10^{-2}\text{ mol}$$

C_P for gold is 25.4 J K^{-1} mol^{-1} (Table 18-1) and ΔT is $49.3 - 16.1$, or 33.2°C, which is 33.2 K

$$\Delta H = q_P = 7.61 \times 10^{-2}\text{ mol}(25.4\text{ J K}^{-1}\text{ mol}^{-1})(33.2\text{ K}) = 64.2\text{ J} ●$$

**18-3
Thermochemistry**

Thermochemistry is the branch of thermodynamics which is concerned with heat liberation and absorption during chemical change. When a reaction takes place at constant pressure and the only work possible is work of expansion, then the heat liberated equals ΔH, where

$$\Delta H = q_P = \text{heat of reaction at constant pressure}$$

Usually when we speak of a heat of reaction, we mean the enthalpy change, and so the term *enthalpy* of reaction is often used. (Only when a reaction takes place at *constant volume* is the heat liberated equal to the energy change: $q_V = \Delta E$.)

When carbon is burned in an excess of oxygen at constant pressure, carbon dioxide is formed and 393.5 kJ of heat is liberated per mole of carbon consumed. More succinctly,

$$C(s) + O_2(g) \longrightarrow CO_2(g) \qquad \Delta H = -393.5 \text{ kJ}$$

This reaction, the oxidation of carbon to form carbon dioxide, can be accomplished in two stages: First, carbon can be burned in a limited amount of oxygen.

$$C(s) + \tfrac{1}{2}O_2(g) \longrightarrow CO(g) \qquad \Delta H = -110.5 \text{ kJ}$$

Then the carbon monoxide formed in this reaction can be burned in additional oxygen.

$$CO(g) + \tfrac{1}{2}O_2 (g) \longrightarrow CO_2(g) \qquad \Delta H = -283.0 \text{ kJ}$$

Note that the net result of these two stages is the same as that of the original single reaction. Note also that the sum of the ΔH values for the two stages equals ΔH for the original single reaction.

When chemical equations are added as if they were algebraic equations, the corresponding ΔH values can also be added.

$$
\begin{array}{lll}
C(s) + \tfrac{1}{2}O_2(g) \longrightarrow CO(g) & & \Delta H = -110.5 \text{ kJ} \\
CO(g) \qquad + \tfrac{1}{2}O_2(g) \longrightarrow \qquad CO_2(g) & & \Delta H = -283.0 \text{ kJ} \\
\hline
CO(g) + C(s) + \ O_2(g) \longrightarrow CO(g) + CO_2(g) & & \Delta H = -393.5 \text{ kJ}
\end{array}
$$

This is an illustration of *Hess's law of constant heat summation*, usually known simply as *Hess's law*. It states that *the change in enthalpy for any reaction depends only on the nature of the reactants and products and is independent of the number of steps or the pathway taken between them.* (We have previously made use of this law in determining lattice energy by means of a Born-Haber calculation, Sec. 10-5, and in showing the relationship between lattice energy, hydration energy, and heat of solution, Sec. 12-3.) Figure 18-2 schematically illustrates Hess's law for the $C–CO–CO_2$ reactions described above.

An equation and its corresponding ΔH value can be multiplied or divided by the same number.

$$C(s) + \tfrac{1}{2}O_2(g) \longrightarrow \ CO(g) \qquad \Delta H = -110.5 \text{ kJ}$$

$$2C(s) + \ O_2(g) \longrightarrow 2CO(g) \qquad \Delta H = -221.0 \text{ kJ}$$

(All this really says is that twice as much heat is liberated when two moles of CO are formed from C and O_2 as when one mole is formed.)

ΔH for a reaction can be thought of as the difference between the enthalpy of the products and that of the reactants. In other words,

$$\Delta H_{\text{reaction}} = H_{\text{products}} - H_{\text{reactants}}$$

This means that the sign of ΔH changes when a reaction reverses, because reactants become products and vice versa. For example,

$$C(s) + O_2(g) \longrightarrow CO_2(g) \qquad \Delta H = -393.5 \text{ kJ}$$

$$CO_2(g) \longrightarrow C(s) + O_2(g) \qquad \Delta H = +393.5 \text{ kJ}$$

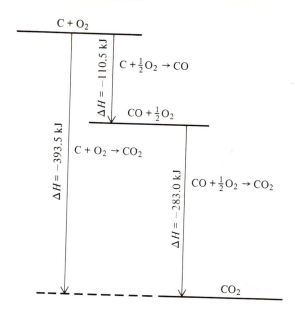

Figure 18-2
Hess's law.

Hess's law greatly simplifies the job of tabulating enthalpy changes for reactions by eliminating the need for a huge list consisting of all possible reactions and their ΔH values. Instead, tabulations of only one specific type of reaction are used to calculate ΔH values for all other reactions. The type of reaction used in the tabulation is called a *formation reaction*. This is one in which *one mole of a single product is formed from its uncombined elements*. In addition, each element is specified as being in its *standard state,* which is the *form which is stable at one atmosphere.* Unless otherwise specified, the product is also in its standard state. When these conditions have been met, the heat of the reaction is called the *standard heat* (or *enthalpy*) *of formation* of the product and is represented by the designation ΔH_f°, where the superscript means *standard*. Although the term standard state does not imply a temperature, and each substance has a standard state at each temperature, 25°C (298 K) is usually specified in tabulations of standard heats of formation. Table 18-2 gives standard heats of formation for a few substances. More are given in App. G.

Table 18-2
Standard heats of formation at 25°C

Substance	ΔH_f°, kJ mol^{-1}	Substance	ΔH_f°, kJ mol^{-1}
$CH_4(g)$	-74.8	$H_2O_2(l)$	-187.6
$CH_3OH(l)$	-239.0	$H_2S(g)$	-20.6
$C_2H_2(g)$	226.8	$H_2SO_4(l)$	-814.0
$C_2H_4(g)$	52.3	$NH_3(g)$	-46.1
$C_2H_6(g)$	-84.6	$NH_4Cl(s)$	-314.4
$CO(g)$	-110.5	$NaCl(s)$	-412.1
$CO_2(g)$	-393.5	$Na_2O(s)$	-415.9
$HCl(g)$	-92.3	$O_3(g)$	143
$H_2O(g)$	-241.8	$SO_2(g)$	-296.8
$H_2O(l)$	-285.8	$SO_3(g)$	-395.7

ADDED COMMENT

Is the following reaction a formation reaction?

$$C_2H_4(g) + H_2(g) \longrightarrow C_2H_6(g)$$

No. One mole of a single product is formed in this reaction, but one of the reactants is a compound. All reactants must be *elements* in a formation reaction.

● **EXAMPLE 18-5**

Problem   Using data from Table 18-2, calculate $\Delta H°$ for the following reaction at 25°C:

$$NH_3(g) + HCl(g) \longrightarrow NH_4Cl(s)$$

Solution   From the table, we find

$$\tfrac{1}{2}N_2(g) + \tfrac{3}{2}H_2(g) \longrightarrow NH_3(g) \qquad \Delta H_f° = -46.1 \text{ kJ mol}^{-1}$$
$$\tfrac{1}{2}H_2(g) + \tfrac{1}{2}Cl_2(g) \longrightarrow HCl(g) \qquad \Delta H_f° = -92.3 \text{ kJ mol}^{-1}$$
$$\tfrac{1}{2}N_2(g) + 2H_2(g) + \tfrac{1}{2}Cl_2(g) \longrightarrow NH_4Cl(s) \qquad \Delta H_f° = -314.4 \text{ kJ mol}^{-1}$$

If we reverse the first two equations, changing the sign on each corresponding ΔH, and then add them to the third equation, we get

$$NH_3(g) \longrightarrow \tfrac{1}{2}N_2(g) + \tfrac{3}{2}H_2(g) \qquad \Delta H° = +46.1 \text{ kJ mol}^{-1}$$
$$HCl(g) \longrightarrow \tfrac{1}{2}H_2(g) + \tfrac{1}{2}Cl_2(g) \qquad \Delta H° = +392.3 \text{ kJ mol}^{-1}$$
$$\underline{\tfrac{1}{2}N_2(g) + 2H_2(g) + \tfrac{1}{2}Cl_2(g) \longrightarrow NH_4Cl(s) \qquad \Delta H_f° = -314.4 \text{ kJ mol}^{-1}}$$
$$NH_3(g) + HCl(g) \longrightarrow NH_4Cl(s) \qquad \Delta H° = -176.0 \text{ kJ mol}^{-1} \;●$$

If you look at Example 18-5 closely, you will see that what we have done to find $\Delta H°$ for the reaction

$$NH_3(g) + HCl(g) \longrightarrow NH_4Cl(s)$$

is the following:

$$\begin{aligned}
\Delta H_{\text{reaction}}° &= (\Delta H_f°)_{NH_4Cl} - [(\Delta H_f°)_{NH_3} + (\Delta H_f°)_{HCl}] \\
&= 314.4 \text{ kJ mol}^{-1} - [-46.1 \text{ kJ mol}^{-1} + (-92.3 \text{ kJ mol}^{-1})] \\
&= -176.0 \text{ kJ mol}^{-1}
\end{aligned}$$

The heat of reaction is equal to the sum of the heats of formation of the products minus the sum of the heats of formation of the reactants, or

$$\Delta H_{\text{reaction}}° = \Sigma\,(\Delta H_f°)_{\text{products}} - \Sigma\,(\Delta H_f°)_{\text{reactants}}$$

[The Greek letter Σ (sigma) means "sum."]

    The standard heat of formation of an element is zero because the formation of an element *from itself* is not really a reaction at all; so for any element $\Delta H_f° = 0$.

● **EXAMPLE 18-6**

Problem   Find the heat of combustion of methanol, CH_3OH, to form carbon dioxide and water vapor.

Solution   The (balanced) equation for the reaction is

$$CH_3OH(l) + \tfrac{3}{2}O_2(g) \longrightarrow CO_2(g) + 2H_2O(g)$$

$$
\begin{aligned}
\Delta H^\circ &= \Sigma \, (\Delta H_f^\circ)_{products} - \Sigma \, (\Delta H_f^\circ)_{reactants} \\
&= [(\Delta H_f^\circ)_{CO_2} + 2(\Delta H_f^\circ)_{H_2O}] - [(\Delta H_f^\circ)_{CH_3OH} + \tfrac{3}{2}(\Delta H_f^\circ)_{O_2}] \\
&= [-393.5 + 2(-241.8)] - [-239.0 + \tfrac{3}{2}(0)] = -638.1 \text{ kJ mol}^{-1}
\end{aligned}
$$

Methanol (wood alcohol) gives off considerable heat when it burns! ●

18-4
The second law

The first law of thermodynamics states that changes in which energy is conserved can occur. For example, a book on a table (higher energy state) can fall to the floor (lower energy state), and the energy lost by the book is converted into heat and sound energy at the moment of impact with the floor, so that the total energy is conserved. (The floor gets a little warmer.) However, the first law is quite inadequate for predicting whether or not a change is spontaneous. Consider the following process: A book rests on the floor next to a table. Then, some of the kinetic energy of the random motion of the molecules in the floor is converted into the energy needed to lift the book to the table. The book rises to the table, and its energy increase equals the decrease in energy of the floor. (The floor gets a little cooler.) Note that in this change the first law is not violated; energy is conserved. Still, experience tells us that the change will not occur. It is evident that in order to make accurate predictions of change we need more than just the first law.

Change, energy,
and disorder

Spontaneous changes are often accompanied by a loss of energy, but energy decrease alone cannot be used as a criterion for spontaneous change. For example, consider an ideal gas which is allowed to expand into a vacuum. Such an expansion is spontaneous, but no work is done ($P_{ext} = 0$), and if no heat is added, ΔE equals 0 by the first law.

Some spontaneous processes absorb heat, and each of these represents a change to a state of higher energy. An example of such an endothermic change is the dissolving of calcium chloride in water. The test tube or beaker becomes cooler as the $CaCl_2$ dissolves and the system absorbs heat from its surroundings. A similar example is the melting of ice above $0°C$. Heat is absorbed in this process, as the system goes to a higher energy state.

Do those spontaneous processes which are accompanied by an *increase* in energy have any other common characteristic? Yes, in each case the system goes from a *more ordered* state to one which is *less ordered*. This *disordering tendency* is just as important as the tendency to go to a state of lower energy. The trouble is, neither tendency *always* dominates the other, and so neither can be used as an absolute criterion for spontaneous change. Above $0°C$ ice melts and goes to a state of higher energy and lower molecular order, but when water freezes below $0°C$, it goes to a state of lower energy and greater molecular order. Thus above $0°C$ the disordering tendency dominates, but below $0°C$ the tendency to decrease energy is more important. Exactly at $0°C$ the two tendencies are exactly balanced against each other, and so the system is at equilibrium, with neither melting nor freezing being spontaneous.

Let us take a closer look at the disordering tendency. Systems tend to change toward states which are more disordered, because the *probability* of such a state is higher than that of a comparatively more ordered state. How are disorder and probability related? Consider a pair of bulbs, designated 1 and 2, respectively, connected by means of a tube, as is shown in Fig. 18-3. If we place a single gas molecule in this double container, the chance that at any instant the molecule occupies bulb 1 is one out of two, and so we say that the probability that it is in bulb 1 is $\frac{1}{2}$. Of course, the probability that it is in bulb 2 is also $\frac{1}{2}$. Now consider two molecules, designated A and B, respectively, in the two-bulb container. Now there are *four* possible arrangements, as is shown in the illustration. Each specific arrangement is equally likely, and so we see that the probability that both molecules are in bulb 1 is $\frac{1}{2} \times \frac{1}{2}$, or $\frac{1}{4}$, and that both are in bulb 2 is also $\frac{1}{4}$. But the probability that one molecule is in each bulb is $\frac{1}{4} + \frac{1}{4}$, or, $\frac{1}{2}$, because there are *two* ways of achieving this even distribution of molecules. Note that the more random or disordered distribution (one molecule in each bulb) is more probable because there are more ways of achieving it—two, instead of one.

If we put four molecules in the double-bulb container of Fig. 18-3, the probability that all are found in bulb 1 is $\frac{1}{2} \times \frac{1}{2} \times \frac{1}{2} \times \frac{1}{2}$, or $\frac{1}{16}$. Also, the probability that the four are evenly distributed, two in each bulb, is $6 \times \frac{1}{16}$, or $\frac{3}{8}$, because there are six ways of distributing the four molecules evenly between the two bulbs.

As we go to more and more molecules in the system, two facts become apparent: the probability that all molecules are in one bulb becomes vanishingly small, and the probability of an even distribution is always highest. To illustrate the first point, consider that our system consists of Avogadro's number of molecules. The probability that all are in one bulb is

$$\left(\frac{1}{2}\right)^{6.02 \times 10^{23}}$$

This is an incomprehensibly small number. Written as a decimal fraction it would require more printer's ink than is available in the entire world to write all the zeros after the decimal point! We can safely say that for practical purposes all the molecules can never be in one bulb.

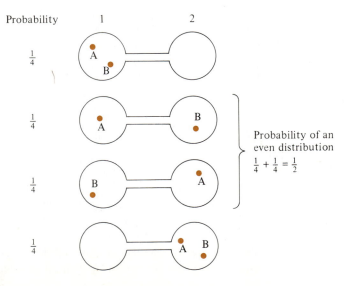

Figure 18-3
Distribution of gas molecules between two bulbs.

Each system tends to go from a state of lower probability to one of higher probability, and since the high-probability states are ones of greater disorder, we conclude that *there is a natural tendency for a system to become more disordered.* This is one way to state the *second law of thermodynamics.*

Probability, entropy, and the second law

The disorder or randomness of a system in a given state can be expressed quantitatively by stating the number of alternate ways its particles can be arranged to constitute that state. This number, W, called the *thermodynamic probability,* is high when randomness is high. In order to see the significance of W consider a tiny crystal consisting of only eight atoms. If this crystal is at absolute zero, all eight are vibrating minimally (all are in their ground vibrational states), and so no transfer of energy from one to another is possible. This state is thus completely ordered, and since there is only one way of constituting it, $W = 1$. Now let us add just enough energy to raise one atom to its first excited vibrational state. Energy exchange among the eight atoms now occurs, so that at any instant any one of the atoms might be the excited one. There are thus eight ways of having one excited atom in our microcrystal, and so W for this state equals 8. A little thought reveals that if considerable energy is added to the crystal, W becomes quite large, as there are many ways of distributing this energy among the atoms. Also, if the number of atoms is large, then W for any amount of energy added is likewise very large.

We now define a new thermodynamic quantity called *entropy* (from the Greek, meaning "a change within"), represented by the letter S. Entropy is related to thermodynamic probability by the equation

$$S = k \ln W$$

Here k is *Boltzmann's constant,* named after Ludwig Boltzmann, the Austrian physicist who first proposed the relationship. k is just the ideal gas constant *per molecule,* rather than per mole, that is,

$$k = \frac{R}{6.02 \times 10^{23}}$$

It is clear that a highly disordered system is one of high W and hence high S. Gases generally have higher entropies than solids do, because they are less ordered. Also, a substance at a high temperature has a higher entropy than when at a low temperature.

In an *isolated system,* one whose boundaries are impenetrable to all forms of energy and matter, entropy increases in every spontaneous change, because the system tends to go toward a more probable state. In other words,

$$\Delta S > 0 \qquad \text{(spontaneous change; isolated system)}$$

If a system is not isolated, if it can exchange energy with its surroundings, then the system plus its surroundings can be considered a single larger isolated system. In this case the total entropy change ΔS_{total} equals the sum of the entropy change of the system, ΔS_{system}, plus that of the surroundings, $\Delta S_{surroundings}$. That is,

$$\Delta S_{total} = \Delta S_{system} + \Delta S_{surrounding} > 0 \qquad \text{(spontaneous change)}$$

The entire surroundings of any system can be considered to be the universe. Therefore, we can say that the entropy of the universe tends to increase. Rudolf

Clausius (1822–1888) summarized the first and second laws of thermodynamics as follows:

First law: The energy of the universe is constant.
Second law: The entropy of the universe is constantly increasing.[2]

Entropy, like energy and enthalpy, is a *state function*. This means that it is a property of the state of a system and is independent of the system's past history. This means that the change in entropy which accompanies a given process depends only on the initial and final states and is independent of the pathway leading from one to the other. In other words,

$$\Delta S = S_2 - S_1$$

where the subscripts 1 and 2 refer to the initial and final states, respectively.

Free energy and spontaneous change   The second law is often difficult to apply directly when trying to determine whether or not a process is spontaneous. The trouble is that spontaneity depends upon the total entropy change, that of the system and its surroundings, not just that of the system alone. And it is often impractical to estimate the entropy change of the surroundings, especially if they are considered to be the universe! However, there is a way around this problem, although in using it we must restrict ourselves to a consideration of changes which occur at constant temperature and pressure. (It turns out that in actual practice the restriction is not very serious.)

For a process which takes place at constant temperature the entropy change *of the surroundings* depends only on the amount of heat absorbed by the surroundings from the system undergoing change and on the temperature at which the heat is transferred. Specifically,

$$\Delta S_{\text{surroundings}} = \frac{\text{heat absorbed by surroundings}}{T}$$

But the heat absorbed by the surroundings equals $-q$, where q is the heat absorbed by the system. (If the system *loses* heat, q is a negative number, which makes $-q$ positive.) At constant pressure $q = \Delta H_{\text{system}}$ (Sec. 18-2); so at constant temperature and pressure

$$\Delta S_{\text{surroundings}} = \frac{-\Delta H_{\text{system}}}{T}$$

(In an exothermic process q is negative, ΔH is negative, $-\Delta H$ is positive, and so ΔS is positive; that is, the entropy of the surroundings increases.) But since

$$\Delta S_{\text{total}} = \Delta S_{\text{system}} + \Delta S_{\text{surroundings}}$$

then

$$\Delta S_{\text{total}} = \Delta S_{\text{system}} - \frac{\Delta H_{\text{system}}}{T}$$

Multiplying through by $-T$ gives us

$$-T\,\Delta S_{\text{total}} = \Delta H_{\text{system}} - T\,\Delta S_{\text{system}}$$

[2]This is actually a rather liberal translation. Clausius spoke of the world, not the universe.

We now define a new thermodynamic function, G, called the *Gibbs free energy*, or simply, the *free energy*.[3]

$$G = H - TS$$

Like E, H, and S, G depends only on the state of a system, and so for any change from state 1 to state 2

$$G_2 - G_1 = H_2 - H_1 - (T_2 S_2 - T_1 S_1)$$

or

$$\Delta G = \Delta H - \Delta(TS)$$

For a change at constant temperature

$$\Delta G = \Delta H - T\Delta S$$

Here ΔG, ΔH, and ΔS all refer to the system; so comparing this equation to the equation

$$-T\Delta S_{\text{total}} = \Delta H_{\text{system}} - T\Delta S_{\text{system}}$$

we see that

$$\Delta G_{\text{system}} = -T\Delta S_{\text{total}} \qquad \text{(constant } T, P\text{)}$$

We are now ready to state a rigorous criterion for spontaneous change in a system held at constant temperature and pressure. According to the second law, $\Delta S_{\text{total}} > 0$ for a spontaneous change. This means that for such a change $-T\Delta S_{\text{total}} < 0$, and so

$$\Delta G_{\text{system}} < 0 \qquad \text{(spontaneous change, constant } T, P\text{)}$$

In other words, *when a system undergoes a spontaneous change at constant temperature and pressure, its free energy decreases.* Note that we now have a *single* criterion for spontaneity, rather than two (ΔH and ΔS).

The opposite of a spontaneous change is a nonspontaneous change. If such a change could occur, it would result in a decrease in the total entropy of system and surroundings, that is,

$$\Delta S_{\text{total}} = \Delta S_{\text{system}} + \Delta S_{\text{surroundings}} < 0 \qquad \text{(nonspontaneous change)}$$

For such a change $-T\Delta S_{\text{total}} > 0$, and so

$$\Delta G_{\text{system}} > 0 \qquad \text{(nonspontaneous change, constant } T, P\text{)}$$

An example of a nonspontaneous change is the *freezing* of water *above* 0°C. For this change $\Delta G_{\text{water}} > 0$. The process which is the reverse of a nonspontaneous change is spontaneous. For the melting of ice above 0°C, $\Delta G_{\text{water}} < 0$.

For a system *at equilibrium*, $\Delta S_{\text{total}} = 0$, and so $\Delta G_{\text{system}} = 0$.

From now on we will drop the subscript *system* and simply write ΔG, ΔH, ΔS, etc., although these will represent thermodynamic changes *of the system*. The relationship between the sign of ΔG and the spontaneity of change is summarized on the next page.

---

[3] G is also sometimes called the *Gibbs energy* or *Gibbs function*. It has also been represented by the letter F.

ΔG (constant T, P)	Change
<0	Spontaneous
$=0$	No net change; system is at equilibrium
>0	Nonspontaneous; reverse change is spontaneous

We have previously referred to the apparently independent tendencies for a system to go to a state of (1) lower energy and (2) greater disorder or randomness. We can now see that when these tendencies oppose each other a third factor determines which tendency dominates. Look closely at the relationship

$$\Delta G = \Delta H - T\Delta S$$

When ΔH is negative (the reaction is exothermic) and ΔS is positive (the system becomes more disordered), ΔG must be negative (the process is spontaneous). (Remember that T is always positive.) When ΔH is positive and ΔS is negative, ΔG must be positive (the process is nonspontaneous). When ΔH and ΔS have the same sign, the relative magnitudes of ΔH and $T\Delta S$ determine the sign of ΔG. These relationships are summarized below.

ΔH	ΔS	$\Delta G(= \Delta H - T\Delta S)$	Change
−	+	−	Spontaneous
+	−	+	Nonspontaneous
−	−	? $\begin{cases} - \text{ at low T} \\ + \text{ at high T} \end{cases}$	Spontaneous / Nonspontaneous
+	+	? $\begin{cases} + \text{ at low T} \\ - \text{ at high T} \end{cases}$	Nonspontaneous / Spontaneous

It can be seen that when ΔH and ΔS have the same sign, it is the *temperature* which determines the sign of ΔG and hence the spontaneity of the reaction.

The freezing and melting of water are changes which provide a good illustration of how temperature determines which process takes place. For the melting process

$$H_2O(s) \longrightarrow H_2O(l)$$

$\Delta H = 6.008$ kJ mol^{-1}, and $\Delta S = 21.99$ J K^{-1} mol^{-1}. The calculation of ΔG for this process at three different temperatures is summarized below.

Temperature		ΔH, J mol^{-1}	ΔS, J K^{-1} mol^{-1}	$-T\Delta S$, J mol^{-1}	$-\Delta G$, J mol^{-1}	Spontaneous melting?
°C	K					
1.0	274.2	6008	21.99	−6030	−22	Yes
0.0	273.2	6008	21.99	−6008	0	Equilibrium
−1.0	272.2	6008	21.99	−5986	+22	No

(In these calculations we have neglected the small changes of ΔH and ΔS with temperature. The resulting numerical errors are small, however, and do not affect the final conclusions.)

**18-5
Entropy and
free-energy changes**

It is possible to determine entropy changes (that is, calculate values of ΔS) for many kinds of processes. Two of the most important are changes of state and chemical reactions.

*Entropy and change
of state*

Consider a substance undergoing a change of state (melting, boiling, etc.) at its equilibrium temperature. For such a change

$$\Delta G = \Delta H - T\,\Delta S = 0$$

and so

$$\Delta S = \frac{\Delta H}{T}$$

For example, if the melting point of a substance is T_m and its molar heat of fusion is ΔH_{fus}, then the increase in entropy which accompanies the melting of one mole of the substance is

$$\Delta S_{\text{fus}} = \frac{\Delta H_{\text{fus}}}{T_m}$$

● **EXAMPLE 18-7**

Problem

The heat of fusion of gold is 12.36 kJ mol^{-1}, and its entropy of fusion is 9.250 J K^{-1} mol^{-1}. What is the melting point of gold?

Solution

Rearranging the relationship

$$\Delta S_{\text{fus}} = \frac{\Delta H_{\text{fus}}}{T_m}$$

we have

$$T = \frac{\Delta H_{\text{fus}}}{\Delta S_{\text{fus}}}$$

$$= \frac{12.36 \times 10^3 \text{ J mol}^{-1}}{9.250 \text{ J K}^{-1} \text{ mol}^{-1}} = 1336 \text{ K}$$

This is $1336 - 273$, or $1063°C$. ●

*The third law
and absolute
entropies*

A state of perfect order is a state of minimum entropy. Such a state can exist only in a perfect crystal at absolute zero. Each atom in such a crystal is locked at a fixed position in a perfect crystal lattice, and each has the same (minimum) energy. There is thus minimum randomness or disorder in both position and energy. The *third law of thermodynamics* states that *the entropy of a pure, perfect crystalline solid is zero at absolute zero.* The entropy of a crystal containing defects, an amorphous solid (glass), or a solid solution is greater than zero and is a measure of the disorder in the substance.

Adding heat to a pure, perfect crystal at absolute zero causes its temperature to rise, and the increased molecular motion increases the randomness of its structure, so that its entropy increases. By determining the heat capacity of a substance over a range of temperatures from 0 K to some higher temperature, it is possible to calculate ΔS for the temperature change. Using the third law, we can write for the temperature change from 0 to T

$$\Delta S_{0 \to T} = S_T - S_0 = S_T - 0 = S_T$$

Here S_T is the *absolute entropy* of the substance at temperature T. *Standard absolute entropies*, $S°$, are absolute entropies of substances in their standard states (Sec. 18-3). These have been determined for many substances. A few values are given in Table 18-3, and more are found in App. G.

Entropy changes in chemical reactions

Standard absolute entropies can be used to calculate standard entropy changes for chemical reactions. For example, for the reaction

$$k\text{A} + l\text{B} \longrightarrow m\text{C} + n\text{D}$$

$$\Delta S° = (mS_\text{C}° + nS_\text{D}°) - (kS_\text{A}° + lS_\text{B}°)$$

● **EXAMPLE 18-8**

Problem   Calculate $\Delta S°$ for the reaction

$$2\text{CH}_3\text{OH}(l) + 3\text{O}_2(g) \longrightarrow 2\text{CO}_2(g) + 4\text{H}_2\text{O}(g)$$

Solution $\Delta S° = (4S_{\text{H}_2\text{O}}° + 2S_{\text{CO}_2}°) - (2S_{\text{CH}_3\text{OH}}° + 3S_{\text{O}_2}°)$
$\quad = [4(188.7) + 2(213.6)] - [2(126.3) + 3(205.1)] = 314.1 \text{ J K}^{-1}$ ●

Standard free energies of formation

In Sec. 18-3 we showed how standard heats of formation are tabulated and used in calculating $\Delta H°$ values for chemical reactions. Standard free energies of formation can be similarly tabulated. Several methods are used for determining standard free energies of formation, $\Delta G_f°$, but one is particularly straightforward. It depends simply on the relationship $\Delta G = \Delta H - T\Delta S$, as is shown in the following example.

● **EXAMPLE 18-9**

Problem   From the data in Tables 18-2 and 18-3 calculate the standard free energy of formation of ammonium chloride, NH_4Cl, at 25°C.

Solution   The equation showing the formation of NH_4Cl from its elements is

$$\tfrac{1}{2}\text{N}_2(g) + 2\text{H}_2(g) + \tfrac{1}{2}\text{Cl}_2(g) \longrightarrow \text{NH}_4\text{Cl}(s)$$

From Table 18-2 we find that $\Delta H_f°$ is $-314.4 \text{ kJ mol}^{-1}$. From Table 18-3 we find that

$$S_{\text{N}_2}° = 191.5 \text{ J K}^{-1}\text{ mol}^{-1}$$
$$S_{\text{H}_2}° = 130.6 \text{ J K}^{-1}\text{ mol}^{-1}$$
$$S_{\text{Cl}_2}° = 223.0 \text{ J K}^{-1}\text{ mol}^{-1}$$
$$S_{\text{NH}_4\text{Cl}}° = 94.6 \text{ J K}^{-1}\text{ mol}^{-1}$$

Therefore, for the formation reaction above,

$$\Delta S_f° = S_{\text{NH}_4\text{Cl}}° - (\tfrac{1}{2}S_{\text{N}_2}° + 2S_{\text{H}_2}° + \tfrac{1}{2}S_{\text{Cl}_2}°)$$
$$= 94.6 - [\tfrac{1}{2}(191.5) + 2(130.6) + \tfrac{1}{2}(223.0)] = -373.8 \text{ J K}^{-1}\text{ mol}^{-1}$$

Now that we have ΔH and ΔS, we can find ΔG.

$$\Delta G_f° = \Delta H_f° - T\Delta S_f°$$
$$= -314.4 \times 10^3 \text{ J mol}^{-1} - (298.2 \text{ K})(-373.8 \text{ J K}^{-1}\text{ mol}^{-1})$$
$$= -2.030 \times 10^5 \text{ J mol}^{-1} \quad \text{or} \quad -203.0 \text{ kJ mol}^{-1}$$

**Table 18-3
Standard absolute
entropies at 25°C**

Substance	$S°$, J K^{-1} mol^{-1}	Substance	$S°$, J K^{-1} mol^{-1}
C(diamond)	2.38	$H_2O(g)$	188.7
C(graphite)	5.74	$H_2O(l)$	69.9
$CH_4(g)$	187.9	$H_2S(g)$	205.7
$CH_3OH(g)$	126.3	$H_2SO_4(l)$	156.9
$C_2H_2(g)$	200.8	$N_2(g)$	191.5
$C_2H_4(g)$	219.5	$NH_3(g)$	192.3
$C_2H_6(g)$	229.5	$NH_4Cl(s)$	94.6
$CO(g)$	197.6	$NaCl(s)$	72.4
$CO_2(g)$	213.6	$Na_2O(s)$	72.8
$Cl_2(g)$	223.0	$O_2(g)$	205.1
$H_2(g)$	130.6	$SO_2(g)$	248.1
$HCl(g)$	186.8	$SO_3(g)$	256.6

Note: Here the entropy change is negative, not favorable to the reaction, but the enthalpy change is also negative, favoring the reaction. The temperature is low enough to prevent ΔS from "overruling" ΔH; so ΔG is negative. ●

**Free energies
of reaction**

Once $\Delta G_f°$ values are available, they can be used to calculate $\Delta G°$ values for many reactions. A brief tabulation of some $\Delta G_f°$ values at 25°C is given in Table 18-4, and a more extensive listing is given in App. G. Note that as is the case with $\Delta H_f°$ values, $\Delta G_f°$ *of an element is zero.*

● **EXAMPLE 18-10**

Problem   Calculate the standard free-energy change for the reaction

$$C_2H_6(g) + \tfrac{7}{2}O_2(g) \longrightarrow 2CO_2(g) + 3H_2O(g)$$

Solution

$$\Delta G°_{reaction} = \Sigma(\Delta G_f°)_{products} - \Sigma(\Delta G_f°)_{reactants}$$
$$\Delta G° = [2(\Delta G_f°)_{CO_2} + 3(\Delta G_f°)_{H_2O}] - [(\Delta G_f°)_{C_2H_6} + \tfrac{7}{2}(\Delta G_f°)_{O_2}]$$
$$= [2(-394.4) + 3(-228.6)] - [(-32.9) + \tfrac{7}{2}(0)] = -1441.7 \text{ kJ} ●$$

18-6
Thermodynamics
and equilibrium

One of the greatest strengths of chemical thermodynamics lies in its ability to describe the approach to equilibrium and the properties of the equilibrium state.

**Table 18-4
Standard free energies
of formation at 25°C**

Substance	$\Delta G_f°$, kJ mol^{-1}	Substance	$\Delta G_f°$, kJ mol^{-1}
$CH_4(g)$	−50.8	$H_2O_2(l)$	−120.4
$CH_3OH(l)$	−166.3	$H_2S(g)$	−33.6
$C_2H_2(g)$	209.2	$H_2SO_4(l)$	−690.1
$C_2H_4(g)$	68.1	$NH_3(g)$	−16.5
$C_2H_6(g)$	−32.9	$NH_4Cl(s)$	−202.7
$CO(g)$	−137.2	$NaCl(s)$	−384.0
$CO_2(g)$	−394.4	$Na_2O(s)$	−376.6
$HCl(g)$	−95.3	$O_3(g)$	163.2
$H_2O(g)$	−228.6	$SO_2(g)$	−300.2
$H_2O(l)$	−237.2	$SO_3(g)$	−371.1

Free energy and position of equilibrium

Free energy decreases during spontaneous change. But even if $\Delta G°$ for a reaction is negative, complete conversion of reactants to products does not occur. Instead, the system reaches equilibrium. Failure of a reaction to go to completion means that the free energy of a certain mixture of reactants and products is less than that of either reactants or products alone. A reaction takes place only as long as the free energy of the system continues to decrease. When a reaction has taken place to the extent that further formation of products will bring about an increase in the free energy of the system, no further net change takes place, that is, the system is at equilibrium.

Figure 18-4 shows how several thermodynamics quantities vary with the extent of a hypothetical reaction. Figure 18-4a shows the change in H, ΔH, when a system goes from pure reactants, each in its standard state, to various mixtures of reactants and products. Figure 18-4b similarly shows ΔS; Fig. 18-4c, $T\Delta S$; and Fig. 18-4d, ΔG ($= \Delta H - T\Delta S$). For this reaction $\Delta H° = -10$ kJ, $\Delta S° = +15$ J K^{-1}, and the temperature is 25°C. Figure 18-4a shows that H decreases essentially linearly with the extent of the reaction. If decrease in enthalpy were all that mattered, the reactants would be completely transformed into products, the state of lowest enthalpy. But the entropy is important, too. The relationship between ΔS and extent of reaction is not a simple one. In Fig. 18-4b we see that the entropy goes through a maximum because of the fact that partway through the reaction a mixture of reactants and products exists. Since T is constant, $T\Delta S$ (Fig.

Figure 18-4
Thermodynamic changes and extent of reaction: (a) ΔH, (b) ΔS, (c) $T\Delta S$, (d) ΔG.

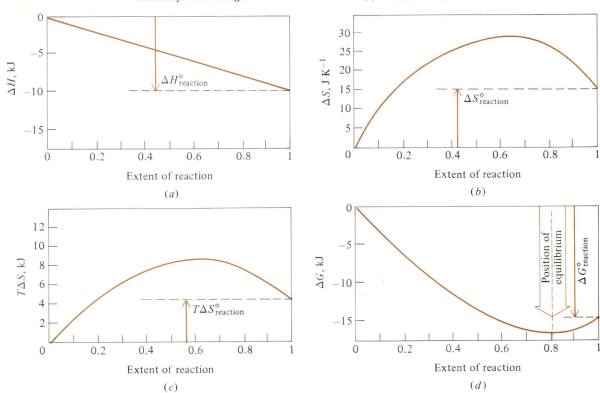

18-4c) goes through a similar maximum.   Because of this, the free energy goes through a minimum as is shown in Fig. 18-4d.   In our hypothetical reaction this minimum occurs when the reaction is 0.8 complete.   This is the equilibrium point.   Whether we start with reactants or products, we will end up with this mixture because the free-energy curve is "downhill" from both the far left and far right of the ΔG graph.   In summary, then, each chemical reaction proceeds spontaneously so that the free energy of the reaction mixture is lowered, until a minimum in the free energy has been reached.   This minimum occurs when the system is at equilibrium.   Once a system has reached equilibrium further net change is impossible, because $G_{reactants} = G_{products}$.

Free energies and equilibrium constants

It is possible to show that the free energy G and partial pressure P of one mole of an ideal gas are related by the expression

$$G = G^\circ + RT \ln P$$

where G° is the standard free energy of the gas, R is the ideal-gas constant, and T is the absolute temperature.   For the gas-phase reaction

$$A_2(g) + B_2(g) \longrightarrow 2AB(g)$$

the free-energy change is

$$\Delta G = 2G_{AB} - (G_{A_2} + G_{B_2}) = 2G_{AB} - G_{A_2} - G_{B_2}$$

But the molar free energies in this relationship can be expressed as follows:

$$G_{AB} = G_{AB}^\circ + RT \ln P_{AB}$$

$$G_{A_2} = G_{A_2}^\circ + RT \ln P_{A_2}$$

$$G_{B_2} = G_{B_2}^\circ + RT \ln P_{B_2}$$

Therefore,

$$\begin{aligned}
\Delta G &= 2G_{AB} - G_{A_2} - G_{B_2} \\
&= 2(G_{AB}^\circ + RT \ln P_{AB}) - (G_{A_2}^\circ + RT \ln P_{A_2}) - (G_{B_2}^\circ + RT \ln P_{B_2}) \\
&= 2G_{AB}^\circ - (G_{A_2}^\circ + G_{B_2}^\circ) + RT(2 \ln P_{AB} - \ln P_{A_2} - \ln P_{B_2}) \\
&= \Delta G^\circ + RT \ln \frac{P_{AB}^2}{P_{A_2}P_{B_2}}
\end{aligned}$$

But $P_{AB}^2/P_{A_2}P_{B_2}$ is just the mass-action expression Q_P for the reaction, expressed in terms of pressures.   Therefore,

$$\Delta G = \Delta G^\circ + RT \ln Q_P$$

At equilibrium ΔG has a fixed value, *zero*.   This means that Q must be a constant. At equilibrium $Q_P = K_P$, the equilibrium constant.   Finally, we have,

$$0 = \Delta G^\circ + RT \ln K_P$$

or

$$\Delta G^\circ = -RT \ln K_P$$

This constitutes the thermodynamic proof that the mass-action expression is a constant at equilibrium at any given temperature.

When gases are not ideal, the relationship between G and G° is given by

$$G = G^\circ + RT \ln a$$

where a is the *chemical*, or *thermodynamic, activity*. Thus activity is a quantity which relates G to G° for a real gas in the same way that partial pressure relates the two for an ideal gas. This leads to the relationship at equilibrium

$$\Delta G^\circ = -RT \ln K_a$$

where K_a, the *thermodynamic equilibrium constant,* is a function of activities, not pressures. For solution reactions a similar relationship can be derived in which the equilibrium constant is a function of concentrations (K_c) if the solution is ideal, or of activities (K_a) if it is not.

The equation

$$\Delta G^\circ = -RT \ln K$$

is useful, because it relates knowledge about standard states to knowledge about the equilibrium state. Written in terms of a common logarithm (log to the base 10) this relationship becomes

$$\Delta G^\circ = -2.303 \, RT \log K$$

● EXAMPLE 18-11

Problem   The standard free energy of formation of ammonia at 25°C is -16.5 kJ mol^{-1}. Calculate the equilibrium constant at this temperature for the reaction

$$N_2(g) + 3H_2(g) \rightleftharpoons 2NH_3(g)$$

Solution   For the reaction as written ΔG° is $2(-16.5)$, or -33.0 kJ, because 2 mol of NH$_3$ are shown to be formed.

$$\Delta G^\circ = -RT \ln K_P$$
$$= -2.303 \, RT \log K_P$$

$$\log K_P = -\frac{\Delta G^\circ}{2.303 \, RT}$$

$$= -\frac{-33.0 \times 10^3 \text{ J mol}^{-1}}{(2.303)(8.314 \text{ J K}^{-1} \text{mol}^{-1})(298 \text{ K})} = 5.78$$

$$K_P = 6.1 \times 10^5 \qquad ●$$

SUMMARY

In this chapter we have introduced the fundamentals of chemical thermodynamics. Our discussion began with a consideration of the way the energy E of a system changes when (1) the system absorbs heat q and (2) the system does work w on its surroundings. The change in energy of the system is related to these by the equation

$$\Delta E = q - w$$

which is a statement of the *first law of thermodynamics*.

Special attention was paid to the work done when a system expands against a constant external pressure, P_{ext}. This work is equal to $P_{ext} \Delta V$, where ΔV is the increase in volume of the system.

The *enthalpy* H of a system is defined as

$$H = E + PV$$

When a process is conducted at constant pressure the change in enthalpy, ΔH, of a system is equal to the heat absorbed by the system. Heats of chemical reactions are usually specified as values of ΔH. The enthalpy change accompanying any reaction depends only on the nature of reactants and products and is independent on the number or nature of the steps between them. This is a statement of *Hess's law*. The *standard heat (enthalpy) of formation* (ΔH_f°) of a compound is ΔH for the reaction in which the compound is formed from its elements, all in their standard states. The *standard state* of a substance is that form which is stable at 1 atm. ΔH_f° values can be used to calculate standard heats of reaction.

The *second law of thermodynamics* states that *the entropy of the universe is constantly increasing*. Entropy measures disorder or randomness in a system; since random states are more probable than ordered states, isolated systems tend to experience an increase in entropy. *Standard absolute entropies* have been determined. These are based on the *third law of thermodynamics* which states that *the entropy of a pure, perfect crystalline solid is zero at absolute zero*.

For a system undergoing changes at constant temperature and pressure, the *free-energy change* serves as a predictor of spontaneous change. Free energy, G, is defined as

$$G = H - TS$$

If ΔG for such a process is negative, it is spontaneous; if ΔG is positive, the process is nonspontaneous, that is, it can not occur. $\Delta G = 0$ corresponds to an equilibrium state. At constant temperature and pressure the relationship

$$\Delta G = \Delta H - T\,\Delta S$$

shows how the interplay of enthalpy, entropy, and temperature determines whether or not a process is spontaneous. Standard free energies of formation of compounds can be used to calculate standard free-energy changes for chemical reactions.

A reaction tends to proceed until the reacting system reaches a free-energy minimum. At this point $\Delta G = 0$; that is, the free energies of reactants and products are equal. The *standard free-energy change* ΔG° for a reaction is related to the equilibrium constant K by the equation

$$\Delta G^\circ = -RT \ln K$$

KEY TERMS

Absolute entropy (Sec. 18-5)
Activity (Sec. 18-6)
Enthalpy (Sec. 18-2)
Entropy (Secs. 18-4, 18-5)
First law (Sec. 18-1)
Formation reaction (Sec. 18-3)
Free energy (Secs. 18-4, 18-5)
Heat capacity (Sec. 18-2)
Isolated system (Sec. 18-4)

Second law (Sec. 18-4)
Spontaneous change (Sec. 18-4)
Standard state (Sec. 18-3)
Surroundings (Sec. 18-4)
Thermochemistry (Sec. 18-3)
Thermodynamic probability (Sec. 18-4)
Third law (Sec. 18-5)
Work of expansion (Sec. 18-1)

QUESTIONS AND PROBLEMS

The first law

18-1 Calculate the work in joules done by an ideal gas when 1.00 mol expands from a volume of 10.0 liters to one of 100.0 liters at a constant 25°C, if the expansion is performed   (a) into a vacuum   (b) against a constant opposing pressure of 0.100 atm.

18-2 Calculate w and ΔE for the expansion of 10.0 g of N_2 from a volume of 10.0 liters to one of 20.0 liters at 300°C against a constant opposing pressure of 0.800 atm, if 125 J of heat is absorbed.

18-3 Discuss the validity of the statement
$$\Delta E = q - w$$
18-4 Discuss the suggestion that heat and work are not forms of energy, but are measures of *change* in energy.

18-5 A certain gas expands into a vacuum. If the system is thermally insulated from its surroundings, how does its energy change? Does your answer depend on whether the gas is ideal or real? Explain.

18-6 Comment on the statement: "Heat stimulates

random motion, but work stimulates directed motion."

18-7 How can a gas be expanded so that it does no work? How can it be expanded so that it does maximum work?

18-8 A system expands from a volume of 2.50 liters to one of 4.50 liters against a constant opposing pressure of 4.80 atm. Calculate **(a)** the work done by the system on its surroundings **(b)** the work done on the system by its surroundings. Express your answer in liter atmospheres.

18-9 A system is compressed by an external pressure of 8.00 atm from an initial volume of 450 ml to a final volume of 250 ml. Calculate **(a)** the work done by the system on its surroundings **(b)** the work done by the surroundings on the system. Express your answer in joules.

18-10 An ideal gas is expanded isothermally ($\Delta E = 0$) from a volume of 2.40 liters to one of 6.70 liters against a constant opposing pressure of 8.00 atm. How many kilojoules of heat are absorbed by the gas?

18-11 Calculate ΔE for a substance which absorbs 1.48 kJ of heat while expanding from a volume of 1.00 liter to one of 5.00 liters against an opposing pressure of 5.00 atm.

18-12 1.00 mol of an ideal gas at STP expands isothermally ($\Delta E = 0$) against a constant opposing pressure of 0.100 atm. How much heat is absorbed by the gas if its final pressure is **(a)** 0.500 atm **(b)** 0.100 atm?

18-13 1.00 mol of an ideal gas at STP expands isothermally ($\Delta E = 0$) to a volume of 60.0 liters. What is the greatest possible amount of work (kJ) the gas can do on its surroundings if the gas expands against a constant opposing pressure?

Energy and enthalpy

18-14 A system expands at a constant 1.00 atm pressure from a volume of 2.00 liters to one of 9.00 liters while absorbing 14.0 kJ of heat. What are ΔH and ΔE for the process?

18-15 A system undergoing reaction liberates 128 kJ of heat while expanding from a volume of 16.2 liters to one of 28.4 liters at a constant 1.00 atm pressure. Calculate ΔH and ΔE.

18-16 For which of the following processes will ΔH be significantly different in magnitude from ΔE? **(a)** Liquid ethanol, C_2H_5OH, burns in oxygen to form CO_2 and H_2O. **(b)** Solid carbon dioxide sublimes. **(c)** Methane gas (CH_4) burns in O_2 to form

CO_2 and H_2O. **(d)** Solid CaO and CO_2 gas combine to form solid $CaCO_3$.

18-17 The density of ice at 0°C is 0.917 g ml^{-1} and of liquid water at the same temperature is 0.9998 g ml^{-1}. If ΔH_{fus} of ice is 6.009 kJ mol^{-1}, what is ΔE_{fus}?

18-18 At 25°C $\Delta H°$ for the combustion of liquid benzene, C_6H_6, to form CO_2 and H_2O, both gaseous, is -3135 kJ mol^{-1}. Calculate $\Delta E°$ at 25°C.

Hess's law

18-19 Calculate the standard heat of combustion of acetylene, C_2H_2, at 25°C. (Assume that gaseous CO_2 and liquid H_2O are formed.)

18-20 Calculate the standard heat of combustion at 25°C for methane, CH_4, to form gaseous CO_2 and gaseous H_2O.

18-21 The standard heat of combustion of sucrose (cane sugar, $C_{12}H_{22}O_{11}$) to form gaseous CO_2 and liquid H_2O is -5647 kJ mol^{-1}. What is the standard heat of formation of sucrose at 25°C?

18-22 Calculate $\Delta H°$ at 25°C for the reaction

$$H_2S(g) + \tfrac{3}{2}O_2(g) \longrightarrow SO_2(g) + H_2O(l)$$

18-23 Calculate $\Delta H°$ at 25°C for the reaction

$$2H_2O_2(l) \longrightarrow 2H_2O(l) + O_2(g)$$

18-24 From the standard heats of formation of gaseous and liquid water at 25°C calculate the heat of vaporization of water at 1 atm and 25°C.

18-25 Calculate $\Delta H°$ at 25°C for the reaction

$$Na_2O(s) + 2HCl(g) \longrightarrow 2NaCl(s) + H_2O(g)$$

Heat capacity

18-26 Why do C_P and C_V have different values for a given substance? In general for what kinds of substances will C_P be numerically close to C_V?

18-27 How much heat is needed to raise the temperature of 146 g of copper from 46.1 to 98.2°C?

18-28 Calculate the amount of heat liberated from a piece of silver weighing 42.1 g when it cools from 14.0 to -32.1°C.

18-29 A piece of iron weighing 25.0 g at a temperature of 14.9°C is placed in contact with a piece of gold weighing 35.0 g at a temperature of 64.1°C. Assuming that no heat is lost from the combined system, what is the final temperature?

18-30 A gold ring at body temperature (say, 37.00°C) is dropped into 20.0 g of water at 10.00°C. If the final temperature is 10.99°C, how much does the ring weigh?

18-31 Calculate the difference $C_P - C_V$ for an ideal gas.

Entropy and the second law

18-32 Indicate whether there is an increase or decrease in the entropy of the substance undergoing each of the following changes: **(a)** water freezes **(b)** chloroform evaporates **(c)** carbon dioxide sublimes **(d)** a gas is compressed at constant temperature **(e)** two different gases are mixed **(f)** two different liquids are mixed **(g)** an automobile rusts.

18-33 Is it theoretically possible to come up with a perfectly ordered deck of cards after shuffling them? Is it probable? Compare the entropy of a perfectly ordered deck of cards with that of one in which each card is randomly located in the deck.

18-34 For each of the following pairs at 1 atm indicate which substance has the higher entropy: **(a)** 1 g of $NaCl(s)$ or 1 g of $HCl(g)$, each at 25°C **(b)** 1 g of $HCl(g)$ at 25°C or 1 g of $HCl(g)$ at 50°C **(c)** 1 g of $HCl(g)$ or 1 g of $Ne(g)$, each at 25°C.

18-35 The molar heat of vaporization of benzene, C_6H_6, at 1 atm is 30.8 kJ mol^{-1}, and its normal boiling point is 80.1°C. Calculate the entropy change when 10.0 g of benzene boils at this temperature.

18-36 The molar heat of vaporization of ethanol, C_2H_5OH, is 39.4 kJ mol^{-1}, and its normal boiling point is 78.3°C. Calculate the entropy change when 10.0 g of ethanol vapor condenses to liquid at this temperature.

18-37 The molar heat of fusion of potassium bromide, KBr, is 20.9 kJ mol^{-1}. If the molar entropy of fusion of KBr is 20.5 J K^{-1} mol^{-1}, what is the melting point of this compound?

18-38 Twenty-five strangers enter a classroom with many seats and a single aisle down the middle. What is the probability that all choose seats on one side of the aisle?

18-39 What is an isolated system? Does any such system really exist?

18-40 It has been said that living organisms violate the second law because they grow and mature creating highly ordered structures as they do. Does life violate the second law? Explain.

18-41 Predict the sign of ΔS for each of the following reactions:

(a) $C_2H_4(g) + 2O_2(g) \longrightarrow 2CO(g) + 2H_2O(g)$
(b) $CO(g) + 2H_2(g) \longrightarrow CH_3OH(l)$
(c) $Hg_2Cl_2(s) \longrightarrow 2Hg(l) + Cl_2(g)$
(d) $Mg(s) + H_2O(l) \longrightarrow MgO(s) + H_2(g)$
(e) $Fe_2O_3(s) + CO(g) \longrightarrow 2FeO(s) + CO_2(g)$

18-42 Calculate ΔS_{298}° for the reaction

$$CO_2(g) + C_{(graphite)} \longrightarrow 2CO(g)$$

Comment on the sign of $\Delta S°$ for this reaction.

18-43 Calculate ΔS_{298}° for the reaction

$$NH_4Cl(s) \longrightarrow NH_3(g) + HCl(g)$$

18-44 Calculate the standard entropy of formation at 298 K of methanol, CH_3OH.

Free energy

18-45 State how ΔG can be used to predict the spontaneity of a reaction.

18-46 Explain in terms of the entropy and enthalpy changes why the *sign* of ΔG_{fus} of a substance changes at the melting point.

18-47 ΔS for a certain reaction is -100 J K^{-1} mol^{-1}. If the reaction occurs spontaneously, what must be the sign of ΔH for the reaction?

18-48 Calculate the standard free energy of combustion of acetylene, C_2H_2, to form $CO_2(g)$ and $H_2O(l)$ at 25°C.

18-49 Calculate the standard free energy of combustion of acetylene, C_2H_2, to form $CO(g)$ and $H_2O(l)$ at 25°C.

18-50 Repeat the calculation of Prob. 18-49 assuming that $H_2O(g)$ is formed.

18-51 Distinguish clearly between ΔG for a reaction and $\Delta G°$ for the same reaction. Which quantity should be used when predicting the direction of spontaneous change? Why?

18-52 Suppose that for a given reaction ΔH is 50 kJ and ΔS is 120 J K^{-1}. Will the reaction be spontaneous at 25°C?

18-53 A given reaction has $\Delta H = 92$ kJ and $\Delta S = 85$ J K^{-1}. Above what temperature will the reaction be spontaneous?

Equilibrium

18-54 The standard free energies of formation of SO_2 and SO_3 at 25°C are -300.2 and -371.1 kJ mol^{-1}, respectively. Calculate the value of the equilibrium constant at 25°C for the reaction

$$2SO_2(g) + O_2(g) \rightleftharpoons 2SO_3(g)$$

18-55 Calculate the value of the equilibrium constant at 25°C for

$$C_{(graphite)} + 2H_2(g) \rightleftharpoons CH_4(g)$$

18-56 The dirty brown color of photochemical smog is largely caused by nitrogen dioxide, NO_2, which can be formed by the oxidation of nitric oxide, NO. If the standard free energies of formation of NO_2 and NO are 240.5 and 210.6 kJ mol^{-1} at 25°C, respectively, what is the value of the equilibrium constant for

$$2NO(g) + O_2(g) \rightleftharpoons 2NO_2(g)$$

Thermodynamics—general

18-57 Accurate experimental measurements show that solid carbon monoxide has a small but significant residual entropy near absolute zero. Suggest a reason for this apparent violation of the third law. (*Hint:* CO has a very low dipole moment.)

18-58 The vapor pressure of water is 23.8 mmHg at 25°C. Calculate ΔG per mole for each of the following processes at 25°C:

(a) $H_2O(l) \longrightarrow H_2O(g, 23.8 \text{ mmHg})$

(b) $H_2O(g, 23.8 \text{ mmHg}) \longrightarrow H_2O(g, 760 \text{ mmHg})$

(c) $H_2O(l) \longrightarrow H_2O(g, 760 \text{ mmHg})$

18-59 The normal boiling point of toluene is 111°C, and its molar heat of vaporization is 33.5 kJ mol^{-1}. Calculate w, q, ΔH, ΔE, ΔG, and ΔS for the vaporization of one mole of toluene at its boiling point.

18-60 For water $\Delta H_{fus} = 6.01$ kJ mol^{-1} and ΔH_{vap} is 40.66 kJ mol^{-1}. Assuming that C_P for liquid water stays constant at 75.6 J K^{-1} mol^{-1}, calculate ΔH for the process

$$H_2O(s, 0°C, 1 \text{ atm}) \longrightarrow H_2O(g, 100°C, 1 \text{ atm})$$

18-61 For the reaction

$$2SO_2(g) + O_2(g) \longrightarrow 2SO_3(g)$$

find ΔG at 25°C if the partial pressure of each gas is 0.0100 atm.

18-62 The activity of a species is 0.50. The free energy of the species is how much less than its standard free energy? (Assume a 25°C temperature.)

18-63 For each of the following changes indicate the algebraic sign of ΔE, ΔH, ΔS, and ΔG. (Consider the system only, not its surroundings.)

(a) The expansion of an ideal gas into a vacuum

(b) The boiling of water at 100°C and 1 atm

(c) The burning of hydrogen in oxygen

18-64 The molar heat of fusion of ice is 6.008 kJ mol^{-1}. (a) Calculate the molar entropy of fusion of ice at 0.00°C (273.15 K). Assuming that ΔS_{fus} does not vary with temperature, calculate ΔG for the melting of one mole of ice at (b) 10.00°C (c) 0.00°C (d) -10.00°C.

19

ELECTROCHEMISTRY

TO THE STUDENT

Matter is composed of electrically charged particles, and so it is not surprising that it is possible to convert chemical energy into electrical energy and vice versa. The study of these interconversion processes is an important part of *electrochemistry,* which is broadly concerned with *the relationship between electrical energy and chemical change.*

In Sec. 13-3 we examined ways of predicting what kinds of reactions can occur when various substances are mixed. At that time we considered precipitation reactions, reactions which form weak electrolytes, and complexation reactions, but we postponed consideration of how to predict redox reactions. In this chapter we will pick up this loose end by applying a little thermodynamics to electron-transfer reactions and obtaining a practical way of predicting their spontaneity.

**19-1
Galvanic cells**

An *electrochemical cell* is a device which permits the interconversion of chemical and electrical energy. There are two kinds of electrochemical cells: *galvanic cells,* in which chemical energy is converted into electrical energy and *electrolytic cells,* in which electrical energy is converted into chemical energy. We will first consider the operation of galvanic cells.

Spontaneous reaction and the galvanic cell

Consider the simple redox reaction

$$Zn(s) + Cu^{2+} \longrightarrow Zn^{2+} + Cu(s)$$

This is a spontaneous reaction, as we can easily show by placing a rod of zinc metal in a solution of cupric sulfate, as pictured in Fig. 19-1. We immediately notice a dark deposit forming on the zinc and obscuring its bright surface. This deposit consists of finely divided particles of copper metal and grows to form a thick, spongy layer. Gradually the characteristic blue color of the $CuSO_4$ solution becomes paler showing that hydrated cupric ions, $Cu(H_2O)_4^{2+}$, are used up in the reaction. The reaction is spontaneous; zinc is oxidized, and cupric ions are reduced:

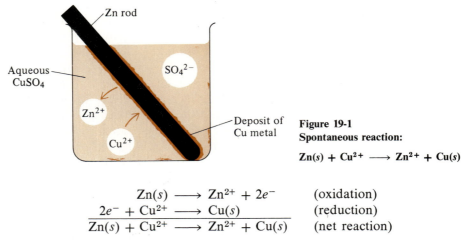

Figure 19-1
Spontaneous reaction:

$$Zn(s) + Cu^{2+} \longrightarrow Zn^{2+} + Cu(s)$$

$$
\begin{array}{ll}
Zn(s) \longrightarrow Zn^{2+} + 2e^- & \text{(oxidation)} \\
\underline{2e^- + Cu^{2+} \longrightarrow Cu(s)} & \text{(reduction)} \\
Zn(s) + Cu^{2+} \longrightarrow Zn^{2+} + Cu(s) & \text{(net reaction)}
\end{array}
$$

For this reaction $\Delta G°$ equals -212 kJ, this large negative value indicating a strong tendency for electrons to be transferred from Zn metal to Cu^{2+} ions, at least when reactants and products are in their standard states (pure metals and 1 M ionic concentrations).

The free-energy change for a reaction depends only on the nature and state of the reactants and products, and not on how the reaction takes place. In other words, as long as we provide the means for electrons to be transferred from Zn to Cu^{2+}, the transfer will occur. Suppose, for instance, that we physically separate the zinc rod from the solution of cupric sulfate, as is shown in Fig. 19-2a. Here the zinc rod is immersed in a solution of zinc sulfate, a copper rod is immersed in a solution of cupric sulfate, and the two rods are electrically connected by a wire outside the apparatus. This is one form of a *galvanic cell*, also called a *voltaic*

Figure 19-2
Galvanic cell: two versions.   (a) Cell with porous partition.   (b) Cell with salt bridge.

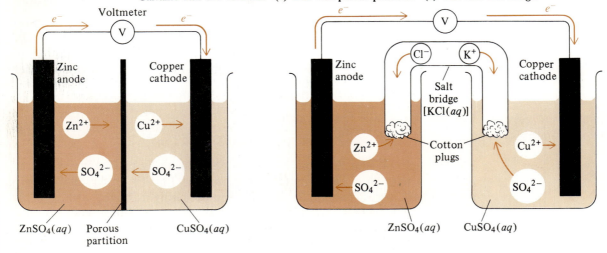

(a)

(b)

cell. The two halves of the cell are called *compartments,* and are separated by a porous partition, perhaps a piece of unglazed clay or porcelain. The zinc and copper rods are called *electrodes* and provide surfaces at which the oxidation and reduction half-reactions can take place.

If the zinc and copper electrodes in Fig. 19-2*a* are connected directly to each other through the external circuit, electrons leave the zinc metal and travel through this circuit to the copper electrode, where they are accepted by Cu^{2+} ions at its surface. These ions are reduced and the resulting copper atoms deposit on the surface of the copper electrode in a process called *plating out.* In this cell the zinc electrode is called an *anode.* An anode is an electrode at which *oxidation* takes place.

$$Zn(s) \longrightarrow Zn^{2+} + 2e^- \qquad \text{(anode half-reaction)}$$

The copper electrode is a *cathode,* an electrode at which *reduction* takes place.

$$2e^- + Cu^{2+} \longrightarrow Cu(s) \qquad \text{(cathode half-reaction)}$$

(Actually, the cathode could be made of any inert but electrically conducting material, such as platinum or graphite.) The porous partition in the cell serves to keep cupric ions away from the zinc anode and thereby prevents electrons from transferring directly from zinc to cupric ions, instead of passing through the external circuit. As the cell reaction occurs, zinc ions migrate away from the zinc anode and toward the copper cathode, as do cupric ions. Positive ions are called *cations* because they migrate toward the *cathode.* Similarly, sulfate ions migrate toward the zinc *anode* and therefore are called *anions.*

Now we place a voltmeter, labeled V in Fig. 19-2*a,* in the external circuit. If this voltmeter has a sufficiently high electrical resistance (a high resistance to electron flow) the passage of electrons through the external circuit essentially stops, as does the oxidation of Zn and the reduction of Cu^{2+}. The voltmeter reads the *electrical potential difference,* or *voltage,* between the two electrodes, usually expressed in *volts.*[1] This gives a measure of the tendency of electrons to flow through the external circuit, and this tendency depends in turn upon the tendencies for anode and cathode half-reactions to occur.

Figure 19-2*b* shows an alternate way of constructing a galvanic cell. In this version the porous partition of Fig. 19-2*a* has been replaced by a salt bridge, a U-tube filled with a solution of potassium chloride. In the salt bridge Cl^- ions migrate toward the anode, and K^+ ions toward the cathode, as the cell discharges.

A salt bridge serves three functions: it physically separates the electrode compartments, it provides electrical continuity (a path for migrating anions and cations) within the cell, and it reduces the so-called *liquid-junction potential,* a voltage produced where two dissimilar solutions are in contact. This voltage arises because of unequal rates of anion and cation migration across the contact region, or junction. A salt bridge usually contains ions which migrate at almost equal rates, thus minimizing the potential. Interpretation of measured cell voltages is more straightforward if no liquid junction is present.

---

[1]A voltmeter connected in this way measures electrical potential difference in much the same way that water-pressure differences are measured by a pressure meter. Electrical potential may be thought of as a kind of electron pressure.

ADDED COMMENT

Why is it that we cannot eliminate the liquid junction potential altogether by just pulling the salt bridge out of the cell in Fig. 19-2b? The reason is that everything comes to a halt if we do: the electrode reactions stop and the voltmeter reading drops to zero. A salt bridge (or porous partition) is necessary in order to provide a path for ions migrating between electrode compartments, and thus to complete the electrical circuit.

In both versions of the galvanic cell shown in Fig. 19-2 the electrode and overall cell reactions are the same:

$$
\begin{array}{lrcl}
\text{Anode:} & Zn(s) & \longrightarrow & Zn^{2+} + 2e^- \\
\text{Cathode:} & 2e^- + Cu^{2+} & \longrightarrow & Cu(s) \\
\hline
\text{Cell:} & Zn(s) + Cu^{2+} & \longrightarrow & Zn^{2+} + Cu(s)
\end{array}
$$

These cells are versions of the *Daniell cell,* a name given to any galvanic cell which makes use of the above reactions. It can be seen that in the Daniell cell the reaction is exactly the same as the one which occurs when a rod of zinc is placed in a beaker of $CuSO_4$ solution. The big difference is that in the Daniell cell electrons must pass through the external circuit before reaching the Cu^{2+} ions in the cathode compartment.

Cell diagrams   Galvanic cells are commonly represented by a shorthand notation called a *cell diagram.* The cell diagram for the Daniell cell is

$$Zn(s)\,|\,ZnSO_4(aq)\,|\,CuSO_4(aq)\,|\,Cu(s)$$

Here each symbol and formula has its usual meaning, and the short vertical lines represent phase boundaries or junctions. The convention usually followed is to show the *anode* at the *left* of the cell diagram. This means that electrons *leave* the cell from the electrode written at the left. When a salt bridge is present to minimize the liquid-junction potential, it is indicated by a double line:

$$Zn(s)\,|\,ZnSO_4(aq)\,\|\,CuSO_4(aq)\,|\,Cu(s)$$

Sometimes only reacting ions are shown in solution phases:

$$Zn(s)\,|\,Zn^{2+}\,\|\,Cu^{2+}\,|\,Cu(s)$$

Electrodes in galvanic cells   The electrodes in a cell serve as devices for removing electrons from the reducing agent at the anode and providing electrons to the oxidizing agent at the cathode. The five most important types of electrodes are

1 Metal–metal ion electrodes
2 Gas-ion electrodes
3 Metal–insoluble salt–anion electrodes
4 Inert, "oxidation–reduction" electrodes
5 Membrane electrodes

The *metal–metal ion* electrode consists of a metal in contact with its ions in solution. An example is a piece of silver metal immersed in a solution of silver nitrate. The diagram for such an electrode serving as a cathode is

$$Ag^+ \,|\, Ag(s)$$

and the electrode half-reaction is

$$Ag^+ + e^- \longrightarrow Ag(s)$$

where the electron shown in this equation comes from the external circuit. When this electrode serves as an anode, it is diagrammed as

$$Ag(s) \,|\, Ag^+$$

and its half-reaction equation is

$$Ag(s) \longrightarrow Ag^+ + e^-$$

where the electron shown leaves the silver and enters the external circuit. The copper–cupric ion and zinc–zinc ion electrodes of the Daniell cell are also of this type.

The *gas-ion* electrode uses a gas in contact with its anion or cation in solution. The gas is bubbled into the solution and electrical contact is made by means of a piece of inert metal, usually platinum. Figure 19-3 shows one construction of the hydrogen–hydrogen ion electrode (usually called, simply, the hydrogen electrode). The diagram for this electrode functioning as a cathode is

$$H^+ \,|\, H_2(g) \,|\, Pt(s)$$

and the electrode half-reaction is

$$2e^- + 2H^+ \longrightarrow H_2(g)$$

(As before, the electrons come from the external circuit.) In this electrode the platinum must be coated with a thin layer of finely divided platinum (platinum black), which catalyzes the reaction.

In the *metal–insoluble salt–anion electrode,* a metal is in contact with one of its insoluble salts and also with a solution containing the *anion* of the salt. An example is the silver–silver chloride electrode

$$Cl^- \,|\, AgCl(s) \,|\, Ag(s)$$

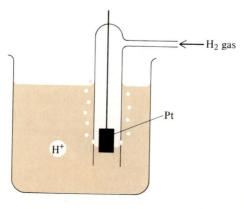

Figure 19-3
Hydrogen electrode.

for which the half-reaction is

$$AgCl(s) + e^- \longrightarrow Ag(s) + Cl^-$$

In this electrode a silver wire is coated with a paste of silver chloride and immersed in the chloride ion–containing solution, as shown in Fig. 19-4.

The *inert, "oxidation–reduction" electrode* is really no more of an oxidation–reduction electrode than any other. It consists of a strip or wire of an inert metal, say, platinum, in contact with a solution which contains ions of a substance in two different oxidation states. Thus the ferric ion, Fe^{3+}, is reduced to the ferrous ion, Fe^{2+}, at such an electrode (acting as a cathode). In this case the diagram is

$$Fe^{3+}, Fe^{2+} | Pt(s)$$

where the comma indicates that both ions are in the solution phase.

We will consider one type of *membrane electrode,* the glass electrode, in Sec. 19-5.

Voltage and spontaneity

When a voltmeter or other voltage-measuring device is connected to a galvanic cell, it shows an *electrical potential difference.* We call this difference *the voltage produced by the cell* and give it a *positive* algebraic sign. The plus sign indicates that the cell reaction proceeds spontaneously, but the problem is, *in which direction?*

To illustrate this problem, consider that we have prepared a cell with an A–A$^+$ ion electrode and a B–B$^+$ ion electrode connected by a salt bridge. Since we have not yet distinguished anode from cathode, we do not know whether to diagram the cell as

$$A(s) | A^+ \| B^+ | B(s)$$

for which the reactions are

Anode:	$A(s) \longrightarrow A^+ + e^-$
Cathode:	$e^- + B^+ \longrightarrow B(s)$
Cell:	$A(s) + B^+ \longrightarrow A^+ + B(s)$

or as

$$B(s) | B^+ \| A^+ | A(s)$$

for which the reactions are

Anode:	$B(s) \longrightarrow B^+ + e^-$
Cathode:	$e^- + A^+ \longrightarrow A(s)$
Cell:	$B(s) + A^+ \longrightarrow B^+ + A(s)$

To determine which electrode is the anode we connect the "−" lead of a voltmeter to the A electrode, the "+" lead to the B electrode, and observe the indicated voltage. If the voltmeter reads a *positive* voltage, it means that we must have connected the voltmeter correctly, and that the A electrode was negatively charged and the B electrode was positively charged. If an electrode appears negatively charged (to the voltmeter), it means that electrons tend to emerge from the cell at that point. Thus we can conclude that the A electrode in our imaginary cell is the anode, and so the cell diagram is

$$A(s) | A^+ \| B^+ | B(s)$$

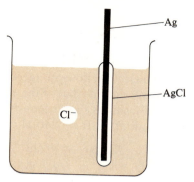

Figure 19-4
Silver-silver chloride electrode.

For this cell we associate a *positive voltage* with the *spontaneous reaction*

$$A(s) + B^+ \longrightarrow A^+ + B(s)$$

The Daniell cell reaction

$$Zn(s) + Cu^{2+} \longrightarrow Zn^{2+} + Cu(s)$$

occurs spontaneously, and at $1\ M$ ionic concentrations and 25°C the measured voltage is $+1.10$ V. This voltage is a direct measure of the tendency for the reaction to take place, that is, its spontaneity, *whether in a galvanic cell or not.* Thus the "Daniell cell reaction" is spontaneous even if we have no cell at all. (See Fig. 19-1.) In addition, we associate a *negative voltage* with a *nonspontaneous reaction.* Thus the voltage for

$$Cu(s) + Zn^{2+} \longrightarrow Cu^{2+} + Zn(s)$$

is -1.10 V ($1\ M$ ionic concentrations, 25°C). This is consistent with the facts: copper metal is *not* oxidized by zinc ions.

ADDED COMMENT

A word is in order about the apparent charges of electrodes. The anode of a galvanic cell appears negative from outside the cell, because electrons, which are negative charges, tend to emerge from it. From *inside* the cell, however, it appears *positive.* In the Daniell cell, for instance, positive zinc ions leave the electrode and enter the solution. Fortunately, we do not sit inside a cell and wonder about electrode signs! A less ambiguous way of designating electrodes is by labeling them simply as anode and cathode.

19-2
Electrolytic cells

The second kind of electrochemical cell is the *electrolytic cell.* In this cell electrical energy from an external source is used to produce chemical reaction.

Nonspontaneous reactions and electrolytic cells

Consider once more the Daniell cell:

$$Zn(s) \,|\, Zn^{2+} \,\|\, Cu^{2+} \,|\, Cu(s)$$

If this is a *standard* Daniell cell (ionic concentration $= 1\ M$), it produces a voltage of 1.10 V, and the zinc electrode is the anode, which means that electrons tend to leave the cell at this electrode. We now connect an external voltage of 1.09 V to the Daniell cell in such a way that it *opposes* the voltage produced by the Daniell cell. (The external voltage is connected negative to negative and positive to positive, so that it tends to pump electrons *into* the zinc anode and *out of* the copper cathode, thus opposing the electron flow produced by the cell.) Since the voltage produced by the Daniell cell exceeds that of the external opposing source (by 0.01 V), the Daniell cell still discharges spontaneously, with electrons leaving the anode and entering the cathode from the external circuit:

$$
\begin{array}{lll}
\text{Anode:} & Zn(s) \longrightarrow Zn^{2+} + 2e^- \\
\text{Cathode:} & 2e^- + Cu^{2+} \longrightarrow Cu(s) \\
\hline
\text{Cell:} & Zn(s) + Cu^{2+} \longrightarrow Cu(s) + Zn^{2+}
\end{array}
$$

Now, if we increase the external opposing voltage by 0.01 V so that it is 1.10 V, it will just counterbalance the cell voltage, so that there is no net electron flow. The effect of this is to stop net production of $Cu(s)$ and Zn^{2+} and to establish a state of equilibrium at each electrode:

$$
\begin{array}{lll}
\text{Zinc electrode:} & Zn(s) \rightleftharpoons Zn^{2+} + 2e^- \\
\text{Copper electrode:} & 2e^- + Cu^{2+} \rightleftharpoons Cu(s) \\
\hline
\text{Cell:} & Zn(s) + Cu^{2+} \rightleftharpoons Cu(s) + Zn^{2+}
\end{array}
$$

Finally, if we increase the external opposing voltage so that it is *larger* than the cell voltage, say 1.11 V, it causes electrons to *enter* the zinc electrode, so that it becomes a *cathode*. At the same time electrons *leave* the copper electrode, so that it becomes an *anode:*

$$
\begin{array}{lll}
\text{Cathode:} & Zn(s) \longleftarrow Zn^{2+} + 2e^- \\
\text{Anode:} & 2e^- + Cu^{2+} \longleftarrow Cu(s) \\
\hline
\text{Cell:} & Zn(s) + Cu^{2+} \longleftarrow Cu(s) + Zn^{2+}
\end{array}
$$

The cell is now an *electrolytic cell*. In an electrolytic cell electrical energy from an external source is used to reverse a thermodynamically spontaneous reaction, that is, *to force a nonspontaneous reaction to occur.*

Electrolysis   In principle any galvanic cell can be converted into an electrolytic cell by connecting an external opposing voltage greater than the voltage produced by the cell. (In practice the reversal of the cell reaction sometimes does not occur due to rate effects.) Consider the galvanic cell

$$Pt(s)\,|\,H_2(g)\,|\,H^+,\ Cl^-\,|\,Cl_2(g)\,|\,Pt(s)$$

If the concentration of HCl is 1 M and if the gases are present at 1 atm pressure, the cell operates as a galvanic cell and produces a voltage of 1.36 V. The processes occurring within the cell are

$$
\begin{array}{lll}
\text{Anode:} & H_2(g) \longrightarrow 2H^+ + 2e^- \\
\text{Cathode:} & 2e^- + Cl_2(g) \longrightarrow 2Cl^- \\
\hline
\text{Cell:} & H_2(g) + Cl_2(g) \longrightarrow 2H^+ + 2Cl^-
\end{array}
$$

This reaction is spontaneous in the direction written, but can be reversed by applying an opposing voltage greater than 1.36 V. This causes the reduction of H^+ to form H_2 and the oxidation of Cl^- to form Cl_2. The reactions are

$$
\begin{array}{lrcl}
\text{Cathode:} & 2e^- + 2H^+ & \longrightarrow & H_2(g) \\
\text{Anode:} & 2Cl^- & \longrightarrow & Cl_2(g) + 2e^- \\
\hline
\text{Cell:} & 2H^+ + 2Cl^- & \longrightarrow & H_2(g) + Cl_2(g)
\end{array}
$$

It can be seen that the overall effect is one of converting the dissolved HCl into H_2 and Cl_2 gases. This is an example of *electrolysis,* a process in which an otherwise nonspontaneous reaction ($\Delta G > 0$) is forced to take place by the application of energy from an external source. An electrolysis reaction often (though not always) results in the decomposition of a compound to form its elements.

Electrolyses are usually accomplished by applying a voltage to a pair of inert (unreactive) electrodes immersed in a liquid. When two platinum electrodes are immersed in hydrochloric acid and a gradually increasing voltage is applied, no appreciable quantity of H_2 or Cl_2 is formed until the applied voltage reaches about 1.36 V. This is because the formation of minute amounts of H_2 and Cl_2 sets up a galvanic cell whose spontaneous reaction opposes the electrolytic decomposition of the HCl. Only when the applied voltage exceeds the galvanic cell's voltage does electrolysis begin. The external voltage necessary to begin electrolysis of a solution is called the *decomposition potential* of the solution. Sometimes the voltage necessary to initiate electrolysis is greater than the voltage produced by the opposing galvanic cell. This is caused by rate effects at the electrodes or in the solution, where slow diffusion of ions toward an electrode may necessitate the application of a higher-than-ideal voltage in order to produce electrolysis.

The electrolysis of molten sodium chloride

Consider the electrolytic cell shown in Fig. 19-5. It consists of a pair of inert electrodes, say, platinum, dipping into molten (liquid) NaCl. Since the melting point of NaCl is about 800°C, the cell must operate above this temperature. The battery connected in the external circuit serves to pump electrons out of the anode and into the cathode. The negatively charged chloride ions in the liquid are attracted to the anode, where each gives up one electron:

$$Cl^- \longrightarrow Cl + e^-$$

Chlorine atoms then pair up to form Cl_2 gas:

$$2Cl \longrightarrow Cl_2(g)$$

so that the net anode half-reaction is

$$2Cl \longrightarrow Cl_2(g) + 2e^-$$

Positively charged sodium ions are attracted to the cathode, where each gains one electron:

$$Na^+ + e^- \longrightarrow Na(l)$$

Since the melting point of sodium is only about 98°C, the sodium is formed as a liquid and runs *up* the surface of the cathode. (*Note:* Sodium metal is less dense than sodium chloride.)

The processes occurring in the cell can be written as

$$
\begin{array}{llr}
\text{Anode:} & 2Cl^- \longrightarrow Cl_2(g) + 2e^- & \text{(oxidation)} \\
\text{Cathode:} & [Na^+ + e^- \longrightarrow Na(l)] \times 2 & \text{(reduction)} \\
\hline
\text{Cell:} & 2Na^+ + 2Cl^- \longrightarrow 2Na(l) + Cl_2(g) &
\end{array}
$$

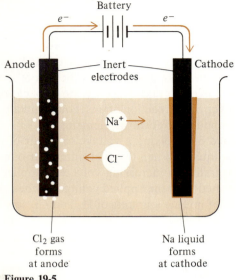

Figure 19-5
Electrolysis of molten NaCl.

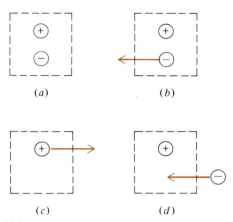

Figure 19-6
Maintenance of electrical neutrality. (*a*) Tiny volume of liquid NaCl. (*b*) Cl^- moves out of volume. (*c*) Na^+ moves out of volume. (*d*) Another Cl^- moves into volume.

As Cl^- ions are removed at the anode, other Cl^- ions move toward this electrode to take their places. Similarly, removal of Na^+ ions at the cathode causes other Na^+ ions to move toward this electrode. The continuous migration of cations toward the cathode and of anions toward the anode is called the *ion current* and should be distinguished from an electron current, as in a metallic conductor. The motion of ions occurs in such a way that there is no positive or negative charge buildup in any part of the liquid. Figure 19-6 shows a tiny volume of liquid NaCl within the cell, a volume so small that it can contain but a single Na^+ ion and a single Cl^- ion, as shown in Fig. 19-6*a*. Now imagine that the Cl^- ion moves out of this tiny volume toward the anode, as shown in Fig. 19-6*b*. In order to maintain the electrical neutrality, two things can happen: the Na^+ ion can move out of the volume (Fig. 19-6*c*), or another Cl^- can move into the volume (Fig. 19-6*d*). In either case, electrical neutrality is preserved. What actually happens is a combination of both Fig. 19-6*c* and *d*, although not to equal extents, because the speeds with which oppositely charged ions migrate through a solution are usually not equal.

The electrolysis of aqueous sodium chloride

Consider now the electrolytic cell shown in Fig. 19-7. This cell contains an aqueous solution of NaCl, instead of pure, liquid NaCl. Because there are so many species present in this cell, several possibilities exist for both anode and cathode reactions:

Possible anode (oxidation) reactions:

$$2Cl^- \longrightarrow Cl_2(g) + 2e^-$$

$$2H_2O \longrightarrow O_2(g) + 4H^+ + 4e^-$$

$$4OH^- \longrightarrow O_2(g) + 2H_2O + 4e^-$$

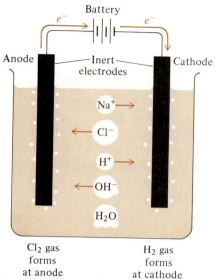

Battery

Anode   —Inert—   Cathode
electrodes

$Na^+ \rightarrow$

$\leftarrow Cl^-$

$H^+ \rightarrow$

$\leftarrow OH^-$

H_2O

Cl_2 gas H_2 gas
forms   forms
at anode   at cathode

Figure 19-7
Electrolysis of aqueous NaCl.

Possible cathode (reduction) reactions:

$$e^- + Na^+ \longrightarrow Na(s)$$

$$2e^- + 2H_2O \longrightarrow H_2(g) + 2OH^-$$

$$2e^- + 2H^+ \longrightarrow H_2(g)$$

It is observed that at the anode chlorine gas is produced; thus the anode reaction is

$$2Cl^- \longrightarrow Cl_2(g) + 2e^-$$

At the cathode hydrogen gas is formed; so we know that either H^+ or H_2O is reduced. Since the concentration of H_2O molecules in aqueous NaCl is so much higher than that of H^+ ions (about 560 million times higher, as calculated from K_w), the cathode reaction is usually written as

$$2e^- + 2H_2O \longrightarrow H_2(g) + 2OH^-$$

In any event, even if H^+ is the actual species reduced at the electrode, the above half-reaction best represents the net change, since it can be considered to be the combination of

$$2e^- + 2H^+ \longrightarrow H_2(g)$$

followed by the shifting of the water equilibrium:

$$H_2O \rightleftharpoons H^+ + OH^-$$

the sum of which leads to the indicated net half-reaction.

In summary then,

Anode:	$2Cl^- \longrightarrow Cl_2(g) + 2e^-$	(oxidation)
Cathode:	$2e^- + 2H_2O \longrightarrow H_2(g) + 2OH^-$	(reduction)
Cell:	$2H_2O + 2Cl^- \longrightarrow H_2(g) + Cl_2(g) + 2OH^-$	

Other electrolyses   HYDROCHLORIC ACID. The electrolysis of a solution of the strong acid HCl is similar to that of aqueous NaCl because the same products are formed, Cl_2 at the anode and H_2 at the cathode. Because of the high concentration of hydrogen ions in this solution, the cathode reaction is usually written to show H^+ being reduced:

$$
\begin{array}{ll}
\text{Anode:} & 2Cl^- \longrightarrow Cl_2(g) + 2e^- \\
\text{Cathode:} & 2e^- + 2H^+ \longrightarrow H_2(g) \\
\hline
\text{Cell:} & 2H^+ + 2Cl^- \longrightarrow H_2(g) + Cl_2(g)
\end{array}
$$

SULFURIC ACID. H_2SO_4 is also a strong acid; so the cathode reaction is the same as with HCl: the reduction of H^+ to H_2. At the anode the bisulfate ion HSO_4^- might be expected to be oxidized, but it turns out that water more readily loses electrons. The reactions are, therefore,

$$
\begin{array}{ll}
\text{Anode:} & 2H_2O \longrightarrow O_2(g) + 4H^+ + 4e^- \\
\text{Cathode:} & [2e^- + 2H^+ \longrightarrow H_2(g)] \times 2 \\
\hline
\text{Cell:} & 2H_2O \longrightarrow 2H_2(g) + O_2(g)
\end{array}
$$

In adding the above two half-reactions we have canceled hydrogen ions from both sides of the cell reaction, because they are *produced* at the anode, but *used up* at the cathode.

Note that the HSO_4^- ion is missing from the above cell reaction, and from the half-reactions, as well. Does this mean that it is not needed? No, it serves two related functions. First, it serves to carry part of the electrical current within the body of the cell, as it moves from cathode to anode. Second, it helps to maintain electrical neutrality in the regions near electrodes. As H^+ ions are used up by the cathode, HSO_4^- ions move away to keep the region electrically neutral. Similarly, as H^+ ions are formed at the anode, more HSO_4^- ions move into that region to preserve electrical neutrality.

SODIUM SULFATE. In a solution of the salt Na_2SO_4, neither sodium ions nor sulfate ions are directly involved in the electrode reactions. At the anode H_2O is more readily oxidized than is SO_4^{2-}, and at the cathode the situation is the same as in the electrolysis of aqueous NaCl, that is, water molecules are easier to reduce than sodium ions. The reactions are, therefore,

$$
\begin{array}{ll}
\text{Anode:} & 2H_2O \longrightarrow O_2(g) + 4H^+ + 4e^- \\
\text{Cathode:} & [2e^- + 2H_2O \longrightarrow H_2(g) + 2OH^-] \times 2 \\
\hline
\text{Cell:} & 6H_2O \longrightarrow 2H_2(g) + O_2(g) + 4H^+ + 4OH^-
\end{array}
$$

If the contents of the cell are stirred during electrolysis, the hydrogen ions produced at the anode react with the hydroxide ions produced at the cathode:

$$H^+ + OH^- \longrightarrow H_2O$$

so that the net equation for the overall process is

$$2H_2O \longrightarrow 2H_2(g) + O_2(g)$$

In this cell, the Na^+ and SO_4^{2-} ions serve to conduct electric current through the cell and in so doing prevent a buildup of net positive or negative charge in the electrode regions.

Faraday's laws   During the early nineteenth century Michael Faraday developed the quantitative statements now known as *Faraday's laws of electrolysis*. These are (1) that *the quantity of a substance produced by electrolysis is proportional to the quantity of electricity used,* and (2) that *for a given quantity of electricity the quantity of substance produced is proportional to its equivalent weight.* (See Sec. 13-5.)

For an illustration of Faraday's first law consider the electrolysis of molten NaCl. At the cathode the half-reaction is

$$Na^+ + e^- \longrightarrow Na(l)$$

The equation itself implies Faraday's first law because it shows that *one* electron is needed to produce *one* sodium atom. This means that *one mole* of electrons are needed to produce *one mole* of sodium atoms. Now, one mole of electrons is a quantity of electricity called *one faraday* (1 $\mathcal{F}$). This is a rather large quantity of electricity. A smaller unit is the coulomb (C); there are 9.65×10^4 coulombs in one faraday. How big is a coulomb? If *one coulomb* of electricity passes through a conductor in *one second,* we say that the conductor carries an electrical current of *one ampere* (A). In other words, one ampere equals one coulomb per second.

An illustration of Faraday's second law can also be found in the electrolysis of molten NaCl. At the anode the half-reaction is

$$2Cl^- \longrightarrow Cl_2(g) + 2e^-$$

Here, *two* electrons must be withdrawn (from two Cl^- ions) in order to produce *one* Cl_2 molecule. Thus *two moles* of electrons are needed to produce *one mole* of Cl_2 molecules. This means that one equivalent of Cl_2 (the amount produced by one mole of electrons) is 0.5 mol. (The equivalent weight is half the molecular weight.) When molten NaCl is electrolyzed, then, *one faraday* of electricity will produce *one equivalent* (1 mol) of Na at the cathode plus *one equivalent* (0.5 mol) of Cl_2 at the anode. (It takes twice as many electrons to produce 1 mol of Cl_2 as it does to produce 1 mol of Na.)

● **EXAMPLE 19-1**

Problem   An aqueous solution of copper sulfate, $CuSO_4$, is electrolyzed using inert electrodes. How many grams of copper metal and of oxygen gas are produced if a current of 5.0 A flows through the cell for 1.5 h?

Solution   Since the electrode half-reactions can be interpreted in terms of faradays of electricity, we need first to find out how many faradays flow through the cell. Since one ampere is one coulomb per second, the total number of coulombs passed is

$$5.0 \text{ C s}^{-1}(60 \text{ s min}^{-1})(60 \text{ min h}^{-1})(1.5 \text{ h}) = 2.7 \times 10^4 \text{ C}$$

And since there are 9.65×10^4 C in one faraday, this is

$$2.7 \times 10^4 \text{ C} \frac{1\mathcal{F}}{9.65 \times 10^4 \text{ C}} = 0.28 \text{ }\mathcal{F}$$

If we now turn to the electrode half-reactions, we see that at the cathode cupric ions, Cu^{2+}, are reduced to Cu metal:

$$Cu^{2+} + 2e^- \longrightarrow Cu(s)$$

From this we see that *one* mole of Cu is produced from *two* faradays of electricity, and so the number of grams of Cu produced is

$$0.28 \ \mathscr{F} \ \frac{1 \ \text{mol Cu}}{2 \ \mathscr{F}} \ \frac{63.5 \ \text{g Cu}}{1 \ \text{mol Cu}} = 8.9 \ \text{g Cu}$$

At the anode oxygen is produced:

$$2H_2O \longrightarrow O_2(g) + 4H^+ + 4e^-$$

The half-reaction tells us that *one* mole of O_2 requires the passage of four faradays of electricity. Thus the number of grams of O_2 formed is

$$0.28 \ \mathscr{F} \ \frac{1 \ \text{mol } O_2}{4 \ \mathscr{F}} \ \frac{32.0 \ \text{g } O_2}{1 \ \text{mol } O_2} = 2.2 \ \text{g } O_2 \ \bullet$$

● EXAMPLE 19-2

Problem   A solution of sulfuric acid was electrolyzed using inert electrodes for a period of 35 min. The hydrogen produced at the cathode was collected over water at a total pressure of 752 mmHg and a temperature of 28°C. If the volume of H_2 was 145 ml, what was the average current which passed during electrolysis? (The vapor pressure of H_2O at 28°C is 28 mmHg.)

Solution   First we find the number of moles of H_2 gas. Its partial pressure was $752 - 28$, or 724, mmHg, and its temperature was $273 + 28$, or 301, K; so from the ideal-gas law we have

$$n = \frac{PV}{RT} = \frac{724 \ \text{mmHg} \left(\frac{1 \ \text{atm}}{760 \ \text{mmHg}}\right)(0.145 \ \text{liter})}{(0.0821 \ \text{liter atm K}^{-1} \ \text{mol}^{-1})(301 \ \text{K})} = 5.6 \times 10^{-3} \ \text{mol}$$

The cathode half-reaction is

$$2e^- + 2H^+ \longrightarrow H_2(g)$$

Since *two* faradays are required to produce *one* mole of H_2, the number of faradays passed was

$$5.6 \times 10^{-3} \ \text{mol} \ \frac{2 \ \mathscr{F}}{1 \ \text{mol } H_2} = 1.1 \times 10^{-2} \ \mathscr{F}$$

In coulombs this is

$$1.1 \times 10^{-2} \ \mathscr{F} \frac{9.65 \times 10^4 \ \text{C}}{1 \ \mathscr{F}} = 1.1 \times 10^3 \text{C}$$

Because this quantity of electricity passed through the cell in 35 min, the average number of coulombs *per second* was

$$\frac{1.1 \times 10^3 \ \text{C}}{35 \ \text{min}(60 \ \text{s min}^{-1})} = 0.52 \ \text{C s}^{-1}$$

This is 0.52 A, because $1 \ \text{C s}^{-1}$ is 1 A. ●

19-3
Standard electrode
potentials   Because the voltage associated with a given reaction gives a measure of that reaction's tendency to take place, it would be useful to have a set of voltages for different half-reactions. These could then be combined to give the voltages for a large number of redox reactions. Unfortunately, the voltage produced by a single

electrode in a cell cannot be measured directly. The way around the problem is to choose one reference electrode and arbitrarily assign some voltage to it. Then voltages can be assigned to any other electrode merely by measuring the voltage produced by a cell utilizing the electrode in question plus the reference electrode. The reference electrode chosen by international agreement is the *standard hydrogen electrode*.

The standard hydrogen electrode

A *standard* electrode is one in which all reactants and products of the electrode half-reaction are in their standard states (Sec. 18-3). The standard state for an ion in solution is the one for which the activity (Sec. 18-6) of the ion is defined as being unity. This is the ion at a $1\ M$ concentration in an ideal solution. As usual, we will approximate activities by concentrations, so that the standard state of an ion becomes, effectively, the ion at a $1\ M$ concentration. Thus, the *standard hydrogen electrode*, often abbreviated s.h.e., is

$$\text{Pt}(s)\,|\,\text{H}_2(g,\ 1\ \text{atm})\,|\,\text{H}^+(1\ M)$$

as an anode, or

$$\text{H}^+(1\ M)\,|\,\text{H}_2(g,\ 1\ \text{atm})\,|\,\text{Pt}(s)$$

as a cathode. (See Fig. 19-3.)

The potential, or voltage, arbitrarily assigned to the s.h.e. is 0 V, whether it operates as an anode or cathode. Thus, using $\mathcal{E}$ to represent voltage, $\mathcal{E}^\circ_{\text{H}_2} = 0$, where the superscript means "standard." This means that when a cell is constructed with the s.h.e. plus some other standard electrode (all reactants and products in their standard states), the measured potential is assigned solely to the other electrode. This is best illustrated by examples.

● EXAMPLE 19-3

Problem

A standard copper–copper ion electrode is combined with a standard hydrogen electrode to make a cell. The cell voltage is measured as 0.34 V at 25°C, and electrons are found to enter the external circuit from the hydrogen electrode. What is the potential for the standard copper–copper ion electrode?

Solution

Since the electrons leave the cell from the hydrogen electrode, it must be the anode. The cell diagram and reactions are, therefore,

$$\text{Pt}(s)\,|\,\text{H}_2(g)\,|\,\text{H}^+\,\|\,\text{Cu}^{2+}\,|\,\text{Cu}(s)$$

Anode:	$\text{H}_2(g) \longrightarrow 2\text{H}^+ + 2e^-$	$\mathcal{E}^\circ = 0$	(defined)
Cathode:	$2e^- + \text{Cu}^{2+} \longrightarrow \text{Cu}(s)$	$\mathcal{E}^\circ = ?$	
Cell:	$\text{H}_2(g) + \text{Cu}^{2+} \longrightarrow 2\text{H}^+ + \text{Cu}(s)$	$\mathcal{E}^\circ = 0.34\ \text{V}$	(measured)

Since $\mathcal{E}^\circ_{\text{cell}} = \mathcal{E}^\circ_{\text{anode}} + \mathcal{E}^\circ_{\text{cathode}}$, clearly

$$\mathcal{E}^\circ_{\text{Cu}} = \mathcal{E}^\circ_{\text{cell}} - \mathcal{E}^\circ_{\text{H}_2} = 0.34 - 0 = 0.34\ \text{V} \quad ●$$

Since an anode reaction is an oxidation, the potential produced at such an electrode is called an *oxidation potential*. Similarly, the potential produced at a cathode is called a *reduction potential*. Either oxidation potentials or reduction potentials could be assembled in a table, but by international agreement the latter are tabulated, as *standard reduction potentials*.

● **EXAMPLE 19-4**

Problem In a galvanic cell consisting of a standard hydrogen and a standard zinc–zinc ion electrode the measured cell voltage was found to be 0.76 V at 25°C. If the hydrogen electrode is the cathode, find the standard reduction potential for

$$2e^- + Zn^{2+} \longrightarrow Zn(s)$$

Solution Since the cathode in this cell is the hydrogen electrode, the cell diagram and reactions are

$$Zn(s) \,|\, Zn^{2+} \,\|\, H^+ \,|\, H_2(g) \,|\, Pt(s)$$

Anode:	$Zn(s) \longrightarrow Zn^{2+} + 2e^-$	$\mathcal{E}° = ?$
Cathode:	$2e^- + 2H^+ \longrightarrow H_2(g)$	$\mathcal{E}° = 0$
Cell:	$Zn(s) + 2H^+ \longrightarrow Zn^{2+} + H_2(g)$	$\mathcal{E}° = 0.76 \text{ V}$

Thus,

$$\mathcal{E}°_{Zn} = \mathcal{E}°_{cell} - \mathcal{E}°_{H_2} = 0.76 - 0 = 0.76 \text{ V}$$

Note, however, that this is the value for the half-reaction occurring as an *oxidation*. To find the voltage for the reverse reaction, a reduction, we need to change the algebraic sign. (Reversing the direction of a half- or cell reaction always changes the sign of its voltage.) Finally, then,

$$2e^- + Zn^{2+} \longrightarrow Zn(s) \qquad \mathcal{E}° = -0.76 \text{ V} \quad ●$$

After the standard potential for any electrode has been determined, that electrode can be used with another to find its potential. This is how the electrode potentials in Table 19-1 were compiled. (A larger table is in App. I.) Notice that each half-reaction is written as a reduction. To find the potential for the reverse half-reaction, the oxidation, just change the sign of the given voltage.

A table of standard reduction potentials can be used for (1) predicting the

Table 19-1
Standard reduction potentials at 25°C

Half-reaction	$\mathcal{E}°$, V
$e^- + Li^+ \rightleftharpoons Li(s)$	-3.05
$2e^- + Ca^{2+} \rightleftharpoons Ca(s)$	-2.76
$e^- + Na^+ \rightleftharpoons Na(s)$	-2.71
$2e^- + Mg^{2+} \rightleftharpoons Mg(s)$	-2.38
$3e^- + Al^{3+} \rightleftharpoons Al(s)$	-1.67
$2e^- + Zn^{2+} \rightleftharpoons Zn(s)$	-0.76
$3e^- + Cr^{3+} \rightleftharpoons Cr(s)$	-0.74
$2e^- + Fe^{2+} \rightleftharpoons Fe(s)$	-0.44
$e^- + Cr^{3+} \rightleftharpoons Cr^{2+}$	-0.41
$2e^- + Sn^{2+} \rightleftharpoons Sn(s)$	-0.14
$2e^- + 2H^+ \rightleftharpoons H_2(g)$	0.00
$2e^- + Sn^{4+} \rightleftharpoons Sn^{2+}$	$+0.15$
$2e^- + Cu^{2+} \rightleftharpoons Cu(s)$	$+0.34$
$2e^- + I_2 \rightleftharpoons 2I^-$	$+0.54$
$e^- + Fe^{3+} \rightleftharpoons Fe^{2+}$	$+0.77$
$e^- + Ag^+ \rightleftharpoons Ag(s)$	$+0.80$
$3e^- + NO_3^- + 4H^+ \rightleftharpoons NO(g) + 2H_2O$	$+0.96$
$2e^- + Br_2 \rightleftharpoons 2Br^-$	$+1.09$
$4e^- + O_2(g) + 4H^+ \rightleftharpoons 2H_2O$	$+1.23$
$2e^- + Cl_2(g) \rightleftharpoons 2Cl^-$	$+1.36$
$5e^- + MnO_4^- + 8H^+ \rightleftharpoons Mn^{2+} + 4H_2O$	$+1.51$
$2e^- + F_2(g) \rightleftharpoons 2F^-$	$+2.87$

Increasing oxidizing strength (species at left of double arrows)

Increasing reducing strength (species at right of double arrows)

voltage which a given standard galvanic cell would produce, (2) predicting the spontaneity of a given redox reaction, (3) comparing the relative strengths of oxidizing agents, and (4) comparing the relative strengths of reducing agents.

Predicting cell voltages is accomplished by adding the standard potentials for the half-reactions *as they occur in the cell*. It is thus necessary to add the potential for an oxidation half-reaction to that for the reduction half-reaction which is coupled with it.

● **EXAMPLE 19-5**

Problem   Calculate the voltage produced by a galvanic cell with the reaction

$$Ag^+ + Cr^{2+} \longrightarrow Ag(s) + Cr^{3+}$$

assuming that all ionic concentrations are 1 M.

Solution   First we separate the cell reaction into half-reactions:

Anode: $Cr^{2+} \longrightarrow Cr^{3+} + e^-$

Cathode: $e^- + Ag^+ \longrightarrow Ag(s)$

Then we assign a voltage to each using values from Table 19-1, and add:

Anode:	$Cr^{2+} \longrightarrow Cr^{3+} + e^-$	$\mathcal{E}° = +0.41$ V
Cathode:	$e^- + Ag^+ \longrightarrow Ag(s)$	$\mathcal{E}° = +0.80$ V
Cell:	$Ag^+ + Cr^{2+} \longrightarrow Ag(s) + Cr^{3+}$	$\mathcal{E}° = +1.21$ V

Notice that we have changed the sign on the chromium half-reaction because it is an *oxidation*. ●

● **EXAMPLE 19-6**

Problem   Find the standard voltage produced by the cell

$$Pt(s)\,|\,Fe^{2+},\ Fe^{3+}\,\|\,Cl^-\,|\,Cl_2(g)\,|\,Pt(s)$$

Solution   The half-reactions and their voltages (from Table 19-1) are

Anode:	$2 \times [Fe^{2+} \longrightarrow Fe^{3+} + e^-]$	$\mathcal{E}° = -0.77$ V
Cathode:	$2e^- + Cl_2(g) \longrightarrow 2Cl^-$	$\mathcal{E}° = +1.36$ V
Cell:	$2Fe^{2+} + Cl_2(g) \longrightarrow 2Fe^{3+} + 2Cl^-$	$\mathcal{E}° = +0.59$ V

The cell will produce 0.59 V. ●

Notice that although we had to multiply the anode half-reaction by 2 in the above cell, we did *not* similarly multiply its oxidation potential. *Voltages which cells produce are independent of the quantities of reactants and products.* This is true because the tendency for a cell reaction to occur depends on the nature and state of its reactants and products, not on how much reactant or product is present. For example, the copper cathode in a Daniell cell can be immersed in 50 or 500 ml of $CuSO_4$ solution at a given concentration, and the tendency for the cathode half-reaction to occur is unaffected.

As was described in Sec. 19-1, a voltmeter connected properly in the external circuit of a galvanic cell reads a positive voltage. In other words, *a positive voltage is associated with a spontaneous reaction.* Predicting the spontaneity of a redox reaction is accomplished by adding the appropriate oxidation and reduction potentials. A positive total indicates that the reaction is spontaneous; a negative total, that it is nonspontaneous.

● **EXAMPLE 19-7**

Problem   Predict whether or not the following reaction can occur (all reactants and products in their standard states):

$$Sn^{2+} + 2I^- \longrightarrow Sn(s) + I_2$$

Solution   Adding the oxidation potential to the reduction potential (Table 19-1), we have

Oxidation:	$2I^- \longrightarrow I_2 + 2e^-$	$\mathcal{E}° = -0.54$ V
Reduction:	$2e^- + Sn^{2+} \longrightarrow Sn(s)$	$\mathcal{E}° = -0.14$ V
Redox:	$Sn^{2+} + 2I^- \longrightarrow Sn(s) + I_2$	$\mathcal{E}° = -0.68$ V

Since the total is *negative*, the reaction is *nonspontaneous;* it cannot occur. (The reversed reaction

$$I_2 + Sn(s) \longrightarrow 2I^- + Sn^{2+}$$

is predicted to be spontaneous, because for it $\mathcal{E}°$ is $+0.68$ V.) ●

But it is not actually necessary to obtain the value of $\mathcal{E}°$ in order to predict reaction spontaneity. Look carefully at Table 19-1 again. Since each half-reaction is written as a reduction, reversing it changes the sign of its cell potential. In order for the sum of the oxidation and reduction potentials to be *positive,* the oxidation half-reaction (sign *changed*) must be *above* the reduction half-reaction in the table. This rule can be even further simplified. Since all chemical species on the left of the double arrows in the table are oxidizing agents (get reduced) and all those on the right are reducing agents (get oxidized), *in order for any oxidizing agent to react with a reducing agent, the oxidizing agent* (on the left) *must be below the reducing agent* (on the right). More briefly: species on a *lower-left–upper-right diagonal* react with each other.

● **EXAMPLE 19-8**

Problem   Will Sn^{2+} oxidize I^- [to form $Sn(s)$ and I_2]?

Solution Sn^{2+} (on the left in Table 19-1) is *not* below I^- (on the right). The reaction therefore cannot take place. (Compare with Example 19-7.) ●

Relative strengths of oxidizing and reducing agents can be found quickly from a table of standard reduction potentials. Since the voltages become more positive going down the table, the tendency for reduction to occur is stronger at the bottom than at the top. Thus the oxidizing agents (on the left) become *stronger* going *down* the table. Since the tendency for reduction increases going down the table, the tendency for the reverse reaction, an oxidation, decreases. Thus the reducing agents become *weaker* going *down* the table, and *stronger* going *up*. [*Examples:* Cr^{3+} is a stronger oxidizing agent than Zn^{2+}; $Zn(s)$ is a stronger reducing agent than $Cr(s)$.]

19-4
Free energy,
cell voltage,
and equilibrium

In Chap. 18 we showed that at constant temperature and pressure the value of ΔG for a reaction can be used to predict the reaction's spontaneity. In this chapter we have seen that the voltage associated with an oxidation-reduction reaction is a measure of the spontaneity of the reaction. You should suspect that there is an exact relationship between ΔG and $\mathcal{E}$.

Thermodynamics
and electrochemistry

Although we will not prove it here, the free-energy decrease for a process taking place at constant temperature and pressure is equal to the theoretical maximum amount of work, other than work of expansion, which can be done by the process. In the case of a reaction taking place in a galvanic cell the maximum *electrical* work $w_{max, elect}$ which can be done is equal to the voltage $\mathcal{E}$ produced by the cell times the amount of electrical charge Q passed by means of the external circuit through the work-producing device (a motor). In other words,

$$w_{max, elect} = \mathcal{E} \times Q$$

In the case of the Daniell cell reaction

$$Zn(s) + Cu^{2+} \longrightarrow Zn^{2+} + Cu(s)$$

when *1* mol of Cu is formed, *2* $\mathcal{F}$ of electrical charge (two moles of electrons) are transferred from the zinc anode through the external circuit to the cathode. Thus the theoretical maximum work which can be done by a Daniell cell is

$$w_{max, elect} = \mathcal{E} \times 2\,\mathcal{F}$$

In the general case if n equals the number of moles of electrons (faradays) transferred when the reaction is interpreted in terms of moles,

$$w_{max, elect} = \mathcal{E} \times n\mathcal{F}$$

As we stated above, this is equal to the free-energy decrease as the reaction takes place.

$$-\Delta G = w_{max, elect} = \mathcal{E} \times n\mathcal{F}$$

or

$$\Delta G = -n\mathcal{F}\mathcal{E}$$

Here $\mathcal{F}$, called *Faraday's constant* (Sec. 19-2), is the number of coulombs per faraday, $9.65 \times 10^4\,C\,\mathcal{F}^{-1}$.

This equation is the tremendously important "bridge" between the free-energy change of thermodynamics and the cell voltage of electrochemistry. It accounts for the fact that either quantity can be used to predict spontaneity for a redox reaction. In the particular case when reactants and products are in their standard states, the relationship becomes

$$\Delta G° = -n\mathcal{F}\mathcal{E}°$$

The following summarizes the relationships among the algebraic sign of ΔG, the sign of $\mathcal{E}$, and the spontaneity of redox reactions:

Reaction	ΔG	$\mathcal{E}$
Spontaneous	−	+
Equilibrium	0	0
Nonspontaneous	+	−

● EXAMPLE 19-9

Problem   Calculate ΔG° for the reaction

$$8H^+ + MnO_4^- + 5Ag(s) \longrightarrow Mn^{2+} + 5Ag^+ + 4H_2O$$

Solution   From Table 19-1 we have the $\mathcal{E}^\circ$ values for the half-reactions

$$5e^- + MnO_4^- + 8H^+ \longrightarrow Mn^{2+} + 4H_2O \qquad \mathcal{E}^\circ = +1.51\ V$$
$$5 \times [Ag(s) \longrightarrow Ag^+ + e^-] \qquad \mathcal{E}^\circ = -0.80\ V$$
$$\overline{8H^+ + MnO_4^- + 5Ag(s) \longrightarrow Mn^{2+} + 5Ag^+ + 4H_2O \qquad \mathcal{E}^\circ = +0.71\ V}$$

In this case n, the number of moles of electrons (faradays) transferred in the reaction as written, is 5, as can be seen from the half-reaction equations.

$$\Delta G^\circ = -n\mathcal{F}\mathcal{E}^\circ$$
$$= -(5\ \mathcal{F})(9.65 \times 10^4\ C\ \mathcal{F}^{-1})(0.71\ V) = -3.4 \times 10^5\ C\ V$$

Because a coulomb-volt is the same thing as a joule (App. B):

$$1\ C\ V = 1\ J$$

and so

$$\Delta G^\circ = -3.4 \times 10^5\ J \quad \text{or} \quad -3.4 \times 10^2\ kJ$$

(The reaction is spontaneous.) ●

The effect of concentration on cell voltage   The voltage produced by a galvanic cell is dependent upon concentrations of reactants and products, and this dependency can be predicted qualitatively by using Le Châtelier's principle. Consider, for example, the standard Daniell cell:

$$Zn(s)\,|\,Zn^{2+}\,\|\,Cu^{2+}\,|\,Cu(s)$$

At 25°C its voltage is 1.10 V. What would happen if the Zn^{2+}-ion concentration were reduced below 1 M? If we look at the anode reaction

$$Zn(s) \longrightarrow Zn^{2+} + 2e^-$$

or at the cell reaction

$$Zn(s) + Cu^{2+} \longrightarrow Zn^{2+} + Cu(s)$$

we can predict that according to Le Châtelier's principle reducing $[Zn^{2+}]$ will *increase* the tendency for the *forward* reaction to occur, and so we should see an increase in the $\mathcal{E}$ of the cell. This is indeed what happens. Similarly, a decrease in $[Cu^{2+}]$ in the Daniell cell *decreases* the tendency for the cathode reaction

$$2e^- + Cu^{2+} \longrightarrow Cu(s)$$

and hence for the cell reaction

$$Zn(s) + Cu^{2+} \longrightarrow Zn^{2+} + Cu(s)$$

to occur. The voltage observed from such a cell is therefore *lower* than the standard-state value, 1.10 V.

The Nernst equation   The dependence of cell voltage upon concentration can also be quantitatively described. In Sec. 18-6 we showed that the free-energy change, ΔG, of a reaction and its standard free-energy change, ΔG°, are related by

$$\Delta G = \Delta G^\circ + RT \ln Q$$

where Q is the mass-action expression for the reaction. We have seen that

$$\Delta G = -n\mathcal{F}\mathcal{E} \quad \text{and} \quad \Delta G° = -n\mathcal{F}\mathcal{E}°$$

Thus we have for a redox reaction

$$-n\mathcal{F}\mathcal{E} = -n\mathcal{F}\mathcal{E}° + RT \ln Q$$

or

$$\mathcal{E} = \mathcal{E}° - \frac{RT}{n\mathcal{F}} \ln Q$$

This *Nernst equation*, as it is called, is named after the German, Walther Nernst, who developed it in 1889. It can be simplified for use at 25°C by substituting

$$R = 8.314 \text{ J K}^{-1} \text{ mol}^{-1}$$

$$T = 298.2 \text{ K}$$

$$\mathcal{F} = 96{,}485 \text{ C mol}^{-1}$$

and

$$\ln = 2.303 \log$$

so that at 25°C

$$\mathcal{E} = \mathcal{E}° - \frac{0.0592}{n} \log Q$$

With the Nernst equation we can calculate the voltage produced by any cell, given $\mathcal{E}°$ values for its electrodes and the appropriate concentrations (partial pressures, in the case of gases).

● EXAMPLE 19-10

Problem   Calculate the voltage produced by the cell

$$Sn(s) \,|\, Sn^{2+} \,\|\, Ag^+ \,|\, Ag(s)$$

if $[Sn^{2+}] = 0.15 \ M$ and $[Ag^+] = 1.7 \ M$.

Solution   The cell reactions and $\mathcal{E}°$ values (Table 19-1) are

Anode:	$Sn(s) \longrightarrow Sn^{2+} + 2e^-$	$\mathcal{E}° = -0.15$ V
Cathode:	$[e^- + Ag^+ \longrightarrow Ag(s)] \times 2$	$\mathcal{E}° = +0.80$ V
Cell:	$Sn(s) + 2Ag^+ \longrightarrow Sn^{2+} + 2Ag(s)$	$\mathcal{E}° = +0.65$ V

The mass-action expression Q for this reaction is

$$Q = \frac{a_{Sn^{2+}}}{a^2_{Ag^+}}$$

Here the activities of the pure solid Sn and Ag have been omitted because they are unity. Assuming that the activities of the ions may be approximated by their concentrations, we get

$$Q = \frac{[Sn^{2+}]}{[Ag^+]^2}$$

The Nernst equation for this reaction is thus

$$\mathcal{E} = \mathcal{E}° - \frac{0.0592}{2} \log \frac{[Sn^{2+}]}{[Ag^+]^2}$$

Note: In this equation $n = 2$, because in the equation *as written*, the number of faradays of electricity passed when the reactants form the products is 2.

$$\mathcal{E} = +0.65 \text{ V} - \frac{0.0592 \text{ V}}{2} \log \frac{0.15}{(1.7)^2}$$

$$= +0.65 \text{ V} + 0.038 \text{ V} = +0.69 \text{ V}$$

The variation of the ionic concentrations from 1 M did not greatly affect the $\mathcal{E}$ for the cell, which is often the case. The fact that $\mathcal{E}$ is more positive then $\mathcal{E}°$ tells us that with these concentrations the cell reaction has a larger tendency to take place than with 1 M concentrations, which is consistent with a prediction made on the basis of Le Châtelier's principle. ●

Standard potentials and equilibrium constants

When a reacting system is at equilibrium, $\Delta G = 0$. Furthermore, when such a system is part of a galvanic cell, $\mathcal{E}$ for the cell is zero, since there is no net tendency for the reaction to proceed one way or the other. At equilibrium the Nernst equation becomes

$$0 = \mathcal{E}° - \frac{RT}{N\mathcal{F}} \ln K$$

since at equilibrium $Q = K$, the equilibrium constant.

Thus
$$\mathcal{E}° = \frac{RT}{n\mathcal{F}} \ln K$$

which at 25°C simplifies to $\quad \mathcal{E}° = \frac{0.0592}{n} \log K$

This allows us to calculate $\mathcal{E}°$ from K, or vice versa.

● **EXAMPLE 19-11**

Problem Calculate the value of the equilibrium constant for the reaction given in Example 19-10.

Solution We found that $\mathcal{E}°$ for the reaction was +0.65 V and $n = 2$; so

$$\mathcal{E}° = \frac{0.0592}{n} \log K$$

$$\log K = \frac{n\mathcal{E}°}{0.0592} = \frac{2(0.65)}{0.0592} = 22$$

$$K = 9.1 \times 10^{22} \qquad\qquad ●$$

19-5
The electrochemical measurement of pH

The Nernst equation provides a method of determining ionic concentrations (actually, activities) from galvanic-cell measurements. With the appropriate cell, it is the basis for the determination of the pH of a solution.

Measurement of pH with the hydrogen electrode

One of the earliest methods of precise pH measurement utilized the hydrogen electrode. This electrode was immersed in the solution whose pH was sought, and which was connected by means of a salt bridge to a reference electrode of known potential. The reference electrode often used was the *calomel electrode*[2]

$$Cl^- \,|\, Hg_2Cl_2(s) \,|\, Hg(l)$$

[2] *Calomel* is an old name for mercurous chloride, Hg_2Cl_2.

This is a metal-insoluble salt electrode for which the half-reaction is

$$2e^- + Hg_2Cl_2(s) \longrightarrow 2Hg(l) + 2Cl^-$$

The entire galvanic cell is diagrammed as

$$Pt(s)\,|\,H_2(g)\,|\,H^+(?\ M)\,\|\,Cl^-\,|\,Hg_2Cl_2(s)\,|\,Hg(l)$$

The electrode and cell reactions for this cell are

Anode:	$H_2(g) \longrightarrow 2H^+ + 2e^-$
Cathode:	$2e^- + Hg_2Cl_2(s) \longrightarrow 2Hg(l) + 2Cl^-$
Cell:	$Hg_2Cl_2(s) + H_2(g) \longrightarrow 2H^+ + 2Cl^- + 2Hg(l)$

The measured cell voltage is the sum of the anode and cathode voltages (neglecting liquid-junction potentials):

$$\mathcal{E}_{cell} = \mathcal{E}_{H_2} + \mathcal{E}_{calomel}$$

Now, the Nernst equation written for the hydrogen electrode alone is

$$\mathcal{E}_{H_2} = \mathcal{E}^\circ_{H_2} - \frac{0.0592}{2} \log \frac{[H^+]^2}{P_{H_2}}$$

But $\mathcal{E}^\circ_{H_2} = 0$; so

$$\mathcal{E}_{cell} = -\frac{0.0592}{2} \log \frac{[H^+]^2}{P_{H_2}} + \mathcal{E}_{calomel}$$

which, when solved for $\log\,[H^+]$ gives

$$\log\,[H^+] = -\frac{\mathcal{E}_{cell} - \mathcal{E}_{calomel}}{0.0592} + \frac{1}{2} \log P_{H_2}$$

Using the definition of pH we get

$$-\log\,[H^+] = pH = \frac{\mathcal{E}_{cell} - \mathcal{E}_{calomel}}{0.0592} - \frac{1}{2} \log P_{H_2}$$

Calomel reference electrodes can be prepared with a predictable, reproducible voltage of 0.280 V. With such an electrode the pH of a solution can be determined from the measured cell voltage and the partial pressure of H_2.

● **EXAMPLE 19-12**

Problem A hydrogen electrode ($P_{H_2} = 723$ mmHg) is immersed in a solution of unknown pH. A reference calomel electrode ($\mathcal{E} = 0.280$ V) is connected to the solution by means of a salt bridge, and the voltage produced by the resulting cell is measured as 0.537 V. What is the pH of the solution?

Solution

$$pH = \frac{\mathcal{E}_{cell} - \mathcal{E}_{calomel}}{0.0592} - \frac{1}{2} \log P_{H_2}$$

$$= \frac{0.537\ V - 0.280\ V}{0.0592\ V} - \frac{1}{2} \log \frac{723\ mmHg}{760\ mmHg} = 4.35 \quad ●$$

pH meters and the glass electrode   The relationship between pH and the voltage of the hydrogen electrode–calomel electrode cell can be written as

$$\text{pH} = \frac{\mathcal{E}_{\text{cell}}}{0.0592} - \left(\frac{\mathcal{E}_{\text{calomel}}}{0.0592} + \frac{1}{2}\log P_{\text{H}_2}\right)$$

The voltage of the calomel electrode is constant, because the salt bridge isolates this electrode from the solution of unknown pH, and so at constant P_{H_2}

$$\text{pH} = \frac{\mathcal{E}_{\text{cell}}}{0.0592} - (\text{constant})$$

This says that the pH of a solution is directly proportional to the measured voltage produced by this cell.

It has been found that a thin glass membrane separating two solutions of different pH develops a potential across itself which changes with the pH of one solution in exactly the same way that the potential of a hydrogen electrode changes with pH. Such a glass membrane is incorporated at the tip of a probe called a *glass electrode*, which today is used universally as one electrode of a *pH meter*. The other electrode is often a calomel electrode, and the entire cell is diagrammed

$$\text{Pt}(s)\,|\,\text{Ag}(s)\,|\,\text{AgCl}(s)\,|\,\text{HCl}(aq, 1\ M)\,|\,\text{glass}\,|\,\text{solution, pH} = ?\,\|\,\text{Cl}^-\,|\,\text{Hg}_2\text{Cl}_2(s)\,|\,\text{Hg}(l)$$

The pH of the tested solution is given as

$$\text{pH} = \frac{\mathcal{E}_{\text{cell}}}{0.0592} - (\text{constant})$$

Figure 19-8 shows a glass and a calomel electrode dipping into the solution to be tested. The electrodes are connected to a pH meter, which is an electronic voltmeter designed to draw essentially no current from the cell. The scale on the pH meter reads directly in pH units.

It is not possible to manufacture a glass electrode with a predetermined $\mathcal{E}^\circ$ value; so before use each pH meter and electrode assembly is standardized, that is,

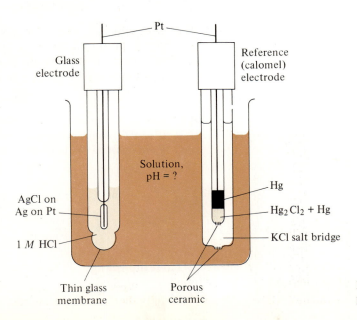

Figure 19-8
pH-meter electrodes.

adjusted so that the meter reads correctly. This is done by immersing the electrodes in a buffer of known pH and then adjusting the pH meter (by turning the appropriate knob) to read that pH. Then the assembly will correctly indicate the pH of any solution into which its electrodes are subsequently dipped. Today some pH meters have both their electrodes incorporated into a single probe for ease of use.

19-6 Commercial galvanic cells

Certain galvanic cells have industrial and domestic applications. These include cells which power portable radios, calculators, hearing aids, and other miniaturized devices. Two familiar galvanic cells are the flashlight battery and the automobile battery. (Originally the term *battery* meant two or more galvanic cells connected in series to produce a total voltage equal to a multiple of the single-cell voltage. Today battery is used to mean one or more cells. A flashlight battery is a single cell.) All of these cells are *energy-storage cells*. A second kind of galvanic cell is the *fuel cell,* in which anode and cathode materials are continuously replaced as they are used up in the production of electrical current.

Cells for energy storage

The common flashlight battery, also called a *dry cell,* or *Leclanché cell,* is an interesting example of a galvanic cell. It is schematically diagrammed in Fig. 19-9 and consists of a zinc can which serves as the anode, a central carbon rod which is the cathode, and a paste of MnO_2, carbon, NH_4Cl, and $ZnCl_2$ moistened with water. (The dry cell is not really dry.) The operation of the dry cell is complex and not entirely understood. Something like this happens, however: zinc is oxidized at the anode

$$Zn(s) \longrightarrow Zn^{2+} + 2e^-$$

At the cathode the MnO_2 is reduced. A suggested reaction is

$$e^- + NH_4^+ + MnO_2(s) \longrightarrow MnO(OH)(s) + NH_3$$

The zinc ions produced at the anode apparently migrate to the cathode where they are complexed by the NH_3 molecules produced there. If too large a current is drawn from a dry cell, it "goes dead" prematurely, possibly due to the formation of an insulating layer of NH_3 gas around the cathode. With "rest" such dry cells will

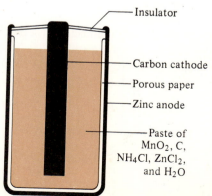

Insulator

Carbon cathode

Porous paper

Zinc anode

Paste of
MnO_2, C,
NH_4Cl, $ZnCl_2$,
and H_2O

Figure 19-9
The dry, or Leclanché, cell.

"rejuvenate" themselves, possibly a result of the migrating Zn^{2+} ions removing the interfering NH_3 molecules.

No electrical energy storage device has been more important than the *lead storage cell*. It is diagrammed as

$$Pb(s)\,|\,PbSO_4(s)\,|\,H^+,\ HSO_4^-\,|\,PbO_2(s)\,|\,Pb(s)$$

The anode of a fully charged cell consists of a frame of ordinary lead filled with some sponge-like lead. When it is oxidized, the product Pb^{2+} ions immediately precipitate as $PbSO_4$ which sticks to the lead frame. The anode half-reaction is

Anode: $Pb(s) + HSO_4^- \longrightarrow PbSO_4(s) + H^+ + 2e^-$

The cathode is another frame of lead, which in this case is filled with lead dioxide, PbO_2. The cathode half-reaction is

Cathode: $2e^- + PbO_2(s) + 3H^+ + HSO_4^- \longrightarrow PbSO_4(s) + 2H_2O$

As current is drawn from the cell, the overall reaction is

Cell: $Pb(s) + PbO_2(s) + 2H^+ + 2HSO_4^- \longrightarrow 2PbSO_4(s) + 2H_2O$

Thus solid $PbSO_4$ is produced at both electrodes, as the cell discharges. Simultaneously H^+ and HSO_4^- (the ions of sulfuric acid, H_2SO_4) are removed from the solution.

Not only can the lead storage cell produce large amounts of current, but it can be recharged. This is done by imposing on the cell a reverse voltage which is a little larger than that produced by the cell, forcing the electrons to flow *into* what was the anode and *out of* what was the cathode. This causes all reactions to reverse as the cell becomes an electrolytic cell, converting $PbSO_4$ to Pb and to PbO_2 at the respective electrodes. If a cell is overcharged, the water becomes electrolyzed, and the hydrogen and oxygen evolution degrade the electrode surfaces. This, together with the addition of impure water to replace that lost by evaporation, causes the $PbSO_4$ to drop from the electrodes. This in turn reduces the capacity of the cell and may eventually produce enough sludge on the bottom of the container to short out the cell internally.

The condition of a lead storage cell may be monitored by measuring the density of its H_2SO_4 solution. (This is what the service-station attendant's "specific-gravity tester" does.) In a completely charged cell the concentration and density of the H_2SO_4 solution are high; discharged, the cell's H_2SO_4 solution is more dilute and hence has a lower density. Since each lead storage cell produces about 2 V, a "12-V battery" has six cells connected in series. Figure 19-10 shows a schematic representation of a lead storage cell.

The *nickel-cadmium cell* has become very common recently. It powers everything from pocket calculators to cordless hedge trimmers. It is diagrammed as

$$Cd(s)\,|\,Cd(OH)_2(s)\,|\,OH^-\,|\,Ni(OH)_2(s)\,|\,NiO_2(s)$$

As the cell discharges the following changes occur:

Anode:	$Cd(s) + 2OH^- \longrightarrow Cd(OH)_2(s) + 2e^-$
Cathode:	$2e^- + NiO_2(s) + 2H_2O \longrightarrow Ni(OH)_2(s) + 2OH^-$
Cell:	$Cd(s) + NiO_2(s) + 2H_2O \longrightarrow Cd(OH)_2(s) + Ni(OH)_2(s)$

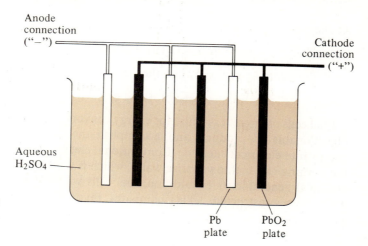

**Figure 19-10
Lead storage cell
(schematic).**

Like the lead storage cell, the nickel-cadmium cell is rechargeable. (Applying an opposing voltage greater than that produced by the cell reverses the above reactions.) Unlike the lead storage cell, it is light in weight and can easily be made in small sizes. The voltage of a nickel-cadmium cell stays essentially constant until the cell is almost discharged. This is true because ionic concentrations inside the cell do not change as the cell discharges.

Fuel cells   A galvanic cell in which the reactants are continuously fed into the cell as the cell produces electrical energy is called a *fuel cell*. Figure 19-11 shows a schematic diagram of a fuel cell which consumes hydrogen and oxygen gases as it operates. Each electrode is made of porous carbon impregnated with a catalyst, such as platinum, silver, or certain transition-metal oxides. At the anode of this fuel cell hydrogen is oxidized:

Anode: $$H_2(g) + 2OH^- \longrightarrow 2H_2O + 2e^-$$

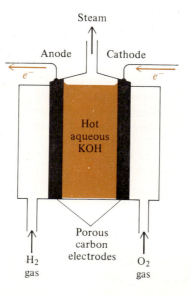

**Figure 19-11
Fuel cell.**

At the cathode oxygen is reduced

Cathode: $4e^- + O_2(g) + 2H_2O \longrightarrow 4OH^-$

Thus the overall reaction is just the conversion of H_2 and O_2 to water:

Cell: $2H_2(g) + O_2(g) \longrightarrow 2H_2O$

Fuel cells offer great promise for energy conversion in the future. They have no objectionable by-products and accomplish the direct conversion of chemical energy into electrical energy. (Conventional electrical power production requires the burning of a fuel to vaporize water, which is used to run turbines connected to electrical generators.) In addition, fuel cells with efficiencies above 80 percent have been constructed. This compares favorably with a high-efficiency (40 percent) steam-turbine generator.

SUMMARY

In this chapter we have considered the interconversion of chemical and electrical energy. A device for accomplishing this interconversion is called an *electrochemical cell,* of which there are two kinds: the *galvanic cell* and the *electrolytic cell.* A galvanic cell takes advantage of the negative free-energy change of a reaction to produce electrical energy. ΔG for an electrolytic cell reaction is positive; so electrical energy must be used to force the reaction to occur. Electrolytic cells are often used to produce decomposition of a compound into its elements. Galvanic and electrolytic cells each employ a redox reaction which is physically "split" so that the oxidation takes place at one electrode, the *anode,* while reduction takes place at the other, the *cathode.*

In the process of electrolysis the products depend upon which cell constituents are most readily and rapidly oxidized (at the anode) and reduced (at the cathode). The quantities of these products formed depend upon their equivalent weights and the quantity of electricity used, a relationship summarized by *Faraday's laws of electrolysis.*

The description of galvanic cells is aided by writing their *cell diagrams.* Because there are several types of electrodes which can be used in galvanic cells, many varieties of cells are possible. Some galvanic cells, *"batteries,"* have practical application as devices for storing energy. Others, *fuel cells,* can accept a continuous feed of reactants, producing electric energy continuously and rejecting products.

Each galvanic cell produces a voltage, or potential, which is the sum of the potentials of its single electrodes, plus, possibly, a *liquid-junction potential.* (The liquid-junction potential can be minimized through the use of a *salt bridge.*) The voltage produced by a cell measures the spontaneity of its cell reaction. *Standard electrode potentials have been tabulated;* in each of these, the potential measures the tendency for a *reduction* reaction to occur relative to the tendency for H^+ to be reduced to H_2. Here, *standard* implies standard states for all reactants and products. The *Nernst equation* permits calculation of cell voltages when reactants and products are not in their standard states.

Galvanic cells may be employed to measure concentrations of ions in solution through the Nernst-equation relationship. pH may be measured by means of the hydrogen electrode, and, more conveniently, the *glass electrode* and *pH meter.*

KEY TERMS

Anode (Sec. 19-1)
Battery (Sec. 19-6)
Cathode (Sec. 19-1)
Cell diagram (Sec. 19-1)
Coulomb (Sec. 19-2)

Daniell cell (Sec. 19-1)
Decomposition potential (Sec. 19-2)
Dry cell (Sec. 19-6)
Electrochemical cell (Sec. 19-1)
Electrochemistry (Sec. 19-1)

Electrode (Sec. 19-1)
Electrolysis (Sec. 19-2)
Electrolytic cell (Sec. 19-2)
Faraday ($\mathcal{F}$) (Sec. 19-2)
Faraday's laws (Sec. 19-2)
Fuel cell (Sec. 19-6)
Galvanic cell (Sec. 19-1)
Gas-ion electrode (Sec. 19-1)
Glass electrode (Sec. 19-5)
Ion current (Sec. 19-2)
Inert, "oxidation-reduction" electrode (Sec. 19-1)
Lead storage cell (Sec. 19-6)
Leclanché cell (Sec. 19-6)
Liquid-junction potential (Sec. 19-1)

Metal-insoluble salt-anion electrode (Sec. 19-1)
Metal–metal ion electrode (Sec. 19-1)
Nernst equation (Sec. 19-4)
Nickel-cadmium cell (Sec. 19-6)
Oxidation potential (Sec. 19-3)
pH meter (Sec. 19-5)
Potential difference (Sec. 19-1)
Reduction potential (Sec. 19-3)
Reference electrode (Sec. 19-5)
Salt bridge (Sec. 19-1)
Standard electrode potential (Sec. 19-3)
Standard hydrogen electrode (s.h.e) (Sec. 19-3)
Voltage (Sec. 19-1)

QUESTIONS AND PROBLEMS

Electrochemical cells

19-1 Distinguish clearly between a galvanic and an electrolytic cell. Comment on the sign of ΔG and $\mathcal{E}$ for the cell reaction for each. Can a given cell serve both as a galvanic and as an electrolytic cell? Explain.

19-2 How is the anode distinguished from the cathode in an electrochemical cell?

19-3 In what way can the electrode designations "+" and "−" be ambiguous in a galvanic cell?

19-4 If you were going to electrolyze water using a lead storage battery as your source of electricity, would you connect the electrode at which you wished hydrogen gas to be formed to the "+" or the "−" terminal of the battery? Explain your reasoning.

Electrolysis

19-5 A source of electricity has one terminal connected to a wire immersed in a solution of H_2SO_4 in a beaker and the other terminal connected to a wire immersed in a solution of H_2SO_4 in a second separate beaker. Explain why no electrolysis takes place.

19-6 Describe all the differences you can think of between metallic and electrolytic conduction.

19-7 Can pure water be electrolyzed? Why is it that an electrolyte is usually added to water before electrolyzing it? What characteristics should such an electrolyte have?

19-8 Contrast the electrolysis of molten NaCl with that of aqueous NaCl. Why are the products different?

19-9 A solution of silver nitrate, $AgNO_3$, is electrolyzed between inert electrodes. Sketch a labeled diagram of the electrolytic cell, and show the direction of electron and ion currents. Write the anode and cathode half-reactions and the cell reaction if $Ag(s)$ and $O_2(g)$ are produced.

19-10 A dilute aqueous solution of NaCl is electrolyzed. At first the gas produced at the anode is pure Cl_2, later it is a mixture of Cl_2 and O_2, and still later it consists of O_2 only. Explain, using equations where appropriate.

19-11 A solution of H_2SO_4 is electrolyzed between inert electrodes. 149 ml of H_2 at 722 mmHg and 29°C are formed at the cathode during 30 min of electrolysis. How many milliliters of O_2 measured at the same temperature and pressure are formed simultaneously at the anode?

19-12 A solution of cupric sulfate, $CuSO_4$, is electrolyzed between inert electrodes. If 4.62 g of Cu is plated out, how many coulombs of electricity have been used?

19-13 How many coulombs of electricity are needed to plate out 35.0 g of gallium from a solution of gallium sulfate, $Ga_2(SO_4)_3$?

19-14 Suppose 4.69×10^3 C were passed through an electrolytic cell containing aqueous $CuSO_4$. How many grams of copper were plated out on the cathode?

19-15 A solution of silver nitrate, $AgNO_3$, is electrolyzed for 45.0 min using a current of 0.335 A. How many grams of silver are plated out?

19-16 How long would 2.25 A be needed to pass through a solution of $CuSO_4$ in order to deposit 50.0 g of copper metal?

19-17 A solution of nickel sulfate, $NiSO_4$, is electrolyzed for 1.50 h between inert electrodes. If 35.0 g of nickel metal is deposited, what was the average current?

19-18 Suppose 645 ml of a solution of $CuSO_4$ was electrolyzed between inert electrodes for 2.00 h using a current of 0.147 A. Assuming that only copper was plated out at the cathode, what was the final pH of the solution?

19-19 Suppose 235 ml of a solution of NaCl was electrolyzed for 75.0 min. If the pH of the final solution was 11.12, what average current was used?

Galvanic cells

19-20 Write the diagram for a cell which utilizes each of the following cell reactions:

(a) $H_2(g) + Cl_2(g) \longrightarrow 2H^+ + 2Cl^-$
(b) $Cl_2(g) + Cd(s) \longrightarrow Cd^{2+} + 2Cl^-$
(c) $2Ag^+ + Cu(s) \longrightarrow 2Ag(s) + Cu^{2+}$
(d) $2Fe^{2+} + Cl_2(g) \longrightarrow 2Fe^{3+} + 2Cl^-$
(e) $Zn(s) + 2AgCl(s) \longrightarrow Zn^{2+} + 2Cl^- + 2Ag(s)$
(f) $Ag^+ + Cl^- \longrightarrow AgCl(s)$

19-21 Sketch a labeled drawing of each of the cells diagrammed for Prob. 19-20. Indicate on your sketch the anode, cathode, and directions of electron and ion flow.

19-22 Write anode and cathode half-reactions and the overall cell reaction for each of the following galvanic cells:

(a) $Pt(s)|H_2(g)|H^+||Ag^+|Ag(s)$
(b) $Cu(s)|Cu^{2+}|Cl^-|Cl_2(g)|Pt(s)$
(c) $Pt(s)|H_2(g)|H^+, Cl^-|Cl_2(g)|Pt(s)$
(d) $Pt(s)|H_2(g)|H^+, Cl^-|AgCl(s)|Ag(s)$
(e) $Zn(s)|Zn^{2+}||Fe^{3+}, Fe^{2+}|Pt(s)$

19-23 What is a salt bridge? What functions can it perform in a galvanic cell? Can a salt bridge in a galvanic cell be replaced by a U-shaped piece of platinum wire? Explain.

19-24 In some modern versions of the Leclanché (dry) cell the locations of the anode and cathode are reversed; that is, the "can" is made of carbon and the rod in the center, of zinc. What advantage do such cells have? What problem do you see in their design?

19-25 In the discharge of the lead storage cell, HSO_4^- ions move to the *cathode* in order to make $PbSO_4$. Considering that they are *anions,* how do they manage to do this?

19-26 Show that as a lead storage cell discharges, its voltage slowly drops. Show that as a nickel-cadmium cell discharges, its voltage remains essentially constant.

19-27 What is a fuel cell? What kinds of substances can be used in fuel cells? Why do fuel cells show great promise for the future for generation of electricity?

Electrode potentials (Use Table 19-1, as needed.)

19-28 What is the standard potential for each of the following half-reactions?

(a) $e^- + Cr^{3+} \longrightarrow Cr^{2+}$
(b) $Cu(s) \longrightarrow Cu^{2+} + 2e^-$
(c) $2Cl^- \longrightarrow Cl_2(g) + 2e^-$
(d) $Cl^- \longrightarrow \frac{1}{2}Cl_2(g) + e^-$

19-29 Calculate the standard potential $\mathcal{E}°$ for each of the following redox reactions:

(a) $3Br_2 + 2Cr(s) \longrightarrow 2Cr^{3+} + 6Br^-$
(b) $Sn^{4+} + Sn(s) \longrightarrow 2Sn^{2+}$
(c) $4MnO_4^- + 12H^+ \longrightarrow 4Mn^{2+} + 5O_2 + 6H_2O$
(d) $8H^+ + 2NO_3^- + 3Cu(s) \longrightarrow$
$$2NO(g) + 3Cu^{2+} + 4H_2O$$

19-30 If all reactants and products are in their standard states, which of the following redox reactions are spontaneous?

(a) $Zn(s) + 2H^+ \longrightarrow Zn^{2+} + H_2(g)$
(b) $Cu(s) + 2H^+ \longrightarrow Cu^{2+} + H_2(g)$
(c) $I_2 + 2Fe^{2+} \longrightarrow 2I^- + 2Fe^{3+}$
(d) $Cr(s) + 3Fe^{3+} \longrightarrow Cr^{3+} + 3Fe^{2+}$
(e) $2Fe^{3+} + Fe(s) \longrightarrow 3Fe^{2+}$

19-31 Of each of the following pairs, choose the better oxidizing agent (assuming standard states for all reactants and products):

(a) Sn^{4+} or Sn^{2+}
(b) Zn^{2+} or Cu^{2+}
(c) $Cl_2(g)$ or Ag^+

19-32 Of each of the following pairs, choose the better reducing agent (assuming standard states for all reactants and products):

(a) Zn or Cu
(b) H_2 or Fe^{2+}
(c) Br^- or Cl^-

19-33 Calculate $\mathcal{E}°$ for each of the following cells:

(a) $Zn(s)|Zn^{2+}||I^-|I_2|Pt(s)$
(b) $Mg(s)|Mg^{2+}||Ag^+|Ag(s)$
(c) $Al(s)|Al^{3+}||H^+|H_2(g)|Pt(s)$
(d) $Ag(s)|Ag^+||Cl^-|Cl_2(g)|Pt(s)$

19-34 Calculate the voltage produced by a Daniell cell in which

$$[Cu^{2+}] = 0.25\ M \text{ and } [Zn^{2+}] = 0.010\ M.$$

19-35 The voltage produced by a Daniell cell is observed to be 1.22 V. If $[Cu^{2+}] = 1.0\ M$, what is $[Zn^{2+}]$ in the cell?

19-36 Calculate the voltage produced by the cell

$$Fe(s)|Fe^{2+}(0.65\ M)||H^+(0.10\ M)|H_2(1.4\ atm)|Pt(s)$$

19-37 Calculate the voltage produced by the cell

$$Cu(s)\,|\,Cu^{2+}(1.2\ M)\,\|\,Cl^-(0.080\ M)\,|\,Cl_2(1.0\ atm)\,|\,Pt(s)$$

19-38 Calculate the voltage produced by the cell

$$Cu(s)\,|\,CuCl_2(aq,1.2\ M)\,|\,Cl_2(1.0\ atm)\,|\,Pt(s)$$

19-39 Calculate the voltage produced by a Daniell cell in which $[Zn^{2+}] = [Cu^{2+}]$.

19-40 A voltage of 1.09 V was obtained from the cell

$$Sn(s)\,|\,Sn^{2+}\,\|\,Ag^+(0.50\ M)\,|\,Ag(s)$$

What was $[Sn^{2+}]$ in the cell?

19-41 Show that the Nernst equation gives the same ε for the reaction:

$$Zn(s) + 2H^+ \longrightarrow Zn^{2+} + H_2(g)$$

as it does for

$$\tfrac{1}{2}Zn(s) + H^+ \longrightarrow \tfrac{1}{2}Zn^{2+} + \tfrac{1}{2}H_2(g)$$

19-42 A hydrogen electrode ($P_{H_2} = 1.0$ atm), a salt bridge, and a calomel electrode ($\varepsilon = 0.280$ V) are used to determine the pH of a solution. If the combination gives a voltage of 0.790 V, what is the pH of the solution?

19-43 The combination of a glass electrode, salt bridge, and calomel electrode are used to determine the pH of a solution. When immersed in a buffer of pH = 7.00, the combination gives a potential of 0.433 V. When immersed in a second solution, the observed voltage is 0.620 V. What is the pH of the second solution?

Electrochemistry and thermodynamics

19-44 Calculate $\Delta G°$ for each of the reactions listed in Prob. 19-29.

19-45 Calculate $\Delta G°$ for each of the reactions listed in Prob. 19-30.

19-46 Calculate $\Delta G°$ for each of the cells listed in Prob. 19-33, assuming that 1 mol of reducing agent is consumed in each case.

19-47 Calculate ΔG for each of the following reactions:

(a) $F_2(1\ atm) + 2Fe^{2+}(0.020\ M) \longrightarrow$
$$2F^-(0.050\ M) + 2Fe^{3+}(0.10\ M)$$

(b) $Cu^{2+}(1.8\ M) + Zn(s) \longrightarrow Cu(s) + Zn^{2+}(0.14\ M)$

(c) $Cl_2(0.050\ atm) + Sn^{2+}(0.10\ M) \longrightarrow$
$$2Cl^-(0.60\ M) + Sn^{4+}(1.0\ M)$$

19-48 $\Delta G°$ for the reaction

$$Hg_2Br_2(s) + Ni(s) \longrightarrow 2Hg(l) + 2Br^- + Ni^{2+}$$

is -72.3 kJ. What voltage will a cell utilizing this reaction produce if the concentration of $NiBr_2$ in the cell is 0.45 M?

19-49 Calculate the value of the equilibrium constant for each of the following:

(a) $Mg^{2+} + H_2(g) \longrightarrow Mg(s) + 2H^+$

(b) $Br_2 + Cu(s) \longrightarrow 2Br^- + Cu^{2+}$

(c) $Fe(s) + 2Fe^{3+} \longrightarrow 3Fe^{2+}$

19-50 What is the value of the equilibrium constant for a reaction for which $\varepsilon° = 0$?

19-51 From $\varepsilon°$ values in Table 19-1, calculate the standard molar free energy of formation of liquid water.

19-52 What is the reduction potential for the hydrogen electrode at pH = 7.00 and 1.0 atm?

19-53 Calculate the standard reduction potential for

$$2e^- + 2H_2O \longrightarrow H_2(g) + 2OH^-$$

19-54 Given the following standard reduction potentials:

$$Ag^+ + e^- \longrightarrow Ag(s) \qquad \varepsilon° = +0.7991\ V$$
$$AgCl(s) + e^- \longrightarrow Ag(s) + Cl^- \qquad \varepsilon° = +0.2225\ V$$

calculate the value of the solubility product for silver chloride.

20

THE NONMETALS

TO THE STUDENT

With this chapter we begin a brief survey of *descriptive chemistry,* the study of the properties and reactions of the elements and their compounds. It would certainly be convenient if we could predict all chemical behavior from theoretical concepts and principles. Then we would never need to memorize that silver chloride is a white solid which is insoluble in water, but soluble in ammonia, for example. But the variations in observed chemical properties are so extensive that it is doubtful that we will ever reach the stage where all chemical behavior can be predicted from so-called first principles; thus the need for this chapter, the first of several on descriptive chemistry.

The changed emphasis in this chapter should not be disconcerting to you. Although you will need to learn (memorize!) more chemical facts than in previous chapters, you will also find that the theoretical foundation laid in earlier chapters will help you organize, remember, and use these facts. And, of course, be ready to *refresh your memory by referring back to these chapters whenever necessary.*

**20-1
Inorganic
nomenclature**

Chemical names and formulas are not new to you; we have been using them all along. Furthermore, we have mentioned a few of the rules of compound naming and formula writing from time to time as it seemed appropriate. At this time we will take a closer look at inorganic chemical nomenclature, and in Chap. 23 we will consider how organic compounds are named.[1]

Elements

The names of the elements vary somewhat from one language to another, but chemical symbols are almost universal. Each symbol consists of one or two letters taken from the name of the element (usually in English or Latin). The first letter of a two-letter symbol is capitalized, but the second is lowercase.

Elements which can exist as two or more molecular allotropes may be system-

---

[1] A somewhat more extensive discussion of nomenclature is found in App. C.

atically named with a prefix indicating the number of atoms per molecule. The prefixes and the numbers they represent are

mono	1	penta	5	nona or ennea	9
di	2	hexa	6	deca	10
tri	3	hepta	7	undeca or hendeca	11
tetra	4	octa	8	dodeca	12

(The alternates given for 9 and 11 are derived from the Latin and Greek, respectively.) Some examples of the use of these prefixes are

Formula	Systematic name	Common name
O_2	Dioxygen	Oxygen
O_3	Trioxygen	Ozone
P_4	Tetraphosphorus	White phosphorus
S_8	Octasulfur	Sulfur

Cations   SIMPLE CATIONS. When an element shows only one simple cationic form, the name of the cation is the same as the name of the element.

Examples:
Na^+ sodium ion
Ca^{2+},   calcium ion
Al^{3+},   aluminum ion

When an element can form two relatively common simple cations (with two different respective oxidation states), each ion should be named so as to differentiate it from the other. There are two ways to do this; these are the *ous-ic* system and the *Stock* system. The *ous-ic* system uses the suffixes *-ous* and *-ic* attached to the root of the element name to indicate, respectively, the *lower* and *higher* oxidation state. The root is usually formed by dropping *-um* or *-ium* from the element name in English or, sometimes, Latin. (If the element name does not end with *-um* or *-ium,* the last syllable is most commonly dropped to form the root.) Some examples of ions named this way are given in Table 20-1.

The second way of naming simple cations is by using the *Stock* system. In this system the oxidation state of the element is indicated by means of a Roman numeral immediately after the name. The ions of Table 20-1 are shown renamed according to the Stock system in Table 20-2.

The Roman numerals of the Stock system are often used in formulas to indicate oxidation state without specifying exact composition. Thus Cr(III) might mean Cr^{3+}, $CrCl^{2+}$, $Cr(OH)_2^+$, $CrCl_4^-$, or any combination of these or other ions in which chromium is in the $+3$ oxidation state.

The Stock system is generally preferred over the *ous-ic* system, because it affords less chance for misunderstanding. The *ous-ic* system is, however, commonly accepted and is a good system if used carefully. Neither system is used when a

Table 20-1
The *ous-ic* system of
naming cations

Element		Name and formula of ion	
English name	**Latin name**	**Lower oxidation state**	**Higher oxidation state**
Copper	Cuprum	Cu^+, cuprous ion	Cu^{2+}, cupric ion
Tin	Stannum	Sn^{2+}, stannous ion	Sn^{4+}, stannic ion
Chromium	—	Cr^{2+}, chromous ion	Cr^{3+}, chromic ion
Iron	Ferrum	Fe^{2+}, ferrous ion	Fe^{3+}, ferric ion
Cobalt	—	Co^{2+}, cobaltous ion	Co^{3+}, cobaltic ion

simple name is completely unambiguous: "the sodium ion" means Na^+ to everybody, for example.

POLYATOMIC CATIONS. Cations with more than one atom are named variously, depending on the type of cation concerned. For example, the diatomic cation Hg_2^{2+} is called either the *mercurous ion* or the *mercury(I) ion,* just as if it were monatomic. Cations which consist of one or more oxygens bonded to another atom are named with a *-yl* suffix.

Examples:
UO_2^{2+}, the uranyl(VI) ion
NO^+, the nitrosyl ion

Some complex cations are universally named (in English) in a traditional, nonsystematic way. The ammonium ion NH_4^+ is a good example of this. The IUPAC name for H_3O^+ is the *oxonium ion,* but in the United States abandonment of the traditional *hydronium ion* has been slow.

Cations (and anions) which are typical complexes consisting of a central atom surrounded by ligands will be discussed in Chap. 22.

Anions   SIMPLE ANIONS. Monatomic anions are named by adding the suffix *-ide* to the root of the element name, where the root usually consists of the first syllable of the element's name.

Examples:
O^{2-},   oxide ion Cl^-,   chloride ion
N^{3-},   nitride ion S^{2-},   sulfide ion

POLYATOMIC ANIONS. A few polyatomic anions are also named with an *-ide* suffix.

Examples:
CN^-,   cyanide ion OH^-,   hydroxide ion
O_2^{2-},   peroxide ion S_2^{2-},   disulfide ion

Table 20-2
Stock system of
naming cations

Element	Name and formula of ion	
	Lower oxidation state	**Higher oxidation state**
Copper	Cu^+, copper(I) ion	Cu^{2+}, copper(II) ion
Tin	Sn^{2+}, tin(II) ion	Sn^{4+}, tin(IV) ion
Chromium	Cr^{2+}, chromium(II) ion	Cr^{3+}, chromium(III) ion
Iron	Fe^{2+}, iron(II) ion	Fe^{3+}, iron(III) ion
Cobalt	Co^{2+}, cobalt(II) ion	Co^{3+}, cobalt(III) ion

OXOANIONS. Anions of oxoacids can be named in a systematic way by considering them as typical complex ions in which oxide ions are ligands. In practice, however, traditional names for these ions are most commonly used. When the central, or principal, atom commonly occurs in only one oxidation state in an oxoanion, the suffix *-ate* is used.

Examples:
CO_3^{2-}, carbonate ion
SiO_4^{4-}, silicate ion

The suffixes *-ite* and *-ate* are used to distinguish between two oxoanions having the same central atom but in two different oxidation states. *-ite* signifies the lower and *-ate* the higher state.

Examples:
NO_2^-, nitrite ion NO_3^-, nitrate ion
SO_3^{2-}, sulfite ion SO_4^{2-}, sulfate ion
AsO_3^{3-}, arsenite ion AsO_4^{3-}, arsenate ion

Up to four different oxidation states can be differentiated in oxoanions using the following system: if the oxidation state of the central atom is less than that in the *-ite* anion, use the *hypo——ite* prefix-suffix combination. If the oxidation state is greater than that in the *-ate* anion use the combination *per——ate*. Some examples are

ClO^-, hypochlorite ion ClO_3^-, chlorate ion
ClO_2^-, chlorite ion ClO_4^-, perchlorate ion

Some anions contain one or more "acidic hydrogens." These anions are those formed when a polyprotic acid loses less than all its available protons. These anions are named with the word *hydrogen* and, if necessary, a prefix.

Examples:
HS^-, hydrogen sulfide ion HCO_3^-, hydrogen carbonate ion
HO_2^-, hydrogen peroxide ion $H_2PO_4^-$, dihydrogen phosphate ion

Note: There is an old but still useful system of naming such ions which have only one hydrogen. This is by using the prefix *bi-* to represent the hydrogen. Thus we have HCO_3^-, the *bicarbonate ion,* and HSO_4^-, the *bisulfate ion.* The origins of this system are very old, and the use of *bi* to mean a hydrogen atom is misleading and unfortunate. Nevertheless, the "bicarbonate ion" is an old friend to many chemists.

Compounds   SALTS. The name of a salt consists of the name of its cation followed by that of its anion. Examples of the formulas and names of some salts are given in Table 20-3.

HYDROXIDES. A hydroxide is a compound in which the hydroxo $(-OH)$ group is bonded to another atom, usually that of a metal. (When the hydroxo group is bonded to a nonmetal, the compound is acidic and so is usually named as an acid.) The nomenclature of hydroxides is analogous to that of salts. Some examples are given in Table 20-4.

Table 20-3
Names of some salts

Formula	Name(s)	Formula	Name(s)
KCl	Potassium chloride	NaH_2PO_4	Sodium dihydrogen phosphate
$CaCl_2$	Calcium chloride	Na_2HPO_4	Sodium hydrogen phosphate
$FeCl_2$	Iron(II) chloride	Na_3PO_4	Sodium phosphate
	or ferrous chloride	$Ca(OCl)_2$	Calcium hypochlorite
		Na_2O_2	Sodium peroxide
$FeCl_3$	Iron(III) chloride	Na_2CO_3	Sodium carbonate
	or ferric chloride	$NaHCO_3$	Sodium hydrogen carbonate (sodium bicarbonate)
$CuSO_4$	Copper(II) sulfate	$NaBrO_3$	Sodium bromate
	or cupric sulfate	$Ca(CN)_2$	Calcium cyanide
		$Al(NO_2)_3$	Aluminum nitrite

BINARY ACIDS. Compounds which are *binary Arrhenius acids* may be named by using the word *hydrogen,* followed by the anion name.

Examples:
HCl,   hydrogen chloride
H_2S,   hydrogen sulfide

A few acids which are not binary (and which are not oxoacids) are named similarly.   Thus HCN is called *hydrogen cyanide.*

Binary acids in *aqueous solution* are often named by using the prefix *hydro-* and the suffix *-ic* with the root of the anion name.

Examples:
HCl,   hydrochloric acid
HF,   hydrofluoric acid

This method is most commonly used for the binary halogen acids in solution.

OXOACIDS. In these compounds the available proton is bonded to an oxygen, and so they can be classed as acidic hydroxo compounds. ¦ They are almost never named following the pattern "hydrogen + anion name." (Thus H_2SO_4 is rarely called hydrogen sulfate.)   Instead, they are named by deleting *-ite* or *-ate* from the anion name and adding *-ous acid* or *-ic acid,* respectively.

Examples:
HClO,   hypochlorous acid $HClO_3$,   chloric acid
$HClO_2$,   chlorous acid $HClO_4$,   perchloric acid

Some acids and alternative ways of naming them are given in Table 20-5.

Other inorganic compounds

Other compounds are named by using the name of the more electropositive element(s) first, followed by the names of electronegative atoms, their names

Table 20-4
Names of some hydroxides

Formula	Name(s)
KOH	Potassium hydroxide
$Ca(OH)_2$	Calcium hydroxide
$Al(OH)_3$	Aluminum hydroxide
$Fe(OH)_2$	Iron(II) hydroxide or ferrous hydroxide
$Fe(OH)_3$	Iron(III) hydroxide or ferric hydroxide

Table 20-5
Names of some acids

Formula	Name(s)	Formula	Name(s)
HBr	Hydrogen bromide Hydrobromic acid (aq)	H_2SO_4	Sulfuric acid
HOBr	Hypobromous acid	H_2SO_3	Sulfurous acid
$HBrO_2$	Bromous acid	H_2CO_3	Carbonic acid
	(existence questionable)	H_3PO_4	Phosphoric acid
		H_3PO_3	Phosphorous acid
$HBrO_3$	Bromic acid	HNO_3	Nitric acid
$HBrO_4$	Perbromic acid	HNO_2	Nitrous acid

modified to end in *-ide*. Prefixes (*di-, tri-*, etc.) are used where useful or necessary. (*Mono-* is often omitted.)

Examples

SO_2,	sulfur dioxide	N_2O_5,	dinitrogen pentaoxide
SO_3,	sulfur trioxide	PCl_5,	phosphorous pentachloride
CO,	carbon monoxide	$BiOCl$,	bismuth oxide chloride
SiC,	silicon carbide	$KMgF_3$,	potassium magnesium fluoride

The preceding are the basic rules for naming inorganic compounds. We will augment them with others from time to time in this and the following chapters. Finally, it should be pointed out that certain compounds, water and ammonia, for instance, are customarily referred to only by their common names.

ADDED COMMENT

Learning the names of polyatomic anions seems to give some students considerable trouble. For those of you with that problem the listing in Table 20-6 is provided for future reference. Also: be prepared to refer to App. C (Chemical Nomenclature) as needed.

Table 20-6
Some common anions

Anion	Name	Anion	Name
CN^-	Cyanide	OH^-	Hydroxide
CO_3^{2-}	Carbonate (bicarbonate)	PO_4^{3-}	Phosphate
		HPO_4^{2-}	Monohydrogen phosphate
HCO_3^-	Hydrogen carbonate	$H_2PO_4^-$	Dihydrogen phosphate
$C_2O_4^{2-}$	Oxalate	SCN^-	Thiocyanate
$HC_2O_4^-$	Hydrogen oxalate	SO_3^{2-}	Sulfite
	(binoxalate)	HSO_3^-	Hydrogen sulfite
CrO_4^{2-}	Chromate		(bisulfite)
$Cr_2O_7^{2-}$	Dichromate	SO_4^{2-}	Sulfate
MnO_4^-	Permanganate	HSO_4^-	Hydrogen sulfate
NO_2^-	Nitrite		(bisulfate)
NO_3^-	Nitrate	$S_2O_3^{2-}$	Thiosulfate

Also

XO^-	Hypo*hal*ite	where X represents F, Cl, Br, or I, and *-hal-*
XO_2^-	*Hal*ite	represents *-fluor-, -chlor-, -brom-,* and *-iod-*, re-
XO_3^-	*Hal*ate	spectively. (The existence of some of these ions
XO_4^-	Per*hal*ate	is uncertain.)

20-2
Hydrogen

Hydrogen is a comparatively abundant element. In the crust of the earth, which we mean here to include the lithosphere, hydrosphere, and atmosphere, of all the elements hydrogen ranks third (after oxygen and silicon) in atomic percent and ninth in percent by mass. It is believed that hydrogen is the most abundant element in the universe. On earth virtually all hydrogen is combined, much of it with oxygen as water. The hydrogen molecule is so light, that when it is released, it rapidly rises to the upper levels of the atmosphere, where it gradually bleeds off into space.

The element

Hydrogen exists uncombined as a diatomic molecule. This is a far more stable state than that of the uncombined atoms:

$$2H(g) \longrightarrow H_2(g) \qquad \Delta G^\circ = -407 \text{ kJ}$$

ΔH° for this reaction is also highly negative, -436 kJ, and as a result atomic hydrogen can be produced only at very high temperatures, 2000 to 3000 K, or by means of an electric arc.

Hydrogen occurs naturally as a mixture of the three isotopes *protium* ($_1^1H$), *deuterium* ($_1^2H$), and *tritium* ($_1^3H$). Deuterium is also symbolized as D and has been called *heavy hydrogen,* while tritium (T) is occasionally called *heavy heavy hydrogen.* The masses and abundances of these isotopes are given in Table 20-7. (Hydrogen is the only element whose isotopes have special names and symbols.) Tritium is radioactive and hence undergoes nuclear decay. As we will see in Chap. 24, such decay follows first-order kinetics. For tritium the decay involves the emission of a beta particle to leave a helium ($_2^3He$) nucleus; the half-life (Sec. 14-2) for this process is about 12 years. Tritium is believed to be formed in the upper atmosphere as a result of cosmic-ray activity.

Considerable study has been made of the differences in the properties of protium and deuterium, both as uncombined elements and in compounds. In no other element are the percentage mass differences between isotopes so large, and as a result the differences in physical and chemical properties are uniquely large for the isotopes of hydrogen. Thus the normal boiling point of H_2 is 20.4 K, that of HD is 22.5 K, and that of D_2 is 23.6 K. Compounds of hydrogen show similar *isotope effects.* The normal freezing point of H_2O is 0.0°C, but that of D_2O is 3.8°C, for example. Chemical isotope effects are also pronounced: at 25°C the pD of pure D_2O is 7.35, as compared to a pH of 7.00 for pure H_2O. Isotope effects are much smaller in other elements because percentage mass differences are much smaller.

There is some evidence that at extremely high pressures hydrogen can be converted into a metallic form with unusual properties. (Perhaps hydrogen is the first alkali metal, after all!)

Preparation
of hydrogen

The preparation of H_2 gas is usually a reduction from the +1 state. This reduction can be accomplished either electrolytically or chemically. *Electrolytic hydrogen* is the purest commercially available grade and is made by the electrolysis of water:

$$2H_2O \longrightarrow 2H_2(g) + O_2(g) \qquad \Delta G^\circ = 474 \text{ kJ}$$

Electrolytic hydrogen is relatively expensive because of the cost of the electrical energy necessary to make it.

Table 20-7
Isotopes of hydrogen

Name	Symbol	Mass, amu	Abundance, atomic %
Protium	$^{1}_{1}H$ or H	1.0078	99.985
Deuterium	$^{2}_{1}H$ or D	2.0141	0.015
Tritium	$^{3}_{1}H$ or T	3.0161	1×10^{-17}

Chemical reduction of water can be accomplished by means of a number of reducing agents. In pure water the reduction potential for

$$2e^- + 2H_2O \longrightarrow H_2(g) + 2OH^-$$

is in fact that for

$$2e^- + 2H^+(1.0 \times 10^{-7}\ M) \longrightarrow H_2(g)$$

which at 1 atm pressure is

$$\mathcal{E} = \mathcal{E}° - \frac{0.0592}{n} \log \frac{P_{H_2}}{[H^+]^2}$$

$$= 0 - \frac{0.0592}{2} \log \frac{1}{(1.0 \times 10^{-7})^2} = -0.41\ V$$

so that any reducing agent with an *oxidation* potential greater than 0.41 V should reduce water at a pH = 7.0. Theoretically, then, metals such as aluminum, zinc, and chromium should reduce water at a pH = 7.0:

$$Al(s) \longrightarrow Al^{3+} + 3e^- \qquad \mathcal{E}° = +1.67\ V$$

$$Zn(s) \longrightarrow Zn^{2+} + 2e^- \qquad \mathcal{E}° = +0.76\ V$$

$$Cr(s) \longrightarrow Cr^{3+} + 3e^- \qquad \mathcal{E}° = +0.74\ V$$

In practice none of the above metals is very effective at producing hydrogen from water, at least not at a pH of 7.0, because the rate is far too slow.

Some metals, such as sodium and calcium are effective at producing hydrogen from water:

$$Na(s) \longrightarrow Na^+ + e^- \qquad \mathcal{E}° = +2.71\ V$$

$$Ca(s) \longrightarrow Ca^{2+} + 2e^- \qquad \mathcal{E}° = +2.76\ V$$

The reaction of sodium (and of the other alkali metals, except for lithium) is rapid and often spectacular:

$$2Na(s) + 2H_2O \longrightarrow H_2(g) + 2Na^+ + 2OH^-$$

When a small piece of sodium is added to water, it floats (density = 0.97 g cm^{-3}), and the heat of reaction causes it to melt ($T_m = 98°C$), so that a little spherical globule of sodium can be seen to skitter and sputter its way across the surface of the water. The hydrogen formed often ignites with a pop. (Using a large lump of sodium sometimes results in a dangerous explosion.)

If the pH of water is lowered below 7.0 by the addition of an acid, the rate of reduction by a metal increases. Zinc, for example, reacts readily with a solution of hydrochloric or sulfuric acid:

$$Zn(s) + 2H^+ \longrightarrow Zn^{2+} + H_2(g)$$

Interestingly enough, some metals, such as zinc or aluminum, will produce hydrogen from basic solutions, as well:

$$Zn(s) + 2OH^- + 2H_2O \longrightarrow H_2(g) + Zn(OH)_4{}^{2-}$$

$$2Al(s) + 2OH^- + 6H_2O \longrightarrow 3H_2(g) + 2Al(OH)_4{}^-$$

These reactions take place because both zinc and aluminum can form hydroxo anions. (Another way of putting it is to say that $Zn(OH)_2$ and $Al(OH)_3$ are *amphoteric*.)

Gaseous water (steam) can be reduced at high temperatures by carbon or low-molecular-weight hydrocarbons such as methane, CH_4. With carbon the reaction is

$$C(s) + H_2O(g) \longrightarrow H_2(g) + CO(g)$$

The H_2–CO mixture is useful as an industrial fuel known as *water gas*. The CO can be removed from it by catalytically oxidizing the CO to CO_2 (using more steam) and then dissolving the CO_2 in water or a basic solution.

Hydrogen is also produced as a useful byproduct of a number of petroleum refining reactions.

Compounds of hydrogen

Hydrogen forms more compounds than does any other element. In these compounds it is assigned oxidation numbers of $+1$, or -1.

OXIDATION STATE $+1$. Most compounds of hydrogen are in the $+1$ oxidation state, but in none of these does the hydrogen exist as a simple $1+$ ion. Rather, it is covalently bonded to a more electronegative atom. This is illustrated in binary compounds such as HF, HCl, HBr, HI, H_2O, H_2S, NH_3, CH_4, etc. Each of these can be formed by direct combination of the elements.

In addition to binary compounds, ternary and other compounds of hydrogen are common. Many of these are hydroxo compounds, such as the acids HOCl, H_2SO_4, etc., the basic hydroxides such as NaOH, $Ca(OH)_2$, etc., and the acid salts such as NaH_2PO_4, $NaHCO_3$, etc. In each of these hydrogen is bonded to oxygen and so is considered to be in the $+1$ oxidation state.

Most of the countless compounds of carbon also have hydrogen in the $+1$ state. These include the hydrocarbons (Chap. 23) such as methane CH_4, ethane C_2H_6, propane C_3H_8, etc., as well as their derivatives.

OXIDATION STATE -1. Compounds in which hydrogen is bonded to a more electropositive element are known as *hydrides*.[2] A hydrogen atom can gain an electron to fill its $1s$ orbital, and the result is a *hydride ion*, H^-. This occurs with highly electropositive elements such as the alkali and alkaline-earth metals, except for Be and Mg. These hydrides can be prepared by direct combination of the elements at elevated temperatures. The reaction for the production of sodium hydride is

$$2Na(l) + H_2(g) \xrightarrow{400°C} 2NaH(s)$$

---

[2] There is, unfortunately, some disagreement over the definition of a hydride. Some books use *hydride* to mean any, or any binary, hydrogen compound, such as H_2O, NH_3, CH_4, etc. We will restrict our use of the word to those compounds in which H has a *negative* oxidation state.

Ionic hydrides have typical salt-like characteristics. They are white crystalline solids with crystal lattices composed of metal ions plus hydride ions. All the alkali–metal hydrides have the NaCl structure. When an ionic hydride is melted or dissolved in a molten alkali–metal halide and then electrolyzed, hydrogen gas is produced at the *anode:*

$$2H^- \longrightarrow H_2 + 2e^-$$

The hydride ion is an excellent reducing agent; it reduces water, for instance, to form H_2:

$$H^- + H_2O \longrightarrow H_2(g) + OH^-$$

Hydrogen can also achieve the -1 state by *sharing* an electron pair with a more electropositive atom. This occurs, for example, in binary covalent hydrides such as SiH_4, AsH_3, and SbH_3. Unlike the ionic hydrides these are volatile substances, because intermolecular attractions are only weak London forces. The covalent hydrides are often given common names: SiH_4 is called *silane,* AsH_3 is *arsine,* and SbH_3 is *stibine,* for example. These hydrides are poorer reducing agents than the ionic hydrides.

METALLIC HYDRIDES. Hydrogen forms an interesting series of compounds called the *metallic hydrides.* These are combinations of a transition metal and hydrogen, in which it is difficult to assign oxidation numbers at all. Some metallic hydrides, such as LaH_3, TiH_2, and AgH, are well-defined compounds with fixed stoichiometries; others seem to have variable compositions, giving rise to formulas such as $TaH_{2.76}$, $VH_{0.56}$, and $PdH_{0.62}$. In some of these the metal seems to merely *dissolve* hydrogen, with hydrogen *atoms* (probably) occupying interstitial positions in the host lattice. This is accompanied by a considerable expansion of the solid. In the formation of other interstitial hydrides an entirely new crystal structure is formed. The metallic hydrides have metallic luster and show either metallic conduction or semiconduction, sometimes depending upon the concentration of hydrogen atoms in the compound. They are often quite weak and brittle, a property thought to be caused by internal strains in the host crystal lattice. Such *hydrogen embrittlement* can be produced at the cathode of an electrolytic cell employing an aqueous medium. When metals are rapidly plated out of solution, small amounts of hydrogen from the reduction of water are produced. Some of this hydrogen apparently dissolves in the base metal, considerably weakening it. (Chrome-plated wire wheels for sports cars were once considered dangerous for this reason.)

**20-3
Oxygen**

Oxygen is the most abundant element on earth as measured either by atomic percent (55 percent) or mass percent (49 percent). It occurs uncombined in the atmosphere, combined with hydrogen in the oceans and combined with silicon, aluminum, and other elements in various rocks and minerals.

The element

Oxygen occurs naturally as a mixture of three stable isotopes: $^{16}_{8}O$, $^{17}_{8}O$, and $^{18}_{8}O$, of which the first is by far the most abundant (99.8 percent). Uncombined it occurs most commonly as the diatomic O_2 (dioxygen) molecule. The bond energy and bond length in this molecule suggest double bonding, as can be seen by the comparison with N_2 (triple-bonded) and F_2 (single-bonded) shown in Table 20-8. A simple valence-bond approach leads to the *incorrect* Lewis structure

$$:\ddot{O}::\ddot{O}:$$

which shows the double bond and conforms to the octet rule, but which fails to account for the *paramagnetism* of oxygen. Magnetic measurements indicate the presence of two unpaired electrons per molecule. The apparent difficulty in assigning electrons to the O_2 molecule does not arise in molecular-orbital (MO) theory, however. The MO electronic configuration can be written

$$O_2: KK(\sigma_s)^2(\sigma_s^*)^2(\sigma_x)^2(\pi_y)^2(\pi_z)^2(\pi_y^*)^1(\pi_z^*)^1$$

and is diagrammed in Fig. 9-39. It can be seen that this description satisfactorily accounts for both the paramagnetism and the double bond. (The latter can be described as a sigma plus two half-pi bonds.) Oxygen gas condenses at 90 K (at 1 atm) to form a pale blue liquid and freezes at 55 K to form a similarly colored solid.

A second allotrope of oxygen is trioxygen, usually called *ozone*, O_3. This is a bent molecule with a bond angle of 117°. The distance between bonded oxygens in ozone is 0.128 nm, longer than that in the double-bonded O_2 molecule (0.121 nm), but shorter than that in hydrogen peroxide (0.149 nm), in which the oxygens are singly bonded. This indicates a bond order which is intermediate between 1 and 2, a view supported by the MO model of the molecule. Using VB theory the O_3 molecule is shown as a resonance hybrid of the two forms:

consistent with a bond order of 1.5.

Ozone can be prepared by passing a silent electrical discharge through oxygen gas:

$$3O_2(g) \longrightarrow 2O_3(g) \qquad \Delta G° = 326 \text{ kJ}$$

(The reverse reaction, though spontaneous, is slow at ordinary temperatures and in the absence of catalysts.) The gas is pale blue and has a peculiar pungent odor which is extremely irritating at higher concentrations.

Ozone is produced in the upper atmosphere through absorption of a photon, $h\nu$, of ultraviolet light by an O_2 molecule:

$$O_2(g) + h\nu \longrightarrow 2O(g)$$

$$O(g) + O_2(g) \longrightarrow O_3^*(g)$$

The ozone molecule formed has an excess of energy (*) and will soon dissociate back to O_2 and O unless it collides with another molecule (M) such as CO_2, N_2, or

Table 20-8 Some diatomic molecules	Molecule	N_2	O_2	F_2
	Bond length, nm	0.110	0.121	0.142
	Bond energy, kJ mol⁻¹	941	493	158
	Bond order	3	2	1

O_2, which can carry off the excess energy, thus stabilizing the ozone molecule:

$$O_3^*(g) + M(g) \longrightarrow O_3(g) + M^*(g)$$

Some ozone is also produced by lightning and, to a smaller extent, by manufactured devices which arc, spark, or give off ultraviolet light.

The ozone in the upper atmosphere is important in shielding us from the intense ultraviolet radiation coming from the sun. The so-called ozone shield is a shell about 30 km thick which contains enough ozone to absorb much of the ultraviolet radiation of wavelengths less than 310 nm. The absorption causes dissociation of O_3 to re-form O_2:

$$O_3(g) + h\nu \longrightarrow O_2(g) + O(g)$$

$$2O(g) \longrightarrow O_2^*(g)$$

$$O_2^*(g) + M(g) \longrightarrow O_2(g) + M^*(g)$$

Since living tissue is very sensitive to the wavelengths of ultraviolet absorbed by the ozone, life on earth owes much to the existence of the ozone shield. In recent years the shield may have been placed in jeopardy from two diverse sources: supersonic airplanes and chlorofluorocarbons. Products in the jet exhausts of supersonic aircraft undoubtedly reduce ozone, thus decreasing its concentration in the shield. Similarly, chlorofluorocarbons, compounds composed of carbon, chlorine, and fluorine, undergo photodecomposition to release chlorine atoms, which react readily with ozone. Chlorofluorocarbons have been used as refrigerants and as propellants in some "aerosol sprays." Atmospheric chemistry is extremely complex, and so it is difficult to determine how seriously the ozone shield has been damaged. In addition, the lifetimes of the chlorofluorocarbons are so long that it may be decades before we can be sure of the extent of ozone depletion in the upper atmosphere.

A third allotropic form of oxygen has been detected in both liquid and gaseous oxygen. It has the formula O_4 and thus appears to be a dimer of O_2. One might predict the formation of this molecule because of the two unpaired electrons in O_2, but the dissociation energy of O_4 appears to be unusually low.

Preparation of oxygen   We have already mentioned the preparation of ozone. Ordinary O_2 can be prepared in the laboratory by the electrolysis of water,

$$2H_2O \longrightarrow 2H_2(g) + O_2(g)$$

or by the thermal decomposition of certain oxides and oxosalts such as potassium chlorate:

$$2KClO_3(l) \longrightarrow 2KCl(s) + 3O_2(g)$$

This decomposition is catalyzed by manganese dioxide, MnO_2, and by many other oxides of elements which show more than one common oxidation state.

Industrially oxygen is also produced by electrolysis, but the cost (of energy) makes this method practical only for production of very pure O_2. Most industrial oxygen is produced by the fractional distillation (Sec. 12-4) of liquid air. This process takes advantage of the fact that the normal boiling point of oxygen (90 K) is higher than that of nitrogen (77 K).

Compounds
of oxygen

The most common oxidation state shown by oxygen in its compounds is the -2 state, but it also shows the -1, $-\frac{1}{2}$, $+\frac{1}{2}$, $+1$, and $+2$ states.

OXIDATION STATE -2. This state is shown by oxygen in oxides and many other compounds. It is achieved when an oxygen atom completes its octet, either by gaining a pair of electrons to form the oxide ion, O^{2-} or by gaining a share of two electrons to form a covalent bond with a less electronegative element. *Ionic oxides* include those of the alkali and alkaline-earth metals, except for BeO, which is covalent. These oxides are basic; they react with water to form hydroxides, in which water can break the metal-hydroxide bond. For sodium oxide the reactions are

$$Na_2O(s) + H_2O \longrightarrow 2NaOH(s)$$

$$NaOH(s) \longrightarrow Na^+ + OH^-$$

Many ionic oxides are very *refractory,* that is, they can be heated to high temperatures without melting or decomposition. Calcium oxide, CaO, "quicklime," has a melting point above 2500°C, for instance. In the early days of the theater quicklime was used for stage illumination, because at high temperatures it gives off a brilliant, white light, the "limelight."

Covalent molecular oxides are formed when oxygen bonds to other nonmetals. Examples include CO_2, SO_2, NO_2, and ClO_2. These oxides are acidic; reaction of each with water produces a hydroxo compound which dissociates as an acid. With CO_2 the reactions are

$$CO_2(g) + H_2O \longrightarrow H_2CO_3$$

$$H_2CO_3 \longrightarrow H^+ + HCO_3^-$$

The oxides of elements of intermediate electronegativity show intermediate bond character, as expected. Many of these oxides are amphoteric, being insoluble in water, but dissolving in either acid or base. With zinc oxide the reactions are

$$ZnO(s) + 2H^+ \longrightarrow Zn^{2+} + H_2O$$

$$ZnO(s) + 2OH^- + H_2O \longrightarrow Zn(OH)_4^{2-}$$

Some elements form several oxides. Chromium, for example, forms chromous oxide (CrO, basic), chromic oxide (Cr_2O_3, amphoteric), and chromium trioxide (CrO_3, acidic). In the corresponding hydroxo compounds the Cr—O bond becomes stronger while the O—H bond becomes weaker, as the oxidation state of the chromium increases.

Hydroxides are formed, at least in principle, when an oxide reacts with water:

$$CaO(s) + H_2O \longrightarrow Ca(OH)_2(s)$$

As has been already mentioned, the term *hydroxide* is usually limited to hydroxo compounds of metals or near-metals. Some hydroxides contain discrete OH^- ions; alkali and alkaline-earth metal hydroxides are in this category. Other elements, notably the transition metals form "hydroxides" which are really hydrated oxides, also called *hydrous oxides*. Thus ferric hydroxide, though usually formulated

Fe(OH)$_3$, is probably better represented as Fe$_2$O$_3$ · xH$_2$O, and called *hydrous ferric oxide*.

Oxoacids include the hydroxo compounds of the nonmetals. Here the hydroxide group does not exist as a discrete ion, but rather it is bonded covalently to an atom such as N, Cl, etc. Derived from oxoacids are the *oxosalts*, such as Na$_2$SO$_4$ (derived from H$_2$SO$_4$), NaOCl (derived from HOCl), etc. Many of these, such as potassium permanganate (KMnO$_4$) and sodium chromate (Na$_2$CrO$_4$), are better known than their parent acids.

OXIDATION STATE -1. Compounds in which oxygen is in the -1 state are called peroxides (or peroxo compounds). The alkali metals and Ca, Sr, and Ba of the alkaline-earth metals form *ionic peroxides*. In these the anion is the discrete O$_2{}^{2-}$ (peroxide) ion. This can be thought of as an oxygen molecule to which two electrons have been added. These electrons fill the two antibonding orbitals which are half-filled in O$_2$ (Fig. 9-39), so that the peroxide ion is not paramagnetic and has a bond order of 1. Its Lewis structure is

$$\left[\; \overset{\displaystyle ..}{\underset{\displaystyle ..}{:\!\text{O}}}:\overset{\displaystyle ..}{\underset{\displaystyle ..}{\text{O}\!:}}\; \right]^{2-}$$

Sodium peroxide can be prepared by heating sodium metal in oxygen:

$$2\text{Na}(s) + \text{O}_2(g) \longrightarrow \text{Na}_2\text{O}_2(s)$$

The O—O single bond is present also in *covalent* peroxides, the most important of which is hydrogen peroxide, H$_2$O$_2$:

$$\begin{array}{c} \text{H} \\ \overset{\displaystyle ..}{\underset{\displaystyle ..}{:\!\text{O}}}:\overset{\displaystyle ..}{\underset{\displaystyle ..}{\text{O}\!:}} \\ \text{H} \end{array}$$

This molecule has the skewed structure shown in Fig. 20-1. As can be seen, the four atoms are not all in the same plane, the two H atoms being separated by a dihedral angle of 94°.

Hydrogen peroxide may be prepared by adding an ionic peroxide to a solution of a strong acid:

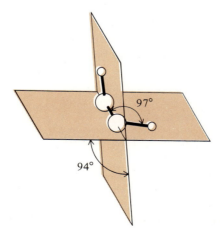

Figure 20-1
Structure of H$_2$O$_2$.

$$BaO_2(s) + 2H^+ \longrightarrow H_2O_2 + Ba^{2+}$$

Commercially, H_2O_2 is made by several methods, one of which starts with the electrolysis of a solution of sulfuric acid or a mixture of H_2SO_4 and $(NH_4)_2SO_4$. Another covalent peroxide, peroxodisulfuric acid, $H_2S_2O_8$, is formed at the anode:

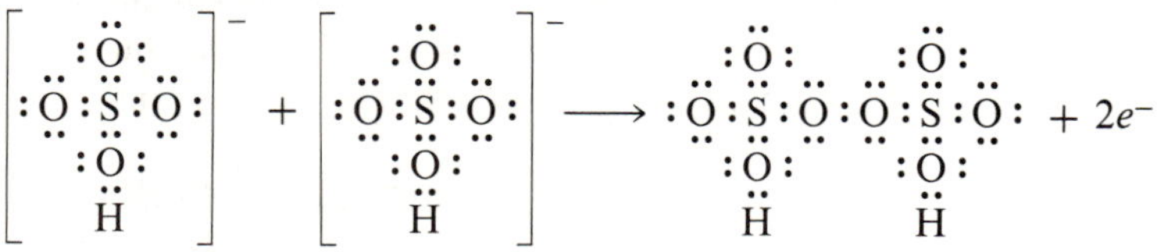

This is then reacted with water, with which it forms peroxomonosulfuric acid, H_2SO_5, which decomposes to form hydrogen peroxide:

$$H_2S_2O_8 + H_2O \longrightarrow H_2SO_5 + H_2SO_4$$

$$H_2SO_5 + H_2O \longrightarrow H_2O_2 + H_2SO_4$$

Fractional distillation is then used to concentrate the solution.

H_2O_2 is readily available as 3% and 30% (by mass) solutions. It is a very weak acid ($K_1 = 2 \times 10^{-12}$) and when pure is a colorless liquid which shows little tendency to decompose at room temperature. Its dielectric constant is 84 at 25°C and its dipole moment is 2.2 D, both values being higher than those of water. The high dielectric constant is partly a consequence of the fact that H_2O_2 is even more hydrogen-bonded (Sec. 12-3) than water. This also in part accounts for its high density, 1.46 g cm^{-3} at 0°C.

Hydrogen peroxide is an even better polar protonic solvent than is water, but its strong oxidizing ability limits its use as such. The reduction potentials for H_2O_2 in acidic and basic solution are, respectively,

$$2e^- + H_2O_2 + 2H^+ \longrightarrow 2H_2O \qquad \mathcal{E}° = 1.78 \text{ V}$$

$$2e^- + HO_2^- + H_2O \longrightarrow 3OH^- \qquad \mathcal{E}° = 0.88 \text{ V}$$

In spite of the fact that the $\mathcal{E}°$ value is higher in acidic solution, most oxidations by H_2O_2 occur more *rapidly* in basic solution.

When H_2O_2 is reduced, it gains two electrons per molecule, and the oxygen atoms go to the -2 state. H_2O_2 can also be oxidized, in which case the oxygen atoms gain one electron each and end up in the zero state. In acidic and basic solutions the respective appropriate reduction potentials are:

$$2e^- + O_2(g) + 2H^+ \longrightarrow H_2O_2 \qquad \mathcal{E}° = 0.68 \text{ V}$$

$$2e^- + O_2(g) + H_2O \longrightarrow HO_2^- + OH^- \qquad \mathcal{E}° = -0.08 \text{ V}$$

In order for H_2O_2 to be oxidized in acidic solution, then, the oxidizing agent must be strong enough to reverse the first of the above reactions. This means that its reduction potential must be greater than 0.68 V. Since the *reduction* potential of H_2O_2 itself is 1.78 V, H_2O_2 ought to be able to oxidize (and reduce) itself:

Reduction:	$2e^- + H_2O_2 + 2H^+ \longrightarrow 2H_2O$	$\mathcal{E}^\circ =$	1.78 V
Oxidation:	$H_2O_2 \longrightarrow 2H^+ + O_2(g) + 2e^-$	$\mathcal{E}^\circ =$	−0.68 V
Overall:	$2H_2O_2 \longrightarrow 2H_2O + O_2(g)$	$\mathcal{E}^\circ =$	1.10 V

Indeed, hydrogen peroxide does undergo such *auto-oxidation-reduction*, or *disproportionation*. Commercial solutions invariably contain inhibitors to tie up trace quantities of transition-metal ions which catalyze the disproportionation. Pure (100 percent) H_2O_2 is quite easy to handle, but small quantities of almost anything, even water, tend to make it explosively unstable.

OXIDATION STATE $-\frac{1}{2}$. If only one electron is added to an oxygen molecule, the result is the superoxide ion O_2^-. The superoxides of K, Rb, and Cs are the easiest to form. They require only heating in O_2:

$$K(s) + O_2(g) \longrightarrow KO_2(s)$$

Superoxides are orange crystalline solids which are very powerful oxidizing agents. They rapidly oxidize even water:

$$2KO_2(s) + H_2O \longrightarrow O_2(g) + HO_2^- + OH^- + 2K^+$$

POSITIVE OXIDATION STATES. Oxygen can form a cation, O_2^+, called the *dioxygenyl ion,* in which a $+\frac{1}{2}$ oxidation state must be assigned to each oxygen atom. This ion occurs in a few known complex salts, such as O_2PtF_6. The electronic configurations and bond lengths of the known dioxygen species are shown in Table 20-9. Notice that as the number of antibonding π electrons increases from O_2^+ to O_2^{2-}, the bond order decreases, and so the bond length increases. The species O_2^{2+} (bond order = 3) and O_2^{3-} (bond order = 0.5) might be expected to exist, but have so far not been confirmed.

In its binary compounds with fluorine oxygen is also assigned a positive oxidation number, since fluorine is more electronegative than oxygen. A series of oxygen fluorides has been observed; of these only OF_2 and O_2F_2 are reasonably stable at room temperature. OF_2 is a bent molecule resembling H_2O and having a bond angle of 103°. O_2F_2 geometrically resembles the H_2O_2 molecule, but the dihedral angle, the angle between the two O—O—F planes, is 87°.

Table 20-9
Bonding in dioxygen species

Species	O_2^+		O_2		O_2^-		O_2^{2-}	
Name	Dioxygenyl ion		Dioxygen		Superoxide ion		Peroxide ion	
Bond length, nm	0.112		0.121		0.133		0.149	
Unpaired electrons	1		2		1		0	
Bond order	2.5		2		1.5		1	
Electronic configuration								
σ_x^*	—		—		—		—	
$\pi_y^* \ \pi_z^*$	1		1	1	1↓	1	1↓	1↓
$\pi_y \ \pi_z$	1↓	1↓	1↓	1↓	1↓	1↓	1↓	1↓
σ_x	1↓		1↓		1↓		1↓	
σ_s^*	1↓		1↓		1↓		1↓	
σ_s	1↓		1↓		1↓		1↓	
KK	1↓	1↓	1↓	1↓	1↓	1↓	1↓	1↓

20-4 Water   We have already described many of the properties of water (see Sec. 12-3, for instance). You should recall that because of hydrogen bonding liquid water is highly associated. In the liquid there are small regions, *flickering clusters,* in which H_2O molecules come together to form the ice I (ordinary ice) structure, only to move apart an instant later. Because of the numbers of these "almost-ice" regions the liquid is less dense than it otherwise might be. Ice I has an open structure with hexagonal channels running through it in three directions as is shown in Fig. 20-2. (The locations of the hydrogen atoms are not shown in this drawing.) When liquid water is cooled, the number and size of the flickering clusters increase. Below 3.98°C their buildup is so pronounced that the liquid expands with further cooling. At 0°C the clusters cease their flickering as the rigid ice structure forms.

Hydrates   Water can become incorporated in solid compounds in a number of different ways. In some cases the water molecule loses its identity as when a hydroxide is formed:

$$CaO(s) + H_2O \longrightarrow Ca(OH)_2(s)$$

In others it retains its identity as a molecule either by serving as a ligand in an aquo complex, such as in $Cr(H_2O)_6Cl_3$, or by occupying specific lattice sites in the solid, without being strongly bonded to a specific atom, as in $CaSO_4 \cdot 2H_2O$. The latter solids are called *hydrates,* and their water molecules are known as *waters of hydration,* or *waters of crystallization.* Many hydrates lose water upon heating. $CaCl_2 \cdot 6H_2O$, for example, undergoes progressive dehydration to form compounds with four, two, one, and, finally, no H_2O molecules per $CaCl_2$ formula unit. (The last is called *anhydrous* $CaCl_2$.) Many anhydrous compounds find application as drying agents, since they tend to recover the lost water. It should be noted that each hydrate is a separate compound with its own crystal structure.

Clathrates   An interesting kind of "compound" is formed when water is frozen after being saturated with a gas composed of small molecules such as Cl_2, Br_2, $CHCl_3$, H_2S, and even the noble gases Ar, Kr, and Xe. In these substances the small molecule

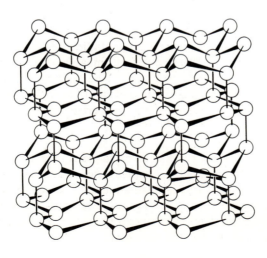

Figure 20-2
Ice I structure (hydrogen atoms are not shown).

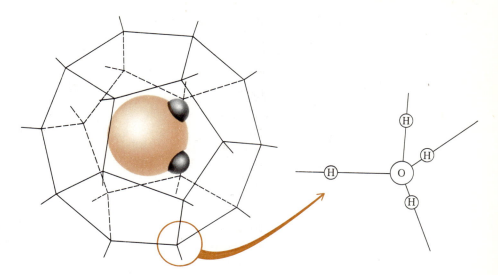

Figure 20-3
H$_2$S molecule in the clathrate (H$_2$S)$_4$(H$_2$O)$_{24}$.

is trapped in a cavity or cage of water molecules. These substances, called *clathrates* (from a Latin word meaning "caged in" or "imprisoned by bars") are not compounds in the usual sense because there is no direct bonding (other than London forces) between the water and the trapped molecule.

The cavity in which the trapped molecule is enclosed varies from one clathrate to another, but always approximates a regular polyhedron. Figure 20-3 shows an almost-regular pentagonal dodecahedron having an oxygen atom at each corner. Many such dodecahedra are linked via hydrogen bonds to form a three-dimensional structure with one trapped molecule per dodecahedron. A number of these structures have been observed. In each, the cavities are somewhat larger than the channels in ordinary ice and owe their stability to hydrogen bonding and the fact that they are kept "propped open" or "inflated" by the enclosed molecule.

20-5
The halogens

The elements of periodic group VIIA include fluorine, chlorine, bromine, iodine, and astatine. Of these we will discuss the first four, since astatine does not occur naturally, and its longest-lived isotope has a half-life of only 8 h. (What little is known of the chemical behavior of astatine can be summed up by saying that it resembles iodine, but is a little more metallic.) The elements of this group are known as the *halogens* (from Greek words meaning "salt formers"). Uncombined, they all exist as *diatomic molecules*. The halogens combine with nearly all elements in the periodic table, most commonly as *halides* in which they exhibit the -1 oxidation state. Some of the properties of the halogens are shown in Table 20-10.

Fluorine

The name of this element comes from *fluorspar,* a mineral now more commonly called *fluorite.* (Fluorspar was named after a Latin word meaning "to flow"; fluorspar was used as a flux in brazing metals.) Fluorine occurs in minerals such as fluorite, CaF_2; cryolite, Na_2AlF_6; and fluorapatite, $Ca_5F(PO_4)_3$.

Elemental fluorine can not be prepared by the chemical oxidation of fluorides, since F_2 is the strongest oxidizing agent:

Table 20-10
Halogens

	F	Cl	Br	I
Valence-shell configuration of atom	$2s^22p^5$	$3s^23p^5$	$4s^24p^5$	$5s^25p^5$
Physical appearance at STP	Pale yellow gas	Greenish-yellow gas	Reddish-brown liquid	Dark gray solid
Color of gas	Pale yellow	Greenish yellow	Orange	Violet
Normal melting point, °C	−223	−102	−7.3	114
Normal boiling point, °C	−187	−35	59	183
Electronegativity	4.0	3.0	2.8	2.5

$$F_2(g) + 2H^+ + 2e^- \longrightarrow 2HF \qquad \mathcal{E}° = 3.06 \text{ V} \qquad \text{(acidic solution)}$$

$$F_2(g) + 2e^- \longrightarrow 2F^- \qquad \mathcal{E}° = 2.87 \text{ V} \qquad \text{(basic solution)}$$

This means that in order to produce the uncombined element, electrolytic methods must be used. Usually a molten mixture of HF and KF is electrolyzed using a carbon anode, at which F_2 is produced, and a silver or stainless-steel cathode, at which H_2 is formed. The products must be kept separated, because they react explosively.

Fluorine's powerful oxidizing ability causes it to react with most substances, often vigorously. It can be stored in containers made of steel, copper, or certain alloys only because it quickly forms a fluoride coating on the metal surface, a coating in the form of a tough, impervious layer which effectively prevents further reaction. It also can be stored and handled in glass, provided that no HF is present as an impurity.

Fluorine reacts violently with hydrogen to form hydrogen fluoride:

$$H_2(g) + F_2(g) \longrightarrow 2HF(g) \qquad \Delta G° = -541 \text{ kJ}$$

Although HF is only a weak acid ($K_a = 6.7 \times 10^{-4}$), it has some properties which make it tricky to handle. Both it and its aqueous solutions (hydrofluoric acid) attack glass. If, for simplicity's sake, we represent glass (a silicate; see Sec. 21-5) as silicon dioxide, the reaction may be written as

$$SiO_2(s) + 6HF \longrightarrow SiF_6^{2-} + 2H^+ + 2H_2O$$

Many fluoro complexes, such as the hexafluorosilicate ion, above, are very stable.

HF has another property which makes it very dangerous to handle. It causes chemical burns which are exceedingly painful and which often take many months to heal. Gaseous HF and acidified fluoride solutions should thus be handled with extreme caution, and HF experiments should always be performed in a fume hood. HF solutions can be stored in containers made of polyethylene plastic or paraffin wax.

Highly electropositive metals form ionic fluorides. Those with formulas MF tend to be soluble in water, but MF_2 compounds do not. On the other hand, nonmetals and some metals in high positive oxidation states form molecular fluorides. These include SF_2, CF_4, SiF_4, TiF_4, and UF_6. Many complex fluorides exist; these include Na_2SiF_6, $SrPbF_6$, $(NH_4)_3ZF_7$, and Na_3TaF_8. The small size of the fluoride ion permits more than six to be coordinated to a central atom, a fairly unusual situation.

The other halogens form a series of oxoacids and oxosalts, a property not shared by fluorine. The only such compound yet observed is HOF, the fluorine analogue of hypochlorous acid, HOCl, but it is extremely unstable.

Chlorine The name of this element comes from a Greek word meaning "green"; uncombined chlorine is a greenish-yellow, poisonous gas. It occurs naturally combined in chlorides such as the minerals halite (NaCl) and sylvite (KCl) in underground deposits, and, of course, in the ocean. Chlorine is a powerful germicide and is used in the purification of drinking water throughout the world. It is also used as an industrial bleach.

Most chlorine is prepared industrially by electrolysis of molten NaCl, giving sodium as a by-product, or of aqueous NaCl. In the latter case when an iron cathode and carbon anode are used, hydrogen is produced at the cathode. When a mercury cathode is used, however, sodium ions are reduced to sodium metal, rather than water to H_2 gas. This happens for two reasons: (1) sodium metal is soluble in mercury (the solution is called an *amalgam*), so that the chemical activity of the sodium is reduced far below that of pure sodium ($a = 1$), and a lower voltage is consequently required for the electrolysis; and (2) an abnormally high voltage (*overvoltage*) is required to produce H_2 on a mercury surface. The reactions are

$$\begin{array}{ll} \text{Anode:} & 2Cl^- \longrightarrow Cl_2(g) + 2e^- \\ \text{Cathode:} & 2 \times [e^- + Na^+ \longrightarrow Na(Hg)] \\ \hline \text{Cell:} & 2Cl^- + 2Na^+ \longrightarrow Cl_2(g) + 2Na(Hg) \end{array}$$

The sodium amalgam is then decomposed by reaction with water, yielding very pure sodium hydroxide:

$$2Na(Hg) + 2H_2O \longrightarrow 2Na^+ + 2OH^- + H_2(g)$$

Unfortunately, in the past, at least, this process has produced industrial wastewater which was badly contaminated with mercury. Today many bodies of water all over the world are loaded with waste mercury from this reaction, although most industrial plants have modified their processes to reduce the mercury lost to low levels. (We will discuss mercury further in Sec. 22-6.)

In the laboratory chlorine is easily produced by chemical oxidation of acidified Cl^- solutions. Manganese dioxide, MnO_2, is often used:

$$MnO_2(s) + 2Cl^- + 4H^+ \longrightarrow Cl_2(g) + Mn^{2+} + 2H_2O$$

CHLORIDES. Chlorides are compounds with chlorine in the -1 oxidation state. They include both the ionic chlorides, such as NaCl and $CaCl_2$ and the covalent chlorides, such as HCl, CCl_4 and SCl_2. Hydrogen chloride, HCl, can be produced by adding concentrated H_2SO_4 to solid NaCl:

$$H_2SO_4(l) + NaCl(s) \longrightarrow HCl(g) + NaHSO_4(s)$$

The bond energy in HCl is less than that in HF, which partially accounts for the fact that HCl is a strong acid in water, while HF is weak.

CHLORINE OXIDES. Chlorine forms a series of oxides, all of which are unstable and potentially explosive: Cl_2O, Cl_2O_3, ClO_2, Cl_2O_4, Cl_2O_6, and Cl_2O_7. Of these, chlorine dioxide, ClO_2, is industrially the most important; it is a powerful oxidizing agent and is used as a bleach of wood pulp, for example. (The process of bleaching generally consists of oxidizing colored organic compounds to form products which are either colorless or less intensely colored.)

ClO_2 is called an *odd molecule,* because it has an odd number of valence electrons. Its Lewis structure is

$$\cdot \ddot{C}l : \ddot{O} :$$
$$: \ddot{O} :$$

According to VSEPR theory (Sec. 9-1), this molecule should be *bent,* the three electron pairs and the single electron occupying approximately tetrahedral positions around the Cl, similar to the orientation of electron pairs around the oxygen in H_2O. But the lone electron should repel the bonding pairs less than a lone pair of electrons; so we predict a bond angle in ClO_2 larger than that in water. Indeed, the observed bond angle turns out to be 116°, greater than that in H_2O (105°) and greater even than the tetrahedral angle (109°).

In basic solution ClO_2 disproportionates to form the chlorate, ClO_3^-, and chlorite, ClO_2^-, ions:

$$2ClO_2(g) + 2OH^- \longrightarrow ClO_3^- + ClO_2^- + H_2O$$

OXOACIDS AND SALTS. Unlike fluorine, chlorine forms a series of oxoacids and corresponding oxosalts. The acids are listed in Table 20-11, along with their Lewis structures and dissociation constants.

Hypochlorous acid, HOCl, which has never been isolated pure, is a weak acid and powerful oxidizing agent and can be prepared by dissolving chlorine gas in water:

$$Cl_2 + H_2O \longrightarrow HOCl + H^+ + Cl^- \qquad K' = 4.7 \times 10^{-4}$$

The product of this disproportionation may be considered to be a mixture of the weak hypochlorous acid plus the strong hydrochloric acid, but the value of the equilibrium constant is small, indicating that not much of either acid is present at equilibrium. However, since the products include H^+ and HOCl, addition of OH^- will shift the equilibrium to the right, forming H_2O and the *hypochlorite ion,* OCl^-:

Table 20-11 Oxoacids of chlorine	Acid	Formula	Chlorine oxidation state	Lewis structure	K_a	Exists pure?
	Hypochlorous	HOCl	+1	$H:\ddot{O}:\ddot{C}l:$	3.2×10^{-8}	No
	Chlorous	$HClO_2$ (HOClO)	+3	$H:\ddot{O}:\ddot{C}l:$ $:\ddot{O}:$	1.1×10^{-2}	No
	Chloric	$HClO_3$ (HOClO$_2$)	+5	$H:\ddot{O}:\ddot{C}l:\ddot{O}:$ $:\ddot{O}:$	(Strong acid)	No
	Perchloric	$HClO_4$ (HOClO$_3$)	+7	$H:\ddot{O}:\ddot{C}l:\ddot{O}:$ $:\ddot{O}:$	(Strong acid)	Yes

$$Cl_2 + H_2O \longrightarrow HOCl + H^+ + Cl^- \qquad \text{(disproportionation)}$$
$$OH^- + HOCl \longrightarrow OCl^- + H_2O \qquad \text{(neutralization)}$$
$$\underline{OH^- + H^+ \longrightarrow H_2O \qquad \text{(neutralization)}}$$
$$Cl_2 + 2OH^- \longrightarrow OCl^- + Cl^- + H_2O \qquad \text{(overall)}$$

The equilibrium constant for the overall process can be found as follows:

$$K = \frac{[OCl^-][Cl^-]}{[Cl_2][OH^-]^2}$$

$$= \frac{[OCl^-][H^+]}{[HOCl]} \times \frac{[HOCl][H^+][Cl^-]}{[Cl_2]} \times \frac{1}{[H^+]^2[OH^-]^2}$$

$$= K_{HOCl}K'\frac{1}{K_w^2}$$

$$= (3.2 \times 10^{-8})(4.7 \times 10^{-4})\left[\frac{1}{(1.0 \times 10^{-14})^2}\right] = 1.5 \times 10^{17}$$

The disproportionation of Cl_2 thus proceeds essentially to completion in basic solution. Such solutions are commonly used as household and commercial laundry bleaches, often called "chlorine bleaches." Mixing these with acidic substances such as vinegar (acetic acid) or some toilet-bowl cleaners (which contain sodium bisulfate, $NaHSO_4$) reverses the above equilibrium and produces quantities of the very poisonous chlorine gas. (Mixing such bleaches with ammonia, a common household cleaner, is also dangerous, forming a variety of toxic products.)

ClO^- undergoes further disproportionation to form the chlorate ion, ClO_3^-:

$$3ClO^- \longrightarrow ClO_3^- + 2Cl^- \qquad K = 10^{27}$$

Although this reaction proceeds essentially to completion, it is slow at room temperature.

Chlorous acid, $HClO_2$, though still weak, is a stronger acid than hypochlorous acid. It can not be isolated pure, but can be made by acidifying a solution containing the *chlorite ion,* ClO_2^-, obtained from the disproportionation of ClO_2 (above). Chlorous acid is unstable, decomposing as follows:

$$5HClO_2 \longrightarrow 4ClO_2(g) + H^+ + Cl^- + 2H_2O$$

The chlorite ion is kinetically quite stable in basic solution and is used as an industrial bleach.

Chloric acid, $HClO_3$, is a strong acid and a powerful oxidizing agent, but can be obtained only in solution. Its anion, the *chlorate ion,* ClO_3^-, is found in many salts, the best known of which is potassium chlorate, $KClO_3$. This can be prepared by heating solutions containing the hypochlorite ion and then adding KCl to precipitate $KClO_3$, which is not very soluble in water. Industrially, $KClO_3$ is prepared by electrolyzing a hot, stirred solution of KCl. The chlorine produced at the anode

Anode:
$$2Cl^- \longrightarrow Cl_2 + 2e^-$$

mixes with hydroxide ions produced at the cathode:

Cathode:
$$2e^- + 2H_2O \longrightarrow H_2(g) + 2OH^-$$

to form hypochlorite ions:

$$Cl_2 + 2OH^- \longrightarrow OCl^- + Cl^- + 2H_2O$$

which, because the solution is hot, undergo disproportionation to chlorate ions:

$$3OCl^- \longrightarrow ClO_3^- + 2Cl^-$$

which then precipitate as $KClO_3$:

$$ClO_3^- + K^+ \longrightarrow KClO_3(s)$$

When solid $KClO_3$ is heated, it first melts and then decomposes. Parallel reactions are possible:

$$2KClO_3(l) \longrightarrow 2KCl(s) + 3O_2(g)$$

$$4KClO_3(l) \longrightarrow 3KClO_4(s) + KCl(s)$$

The first of these reactions can be speeded up by catalysts such as MnO_2 and Fe_2O_3, so that essentially no $KClO_4$ is formed.

Unlike the other oxoacids of chlorine, *perchloric acid,* $HClO_4$, can be isolated as a pure substance. It is a colorless, oily liquid at room temperature and a stronger acid even than H_2SO_4, although in water this property is hidden by the leveling effect (Sec. 13-1). $HClO_4$ is a powerful oxidizing agent and, when pure, is quite unstable, often decomposing explosively. Commercially, perchloric acid is available as 60 to 70% (by mass) aqueous solutions. Even these should be handled with caution and kept out of contact with good reducing agents.

The perchlorate ion is a good oxidizing agent, but is often a slow one. It is useful in chemical research for adjusting the overall concentration of ions in solution without causing any unwanted reactions. (The perchlorate ion has been rarely observed to serve as a ligand in any complex.) The ClO_4^- ion has been used in qualitative testing for the potassium ion in solution, because $KClO_4$ is somewhat insoluble.

STRUCTURE OF THE CHLORINE OXOANIONS. The OCl^-, ClO_2^-, ClO_3^- and ClO_4^- anions show geometrical structures as predicted by VSEPR theory (Sec. 9-1). The hypochlorite ion

$$\left[:\overset{..}{\underset{..}{O}} : \overset{..}{\underset{..}{Cl}} : \right]^-$$

is of course linear. The chlorite ion

$$\left[:\overset{..}{\underset{..}{O}} : \overset{..}{\underset{..}{Cl}} : \\ :\overset{..}{\underset{..}{O}} : \right]^-$$

is bent with two lone pairs on the chlorine. The chlorate ion

$$\left[:\overset{..}{\underset{..}{O}} : \overset{..}{Cl} : \overset{..}{\underset{..}{O}} : \\ :\overset{..}{\underset{..}{O}} : \right]^-$$

is trigonal pyramidal with a lone pair on the chlorine. The perchlorate ion

$$\left[\begin{array}{c} :\overset{..}{O}: \\ :\overset{..}{O}:\overset{..}{\underset{..}{Cl}}:\overset{..}{O}: \\ :\overset{..}{O}: \end{array} \right]^{-}$$

is tetrahedral with all of chlorine's valence-shell electrons used in bonding.

ACIDITY OF THE CHLORINE OXOACIDS. As is shown in Table 20-11, the oxoacids of chlorine become more acidic as the oxidation number of the chlorine increases. This is a result of the high electronegativity of oxygen. As more oxygen atoms are added to the chlorine, they tend to withdraw electrons from the O—H bond making it easier to break by water molecules. Thus the higher the oxidation state, the stronger is the acid.

Bromine This element, named after a Greek word meaning "stench," occurs naturally throughout the world combined as bromides which are associated in small quantities with chlorides. Uncombined, it is a reddish-brown liquid with a high vapor pressure at room temperature. Br_2 is obtained industrially by oxiding the Br^- in acidified seawater using chlorine as the oxidizing agent:

$$Cl_2(g) + 2Br^- \longrightarrow 2Cl^- + Br_2(g)$$

The bromine is swept out of the reaction mixture by a stream of air and later condensed to form the liquid. In the laboratory bromine can be prepared by the oxidation of Br^- in acid solution using MnO_2 or a similar oxidizing agent.

Bromine should be handled with considerable caution; it forms painful chemical burns on the skin which are extremely slow to heal.

POSITIVE OXIDATION STATES OF BROMINE. Bromine forms several oxides, of which Br_2O and BrO_2 have been best characterized. In addition, it forms oxoacids and oxosalts which are analogous to those formed by chlorine, except for the fact that neither bromous acid, $HBrO_2$, nor any bromite, BrO_2^-, have ever been identified, either pure or in solution. Actually, none of the oxoacids of bromine has been isolated in the pure state.

Hypobromous acid ($K_a = 2 \times 10^{-9}$) is slightly weaker than hypochlorous acid. The hypobromite ion OBr^- can be prepared as the hypochlorite ion is, by dissolving the halogen in basic solution:

$$Br_2(l) + 2OH^- \longrightarrow OBr^- + Br^- + H_2O$$

In this case, however, the rate of subsequent disproportionation is significant at room temperature, leading to the formation of the bromate ion:

$$3OBr^- \longrightarrow BrO_3^- + 2Br^-$$

Bromic acid, $HBrO_3$, is known only in solution. It is a strong acid, and attempts to concentrate it in solution by evaporation lead to its decomposition:

$$4H^+ + 4BrO_3^- \longrightarrow 2Br_2(g) + 5O_2(g) + 2H_2O(g)$$

Perbromates, BrO_4^-, can be prepared by oxidation of BrO_3^-, either electrolytically or by using F_2 or XeF_2 as the oxidizing agent. Perbromic acid is strong and decomposes when attempts are made to concentrate it.

The bonding and structures of the bromine oxoacids and anions parallel those of the corresponding chlorine compounds.

Iodine   The origin of this name is a Greek word meaning "violet." Iodine occurs naturally as the iodide ion in the ocean, especially in certain marine organisms such as seaweeds, which concentrate it. It also is found as the iodate ion IO_3^- mixed in small quantities with $NaNO_3$ (Chile saltpeter) in deposits in South America. The uncombined element is prepared in industry either by oxidizing I^- with Cl_2 gas or by reducing IO_3^- to I^-,

$$IO_3^- + 3SO_2(g) + 3H_2O \longrightarrow I^- + 3HSO_4^- + 3H^+$$

followed by an oxidation by IO_3^-:

$$IO_3^- + 5I^- + 6H^+ \longrightarrow 3I_2(s) + 3H_2O$$

Iodine is a steel-gray solid with a semimetallic luster. The odor of its vapor is easily detectable at room temperature. I_2 has a normal melting point at 114°C and a normal boiling point at 183°C. Its vapor is a striking violet, a color which is duplicated in its solutions in nonpolar solvents such as CCl_4 and CS_2. In polar solvents such as water and ethanol, the color of iodine solutions is brown, believed to be due to the formation of *charge-transfer complexes,* in which some electronic charge is transferred from the solvent molecule(s) to the iodine molecule. Iodine forms a deep blue-black complex with starch.

A solution of iodine in ethanol (*tincture of iodine*) has been widely used as a disinfectant and antiseptic; I_2 oxidizes and destroys most microorganisms. To some extent it does the same to living tissue, thus delaying the healing process, and so tincture of iodine has fallen into some disfavor. In 1955, however, an iodine complex with an organic polymer called *povidone* was prepared; in this complex the activity of the I_2 is reduced so that it is less destructive to tissue, but still functions as an antiseptic.

Iodine shows only a limited solubility in water, but is much more soluble in solutions containing the iodide ion. This has been attributed to the formation of the brown triiodide ion, I_3^-:

$$I_2 + I^- \longrightarrow I_3^- \qquad K = 1.4 \times 10^{-3}$$

This ion illustrates clear violation of the octet rule:

$$\left[:\ddot{\underset{..}{I}}: \ddot{\underset{.\,.}{I}} :\ddot{\underset{..}{I}}: \right]^-$$

According to VSEPR theory the five electron pairs on the central iodine should be arrayed at the corners of a trigonal pyramid. Electron repulsion is minimized with the lone pairs occupying the equatorial positions, leaving the bonding pairs at the axial positions. This leads to the prediction of a linear shape for the I_3^- ion, a prediction born out by x-ray diffraction measurements.

POSITIVE OXIDATION STATES OF IODINE. The most important *oxide* of iodine is diiodine pentaoxide, I_2O_5. It is the anhydride of *iodic acid* and can be made by dehydrating this acid:

$$2HIO_3(s) \longrightarrow I_2O_5(s) + H_2O(g)$$

It is an oxidizing agent perhaps best known for its use in the quantitative analysis for carbon monoxide. It oxidizes the CO,

$$I_2O_5(s) + 5CO(g) \longrightarrow I_2 + 5CO_2(g)$$

and the amount of iodine liberated can be determined by adding starch as an indicator and then titrating with sodium thiosulfate, $Na_2S_2O_3$, of known concentration. The thiosulfate ion reduces the iodine,

$$I_2 + 2\underset{\text{Thiosulfate}}{2S_2O_3^{2-}} \longrightarrow 2I^- + \underset{\text{Tetrathionate}}{S_4O_6^{2-}}$$

and the end point of the titration is observed when the blue starch-iodine complex disappears.

Of the *oxoacids* of iodine, only the one with iodine in the $+3$ state is missing, as it is with bromine. Hypoiodous acid, HOI, is an extremely weak acid ($K_a = 4 \times 10^{-13}$). The decrease in acid strength from HOCl to HOI reflects the increase in metallic character going down the group. As the electronegativity of the halogen decreases, the O—H bond becomes stronger. In fact, the *basic* dissociation constant of HOI (to form I^+ and OH^-) has been estimated as 3×10^{-10}!

Hypoiodous acid can be prepared in low concentrations by adding I_2 to water:

$$I_2 + H_2O \longrightarrow HOI + H^+ + I^-$$

But in a basic solution

$$I_2 + 2OH^- \longrightarrow OI^- + H_2O$$

the hypoiodite ion immediately disproportionates:

$$3OI^- \longrightarrow IO_3^- + 2I^-$$

Iodic acid is strong and, unlike $HClO_3$ and $HBrO_3$, can be isolated as a pure, white, crystalline solid. The iodate ion is a good oxidizing agent, though poorer than chlorate or bromate.

In the $+7$ oxidation state iodine forms at least three acids or their salts. The acids are named *metaperiodic acid* (HIO_4), *mesoperiodic acid* (H_3IO_5), and *paraperiodic acid* (H_5IO_6). The best characterized of these is the last, H_5IO_6, which seems to be triprotic ($K_1 = 2 \times 10^{-2}$; $K_2 = 1 \times 10^{-6}$; $K_3 = 2 \times 10^{-13}$).

Interhalogen compounds

The halogens form an impressive series of binary compounds among themselves. All possible XY combinations have been observed, where X and Y represent different halogens. In addition, compounds of the types XY_3, XY_5, and XY_7 are known, where X is always the larger atom.

Some interhalogens can be prepared by direct combination:

$$Br_2(l) + 3F_2(g) \longrightarrow 2BrF_3(l)$$

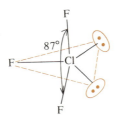

Figure 20-4
Structures of two inter-halogen compounds.

Chlorine trifluoride

Bromine pentafluoride

Others require more indirect routes:

$$3I_2(g) + 5AgF(s) \longrightarrow 5AgI(s) + IF_5(l)$$

Many of the interhalogens are quite unstable, some violently so.

The structures of the interhalogens, at least of those whose stabilities have permitted structure determinations, generally support predictions made on the basis of VSEPR theory. All of the interhalogens except those of the XY type show violation of the octet rule.

Structures of two interhalogens are shown in Fig. 20-4.

20-6
The chalcogens, especially sulfur

The elements of periodic group VIA are known as the *chalcogens*. This word is derived from Greek words meaning "copper former" and seems like a strange name, until one learns that the ores from which copper is obtained have formulas like Cu_2S, $CuFeS_2$, Cu_2O, and $Cu_2CO_3(OH)_2$. Members of group VIA other than oxygen and sulfur are also often found as impurities in such ores.

In the chalcogens we see more clearly than in the halogens the trend from nonmetallic to metallic properties going down a group. Oxygen is clearly a nonmetal. Sulfur is also a nonmetal; none of its allotropic forms has any metallic appearance. The sulfur atom shows the typical nonmetallic tendency to gain electrons; the only simple ion it forms is the sulfide ion, S^{2-}. In most of its compounds in which it shows positive oxidation states, it *shares* electrons with more electronegative oxygen atoms, and its hydroxo compounds are all acidic. Selenium also exists as several allotropes, the most stable of which at room temperature has a subdued metallic luster, is a semiconductor, and is best classed as a semimetal, or metalloid. It does not form cations, and its hydroxo compounds are acidic, although less so than those of sulfur. Tellurium is also a metalloid, but its hydroxo compounds are less acidic than those of selenium. Little is known of polonium; its radioactivity (it is an alpha emitter) has restricted study of it. It appears to be a metal, forming a basic hydroxide. As we look at the various main groups (A groups) of the periodic table we will see further illustration of this nonmetal-to-metal trend. Some of the properties of the chalcogens are shown in Table 20-12.

In this section we will concentrate only on sulfur, since we have already described oxygen (Sec. 20-3), and since we will describe selenium and tellurium with the metalloids (Sec. 21-5).

Sulfur

Sulfur is an element known since antiquity; its name comes from Sanskrit. For a while it was known as *brimstone* (of "fire-and-brimstone"), a corruption of a

Table 20-12
Chalogens

	O	S	Se	Te	Po
Valence-shell configuration	$2s^2 2p^4$	$3s^2 3p^4$	$4s^2 4p^4$	$5s^2 5p^4$	$6s^2 6p^4$
Physical appearance (most stable allotrope at STP)	Colorless gas	Yellow solid	Dark gray solid, metallic luster	Grayish-white metallic solid	Metallic solid
Normal melting point, °C	−218	119	217	450	254
Normal boiling point, °C	−183	445	685	1390	962
Electronegativity	3.5	2.5	2.4	2.1	2.0

German word meaning "burning stone." The alchemists tried vainly to infuse the yellow color of sulfur into lead in order to make gold.

Sulfur is found widely distributed in the earth's crust. It is found in vast underground beds of 99.8 percent pure uncombined sulfur and in many sulfide minerals such as galena (PbS), pyrite (FeS_2), sphalerite (ZnS), and various calcium, magnesium, and other sulfates. The traditional way of obtaining sulfur has been by means of the *Frasch process,* in which superheated water (at about 170° C and under pressure) and compressed air are forced down through pipes to underground sulfur deposits. The sulfur melts and is forced up to the surface as an air-water-sulfur froth.

Today an increasing amount of sulfur is being obtained from sulfide minerals and from petroleum, in which it is an undesirable impurity. The reasons for this are both economic and environmental. Excess sulfur from metallurgical smelting and oil-refining operations was traditionally burned off as sulfur dioxide, SO_2, which forms sulfurous acid, H_2SO_3, in the atmosphere. Slow subsequent oxidation produces sulfuric acid, H_2SO_4, which is destructive to almost everything from lungs to limestone. Recovery of by-product sulfur compounds from various industrial processes has fortunately been on the increase during recent years.

Sulfur exhibits a variety of allotropic forms. Stable at room temperature is a yellow solid composed of orthorhombic crystals. This is known as *rhombic sulfur,* or α-sulfur, and is soluble in nonpolar solvents such as CS_2 and CCl_4. In this form the molecule is cyclic: an S_8 puckered ring (Fig. 20-5). These rings are packed together in an orthorhombic crystal lattice.

When rhombic sulfur is slowly heated, it changes to a *monoclinic* crystalline

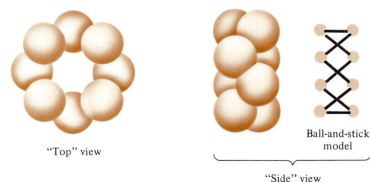

"Top" view

Ball-and-stick model

"Side" view

Figure 20-5
Sulfur molecule.

form at 96°C (at 1 atm). The change is very slow, as are most solid-solid phase transformations. The structure of monoclinic (or β) sulfur is not known, but evidence indicates that the same S_8 cyclic molecules are present, packed together in a monoclinic crystal lattice. Heating of monoclinic sulfur eventually produces melting at 119°C to form a thin, light, straw-colored liquid. This liquid consists primarily of S_8 rings, but the evidence indicates that some larger rings, perhaps as large as S_{20} are also present. Above 160°C a striking change takes place: the liquid darkens to a deep red-brown and becomes very viscous. Apparently thermal motion causes most of the rings to break open, and their ends reconnect to form very long chains of S atoms, which then become entangled with each other. With continued heating the liquid becomes less viscous, as the average chain length decreases, until at 445°C it boils. The gas consists of a mixture of fragments, which dissociate to form still smaller fragments at higher temperatures. At 800°C, S_2 molecules predominate; these are like O_2 in electronic configuration. Above 2000°C monatomic sulfur is present.

When liquid sulfur at about 350° is suddenly quenched by pouring it into water, a strange elastic substance called *plastic,* or *amorphous, sulfur* is formed. It looks and feels much like gum rubber and apparently consists of many intertwined long sulfur chains. After standing for several days, it gradually reverts to the stable orthorhombic crystals. Figure 20-6 schematically shows the phase diagram for sulfur. As can be seen, there are three triple points: rhombic-monoclinic-gas, monoclinic-liquid-gas, and rhombic-monoclinic-liquid.

The common oxidation states of sulfur in its compounds are -2, $+4$, and $+6$. The -2 state corresponds to the gain of two electrons to form the sulfide ion S^{2-} or the sharing of two electrons with one or two more electropositive atoms. The $+4$ state corresponds to the sharing of sulfur's four $3p$ electrons with more electronegative atoms, while the $+6$ state indicates that all six of the valence electrons are similarly used in bonding. The existence of a positive oxidation state corresponding to the group number and of another state two units lower is an illustration of the *inert-pair effect,* the term referring to the valence-shell *s* electrons, used in bonding in the higher oxidation state but not in the lower.

SULFIDES. Sulfides can generally be prepared by direct combination of the elements. Their solubilities in water vary tremendously: sodium sulfide, Na_2S, is extremely soluble, while the solubility product for platinum sulfide, PtS, is 1×10^{-72}. (How many Pt^{2+} ions are present in 1 liter of a saturated PtS solution?) The parent of the metal sulfides is hydrogen sulfide, H_2S, a gas with "the odor of rotten eggs." (Actually, when eggs go rotten, they do give off H_2S, among other things.) Hydrogen sulfide is extremely poisonous, more so than either hydrogen cyanide or carbon monoxide. Fortunately, the human nose can detect it at levels far below the danger point; unfortunately, it has the property of gradually deadening the sense of smell, making the nose a rather imperfect H_2S detector.

Hydrogen sulfide can be made by reacting hydrogen with molten sulfur, heating sulfur with various hydrocarbons such as paraffin, or by reacting the more soluble sulfides with an acid:

$$2H^+ + FeS(s) \longrightarrow H_2S(g) + Fe^{2+}$$

This reaction was used for years for the laboratory preparation of H_2S, but today it

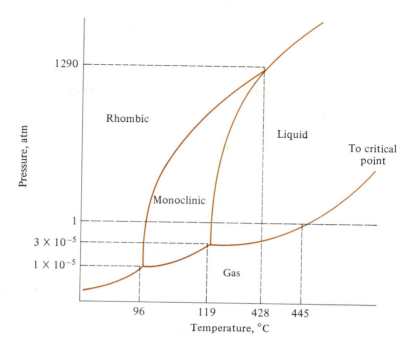

Figure 20-6
Phase diagram for
sulfur (both axes
distorted).

has been largely replaced by the hydrolysis of thioacetamide. This reaction is fairly slow in neutral solution, but faster in the presence of acid or base. In acidic solution the reaction is

$$H-\underset{\underset{H}{|}}{\overset{\overset{H}{|}}{C}}-C\overset{S}{\underset{N-H}{\diagdown}} + H^+ + 2H_2O \longrightarrow H-\underset{\underset{H}{|}}{\overset{\overset{H}{|}}{C}}-C\overset{O}{\underset{O-H}{\diagdown}} + NH_4^+ + H_2S$$

Thioacetamide   Acetic acid

Hydrogen sulfide is a gas under ordinary conditions, its low boiling point ($-61°C$) as compared with that of water indicating that there must be little hydrogen bonding in the liquid. H_2S is a bent molecule with a bond angle of $92°$. This is less than in H_2O ($105°$), probably because of the lower electronegativity of sulfur. If the bonding pairs are withdrawn from the sulfur in H_2S more than from the oxygen in H_2O, this will serve to reduce bonding pair–bonding pair repulsion, thus decreasing the bond angle in H_2S.

Hydrogen sulfide is a weak diprotic acid with $K_1 = 1.1 \times 10^{-7}$ and $K_2 = 1.0 \times 10^{-14}$. We have already discussed the role of hydrogen sulfide in precipitating sulfides of varying solubility in Sec. 17-4. Notice the small value of K_2 for H_2S; this means that the sulfide ion is extremely well hydrolyzed, and so solutions of alkali-metal sulfides—all of which are quite soluble—are extremely basic.

The sulfide ion will react with sulfur to form polysulfide ions. When solutions of alkali metal or ammonium sulfide are boiled with sulfur these ions are formed:

$$S^{2-} + S(s) \longrightarrow S_2^{2-}$$

$$S_2^{2-} + S(s) \longrightarrow S_3^{2-}$$

etc., where we have for simplicity represented sulfur by its symbol, rather than by S_8. These ions are really only sulfide ions with varying numbers of sulfurs tacked on; S_4^{2-} is, for example,

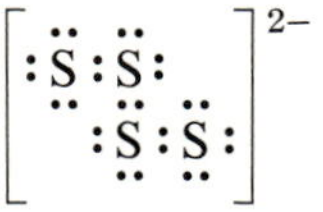

$$\left[\begin{array}{cc} \ddot{S} : \ddot{S} \\ \ddot{S} : \ddot{S} \end{array} \right]^{2-}$$

Old solutions of sulfides usually contain polysulfides, because the sulfide ion is oxidized to sulfur by the oxygen of the air:

$$O_2(g) + 2S^{2-} + 2H_2O \longrightarrow 2S(s) + 4OH^-$$

Some of the sulfur then combines (slowly) with S^{2-} to form a mixture of poly-sulfides (sometimes formulated S_x^{2-}).

SULFUR DIOXIDE AND THE SULFITES. When sulfur is burned in air, sulfur dioxide, SO_2, is produced:

$$S(s) + O_2(g) \longrightarrow SO_2(g)$$

This is a colorless gas which condenses to a liquid at 10°C (at 1 atm). It is toxic, but its extremely sharp, irritating odor makes it virtually impossible to inhale a fatal dose. It was once used as a commercial and domestic refrigerant, but now is produced primarily for further oxidation in the manufacture of sulfuric acid (see below).

SO_2 is a bent molecule with a resonance hybrid structure analagous to that of ozone:

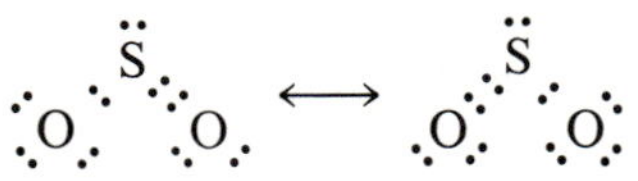

Sulfur dioxide is the anhydride of sulfurous acid, H_2SO_3, and aqueous solutions of SO_2 are usually referred to as such. The concentration of H_2SO_3 in such solutions is apparently quite low, however; so the first dissociation is often represented as

$$SO_2 + H_2O \longrightarrow H^+ + HSO_3^- \qquad K_1 = 1.3 \times 10^{-2}$$

The salts of sulfurous acid are the *hydrogen sulfites* which contain the HSO_3^- ion, and the *sulfites* which contain the trigonal pyramidal SO_3^{2-} ion. Any sulfite or bisulfite salt will evolve SO_2 gas when acidified:

$$SO_3^{2-} + 2H^+ \longrightarrow H_2O + SO_2(g)$$

Sulfites are reducing agents; their aqueous solutions are not stable, being gradually transformed to sulfate solutions due to oxidation by O_2:

$$2SO_3^{2-} + O_2(g) \longrightarrow 2SO_4^{2-}$$

Sulfites are also oxidizing agents. They will, for example, oxidize the sulfide ion in solution:

$$3H_2O + SO_3^{2-} + 2S^{2-} \longrightarrow 3S(s) + 6OH^-$$

Thus SO_3^{2-} and S^{2-} cannot coexist stably in solution.

SULFUR TRIOXIDE AND THE SULFATES. The oxidation of SO_2 to SO_3 by oxygen is spontaneous:

$$SO_2(g) + \tfrac{1}{2}O_2(g) \longrightarrow SO_3(g) \qquad \Delta G° = -70.9 \text{ kJ}$$

but very slow. This reaction is a key step in the manufacture of sulfuric acid by the *contact process*. In order to speed up the reaction, a catalyst, usually vanadium pentoxide, V_2O_5, is used.

The SO_3 molecule is trigonal planar and may be represented as a resonance hybrid of the three forms:

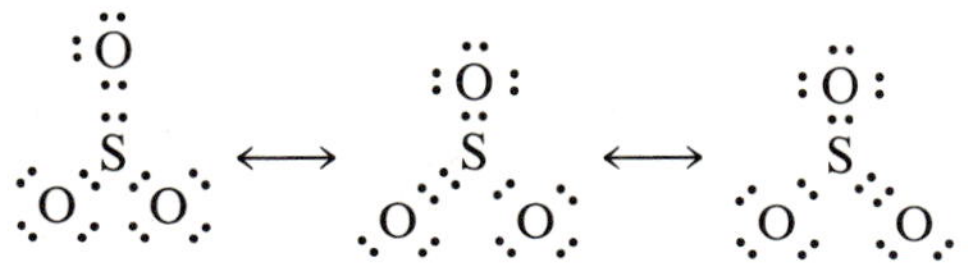

SO_3 is the anhydride of *sulfuric acid,* H_2SO_4. This acid is the most important industrial chemical with the exception, of course, of water. Nearly everything manufactured on earth depends in one way or another on sulfuric acid. (In 1978, 40 million tons were produced in the United States.) It is made today almost exclusively by the *contact process*. As has been described, this process involves the catalytic oxidation of SO_2 to SO_3. In theory one should then be able to add water to get H_2SO_4:

$$H_2O(l) + SO_3(g) \longrightarrow H_2SO_4(l)$$

In practice this reaction is slow; the H_2SO_4 tends to form as a fog whose particles refuse to coalesce. The problem is handled industrially by dissolving the SO_3 in H_2SO_4:

$$SO_3(g) + H_2SO_4(l) \longrightarrow H_2S_2O_7(l)$$

$H_2S_2O_7$ is known as *disulfuric acid,* or *pyrosulfuric acid.* (Older names which are sometimes still heard are *fuming sulfuric acid* and *oleum.*) Addition of water to disulfuric acid produces the desired sulfuric acid:

$$H_2S_2O_7(l) + H_2O(l) \longrightarrow 2H_2SO_4(l)$$

In each of these sulfur compounds each sulfur is surrounded tetrahedrally by four oxygens, so that the above reaction can be shown as

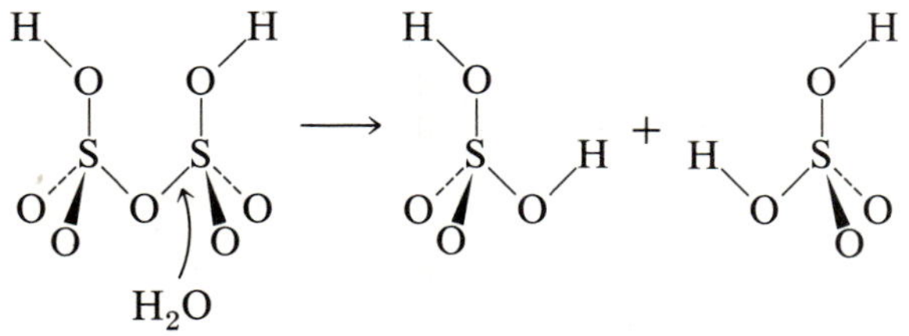

(Here the wedges indicate bonds to oxygens above the plane of the paper, and the dashed lines to oxygens below this plane.)

Sulfuric acid is commercially available as an approximately 95 percent by mass, or 18 M, solution. Its boiling point is about 330°C, although at this temperature it begins to decompose into SO_3 and H_2O. The pure liquid is even more self-dissociated than water:

$$H_2SO_4 \rightleftharpoons H^+ + HSO_4^- \qquad K = [H^+][HSO_4^-] = 2 \times 10^{-4}$$

but this dissociation is usually written as

$$2H_2SO_4 \rightleftharpoons H_3SO_4^+ + HSO_4^- \qquad K = [H_3SO_4^+][HSO_4^-] = 2 \times 10^{-4}$$

Eighteen-molar or "concentrated," H_2SO_4 is a powerful dehydrating agent and finds use as such in the laboratory. It owes its dehydrating ability to the tendency to form hydrates such as $H_2SO_4 \cdot H_2O$, $H_2SO_4 \cdot 2H_2O$, etc. It will even "suck" water out of carbohydrates and some other organic compounds which contain hydrogen and oxygen. With sucrose (cane sugar) the reaction can be written as

$$C_{12}H_{22}O_{11}(s) + 11H_2SO_4 \longrightarrow 12C(s) + 11H_2SO_4 \cdot H_2O$$

from which the carbon is produced as a spongy mass.

The oft-repeated rule, "Always add acid to water, not vice versa," is most important when diluting concentrated sulfuric acid. If water is added to the acid, the less-dense water tends to float on the surface of the H_2SO_4 (density about 1.8 g cm^{-3}). The mixing of the two liquids is thus limited to the area of contact of the two layers, so that the large heat of hydration is all generated here. This can cause violent spattering of hot H_2SO_4. The recommended procedure for diluting concentrated H_2SO_4 is to add the acid slowly in a thin stream to the water, while stirring the mixture briskly. This permits the heat to be generated more or less evenly throughout the solution. (Even so, the increase in temperature is remarkable, and so cooling the outside of the container is a wise procedure.)

Sulfuric acid is an oxidizing agent, but not a very strong or rapid one unless it is hot and concentrated. When it is reduced, the product is usually SO_2:

$$3H_2SO_4(l) + 2Br^- \longrightarrow Br_2 + SO_2(g) + 2HSO_4^- + 2H_2O$$

Copper metal, not a strong enough reducing agent to be dissolved in so-called nonoxidizing acids like HCl, will dissolve in hot, concentrated H_2SO_4:

$$3H_2SO_4 + Cu(s) \longrightarrow Cu^{2+} + SO_2(g) + 2HSO_4^- + 2H_2O$$

(In the last two equations we have represented the acid as H_2SO_4, because this is the predominant species present; significant dissociation to form hydronium and bisulfate ions requires the presence of water.)

Sulfuric acid is a strong diprotic acid. Even its second dissociation constant is quite large, 1.2×10^{-2}. It forms two series of salts, the *hydrogen sulfates,* or *bisulfates,* which contain the HSO_4^- ion, and the *sulfates* which contain the SO_4^{2-} ion. Bisulfates are acidic, due to the high value of K_2 for H_2SO_4. Sulfates might be expected to be basic, but the extent of hydrolysis is so slight that the effect is very small: a 1 M Na_2SO_4 solution has pH = 7.9.

THIOSULFATES. Sulfur atoms can substitute for oxygen atoms in oxoacids and oxoanions. An example is the *thiosulfate ion* $S_2O_3^{2-}$. (The *thio-* prefix indicates the S-for-O substitution.)

$$\left[\begin{array}{c} :\ddot{O}: \\ :\ddot{O}:S:\ddot{O}: \\ :\ddot{O}: \end{array}\right]^{2-} \quad \left[\begin{array}{c} :\ddot{S}: \\ :\ddot{O}:S:\ddot{O}: \\ :\ddot{O}: \end{array}\right]^{2-}$$

Sulfate   Thiosulfate

This ion can be prepared by boiling sulfur in a sulfite solution:

$$S(s) + SO_3^{2-} \longrightarrow S_2O_3^{2-}$$

The parent acid, *thiosulfuric acid,* is unstable, even in solution. Acidifying a thiosulfate solution forms sulfur, first as a colloid, detectable only by means of the Tyndall effect (Sec. 12-1), and later as a milky suspension, which gradually coagulates and settles:

$$S_2O_3^{2-} + 2H^+ \longrightarrow S(s) + SO_2 + H_2O$$

The thiosulfate ion is an excellent complexing agent, especially for the silver ion:

$$Ag^+ + 2S_2O_3^{2-} \longrightarrow Ag(S_2O_3)_2^{3-}$$

This complex is so stable ($K_{diss} = 6 \times 10^{-14}$) that thiosulfate solutions will dissolve the insoluble silver salts AgCl, AgBr, and AgI.

SULFUR HALIDES. Sulfur forms a varied series of compounds with the halogens. These include compounds with the general formulas S_2X_2, SX_2, SX_4, SX_6, and S_2X_{10}, although not all halogens (X) are represented in each formulation. All of these are low-boiling, molecular compounds. The structures of the molecules can be predicted from VSEPR theory: S_2X_2 molecules have a skewed structure resembling that of H_2O_2; SX_2 molecules are bent; SX_4 molecules have the seesaw shape derived from a trigonal bipyramid with a single equatorial lone pair; SF_6, the only well-characterized hexafluoride, is octahedral; finally, S_2F_{10} may be considered to be two SF_6 octahedra, each of which has lost an F atom so that the S atoms can bond directly. The structures of the sulfur halides are summarized in Fig. 20-7.

Sulfur hexafluoride, SF_6, deserves special mention. At room temperature it is a gas with such a high molecular weight that it is more than 4 times as dense as oxygen gas. It is remarkably inert, probably due to its symmetry; the "shell" of

Figure 20-7
Structures of the sulfur halides.

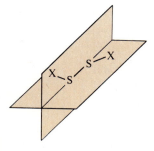   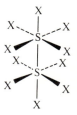

Disulfur dihalide   Sulfur dihalide   Sulfur tetrahalide   Sulfur hexahalide   Disulfur decahalide

fluorines protects the $+6$ sulfur from attack by reactants. This leads us to predict that its inertness might be kinetic, not thermodynamic. Consider, for instance, the reaction

$$SF_6(g) + 3H_2O(g) \longrightarrow SO_3(g) + 6HF(g)$$

Using standard free energies of formation, we can calculate $\Delta G°$ for this reaction:

$$\Delta G° = \Delta G°_{SO_3} + 6\,\Delta G°_{HF} - (\Delta G°_{SF_6} + 3\,\Delta G°_{H_2O})$$
$$= -371.1 + 6(-270.7) - [(-1105.4) + 3(-228.6)] = -204.1 \text{ kJ}$$

Thermodynamically the above reaction is spontaneous, but it is too slow to be observed. Sulfur hexafluoride finds application as a gaseous insulator in transformers, high-voltage switches, and electrostatic loudspeakers.

20-7
The group-VA
nonmetals:
nitrogen
and phosphorus

The elements of group VA are nitrogen, phosphorus, arsenic, antimony, and bismuth. These are occasionally called the *pnicogens,* or *pnigogens,* from a Greek word meaning "to suffocate." (In German the name for nitrogen is *Stickstoff,* literally, "suffocating stuff.") As in group VIA there is a clear transition from nonmetallic to metallic character going down the group. Chemically, nitrogen and phosphorus are nonmetals; their hydroxo compounds are acidic. Arsenic and antimony are best classed as metalloids, and their hydroxides are amphoteric. Bismuth is predominately metallic, forming hydroxides which are basic.

The physical properties of the group VA elements reflect a similar decrease in the tightness with which electrons are held. Nitrogen is a nonmetal. Phosphorus exhibits a bewildering collection of allotropes, but one of them is black and conducts electricity fairly well. Arsenic, antimony, and bismuth each have at least one metallic-appearing form and the conductivity of these increases in the expected order: arsenic < antimony < bismuth.

In this section we will focus on nitrogen and phosphorus. (We will discuss arsenic and antimony in Sec. 21-5 and bismuth in Sec. 21-4.) Some of the properties of the group VA elements are summarized in Table 20-13.

Nitrogen

The name of this element comes from Greek words meaning "nitron-forming." *Nitron* referred to potassium nitrate, KNO_3. In French the word for nitrogen is *azote,* coined by Lavoisier and meaning "no life." (Lavoisier observed that a

Table 20-13
Group VA elements

	N	P	As	Sb	Bi
Valence shell configuration	$2s^2 2p^3$	$3s^2 3p^3$	$4s^2 4p^3$	$5s^2 5p^3$	$6s^2 6p^3$
Physical appearance (most stable allotrope at STP)	Colorless gas	Black solid*	Metallic solid	Metallic solid	Metallic solid
Normal melting point, °C	-210	416 (red sublimes)	615 (sublimes)	630	271
Normal boiling point, °C	-196	—	—	1635	1420
Electronegativity	3.0	2.1	2.0	1.9	1.9

*Other allotropes are more commonly observed.

mouse died when given an atmosphere of nitrogen to breathe). Ammonium salts and nitrates were known to the earliest alchemists, who also prepared what must have been nitric acid.

Nitrogen occurs on earth as the major constituent of air ($\sim$78 percent by volume). Inorganic compounds of nitrogen are not commonly found as minerals, because most of them are soluble in water. In some locations with dry climates (or which in the past have had dry climates) there are beds of nitrates, usually impure sodium or potassium nitrate.

Nitrogen is found combined in organic compounds in all living matter, animal and plant. *Proteins,* for example, are giant molecules in which the building blocks are nitrogen-containing compounds called *amino acids* (Chap. 23). A simple amino acid is aminoacetic acid, or glycine,

The source for the nitrogen combined in living matter is ultimately the atmosphere, although N_2 gas itself is not useable for protein synthesis. One consequence of electrical storms is, however, the production of small quantities of nitrates (NO_3^-) and nitrites (NO_2^-) which are carried to earth in rain. In addition certain bacteria in the soil and in the roots of some plants, chiefly the legumes, convert atmospheric nitrogen into organic nitrogen-containing compounds which are then converted by other bacteria into nitrates, the form of combined nitrogen most useable by plants in their synthesis of protein. The drain on atmospheric nitrogen is compensated by the production of N_2 by certain soil bacteria and from the decay of plant and animal protein material. A complex interrelationship exists involving nitrogen in the atmosphere, NO_3^-, NO_2^-, and NH_4^+ in the soil, and organic nitrogen compounds in bacteria and higher organisms, plant and animal; this relationship, the *nitrogen cycle,* keeps the nitrogen content of the atmosphere constant.

Elemental nitrogen is prepared industrially by the fractional distillation of liquid air (Sec. 12-4). In the laboratory it can be formed from the thermal decomposition of certain compounds. Very pure N_2 can be prepared, for example, when sodium azide, NaN_3, is *cautiously* heated under vacuum:

$$2NaN_3(s) \longrightarrow 2Na(l) + 3N_2(g)$$

Nitrogen is also produced by heating ammonium nitrite or a mixture of an ammonium salt and a nitrite:

$$NH_4NO_2(s) \longrightarrow N_2(g) + 2H_2O(g)$$

Again, the heating must be done with care; nitrites (and also nitrates) can explode if heated too vigorously or if trace quantities of reducing agents are present. The ammonium nitrite decomposition has had an interesting application in the internal pressurization of tennis balls. Before the two halves of the ball are sealed together, if small quantities of NH_4Cl and $NaNO_2$ are placed inside, the heat used in the sealing process produces nitrogen, which pressurizes the ball.

The nitrogen molecule can be represented as

$$:N:::N:$$

while with the molecular-orbital model the triple bond is described as $(\sigma_x)^2(\pi_y)^2(\pi_z)^2$. This bond is very strong, having a bond energy of 945 kJ mol^{-1}. The great strength of the N_2 triple bond is responsible for the slow rate with which nitrogen reacts with most other substances. The earth would be a much different place if this bond were more easily broken. Consider, for example, the reaction between N_2 and O_2 in the air and H_2O in the oceans to form nitric acid:

$$2N_2(g) + 5O_2(g) + 2H_2O(l) \longrightarrow 4H^+ + 4NO_3^-$$

ΔG° for this reaction at 25°C is $+29.2$ kJ; so the reaction appears to be nonspontaneous. But 29.2 kJ represents the *standard* free-energy change. If we set $P_{N_2} = 0.78$ atm, $P_{O_2} = 0.21$ atm, and $[H^+] = [NO_3^-] = 0.010\ M$, then ΔG for the reaction turns out to be (see Sec. 18-6):

$$\Delta G = \Delta G^\circ + RT \ln \frac{[H^+]^4[NO_3^-]^4}{P_{N_2}{}^2 P_{O_2}{}^5}$$

$$= 29.2 \times 10^3\ J + 2.303(8.314)(298) \log \frac{(0.010)^4(0.010)^4}{(0.78)^2(0.21)^5}\ J$$

$$= -41.5 \times 10^3\ J \quad \text{or} \quad -41.5\ kJ$$

Thus the air and oceans are thermodynamically unstable, tending to change into a dilute solution of nitric acid! However, the strength of the N_2 triple bond provides for a very high activation energy for this reaction, and so it is very slow, indeed. (Science-fiction writers have "invented" a catalyst which would allow this reaction to occur rapidly at room temperature!)

In its compounds nitrogen shows all the negative and positive oxidation states from -3 to $+5$. In most of these compounds nitrogen *shares* electron pairs with other atoms. Only with highly electropositive metals such as lithium and sodium does nitrogen form simple *nitride* ions, N^{3-}. The nitrides of less electropositive elements, aluminum, for example, consist of giant covalent networks. And the so-called transition-metal nitrides do not contain -3 nitrogen at all. They consist of N atoms in the interstices of a metal-atom lattice. *Case-hardened steel* is made by dipping hot steel in molten sodium cyanide, NaCN, to form an extremely tough, hard, surface layer of iron with interstitial N and C atoms.

AMMONIA. One of the most important industrial chemicals is ammonia, NH_3, which is produced by means of the *Haber process*:

$$N_2(g) + 3H_2(g) \longrightarrow 2NH_3(g) \qquad \Delta H^\circ = -92.2\ kJ$$

A fairly high pressure (500 to 1000 atm) is used to favor the formation of the product, together with a temperature of 400 to 500°C, high enough to permit the reaction to take place at a reasonable rate, but not high enough to shift the equilibrium far to the left. An iron catalyst is also used to speed the rate. Ammonia can be produced in the laboratory by adding a strong base to an ammonium salt and warming, if necessary:

$$NH_4^+ + OH^- \longrightarrow NH_3(g) + H_2O$$

Ammonia is a colorless gas (normal boiling point $-33.4°C$) with a characteristic pungent odor. It is not very toxic, although inhalation of the gas in high concentrations causes severe respiratory problems. Ammonia is usually considered to be nonflammable, but it can be made to burn in air:

$$4NH_3(g) + 3O_2(g) \longrightarrow 2N_2(g) + 6H_2O(g)$$

and some mixtures of NH_3 and air will even explode.

Liquid ammonia is colorless and resembles water in many respects. It has a fairly high dielectric constant (about 22 at its normal boiling point), is hydrogen bonded, and self-ionized

$$2NH_3 \longrightarrow NH_4^+ + NH_2^- \qquad K = 1 \times 10^{-28}$$

Liquid ammonia is consequently a good solvent for ionic solids, although solubilities are generally lower than in water. In one respect, however, liquid ammonia is quite unlike water in solvent properties; whereas water reacts vigorously with highly electropositive metals such as sodium,

$$H_2O + Na(s) \longrightarrow Na^+ + OH^- + \tfrac{1}{2}H_2(g)$$

the corresponding reaction in ammonia

$$NH_3 + Na(s) \longrightarrow Na^+ + NH_2^- + \tfrac{1}{2}H_2(g)$$

is very slow, especially if the reactants are pure. Instead, sodium *dissolves* in ammonia to form a strikingly blue solution. Magnetic and electrical conductivity measurements show that the solution contains unpaired electrons, apparently *solvated* (*ammoniated*, in this case) as if they were anions. The dissolving process can thus be represented as

$$Na(s) \longrightarrow Na^+(NH_3) + e^-(NH_3)$$

The mobility of the electrons is very high, giving rise to very high conductivities of these solutions. As the concentration of the solution increases its color changes to a brilliant bronze, and its conductivity becomes as high as that of some metals. In dilute solution the electron is apparently weakly trapped in a cavity among the surrounding NH_3 molecules. In concentrated solution the electron is essentially delocalized, as in a liquid metal. Solutions of all the alkali metals, of some of the alkaline-earth metals (Ca, Sr, and Ba) and of some of the more electropositive transition metals can be prepared. These solutions are not stable indefinitely; the expected decomposition, which can be written

$$e^-(NH_3) + NH_3 \longrightarrow NH_2^-(NH_3) + \tfrac{1}{2}H_2(g)$$

takes place at a rate which depends on the metal ion present and on the presence of catalysts such as OH^- and NH_2^-.

Ammonia is extremely soluble in water. Historically these solutions have been called *ammonium hydroxide,* but this compound has never been isolated. Furthermore, the evidence indicates that discrete NH_4OH molecules do not exist even in solution. What actually is present? The system is evidently an exceedingly complex one involving proton interchanges among the following hydrated species:

H_2O, NH_3, NH_4^+, OH^-, and H^+. The basic dissociation of ammonia can be represented simply as either

$$NH_3 + H_2O \rightleftharpoons NH_4^+ + OH^-$$

or

$$NH_4OH \rightleftharpoons NH_4^+ + OH^-$$

for which the dissociation constant K_b at 25°C has been determined to be 1.81×10^{-5}.

The ammonia molecule is a trigonal pyramid with the H—N—H angle equal to 108°. It undergoes an interesting, rapid, oscillatory motion known as *inversion,* in which the nitrogen atom can be pictured as diving back and forth through the plane of the three hydrogens in the molecule. Actually the hydrogens do most of the moving since they are much lighter than the nitrogen. The inversion of NH_3 is shown schematically in Fig. 20-8.

Salts of ammonia contain the tetrahedral *ammonium ion,* NH_4^+. Solutions of ammonium salts of strong acids are acidic due to hydrolysis, which can be represented as

$$NH_4^+ + H_2O \rightleftharpoons NH_4OH + H^+$$

or as

$$NH_4^+ + H_2O \rightleftharpoons NH_3 + H_3O^+$$

or even as

$$NH_4^+ \rightleftharpoons NH_3 + H^+$$

Solid NH_4Cl sublimes with decomposition at 340°C:

$$NH_4Cl(s) \rightleftharpoons NH_3(g) + HCl(g).$$

The two product gases immediately recombine to form a dense smoke of NH_4Cl. Since H_2SO_4 is much less volatile than HCl, heating ammonium sulfate produces only ammonia as the gaseous product:

$$(NH_4)_2SO_4(s) \longrightarrow NH_4HSO_4(s) + NH_3(g)$$

When ammonium nitrate is heated, nitrous oxide, N_2O is formed:

$$NH_4NO_3(s) \longrightarrow N_2O(g) + 2H_2O(g)$$

(This reaction can be difficult to control; it sometimes takes off explosively.)

OTHER NEGATIVE OXIDATION STATES OF NITROGEN. The -2 state of nitrogen is represented by hydrazine, N_2H_4, which contains an N—N single bond:

$$H\!:\!\overset{\cdot\cdot}{\underset{H}{N}}\!:\!\overset{\cdot\cdot}{\underset{H}{N}}\!:\!H$$

Hydrazine is a colorless liquid, which resembles ammonia in some ways although its reducing action is powerful and often violent. It burns readily in air and has

Figure 20-8
Inversion of ammonia.

been used in rocket propulsion systems in which it is oxidized by various oxidants, including H_2O_2 and N_2O_4.

Nitrogen exhibits the -1 oxidation state in *hydroxylamine*, NH_2OH:

$$H:\overset{\displaystyle ..}{\underset{\displaystyle H}{N}}:\overset{\displaystyle H}{\overset{\displaystyle ..}{\underset{\displaystyle ..}{O}}}:$$

It is a weak base, dissociating to give *hydroxylammonium ions:*

$$NH_2OH + H_2O \longrightarrow NH_3OH^+ + OH^- \qquad K_b = 9.1 \times 10^{-9}$$

Pure hydroxylamine is a highly unstable solid at room temperature.

OXIDES OF NITROGEN. Nitrogen combines with oxygen to form a series of oxides. *Nitrous oxide*, N_2O, is a linear molecule containing nitrogen in the $+1$ state. It may be described as a resonance hybrid of the following principal forms:

$$:N::N::\underset{\displaystyle ..}{O}: \longleftrightarrow :N:::N:\overset{\displaystyle ..}{\underset{\displaystyle ..}{O}}:$$

Compared with the other oxides of nitrogen, nitrous oxide is surprisingly stable and unreactive, at least at room temperature. It will undergo thermal decomposition to form its elements, which may account for its ability to support the combustion of most things that will burn in oxygen gas. N_2O has been known as *laughing gas* since the early 1800s, when it was discovered that inhalation of it produced happy feelings. It has been used for many years mixed with oxygen as a mild general anesthetic. (It is the dentist's "gas.")

The $+2$ state of nitrogen is represented by another oxide, NO, usually called *nitric oxide*. It is formed from N_2 and O_2 as a result of lightning discharges in the lower atmosphere and ultraviolet radiation in the upper atmosphere. It is also produced in high-temperature fires, including those which occur in automobile engines. It can be formed from the reduction of nitric acid:

$$NO_3^- + 3Fe^{2+} + 4H^+ \longrightarrow NO(g) + 3Fe^{3+} + 2H_2O$$

NO is an odd molecule and so is paramagnetic. Its structure may be represented as a resonance hybrid of the forms:

$$:\underset{\displaystyle .}{N}::\underset{\displaystyle ..}{O}: \longleftrightarrow :\underset{\displaystyle ..}{N}::\underset{\displaystyle .}{O}:$$

The MO picture of this molecule indicates a bond order of 2.5, in agreement with the observed bond length: $KK(\sigma_s)^2(\sigma_s^*)^2(\sigma_x)^2(\pi_y)^2(\pi_z)^2(\pi_y^*)^1$.

Nitric oxide is a colorless gas which reacts readily with oxygen to form brown NO_2, nitrogen dioxide. NO shows some tendency to dimerize in the liquid state. The solid may be composed entirely of N_2O_2 dimers.

Nitrogen dioxide, NO_2, is a brown gas formed by the oxidation of NO, as mentioned above. It can also be produced by the reduction of concentrated nitric acid by some metals and by the thermal decomposition of nitrates:

$$2Pb(NO_3)_2(s) \longrightarrow 2PbO(s) + 4NO_2(g) + O_2(g)$$

NO_2 is another odd molecule, but shows a much greater tendency to dimerize than does NO:

$$2NO_2(g) \longrightarrow N_2O_4(g) \qquad \Delta H^\circ = -61 \text{ kJ}$$

As nitrogen dioxide is cooled, its brown color fades and its paramagnetism decreases, as the percentage of NO_2 in the mixture decreases. In the liquid at its normal boiling point (21°C) the percentage of NO_2 is only 0.1 percent. For this bent molecule the forms which contribute primarily to the resonance hybrid are:

OXOACIDS AND ANIONS OF NITROGEN. Nitrogen forms two important oxoacids. The first is *nitrous acid*, HNO_2, a weak acid ($K_a = 5.0 \times 10^{-4}$) which can be prepared by acidifying solutions of nitrites, but which cannot be isolated pure. Alkali-metal nitrites can be prepared by the thermal decomposition of nitrates:

$$2NaNO_3(l) \longrightarrow 2NaNO_2(l) + O_2(g)$$

In solution nitrous acid slowly disproportionates to form NO and nitric acid:

$$3HNO_2 \longrightarrow 2NO(g) + H^+ + NO_3^- + H_2O$$

The bonding in HNO_2 is described by the Lewis structure:

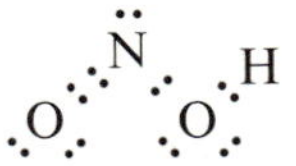

while the NO_2^- ion is best represented as a resonance hybrid:

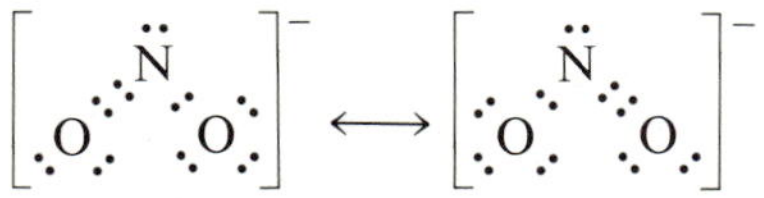

Nitrites are poisonous, but have been widely used at low concentrations to preserve pork and other meats. There is evidence, however, that these nitrites react with proteins to form organic compounds called *nitrosoamines* which are carcinogens (cancer-producing agents).

Nitric acid, HNO_3, is an important industrial chemical, most of which is produced by the *Ostwald process*. In this process ammonia is first oxidized using a platinum or platinum-rhodium catalyst:

$$4NH_3(g) + 5O_2(g) \xrightarrow{\text{Pt}} 4NO(g) + 6H_2O(g)$$

The nitric oxide is then mixed with more oxygen to produce nitrogen dioxide,

$$2NO_2(g) + O_2(g) \longrightarrow 2NO_2(g)$$

which is then dissolved in water where it disproportionates:

$$3NO_2(g) + H_2O \longrightarrow 2H^+ + 2NO_3^- + NO(g)$$

The resulting solution is distilled to yield pure HNO_3, and the byproduct NO is recycled.

The forms which contribute to the HNO_3 resonance hybrid are

Nitric acid is a strong acid which, when either pure or in solution, undergoes slow photochemical decomposition:

$$4H^+ + 4NO_3^- \longrightarrow 4NO_2 + 2H_2O + O_2(g)$$

The dissolved NO_2 colors the solution yellow or brown, but the decomposition can be greatly slowed by keeping the solution in a brown bottle.

Nitric acid is a powerful oxidizing agent when concentrated. Its reduction products vary greatly depending upon the reducing agent and the concentrations involved. Possible products include NO_2, HNO_2, NO, N_2O, N_2, NH_3OH^+, $N_2H_5^+$, and NH_4^+. Some typical reactions are

1    $Cu(s) + 2NO_3^- + 4H^+ \longrightarrow Cu^{2+} + 2NO_2(g) + 2H_2O$
2    $3Cu(s) + 2NO_3^- + 8H^+ \longrightarrow 3Cu^{2+} + 2NO(g) + 4H_2O$
3    $4Zn(s) + 2NO_3^- + 10H^+ \longrightarrow 4Zn^{2+} + N_2O(g) + 5H_2O$
4    $4Zn(s) + NO_3^- + 10H^+ \longrightarrow 4Zn^{2+} + NH_4^+ + 3H_2O$

Of the reactions with copper, reaction 1 is favored with concentrated HNO_3 and reaction 2 with dilute. Of the reactions with zinc, reaction 3 takes place with dilute HNO_3, and reaction 4 with dilute HNO_3 to which H_2SO_4 has been added to increase [H^+]. With many metals a mixture of products is obtained.

Concentrated nitric acid will dissolve nearly all metals. The few that resist its oxidizing ability include the noble metals platinum, gold, rhodium and iridium. Some, iron and aluminum, for example, react to form an impervious oxide film which essentially stops the reaction. (See passivation, Sec. 22-6.) Even metals like platinum succumb to a powerful mixture known as *aqua regia*, which is a mixture of two or three parts concentrated HCl to one part concentrated HNO_3, by volume. Aqua regia will dissolve metals and many compounds which are insoluble in either acid alone. It owes its solvent power to the oxidizing ability of NO_3^- and the complexing ability of Cl^-. Typical reactions are

$$3Pt(s) + 4NO_3^- + 16H^+ + 18Cl^- \longrightarrow 3PtCl_6^{2-} + 4NO(g) + 8H_2O$$

$$3HgS(s) + 2NO_3^- + 8H^+ + 12Cl^- \longrightarrow 3HgCl_4^{2-} + 2NO(g) + 3S(s) + 4H_2O$$

The salts of nitric acid are the nitrates, most of which are quite soluble in water. The nitrate ion is completely unhydrolyzed and is a poor complexing agent. It can be detected in qualitative analysis by adding ferrous sulfate, $FeSO_4$, and then carefully pouring concentrated H_2SO_4 down the inside wall of the test tube. At the top of the more dense H_2SO_4 layer the NO_3^- is reduced to NO by Fe^{2+}. The NO then complexes with more Fe^{2+} to form the nitrosyl ferrous complex, $FeNO^{2+}$, which is visible as a brown ring or layer. This is the *brown-ring test* for nitrates.

The nitrate ion is a trigonal planar ion which can be represented as a resonance hybrid of the three equivalent contributing forms:

Phosphorus (the name comes from a Greek word meaning "light-bearing") is the second member of group VA. Unlike nitrogen it does not occur uncombined in nature; most phosphorus is found in deposits of phosphate rock, impure $Ca_3(PO_4)_2$; apatite, $Ca_5F(PO_4)_3$; and similar calcium phosphate compounds. Elemental phosphorus is obtained industrially by heating the phosphorus-containing rock with carbon (in the form of coke) and silicon dioxide (sand) in an electric furnace:

$$2Ca_3(PO_4)_2(s) + 6SiO_2(s) + 10C(s) \longrightarrow P_4(g) + 6CaSiO_3 + 10CO(g)$$

The phosphorus is formed first as P_2 molecules which dimerize and are condensed under water to form solid P_4, white phosphorus.

Phosphorus can exist in at least six different solid allotropic forms, of which we will mention only three. *White phosphorus,* mentioned above, is a very reactive, poisonous, volatile, waxy, yellow-white substance which is very soluble in nonpolar solvents such as benzene, C_6H_6, and carbon disulfide, CS_2. It has a cubic structure (although a hexagonal variety also exists) in which P_4 molecules occupy the lattice points. The P_4 molecule has the compact, nonpolar, tetrahedral structure shown in Fig. 20-9. White phosphorus must be kept stored under water in order to prevent it from igniting spontaneously in air:

$$P_4(s) + 5O_2(g) \longrightarrow P_4O_{10}(s) \qquad \Delta H° = -2940 \text{ kJ}$$

White phosphorus boils at 280°C to form P_4 vapor which dissociates above 700°C to form P_2, analogous to N_2. Heating white phosphorus to about 260°C using iodine or sulfur as a catalyst produces a red amorphous form of phosphorus which subsequently crystallizes at higher temperatures to make the form known as *red phosphorus*. This is a much less reactive, less poisonous form of phosphorus, which can be stored in contact with the air and which is insoluble in most solvents. It has a structure composed of "tubes" of pentagonal cross section arranged in layers, such that the tubes in adjacent layers are at right angles to each other. The structure of one such tube is shown in Fig. 20-10.

White and red phosphorus are the most common forms of the element, but the form which is most thermodynamically stable under ordinary conditions is believed to be *black phosphorus*. Its structure consists of interconnected, corrugated layers of phosphorus atoms.

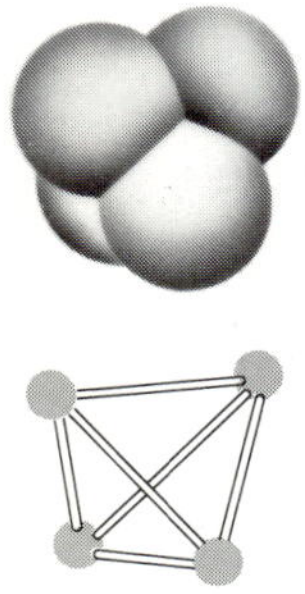

PHOSPHORUS OXIDES. When phosphorus is burned in a limited supply of oxygen, the product is tetraphosphorus hexaoxide, P_4O_6. It is also known as phosphor*ous* oxide (note the suffix) or *phosphorus trioxide,* an old name still justifiable on the basis of its empirical formula. P_4O_6 is a volatile liquid freezing at about room temperature (23.8°C). Its molecule is shown in Fig. 20-11 with the P_4 molecule shown for comparison. As can be seen, the white phosphorus tetrahedron is expanded to accommodate an oxygen atom covalently bonded between each pair of phosphorus atoms.

When phosphorus is burned in an excess of oxygen, or simply in the open air, it forms tetraphosphorus decaoxide, P_4O_{10}, also called phosphor*ic* oxide or *phosphorus pentoxide*. This is a volatile, white solid which sublimes (at 1 atm) at 360°C. Its molecule is like that of P_4O_6, except that each phosphorus atom

Figure 20-9
White phosphorus molecule.

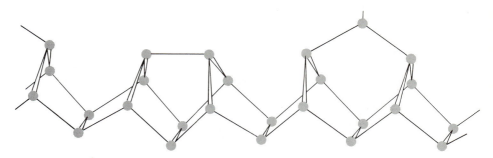

Figure 20-10
Red phosphorus struc-
ture.

accommodates one more oxygen in addition to the bridging oxygens. (See Fig. 20-11.)

P_4O_{10} is one of the strongest dehydrating agents known. When exposed to water vapor it becomes hydrated to produce a series of phosphoric acids (see below) which form a gummy layer on the surface. When P_4O_{10} is hydrated rapidly, as when liquid water is added to it, large amounts of heat are suddenly given off, and violent spattering can occur.

OXOACIDS AND SALTS OF PHOSPHORUS. When water is added to P_4O_6, *phosphorous acid* is formed:

$$P_4O_6(s) + 6H_2O \longrightarrow 4H_3PO_3$$

This acid, a solid at room temperature, might be better formulated as H_2PHO_3 to

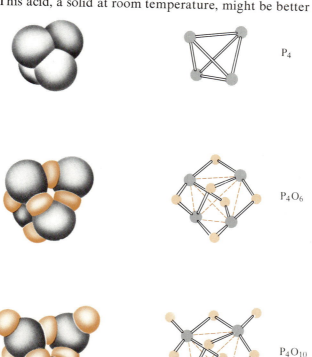

P_4

P_4O_6

P_4O_{10}

Figure 20-11
P_4, P_4O_6, and P_4O_{10} structures.

show that it is only a *diprotic* acid. ($K_1 = 1.6 \times 10^{-2}$ and $K_2 = 7 \times 10^{-7}$.) Its Lewis structure is drawn as

$$
\begin{array}{c}
\text{H} \\
:\!\ddot{\text{O}}\!:\!\text{P}\!:\!\ddot{\text{O}}\!:\!\text{H} \\
:\!\ddot{\text{O}}\!: \\
\text{H}
\end{array}
$$

Since the H—P bond is not broken by water, the acid can form only two series of salts: the *dihydrogen phosphites,* containing the $H_2PO_3^-$ ion, and the *monohydrogen phosphites,* containing the HPO_3^{2-} ion. The latter are often called, simply, the *phosphites.*

When water is added to P_4O_{10} a series of *phosphoric acids* is formed. The equations for the P_4O_{10} hydration are

$$P_4O_{10}(s) + 2H_2O \longrightarrow \underset{\text{Metaphosphoric acid}}{4HPO_3(s)}$$

$$2HPO_3(s) + H_2O \longrightarrow \underset{\text{Pyrophosphoric acid}}{H_4P_2O_7(s)}$$

$$H_4P_2O_7(s) + H_2O \longrightarrow \underset{\text{Orthophosphoric acid}}{2H_3PO_4(s)}$$

The name *phosphoric acid* always refers to the ortho acid. It is a poor oxidizing agent and a triprotic acid with $K_1 = 7.6 \times 10^{-3}$, $K_2 = 6.3 \times 10^{-8}$, and $K_3 = 4.4 \times 10^{-13}$. Its Lewis structure is

$$
\begin{array}{c}
:\!\ddot{\text{O}}\!: \\
\text{H}:\!\ddot{\text{O}}\!:\!\text{P}\!:\!\ddot{\text{O}}\!:\!\text{H} \\
:\!\ddot{\text{O}}\!: \\
\text{H}
\end{array}
$$

H_3PO_4 forms three series of salts: the *dihydrogen phosphates,* containing $H_2PO_4^-$, the *monohydrogen phosphates,* containing HPO_4^{2-}, and the *phosphates,* containing PO_4^{3-}. K_3 for H_3PO_4 is so small that solutions of Na_3PO_4, trisodium phosphate, are quite basic. This compound was once widely used in household and other detergent formulations, but concern over freshwater pollution by phosphates has limited this application in recent years.

20-8 Carbon   In group IVA of the periodic table there is only one good nonmetal for us to discuss: *carbon.* (The name comes from a Latin word meaning "charcoal.") Silicon, directly below carbon in the group, is best classed as a metalloid, as it shows considerable semimetallic character. As we have moved from right to left across the periodic table, the groups have become increasingly metallic, with nonmetallic elements becoming limited to the top of the group. Carbon has one allotrope which is electrically conductive; so we see a hint of metallic character even here. The hydroxo compounds of carbon are clearly acidic, however.

Elemental carbon   *Diamond* and *graphite* are the two common allotropic forms of carbon. Diamond crystallizes in cubic or octahedral crystals which are colorless when pure and which often show curved faces and edges due to lattice defects. It is the hardest naturally occurring substance and is used for grinding, machining, and engraving hard metals and other substances. Diamond's hardness is accounted for by its closely interlocked, three-dimensional structure, shown in Fig. 20-12. What may appear to be separated layers in this structure are not really such, since all four bonds of each carbon are equivalent in all respects. The four valence electrons of each carbon are shared with electrons of four adjacent carbon atoms in sp^3 (tetrahedral) hybrid orbitals.

Diamond is an electrical insulator since all valence electrons are fully involved in single (sigma) bond formation. It has a high *refractive index,* which means that light entering it from the air is bent strongly toward the interior of the diamond. It also shows great *dispersion* which means that the angle of bending of light varies with its wavelength (color). These two properties are exploited by diamond cutters, who make facets on gem-quality diamonds, producing great "fire" and sparkle.

The second allotrope of carbon is *graphite,* whose properties are far different from those of diamond. Graphite is a soft, black solid with a semimetallic luster that hints of loosely held electrons. It has a layer-type structure in which each layer is a "chicken-wire" net of carbon atoms, as shown in Fig. 20-13a. In each of these nets each carbon is bonded to three others at a distance of 0.142 nm, which is shorter than the C—C distance in diamond (0.154 nm). Each net is one layer in a stack with an interlayer distance of 0.334 nm. (See Fig. 20-13b.)

Within each layer in the graphite structure each carbon bonds to three adjacent carbons using three of its four valence electrons in sp^2 hybrid orbitals. This gives rise to a σ skeleton which is then supplemented by π bonding as follows: The fourth valence electron of each carbon is in a p orbital whose axis is perpendicular to the layer. These p orbitals overlap side to side to form a huge delocalized π orbital extending over and under the whole layer. Each carbon contributes one electron to this delocalized π cloud, and these electrons are responsible for the semimetallic luster of graphite. They also account for the observed electrical conductivity of graphite, which is about 10^5 times greater in directions parallel to

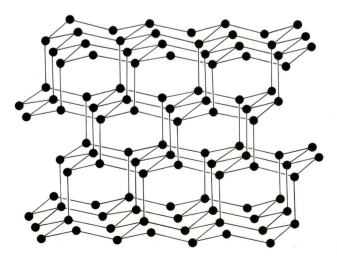

Figure 20-12
Diamond lattice.

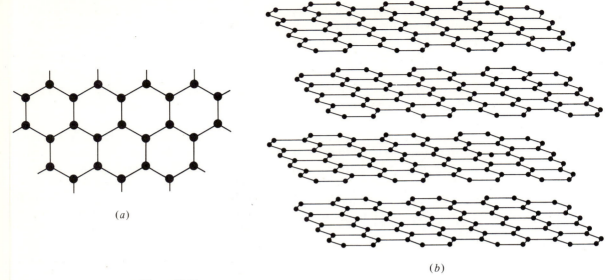

Figure 20-13
Structure of graphite.   (*a*) **A single layer of carbon atoms.** (*b*) **The stacking of layers.**

the planes than in the perpendicular direction. This delocalized π MO model of graphite is shown schematically in Fig. 20-14.

The structure of a graphite layer may alternatively be pictured as a resonance hybrid of the three equivalent forms shown in Fig. 20-15.

Bonding *between* layers in graphite is weak, consisting of London forces which can hold a graphite crystal together only because the layers are so large. The layers can easily slip over each other, which accounts for the successful use of graphite as a lubricant. During World War II, however, it was found that the graphite used to lubricate various components of high-flying aircraft lost its lubricity. After several planes were lost, it was determined that graphite layers do not slip easily under vacuum. The presence of some small "ball-bearing" molecules like H_2O between the layers is believed to facilitate slippage.

At 25°C and 1 atm graphite is the more stable form of carbon:

$$C(\text{diamond}) \longrightarrow C(\text{graphite}) \qquad \Delta G° = -2.9 \text{ kJ}$$

but the transformation is too slow to be observed under these conditions. Since diamond is more dense than graphite, it becomes stable at high pressures. Thus at

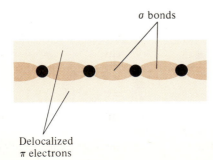

σ bonds

Delocalized
π electrons

Figure 20-14
Bonding within a graphite layer (schematic).

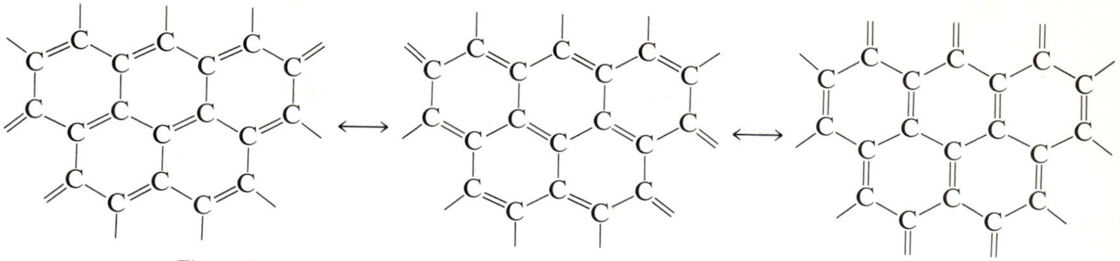

Figure 20-15
Bonding in graphite: resonance model.

about 10^5 atm and 3000°C the graphite-diamond transformation can be accomplished with the help of transition-metal catalysts. The process is used for making small, industrial diamonds and has been successful in producing gem-quality diamonds of various sizes, but such diamonds are more costly to make than to mine.

Several so-called amorphous forms of carbon exist; these include charcoal, soot, and lampblack and are in reality microcrystalline forms of graphite. The tiny graphite fragments in these may be so small that the total surface area per gram of carbon is enormous. Since the electrons at the surfaces of these particles are not all needed for bonding within the particles, some are free to bond to other atoms. Such *activated carbons* can adsorb large quantities of gases and of components from liquid solutions and are used for such diverse purposes as controlling hydrocarbon emissions from fuel tanks in automobiles to purifying water for municipal supplies.

Inorganic compounds of carbon
    The chemical derivatives of the *hydrocarbons* (compounds of hydrogen and carbon only) are usually classed as *organic compounds*. We will discuss these in Chap. 23. At this time we will describe some of the important *inorganic carbon compounds*. (The distinction between organic and inorganic is not sharp, however.)

Carbon monoxide, CO, is a colorless, odorless gas. It is produced when a carbon-containing substance such as wood, coal, gasoline, charcoal, or tobacco is burned, especially when the supply of air is limited. The carbon monoxide molecule has a triple-bonded structure:

$$:C:::O:$$

much like that of the N_2 molecule. (CO and N_2 are *isoelectronic;* that is, they have the same electronic configuration.)

CO serves as a ligand in stable uncharged complexes called *carbonyls*. $Ni(CO)_4$ and $Fe(CO)_5$ can each be formed by direct combination of the corresponding element with carbon monoxide. Nickel carbonyl is formed at room temperature and 1 atm:

$$Ni(s) + 4CO(g) \rightleftharpoons Ni(CO)_4(g)$$

The reaction reverses at higher temperatures (or lower pressures) and is used to purify nickel metal (the *Mond process*).

The well-known toxic nature of carbon monoxide is related to its ability to complex metals. The giant protein molecule of *hemoglobin* in the blood contains 4 ferrous ions, Fe^{2+}, which are complexed in such a way that each can accept one more ligand. When that ligand is oxygen (O_2), hemoglobin is converted to *oxyhemoglobin* by a reversible process:

$$\text{Hemoglobin} + 4O_2 \rightleftharpoons \text{oxyhemoglobin} \quad \text{(4 steps)}$$

Oxyhemoglobin is formed in the lungs and carried to the cells, where it gives up its oxygen. Hemoglobin bonds much more strongly to carbon monoxide than to oxygen, however, so that inhalation of even small amounts of CO can form enough of the too-stable *carboxyhemoglobin* to limit the oxygen-carrying ability of the blood.

Combustion of carbon or of carbon-containing compounds in an excess of oxygen yields the other important oxide of carbon, *carbon dioxide*, CO_2. It is produced industrially in large quantities in fermentation processes and also in the manufacture of calcium oxide, CaO, from limestone, $CaCO_3$. And, of course, the respiration of animals also produces CO_2.

Carbon dioxide is the anhydride of carbonic acid, H_2CO_3. This acid can not be isolated, but exists at low concentrations in solutions of CO_2:

$$CO_2(g) + H_2O \rightleftharpoons H_2CO_3 \quad K \approx 2 \times 10^{-3}$$

As has been noted in Sec. 16-6, all the dissolved CO_2 is usually considered as "carbonic acid," so that the published value of K_1 for its first dissociation, while usually written as

$$K_1 = \frac{[H^+][HCO_3^-]}{[H_2CO_3]}$$

or as

$$K_1 = \frac{[H^+][HCO_3^-]}{[CO_2]}$$

is in reality

$$K_1 = \frac{[H^+][HCO_3^-]}{[\text{carbonic acid}]}$$

where [carbonic acid] refers to the total molar concentration of all neutral carbon-containing species. Using this convention, then, $K_1 = 4.2 \times 10^{-7}$ and $K_2 = 5.6 \times 10^{-11}$.

Because carbonic acid is diprotic, it forms two series of salts, the *bicarbonates* (hydrogen carbonates), containing the HCO_3^- anion, and the *carbonates*, containing the CO_3^{2-} anion. Sodium carbonate, Na_2CO_3, gives a basic reaction in water due to the hydrolysis of the carbonate ion:

$$CO_3^{2-} + H_2O \rightleftharpoons HCO_3^- + OH^- \quad K_h = 1.8 \times 10^{-4}$$

$Na_2CO_3 \cdot 10H_2O$ is known as *washing soda;* it was at one time commonly used in the home to augment the cleaning action of soap. (Oils and greases can be more readily removed from clothing in basic solution.) Sodium carbonate is used today as an additive in detergents.

Sodium bicarbonate, $NaHCO_3$, produces an only slightly basic solution because of the competing equilibria:

$$HCO_3^- + H_2O \rightleftharpoons H_2CO_3 + OH^- \qquad K_h = 2.4 \times 10^{-8}$$

$$HCO_3^- \rightleftharpoons CO_3^{2-} + H^+ \qquad K_2 = 5.6 \times 10^{-11}$$

the first of which predominates slightly over the second. $NaHCO_3$ is known as *baking soda*. The rising action of baking cakes, etc., is produced by the reaction of HCO_3^- with acidic substances:

$$HCO_3^- + H^+ \longrightarrow H_2O + CO_2(g)$$

With nitrogen, carbon forms compounds which can be considered derivatives of *cyanogen*, $(CN)_2$, a highly toxic gas:

$$:N:::C:C:::N:$$

It is often called a *pseudohalogen*, because it mimics the behavior of the halogens. Compare, for example, the disproportionations of Cl_2 and $(CN)_2$ in basic solution:

$$Cl_2 + 2OH^- \longrightarrow OCl^- + Cl^- + H_2O$$

$$(CN)_2 + 2OH^- \longrightarrow OCN^- + CN^- + H_2O$$

The *cyanide ion* CN^- is the anion of *hydrogen cyanide,* HCN, sometimes named as an acid as *hydrocyanic acid,* a very weak acid ($K = 4.0 \times 10^{-10}$) in aqueous solution. The gas has the odor of bitter almonds and is very toxic.

The other ion produced by the disproportionation of cyanogen is the cyanate ion OCN^-. Ammonium cyanate, NH_4OCN, is of historical interest. In 1828 Wöhler prepared urea,

by heating ammonium cyanate, NH_4OCN. Until that time it was thought that urea could be produced only as a by-product of living organisms and that chemicals such as urea contained some kind of "vital force," which was absent from "inorganic" chemicals such as ammonium cyanate.

Related to the cyanate ion is the *thiocyanate* ion SCN^-. This is a useful complexing agent; the red $FeNCS^{2+}$ complex serves as a virtually unambiguous indicator of the presence of Fe^{3+} in solution.

Carbon forms other compounds with nonmetals; these include carbon tetrachloride, CCl_4; carbon disulfide, CS_2; phosgene, $COCl_2$, a very poisonous gas; and fluorocarbons, such as C_2F_6.

With metals carbon forms several types of *carbides*. The carbides of the highly electropositive metals contain discrete carbide anions such as C^{4-} (in Al_4C_3) and C_2^{2-} (in CaC_2). Calcium carbide reacts with water to form the hydrocarbon acetylene, C_2H_2, which was used in miners' lamps and early automobile headlights.

$$CaC_2(s) + 2H_2O \longrightarrow C_2H_2(g) + 2OH^- + Ca^{2+}$$

Carbides of the transition metals form nonstoichiometric interstitial carbides.

Some of these are very hard and high-melting; tungsten and tantalum carbides are used for cutting tools in the machining of metals.

With metalloids carbon forms very hard, covalent solids. Silicon carbide, SiC, or *carborundum,* is used as an abrasive for sharpening and grinding metals and other substances. It has the diamond structure (Fig. 20-12), except that silicon and carbon atoms alternate throughout the lattice.

20-9
The noble gases

The noble gases helium, neon, argon, krypton, xenon, and radon occur in the atmosphere in trace to small quantities. Helium also occurs in some natural gases in the United States. It is obtained by liquefaction and subsequent fractional distillation. Neon, argon, krypton and xenon are all obtained by fractional distillation of liquid air. Radon is intensely radioactive; it is a product of radioactive decay of radium (Chap. 24). Some of the properties of the noble gases are shown in Table 20-14.

Until 1962 few chemists gave serious thought to the possibility of any of the noble gases forming any real compounds. To be sure, noble-gas clathrates (Sec. 20-4) were known, but these were considered "pseudo-compounds." In addition, each of these elements has an outer-shell configuration which implied such stability that the term *inert gas* was in common use. Then Neil Bartlett, at the University of British Columbia, refusing to be intimidated by tradition, tried to make a noble-gas compound, the first such attempt recorded in many years. He had just prepared the compound O_2PtF_6 which contains the dioxygenyl cation O_2^+, and realizing that the first ionization energy of O_2 is very close to that of Xe, he tried to make $XePtF_6$ by similar methods—and succeeded! Then the dam broke; once it was realized that the inertness of these elements was not absolute, many other compounds were soon prepared.

Most of the well-characterized compounds of the noble gases contain xenon. A few krypton compounds have been prepared, and there is evidence for at least one argon compound. A detailed discussion of the chemistry of these compounds is out of place in this book, but we will mention a few of the compounds of xenon.

Compounds
of xenon

Xenon shows the $+2$, $+4$, $+6$, and $+8$ oxidation states in its compounds. Some of these are explosively unstable, while others are surprisingly stable. *Xenon difluoride,* XeF_2, is a white solid made by reacting xenon with a limited supply of fluorine at high pressure. The XeF_2 molecule has a linear structure. *Xenon tetrafluoride,* XeF_4, is a white solid also prepared by direct reaction of the elements, but with a higher partial pressure of fluorine. It is a good oxidizing agent, and its

Table 20-14
Noble gases

	He	Ne	Ar	Kr	Xe	Rn
Valence-shell configuration	$1s^2$	$2s^22p^6$	$3s^23p^6$	$4s^24p^6$	$5s^25p^6$	$6s^26p^6$
Normal melting point, °C	-272	-249	-189	-157	-112	-71
Normal boiling point, °C	-269	-246	-186	-152	-107	-62
Volume percentage in atmosphere	0.0005	0.0016	0.93	0.0001	8×10^{-6}	6×10^{-18}

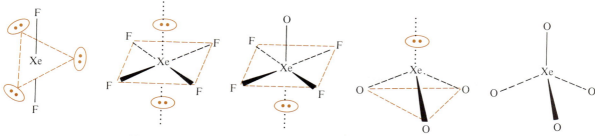

Figure 20-16
Some xenon compounds (lone pairs shown).

molecule has a square planar structure. The +6 state of xenon is represented by numerous compounds including XeF_6 (complex structure), $XeOF_4$ (tetragonal pyramidal structure), and XeO_3 (trigonal pyramidal structure). *Xenon tetraoxide,* XeO_4, has been prepared and has been shown to have a tetrahedral structure. In addition a number of *perxenate* salts, such as $Na_4XeO_6 \cdot 8H_2O$, have been isolated.

The structures of these unusual compounds agree remarkably with predictions made on the basis of VSEPR theory. (See Fig. 20-16.)

SUMMARY

We have started a brief but systematic study of the elements and their compounds, but we have prefaced this study with a brief summary of the most important aspects of *inorganic nomenclature.* The rules given included those for naming elements, cations, anions, salts, hydroxides, binary acids, oxoacids, and other compounds. (A more extensive discussion is found in App. C.)

Hydrogen exists uncombined as H_2 molecules. Three isotopes of hydrogen are known. The mass differences among these give rise to differences in physical and chemical properties known as *isotope effects,* which are greater for hydrogen than for any other element. Elemental hydrogen is usually prepared by the reduction of compounds in which the element is in the +1 oxidation state. This can be accomplished with chemical reducing agents or by electrolysis. Hydrogen shows the +1 oxidation state in most of its compounds, but adopts the −1 state in the *hydrides.*

Oxygen is the most abundant element on earth and is found in the atmosphere uncombined as O_2 molecules and at low concentrations as *ozone,* O_3. Elemental oxygen can be prepared by the electrolysis of water, the thermal decomposition of certain oxides and oxosalts, or the fractional distillation of liquid air. In its compounds oxygen exhibits oxidation states ranging from −2 to positive states in fluorine compounds. The −2 state is the most common and includes *oxides, hydrox-ides, oxoacids,* and *oxosalts. Peroxides* contain oxygen in the −1 state.

The most important hydrogen-oxygen compound is water. Water molecules are incorporated into the crystal lattices of *hydrates.* They can also trap small molecules in weak structures called *clathrates.*

The elements of periodic group VIIA, the *halogens,* are fluorine, chlorine, bromine, iodine, and astatine. These elements are nonmetallic, but become slightly less so going down the group. The elemental halogens can be prepared by electrolytic or chemical reduction, except for fluorine, which can only be prepared electrolytically. In its compounds, fluorine shows only the −1 oxidation state, but the other halogens show states up to a maximum of +7.

The *chalcogens,* the members of group VIA, include oxygen, sulfur, selenium, tellurium, and polonium, of which only the first two are typical nonmetals. Elemental *sulfur* exists in a variety of forms of which *rhombic sulfur* is stable under ordinary conditions. The most important oxidation states of sulfur are −2 (in the *sulfides*), +4 (in *sulfur dioxide* and its derivatives), and +6 (in *sulfur trioxide* and its derivatives).

The group VA elements, sometimes called the *pnicogens,* consist of nitrogen, phosphorous, arsenic, antimony, and bismuth. *Nitrogen* occurs naturally as a triple-bonded diatomic molecule. In its compounds it shows oxidation states from −3 to +5. Its most important compounds are *ammonia,* the *nitrites,* and the *nitrates. Phosphorus* exists as several allotropes, of which the white and red varieties are most common. The most important oxidation states shown by phosphorus in its compounds are the $+3$ and $+5$ states.

Carbon is the only typical nonmetal in group IVA. It occurs uncombined as diamond and graphite. The properties of these allotropes are dramatically different and reflect fundamental differences in structure. Inorganic carbon compounds include the oxides and their derivatives and cyanogen and its derivatives.

The *noble gases* (group 0) are generally quite unreactive, but xenon, and to a smaller extent, radon and krypton, do form some compounds. The element fluorine is commonly combined in these compounds.

KEY TERMS

Activated carbon (Sec. 20-8)
Allotrope (Sec. 20-3)
Aqua regia (Sec. 20-7)
Case hardening (Sec. 20-7)
Chalcogen (Sec. 20-6)
Clathrate (Sec. 20-4)
Contact process (Sec. 20-6)
Deuterium (Sec. 20-2)
Flickering cluster (Sec. 20-4)
Haber process (Sec. 20-7)
Halogen (Sec. 20-5)
Hydrate (Sec. 20-4)
Hydride (Sec. 20-2)
Hydrous oxide (Sec. 20-4)
Inert-pair effect (Sec. 20-6)

Interhalogen (Sec. 20-5)
Inversion (of ammonia) (Sec. 20-7)
Isotope effects (Sec. 20-2)
Noble gas (Sec. 20-9)
Ostwald process (Sec. 20-7)
ous-ic system (Sec. 20-1)
Ozone (Sec. 20-3)
Peroxo compound (Sec. 20-3)
Pnicogen (Sec. 20-7)
Protium (Sec. 20-2)
Pseudohalogen (Sec. 20-8)
Refractory (Sec. 20-3)
Stock system (Sec. 20-1)
Tritium (Sec. 20-2)
Water of hydration (or crystallization) (Sec. 20-4)

QUESTIONS AND PROBLEMS

Nomenclature

20-1 Give the prefixes which are used in chemical nomenclature to represent each of the numbers from 1 through 8.

20-2 Give an acceptable alternate name for **(a)** the chromic ion **(b)** the tin(II) ion **(c)** the ferrous ion.

20-3 Explain the meaning of each of the following suffixes: **(a)** *-ide* **(b)** *-ite* **(c)** *-ate* **(d)** *-yl.*

20-4 Explain the meaning of each of the following prefixes: **(a)** *hepta-* **(b)** *hypo-* **(c)** *tetra-* **(d)** *per-.*

20-5 Give a systematic name for **(a)** calcium bicarbonate, $Ca(HCO_3)_2$ **(b)** potassium bisulfite, $KHSO_3$.

20-6 Name each of the following as an acid: **(a)** hydrogen bromide (HBr) **(b)** hydrogen cyanide (HCN) **(c)** hydrogen sulfide (H_2S) **(d)** hydrogen iodide (HI).

20-7 Name each acid in the series $HOBr$, $HBrO_2$, $HBrO_3$, $HBrO_4$.

20-8 Name each anion in the series IO^-, IO_2^-, IO_3^-, IO_4^-.

20-9 Name each of the following ions: **(a)** CrO_4^{2-} **(b)** SCN^- **(c)** PO_4^{3-} **(d)** SO_3^{2-} **(e)** S^{2-} **(f)** SO_4^{2-} **(g)** HSO_3^- **(h)** $S_2O_3^{2-}$ **(i)** $Cr_2O_7^{2-}$.

20-10 Name each of the following compounds: **(a)** $AlCl_3$ **(b)** V_2O_5 **(c)** MnO **(d)** $MgSO_3$ **(e)** $Cd(CN)_2$ **(f)** $AgClO_4$ **(g)** $(NH_4)_2Cr_2O_7$ **(h)** Ba_3P_2 **(i)** $Mg(HSO_3)_2$ **(j)** HNO_2 **(k)** BaO_2 **(l)** PH_3 **(m)** $AuCl_3$ **(n)** CaH_2 **(o)** HOBr.

20-11 Write the formula for each of the following substances: **(a)** rubidium chlorate **(b)** calcium sulfite **(c)** silicon carbide **(d)** calcium nitrite **(e)** perbromic acid **(f)** copper(I) iodide **(g)** gold(I) chloride **(h)** nitrous acid **(i)** zinc bisulfate **(j)** calcium hydrogen peroxide **(k)** calcium telluride **(l)** gallium oxide **(m)** tetrasulfur **(n)** cobalt(III) sulfate **(o)** barium hydrogen carbonate.

Hydrogen

20-12 Name and characterize the isotopes of hydrogen. Why is the isotope effect greater for hydrogen than for any other element?

20-13 The vibrational zero-point energy (energy in the lowest vibrational state) is less for D_2 than for H_2. How will this affect the rate of a D_2 reaction as compared with one using H_2? Explain.

20-14 Calculate the ratio of the rate of diffusion of H_2 to that of D_2 at the same temperature. Repeat the calculation for $^{16}O_2$ and $^{17}O_2$. (The mass of ^{16}O is 15.9949 amu and that of ^{17}O is 16.9991 amu.)

20-15 Write an equation for the reaction which occurs when sodium hydride is added to **(a)** water **(b)** liquid ammonia.

20-16 If pure D_2O has a pD = 7.35, what is K_w for D_2O?

20-17 Which of the following metals are thermodynamically capable of reducing H_2O to H_2? **(a)** Sn **(b)** Ag **(c)** Cd **(d)** Au.

20-18 Write a balanced equation for **(a)** reduction of H_2O by Ba **(b)** oxidation of methane, CH_4, by water vapor to produce carbon monoxide **(c)** adding NaH to $1\,M$ HCl **(d)** burning H_2 in Cl_2.

20-19 Justify the fact that beryllium does not form an ionic hydride.

Oxygen

20-20 Draw a Lewis structure which is consistent with the fact that O_2 is paramagnetic and has a double bond.

20-21 Which of the following would be expected to be paramagnetic? **(a)** Na_2O **(b)** Na_2O_2 **(c)** KO_2 **(d)** O_2F_2.

20-22 What is ozone? Would you expect its molecule to be polar? Explain.

20-23 Why is it that at room temperature oxides of metals are solids, while those of nonmetals tend to be gases?

20-24 Write equations illustrating the amphoterism of **(a)** $Zn(OH)_2$ **(b)** $Cr(OH)_3$.

20-25 Justify the fact that calcium oxide is $Ca^{2+}O^{2-}$, not Ca^+O^-. How could you show experimentally that this is true?

20-26 Would you expect the species O_2^{4-} to be stable? Explain.

20-27 Potassium superoxide forms tetragonal crystals. Predict the nature of the KO_2 unit cell.

20-28 BaO_2 is a peroxide, while PbO_2 is not, yet their formulas are similar. Explain.

20-29 Write an equation for **(a)** thermal decomposition of $NaNO_3$ to form $NaNO_2$ and O_2 **(b)** thermal dehydration of NaOH **(c)** burning Na in O_2 **(d)** burning Li in O_2 **(e)** dissolving Na_2O_2 in water **(f)** neutralization of H_2O_2 by a strong base.

Water

20-30 At what temperature is the density of liquid water at maximum? Why does this maximum occur?

20-31 What is a clathrate? Why is it that clathrates are not considered to be "true" compounds?

20-32 Write an equation for the reaction which occurs when each of the following is added to water: **(a)** K_2O **(b)** K_2O_2 **(c)** KO_2 **(d)** SO_2 **(e)** Ca **(f)** Cl_2O.

Halogens

20-33 Although fluorine oxidizes all metals, it can be stored in some metallic containers. Explain.

20-34 Which would you expect to have the higher melting point CaF_2 or SF_2? Justify your choice.

20-35 The unstable compound HOF has been called *hypofluorous acid*. In what way is this name inappropriate? Suggest an alternate name for this compound.

20-36 From data given in the chapter calculate ΔG° for the disproportionation of Cl_2 in basic solution at $25°C$.

20-37 Account for the decrease in acid strength in each series: **(a)** $HClO_4$, $HClO_3$, $HClO_2$, HOCl **(b)** HOCl, HOBr, HOI.

20-38 The dissociation energy of F_2 and the electron affinity of F are both less than the corresponding quantities for chlorine. Why, then, is F_2 a better oxidizing agent than Cl_2?

20-39 Suggest a reason for the fact that crystalline sodium iodide often has a yellow tinge, in spite of the fact that both Na^+ and I^- are colorless.

20-40 Both Br_2 and I_2 are yellow in dilute aqueous solution. How can you distinguish between these two solutions in the laboratory?

20-41 Using VSEPR theory predict the shape of **(a)** Cl_2O **(b)** ClO_2 **(c)** ClO_2^- **(d)** ClO_3^- **(e)** Cl_2O_6 **(f)** Cl_2O_7 **(g)** IFO_2.

20-42 Draw a Lewis structure for **(a)** HOCl **(b)** I_2O_7 **(c)** BrO_2 **(d)** I_3^- **(e)** OCl_2^-.

20-43 Liquid iodine conducts electricity by a predominately electrolytic mechanism. Given the stability of the I_3^- ion, write an equation for the self-dissociation of I_2.

20-44 Predict the structure of each of the following interhalogens by means of VSEPR theory: **(a)** ClF_3 **(b)** IF_5 **(c)** ClF_5 **(d)** ICl_2^- **(e)** BrF_4^- **(f)** IF_4^+ **(g)** BrF_2^+.

20-45 Write an equation for each of the following: **(a)** acidification of an NaF solution **(b)** oxidation of Cl^- to Cl_2 by MnO_4^- in acidic solution **(c)** disproportionation of OBr^- in basic solution **(d)** disproportionation of ClO_2 in basic solution.

20-46 Write an equation for each of the following: **(a)** oxidation of aqueous NaI by O_2 **(b)** reduction of IO_3^- by I^- in acidic solution **(c)** dehydration of HIO_3 **(d)** oxidation of solid NaI by hot, concentrated H_2SO_4 to form IO_3^-.

20-47 Write an equation for each of the following: **(a)** oxidation of bromate to perbromate by fluorine **(b)** oxidation of water by fluorine to produce oxygen **(c)** burning of CH_4 in fluorine.

Sulfur

20-48 Describe the changes which occur when sulfur is heated from room temperature to 2000°C. Correlate changes in appearance with changes in structure.

20-49 Why is the S_8 ring puckered?

20-50 Solutions of sulfides often slowly deteriorate in the laboratory. Why?

20-51 Write a Lewis structure for **(a)** SO_2 **(b)** SO_3 **(c)** $S_2O_3^{2-}$ **(d)** $S_4O_6^{2-}$ **(e)** $S_2O_7^{2-}$ **(f)** $S_2O_8^{2-}$.

20-52 Outline the chemistry of the contact process for the manufacture of H_2SO_4.

20-53 Show how VSEPR theory predicts the structures of **(a)** SCl_2 **(b)** $SOCl_2$ **(c)** SF_6 **(d)** SF_4.

20-54 A solution is known to contain either SO_3^{2-} or SO_4^{2-} (but not both). Suggest two ways by which the anion could be identified.

20-55 Is ΔS for the rhombic-monoclinic sulfur transformation positive or negative? Explain.

20-56 Write an equation for each of the following: **(a)** dissolving of ZnS in dilute HCl **(b)** dissolving of CuS in aqueous HNO_3 to form NO gas **(c)** dissolving of PtS in aqua regia **(d)** addition of HCl to aqueous Na_2SO_3 **(e)** oxidation of S^{2-} in basic solution by CrO_4^{2-} to form $Cr(OH)_4^-$ **(f)** oxidation of SO_3^{2-} by HO_2^- in basic solution **(g)** dissolving of AgBr in aqueous $Na_2S_2O_3$ **(h)** burning of H_2S in air.

Nitrogen and phosphorus

20-57 What conditions are chosen for the Haber process? Why?

20-58 What is the relation between the blue color of a solution of sodium in liquid NH_3 and color centers in crystals?

20-59 The insides of windows in poorly ventilated laboratories often develop a white film. How does the use of ammonia contribute to this?

20-60 Justify the fact that hydroxylamine is a weaker base than ammonia.

20-61 Draw Lewis structures for the nitrogen oxides **(a)** N_2O **(b)** NO **(c)** NO_2.

20-62 Why does the paramagnetism of NO_2 decrease as the temperature is lowered?

20-63 Suggest a reason for the fact that when Cu reduces concentrated HNO_3, NO_2 is formed, while with dilute HNO_3, NO is the product.

20-64 Explain in terms of Le Châtelier's principle why HNO_3 will not dissolve platinum metal, but aqua regia will.

20-65 Account for the fact that at room temperature nitrogen exists as diatomic molecules, but phosphorus does not.

20-66 Contrast the structures and properties of white and red phosphorus.

20-67 How are the structures of P_4, P_4O_6, and P_4O_{10} related?

20-68 Liquid PCl_5 exhibits electrolytic conductivity. Suggest what species might be present in the liquid.

20-69 Write equations showing the progressive hydration of P_4O_{10}.

20-70 The standard heat of formation of N_2O is $+82$ kJ mol^{-1}. When a substance burns in N_2O, is its heat of combustion likely to be higher or lower than that when it burns in O_2? Explain.

20-71 Account for the fact that the ideal-gas law seems to fail completely for NO_2.

20-72 List in order of decreasing concentration all the solute species present in **(a)** 0.1 M H_3PO_4 solution **(b)** 0.1 M Na_3PO_4 solution.

20-73 Account for the fact that the dipole moment of NH_3 (1.50) is *larger* than that of NF_3 (0.20).

20-74 Write an equation for each of the following: **(a)** thermal decomposition of sodium azide **(b)** heating of magnesium in nitrogen **(c)** addition of water to magnesium nitride **(d)** dissolving of calcium in liquid ammonia **(e)** combustion of methane in nitrous oxide.

20-75 Write an equation for each of the following: **(a)** oxidation of phosphorus in excess O_2 **(b)** mixing of a solution of phosphorous acid with an excess of base **(c)** reaction of solutions of $NaHCO_3$ and NaH_2PO_4.

Carbon

20-76 Contrast the differences in properties and structure of diamond and graphite.

20-77 How is carbon monoxide formed? Why is carbon monoxide poisonous?

20-78 Why is the carbon-carbon bond length shorter in

graphite than in diamond? Why is diamond more dense than graphite?

20-79 Although calcium carbonate is quite insoluble in water, calcium bicarbonate is not. Why is it that calcium carbonate can be dissolved in an aqueous CO_2 solution?

20-80 Draw a Lewis structure for (a) CO_2 (b) CO (c) $HCHO_2$ (d) CO_3^{2-} (e) HCN (f) CaC_2 (g) $COCl_2$.

20-81 List the solute species present in order of decreasing concentration in a saturated aqueous CO_2 solution.

20-82 Compare the bonding in HCN with that in N_2. Why is the normal boiling point of HCN (26°C) so much higher than that of N_2 (−196°C).

20-83 Write an equation for each of the following: (a) reaction of calcium carbide with water (b) disproportionation of cyanogen in basic solution (c) thermal decomposition of nickel tetracarbonyl (d) precipitation of barium carbonate by bubbling CO_2 into saturated $Ba(OH)_2$ (e) reduction of Fe_2O_3 by CO to form iron.

Noble gases

20-84 Why is it that so many of the noble-gas compounds are compounds of fluorine?

20-85 Predict the structure of each of the following on the basis of VSEPR theory: (a) KrF_2 (b) XeF_4 (c) XeF_2^+ (d) XeO_3 (e) XeF_5^+ (f) XeF_2^{2+}.

21

THE REPRESENTATIVE
METALS AND METALLOIDS

TO THE STUDENT

Now that we have concluded discussion of the nonmetals, we turn our attention to the metals. In this chapter we look at the representative metals (the main-group or A-group metals), and in Chap. 22 we discuss the transition (B-group) metals. But before we start, we should review the meaning of *metal*. A metal is a substance which shows a characteristic shiny appearance (metallic luster), and which is a good conductor of heat and electricity. As we saw in Sec. 10-5, a metallic crystal is permeated by a cloud of delocalized (free) electrons, which accounts for these physical properties. This model of a metallic crystal is also consistent with the notion that a metal atom has only a weak hold on its valence electron(s), a view supported by observed low ionization energies (Sec. 7-3).

The chemical properties of the metals are also related to the ease of removal of electrons and the accompanying small tendency to gain electrons. Thus we find metals existing typically as positive ions in both solid compounds and in aqueous solutions. Metals have low electronegativities, indicating a small electron-attracting tendency. The metal-oxygen bond in a hydroxide is, as a result, typically quite ionic, so that metal hydroxides are basic.

The trend from metallic to nonmetallic going from left to right across a period or from bottom to top in a group in the periodic table is a gradual one, so that there are some elements which "straddle the fence" and are best classed as *metalloids* or *semimetals*. These elements are given special treatment in Sec. 21-5.

21-1
The alkali metals

The elements of periodic group IA are known as the *alkali metals*. (The word *alkali* is derived from an ancient Arabic word meaning "plant ashes"; potassium and sodium are found in the ashes of burned plant material.) These are all soft metals with low melting points, a consequence of the fact that the bonding in these elements is almost purely metallic. The densities of the alkali metals are low, primarily a consequence of their large atomic radii. Their electrical conductivities are very high, although not as high as that of silver, the best electrical conductor at room temperature. All of the alkali metals crystallize in a body-centered cubic structure. The melting points and densities of these metals are shown in Table 21-1.

The English names for the alkali metals come from a variety of interesting sources. *Lithium* comes from a Greek word meaning "rock" (lithium is found in rocks and minerals). *Sodium* comes from a Latin name for a headache remedy. *Potassium* is derived from *pot ash,* an old name for impure potassium carbonate, which was obtained as the residue left in an iron pot after the evaporation of an aqueous extraction of wood ashes. *Rubidium* and *cesium* have origins in Latin words meaning "red" and "blue," respectively. (These elements produce atomic line spectra with characteristic lines in the red and blue regions of the visible spectrum, respectively.) *Francium* was discovered by the French physicist Marguerite Perey.

Sodium and potassium are relatively abundant in the earth's crust (they are the sixth and seventh most abundant elements, respectively). They occur in numerous mineral deposits, such as *halite* (rock salt, NaCl), *sylvite* (KCl), and *carnallite* (KMgCl$_3 \cdot$ 6H$_2$O), as well as in the ocean. Lithium, rubidium, and cesium are much scarcer; lithium is found in a few silicate minerals, while rubidium and cesium are found in trace quantities in many rocks and minerals scattered around the globe. All of the isotopes of francium have very short half-lives, but enough of it has been prepared artificially to verify that its chemical properties are those to be expected of an alkali metal.

Preparation of the elements

The alkali metals can all be prepared by electrolytic reduction, often of a mixture of molten halides or hydroxides. Thus lithium is prepared by the electrolysis of an LiCl–KCl mixture, in which the KCl serves to lower the melting point. A KCl–CaCl$_2$ mixture is used for obtaining potassium. Chemical reduction can be used for the production of potassium, rubidium, and cesium. At high temperatures the hydroxides, chlorides, and carbonates of these elements can be reduced with hydrogen, carbon, or calcium. With rubidium carbonate the reduction by carbon is shown as

$$Rb_2CO_3(s) + 2C(s) \longrightarrow 2Rb(g) + 3CO(g)$$

from which pure rubidium can be obtained by subsequent condensation.

Reactions of the elements

The alkali metals invariably react to form 1+ ions both in crystalline products and in solution. Thus, for example, sodium metal reacts vigorously with water to produce a solution of sodium hydroxide:

$$2Na(s) + 2H_2O \longrightarrow 2Na^+ + 2OH^- + H_2(g)$$

Table 21-1
The alkali metals

	Li	Na	K	Rb	Cs	Fr
Valence-shell electronic configuration	$2s^1$	$3s^1$	$4s^1$	$5s^1$	$6s^1$	$7s^1$
Normal melting point, °C	180	98	64	39	29	—
Density, g cm^{-3}	0.54	0.97	0.86	1.5	1.9	—

Why are 2 + ions not observed? Breaking into the $(n - 1)$ shell requires too much energy, as is shown by the ionization energies of the atoms of these elements (Table 21-2). In each case the *second* ionization energy is so much larger than the first that 2 + ions are an impossibility, except under extreme, energy-rich conditions (in electric arcs, at ultrahigh temperatures, etc.).

All the alkali metals are very strong reducing agents, as can be seen by the standard reduction potentials shown in Table 21-3. Each potential is highly negative, indicating a very small tendency for the half-reaction to occur as written. Written in reverse, each half-reaction, now an *oxidation*,

$$M(s) \longrightarrow M^+ + e^-$$

has an *oxidation potential* which is a large *positive* number, showing the strong tendency for each alkali metal to become oxidized.

An apparent paradox shows up here. $\mathcal{E}°$ for the oxidation of lithium is 3.05 V, the highest of the group. Why is it easier to oxidize lithium than any of the other alkali metals? Lithium has the highest ionization energy, and therefore more energy is needed to remove an electron from its atom; so should not its tendency to form a 1 + ion be lowest among the alkali metals, rather than highest? The seeming contradiction can be resolved by considering that the ionization energy is the energy needed to drive off an electron from a *gaseous* atom:

$$M(g) \longrightarrow M^+(g) + e^-$$

while oxidation potential measures the relative tendency for a different process to take place:

$$M(s) \longrightarrow M^+(aq) + e^-$$

This latter reaction can be broken into steps:

Step 1: $M(s) \longrightarrow M(g)$

Step 2: $M(g) \longrightarrow M^+(g) + e^-$

Step 3: $M^+(g) \longrightarrow M^+(aq)$

The first step requires an amount of energy, the sublimation energy, which is about the same for all the alkali metals. Step 2 requires energy, the ionization energy, which is highest for lithium. Step 3, however, liberates energy, the hydration energy of the ion. For lithium this is much larger than for any of the other group I metals. Since the 1 + charge of the lithium ion is concentrated in such a small volume, it interacts very strongly with water molecules. Thus the energy liberated

Table 21-2
Ionization energies
of alkali-metal atoms

Element	Ionization energy, kJ mol^{-1}	
	First	Second
Lithium	520	7296
Sodium	496	4563
Potassium	419	3069
Rubidium	403	2650
Cesium	376	2420
Francium	370	2170

Table 21-3 Standard reduction potentials for the alkali metals

Reduction	$\mathcal{E}°$, V
$e^- + Li^+ \longrightarrow Li(s)$	-3.05
$e^- + Na^+ \longrightarrow Na(s)$	-2.71
$e^- + K^+ \longrightarrow K(s)$	-2.92
$e^- + Rb^+ \longrightarrow Rb(s)$	-2.92
$e + Cs^+ \longrightarrow Cs(s)$	-2.93

when Li^+ is hydrated more than compensates for the extra energy required to remove an electron from an Li atom.[1]

One might expect the alkali metals to react with oxygen gas to form oxides, but lithium is the only member of the group which readily does this, and then only at elevated temperatures:

$$4Li(l) + O_2(g) \longrightarrow 2Li_2O(s)$$

Sodium reacts with oxygen to form a little Na_2O, but the predominant product is *sodium peroxide:*

$$2Na(s) + O_2(g) \longrightarrow Na_2O_2(s)$$

Under most conditions potassium, rubidium, and cesium form the *superoxides:*

$$K(s) + O_2(g) \longrightarrow KO_2(s)$$

The reactions of sodium, potassium, rubidium, and cesium with the oxygen and water vapor in the air are sufficiently rapid and exothermic to make storage of the metals somewhat of a problem. They are usually kept under a hydrocarbon solvent such as paraffin oil, cyclohexane, or kerosene. Although lithium reacts more slowly with the air, it is usually stored similarly. Surprisingly, lithium reacts faster with the nitrogen of the air than it does with the oxygen:

$$6Li(s) + N_2(g) \longrightarrow 2Li_3N(s)$$

The product, lithium nitride, tends to adhere tightly to the surface of the metal and retard further reaction.

With hydrogen the alkali metals form ionic hydrides and with the halogens, ionic halides, although the rates of reaction vary over a wide range.

As has been mentioned (Sec. 20-7) all of the alkali metals dissolve in liquid ammonia to form solutions which are blue when dilute and metallic-appearing when concentrated. The presence of ammoniated electrons in dilute solutions and delocalized electrons in concentrated ones has been demonstrated, and so these solutions might be referred to as solutions of *sodium electride!* At low temperatures a solution of lithium in liquid NH_3 will precipitate the unusual metallic compound $Li(NH_3)_4$.

Alkali-metal compounds

The alkali metals form a wide variety of compounds. These include the ionic salts of most binary acids and oxoacids, most of which are quite soluble in water. The oxides and hydroxides are basic, as expected. In aqueous solution the alkali-metal

---

[1] We are guilty here of not being very careful in talking about *energy* changes on the one hand and $\mathcal{E}°$ values (which measure *free-energy* changes) on the other. Making a more detailed comparison by including consideration of entropy changes does not alter the conclusion, however.

ions show only a weak tendency to be hydrated, except for the lithium ion, which is the smallest. Unlike the salts of the other alkali metals, lithium salts typically crystallize from aqueous solution as hydrates: $LiCl \cdot 2H_2O$, $LiClO_4 \cdot 3H_2O$, $LiI \cdot 3H_2O$, etc.

The high solubilities of most sodium and potassium compounds make it difficult to test for the presence of these ions in solution by means of precipitation reactions. As a result, *flame tests* are often used. In these, a platinum wire is dipped in the test solution to which some HCl has been added. When the wire is then placed in a flame, the water quickly evaporates, and any metal chloride present melts and evaporates into the flame. Excitation of the atom then occurs, and then electronic transitions from excited to lower-energy states produce spectral lines of characteristic color. The sodium flame is a brilliant, persistent yellow color, and the test is, if anything, too sensitive. (Fainter colors produced by other ions are often obscured by the sodium yellow.) On the other hand, the potassium test is a fleeting pale violet color which is hard for many to see. Sodium will precipitate in a few compounds, however; the yellow, slowly forming, crystalline precipitate of sodium zinc uranyl acetate, $NaZn(UO_2)_3(C_2H_3O_2)_9 \cdot 9H_2O$, can be used to indicate the presence of sodium. Potassium perchlorate, $KClO_4$, and potassium hexanitrocobaltate, $K_3Co(NO_2)_6$, have low enough solubilities to be used as indications of potassium in solution.

Uses of the alkali metals and their compounds

Because of their softness and reactivity the alkali metals cannot be used for structural purposes. Sodium has been used as a heat exchanger in nuclear reactors because of its high thermal conductivity. Sodium-filled exhaust valves are used in gasoline and diesel engines; the sodium in the hollow stem of the valve rapidly conducts heat away from the valve head. Cesium exhibits a strong photoelectric effect because of its very low ionization energy and accordingly finds practical application in photoconductive *photocells*. Such a cell contains a pair of oppositely charged electrodes in an evacuated bulb or cell. The negative electrode is coated with cesium or a cesium alloy and emits electrons into the space between the electrodes when light strikes it. These electrons complete the circuit and allow current to flow through an external circuit. This current can be used to open doors, ring bells, etc.

Many compounds of the alkali metals, particularly those of sodium and potassium, are important industrially. Sodium hydroxide (its common name is *caustic soda*) and potassium hydroxide (*caustic potash*) are used in the manufacture of countless products including soaps, dyes, pigments, greases, and paper products. Lithium salts have been used successfully to treat some forms of mental disease.

21-2 The alkaline-earth metals

The elements of group IIA are known as the *alkaline-earth metals*. The term *earth* is an old alchemical term which referred to any nonmetallic substance which was not very soluble in water and which was stable at high temperatures. Many "earths" were oxides and when it was found that the oxides of the group IIA elements gave alkaline (basic) reactions, they were called the *alkaline earths*.

Except for beryllium, the elements of this group are all typical metals. They are good conductors of heat and electricity, but are harder, more dense, and higher melting than the alkali metals (Table 21-4). Evidently the additional valence

Table 21-4
Alkaline-earth metals

	Be	**Mg**	**Ca**	**Sr**	**Ba**	**Ra**
Valence-shell electronic configuration	$2s^2$	$3s^2$	$4s^2$	$5s^2$	$6s^2$	$7s^2$
Normal melting point, °C	1280	651	851	800	850	960
Density, g cm^{-3}	1.86	1.75	1.55	2.6	3.6	5.0

electron per atom makes the crystal lattices of these metals more rigid than those of the alkali metals. Beryllium and magnesium form hexagonal close-packed lattices, calcium and strontium form face-centered cubic structures at room temperature, and barium crystallizes in a body-centered cubic structure. All these elements appear metallic, although beryllium is a dark gray in color.

The names of the alkaline-earth metals have the following origins: *Beryllium* comes from a Latin word meaning "sweet" (yes, but poisonous!); *magnesium* from Magnesia, a region in Greece; *calcium* from an ancient word originally meaning "corrosion product," and later, "lime"; *strontium* from Strontian, a town in Scotland; *barium* from a Greek word meaning "heavy"; *radium* from a Latin word meaning "ray."

The alkaline-earth metals occur widely distributed in the earth's crust as carbonates, silicates, phosphates, and sulfates. Magnesium and calcium are the most abundant; entire mountains are made of *limestone*, $CaCO_3$, and *dolomite*, $CaMg(CO_3)_2$. Magnesium is also found in the ocean. Beryllium is relatively scarce, its most common mineral being *beryl*, $Be_3Al_2Si_6O_{18}$, which sometimes is found in gem quality as *emerald*. (The characteristic green coloration is caused by chromium present as an impurity.) Strontium and barium are comparatively rare, occurring primarily as the carbonate and sulfate, respectively. Radium is extremely rare and is found in uranium ores such as *pitchblende,* in which it is produced as a result of the radioactive decay of uranium (Chap. 24).

Preparation of the elements

The alkaline-earth metals are prepared by the electrolytic or chemical reduction from the +2 state. Molten beryllium fluoride is reduced by magnesium to yield the element:

$$BeF_2(l) + Mg(s) \longrightarrow Be(s) + MgF_2(s)$$

Alternatively, electrolysis of molten BeF_2 yields beryllium at the cathode and fluorine at the anode.

Much magnesium is obtained from the ocean. Calcium oxide is added to seawater:

$$CaO(s) + H_2O + Mg^{2+} \longrightarrow Mg(OH)_2(s) + Ca^{2+}$$

The magnesium hydroxide is then filtered off, washed, and dissolved in hydrochloric acid:

$$Mg(OH)_2(s) + 2H^+ \longrightarrow Mg^{2+} + 2H_2O$$

and the resulting solution evaporated:

$$Mg^{2+} + 2Cl^- \longrightarrow MgCl_2(s)$$

Molten $MgCl_2$ is then electrolyzed to produce magnesium and chlorine at cathode and anode, respectively.

A second industrial process for producing magnesium involves thermally dehydrating the hydroxide

$$Mg(OH)_2(s) \longrightarrow MgO(s) + H_2O(g)$$

and then reducing the magnesium oxide with carbon at 2000°C:

$$MgO(s) + C(s) \rightleftharpoons Mg(g) + CO(g)$$

Cooling the gaseous products condenses the magnesium and favors further production of products, according to Le Châtelier's principle.

Metallic calcium and strontium are prepared industrially by the electrolysis of their chlorides, to which some KCl has been added to lower the melting point. Barium can also be obtained by electrolysis, but is usually prepared by the high-temperature reduction of barium oxide by aluminum under vacuum:

$$2Al(s) + 3BaO(s) \longrightarrow 3Ba(l) + Al_2O_3(s)$$

Reactions of the elements

The alkaline-earth metals nearly always react to form compounds in which the metal shows the +2 oxidation state. Although beryllium shows a tendency to form covalent bonds, the members of this group typically form 2+ ions, both in solid compounds and in aqueous solution. It is, unfortunately, very tempting to explain the stability of ions with this charge by merely noting the s^2 configuration of each valence shell, without really justifying the absence of 1+ and 3+ ions of these elements. Consider their ionization energies, for example. In Table 21-5 we see the first, second, and third ionization energies of the alkaline-earth metal atoms. In each case the removal of the second electron requires about twice the energy needed to remove the first. This extra energy is considerable; so why is it that we do not see 1+ ions of these elements?[2] Under usual conditions the 2+ ions of the alkaline-earth metals are more stable than their 1+ counterparts because of the increased hydration energies (for ions in solution) and lattice energies (for ions in a crystal) which accompany the more highly charged ions. Since a 2+ ion is not only more highly charged but also smaller (remember, the valence shell is gone now), it will interact much more strongly with water molecules in solution or anions in a crystal than will the corresponding 1+ ion. This stabilizes the 2+ ion *more than* the 1+ ion, and so 1+ ions of these elements are not normally observed.

If hydration and lattice energies provide stabilization for 2+ ions, should the 3+ ions be even more stable? No, because of two effects. First, the energy necessary to remove a third electron, that is, to break into the $(n-1)$ shell, is extremely large, as can be seen from Table 21-5. Second, the size of each 3+ ion is not much smaller than that of the corresponding 2+ ion, and so the increases in hydration and lattice energy are not enough to compensate for the large third ionization energies for these elements. Therefore, the alkaline-earth metal ions almost invariably carry a 2+ charge.

[2] Actually, 1+ ions of some of the alkaline-earth metals can be formed, but only under unusual conditions, such as in the gaseous state at very high temperatures or in certain solutions as short-lived electrolysis products.

Table 21-5
Ionization energies
of alkaline-earth-
metal atoms

Element	Ionization energy, kJ mol^{-1}		
	First	Second	Third
Beryllium	899	1,757	14,849
Magnesium	738	1,450	7,730
Calcium	590	1,145	4,941
Strontium	549	1,064	4,207
Barium	503	965	3,420
Radium	509	978	

Table 21-5 Ionization energies of alkaline-earth-metal atoms

The alkaline-earth metals are powerful reducing agents, as can be seen by their reduction potentials (Table 21-6). As a matter of fact, except for beryllium and magnesium, these elements are about as good reducing agents as the alkali metals. This might seem surprising in view of the high second ionization energy of the alkaline-earth-metal atoms. The higher hydration energy of the 2+ ions almost exactly compensates for this, however; so the reduction potentials of Ca, Sr, and Ba are almost the same as those of their group IA counterparts K, Rb, and Cs. The lower values for the lighter members of group IIA, Be and Mg, are a result of their higher ionization energies.

Except for beryllium, the alkaline-earth metals (M) react with water to form hydrogen:

$$M(s) + 2H_2O \longrightarrow M^{2+} + H_2(g) + 2OH^-$$

The reaction with magnesium is usually very slow, however, because of the fact that this element rapidly forms a thin protective surface layer of MgO, which effectively retards reaction with many substances, especially at room temperature.

All the alkaline-earth metals react with oxygen at ordinary pressures to form oxides, MO. At high pressures and temperatures barium forms the peroxide:

$$Ba(s) + O_2(g) \longrightarrow BaO_2(s)$$

Barium peroxide serves as a convenient source for hydrogen peroxide, which is produced by adding acid:

$$BaO_2(s) + 2H^+ \longrightarrow Ba^{2+} + H_2O_2$$

The group IIA elements all react with the halogens to form halides, MX_2, and with nitrogen to form nitrides, M_3N_2. Calcium, strontium, and barium react with hydrogen at elevated temperatures to form ionic hydrides, MH_2.

Alkaline-earth-metal compounds Except for those of beryllium, the compounds of the alkaline-earth metals exhibit typical ionic crystal lattices. The oxides of Mg, Ca, Sr, and Ba all show the NaCl structure, while $CaCl_2$, $SrCl_2$, and $BaCl_2$ all crystallize in the fluorite (CaF_2)

Table 21-6
Standard reduction
potentials for the
alkaline-earth metals

Reduction	$\mathcal{E}°$, V
$2e^- + Be^{2+} \longrightarrow Be(s)$	1.69
$2e^- + Mg^{2+} \longrightarrow Mg(s)$	2.38
$2e^- + Ca^{2+} \longrightarrow Ca(s)$	2.76
$2e^- + Sr^{2+} \longrightarrow Sr(s)$	2.89
$2e^- + Ba^{2+} \longrightarrow Ba(s)$	2.90
$2e^- + Ra^{2+} \longrightarrow Ra(s)$	2.92

Table 21-6 Standard reduction potentials for the alkaline-earth metals

structure, for example. Beryllium, on the other hand, has an extremely small atomic radius (about 0.09 nm), and so its electronegativity is high enough (1.5) to result in considerable covalent bond formation by this element. Thus $BeCl_2$ is essentially a covalent salt, as shown by its low melting point (430°C) and low electrical conductivity in the molten state, an indication of the presence of $BeCl_2$ molecules, instead of Be^{2+} and Cl^- ions. Solid $BeCl_2$ has been shown by x-ray diffraction to consist of long chains in which pairs of Cl atoms serve as bridges between Be atoms (Fig. 21-1). Each Be atom thus bonds approximately tetrahedrally (sp^3 hybridization) to four Cl atoms. Gaseous $BeCl_2$ consists of linear molecules (sp hybridization).

Whereas most alkali-metal compounds are soluble in water, many alkaline-earth-metal compounds are not, especially those in which the anion charge is greater (more negative) than 1−. The high lattice energy is apparently reponsible for this, even though hydration energies are also high. Most of the halides are soluble, although fluorides tend to be insoluble.

Although beryllium oxide, BeO, is essentially inert in contact with water, the other alkaline-earth-metal oxides react readily with water in an exothermic process known as *slaking*. *Slaked lime* is calcium hydroxide and is prepared from *lime*, or *quicklime*, calcium oxide:

$$CaO(s) + H_2O \longrightarrow Ca(OH)_2(s) \qquad \Delta H = -66 \text{ kJ}$$

The hydroxides of the alkaline-earth metals are strong bases, except for $Be(OH)_2$, which is weak and amphoteric. They are all much less soluble than the alkali-metal hydroxides.

The small size of the beryllium atom is responsible for the fact that this element is the least metallic in group IIA. Its higher electronegativity is shown in the amphoterism of $Be(OH)_2$. This can be illustrated by the equations:

$$Be(OH)_2(s) + 2H^+ \longrightarrow Be^{2+} + 2H_2O$$

$$Be(OH)_2(s) + 2OH^- \longrightarrow Be(OH)_4^{2-}$$

where $Be(OH)_4^{2-}$ is called the *beryllate ion*. Actually, in most cases the reactions are more complicated than the above, because beryllium shows a great tendency to form many cationic and anionic complexes.

The *nitrates* of the alkaline-earth metals are all soluble, but the *sulfates* $CaSO_4$, $SrSO_4$, and $BaSO_4$ are not, the solubility decreasing going down the group. Calcium sulfate crystallizes as a dihydrate, $CaSO_4 \cdot 2H_2O$, which occurs naturally as the rock *gypsum*.

The carbonates of the group IIA metals are all insoluble in water, but soluble in dilute acids because of the formation of the bicarbonate ion. For calcium carbonate the reaction is

$$CaCO_3(s) + H^+ \longrightarrow Ca^{2+} + HCO_3^-$$

In qualitative analysis Mg^{2+}, Ca^{2+}, Sr^{2+}, and Ba^{2+} can be precipitated as white,

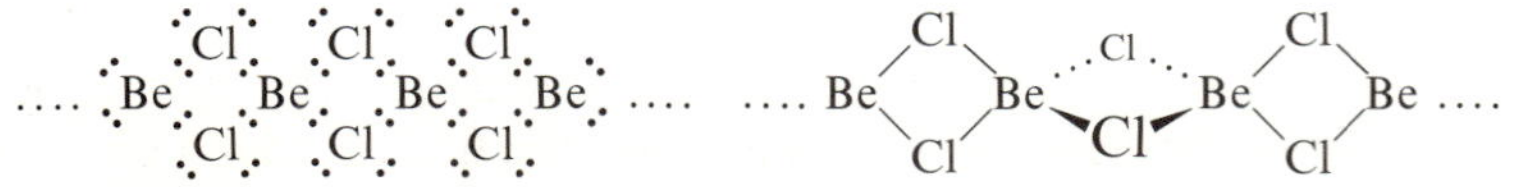

Figure 21-1
The $BeCl_2$ structure.

insoluble carbonates from basic solutions. Magnesium will precipitate as white magnesium ammonium phosphate, $MgNH_4PO_4 \cdot 6H_2O$, or as the hydroxide. In the latter case if *magnesium reagent,* or *magneson,* (*p*-nitrobenzeneazoresorcinol) is present, a blue *lake* is formed (the magnesium hydroxide is dyed blue). Calcium precipitates as an insoluble oxalate, CaC_2O_4. Strontium and barium both form insoluble chromates, $SrCrO_4$ and $BaCrO_4$. They can be selectively precipitated by controlling the pH. In slightly acidic solution the chromate concentration is low because of the equilibrium

$$2CrO_4{}^{2-} + 2H^+ \rightleftharpoons Cr_2O_7{}^{2-} + H_2O$$

Under these conditions K_{sp} for $BaCrO_4$ (8×10^{-11}) is usually exceeded, but that of $SrCrO_4$ (4×10^{-5}) is usually not. Under more basic conditions the $SrCrO_4$ will precipitate. Flame tests are often used to confirm these elements. Although magnesium gives no flame color, Ca^{2+} gives an orange-red, Sr^{2+} a crimson, and Ba^{2+} an apple-green color.

Water hardness Natural waters in much of the world contain cations which interfere with the cleaning action of soaps and, to a smaller extent, detergents. Such water is said to be *hard.* Soaps are salts of *fatty acids,* organic acids whose molecules have long chains of carbon atoms. A typical soap is sodium stearate, $NaC_{18}H_{35}O_2$, which dissolves to release stearate ions:

$$NaC_{18}H_{35}O_2(s) \longrightarrow Na^+ + C_{18}H_{35}O_2{}^-$$

The stearate ions are responsible for the cleansing action of the soap, but are unfortunately precipitated by certain cations, the most important of which are Ca^{2+}, Mg^{2+}, and Fe^{2+}. The product is an insoluble soap. With calcium ions the reaction is

$$Ca^{2+} + 2C_{18}H_{35}O_2{}^- \longrightarrow Ca(C_{18}H_{35}O_2)_2(s)$$

Not only is this inefficient, since some soap is wasted by being used in the precipitation reaction, but the insoluble calcium soap forms a scum in the soapy water.

Groundwater is usually hard in regions where limestone deposits are found. Because of the CO_2 in the atmosphere, falling rain can be considered to be a dilute solution of carbonic acid:

$$CO_2(g) + H_2O \rightleftharpoons H_2CO_3$$

$$H_2CO_3 \rightleftharpoons H^+ + HCO_3{}^-$$

which slowly dissolves limestone:

$$H^+ + CaCO_3(s) \longrightarrow HCO_3{}^- + Ca^{2+}$$

so that the overall change is

$$CO_2(g) + H_2O + CaCO_3(s) \longrightarrow Ca^{2+} + 2HCO_3{}^-$$

The result is hard water (and limestone caves). This type of water hardness is known as *temporary hardness,* because the unwanted Ca^{2+} (or Mg^{2+}, or Fe^{2+}) can be easily removed by boiling. This reverses all the above reactions, since CO_2 is volatile:

$$Ca^{2+} + 2HCO_3^- \longrightarrow CaCO_3(s) + CO_2(g) + H_2O$$

The calcium carbonate forms a deposit which may slowly build up (on the inside of a teakettle, for instance) and which reduces the efficiency of heat transfer into the water. When this occurs in large water boilers, the deposit, known as *boiler scale,* causes local overheating in the metal of the boiler, resulting in boiler failure.

When the anion present in hard water is not HCO_3^- (when it is SO_4^{2-}, NO_3^-, or Cl^-, for example), the hardness is known as *permanent hardness.* In this case the unwanted 2+ ions cannot normally be removed by boiling. (When the sulfate ion is present, $CaSO_4$ may precipitate as a particularly destructive form of boiler scale. This may occur in high-temperature, high-pressure boilers in which the rate of evaporation of water is very high, and K_{sp} for $CaSO_4$ is consequently exceeded.)

How can hard water be softened? One set of methods involves the precipitation of the unwanted ions. Boiling water with temporary hardness accomplishes this, but is not practical for large quantities of water, because of the energy required and the boiler-scale problem. Similarly impractical is the addition of enough soap to precipitate Ca^{2+} and do the cleaning besides; this is wasteful, and the soap scum formed is often a nuisance.

On a large scale the *lime-soda method* is useful for softening water. In this method the water is first analyzed for temporary and permanent hardness. Then enough slaked lime, $Ca(OH)_2$, is added to remove the temporary hardness. This makes the solution basic,

$$Ca(OH)_2(s) \longrightarrow Ca^{2+} + 2OH^-$$

and converts HCO_3^- to CO_3^{2-}:

$$OH^- + HCO_3^- \longrightarrow H_2O + CO_3^{2-}$$

If 1 mol of $Ca(OH)_2$ is added per mole of dissolved HCO_3^-, then enough CO_3^{2-} is formed to precipitate not only the dissolved Ca^{2+} which originally made the water hard, but also the Ca^{2+} from the slaked lime:

$$Ca^{2+} + CO_3^{2-} \longrightarrow CaCO_3(s)$$

(*Remember: Two* HCO_3^- ions are originally present for each Ca^{2+} ion.) Finally, any residual permanent hardness (or excess added Ca^{2+}) can be removed by adding washing soda, Na_2CO_3. [Although other bases such as $NaOH$ or NH_3 will work satisfactorily for the softening of water with temporary hardness, $Ca(OH)_2$ is commonly used because it is cheap.]

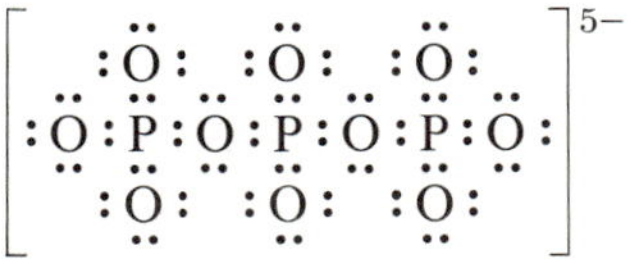

A second general method of softening water is to complex the unwanted ions, instead of precipitating them. Adding sodium triphosphate (also called sodium tripolyphosphate), $Na_5P_3O_{10}$, releases the triphosphate ion which complexes ions such as Ca^{2+}, reducing their concentrations so that they will not interfere with the action of soaps.

Other, more sophisticated methods for softening water are also used. These include general purification methods such as reverse osmosis (Sec. 12-4). For most industrial and domestic water softening, an elegant method known as *ion exchange* is used. The first ion-exchange media were minerals known as *zeolites*. These are complex aluminosilicates, structures in which aluminum, silicon, and oxygen atoms are bonded in rigid covalent networks containing long channels. Sodium ions in these channels are bonded (ionically) to the aluminosilicate framework, but can be displaced by cations of greater charge, such as Ca^{2+}. When hard water is passed through a column packed with this sodium zeolite, NaZ, the dipositive ions displace the sodium ions:

$$2NaZ(s) + Ca^{2+} \rightleftharpoons CaZ_2(s) + 2Na^+$$

Since the sodium ions added to the water will not precipitate soap, the water has been softened. The zeolite must be recharged periodically, after its ability to exchange sodium ions for calcium ions has been exhausted. This is done by passing a concentrated NaCl solution through it, reversing the above reaction (Le Châtelier's principle) and replacing the calcium zeolite by the sodium form.

Today many water softeners make use of synthetic ion exchangers. In most cases these are organic resins which have greater capacity than the natural zeolites. Furthermore, cation-exchange resins have been developed which will replace all metal cations in solution by H^+, and anion resins which will replace all anions by OH^-. Running impure water through a combination of these yields water with very low total ionic concentrations. This so-called deionized water is similar in properties to distilled water, except that no molecular impurities have been removed.

Uses of the alkaline-earth metals and their compounds

Beryllium is rare, expensive, and toxic; so it and its compounds find limited use. Pure beryllium is transparent to neutrons and x rays, and so it is used for structural parts in nuclear reactors and for windows in some x-ray tubes. Beryllium-copper alloys are as hard and tough as some steels and are used in the manufacture of tools for use when danger of fire or explosion exists. (Copper-beryllium alloys will not give off sparks when struck.) Beryllium oxide is very refractory (melting point $= 2670°C$) and is used to make high temperature electrical insulators.

Magnesium finds extensive use in the manufacture of light, strong alloys which are used particularly in the aircraft industry. Some photographic flashbulbs contain magnesium in an atmosphere of oxygen; the magnesium is ignited electrically.

With minor exceptions calcium, strontium, and barium are not used as uncombined metals, since their great reactivity with oxygen and water makes this impractical. Calcium hydroxide is a very important industrial base, and calcium oxide is used in the manufacture of cement. *Plaster of paris* is $CaSO_4 \cdot \frac{1}{2}H_2O$ and is made by heating gypsum:

$$CaSO_4 \cdot 2H_2O(s) \rightleftharpoons CaSO_4 \cdot \tfrac{1}{2}H_2O(s) + \tfrac{3}{2}H_2O$$

When water is added to the plaster of paris, the above reaction is reversed and gypsum is reformed. There is a slight volume expansion when this happens, and so fine details of the inside of a mold are reproduced. Strontium compounds find use in fireworks and red highway flares. Barium sulfate is so insoluble ($K_{sp} = 1.5 \times 10^{-9}$) that although Ba^{2+} is poisonous, $BaSO_4$ can be safely ingested. Because of this and the fact that Ba^{2+} ions strongly scatter x rays, $BaSO_4$ finds medical use in making x-ray pictures; interior structures in the body are clearly characterized because of the x ray–opaque $BaSO_4$. Radium finds use as an alpha emitter in the radiation treatment of cancer.

21-3
The group-IIIA
metals

Group IIIA has no special name; it consists of the elements boron, aluminum, gallium, indium, and thallium. These elements are generally less metallic than the corresponding alkaline-earth metals, but, as usual, metallic character increases going down the group. Boron is actually more nonmetallic than metallic (its oxide and hydroxide are acidic), a result of the small size of the boron atom. We postpone discussion of boron until we consider the metalloids in Sec. 21-5.

Each of the group IIIA atoms has a single valence-shell p electron in addition to its two s electrons. We might thus expect the *inert-pair effect* (Sec. 20-6) to be evident in the observed oxidation states of these elements, but although $+1$ compounds of all of them are known, under normal conditions the $+3$ state is observed, except for thallium which also commonly shows a $+1$ state. Some of the properties of the group IIIA elements are shown in Table 21-7.

Aluminum was named after one of its compounds, *alum,* from a Latin word meaning "astringent taste." *Gallium* comes to us from the Latin, *Gallia* (France). *Indium* was named for the color *indigo;* its spectral lines are that color. *Thallium* comes from a Greek word meaning "green shoot" or "twig"; spectral lines of thallium are green.

Aluminum

Although aluminum ranks third in abundance in the lithosphere, it is not practical to extract the aluminum from most rocks and minerals which contain it. Most aluminum occurs combined as aluminosilicates such as the clays, micas, and feldspars.

Aluminum ore is the mineral bauxite (the original, French, pronunciation is boe'-zite; it is more commonly pronounced box'-ite), which is impure hydrated aluminum oxide, $Al_2O_3 \cdot xH_2O$. This is first freed from its impurities, primarily ferric oxide and silicon dioxide, by the *Bayer process,* in which the aluminum oxide is first dissolved in hot, concentrated sodium hydroxide. The ferric oxide, being basic, is insoluble, but silicon dioxide (acidic) and aluminum oxide (amphoteric) dissolve. The aluminum oxide dissolves to form the *aluminate ion:*

$$Al_2O_3(s) + 2OH^- + 3H_2O \longrightarrow 2Al(OH)_4^-$$

Table 21-7
Group IIIA elements

	B	Al	Ga	In	Tl
Valence-shell electronic configuration	$2s^22p^1$	$3s^23p^1$	$4s^24p^1$	$5s^25p^1$	$6s^26p^1$
Normal melting point, °C	2300	660	30	156	449
Density, g cm^{-3}	2.4	2.7	5.9	7.3	11.8

This solution is then cooled, agitated with air, and seeded with some $Al(OH)_3$, resulting in precipitation of $Al(OH)_3$:

$$Al(OH)_4^- \longrightarrow Al(OH)_3(s) + OH^-$$

The air provides CO_2 which, being acidic, aids in the precipitation:

$$Al(OH)_4^- + CO_2(s) \longrightarrow Al(OH)_3(s) + HCO_3^-$$

The hydroxide is then heated to form aluminum oxide, *alumina:*

$$2Al(OH)_3(s) \longrightarrow Al_2O_3(s) + 3H_2O$$

Aluminum metal is prepared electrolytically by the *Hall process.* In the late ninteenth century Charles Martin Hall was an undergraduate student at Oberlin College, when one of his chemistry professors interested him in trying to find a method for producing aluminum which would be practical for use on an industrial scale. (In the middle 1800s aluminum had sold for more than $500 per pound and was used largely as a precious metal in jewelry, etc.) Hall had a hunch that electrolysis would work and although he knew nothing of standard oxidation potentials, or even of the existence of ions (they had not yet been shown to exist), he did know of Faraday's work on electrolysis. But Hall soon found out that electrolysis of solutions of aluminum salts produced only hydrogen at the cathode. [The standard electrode potential for the reduction of Al^{3+} to $Al(s)$ is -1.67 V, and so H^+, or H_2O, is reduced, rather than Al^{3+}.]

A second possibility, the electrolysis of molten alumina, Al_2O_3, was rejected because of the high melting point of this compound (2045°C). Hall then sought a nonaqueous solvent for Al_2O_3 and eventually hit upon *cryolite* (from the Greek: "frost stone"), a mineral found in Greenland which looks like ice, but melts at 950°C (hence, its name). Cryolite is sodium hexafluoroaluminate, Na_3AlF_6, and when molten, was found to dissolve Al_2O_3. Hall had set up a sort of laboratory in the family woodshed, and in 1886, less than a year after graduating from college, he successfully electrolyzed a solution of alumina in molten cryolite to form a few shiny globules of aluminum.[3]

The electrolytic cell used in the Hall process is shown schematically in Fig. 21-2. At the graphite anodes a mixture of products, including O_2 and F_2, is formed. These gases are very reactive, especially at elevated temperatures, and so the anodes gradually corrode and must be replaced periodically. The molten aluminum is formed at the graphite cathode, drawn off at the bottom of the cell, and cast into ingots. Today most of the cryolite used in the Hall process is produced synthetically.

Although aluminum is far cheaper now than it was 100 years ago, vast quantities of electrical energy are needed to produce it. One equivalent of aluminum weighs $\frac{27}{3}$, or 9 g; this means that one faraday of electricity is necessary to produce only 9 g of the metal. This is why the recycling of aluminum beverage cans (and other objects) is so important. Of course, the energy put into the production of aluminum is not totally lost; we may someday discover how to use aluminum in a commercially practical battery which will produce electrical energy from the

---

[3]Coincidentally, the same process was discovered one week later by Paul Héroult in France. In Europe the process is usually referred to as the Héroult process.

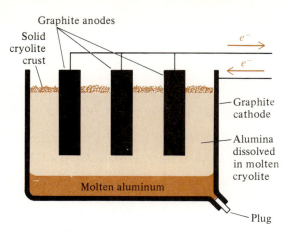

Figure 21-2
Hall cell for production of aluminum.

anodic oxidation of aluminum. (It has been suggested that the ultimate propulsion system for an automobile is an electric motor powered by a fuel cell which consumes beer cans collected along the highway.)

Aluminum is an extremely versatile metal. It can be rolled, pressed, cast, bent, extruded, or drawn into almost any imaginable shape. Its low density makes it useful in aircraft construction, and recently the automotive industry has been using more aluminum each year in order to help lighten vehicles. Actually, pure aluminum is too soft to be used structurally, but alloys which incorporate small quantities of copper, silicon, manganese, or magnesium have strengths and hardnesses which approach those of some steels. Pure aluminum is an excellent electrical conductor and finds application in electrical wiring, replacing increasingly scarce copper.

Aluminum is thermodynamically highly reactive, as shown by its standard reduction potential (above) and the free-energy changes for reactions such as

$$2Al(s) + \tfrac{3}{2}O_2(g) \longrightarrow Al_2O_3(s) \qquad \Delta G^\circ = -1576 \text{ kJ mol}^{-1}$$

Reactions of metallic aluminum at room temperature are slowed considerably by the formation of a surface skin of Al_2O_3. This surface layer is smooth and very tough and hard, and so aluminum and most of its alloys are self-protected against attack by the atmosphere. An extra-thick layer of alumina can be deposited electrolytically on the surface of aluminum in a process known as *anodizing*. The alumina can be colored by the addition of various dyes to the anodizing solution, and the resulting color coat is far more durable than any paint.

Aluminum forms the previously mentioned oxide and hydroxide and also a wide range of salts. The aluminum ion, being small and highly charged, hydrolyzes extensively in water. This reaction may be written as

$$Al^{3+} + H_2O \rightleftharpoons AlOH^{2+} + 2H^+$$

or as $\qquad Al(H_2O)_6^{3+} + H_2O \rightleftharpoons Al(OH)(H_2O)_5^{2+} + H_3O^+$

(The above represent only the first step in the hydrolysis.) Addition of a limited amount of base to a solution containing Al^{3+} precipitates aluminum hydroxide, $Al(OH)_3$. This is a white or colorless precipitate which has almost the same refractive index as water, making it hard to see. In qualitative analysis aluminum hydroxide is made more visible by adding the dye *aluminon* (the ammonium salt of aurintricarboxylic acid), which forms a red lake with the hydroxide. Actually, freshly precipitated "aluminum hydroxide" is not stoichiometrically well characterized, and it is more correctly referred to as *hydrous aluminum oxide,* $Al_2O_3 \cdot xH_2O$.

Aluminum hydroxide is amphoteric, and so it will dissolve in excess base:

$$Al(OH)_3(s) + OH^- \longrightarrow Al(OH)_4^-$$

The stability of the aluminate ion, more completely formulated as $Al(OH)_4(H_2O)_2^-$, accounts for the fact that aluminum metal dissolves not only in acids,

$$2Al(s) + 6H^+ \longrightarrow 2Al^{3+} + 3H_2(g)$$

but also in bases,

$$2Al(s) + 2OH^- + 6H_2O \longrightarrow 2Al(OH)_4^- + 3H_2(g)$$

Anhydrous aluminum oxide occurs naturally as the mineral *corundum.* Although the Al—O bond has considerable covalent character, the structure of Al_2O_3 may be imagined as a hexagonal close-packed array of oxide ions in which two-thirds of the octahedral holes are occupied by aluminum ions. Pure Al_2O_3 is colorless (or white, if powdered), but when Cr^{3+} ions substitute for some of the Al^{3+} ions, a red color is produced, as in the gemstone *ruby.* Substitution by Fe^{3+} and Ti^{3+} ions produces the blue color in *sapphire.*

The extreme hardness of alumina is due in part to the strength of the Al—O bond. This is taken advantage of in the *thermite reaction* is which ferric oxide reacts with powdered aluminum:

$$Fe_2O_3(s) \longrightarrow 2Fe(l) + \tfrac{3}{2}O_2(g) \qquad \Delta H^\circ = +\ 824\ kJ$$
$$\underline{2Al(s) + \tfrac{3}{2}O_2(g) \longrightarrow Al_2O_3(s) \qquad \Delta H^\circ = -1676\ kJ}$$
$$Net:\ 2Al(s) + Fe_2O_3(s) \longrightarrow 2Fe(l) + Al_2O_3(s) \qquad \Delta H^\circ = -\ 852\ kJ$$

The large amount of heat evolved results in temperatures approaching 3000°C and produces the product iron in molten form. Thermite is used for welding in locations where neither electricity nor welding gases are available.

Aluminum was first discovered in a series of salts known as the *alums.* These are hydrated sulfates containing both uni- and tripositive ions: $A^+B^{3+}(SO_4)_2 \cdot 12H_2O$. Ordinary alum is $KAl(SO_4)_2 \cdot 12H_2O$. Unipositive ions such as Li^+, Na^+, and even NH_4^+ may substitute for K^+, and tripositive ions such as Cr^{3+} and Fe^{3+} may substitute for Al^{3+} to form a series of different alums, all of which crystallize in the same lattice and which produce large, easy-to-grow, octahedral crystals. Figure 10-2 shows a large, laboratory-grown crystal of an alum. (Beginners in the art of crystal growing often start with the alums and soon find that layers of one alum can be easily built up on a crystal of another, when the crystal is grown by evaporation at approximately constant temperature.) Vast quantities of alum are used in the paper and dye industries.

Aluminum chloride crystallizes from aqueous solutions as a hydrate, $AlCl_3 \cdot 6H_2O$, or better, $Al(H_2O)_6Cl_3$. This is a typical ionic salt containing the hydrated Al^{3+} ion. Anhydrous aluminum chloride, on the other hand, consists of a three-dimensional covalent network in which chlorine atoms form bridges

Figure 21-3
The Al_2Cl_6 structure.

between adjacent aluminum atoms. $AlCl_3$ sublimes at 178°C (1 atm pressure) to form a gas which consists of Al_2Cl_6 dimers as shown in Fig. 21-3. As can be seen, each molecule uses two of its six Cl atoms to "bridge bond" the two Al atoms. Above about 400°C dissociation of Al_2Cl_6 is significant:

$$Al_2Cl_6(g) \rightleftharpoons 2AlCl_3$$

The monomer $AlCl_3$ has been shown by electron diffraction to have a trigonal-planar structure (sp^2 hybridization).

Gallium, indium, and thallium

The heavier three members of group IIIA are all rare and are found primarily as impurities in zinc, lead, and cadmium ores, as well as with many aluminum-containing minerals. Because of their scarcity they are used only in small quantities. Gallium has been used as a dopant in the manufacture of semiconductors and as a component of alloys. Gallium arsenide is used in solid-state lasers. Indium has been used in the manufacture of bearings and solid-state devices. Thallium is used in photocells, infrared detectors, and in the manufacture of special glasses. Thallium(I) sulfate is extremely toxic and is used as a rodent and ant killer.

Gallium, indium, and thallium are not as metallic as one might expect. This is largely a consequence of the transition-metal series which precedes each in the periodic table. Since the $(n - 1)d$ subshell is filled, the resulting high nuclear charge and small size give the atoms of each of these elements a tighter hold on their electrons than would otherwise be the case. Gallium hydroxide, $Ga(OH)_3$, is consequently very much like aluminum hydroxide in its amphoterism and even indium hydroxide dissolves slightly in base, though more readily in acid. Thallium is the only element in the group which illustrates the inert pair effect: it forms stable Tl(I) (thallous) and Tl(III) (thallic) compounds.

21-4
Other representative metals

Other important nontransition metals are tin and lead (periodic group IVA) and bismuth (group VA).

Tin

The element *tin* (the name comes from the Anglo-Saxon) has been known since prehistoric times. Its most important source is the mineral *cassiterite*, SnO_2, from which it may be obtained by reduction by carbon,

$$SnO_2(s) + 2C(s) \longrightarrow Sn(l) + 2CO(g)$$

followed by electrolytic purification.

Tin shows three solid allotropic forms: α-tin, or *gray tin*, is a nonmetallic form stable below 13°C. In it tin atoms are covalently bonded in the diamond lattice. From 13 to 161°C β-tin, or *white tin*, is the stable form. It is ordinary, metallic tin and crystallizes in a tetragonal lattice. Above 161°C (up to 232°C, the melting point) γ-tin, or *rhombic tin*, is stable. It crystallizes in an orthorhombic lattice and is quite brittle. The $\beta \longrightarrow \alpha$ transition is slow, and when it occurred in old tin organ pipes in cold European cathedrals, it was called *tin disease*. (The evil looking "growths" of gray tin were thought to be the work of the devil.)

Tin is commonly used in plating ("tin cans" are tin-plated iron) and in alloys such as *bronze* (with copper), *pewter* and *solder* (with lead), and some bearing metals.

The valence-shell configuration of the tin atom is $5s^2 5p^2$, and the inert-pair effect is evident in the existence of compounds with tin in the $+2$ state (*stannous* compounds) and the $+4$ state (*stannic* compounds). The standard reduction potential for $Sn^{2+} + 2e^- \longrightarrow Sn(s)$ is -0.14 V, and so tin slowly dissolves in dilute acids:

$$Sn(s) + 2H^+ \longrightarrow Sn^{2+} + H_2(g)$$

The stannous ion forms a wide variety of compounds such as the hydroxide, oxide, sulfide, halides, etc. The stannous ion is moderately hydrolyzed in water:

$$Sn^{2+} + H_2O \rightleftharpoons SnOH^+ + H^+$$

Addition of a base to the stannous ion precipitates stannous hydroxide:

$$Sn^{2+} + 2OH^- \longrightarrow Sn(OH)_2(s)$$

This is amphoteric and dissolves in excess base to give the *stannite ion:*

$$Sn(OH)_2(s) + OH^- \longrightarrow Sn(OH)_3^-$$

The stannous ion is a useful mild reducing agent. For example, it reduces mercuric chloride first to insoluble mercurous chloride and then, if present in excess, to metallic mercury:

$$2HgCl_2 + Sn^{2+} \longrightarrow Sn^{4+} + Hg_2Cl_2(s) + 2Cl^-$$

$$Hg_2Cl_2(s) + Sn^{2+} \longrightarrow 2Hg(l) + Sn^{4+} + 2Cl^-$$

(Here we have written mercuric chloride as a molecule, since it is one of the uncommon *weak* salts.)

In basic solution Sn(II) is an even better reducing agent than in acidic solution. Thus the stannite ion is readily oxidized to the *stannate ion,* $Sn(OH)_6^{2-}$.

Sn(IV) forms an insoluble hydrous oxide, often represented for simplicity as stannic hydroxide, $Sn(OH)_4$. It dissolves in excess base to form the stannate ion. Actually, two different hydrous oxides of Sn(IV) are known: one is soluble in acid, while the other, formed by the oxidation of tin metal or the stannous ion by nitric acid, is insoluble in acid. The differences between these forms are presently not well understood.

In qualitative analysis tin is usually precipitated either as brown stannous sulfide, SnS, or as yellow stannic sulfide, SnS_2. The former is insoluble in excess sulfide, while the latter dissolves to form the *thiostannate ion:*

$$SnS_2(s) + S^{2-} \longrightarrow SnS_3^{2-}$$

Stannous sulfide will dissolve, however, in a solution containing polysulfide ions, here represented as S_2^{2-}:

$$SnS(s) + S_2^{2-} \longrightarrow SnS_3^{2-}$$

Conversion of this complex to a chloro complex is accomplished by adding excess HCl. After boiling to remove all sulfide, the tin(IV) can be reduced to tin(II) by iron and then reoxidized by mercuric chloride to form Hg_2Cl_2 and/or Hg.

Lead   *Lead* (another old Anglo-Saxon word) was also known in ancient times. Its only important ore is the mineral *galena*, PbS, from which it is obtained by several methods. One of these involves *roasting* the ore, heating it in the presence of air to convert it to the oxide:

$$2PbS(s) + 3O_2(g) \longrightarrow 2PbO(s) + 2SO_2(g)$$

Some of the PbO is converted to lead sulfate in this process:

$$PbS(s) + 2O_2(g) \longrightarrow PbSO_4(s)$$

The product (PbO and $PbSO_4$) is then mixed with additional PbS and reheated without air:

$$PbS(s) + 2PbO(s) \longrightarrow 3Pb(l) + SO_2(s)$$

$$PbS(s) + PbSO_4(s) \longrightarrow 2Pb(l) + 2SO_2(g)$$

The resulting lead is quite impure, however, and must be further purified to remove many contaminating metals.

Lead exists in only one common allotropic form, the familiar, soft, low-melting, gray metal. Its primary use is in the manufacture of lead storage batteries (Sec. 19-6). Large quantities of lead have been used in the synthesis of tetraethyllead, $(C_2H_5)_4Pb$, used as an antiknock additive in gasoline. Small quantities of this compound raise the octane number of gasoline, which means it prevents detonation ("knock") in the engine cylinders and promotes smooth burning of the gasoline. Most of the lead gets blown out the exhaust pipe, contributing to environmental pollution, however. Also, lead compounds poison the catalysts in the catalytic converters of newer automobiles, and so alternate ways of increasing the octane numbers of gasolines are being introduced. Lead is also used in various alloys and in *red lead* corrosion-resistant paint, which contains Pb_3O_4.

In its compounds lead more commonly shows the $+2$ oxidation state than the $+4$. The lead(II) ion, sometimes called the *plumbous ion*, precipitates as a number of insoluble salts, including $PbCl_2$, PbS, $PbSO_4$, $PbCrO_4$, and $PbCO_3$. The lead(II) halides will dissolve in excess halide ion to form complexes such as $PbCl_3^-$ and $PbBr_4^{2-}$. Lead(II) hydroxide precipitates from a Pb^{2+} solution when a base is added;

$$Pb^{2+} + 2OH^- \longrightarrow Pb(OH)_2(s)$$

Plumbous hydroxide is amphoteric, dissolving in excess base to form the *plumbite ion:*

$$Pb(OH)_2(s) + OH^- \longrightarrow Pb(OH)_3^-$$

Lead(II) forms some salts which are weak electrolytes. Insoluble lead sulfate, $PbSO_4$, dissolves in a solution of ammonium acetate, for example, because of the formation of the soluble but slightly dissociated lead acetate:

$$PbSO_4(s) + 2C_2H_3O_2^- \longrightarrow Pb(C_2H_3O_2)_2 + SO_4^{2-}$$

In the $+4$, or *plumbic,* state the most important compound is plumbic oxide, usually called lead dioxide, PbO_2. This is a powerful oxidizing agent and is used in the lead storage cell. It can be made by the oxidation of the plumbite ion by hypochlorite:

$$Pb(OH)_3^- + OCl^- \longrightarrow PbO_2(s) + OH^- + Cl^- + H_2O$$

Pb_3O_4, *red lead,* is in reality a complex covalent compound with both Pb(II) and Pb(IV) atoms bonded to oxygens. It is made by heating lead(II) oxide in air at 500°C;

$$6PbO(s) + O_2(g) \longrightarrow 2Pb_3O_4(s)$$

Lead, like most of the heavy metals, is a toxic element. Some historians feel that the decline of the Roman Empire was due to illness, infertility, and death brought about by lead poisoning. It seems that the Roman aristocracy made great use of lead in water pipes and cooking vessels. More recently, *white lead,* $Pb_3(OH)_2(CO_3)_2$, was used as a pigment in paints, and today young children sometimes experience its toxicity after eating chips of old paint from woodwork, toys, etc. Heavy metals are usually inhibitors of many enzyme-catalyzed reactions.

In qualitative analysis lead is usually separated and/or confirmed by precipitation as the chloride, sulfide, chromate, or sulfate.

Bismuth *Bismuth* (the name is thought to be a corruption of German words meaning "white mass") occurs naturally as Bi_2O_3 and Bi_2S_3. The sulfide is converted to the oxide by roasting in air, and the oxide is reduced with carbon.

Bismuth is a dense metal with a dull metallic luster. It is used in various alloys including some which are quite low melting and are used as sensors in fire detection and sprinkler systems. Like water, bismuth is one of the few substances which expand on freezing; it contributes this property to some of its alloys, which are used for type metal. These expand as they are cast to produce faithful reproduction of fine detail in the mold.

The chemistry of bismuth is almost exclusively one of the +3 state. Bismuth burns in oxygen to give the trioxide:

$$4Bi(s) + 3O_2(g) \longrightarrow 2Bi_2O_3(s)$$

This and the hydroxide, $Bi(OH)_3$, show no amphoterism, being soluble in acids, but not in bases.

The bismuth ion, Bi^{3+}, shows such a strong tendency to hydrolyze that addition of bismuth salts to water produces a cloudy precipitate of the hydroxide:

$$Bi^{3+} + 3H_2O \longrightarrow Bi(OH)_3(s) + 3H^+$$

If the chloride ion is present, as when $BiCl_3$ is added to water, the precipitate contains some chloride and is usually formulated $Bi(OH)_2Cl$ or $BiOCl$. This hydrolysis can be repressed only by keeping the solution strongly acidic.

"Sodium bismuthate" is a poorly characterized substance prepared by fusing Bi_2O_3 with Na_2O_2 and NaOH. It is often represented as $NaBiO_3$, but is probably very impure Bi_2O_5; it is a powerful oxidizing agent and will oxidize Mn^{2+} to MnO_4^-.

In qualitative analysis bismuth is precipitated as a brownish-black sulfide, Bi_2S_3, which is insoluble in both excess sulfide and polysulfide. Confirmation of bismuth usually involves observing the above-mentioned hydrolysis of Bi^{3+} or the reduction of $Bi(OH)_3$ by the stannite ion to form black, finely divided bismuth metal:

$$2Bi(OH)_3(s) + 3Sn(OH)_3^- + 3OH^- \longrightarrow 2Bi(s) + 3Sn(OH)_6^{2-}$$

21-5
The metalloids

The word *metalloid* means "like a metal." The metalloids are neither typical metals nor typical nonmetals, but show properties of both. They are often called *semimetals*. The metalloids include the elements boron (group IIIA), silicon and germanium (group IVA), arsenic and antimony (group VA), and selenium and tellurium (group VIA). These elements are found in a narrow region running diagonally across the periodic table, as shown in Fig. 21-4. (Sometimes selenium is classed as a nonmetal, and astatine, polonium, and even bismuth have been classed as metalloids, but these decisions are somewhat arbitrary.)

The uncombined metalloids tend to be hard and brittle and to be poor-to-fair conductors of heat and electricity. They often show electrical semiconduction or can be turned into semiconductors by doping. They typically show a semimetallic luster. These properties indicate that in the solids the valence electrons are not as freely delocalized as in the metals and that there is considerable covalent bonding present. As expected, these elements have intermediate electronegativities; on the Pauling scale they range from 1.8 to 2.4. Their hydroxo compounds are either weakly acidic or amphoteric.

Boron

The name *boron* comes from an Arabic word meaning "white." (Borax, its most common source, is white.) The element is in periodic group IIIA and is always found combined in nature, usually as borate salts, such as *borax*, $Na_2B_4O_5(OH)_4 \cdot 8H_2O$. Boron compounds find a wide variety of applications: borax is used in some detergent formulations, where it assists in the cleaning process and also leaves a slight sheen on ironed linens. Boron is incorporated into *borosilicate glasses,* such as Pyrex; these have much lower coefficients of thermal expansion than ordinary soda-lime ("soft") glasses and therefore can better withstand sudden changes in temperature. Fibers of elemental boron are com-

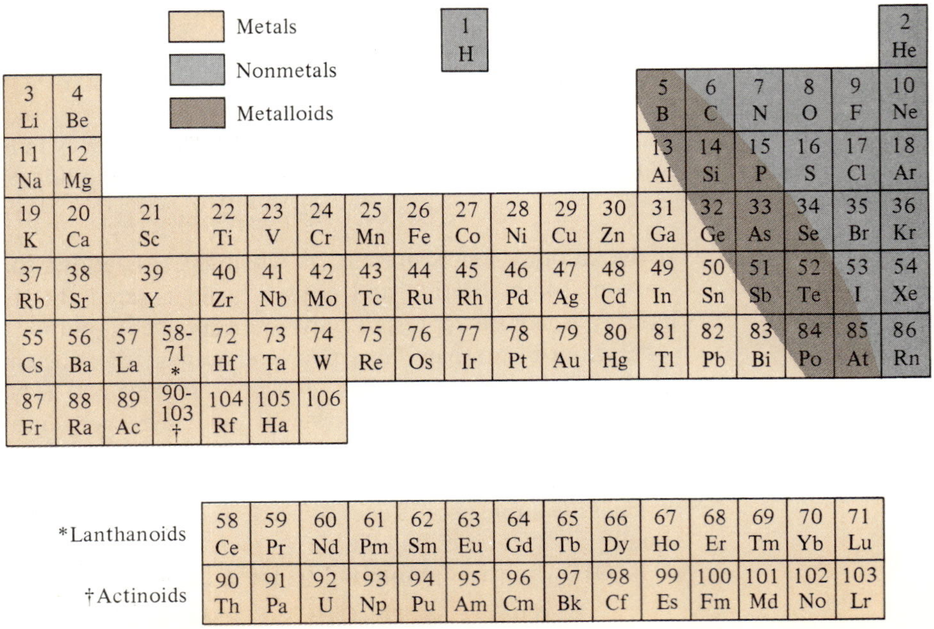

Figure 21-4
The metalloids.

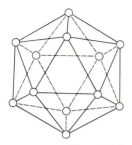

Figure 21-5
The B_{12} icosahedral
unit of structure.

bined with certain epoxy plastics to produce composite materials with great strength and light weight.

Uncombined boron is obtained by dissolving borax in acid and crystallizing out boric acid, H_3BO_3. Dehydration to form B_2O_3 is then followed by reduction with magnesium:

$$B_2O_3(s) + 3Mg(s) \longrightarrow 2B(s) + 3MgO(s)$$

Purer boron may be prepared by reducing boron trichloride with hydrogen gas at high temperature,

$$2BCl_3(g) + 3H_2(g) \longrightarrow 2B(s) + 6HCl(g)$$

or by the thermal decomposition of boron triiodide at the surface of an electrically heated tungsten or tantalum wire,

$$2BI_3(g) \longrightarrow 2B(s) + 3I_2(g)$$

Elemental boron has a dull, semimetallic luster and is an electrical semiconductor. It is the second-hardest element (after diamond) and exists in a number of crystalline modifications, all of which contain B_{12} structural units, and which differ from each other in the way these units are linked in the crystal lattice. In each of these B_{12} units boron atoms are at the 12 corners of an icosahedron (a regular 20-sided polyhedron), as shown in Fig. 21-5. This icosahedral structure shows up in some boron compounds, as well.

Boron forms a series of trihalides, BX_3, all of which are low-melting, molecular compounds. Since the valence-shell electronic configuration of the boron atom is $2s^2 2p^1$, the Lewis structure of BF_3, for example, can be written

$$\ddot{\ddot{F}} : \\ B : \ddot{F} : \\ \ddot{\ddot{F}} :$$

The molecule is trigonal planar, as predicted from VSEPR theory, and so the boron atom presumably uses sp^2 hybridization to form the three equivalent bonds. The bond distances in these molecules are somewhat shorter than expected, however, indicating the possibility of some double-bond character in each B—X bond. This can be accounted for by assuming overlap of one of the filled p orbitals of the halogen atom with the empty, unhybridized p orbital of the boron. The boron halides are Lewis acids; in reactions the boron atom accepts a pair of electrons to complete its octet. Some examples are

$$\left[: \ddot{F} : \right]^- + \begin{array}{c} \ddot{F} : \\ B : \ddot{F} : \\ \ddot{F} : \end{array} \longrightarrow \left[\begin{array}{c} \ddot{F} : \\ : \ddot{F} : B : \ddot{F} : \\ \ddot{F} : \end{array} \right]^-$$

$$\begin{array}{c} H \\ H : N : \\ H \end{array} + \begin{array}{c} \ddot{F} : \\ B : \ddot{F} : \\ \ddot{F} : \end{array} \longrightarrow \begin{array}{c} H \;\; \ddot{F} : \\ H : N : B : \ddot{F} : \\ H \;\; \ddot{F} : \end{array}$$

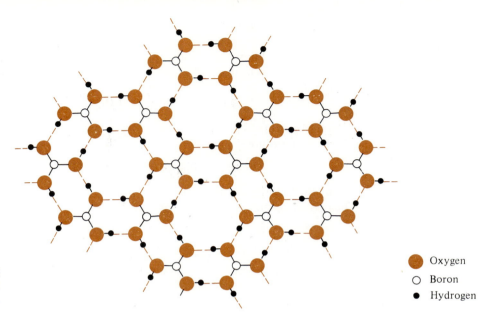

**Figure 21-6
Structure of solid
H_3BO_3.**

Oxygen
Boron
Hydrogen

The latter compound is an example of an *addition compound,* or *adduct.*

The hydroxo compound of boron is an acid, *boric acid,* H_3BO_3. It is a sparingly soluble solid in which no discrete H_3BO_3 molecules exist. Rather, flat BO_3 units are hydrogen-bonded together in an infinite, two-dimensional layer, as shown in Fig. 21-6. Boric acid is a weak, *monoprotic* acid with $K_a = 6 \times 10^{-10}$. At low concentrations its dissociation is represented as

$$H_3BO_3 + H_2O \rightleftharpoons B(OH)_4^- + H^+$$

In more concentrated solutions the simple H_3BO_3 molecule is apparently replaced by more complex species.

Boron forms an extensive series of salts, the *borates,* in which the boron is present in various oxoanions. Some of the borate anions are discrete, such as the trigonal planar orthoborate ion, BO_3^{3-}, while others are extended chains. A few borate anions are illustrated in Fig. 21-7. In each of the ions pictured, each boron atom is surrounded by three oxygen atoms, some of which bridge two boron atoms, and the rest of which are terminal. The number of terminal oxygen atoms equals the charge on the ion. In the extended chain borate anion in Fig. 21-7 the empirical formula is found by identifying the *repeat unit* in the chain, shown by broken lines in the illustration.

Boron combines with hydrogen to form an extremely interesting series of compounds, the *boron hydrides,* or *boranes.* Surprisingly, the simplest of these, BH_3, has been detected only as a short-lived, transient species. At least 17 stable boranes are known, each of which is a low-melting, reactive, molecular compound. Representative formulas include B_2H_6, B_4H_8, B_4H_{10}, $B_{10}H_{16}$, and $B_{18}H_{22}$. The lighter boranes spontaneously ignite in air.

The boranes are classed as *electron-deficient compounds,* because in each case insufficient electrons seem to be present to hold the molecule together. This is illustrated most simply with *diborane,* B_2H_6. In this molecule a total of 12 valence

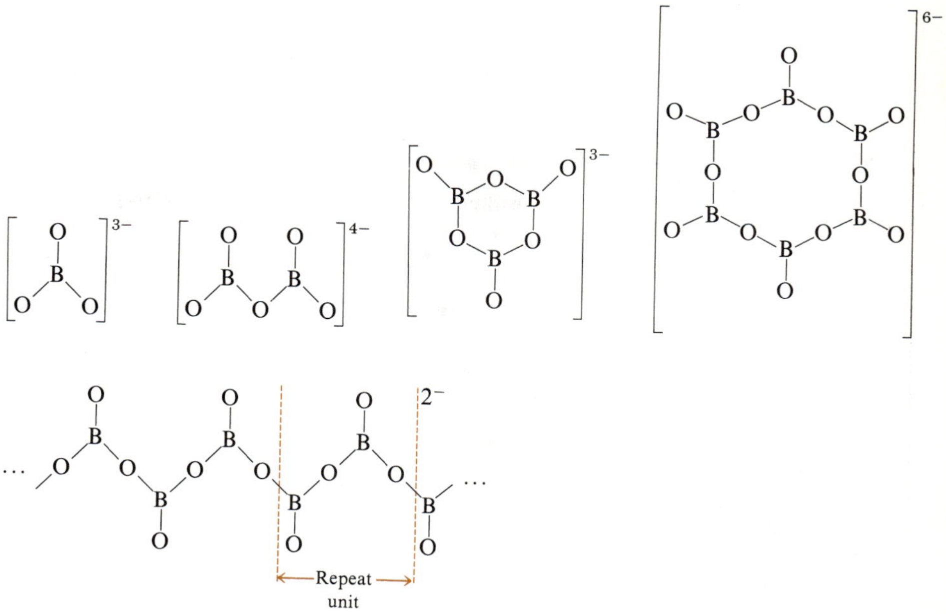

Figure 21-7
Some borate anions.

electrons (6 from the borons and 6 from the hydrogens) is available for bonding. If these are assigned in pairs to bond hydrogen to a boron, we end up with two BH$_3$ fragments which are not bonded to each other:

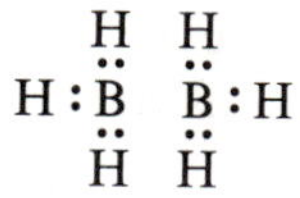

$$\begin{array}{cc} H & H \\ \cdot\cdot & \cdot\cdot \\ H:B & B:H \\ \cdot\cdot & \cdot\cdot \\ H & H \end{array}$$

A clue to the bonding in diborane comes from knowledge of the geometry of the molecule. It has the structure shown in Fig. 21-8, in which two hydrogens act as bridges between the two borons. But how does an H atom bond to two B atoms? It has been shown that in B$_2$H$_6$ two *three-center bonds* are used. Most covalent bonds are two-center bonds, that is, a pair of electrons bonds two atoms together. However, MO theory shows that it is possible to combine *three* atomic orbitals from *three* atoms to form a bonding molecular orbital which will accommodate a pair of electrons and which will bond all three atoms together.

The bonding scheme in diborane is shown in Fig. 21-9. Each boron uses all of its valence-shell orbitals to form four sp^3 hybrid orbitals, two of which are used to bond the terminal H atoms with conventional two-center bonds. The remaining two hybrid orbitals on each boron overlap with the 1s orbitals of two H atoms to form two three-center molecular orbitals. Each of these can be shown to be of lower energy, and therefore increased stability, than the original atomic orbitals. Three-center bonding is used to account for the structures of all the boranes.

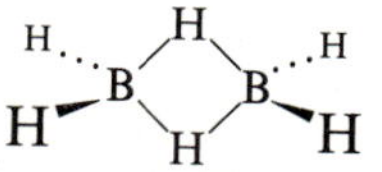

Boron forms an interesting pair of *nitrides,* each of which has the formula BN. *Hexagonal boron nitride* is a slippery white solid with many of the properties of graphite. In fact, it has the graphite structure (Sec. 20-8), except that B and N atoms alternate in the lattice. (Note that the valence electrons contributed by one B and one N atom are equivalent to those from two C atoms.) *Cubic boron nitride,*

Figure 21-8
Structure of diborane.

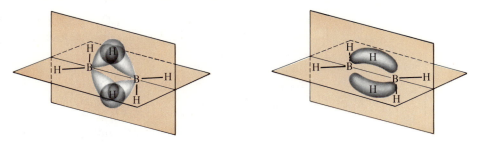

Figure 21-9
Three-center bonds in diborane.

or *borazon,* is formed at high pressure and is about as hard as diamond. It is used in cutting tools and abrasives and is more resistant to high-temperature oxidation than is diamond.

Silicon   In group IVA *silicon* is classed as a metalloid. (The name is derived from a Latin word meaning "flint"; flint is impure silicon dioxide.) In abundance in the earth's crust silicon is exceeded only by oxygen. The vast majority of rocks, soils, sands, and clays are composed of silicon compounds, either as various forms of silica, SiO_2 (quartz, flint, opal, etc.), or as silicates such as the feldspars, micas, and many others.

Uncombined silicon is obtained from the high-temperature reduction of SiO_2 using magnesium or carbon or, for purer silicon, the reduction of $SiCl_4$ by magnesium. The element crystallizes in the diamond lattice to form a very hard, high-melting, metallic-appearing solid. (No silicon analog of graphite has been prepared.) In spite of its metallic luster, it is a poor conductor of electricity at room temperature. It is a semiconductor, however.

The chemistry of silicon is largely the chemistry of its oxo compounds. Although it forms binary compounds with hydrogen (the *silanes:* SiH_4, Si_2H_6, etc.) and with the halogens, such as silicon tetrachloride, $SiCl_4$, these compounds tend to be converted to SiO_2 in the presence of air. Silicon chemistry is largely the chemistry of the $+4$ state in which all four of its $2s^2 2p^2$ electrons are shared with more electronegative elements.

Unlike carbon, silicon does not double-bond with oxygen; instead, most silicon-oxygen compounds show sp^3 hybridization of the valence-shell orbitals of Si with resulting tetrahedral bonding of oxygen atoms around each silicon atom. The simplest such structure is the orthosilicate ion, SiO_4^{4-},

$$\begin{bmatrix} & \ddot{\underset{\cdot\cdot}{O}} & \\ :\ddot{\underset{\cdot\cdot}{O}}: & Si & :\ddot{\underset{\cdot\cdot}{O}}: \\ & :\ddot{\underset{\cdot\cdot}{O}}: & \end{bmatrix}^{4-}$$

This is the anion of orthosilicic acid, $Si(OH)_4$, which can be prepared only in dilute solution. (Attempts to concentrate it result only in complex long-chain silicic acids.) The discrete SiO_4^{4-} ion is found in a few minerals, including *olivine* (Mg_2SiO_4) and *zircon* ($ZrSiO_4$).

The disilicate, or pyrosilicate, ion, $Si_2O_7^{6-}$, is found in thortveitite ($Sc_2Si_2O_7$) and barysilite ($Pb_3Si_2O_7$). In this ion one of the seven oxygens acts as a bridge between two silicons.

Some minerals consist of cyclic silicate anions. These include *benitoite* ($BaTiSi_3O_9$), in which alternating silicon and oxygen atoms form a six-membered $Si_3O_9^{6-}$ ring, and *beryl* ($Be_3Al_2Si_6O_{18}$), in which the $Si_6O_{18}^{12-}$ ion has a 12-membered ring. In each of these each silicon is oxygen-bridged to two other silicons. The structures of the discrete silicate anions are shown schematically in Fig. 21-10. Remember that in each anion each silicon is surrounded tetrahedrally by four oxygen atoms. Further, note that the charge on each ion equals the number of terminal oxygens.

Extended silicate anions are also known. The *infinite chain* SiO_3^{2-} ion is found in the minerals classed as *pyroxenes*. These include *enstatite* ($MgSiO_3$) and *jadeite* [$NaAl(SiO_3)_2$]. The pyroxene chain is shown in Fig. 21-11a. Long chains like this are packed together with cations in the spaces between, and electrostatic attraction between the cations and chains holds the assembly together. Note the broken lines in the illustration; these indicate the repeat unit, which is specified by the empirical formula, SiO_3^{2-}.

When two silicate chains are cross-linked so that half the silicons have three bridge oxygens, and the rest have two, the result is an *infinite band* $Si_4O_{11}^{6-}$ ion, as shown in Fig. 21-11b. This anion occurs in numerous minerals called *amphiboles*. The various asbestos minerals are of this type, their fibrous nature a result of the one-dimensional anion structure.

When all silicons are attached to three oxygen bridges in a silicate, infinite sheet anions, $Si_2O_5^{2-}$, shown in Fig. 21-11c, are formed. These two-dimensional anions are found in *talc*, $Mg_3(Si_2O_5)_2(OH)_2$ and in the *clays* and *micas*. Usually extra cations, hydroxide ions, and sometimes water molecules are sandwiched between the sheets in a complex structure. Aluminum ions may be present, either as cations between the sheets or substituting regularly for silicons in the sheets themselves.

As the number of bridge oxygens attached to each silicon exceeds three, anion structures which are three-dimensional result. These include the feldspars, such as orthoclase ($KAlSi_3O_8$) and the zeolites, such as analcite ($NaAlSi_2O_6 \cdot H_2O$).

The limit of this series of silicates is found in silica, SiO_2, in which each silicon has four oxygens which bridge to four other silicons tetrahedrally located around the first. There are at least six crystalline forms of silica, the best-known being common *quartz*. *Vitreous silica*, or *fused quartz*, is silica which has been super-

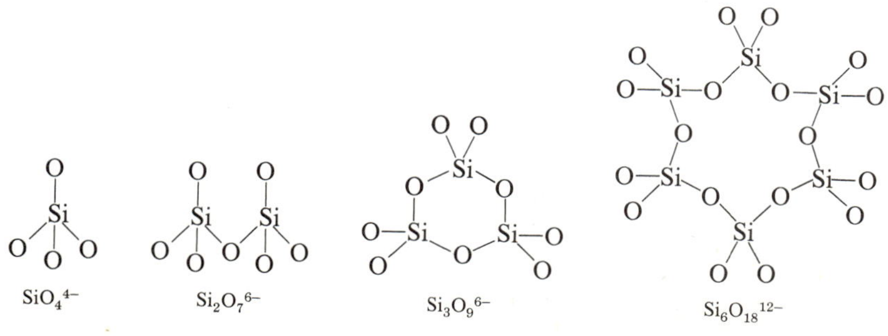

Figure 21-10
Discrete silicate anions (schematic).

SiO_4^{4-} $Si_2O_7^{6-}$ $Si_3O_9^{6-}$ $Si_6O_{18}^{12-}$

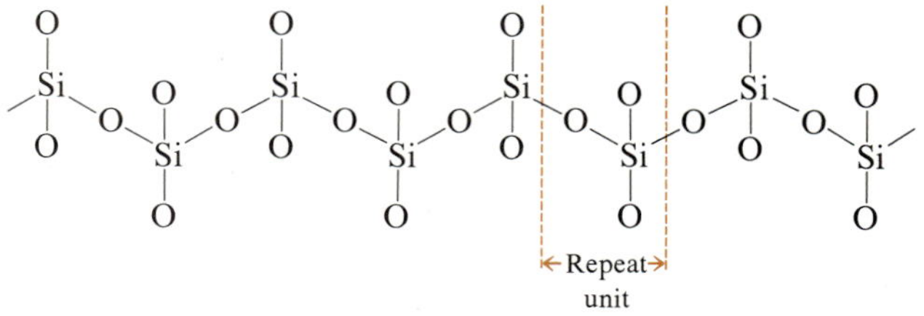

(a) Chain, SiO_3^{2-} (pyroxenes)

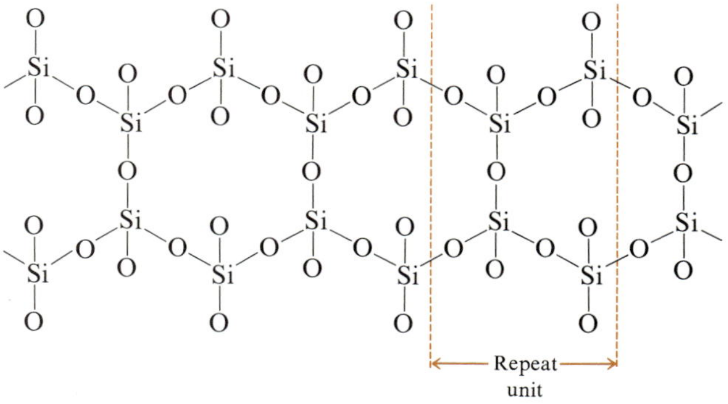

(b) Band, $Si_4O_{11}^{6-}$ (amphiboles)

Figure 21-11
Extended silicate an-
ions (schematic).
(a) Chain, SiO_3^{2-} (py-
roxenes). (b) Band,
$Si_4O_{11}^{6-}$ (amphiboles).
(c) Sheet, $S_2O_5^{2-}$
(micas, talc, clays,
etc.).

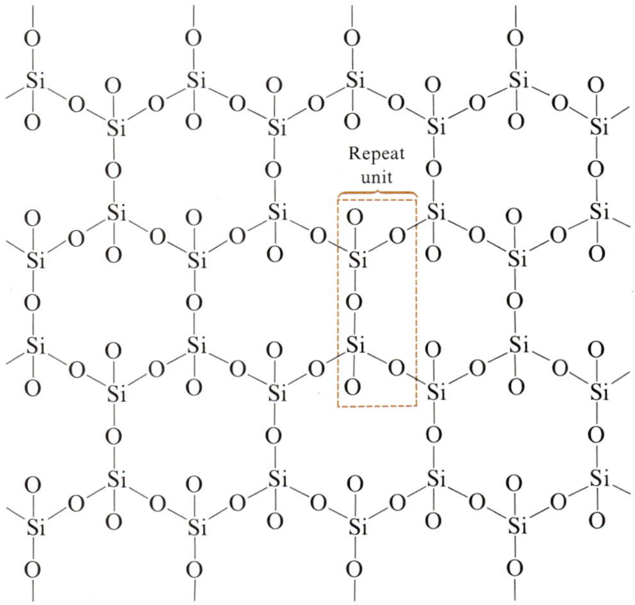

(c) Sheet, $Si_2O_5^{2-}$ (micas, talc, clays, etc.)

cooled from the liquid state. This glassy substance is transparent to near-ultraviolet light and has an extremely low coefficient of thermal expansion. (White-hot vitreous silica can be plunged into cold water with much fuss, but without breaking.)

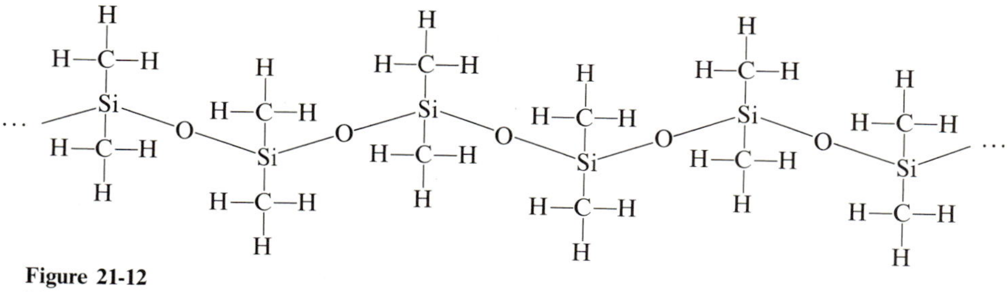

Figure 21-12
Methyl silicone.

Silicon forms some commercially important compounds called the *silicones*. A typical silicone molecule has a long chain of alternating silicon and oxygen atoms as a backbone, with organic substituents, such as the methyl group, $-CH_3$, on the outside of the chain. The structure of methyl silicone is shown in Fig. 21-12. This is a useful lubricant. When such chains are cross-linked, waxy, greasy, or rubbery silicones result. Silicones have much greater thermal stability than ordinary hydrocarbon oils, greases, and rubbers, mostly as a result of the silicate backbones of their molecules.

Germanium   Located below silicon in group IVA is *germanium*. (It is named after *Germany*.) It is a grayish-white, brittle solid which, like silicon, crystallizes in the diamond lattice. Like silicon, also, it is a semiconductor and is used in the fabrication of solid-state electronic devices, such as transistors, integrated circuits, etc. Germanium semiconductor devices are more subject to failure at high temperatures than are corresponding silicon devices, so their use has been somewhat limited in recent years.

The chemistry of germanium is much like that of silicon. However unlike SiO_2, GeO_2 crystallizes in a form in which six oxygens surround each germanium atom, a consequence of the larger atomic radius of Ge. Germanium also forms a series of germanates with both discrete and extended oxoanions.

Arsenic   The element *arsenic* is directly below phosphorus in group VA. (The name comes from the Greek: "yellow pigment," which refers to As_4S_6.) The element exhibits three allotropic forms. *Gray arsenic* is stable at room temperature and is a metallic-appearing solid in which corrugated layers of arsenic atoms are held together by London forces. *Yellow arsenic* consists of As_4 molecules (like those of white phosphorus), as does the gas. *Black arsenic* is a poorly understood and unstable form.

Arsenic has a $4s^2 4p^3$ valence-shell electronic configuration and illustrates the inert-pair effect, most of its compounds being in the $+3$ and $+5$ states. The $+3$ state is represented by As_4O_6, called *arsenious oxide*, or *white arsenic*. It is the anhydride of *arsenious acid,* written either as H_3AsO_3, or as $As(OH)_3$, which exists only in solution. This is an amphoteric hydroxide, reacting with acid to form $As(OH)_2{}^+$ and with base to form $As(OH)_4{}^-$, or $H_2AsO_2{}^-$, ions. Oxo salts such as *sodium arsenite*, Na_3AsO_3, can be prepared.

The $+5$ state of arsenic is represented by As_4O_{10}, *arsenic oxide,* by H_3AsO_4, *arsenic acid,* and by its salts, the *arsenates*. Arsenic acid is a triprotic acid, a shade weaker than phosphoric acid. It forms arsenate salts such as Na_3AsO_4, Na_2HAsO_4, and NaH_2AsO_4.

In qualitative analysis arsenic precipitates as As_2S_3 and As_2S_5, both yellow in color. In solution arsenic can be reduced with aluminum and base, or zinc and acid, to form gaseous AsH_3 (arsine) which forms a black stain of metallic silver on filter paper moistened with a silver nitrate solution.

Antimony

Below arsenic in group VA is antimony, also a metalloid. Its name is derived from the Latin *anti* ("opposite of") and *monium* ("isolated condition"), meaning that it tends to combine readily! Antimony, like arsenic, exhibits *gray, yellow,* and *black* allotropes, of which the gray form is metallic and is stable at room temperature. A fourth black-colored allotrope is called *explosive antimony;* its structure is unknown, as it reverts violently to gray antimony when scratched. Gray antimony is a poor electrical conductor; the other allotropes are nonconductors.

The $+3$ and $+5$ states are most important for antimony, the $+3$ state being represented by Sb_4O_6, *antimonous oxide,* which is an amphoteric anhydride. When added to water Sb_4O_6 becomes hydrated but remains insoluble. It dissolves in either acid, to form $Sb(OH)_2^+$, or base, to form the antimonite ion, $Sb(OH)_4^-$.

The $+5$ state for antimony is represented by Sb_2O_5, *antimonic oxide.* It is an acidic anhydride, dissolving in base to form the *antimonate* ion, $Sb(OH)_6^-$.

In qualitative analysis antimony is precipitated as orange-red Sb_2S_3 or Sb_2S_5. In solution Sb^{3+} shows a great tendency to hydrolyze and precipitate white $Sb(OH)_3$ or $Sb(OH)_2Cl$. Reduction of dissolved antimony by aluminum in basic solution yields black specks of elemental antimony.

Selenium

The element *selenium* (group VIA) shows enough metallic behavior to be classed as a metalloid. (Its name comes from the Greek and means "moon"; it was so named because it resembles tellurium which was named for the earth.) Selenium can exist in six known allotropic modifications; the common form is *gray selenium,* a dark solid which has a semimetallic luster, but which is a poor electrical conductor. However, selenium is a *photoconductor.* This means its conductivity increases in the presence of light, a property which is made use of in photoelectric cells, such as those used in some photographic exposure meters. Much selenium goes into the glass industry. The pink tinge which it imparts to glass compensates for the green color caused by iron impurities, and so the resulting glass appears almost colorless.

Selenium behaves chemically much like sulfur. Hydrogen selenide, H_2Se, is even more vile-smelling than H_2S. SeO_2 is the anhydride of selenious acid, weaker for the first dissociation than H_2SO_3, but unexpectedly stronger for the second. SeO_3 is the anhydride of selenic acid, H_2SeO_4, which is similar in many respects to H_2SO_4, but is a slightly weaker acid and a stronger oxidizing agent.

Tellurium

The element *tellurium* is also in group VIA, just below selenium. As mentioned above, the element was named after the earth. The element exists most stably as a gray allotrope, which is a better electrical conductor than selenium and which shows semiconduction.

Tellurium forms hydrogen telluride, H_2Te, which has an indescribably atrocious odor. [Research chemists who must work with H_2Te suddenly find themselves social outcasts; the compound is apparently absorbed into protein (skin, hair, woolen clothing, etc.) and only slowly released.] The element also forms TeO_2 and TeO_3 and corresponding oxoacids and salts.

SUMMARY

In this chapter we have continued our discussion of descriptive chemistry by examining the properties of the representative metals and the metalloids. The most metallic elements are the *alkali metals* (group IA), which have ns^1 valence-shell populations. These are soft, light, low-melting metals with good electrical conductivity and metallic luster. The uncombined elements can be formed by the electrolysis of their molten salts. They form ionic compounds in which they show the +1 oxidation state and which tend to be quite soluble in water.

The *alkaline-earth metals* (group IIA) are somewhat less electropositive than the alkali metals. They typically form ionic compounds with the metal in the +2 state. (Beryllium forms many covalent compounds, however.) Many of the alkaline-earth-metal compounds have low aqueous solubilities, especially when the anion present carries a −2 or −3 charge. The calcium ion is largely responsible for *hardness* in natural waters. It can be removed by precipitation, complexation, or ion exchange.

The most important metal of periodic group IIIA is *aluminum*, which is produced by the reduction of Al_2O_3 dissolved in cryolite. Aluminum is a highly reactive metal, expecially when it forms bonds with oxygen. In its compounds aluminum shows the +3 oxidation state. Its hydroxide is amphoteric.

Tin and *lead* are important metals of periodic group IVA. These elements show the +2 and +4 oxidation states in their compounds. Their +2 hydroxides are amphoteric. *Bismuth* (group VA) usually shows only the +3 state, and its hydroxide is basic.

The most important *metalloids* (semimetals) are boron (group IIIA) and silicon (group IVA), although arsenic, antimony, selenium, and tellurium fall into this category. The metalloids typically show a semimetallic luster, are semiconductors, and form weakly acidic or amphoteric hydroxides. *Boron* forms molecular halides, a wide variety of borates, as well as the electron-deficient boranes. The most common compounds of *silicon* are the silicates of which many varieties occur as natural minerals. Silicate anions can be discrete, infinite-chain, infinite-band, infinite-sheet, or three-dimensional in character.

KEY TERMS

Alkali metal (Sec. 21-1)
Alkaline-earth metal (Sec. 21-2)
Alum (Sec. 21-3)
Bayer process (Sec. 21-3)
Borane (Sec. 21-5)
Borosilicate glass (Sec. 21-5)
Deionized water (Sec. 21-2)
Electron-deficient compound (Sec. 21-5)
Flame test (Sec. 21-1)
Hall process (Sec. 21-3)
Hard water (Sec. 21-2)
Ion exchange (Sec. 21-2)
Lime (Sec. 21-2)

Lime-soda method (Sec. 21-2)
Metalloid (Sec. 21-5)
Permanent water hardness (Sec. 21-2)
Plaster of paris (Sec. 21-2)
Quicklime (Sec. 21-2)
Roasting (Sec. 21-4)
Silicone (Sec. 21-5)
Slaked lime (Sec. 21-2)
Temporary water hardness (Sec. 21-2)
Thermite reaction (Sec. 21-3)
Three-center bond (Sec. 21-5)
Vitreous silica (Sec. 21-5)
Zeolite (Sec. 21-2)

QUESTIONS AND PROBLEMS

Alkali metals

21-1 Why do the group IA elements **(a)** show metallic properties **(b)** commonly show only the $+1$ oxidation state in compounds?

21-2 Which of the alkali metals is the best reducing agent? How can you account for this?

21-3 Why is it that a *nonaqueous* electrolysis must be used to obtain metallic sodium from its salts?

21-4 In the gas phase at elevated temperatures the alkali metals are observed as diatomic melecules. Why is it that the halogens form diatomic molecules at room temperature, but the alkali metals do not?

21-5 How could you experimentally distinguish between **(a)** Na_2O and Na_2O_2 **(b)** NaCl and KCl **(c)** K_2O and K_2CO_3 **(d)** K_2O and KCl?

21-6 Write a balanced equation for **(a)** the thermal decomposition of sodium bicarbonate to form sodium carbonate **(b)** the synthesis of potassium superoxide **(c)** burning LiH in oxygen **(d)** the reaction of cesium with water **(e)** the reaction of K_2O with CO_2.

21-7 What experimental evidence would you need in order to show that in the compound CsI_3 cesium should not be assigned an oxidation number of $-\frac{1}{3}$?

21-8 Suggest an explanation for the fact that when the alkali metals react with oxygen, the products depend upon the size of the metal atom.

21-9 Describe how you could make each of the following conversions in the laboratory (in each case specify reagents used and write appropriate equations): **(a)** NaCl to NaOH **(b)** NaOH to NaCl **(c)** NaCl to Na_2O_2 **(d)** Li_3N to Li_2SO_4.

21-10 Account for the fact that the electrical conductivity of a $1\ M\ Li_2SO_4$ solution is less than that of a $1\ M\ Cs_2SO_4$ solution.

21-11 Sodium fires pose a special problem for the firefighter. How should such a fire be extinguished?

Alkaline-earth metals

21-12 Account for the fact that the alkaline-earth metals are generally higher melting than the alkali metals.

21-13 Metallic oxides are expected to react with nonmetallic oxides. What product is formed when $BaO(s)$ reacts with $SO_2(g)$? What kind of reaction is this?

21-14 Account for the fact that the group IIA metals commonly show only the $+2$ oxidation state.

21-15 How can it be shown experimentally that CaO is $Ca^{2+}O^{2-}$, and not Ca^+O^-?

21-16 Account for the tendency shown by beryllium to form covalent bonds. Why is the oxidation state of beryllium limited to $+2$ in these compounds?

21-17 Write a balanced equation for **(a)** reacting calcium metal with excess water **(b)** dissolving beryllium in dilute HCl **(c)** dissolving calcium carbonate in dilute HCl **(d)** softening of temporary hard water by heating **(e)** thermal decomposition of $CaCO_3$ **(f)** dissolving $Ba_3(PO_4)_2$ in dilute HCl.

21-18 How could you experimentally distinguish between **(a)** CaO and $Ca(OH)_2$ **(b)** $BaSO_4$ and $BaSO_3$ **(c)** KCl and $CaCl_2$ **(d)** BeO and SrO **(e)** CaH_2 and $CaCl_2$ **(f)** $BaCO_3$ and $BaCl_2$.

21-19 Account for each of the following trends going down group IIA from Be to Ba: **(a)** The hydroxides become more basic. **(b)** The hydroxides become more soluble. **(c)** The carbonates become more stable with respect to thermal decomposition to form the oxide and CO_2. **(d)** The sulfates become less soluble. **(e)** The reduction potentials [for $M^{2+} + 2e^- \longrightarrow M(s)$] become more negative.

21-20 Taking the ionic radius for Li^+ as 0.078 nm and that for Be^{2+} as 0.034 nm, calculate the average charge density (charge per unit volume) in each of these ions. How does this account for the greater tendency of beryllium to hydrolyze?

21-21 State how each of the following conversions could be accomplished in the laboratory: **(a)** $CaCO_3$ to CaH_2 **(b)** $BaCO_3$ to $BaSO_4$ **(c)** BaF_2 to BaO_2 **(d)** $MgCO_3$ to $MgCl_2$.

21-22 What is hard water? Why is hardness undesirable in water? How does temporary hardness differ from permanent hardness?

21-23 How does ground water near large gypsum deposits differ from that near large limestone deposits?

21-24 Stalactites and stalagmites are columns suspended from the ceilings and rising from the floors, respectively, in limestone caves. Suggest a mechanism for their formation.

21-25 Describe how ion exchange can be used to produce a substitute for distilled water. What advantage does ion exchange hold over distillation in the production of pure water? Is distilled water superior to deionized water? Explain.

Group IIIA metals

21-26 Gallium resembles aluminum more than indium in its chemical behavior. Account for this in terms of the electronic configurations of the atoms of these elements.

21-27 Why is it that aluminum can not be electrolytically reduced from aqueous solutions of its compounds?

21-28 How can you demonstrate experimentally that aluminum hydroxide is amphoteric?

21-29 Describe the removal of oxide impurities from Al_2O_3 by means of the Bayer process.

21-30 $\Delta G°$ for the oxidation of aluminum by O_2 is very negative. Why is the reaction so slow?

21-31 Account for the fact that in the vast majority of the compounds of aluminum the simple Al^{3+} ion does not exist.

21-32 Contrast the structures of $AlCl_3 \cdot 6H_2O$ and gaseous aluminum chloride.

21-33 Write a balanced equation for **(a)** the disproportionation of AlCl **(b)** the dissolving of aluminum hydroxide in excess base **(c)** the hydrolysis of aluminum cyanide **(d)** the reduction of MnO_2 by aluminum to form manganese metal **(e)** the oxidation of aluminum by hot, concentrated H_2SO_4 to liberate SO_2 gas **(f)** the reaction of lithium aluminum hydride, $LiAlH_4$, with water to evolve H_2 gas.

21-34 Suggest a means for making artificial cryolite.

Tin, lead, and bismuth

21-35 Lead usually has a dull gray appearance. Why is it not more metallic looking?

21-36 Explain the occurrence of "tin disease" in European cathedrals. Why is it much less common in the United States?

21-37 Write equations illustrating the amphoterism of stannous hydroxide.

21-38 Account for the fact that a fresh precipitate of $PbCl_2$ is soluble in excess HCl.

21-39 Account for the fact that lead chromate, $PbCrO_4$, is soluble in **(a)** concentrated HNO_3 **(b)** ammonium acetate.

21-40 Suggest a reason for the fact that Sn(II) is a better reducing agent in basic solution than in acidic.

21-41 How could you show experimentally that Bi_2O_3 is not amphoteric?

21-42 Write a balanced equation for **(a)** the burning of bismuth in air **(b)** the dissolving of stannic sulfide in aqueous sodium sulfide **(c)** the oxidation of Mn^{2+} by Bi_2O_5 (in "sodium bismuthate") in acidic solution **(d)** the dissolving of tin in dilute HCl **(e)** the dissolving of PbS in concentrated HNO_3 to form S and NO **(f)** the formation of a precipitate from the addition of dilute HCl to a solution of sodium stannite.

Metalloids

21-43 Show that elemental boron must be classed as an electron-deficient substance.

21-44 Account for the fact that the B—F distance in BF_3 is less than that in BF_4^-.

21-45 Draw resonance forms which help account for the "short" B—F distance in BF_3.

21-46 What is a three-center bond? How is it described in terms of MO theory?

21-47 Tetraborane, B_4H_{10}, has a molecule with four BHB three-center bonds. Suggest a structure for this molecule.

21-48 How would you expect the B—N bond length in hexagonal boron nitride to compare with that in the cubic form? Justify your answer.

21-49 Draw the structure of each of the following discrete silicate anions: **(a)** SiO_4^{2-} **(b)** $Si_2O_7^{6-}$ **(c)** $Si_3O_9^{6-}$ **(d)** $Si_4O_{12}^{8-}$ **(e)** $Si_6O_{18}^{12-}$.

21-50 Draw the structure of each of the following extended silicate anions: **(a)** SiO_3^{2-} **(b)** $Si_4O_{11}^{6-}$ **(c)** $Si_2O_5^{2-}$.

21-51 In one of the feldspars, an aluminum atom substitutes for one out of every four silicon atoms in a distorted SiO_2 structure. Electrical neutrality is maintained by the presence of Na^+ ions. What is the formula of this mineral?

21-52 Account for the fact that the mica minerals can be easily cleaved into thin sheets.

21-53 In vitreous silica and in quartz each silicon is surrounded tetrahedrally by four oxygens. How do the structures of these forms of SiO_2 differ from each other?

21-54 Predict the structure of each of the following molecules: **(a)** Si_2Cl_6 **(b)** SbH_3 **(c)** SnH_4 **(d)** As_4O_6 **(e)** AsF_5 **(f)** $SbOF_3$.

21-55 Account for the fact that a graphite-like form of silicon is not observed.

21-56 Account for the fact that arsenic and selenic acids are fairly strong, but antimonic and telluric acids are weak.

21-57 What accounts for the great difference in properties between CO_2 and SiO_2?

21-58 Write a balanced equation for **(a)** burning hydrogen selenide in oxygen **(b)** acidification of SbS_3^{3-} to form a precipitate **(c)** reduction of Ag^+ by AsH_3 **(d)** dissolving As_2O_3 in dilute HCl **(e)** hydrolysis of Ge_2H_6 to form GeO_2 **(f)** dissolving of silica in hydrofluoric acid **(g)** oxidation of B_2H_6 in air **(h)** dehydration of H_3BO_3 **(i)** reaction of magnesium boride, Mg_3B_2, with water to form diborane **(j)** reduction of SiF_4 by potassium metal **(k)** reaction of diborane with water to form hydrogen gas.

22

THE TRANSITION METALS

TO THE STUDENT

Between groups IIA and IIIA in the periodic table is a large block of elements known as the *transition metals,* a term first used by Mendeleev (Sec. 7-1). This block includes the 10 elements scandium ($Z = 21$) through zinc ($Z = 30$) in the fourth period and the corresponding elements below them in succeeding periods. Also included are the lanthanoids ($Z = 58$ through 71) and actinoids ($Z = 90$ through 103). (There is some disagreement over which elements ought to be classified as transition metals. Sometimes Zn, Cd, and Hg are excluded, and sometimes Cu, Ag, and Au, as well. We follow one common practice and include both of these subgroups of elements, however.)

In this chapter we begin by considering the general properties of the transition metals, then move to an examination of bonding and structure in transition-metal complex ions, and finally consider the descriptive chemistry of the more important transition metals.

22-1
Electronic
configurations

When we build up the electronic configurations of the atoms (Sec. 6-1), we find that each transition-metal series results from the belated filling of an inner subshell of electrons. (See Fig. 7-3.) Thus the first series (fourth period) corresponds to the filling of the $3d$ subshell, as is shown in Table 22-1. The "pre-transition" element calcium ($Z = 20$) has an $[Ar]4s^2$ configuration. Scandium ($Z = 21$), the first transition metal in this series, has the configuration $[Ar]3d^14s^2$. Addition of electrons to the $3d$ subshell continues until, at zinc ($Z = 30$), the $3d$ subshell is filled, with the $4s$ subshell still containing two electrons, as it did with calcium: $[Ar]3d^{10}4s^2$. This transition from an empty $3d$ subshell to a filled one occurs with only two irregularities, one at chromium ($Z = 24$), and the other at copper ($Z = 29$). The $3d^54s^1$ configuration of chromium and the $3d^{10}4s^1$ configuration of copper reflect the fact that the $3d$ and $4s$ energies are very close together across the series and that an exactly half-filled (for Cr) or filled (for Cu) d subshell provides enough extra stability to produce these configurations.

Table 22-1
Electronic
configurations: first
transition-metal
series

	Ca		Sc	Ti	V	Cr	Mn	Fe	Co	Ni	Cu	Zn		Ga
			\multicolumn{10}{Transition metals}											
Z	20		21	22	23	24	25	26	27	28	29	30		31
Population:														
K shell	2		2	2	2	2	2	2	2	2	2	2		2
L shell	8		8	8	8	8	8	8	8	8	8	8		8
M shell $\{$	$3s^2$		$3s^2$	$3s^2$	$3s^2$	$3s^2$	$3s^2$	$3s^2$	$3s^2$	$3s^2$	$3s^2$	$3s^2$		$3s^2$
	$3p^6$		$3p^6$	$3p^6$	$3p^6$	$3p^6$	$3p^6$	$3p^6$	$3p^6$	$3p^6$	$3p^6$	$3p^6$		$3p^6$
	—		$3d^1$	$3d^2$	$3d^3$	$3d^5$	$3d^5$	$3d^6$	$3d^7$	$3d^8$	$3d^{10}$	$3d^{10}$		$3d^{10}$
N shell	$4s^2$		$4s^2$	$4s^2$	$4s^2$	$4s^1$	$4s^2$	$4s^2$	$4s^2$	$4s^2$	$4s^1$	$4s^2$		$4s^24p^1$

The second, third, and fourth transition-metal series correspond to the filling of the $4d$, $5d$, and $6d$ subshells, respectively. Throughout each of these series the population of the ns subshell remains almost constant. Note that the third and fourth series include the *lanthanoids* and *actinoids,* respectively, which are often counted with group IIIB. Each of these is sometimes called an *inner transition series* and corresponds to the filling of the $(n - 2)f$ subshell, while the populations of the $(n - 1)d$ and ns subshells remain almost constant. The electronic configurations of these elements can be found in Table 6-2.

22-2
General properties

The properties of the transition elements are typically metallic: a *high reflectivity* and a silvery or golden *metallic luster,* coupled with *good electrical and thermal conductivities.*

Physical properties

Although the *hardnesses* and *melting points* of the transition metals vary widely, these elements generally tend to be harder and higher melting than the alkali and alkaline-earth metals. That it is difficult to distort or break up the crystal lattices of most of these elements suggests the presence of covalent bonding which supplements the metallic bonding in the solid. This covalent bonding makes use of the partly filled d orbitals of adjacent metal atoms in the solid-state structure.

The *densities* of the transition metals vary from the low 3.0 g cm^{-3} of scandium to the high 22.6 g cm^{-3} of iridium and osmium. (A one-gallon bucketful of iridium would weigh about 190 pounds.) The high densities are accounted for by high atomic weights, small atomic volumes, and close packing. Most of the metals crystallize in hexagonal or cubic close-packed structures or body-centered cubic structures. Table 22-2 shows the melting points and densities of the transition metals (except for those of the lanthanoids and actinoids). Also given are the origins of their names.

Chemical properties

The majority of the transition metals do not react rapidly at room temperature with common gases or liquids (hydrogen, oxygen, water, the halogens, etc.). Often this low apparent reactivity is a result of the formation of a thin protective coating of reaction product. For example, when exposed to air many of these metals form a protective oxide or nitride layer which essentially stops further reaction in air and often greatly retards reactions with other substances, as well. Some of the metals react vigorously when in a freshly prepared, finely divided state; *pyrophoric iron,* made by the reduction of Fe_2O_3 by H_2, bursts into flame when exposed to air.

The standard reduction potentials of the transition metals (except for the

Table 22-2
The transition metals

Element	Symbol	Melting point, °C	Density, g cm^{-3}	Origin of English name
FIRST SERIES				
Scandium	Sc	1540	3.0	*Scandinavia*
Titanium	Ti	1680	4.5	Latin: *Titans* (giants)
Vanadium	V	1920	6.1	*Vanadis* (Norse goddess)
Chromium	Cr	1900	7.2	Greek: "color"
Manganese	Mn	1250	7.3	Latin: "magnet"
Iron	Fe	1540	7.9	Anglo-Saxon word
Cobalt	Co	1490	8.9	German: "goblin"
Nickel	Ni	1450	8.9	German: "Satan"
Copper	Cu	1080	8.9	Latin: derived from *Cyprus*
Zinc	Zn	419	7.1	German word (*Zink*)
SECOND SERIES				
Yttrium	Y	1510	4.5	*Ytterby:* Swedish village
Zirconium	Zr	1850	6.5	Arabic: "gold color"
Niobium	Nb	2420	8.6	Greek: *Niobe* (goddess)
Molybdenum	Mo	2620	10.2	Greek: "lead"
Technetium	Tc	2140	11.5	Greek: "artificial"
Ruthenium	Ru	2400	12.5	Latin: "Russia"
Rhodium	Rh	1960	12.4	Greek: "rose"
Palladium	Pd	1550	12.0	*Pallas* (asteroid)
Silver	Ag	961	10.5	From Anglo-Saxon
Cadmium	Cd	321	8.6	Greek: "earth"
THIRD SERIES				
Lanthanum	La	920	6.2	Greek: "concealed"
Hafnium	Hf	2000	13.3	Latin: *Copenhagen*
Tantalum	Ta	3000	16.6	*Tantalus* (Greek god)
Tungsten	W	3390	19.4	Swedish: "heavy stone"
Rhenium	Re	3170	21.0	Latin: *Rhine* (river)
Osmium	Os	2700	22.6	Greek: "odor"
Iridium	Ir	2440	22.6	Latin: "rainbow"
Platinum	Pt	1770	22.4	Spanish: "silver"
Gold	Au	1060	19.3	From Sanskrit
Mercury	Hg	−39	13.6	*Mercury* (planet)

lanthanoids and actinoids) are given in Table 22-3. Many of these are negative, indicating that in each case the metal should dissolve readily in dilute acid. For example, chromium should dissolve in hydrochloric acid,

$$\begin{array}{ll} 2 \times [\text{Cr}(s) \longrightarrow \text{Cr}^{3+} + 3e^-] & \mathcal{E}° = 0.74 \text{ V} \\ 3 \times [2e^- + 2\text{H}^+ \longrightarrow \text{H}_2(g)] & \mathcal{E}° = 0.00 \\ \hline 2\text{Cr}(s) + 6\text{H}^+ \longrightarrow 2\text{Cr}^{3+} + 3\text{H}_2(g) & \mathcal{E}° = 0.74 \text{ V} \end{array}$$

but the reaction is very slow to start unless the protective oxide coating on the chromium is first broken. However, some transition metals have rather positive reduction potentials. This means that they are poorer reducing agents than hydrogen, and so are difficult to dissolve. The term *noble metals* is used for these elements, which include silver, gold, and platinum, as well as ruthenium, rhodium, palladium, osmium, and iridium. (Mercury also qualifies as being noble chemically, but not aristocratically; it has a reduction potential more negative than silver, but makes poor crowns.) Dissolving the noble metals requires the use of an oxidizing acid and sometimes a complexing agent. The most unreactive of these

Table 22-3
Transition metals: standard reduction potentials and common oxidation states in compounds (less common oxidation states shown in parentheses)

First series:	Sc	Ti	V	Cr	Mn	Fe	Co	Ni	Cu	Zn
$\varepsilon°$, V	-2.08	-1.63	-1.2	-0.74	-1.18	-0.44	-0.28	-0.25	$+0.34$	-0.76
To metal, from	Sc^{3+}	Ti^{2+}	V^{2+}	Cr^{3+}	Mn^{2+}	Fe^{2+}	Co^{2+}	Ni^{2+}	Cu^{2+}	Zn^{2+}
Oxidation states	$+3$	$(+2)$	$+2$	$+2$	$+2$	$+2$	$+2$	$+2$	$(+1)$	$+2$
		$+3$	$+3$	$+3$	$(+3)$	$+3$	$+3$		$+2$	
		$+4$	$+4$	$(+4)$	$(+4)$	$(+4)$				
			$+5$	$+6$	$(+6)$	$(+6)$				
					$+7$					

Second series:	Y	Zr	Nb	Mo	Tc	Ru	Rh	Pd	Ag	Cd
$\varepsilon°$, V	-2.37	-1.5	-0.65	-0.2	$+0.4$	$+0.5$	$+0.6$	$+1.2$	$+0.80$	-0.40
To metal, from	Y^{3+}	ZrO^{2+}	Nb_2O_5	Mo^{3+}	Tc^{2+}	Ru^{2+}	Rh^{2+}	Pd^{2+}	Ag^+	Cd^{2+}
Oxidation states	$+3$	$+4$	$(+3)$	$(+2)$	$(+2)$	$(+2)$	$(+1)$	$+2$	$+1$	$+2$
			$(+4)$	$(+3)$	$+4$	$+3$	$(+2)$		$(+2)$	
			$+5$	$(+4)$	$(+5)$	$+4$	$+3$		$(+3)$	
				$+5$	$(+6)$	$+6$	$+4$			
				$+6$	$+7$	$+8$	$(+6)$			

Third series:	La	Hf	Ta	W	Re	Os	Ir	Pt	Au	Hg
$\varepsilon°$, V	-2.37	-1.7	-0.81	-0.12	-0.25	$+0.9$	$+1.0$	$+1.2$	$+1.7$	$+0.78$
To metal, from	La^{3+}	HfO^{2+}	Ta_2O_5	WO_2	ReO_2	Os^{2+}	Ir^{2+}	Pt^{2+}	Au^+	Hg_2^{2+}
Oxidation states	$+3$	$+4$	$(+4)$	$(+2)$	$(+4)$	$(+2)$	$(+1)$	$+2$	$+1$	$+1$
			$+5$	$(+3)$	$(+5)$	$+3$	$(+2)$	$+4$	$+3$	$+2$
				$(+4)$	$(+6)$	$+4$	$+3$			
				$+5$	$+7$	$+8$	$+4$			
				$+6$			$(+6)$			

elements resist attack from any known aqueous solution and can be dissolved only by being heated with certain molten salts.

The transition metals show a wide array of *oxidation states*. (See Table 22-3.) In many cases the lowest oxidation state is the $+2$ state; this generally corresponds to removal of the two ns electrons. Higher oxidation states correspond to the additional loss of the $(n-1)d$ electrons or to sharing them with more electronegative atoms. In the first half of each transition-metal series the maximum observed oxidation state corresponds to removal (or sharing with more electronegative atoms) of all of the ns and $(n-1)d$ electrons. Iridium and osmium even exhibit the $+8$ oxidation state. The decrease in the maximum state after manganese in the first series and after iridium and osmium in the second and third series, respectively, reflects the difficulty of breaking into a half-filled d subshell. Note also that going down any group there is an increase in the stability of the higher oxidation states, as $(n-1)d$ energies become closer to ns energies with increasing atomic size. Variability in oxidation states is a characteristic of most transition metals. (Actually, Table 22-3 shows only the more common oxidation states of these elements; manganese, for example, shows every integral oxidation state from $+1$ through $+7$.)

ADDED COMMENT

Why, when electrons are removed from a transition-metal atom, do the ns electrons come off first? In the Aufbau process they were added *first* and so would seem to be of lower energy; so why do they not come off *last?* Actually, when the ns electrons are first added (as in potassium and calcium of the fourth period), this orbital is of lower energy than the $(n-1)d$ orbitals. But the increase in nuclear charge, which is not compensated by any additional inner-shell shielding, pulls in the $(n-1)d$ orbitals so that their energy drops. It becomes slightly lower than the ns energy at the first element of each transition series and remains lower (though not by much) through the series.

Remember, also, that in the Aufbau procedure electrons are added to the extranuclear region while the nuclear charge is simultaneously increased. On the other hand, when we consider creation of positive oxidation states, we imagine removal (or sharing) of electrons while the *nuclear charge remains constant*. (The two processes are *not* opposites.)

The chemistry of the transition metals is dominated by their tendency to form compounds which contain *complex ions*. Furthermore, many of these compounds show striking *colors*. In contrast, most of the compounds of the representative metals are white (when powdered) or colorless (when in the form of a single crystal). In addition, it is common for transition-metal compounds to be *paramagnetic* (Sec. 6-1). We will discuss the origins of these spectral and magnetic properties after we have considered bonding and structure in complex ions.

**22-3
Complex ions:
general structure
and nomenclature**

As we have previously mentioned, it is difficult to come up with a rigorous definition of a complex ion. Usually the term is taken to mean a polyatomic species which consists of a central metal ion surrounded by several *ligands*, where a ligand is an ion or a molecule bonded to the central ion.

*Coordination
number*

The formation of a complex ion may be considered to be an example of a Lewis acid-base reaction (Sec. 13-1). For example, in the reaction between cupric ions (Cu^{2+}) and ammonia molecules (NH_3)

each NH_3 acts as a Lewis base and contributes *both* of its lone-pair electrons to the formation of a *coordinate* covalent bond (Sec. 8-2) which bonds the nitrogen to the copper. The resulting $Cu(NH_3)_4^{2+}$ complex ion is therefore described as consisting of four NH_3 molecules *coordinated* to a central Cu^{2+} ion. Furthermore, *coordination number* is defined as the number of bonds made by ligands to the central ion in a complex. Thus the coordination number of Cu^{2+} in the above complex is 4. (The whole field of study of complex ion formation is often called *coordination chemistry*.)

Coordination numbers ranging from 2 to more than 8 have been observed, with 2, 4, and 6 being the most common. Coordination number and geometry in a complex are obviously related. In the complex $Ag(NH_3)_2^+$ the coordination number of the silver is 2, and the complex ion is *linear*. In $Zn(NH_3)_4^{2+}$ the coordination number of the zinc is 4, and the complex is *tetrahedral*. A *square-planar* geometry is also possible with a coordination number of 4 and is found in $Cu(NH_3)_4^{2+}$. In $Cr(NH_3)_6^{3+}$ the coordination number of the chromium is 6, and the complex is *octahedral*. Octahedral complexes are by far the most common of transition-metal complexes.

Nomenclature   A systematic set of rules for formulating and naming complex ions has been adopted by IUPAC. At this time we will list and give examples of only the most basic of these rules. (Additional details can be found in App. C.)

FORMULAS OF COMPLEXES

1 *The formula of a complex shows the central atom first, followed by the ligands.* Examples: $Cu(NH_3)_4^{2+}$; $CrCl_6^{3-}$.
2 *When the ligands in a complex are not all alike, anion ligands are written before neutral ligands.* Examples: $CrCl_2(NH_3)_4^+$; $Fe(OH)(H_2O)_5^{2+}$.

NAMES OF COMPLEXES

1 *Complex ions are named by specifying first the ligands and then the central atom.* (Note that this is the reverse of the order in a formula.)
2 *The oxidation state of the central metal atom is indicated with a Roman numeral in parentheses at the end of the name of the metal.*
3 *Anionic ligands have names ending in -o.*
  a When serving as ligands, anions whose names end in *-ide* have this suffix replaced with *-o*.

Anion	Anion name	Ligand name
Cl^-	Chlor*ide*	Chlor*o*
Br^-	Brom*ide*	Brom*o*
OH^-	Hydrox*ide*	Hydrox*o*
CN^-	Cyan*ide*	Cyan*o*

  b When serving as ligands, anions whose names end in *-ate* have this suffix replaced with *-ato*.

Anion	Anion name	Ligand name
SO_4^{2-}	Sulf*ate*	Sulf*ato*
$S_2O_3^{2-}$	Thiosulf*ate*	Thiosulf*ato*
$C_2O_4^{2-}$	Oxal*ate*	Oxal*ato*
SCN^-	Thiocyan*ate*	Thiocyan*ato*

4 *Neutral ligands are named as the neutral molecule.* There are some exceptions to this rule, the most important of which are H_2O, named as the *aquo* ligand; NH_3, named as the *ammine* ligand; and CO, named as the *carbonyl* ligand.
5 *The numbers of ligands in a complex are specified by using the Greek prefixes di-, tri-, tetra-, etc.* The prefixes *bis-* (twice), *tris-* (thrice), and *tetrakis-* (four times) are also used to eliminate ambiguity when the name of a ligand itself contains a Greek prefix. Example: bis(ethylenediamine).
6 *The name of an anionic complex ends in -ate.* This is added to the stem of the English, or sometimes the Latin, name of the metal. Thus in an anionic

Table 22-4
Names of some
complexes

Formula	Name
$Co(H_2O)_6{}^{2+}$	Hexaaquocobalt(II) ion
$CoCl_4{}^{2-}$	Tetrachlorocobaltate(II) ion
$Ni(CN)_4{}^{2-}$	Tetracyanonickelate(II) ion
$Cr(OH)_2(H_2O)_4{}^+$	Dihydroxotetraaquochromium(III) ion
$Cr(OH)_3(H_2O)_3$	Trihydroxotriaquochromium(III)
$Cr(OH)_4(H_2O)_2{}^-$	Tetrahydroxodiaquochromate(III) ion
$CoCl_4(NH_3)_2{}^-$	Tetrachlorodiamminecobaltate(III) ion

complex *chromium* becomes *chromate, aluminum* becomes *aluminate,* and *iron* becomes *ferrate.* (Cationic and neutral complexes are named with the English name of the metal and no suffix.)

Table 22-4 illustrates the use of these rules.

Finally, it should be noted that as usual an ionic *compound* is formulated and written with the cation followed by the anion.

**22-4
Complex ions:
bonding**

What is the nature of the force which bonds a ligand to the central ion in a complex? We might imagine that in the complex $Fe(CN)_6{}^{3-}$ ion six CN^- ions are bonded to the central Fe^{3+} ion by six ionic bonds. And in $Cr(H_2O)_6{}^{3+}$ we might picture the six H_2O molecules as being held to the central Cr^{3+} ion by ion-dipole forces. Such an *electrostatic model* of the bonding in complexes is attractive because of its simplicity, but it is seldom used because it fails to account for many observed properties. Instead, the metal-ligand bond in a complex is usually considered to be predominately covalent, and both valence-bond and molecular-orbital theories have been used to describe its properties.

ADDED COMMENT

If your recollection of VB and MO theories is hazy, it would be wise to reread Secs. 9-3 (Hybrid orbitals) and 9-4 (The molecular-orbital model) at this time.

*Complex ions:
valence-bond theory*

The first reasonably successful description of bonding in complex ions was provided by the valence-bond, hybrid-orbital approach (Sec. 9-3). According to this theory, the orbitals of an uncombined ion become hybridized and the resulting hybrid orbitals overlap with ligand orbitals to form covalent bonds. Consider, for example, the formation of six bonds by Fe^{3+} to form $Fe(CN)_6{}^{3-}$, the hexacyanoferrate(III), or ferricyanide, ion. An isolated iron atom has the following electronic configuration:

Fe ($Z = 26$): [Ar] $\underline{\uparrow\downarrow}\ \underline{\uparrow}\ \underline{\uparrow}\ \underline{\uparrow}\ \underline{\uparrow}$ $\underline{\uparrow\downarrow}$ $\underline{\quad}\ \underline{\quad}\ \underline{\quad}$
$\underbrace{}_{3d}\ \ \underbrace{}_{4s}\ \underbrace{}_{4p}$

Removal of three electrons produces the isolated, ground-state, ferric ion:

Fe^{3+} (ground state): [Ar] $\underline{\uparrow}\ \underline{\uparrow}\ \underline{\uparrow}\ \underline{\uparrow}\ \underline{\uparrow}$ $\underline{\quad}$ $\underline{\quad}\ \underline{\quad}\ \underline{\quad}$
$\phantom{Fe^{3+} (ground state): [Ar] }\underbrace{}_{3d}\ \ \underbrace{}_{4s}\ \underbrace{}_{4p}$

Now consider that the five *d* electrons are bunched up in only three of the 3*d* orbitals (the d_{xy}, d_{yz}, and d_{xz}). This leaves the $3d_{x^2-y^2}$, $3d_{z^2}$, $4s$, and all three $4p$ orbitals vacant. These vacant orbitals are considered to mix to form six equivalent d^2sp^3 hybrid orbitals:

Fe^{3+} (before hybridization): [Ar] $\underset{3d}{\underline{\uparrow\downarrow}\ \underline{\uparrow\downarrow}\ \underline{\uparrow}}$ $\underset{4s}{\underline{\quad}}$ $\underset{4p}{\underline{\quad}\ \underline{\quad}\ \underline{\quad}}$

Mixing ↓

Fe^{3+} (after hybridization): [Ar] $\underset{3d}{\underline{\uparrow\downarrow}\ \underline{\uparrow\downarrow}\ \underline{\uparrow}}$ $\underset{d^2sp^3}{\underline{\quad}\ \underline{\quad}\ \underline{\quad}\ \underline{\quad}\ \underline{\quad}\ \underline{\quad}}$

Now the ferric ion can act as a Lewis acid, accepting six pairs of electrons from six ligands and forming six coordinate covalent bonds. It has been shown that when the cyanide ion

$$[:C:::N:]^-$$

serves as a ligand in $Fe(CN)_6{}^{3-}$, it bonds to the central ferric ion through its carbon atom. Evidently the lone pair of the carbon is donated to form the bond, and so if six CN^- ions donate six electron pairs the resulting configuration in $Fe(CN)_6{}^{3-}$ is

Fe^{3+} (bonded): [Ar] $\underset{3d}{\underline{\uparrow\downarrow}\ \underline{\uparrow\downarrow}\ \underline{\uparrow}}$ $\underset{d^2sp^3}{\boxed{\underline{\uparrow\downarrow}\ \underline{\uparrow\downarrow}\ \underline{\uparrow\downarrow}\ \underline{\uparrow\downarrow}\ \underline{\uparrow\downarrow}\ \underline{\uparrow\downarrow}}}$

where the colored arrows represent the six donated pairs of electrons. The d^2sp^3 orbitals are shown by quantum mechanics to be *octahedral* hybrids, that is, the six major lobes in the set point out toward the corners of a regular octahedron. Figure 22-1 schematically shows the overlap of the six d^2sp^3 hybrid orbitals of

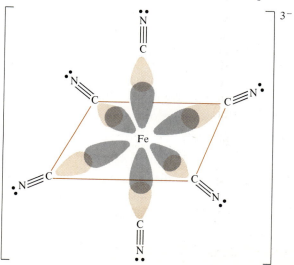

Figure 22-1
The octahedral $Fe(CN)_6{}^{3-}$ ion.

Fe^{3+} with the six CN^- ligand orbitals (containing the lone pairs) to form six bonds and a resulting octahedral structure.

Other geometries in complexes can also be accounted for. It can be shown, for example, that formation of dsp^2 hybrid orbitals leads to a square-planar geometry. Such hybridization is observed in complexes of d^8 ions such as the nickel ion, Ni^{2+}. The eight d electrons can be bunched up into four of the five $3d$ orbitals. This leaves the remaining unoccupied orbital (the $3d_{x^2-y^2}$) to mix with the $4s$, $4p_x$, and $4p_y$ orbitals to form four equivalent dsp^2 hybrid orbitals which have a square-planar geometry. These are then available for occupancy by four electron pairs from four ligands. The electronic configurations of the isolated, ground-state Ni^{2+} ion and of this species in the square-planar $Ni(CN)_4{}^{2-}$ [tetracyanonickelate(II)] ion are shown below:

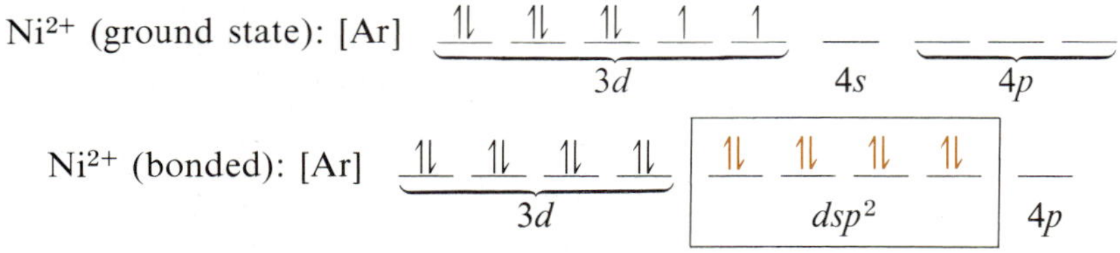

Ni^{2+} (ground state): [Ar] $\underset{3d}{\underline{\uparrow\downarrow}\;\underline{\uparrow\downarrow}\;\underline{\uparrow\downarrow}\;\underline{\uparrow}\;\underline{\uparrow}}\quad \underset{4s}{\underline{}}\quad \underset{4p}{\underline{}\;\underline{}\;\underline{}}$

Ni^{2+} (bonded): [Ar] $\underset{3d}{\underline{\uparrow\downarrow}\;\underline{\uparrow\downarrow}\;\underline{\uparrow\downarrow}\;\underline{\uparrow\downarrow}}\quad \boxed{\underset{dsp^2}{\underline{\uparrow\downarrow}\;\underline{\uparrow\downarrow}\;\underline{\uparrow\downarrow}\;\underline{\uparrow\downarrow}}}\quad \underset{4p}{\underline{}}$

The colored arrows represent the electron pairs donated by the cyanide ligands.

Tetrahedral geometry is explained by assuming that sp^3 hybrid orbitals form. Zn^{2+} is a d^{10} ion and so has no vacant $3d$ orbitals available for hybridization:

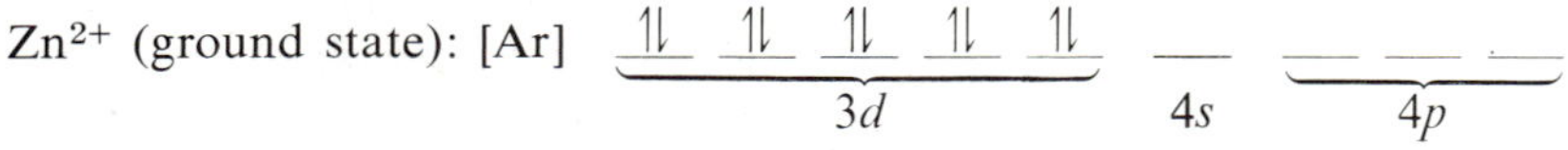

Zn^{2+} (ground state): [Ar] $\underset{3d}{\underline{\uparrow\downarrow}\;\underline{\uparrow\downarrow}\;\underline{\uparrow\downarrow}\;\underline{\uparrow\downarrow}\;\underline{\uparrow\downarrow}}\quad \underset{4s}{\underline{}}\quad \underset{4p}{\underline{}\;\underline{}\;\underline{}}$

In the complex $Zn(CN)_4{}^{2-}$ [tetracyanozincate(II)] ion sp^3 hybrid orbitals (tetrahedral) are used to bond four CN^- ligands:

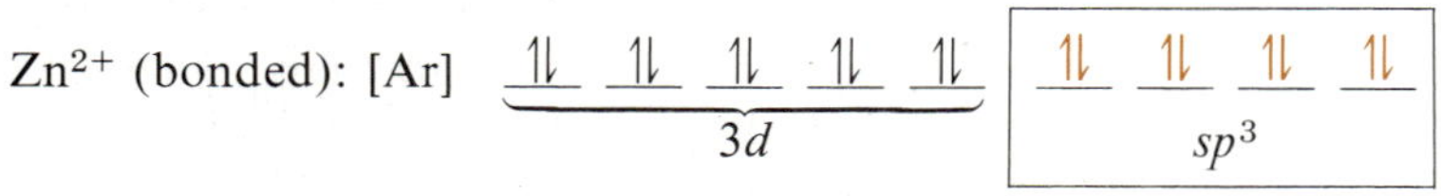

Zn^{2+} (bonded): [Ar] $\underset{3d}{\underline{\uparrow\downarrow}\;\underline{\uparrow\downarrow}\;\underline{\uparrow\downarrow}\;\underline{\uparrow\downarrow}\;\underline{\uparrow\downarrow}}\quad \boxed{\underset{sp^3}{\underline{\uparrow\downarrow}\;\underline{\uparrow\downarrow}\;\underline{\uparrow\downarrow}\;\underline{\uparrow\downarrow}}}$

The valence-bond, hybrid-orbital approach satisfactorily accounts for the existence of various geometries in complexes. Nevertheless, it has certain shortcomings. For example, it can be shown (by means of magnetic measurements; see below) that the octahedral complexes of Fe(III) fall into two classes. The first, *low-spin complexes,* consists of complexes with *one* unpaired electron per ion, such as the above-described $Fe(CN)_6{}^{3-}$ ion. But in a second class of Fe(III) complexes each ion has *five* unpaired electrons. These are called *high-spin complexes* and include $FeF_6{}^{3-}$. The existence of two classes of complexes of iron, and of some other elements, is not readily explained by using a valence-bond approach, and so we will examine other bonding models for complex ions.

Complex ions: ligand-field theory

An elegant model of the bonding in complex ions is found in the application of molecular-orbital theory to these complexes and is usually called *ligand-field theory.* A complete discussion of the way the atomic orbitals of a central ion and six ligands combine to form molecular orbitals in a complex is not appropriate here, but a capsule description of the process is in order.

Imagine that a complex ion is formed as follows: Six ligands approach a metal ion by moving in toward it along the three cartesian coordinate axes. This will form an octahedral complex with one ligand at each of the six corners of the octahedron and with the metal ion at the center. As the ligands move in, an inward-pointing orbital of each, say a p orbital, overlaps certain orbitals of the central ion. These are the ns orbital, the np_x, np_y, and np_z orbitals, and the $(n-1)d_{x^2-y^2}$ and $(n-1)d_{z^2}$ orbitals, as is shown in Fig. 22-2. Only these orbitals can provide favorable overlap with the ligand orbitals. [*Favorable overlap means that there is a net reinforcement of the wave functions in the overlap region* (Sec. 9-2). This means that there is a buildup of electronic charge density there; in other words, a bond.] As can be seen in Fig. 22-2, the inward-pointing lobes of the ligand orbitals favorably overlap with the designated d, s, and p orbitals, but not with the d_{xy}, d_{xz}, and d_{yz} orbitals. The reason for this is that the wave function carries opposite algebraic signs in adjacent lobes of these three orbitals so that no matter what the sign on the ligand orbital lobe is, net reinforcement is impossible. (This just means that reinforcement by one d-orbital lobe is offset by cancellation by the other.)

Favorable overlap of atomic orbitals leads to the formation of molecular orbitals. The six metal orbitals thus combine with the six ligand orbitals to form twelve molecular orbitals, six bonding and six antibonding, as shown in Fig. 22-3. In the diagram the subsets of molecular orbitals are labeled with shorthand designations for the symmetries of the orbitals, but except for the t_{2g} and e_g^* designations we need not concern ourselves with these. Note, however, that the electron pairs from the ligand orbitals all end up in low-energy, *bonding* molecular orbitals. (Each of these started out as a lone pair in, for example, a cyanide ion, a water molecule, a fluoride ion, etc.) These molecular orbitals are all of the σ type because each is symmetrical around the bond axis.

Higher in energy than the bonding orbitals in the complex are the t_{2g} orbitals, which are *nonbonding orbitals*. While electrons in bonding orbitals increase the stability of a molecule (or complex ion), and electrons in antibonding orbitals decrease it, electrons in nonbonding orbitals have no effect on the stability of a collection of atoms. The three t_{2g} orbitals correspond to the d_{xy}, d_{xz}, and d_{yz} orbitals of the original central ion. Figure 22-3 also shows that there are six antibonding orbitals in the complex, each higher in energy than the bonding or nonbonding orbitals. Note that the antibonding orbitals of lowest energy are the two labeled e_g^*.

In Fig. 22-3 we have not shown any of the electrons in the $(n-1)d$ subshell of the metal ion. There may be various numbers of these, depending on the identity and oxidation state of the ion, and these end up distributed between the t_{2g} and e_g^* subsets of molecular orbitals. We will discuss their precise distribution shortly.

When carried out with the fewest possible approximations, calculations based on the ligand-field model yield excellent descriptions of complex ions. But such calculations are not appropriate for this text. Fortunately, however, there is a simplified, yet surprisingly powerful, version of ligand-field theory which is amazingly successful in predicting many of the properties of complex ions. This is called *crystal-field theory,* and for most of the remainder of this chapter we will emphasize this model.[1]

---

[1]Crystal-field theory was developed in 1929 by the great physicist Hans Bethe, who applied it initially to ions in crystals.

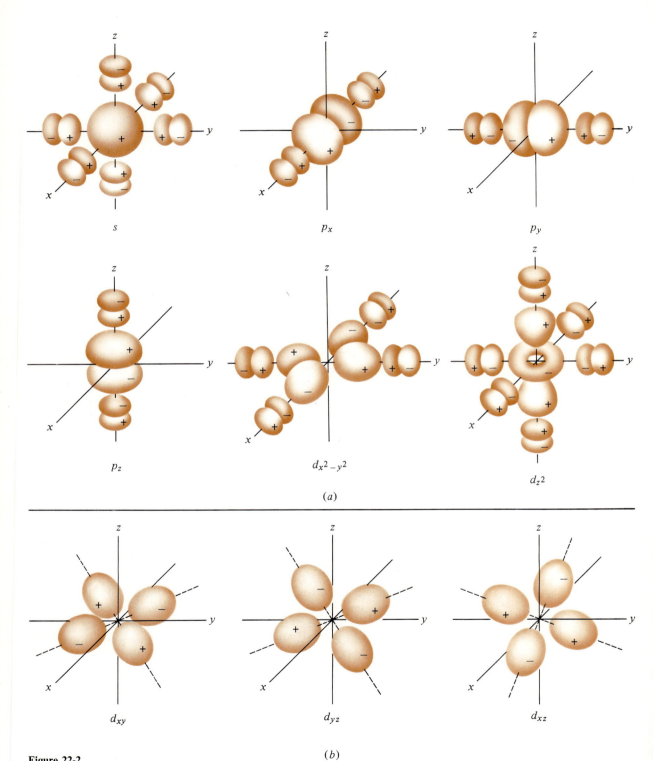

s

p_x

p_y

p_z

$d_{x^2-y^2}$

d_{z^2}

(a)

d_{xy}

d_{yz}

d_{xz}

(b)

Figure 22-2
Overlap of atomic orbitals in an octahedral complex. (*a*) Favorable overlap. (*b*) No net overlap.
Note: Signs are *not* charges, but rather are algebraic signs of the corresponding wave functions.

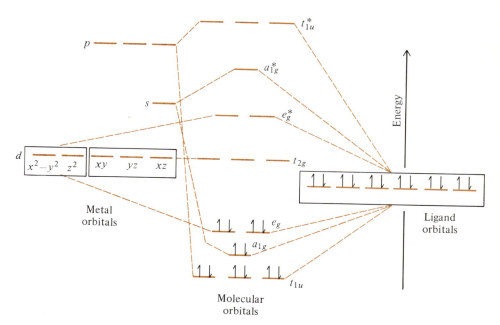

Figure 22-3
Molecular-orbital energy-level diagram for an octahedral complex.

Crystal-field theory

In order to account for the magnetic and spectral properties of octahedral complexes it is necessary to be able to assign electrons to the t_{2g} and e_g^* subsets of orbitals shown in Fig. 22-3. Crystal-field theory ignores all the other orbitals in the complex and pictures these orbitals as arising primarily from the original d orbitals of the central metal ion. In the original uncomplexed ion these orbitals are all of the same energy. *Crystal-field theory makes use of an electrostatic model to predict how the d-orbital energies are split to yield a higher energy (e_g^*) and a lower-energy (t_{2g}) subset of levels.*

Consider, again, the approach of six ligands along the cartesian coordinate axes toward an ion at the origin. The ligands are either anions or dipoles with their negatively charged ends pointing inwards. As they approach the central metal ion, its d orbitals are affected differently, depending on whether their lobes point directly at the approaching ligands or at a 45° angle to them (see Fig. 6-26). According to crystal-field theory, since the $d_{x^2-y^2}$, and d_{z^2} orbitals face the approaching ligands head on, the energies of any electrons in these orbitals will be raised considerably due to electrostatic repulsion. But the approaching ligands can "sneak" partway into the regions between the lobes of the d_{xy}, d_{yz}, and d_{xz} orbitals, so that electrostatic repulsion between the negative charges on the ligands and electrons in these orbitals is minimized. The result is that the d orbitals are split into two subsets: the higher-energy e_g subset and the lower-energy t_{2g} subset.[2] This splitting of the d orbitals is shown schematically in Fig. 22-4.

It should be emphasized that crystal-field theory is a model which provides us with a way to predict the splitting of d-orbital energies in octahedral and other fields. But the forces involved in complex formation are undoubtedly not much

---

[2] These two subsets of orbitals are labeled e_g and t_{2g} in crystal-field theory. They are obviously the e_g^* and t_{2g} subsets, respectively, of the more complete ligand-field theory. We will drop the asterisk on "e_g," however, because the fact that these orbitals are antibonding is not important in crystal-field theory.

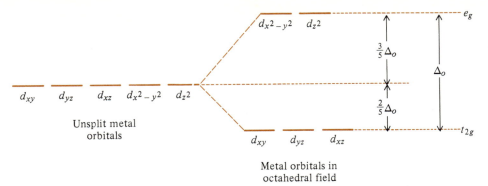

Figure 22-4
Crystal-field theory:
d-orbital splitting in an
octahedral field.

like the simple electrostatic forces considered in crystal-field theory. In spite of this weakness, however, the theory is useful because in many cases it does predict the correct energy splittings.

With crystal-field theory one can predict the nature of the *d*-orbital splitting for many ion geometries. When a *tetrahedral* complex is formed, for example, the ligands can be imagined as approaching along the lines shown in Fig. 22-5. As can be seen, the negative charges of the ligands approach the off-axis t_{2g} orbitals of the central metal ion more closely than they do the e_g orbitals. Electrons in the t_{2g} subset will therefore be repelled more strongly by the approaching negative charges and will thus end up at higher energies. Tetrahedral splitting can thus be thought of as the "inverse" of octahedral splitting, as is shown in Fig. 22-6.

The energy difference between the e_g and t_{2g} orbital subsets is called the *ligand-field* (or *crystal-field*) *splitting energy* and is designated Δ_o in an octahedral complex and Δ_t in a tetrahedral complex. These quantities are indicated in the splitting diagrams of Figs. 22-4 and 22-6, respectively. It can be seen from Fig. 22-4 that in an octahedral complex the energy of each e_g orbital is $\frac{3}{5}\Delta_o$ above the average *d*-orbital energy, while that of each t_{2g} orbital is $\frac{2}{5}\Delta_o$ below. Figure 22-6 shows the corresponding relationship for a tetrahedral complex. The actual magnitude of Δ depends largely upon the nature of the ligands, as we will see shortly.

In most cases the *d* orbitals of the central metal ion are not empty, but contain

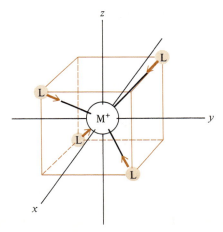

Figure 22-5
Approach of four ligands to form a tetrahedral complex.

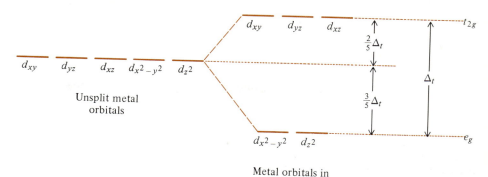

Figure 22-6
d-orbital splitting in a
tetrahedral field.

up to 10 electrons, and in the ground state of an octahedral complex these are distributed between the t_{2g} and e_g subsets. Furthermore, from observations of distributions in many complexes the following tendencies become apparent: (1) the electrons tend to occupy the lowest available energy levels and (2) the electrons tend to spread out and occupy different orbitals in order to minimize interelectronic repulsions (Hund's rule, Sec. 6-1). These two tendencies are not always compatible, however. To be specific, two cases are possible. (1) If the electric field produced by the ligands is *weak*, Δ_o will be small, and the electrons will spread out to occupy the e_g orbitals as well as the t_{2g}. (2) If the electric field of the ligands is *strong*, Δ_o will be large, and electrons will be forced to pair up in the t_{2g} orbitals. In the latter case the energy needed to force the electrons to pair is less than the energy needed to force some of them to populate the e_g orbitals. Which of these two situations actually occurs depends on the strength of the electrostatic field produced by the ligands.

As an example, imagine a d^5 ion such as Fe^{3+} subjected to a weak ligand field. In this case Δ_o is small, and the five d electrons spread out to occupy all five orbitals in the complexed ion:

Weak field:

On the other hand, in a strong ligand field Δ_o is large enough to force the electrons to pair:

Strong field:

In the weak-field case the total spin (sum of the spins of the unpaired electrons) is $5 \times \frac{1}{2}$, or $\frac{5}{2}$, and so the complex is a *high-spin complex*. In the strong-field case the total spin is only $\frac{1}{2}$, and so the complex is a *low-spin complex*.

We see that the strength of the ligand field determines the magnitude of Δ_o, which in turn determines how the electrons are apportioned among the d orbitals of the metal. Figure 22-7 shows d^1 though d^{10} configurations for both the weak- and strong-field cases. Note that the configuration is unaffected by ligand-field

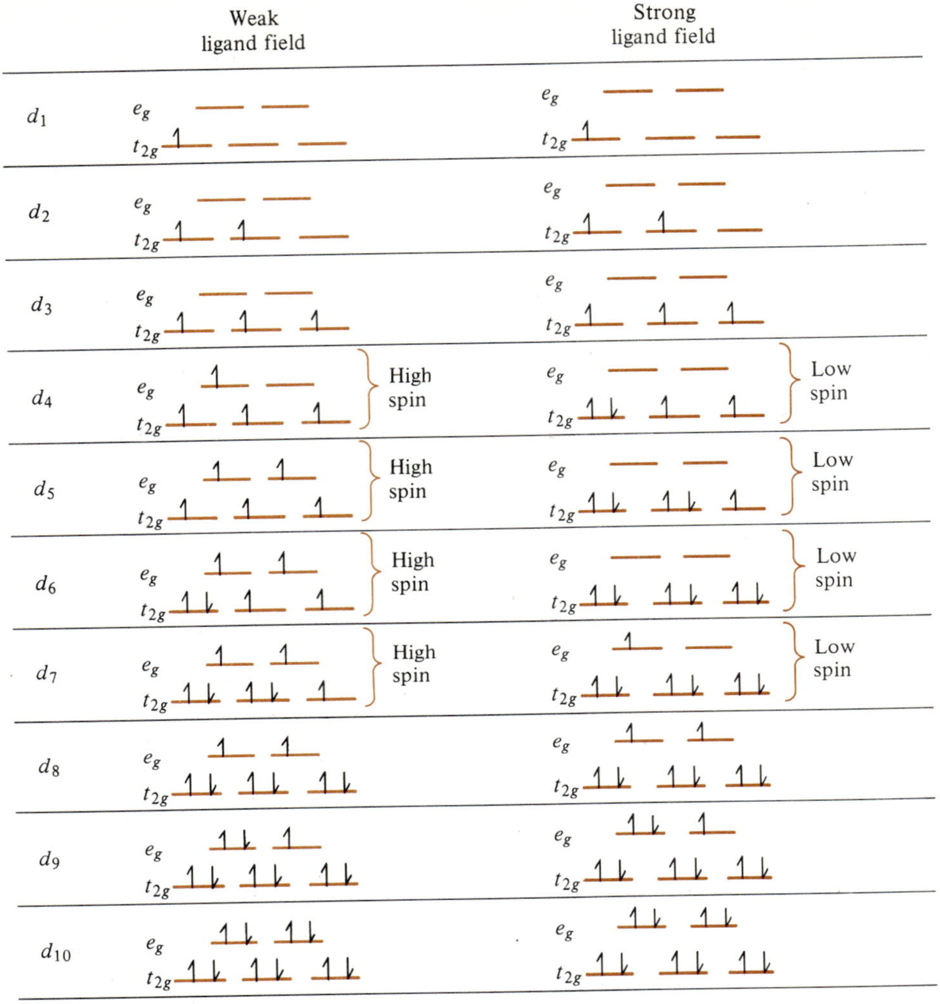

	Weak ligand field	Strong ligand field
d_1	e_g / t_{2g}	e_g / t_{2g}
d_2	e_g / t_{2g}	e_g / t_{2g}
d_3	e_g / t_{2g}	e_g / t_{2g}
d_4	e_g / t_{2g} — High spin	e_g / t_{2g} — Low spin
d_5	e_g / t_{2g} — High spin	e_g / t_{2g} — Low spin
d_6	e_g / t_{2g} — High spin	e_g / t_{2g} — Low spin
d_7	e_g / t_{2g} — High spin	e_g / t_{2g} — Low spin
d_8	e_g / t_{2g}	e_g / t_{2g}
d_9	e_g / t_{2g}	e_g / t_{2g}
d_{10}	e_g / t_{2g}	e_g / t_{2g}

Figure 22-7 d-orbital populations in strong and weak ligand fields.

strength in the d^1, d^2, d^3, d^8, d^9, and d^{10} cases, while high- and low-spin configurations are possible for d^4, d^5, d^6, and d^7 ions.

Magnetic properties Great support for the splitting of d-orbital energies comes from magnetic measurements. *Paramagnetism,* a weak *attraction into* a magnetic field, is a result of the presence of unpaired electrons in a substance (see Sec. 6-1). *Diamagnetism,* an even weaker *repulsion out of* a magnetic field, is caused by a change in motion of all the electrons in a substance so as to reduce an externally imposed magnetic field. All substances show some diamagnetism.

Magnetic measurements can be made on a magnetic balance such as the *Gouy balance,* shown in Fig. 22-8. With this device the sample is weighed, first with an external magnetic field applied (electromagnet turned on) and then with no external field (magnet turned off). A paramagnetic substance appears to weigh more in the presence of the field because magnetic attraction augments gravitational attraction; a substance which is only diamagnetic appears to weigh less.

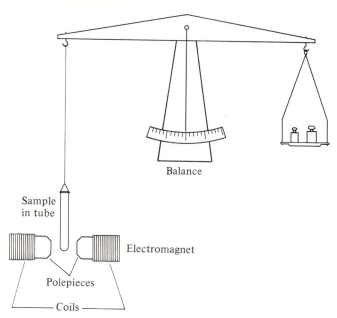

Balance

Sample
in tube

Electromagnet

Polepieces

Coils

Figure 22-8
The Gouy balance for magnetic measurements.

Paramagnetism is generally caused by both the orbital and spin motions of electrons. (Any rotating or revolving charged object generates a magnetic field, that is, "acts like a magnet"). For reasons which we shall not go into here the orbital effect is often not very strong, particularly for the first series of transition elements. The *spin-only* paramagnetic effect of a pair of electrons with antiparallel spins is zero, and so it is only the unpaired electrons whose spins contribute to the measured paramagnetism. For this reason a measurement of the paramagnetism of a substance can often be used to determine the number of unpaired electrons present. Paramagnetic measurements generally support the electronic configurations of transition metal complexes predicted by ligand- and crystal-field theories. Further, they can be used in conjunction with crystal-field theory to classify ligands as strong-field or weak-field.

● **EXAMPLE 22-1**

Problem   Magnetic measurements show that CoF_6^{3-} has four unpaired electrons, while $Co(NH_3)_6^{3+}$ has none. Classify the fluoride ion and the ammonia molecule as strong- or weak-field ligands.

Solution   Each of these complexes contains Co(III), a d^6 species. In a weak-field situation the d-orbital splitting is small, and so both t_{2g} and e_g levels are occupied. The resulting configuration is $t_{2g}^4 e_g^2$, giving a total of four unpaired electrons. (See Fig. 22-7.) When subjected to a strong ligand field the complex becomes low-spin as the d-orbital splitting becomes large enough to force all d electrons to occupy the t_{2g} level. The resulting configuration is $t_{2g}^6 e_g^0$ and is characterized by having all electrons paired. Thus F^- and NH_3 serve as a weak-field and a strong-field ligand, respectively. ●

Color and
d-d transitions   The complexes of the transition metals are noted for the wide variety of colors which they exhibit. When viewed under white light, any substance appears

colored if it absorbs light from some portion of the visible spectrum. It turns out that in many complexes the ligand-field splitting energy corresponds to some energy of visible light. In other words, the energy needed to cause a $t_{2g} \longrightarrow e_g$ electronic transition corresponds to energy found in white light. For example, solutions of the d^1 octahedral complex $Ti(H_2O)_6^{3+}$ are violet. This color results from the transition

$$\text{Energy} \uparrow \quad \begin{array}{c} e_g \\ t_{2g} \end{array} \quad \left| \begin{array}{ccc} \text{—} & \text{—} & \text{—} \\ \underline{1} & \text{—} & \text{—} \end{array} \right. \longrightarrow \left. \begin{array}{ccc} \underline{1} & \text{—} & \text{—} \\ \text{—} & \text{—} & \text{—} \end{array} \right|$$

In $Ti(H_2O)_6^{3+}$ the ligand-field splitting energy (Δ_o) corresponds to the energy of light which has a wavelength of about 500 nm. Thus, when these ions are exposed to white light they absorb light of this wavelength as $e_g \longrightarrow t_{2g}$ transitions occur. (Considered slightly differently, 500-nm light *causes* the transitions to occur.) The light leaving the sample no longer consists of all wavelengths within the visible spectrum and so appears colored, violet in this case. (The energy absorbed during the transitions is later dissipated as heat as the electrons *relax* to their original t_{2g} levels.) Transitions from one d sublevel to another in a complex are known as *d-d transitions*.

Ions with no d electrons are colorless; these include $Sc(H_2O)_6^{3+}$ and $Ti(H_2O)_6^{4+}$. Similarly, ions with d^{10} configurations, such as the hydrated Ag^+ and Zn^{2+} ions, are also colorless, because in these ions d-d transitions are impossible since all d orbitals are filled.

The d^5 species Mn(II) presents an interesting situation: $Mn(H_2O)_6^{2+}$ is extremely pale pink in solution. (In fact, its color is not apparent at all in dilute solutions.) On the other hand, $Mn(CN)_6^{4-}$ is an intense violet. How can we account for the marked difference in appearance between these two ions? Water is a weak-field ligand, and so the following *d-d* transition might be expected to take place in $Mn(H_2O)_6^{2+}$:

$$\text{Energy} \uparrow \quad \begin{array}{c} e_g \\ t_{2g} \end{array} \quad \left| \begin{array}{ccc} \underline{1} & \underline{1} & \\ \underline{1} & \underline{1} & \underline{1} \end{array} \right. \longrightarrow \left. \begin{array}{ccc} \underline{1} & \underline{1\!\downarrow} & \\ \underline{1} & \underline{1} & \text{—} \end{array} \right|$$

But quantum mechanics shows us that transitions in which the total spin changes (from $\frac{5}{2}$ to $\frac{3}{2}$ in this case) are highly improbable. (They are said to be *spin forbidden*.) This means that the transition takes place only occasionally, and the color of a solution containing this complex is accordingly faint.

But what about the vividly colored $Mn(CN)_6^{4-}$ ion? The CN^- ion is a strong-field ligand, and so transitions of the type

$$\text{Energy} \uparrow \quad \begin{array}{c} e_g \\ t_{2g} \end{array} \quad \left| \begin{array}{ccc} \text{—} & \text{—} & \\ \underline{1\!\downarrow} & \underline{1\!\downarrow} & \underline{1} \end{array} \right. \longrightarrow \left. \begin{array}{ccc} \text{—} & \underline{1} & \\ \underline{1\!\downarrow} & \underline{1\!\downarrow} & \text{—} \end{array} \right|$$

which are not spin forbidden, are highly probable. The ion is consequently a deep violet.

The spectrochemical series

We have seen that different ligands have different abilities to split the set of d orbitals into subsets. It is possible to predict this splitting ability from ligand-field theory and also to obtain it from spectral and magnetic measurements. As a result, different ligands can be arranged in order of increasing field strength, in a sequence known as the *spectrochemical series:*

$$I^- < Br^- < Cl^- \approx -SCN^- < F^- < OH^- < H_2O < -NCS^- < NH_3$$

$$< NO_2^- < CO \approx CN^-$$

Weak-field ligands   Strong-field ligands

It should be noted, however, that the dividing line between strong-field and weak-field ligands is not sharp in this series. That is, the effectiveness of many ligands in producing a large t_{2g}–e_g splitting varies somewhat from one ion to another. It is therefore difficult to predict whether a given complex containing ligands near the middle of the series will be high-spin or low-spin. Actually, Δ depends not only on the nature of the ligands in a complex, but also on the identity and oxidation state of the central atom. Still, knowledge of the spectrochemical series is often useful in predicting occupancies of the t_{2g} and e_g levels, especially when all the ligands are the same and are near one of the ends of the series.

ADDED COMMENT

Valence-bond theory, ligand-field theory, crystal-field theory! Do not be awed by what may seem to be an overabundance of descriptions of the bonding in complex ions. Remember that a theory is only a model and that each of these models has its own advantages. For example, valence-bond theory provides the easiest way to predict the spatial geometry of a complex, crystal-field theory supplies the most convenient way to predict approximate magnetic and spectral properties, and ligand-field theory gives the most *accurate* predictions of most properties. Try not to be preoccupied with determining which theory is "right," only with learning the ways in which each can be useful.

22-5 Complex ions: stereochemistry

The spatial geometry of the central metal ion and its ligands in a complex is known as the *stereochemistry* of the complex. Different stereochemistries can be grouped according to the coordination number of the central ion. Here we will mention only the most important coordination numbers: 2, 4, and 6.

Coordination number 2

The most common complexes showing the coordination number 2 are those of Ag(I) and Cu(I). These are *linear* complexes and include $Ag(NH_3)_2^+$, $AgCl_2^-$, $Ag(CN)_2^-$, and $CuCl_2^-$.

Coordination number 4

There are two common geometries associated with this coordination number: *square planar* and *tetrahedral*. Each of these configurations permits a different form of *stereoisomerism*. (Molecules which have the same molecular formula but different structures are called *isomers*. When the molecules also have the same bonds, but differ in the spatial arrangement of the atoms, they are called *stereoisomers*.) Consider the two square-planar stereoisomers of $PtCl_2(NH_3)_2$. In one, the two ammonia molecules (*ammine* ligands) occupy a pair of adjacent corners on the square, while the two chlorides occupy the other pair:

$$\begin{array}{ccc} Cl & & NH_3 \\ & Pt & \\ Cl & & NH_3 \end{array}$$

cis-$PtCl_2(NH_3)_2$

This is called the *cis* isomer, *cis* meaning "adjacent." The other isomer is the *trans* ("opposite") form:

$$\begin{array}{ccc} H_3N & & Cl \\ & Pt & \\ Cl & & NH_3 \end{array}$$

trans-$PtCl_2(NH_3)_2$

Other examples of square-planar complexes are $Ni(CN)_4{}^{2-}$, $PdCl_4{}^{2-}$, and $AgF_4{}^{-}$. This geometry is especially common for ions with d^8 electronic configurations.

In a *tetrahedral complex* the four ligands occupy the corners of a regular tetrahedron. Examples of these complexes include $ZnCl_4{}^{2-}$, $Co(CO)_4{}^{-}$, and $MnO_4{}^{-}$. Cis-trans isomerism is not possible in tetrahedral complexes. (Why not?) But the door is open for a new type of stereoisomerism called *optical isomerism*. Two structures which are *mirror images* of each other, but which are not identical, are called *optical isomers*. Optical isomerism is possible in a tetrahedral complex which has *four different ligands* bonded to the central atom.

But consider first a complex of the type MA_2BC, that is, one in which two of the four ligands are the same, while the rest are different. Such a complex is schematically represented in Fig. 22-9a. The model (a thing-and-stick model?) of this complex is shown at the left and its mirror image at the right. It can be seen that with a slight rotation the mirror image can be superimposed on the structure at the left, showing that the two are really identical.

Now consider the MABCD complex of Fig. 22-9b. Here the mirror image (on

Figure 22-9
Chirality in complexes and hands. (*a*) MA$_2$BC, a nonchiral complex. (*b*) MABCD, a chiral pair. (*c*) A chiral pair of hands.

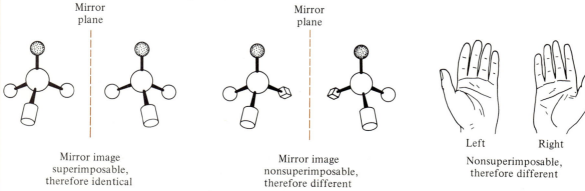

Mirror plane

Mirror image superimposable, therefore identical

Mirror plane

Mirror image nonsuperimposable, therefore different

Left   Right

Nonsuperimposable, therefore different

(*a*) MA$_2$BC, a nonchiral complex   (*b*) MABCD, a chiral pair   (*c*) A chiral pair of hands

the right) is not identical to the original. No amount of twisting around will permit one to be superimposed on the other. The structures are similar to a pair of hands (belonging to the same person), shown in Fig. 22-9c. Your left hand is the mirror image of your right, but the two are not superimposable.

When four different ligands are bonded to the central atom in a complex, the central atom is said to be *asymmetric*, and the entire structure is said to be *chiral*. (*Chiral* means "handed," as in "right-handed" and "left-handed.") Each member of a chiral pair of structures such as those in Fig. 22-9b is called an *optical isomer*, or *enantiomer*.

Optical isomers are so named because, whether pure or in solution, they rotate the plane of plane-polarized light. Figure 22-10 schematically shows an apparatus for measuring this rotation: the *polarimeter*. In ordinary light electric and magnetic fields vibrate in all directions which are perpendicular to the direction of propagation of the light. In the diagram the *polarizer* is a filter which removes all light except that in which the electric field vibrates in one plane. (The magnetic field vibrates in a plane perpendicular to this one.) Such *plane-polarized light* then passes through a tube containing the sample. If all the species present are of the same chirality, or if there are more right-handed complexes than left-handed, or *vice versa,* the plane of polarization rotates as the light passes through the sample. Another polarizing filter, the *analyzer,* can then be used to measure the angle of rotation. Substances which rotate the plane of polarized light are said to be *optically active.*

Synthesis of tetrahedral complexes of the type MABCD is difficult and normally yields a mixture of the two enantiomers. Worse yet, the chiral species in these complexes are in a rapid interconversion equilibrium, as metal-ligand bonds are

Figure 22-10

Rotation of the plane of polarized light by an optically active sample; the polarimeter.

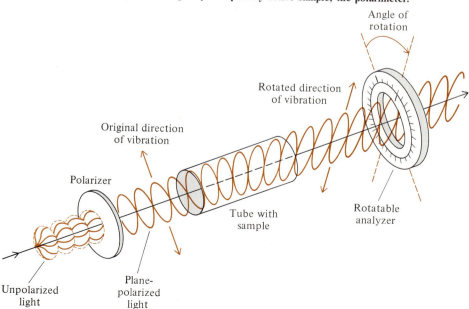

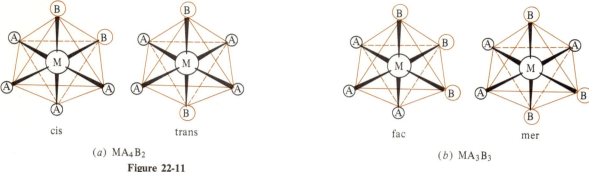

cis   trans   fac   mer

(a) MA_4B_2 (b) MA_3B_3

Figure 22-11
Octahedral stereoisomerism: (a) MA_4B_2 and (b) MA_3B_3.

rapidly broken and reformed. (The complexes are said to be *labile.*) It happens, therefore, that optical isomerism is seldom observed in simple tetrahedral complexes. Optical isomerism is common in octahedral complexes and also in many organic molecules which contain an asymmetric carbon atom (Sec. 23-13).

Coordination — The geometry of this coordination number is almost always based on an *octahe-*
number 6 — *dron.* Octahedral coordination is the most common of all coordinations and permits both cis-trans and optical isomerism. Since the six corners of an octahedron are equivalent, only one structure is possible for MA_6 and MA_5B complexes. For MA_4B_2, however, cis and trans forms can exist. In the cis isomer the two B ligands occupy adjacent corners on the octahedron; in the trans form they are on opposite corners as is shown in Fig. 22-11a. For complexes of the type MA_3B_3 two isomers are again possible; these are the *facial* (abbreviated *fac*), and *meridional* (*mer*) forms, shown in Fig. 22-11b.

Some ligands bond at several places to a central ion in a complex. These are called *polydentate,* or *multidentate,* ("many-toothed") ligands. A common *bidentate* ligand is ethylenediamine, often abbreviated *en:*

$$
\begin{array}{c}
\text{H} \quad \overset{\text{H}}{\underset{}{|}} \quad \overset{\text{H}}{\underset{}{|}} \quad \text{H} \\
\text{H} \diagdown \overset{}{\underset{}{\text{C}}} \text{—} \overset{}{\underset{}{\text{C}}} \diagup \text{H} \\
\overset{}{\underset{}{\text{N}}} \overset{\cdot\cdot}{\quad} \quad \overset{\cdot\cdot}{\underset{}{\text{N}}} \\
\text{H} \qquad\qquad \text{H}
\end{array}
$$

It can bond at two places (note the lone pairs on the nitrogens) to a central atom in a complex, as in the tetrachloroethylenediaminecobaltate(III) ion, as shown in Fig.

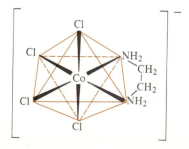

Figure 22-12
The $CoCl_4en^-$ ion.

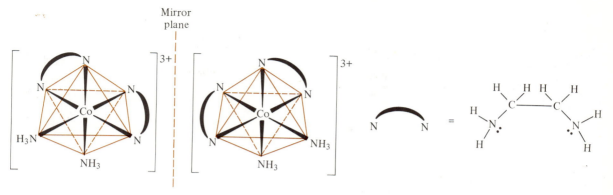

Figure 22-13
Optical isomerism in a chelate.

22-12.   Note that the *en* ligand must attach to cis positions, because the trans positions are too far apart for it to span.   Polydentate ligands which have two, three, four, five, or six points of attachment are known.   These are often called *chelating agents* and the resulting complex, a *chelate* (from the Greek for "claw").

Optical isomerism is common in chelates.   Figure 22-13 shows the two enantiomers of the *cis*-diamminebis(ethylenediamine)cobalt(III) ion.   (In the drawing the two bidentate ligands are shown as arcs.)   Note that one form is not superimposable on its mirror image; they are an enantiomeric pair.

22-6
Descriptive
chemistry of
selected transition
metals

Our discussion of the detailed chemistry of the transition metals centers on those elements which are most important chemically or industrially.   This includes most of the elements of the first transition-metal series plus a few of those in the second and third series.   Second- and third-series elements in a given subgroup generally tend to be much more like each other chemically than do elements of the first and second series.   This is a result of the fact that the second- and third-series elements have very similar atomic radii.   Why should this be so?   During the Aufbau of the lanthanoids 14 electrons are added to an inner shell, while the nuclear charge increases.   This produces a gradual decrease in atomic size as all electrons are pulled in more strongly.   Thus hafnium $(Z = 72)$ is smaller than it would be, had the lanthanoids not just preceded it.   We would normally expect hafnium to be somewhat larger than zirconium $(Z = 40)$, directly above it in the periodic table, but the fact that hafnium has one more shell of electrons than does zirconium is just about balanced by the higher nuclear charge of Hf and the poor shielding ability of the $(n - 2)f$ electrons.   The shrinkage which occurs as the lanthanoids are built up is called the *lanthanoid contraction;* it accounts for the chemical similarity between the second- and third-series transition elements.   The elements zirconium and hafnium, for instance, are invariably found together combined in similar compounds and are difficult to separate.

Titanium

Titanium is a silvery white metal which resists corrosion unusually well.   It is exceptionally hard and strong for its low density (4.5 g cm^{-3}) and finds extensive application, both pure and in alloys, in jet and rocket engines and in aircraft and space vehicles. It also is used where its high resistance to corrosion is important. Titanium is thought to form a self-protective oxide-nitride coat.

Titanium, a d^2 element, forms compounds in the $+2$, $+3$, and $+4$ oxidation states. The $+2$ stated corresponds to the "loss" of the two $4s$ electrons and the higher oxidation states to the "loss" of one or both of the $3d$ electrons. (The quotation marks remind us that, particularly in the case of the higher oxidation states, the electrons are not lost completely, but rather are shared with more electronegative atoms.)

Ti(II). The $+2$ state for titanium is an uncommon one, largely because Ti^{2+} is such a good reducing agent, good enough to reduce water to hydrogen. Titanium dichloride, $TiCl_2$, can be made by reducing $TiCl_4$ with metallic titanium at high temperatures:

$$TiCl_4(g) + Ti(s) \longrightarrow 2TiCl_2(s)$$

Ti(III). The Ti^{3+} ion is called the *titanous ion*. In aqueous solution it is an octahedral aquo complex, $Ti(H_2O)_6^{3+}$. This is a d^1 ion with a violet color, a result of the $t_{2g} \longrightarrow e_g$ electronic transition. It is readily oxidized by oxygen in the air to the $+4$ state.

Ti(IV). This is the most important and stable oxidation state of titanium. It is represented in titanium tetrachloride, $TiCl_4$, a colorless liquid which hydrolyzes rapidly when exposed to the moisture in the air to form a dense smoke, a process used to produce smoke screens in World War I. Two reactions which probably occur are

$$TiCl_4(l) + 2H_2O(g) \longrightarrow TiO_2(s) + 4HCl(g)$$
$$TiCl_4(l) + H_2O \longrightarrow TiOCl_2(s) + 2HCl(g)$$

Titanium dioxide, TiO_2, is a white powder used as a paint pigment. It occurs naturally as several minerals, one of which is *rutile,* which has a higher refractive index than diamond, but which is too soft for use as a gemstone.

The simple *titanic ion,* Ti^{4+}, tends to hydrolyze so strongly that it does not exist in aqueous solution. Instead, a mixture of oxo and hydroxo ions are present and these are usually called, for simplicity's sake, the *titanyl ion,* TiO^{2+}. Actually, salts such as titanyl sulfate $TiOSO_4 \cdot H_2O$ can be prepared, but in these the titanyl ion is an extended-chain cation.

Vanadium

Vanadium is very hard, strong, and corrosion-resistant, but is more dense than titanium. It is rarely produced in pure form, but rather as an alloy with iron, *ferrovanadium.* Much vanadium is used to make alloy steels, to which it imparts strength and ductility.

In its compounds vanadium, a d^3 element, exhibits oxidation states of $+2$, $+3$, $+4$, and $+5$. Oxides, halides, and anionic, neutral, and cationic complexes are represented in most of these states.

V(II). The *vanadous ion,* V^{2+}, exists as a violet hexaaquo complex in aqueous solution. It is readily oxidized by O_2 and, like Ti^{2+}, by water, as well. It can be formed from the reduction of higher states in solution, either electrolytically or by using a reducing agent such as zinc in acidic solution.

V(III). In the $+3$ state vanadium forms an extensive series of complexes, such as the blue cationic $V(H_2O)_6{}^{3+}$, called the *vanadic ion*, the anionic $VF_6{}^{3-}$, and the neutral $VF_3(H_2O)_3$. Addition of base to a solution containing the vanadic ion precipitates a green compound represented either as the hydroxide, $V(OH)_3$, or as a hydrous oxide, $V_2O_3 \cdot nH_2O$. It is a basic oxide and is slowly oxidized in air to the $+4$ state.

V(IV). Heating a mixture of V_2O_3 and V_2O_5 produces the dark blue dioxide, VO_2. It is a basic oxide, dissolving in acids to yield the blue-green oxopentaaquo-vanadium(IV) ion, often called the vanadyl(IV) ion and formulated as VO^{2+}. This ion forms a series of vanadyl(IV) salts such as $VOCl_2$ and $VOSO_4$.

V(V). V_2O_5, vanadium pentoxide, is formed by heating ammonium metavanadate:

$$2NH_4VO_3(s) \longrightarrow V_2O_5(s) + 2NH_3(g) + H_2O(g)$$

or by acidifying its solution with H_2SO_4:

$$2H^+ + 2VO_3{}^- \longrightarrow V_2O_5(s) + H_2O$$

V_2O_5 is an amphoteric oxide dissolving in base to give the *vanadate ion*, $VO_4{}^{3-}$:

$$V_2O_5 + 6OH^- \longrightarrow 2VO_4{}^{3-} + 3H_2O$$

and in acid to give a complicated mixture of hydroxo and oxo species. In strongly acidic solutions the *dioxovanadium(V) ion*, $VO_2{}^+$, is believed to be present.

The only binary halide formed by V(V) is VF_5; VOF_3 and $VOCl_3$ are also known, however. $VOCl_3$ is an unusual compound: it strongly resists reduction, being kinetically inert even to sodium metal. It forms solutions with nonmetals and nonpolar organic solvents, but violently decomposes when added to water, forming V_2O_5.

Chromium   Chromium metal is a silvery white, corrosion-resistant metal which is very hard but somewhat brittle when pure. It is used for alloying steel and for plating iron and other metals. The reduction potential for

$$Cr^{3+} + 3e^- \longrightarrow Cr(s) \qquad \mathcal{E}° = -0.74 \text{ V}$$

is sufficiently negative that chromium might be expected to reduce water and become oxidized in the process. The reaction is normally slow, however, and the permanence of the shine on "chrome-plated" bumpers is a consequence of the formation of a tough, smooth, invisible oxide coat. (Those who live where winters are "salty" might wish to dispute that point!) Chromium dissolves in dilute HCl and H_2SO_4, but becomes passive in HNO_3. (See *Iron*, this section.)

Chromium $(3d^54s^1)$ shows oxidation states from $+1$ to $+6$, but the most common are the $+3$ and $+6$ states, with the $+2$ and $+4$ states being of secondary importance.

Cr(II). The *chromous ion*, Cr^{2+}, is a blue hexaaquo ion which can be formed by reduction of Cr^{3+} or $Cr_2O_7{}^{2-}$, either electrolytically or by zinc metal. It precipitates the yellow, nonamphoteric chromous hydroxide, $Cr(OH)_2$, when a base is added to its solution. Both Cr^{2+} and $Cr(OH)_2$ will reduce water to hydrogen, but the reaction is slow. Oxidation by O_2 yields Cr^{3+} and $Cr(OH)_3$, respectively.

Cr(III). In the $+3$ state chromium is found in countless complexes, nearly all of which are octahedral. The *chromic ion*, Cr^{3+}, is a violet hexaaquo complex in aqueous solution and in solid salts such as $Cr(H_2O)_6Cl_3$ and the chromium-containing alums (Sec. 21-3). Addition of base to Cr^{3+} yields grey-green, gelatinous, chromic hydroxide, $Cr(OH)_3$, or $Cr_2O_3 \cdot xH_2O$. Chromic hydroxide is amphoteric; dissolving it in base yields the green *chromite ion* formulated as $Cr(OH)_4(H_2O)_2^-$, $Cr(OH)_4^-$, or even CrO_2^-. (Note that the last two formulations differ only by two H_2O molecules.) Actually, the ion is probably more complicated than any of these.

Cr(IV). The most important Cr(IV) compound is the oxide, CrO_2. It can be prepared by oxidizing $Cr(OH)_3$ by O_2 at elevated temperatures:

$$4Cr(OH)_3(s) + O_2(g) \longrightarrow 4CrO_2(s) + 6H_2O(g)$$

CrO_2 is a metallic conductor and is ferromagnetic (see *Iron*, this section); its magnetic behavior accounts for its use as a coating in high-quality recording tape.

Cr(VI). When the chromite ion, $Cr(OH)_4^-$, is oxidized by hydrogen peroxide (which exists as HO_2^- in basic solution) the product is the yellow, tetrahedral *chromate ion*, CrO_4^{2-}:

$$2Cr(OH)_4^- + 3HO_2^- \longrightarrow 2CrO_4^{2-} + OH^- + 5H_2O$$

In dilute solution this ion reacts with acid to form the orange *hydrogen chromate ion:*

$$\underset{\text{Yellow}}{CrO_4^{2-}} + H^+ \longrightarrow \underset{\text{Orange}}{HCrO_4^-}$$

In concentrated acid H_2CrO_4, *chromic acid,* is formed. Since K_2 for this acid is 1.3×10^{-6}, the chromate-hydrogen chromate system serves as an inorganic acid-base indicator. Actually, in the range of concentrations usually encountered, say, about 0.1 M, the hydrogen chromate ion is largely dimerized to form the orange *dichromate ion:*

$$\underset{\text{Orange}}{2HCrO_4^-} \rightleftharpoons \underset{\text{Orange}}{Cr_2O_7^{2-}} + H_2O \qquad K = 160$$

so that the acid-base reaction is usually written as:

$$\underset{\text{Yellow}}{2CrO_4^{2-}} + 2H^+ \longrightarrow \underset{\text{Orange}}{Cr_2O_7^{2-}} + H_2O$$

The dichromate ion may be thought of as two CrO_4 tetrahedra sharing a corner oxygen, as shown in Fig. 22-14. The Cr(III) $\longrightarrow$ Cr(VI) conversion by hydrogen peroxide is accomplished in basic solution, as described above. In acidic solutions, however, the conversion goes the other way, Cr(VI) $\longrightarrow$ Cr(III):

$$8H^+ + Cr_2O_7^{2-} + 3H_2O_2 \longrightarrow 2Cr^{3+} + 3O_2(g) + 7H_2O$$

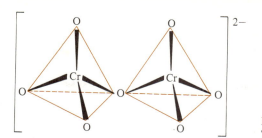

Figure 22-14
The dichromate ion.

Why does H_2O_2 oxidize Cr(III) to Cr(VI) in basic solution, but reduce Cr(VI) to Cr(III) in acidic? The answer is suggested by the reduction potentials:

Acidic: $6e^- + 14H^+ + Cr_2O_7^{2-} \longrightarrow 2Cr^{3+} + 7H_2O$ $\mathcal{E}° = 1.33$ V

Basic: $3e^- + 4H_2O + CrO_4^{2-} \longrightarrow Cr(OH)_4^- + 4OH^-$ $\mathcal{E}° = -0.23$ V

It can be seen, first of all, that the acidic reduction is more spontaneous than the basic. Second, each reduction is favored by increasing $[H^+]$. The reason is that reduction of Cr(VI) involves adding H^+ ions to the complex, as $Cr_2O_7^{2-}$ changes to $Cr(H_2O)_6^{3+}$ or CrO_4^{2-} to $Cr(OH)_4(H_2O)_2^-$. Thus reduction is favored in acidic solution, oxidation in basic. (This generalization is useful for other similar redox systems.)

The reduction of $Cr_2O_7^{2-}$ by H_2O_2 occurs via the unstable intermediate, chromium peroxide, CrO_5. This is a blue substance which lasts for up to a minute or two in water, but which can be somewhat stabilized in ether. It eventually decomposes to form Cr^{3+}. The blue CrO_5 has been used as a confirmatory test for chromium in qualitative analysis.

The anhydride of chromic acid is chromium trioxide, CrO_3. This is a powerful oxidizing agent and is an ingredient in the chemist's *cleaning solution*. (The mixture of CrO_3 with a little water and some concentrated H_2SO_4 is used for cleaning glassware. Find the recipe in a handbook before you make it, though.) Sodium dichromate, $Na_2Cr_2O_7$, is a good substitute for CrO_3 for this purpose and is sometimes used, instead.

Manganese   Manganese is a white, brittle metal, which is considerably more reactive (kinetically) than titanium, vanadium, or chromium. Its principal use is in the alloying of steels.

In manganese (d^5) the maximum oxidation state is $+7$. Also important are the $+2$ and $+4$ states and, to a lesser extent, the $+3$ and $+6$ states.

Mn(II). The $+2$ state of manganese is an important one. Unlike Ti^{2+}, V^{2+}, and Cr^{2+}, the *manganous ion*, Mn^{2+}, in aqueous solution is not oxidized by either oxygen or water. Manganous salts such as the sulfate, $MnSO_4 \cdot 5H_2O$, are a pale pink, but in water $Mn(H_2O)_6^{2+}$ shows this color only in very concentrated solutions.

Addition of base to Mn^{2+} forms white (if pure), insoluble manganous hydroxide, $Mn(OH)_2$. Unlike its parent Mn^{2+}, this compound is oxidized by O_2 to form dark brown MnOOH.

$$4Mn(OH)_2(s) + O_2(g) \longrightarrow 4MnOOH(s) + 2H_2O$$

Mn(III). The $+3$ state of manganese is quite unstable. (Compare with chromium.) The *manganic ion*, Mn^{3+}, a hexaaquo complex in water, is unstable because:

1 It is a powerful oxidizing agent and will oxidize water:

$$4 \times [e^- + Mn^{3+} \longrightarrow Mn^{2+}] \qquad\qquad \mathcal{E}° = 1.51 \text{ V}$$
$$2H_2O \longrightarrow O_2(g) + 4H^+ + 4e^- \qquad\qquad \mathcal{E}° = -1.23 \text{ V}$$

$$\overline{4Mn^{3+} + 2H_2O \longrightarrow 4Mn^{2+} + O_2(g) + 4H^+ \qquad \mathcal{E}° = 0.28 \text{ V}}$$

2 It will even oxidize itself, a good example of *disproportionation:*

$$Mn^{3+} + e^- \longrightarrow Mn^{2+} \qquad\qquad \mathcal{E}° = 1.51 \text{ V}$$
$$Mn^{3+} + 2H_2O \longrightarrow MnO_2(s) + 4H^+ + e^- \qquad\qquad \mathcal{E}° = -0.95 \text{ V}$$

$$\overline{2Mn^{3+} + 2H_2O \longrightarrow Mn^{2+} + MnO_2(s) + 4H^+ \qquad \mathcal{E}° = 0.56 \text{ V}}$$

Mn(III) is relatively stable in certain solid compounds, such as $Mn_2(SO_4)_3$ and MnF_3 and in complexes such as the hexacyanomanganate(III) ion, $Mn(CN)_6^{3-}$.

Mn(IV). The most important compound of manganese in the $+4$ state is *manganese dioxide*, MnO_2, a brown-to-black substance which occurs naturally as the mineral *pyrolusite*. MnO_2 is a good oxidizing agent and has been used in the laboratory preparation of chlorine from HCl:

$$MnO_2(s) + 4H^+ + 2Cl^- \longrightarrow Mn^{2+} + Cl_2(g) + 2H_2O$$

Much manganese dioxide is nonstoichiometric (Sec. 11-3).

Mn(VI). The only important representative of the $+6$ state for manganese is the *manganate ion*, MnO_4^{2-}. It has a deep green color and may be formed by the reduction of MnO_4^- in *strongly* basic solution.

Mn(VII). The most important Mn(VII) species is the *permanganate ion*, MnO_4^-. This ion has an extremely intense purple color which can be clearly seen even in 10^{-4} M solutions. The most common source of this ion is the salt potassium permanganate, $KMnO_4$. The MnO_4^- ion is a useful, powerful oxidizing agent and is commonly used in redox titrations. In these the MnO_4^- solution is added to the solution of reducing agent (*not* vice versa), so that at the end point the sudden increase in MnO_4^- can easily be seen. The titration is run in acidic solution, in which the reduction half-reaction is:

$$5e^- + MnO_4^- + 8H^+ \longrightarrow Mn^{2+} + 4H_2O \qquad \mathcal{E}° = 1.51 \text{ V}$$

If the reducing agent is added to MnO_4^-, the Mn^{2+} formed is oxidized by the MnO_4^- which is in excess (until the end point):

$$2MnO_4^- + 3Mn^{2+} + 2H_2O \longrightarrow 5MnO_2(s) + 4H^+$$

and the brown-black MnO_2 makes the end point impossible to see. If the titration is run in neutral or slightly basic solution, the product is again MnO_2:

$$3e^- + MnO_4^- + 2H_2O \longrightarrow MnO_2(s) + 4OH^- \qquad \mathcal{E}° = 0.59 \text{ V}$$

Reduction of MnO_4^- in strongly basic solution forms the manganate ion:

$$e^- + MnO_4^- \longrightarrow MnO_4^{2-} \qquad \mathcal{E}° = 0.56 \text{ V}$$

Iron   Iron is the most used of all metals. Being very abundant (about 5 percent of the earth's crust) and easy to obtain from its ores, iron has become indispensable for making everything from automobiles to zither strings. It occurs naturally as *hematite* (Fe_2O_3), *limonite* ($Fe_2O_3 \cdot H_2O$), *magnetite* (Fe_3O_4), *siderite* ($FeCO_3$), *pyrite* (FeS_2), and as an impurity in many other minerals. All of the above-named minerals serve as iron ores, except for pyrite (*iron pyrites,* or *fool's gold*), from which it is difficult and costly to remove all the sulfide.

The reduction of iron ore, known as *smelting,* takes place in a *blast furnace,* a huge affair built roughly like a cylinder standing on one end. It is run continuously and charged periodically from the top with iron ore, limestone ($CaCO_3$), and coke (carbon). Molten iron and *slag,* a semimolten material consisting mostly of silicates, are drawn from separate openings at the bottom. Hot air is blown into the bottom of the blast furnace, where carbon burns to form carbon monoxide,

$$2C(s) + O_2(g) \longrightarrow 2CO(g)$$

which is the major reducing agent in the furnace. If the ore is Fe_2O_3, it is reduced by the CO to Fe_3O_4 at the top of the furnace at a temperature of about 300°C:

$$CO(g) + 3Fe_2O_3(s) \longrightarrow 2Fe_3O_4(s) + CO_2(g)$$

The Fe_3O_4 gradually drops down and is reduced (at about 600°C) to FeO:

$$CO(g) + Fe_3O_4(s) \longrightarrow 3FeO(s) + CO_2(g)$$

Still lower in the furnace, the FeO is reduced to iron (at 800 to 1600°C):

$$CO(g) + FeO(s) \longrightarrow Fe(l) + CO_2(g)$$

The limestone fed into the furnace undergoes thermal decomposition,

$$CaCO_3(s) \longrightarrow CaO(s) + CO_2(g)$$

and the calcium oxide reacts with silica and silicate impurities to form a silicate slag. The reaction can be approximated by

$$CaO(s) + SiO_2(s) \longrightarrow CaSiO_3(l)$$

The slag is much less dense than the molten iron, and so it floats and can be drawn off separately. Some slag is used for making roads, building blocks, and artificial stone.

The product of the blast furnace is *pig iron,* which contains up to 5 percent C and up to another 5 percent (total) Si, P, Mn, and S. These impurities are usually oxidized by air and are then removed from the iron. Partial purification yields *cast iron,* which still contains considerable carbon. When freshly solidified cast iron is cooled slowly the carbon comes out of solid solution as graphite and the product is *gray cast iron,* a relatively soft, tough product. Rapid chilling, on the other hand, forms *white cast iron,* in which the carbon is combined in grains of iron carbide, or *cementite,* Fe_3C. White cast iron is very hard, but very brittle. *Wrought iron,* iron from which most of the carbon has been removed, is quite soft and malleable. *Mild steel* has about 0.1 percent carbon, while other steels have up to 1.5 percent. Hundreds of *stainless steels* are known. These corrosion-resistant alloys often contain chromium and/or nickel. Vanadium, titanium, manganese, tungsten, and other metals are also used in steels.

Iron dissolves in nonoxidizing acids such as HCl and dilute H_2SO_4 to form the ferrous ion:

$$Fe(s) + 2H^+ \longrightarrow Fe^{2+} + H_2(g)$$

In dilute HNO_3 the iron is oxidized to the +3 state:

$$Fe(s) + NO_3^- + 4H^+ \longrightarrow Fe^{3+} + NO(g) + 2H_2O$$

In concentrated HNO_3, however, little or no reaction is observed. In this case the iron has been made *passive* by the strong oxidizing agent and is now unreactive even with dilute nonoxidizing acids. Passive iron is believed to be protected by a thin oxide film. Breaking this film by scratching it restores the iron's reactivity.

Metallic iron shows *ferromagnetism,* a very strong attraction into a magnetic field. Like paramagnetism, ferromagnetism has its origin in unpaired electron spins. But it is a much stronger effect than paramagnetism. Ferromagnetism occurs when atoms with unpaired electron spins are just the right distance apart to permit the individual spins to align with each other within a relatively large region (10^6 atoms, or greater) called a *domain.* Because the individual spins act cooperatively within a domain, the result is a large magnetic effect. Ferromagnetic domains are large enough so that with proper technique domain boundaries can be seen under the microscope. Only solids can show ferromagnetism. The only elements which are ferromagnetic at room temperature are iron, cobalt, and nickel although some of the lanthanoids are ferromagnetic at low temperatures. Ferromagnetic compounds, such as CrO_2, and Fe_3O_4 exist, as do alloys, such as *alnico* (Al, Ni, Co, Fe, and Cu).

CORROSION OF IRON. Iron rusts when exposed to damp air or to air-saturated water. *Rust* is primarily a hydrous ferric oxide of variable composition, $Fe_2O_3 \cdot xH_2O$. It does not form as an adherent film, but rather flakes off, exposing more iron to corrosion. The mechanism of rusting is complex and apparently varies, depending on conditions. Rusting is accelerated by the presence of acids, salts, and less reactive metals, and by elevated temperatures.

Water and oxygen are both necessary for rust formation, which is thought to begin by the oxidation of iron to the +2 state:

$$Fe(s) \longrightarrow Fe^{2+} + 2e^- \tag{22-1}$$

The ferrous ion produced then migrates away from the iron through the water, which may be only a thin surface layer. Oxygen is the oxidizing agent in this reaction:

$$O_2(g) + 4H^+ + 4e^- \longrightarrow 2H_2O \tag{22-2}$$

O_2 also performs the second oxidation, which accomplishes the Fe(II) $\longrightarrow$ Fe(III) conversion:

$$Fe^{2+} \longrightarrow Fe^{3+} + e^- \tag{22-3}$$

The ferric ion immediately hydrolyzes and precipitates rust:

$$2Fe^{3+} + (3 + x)H_2O \longrightarrow Fe_2O_3 \cdot xH_2O(s) + 6H^+ \tag{22-4}$$

This reaction is aided by the consumption of H^+ by reaction (22-2). Note that reaction (22-1), which forms the surface corrosion pits on the iron, can occur at some distance from reaction (22-2). This is possible because the electrons released in (22-1) can be conducted through the metal to another point on its surface, where they reduce O_2. Similarly, diffusion of Fe^{2+} from the pitting region can result in the deposition of rust wherever O_2 is available. The system behaves as a galvanic cell with anode and cathode sites removed from each other, as is shown in Fig. 22-15.

Iron can be protected from corrosion by a number of techniques: Covering the surface with a layer of paint, grease, or oil offers some protection. Iron can be rendered passive by the use of a strong oxidizing agent incorporated in paint. (Red lead, Pb_3O_4, and zinc chromate, $ZnCrO_4$, are used for this purpose.) Plating with another metal is a common method of protection, tin plating and *galvanizing* (coating with zinc) being common examples. Unless the surface layer of plated metal is broken, tin protects iron as well as zinc does. But when tin plating becomes scratched or worn through, the exposed iron can rust. With galvanizing, however, the corrosion product of zinc is insoluble zinc hydroxide carbonate, $Zn_2(OH)_2CO_3$, which tends to plug holes and scratches in the zinc.

Galvanizing is an efficient method of protecting iron from corrosion for another reason: it is an example of *cathodic protection,* in which advantage is taken of the fact that the reduction potential of zinc is more negative than that of iron:

$$Zn^{2+} + 2e^- \longrightarrow Zn(s) \qquad \mathcal{E}° = -0.76 \text{ V}$$

$$Fe^{2+} + 2e^- \longrightarrow Fe(s) \qquad \mathcal{E}° = -0.44 \text{ V}$$

Since zinc is a better reducing agent than iron, electrically connecting a piece of zinc to the iron to be protected (a fence, pipeline, ship's hull, etc.) allows the zinc to be oxidized preferentially and thus to supply electrons to the iron for the reduction

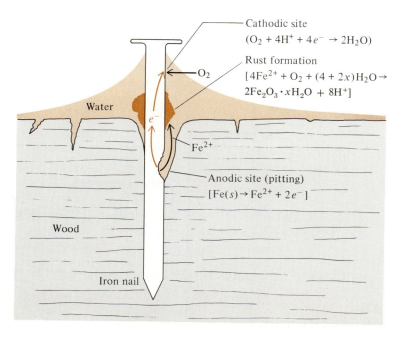

Cathodic site
$(O_2 + 4H^+ + 4e^- \rightarrow 2H_2O)$

Rust formation
$[4Fe^{2+} + O_2 + (4 + 2x)H_2O \rightarrow$
$2Fe_2O_3 \cdot xH_2O + 8H^+]$

O_2

Water

e^-

Fe^{2+}

Anodic site (pitting)
$[Fe(s) \rightarrow Fe^{2+} + 2e^-]$

Wood

Iron nail

**Figure 22-15
A nail rusts.**

of O_2. (See Fig. 22-16.) The zinc is the anode in a galvanic cell and forces the iron to become a cathode, reducing O_2 (or H^+, if the solution is acidic enough). As long as the iron is a cathode, it cannot corrode.

Galvanizing protects because it offers cathodic protection to iron. On the other hand, since tin's reduction potential is *less* negative than that of iron:

$$Sn^{2+} + 2e^- \longrightarrow Sn(s) \qquad \mathcal{E}° = -0.14 \text{ V}$$

tin plating affords no cathodic protection at all; actually, iron exposed to a corroding environment will corrode *faster* when electrically connected to tin than when not, since the iron cathodically protects the tin. This is why "tin cans" rust so rapidly when left outdoors.

In its compounds iron (d^6) commonly shows the $+2$ and $+3$ oxidation states. (The $+4$ and $+6$ states are known, but less common.) Note that when we move from left to right across the first transition metal series, iron is the first element not to exhibit an oxidation state which corresponds to an apparent loss of all ns and $(n-1)d$ electrons. As we move farther to the right beyond iron the higher states become increasingly less stable.

Fe(II). The pale green *ferrous ion*, Fe^{2+}, is actually an octahedral hexaaquo complex in water. It precipitates ferrous hydroxide, $Fe(OH)_2$, when treated with base. This is a white precipitate when pure, but is almost always observed as a light green, thought to be caused by an intermediate in the fairly rapid oxidation of $Fe(OH)_2$ to $Fe(OH)_3$ by O_2. The ferrous ion itself is very slowly oxidized to Fe^{3+}, even in acidic solution. Ferrous hydroxide is slightly amphoteric, dissolving in hot, concentrated NaOH.

Fe(III). The *ferric ion*, Fe^{3+}, is essentially colorless in water, although solutions of ferric salts are usually seen as yellow or yellow-brown because of hydrolysis to form complexes such as $Fe(OH)^{2+}$ and $Fe(OH)_2^+$, which are actually $Fe(OH)(H_2O)_5^{2+}$ and $Fe(OH)_2(H_2O)_4^+$, respectively. Addition of base to the ferric ion precipitates a reddish-brown, gelatinous material commonly called ferric hydroxide, $Fe(OH)_3$, in reality a hydrous oxide, $Fe_2O_3 \cdot xH_2O$. This is much less soluble than ferrous hydroxide and is not amphoteric.

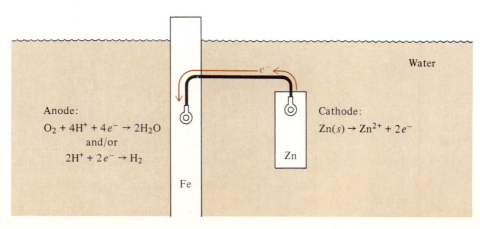

Figure 22-16
Cathodic protection.

Fe(III) forms many complexes. The most common of these are anionic and include FeF_6^{3-}, $Fe(CN)_6^{3-}$, and $Fe(C_2O_4)_3^{3-}$. The last, the trisoxalatoferrate(III) ion, is a chelate in which each of three bidentate oxalate ligands bridge two adjacent octahedral positions around the iron. Actually, this formula represents a chiral pair. (See if you can draw them.)

Both the ferric and ferrous ions form very stable complexes with cyanide. These are the hexacyanoferrate(II) ion, $Fe(CN)_6^{4-}$, commonly called the *ferrocyanide ion,* and the hexacyanoferrate(III) ion, $Fe(CN)_6^{3-}$, the *ferricyanide ion.* Mixing of either (a) a solution of potassium ferrocyanide, $K_4Fe(CN)_6$, with one containing the ferric ion or (b) a solution of potassium ferricyanide, $K_3Fe(CN)_6$, with one containing the ferrous ion produces a deep blue precipitate of $KFeFe(CN)_6 \cdot H_2O$. Historically, the product of case (a) has been called *Prussian blue,* and that of case (b), *Turnbull's blue,* but it is now known that the compounds are identical. This substance has a three-dimensional giant-molecule lattice with alternating Fe(II) and Fe(III) at the anion and cation sites in an NaCl lattice. Between each adjacent pair of iron ions is a $-C\equiv N-$ group bridging the two, and K^+ ions and H_2O molecules alternately occupy the cubic holes in the structure.

Fe(III) is commonly detected in solution by reacting it with the thiocyanate ion, SCN^-, to form the red $FeNCS^{2+}$, actually $Fe(NCS)(H_2O)_5^{2+}$, complex ion.

Cobalt

Cobalt is a hard, relatively unreactive metal with a blue-silver luster. Like iron, it is ferromagnetic and is made passive by strong oxidizing agents. It is used widely in alloys with iron, nickel, aluminum, and others.

In its compounds cobalt (d^7) exhibits primarily the $+2$ and $+3$ states; in this respect cobalt resembles its neighbor, iron.

Co(II). The pink cobaltous ion, Co^{2+}, is too poor a reducing agent to reduce oxygen (contrast with Fe^{2+}), so cobaltous solutions keep indefinitely. When a solution containing the cobaltous ion is added to one of a strong base, blue cobaltous hydroxide, $Co(OH)_2$ is formed; when base is added to a solution of Co^{2+} at $0°C$, pink $Co(OH)_2$ is formed. Surprisingly, no one has yet been able to determine exactly what causes the color difference.

Co(II) forms both tetrahedral and octahedral complexes; the former include $CoCl_4^{2-}$, $Co(OH)_4^{2-}$, and $Co(NO_2)_4^{2-}$. The hexaaquocobalt(II) ion is believed to be in equilibrium with a small concentration of the tetraaquocobalt(II) ion:

$$Co(H_2O)_6^{2+} \rightleftharpoons Co(H_2O)_4^{2+} + 2H_2O$$

Co(III). As an aquo complex, Co(III) is unstable in solution. It is an extremely powerful oxidizing agent:

$$Co^{3+} + e^- \longrightarrow Co^{2+} \qquad \mathcal{E}° = 1.81 \text{ V}$$

and oxidizes water to oxygen. On the other hand, most ligands stabilize Co(III) more than Co(II), so that the higher state can be observed in countless complex ions. In the presence of ammonia, for example, the above reduction potential becomes

$$Co(NH_3)_6^{3+} + e^- \longrightarrow Co(NH_3)_6^{2+} \qquad \mathcal{E}° = 0.1 \text{ V}$$

Many Co(III) complexes can be made by adding the desired ligands to a Co(II) solution and oxidizing with O_2 or H_2O_2. Anionic, neutral, and cationic Co(III) complexes are all common as are *polynuclear* (or *polycentric*) *complexes* in which two or more cobalt atoms are bridged by bidentate ligands:

$$\left[\begin{array}{c} H_3N \quad NH_3 \quad H_2 \quad NH_3 \quad NH_3 \\ \quad \quad | \quad \quad N \quad \quad | \\ \quad \quad Co \quad \quad \quad Co \\ H_3N \quad | \quad O-O \quad | \quad NH_3 \\ \quad \quad NH_3 \quad \quad NH_3 \end{array} \right]^{3+}$$

Nickel   Nickel is the third of the three elements iron, cobalt, and nickel, which are sometimes called the *iron triad;* chemical similarities among these three elements are greater than the similarities found going down each B group, nickel-palladium-platinum, for example. Nickel is a fairly hard metal with a faintly yellowish luster, perhaps in part due to a self-protective oxide coating. Nickel is used as one of the base layers in chromium plating. (In order to get chromium to adhere to iron, the iron is first plated with copper, then nickel, and finally chromium, in a process known as *triple-plating*.) Nickel is also used as a catalyst in certain hydrogenation reactions, such as in the manufacture of margarine and shortenings from liquid fats. (See Sec. 23-10.)

The trend toward decreased stability of the higher oxidation states continues with nickel. For this element the $+2$ state is the rule with $+3$ being quite uncommon and $+4$ extremely rare.

Ni(II). The *nickelous ion,* Ni^{2+}, is a d^8 ion, which is green as a hexaaquo complex in aqueous solution. It precipitates as green nickelous hydroxide, $Ni(OH)_2$, when a strong base is added to an Ni^{2+} solution. $Ni(OH)_2$ is not amphoteric, but will dissolve in ammonia to form ammine complexes, such as $Ni(NH_3)_6^{2+}$.

Ni(II) forms anionic, neutral, and cationic complexes. Geometries include octahedral, tetrahedral, square planar, and even tetragonal pyramidal. Square-planar complexes are common for d^8 ions, because the ligand-field splitting produces one high-energy orbital, as shown in Fig. 22-17. The orbital subsets in a square-planar complex can be imagined as resulting from pulling the ligands on the z axis out away from the central metal in an octahedral complex. This is

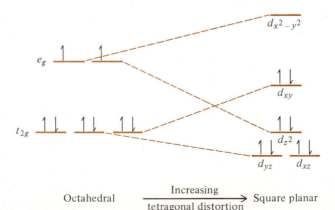

Octahedral $\xrightarrow[\text{tetragonal distortion}]{\text{Increasing}}$ Square planar

Figure 22-17
Ligand-field splitting in d^8 square-planar complexes.

called *tetragonal distortion,* and when the z ligands have been essentially removed, a square-planar complex is left. Removal of the z-axis ligands reduces the energy of the d_{z^2} and, to a lesser extent, the d_{yz} and d_{xz} orbitals. The resulting set of energy levels permits pairing of all electrons in a d^8 complex, an arrangement which leaves the high-energy $d_{x^2-y^2}$ orbital vacant and lowers the total energy of the complex, thus stabilizing it. Sometimes the z ligands move only partway out, producing a distorted octahedral, or, better, *tetragonal bipyramidal,* complex, as in $Ni(H_2O)_2(NH_3)_4^{2+}$.

An important square-planar nickel complex is bis(dimethylglyoximato)-nickel(II):

This is a red solid used to confirm nickel in qualitative analysis and to precipitate it in quantitative analysis.

Nickel metal combines with carbon monoxide at 50°C to form a neutral tetrahedral complex, tetracarbonylnickel(0), or nickel tetracarbonyl:

$$Ni(s) + 4CO(g) \rightleftharpoons Ni(CO)_4(g)$$

This is a volatile, highly poisonous compound which is used in the *Mond process* to purify nickel, since the reaction reverses at about 200°C to yield the high-purity metal.

Ni(III). The simple *nickelic ion,* Ni^{3+}, does not exist. Oxidation of $Ni(OH)_2$ by hypobromite or hypochlorite yields a black hydrous oxide, often formulated $NiO(OH)$. This has been used as the cathode of the *Edison cell* storage battery, which has an iron anode and uses potassium hydroxide as an electrolyte. The cell reactions for discharge are

Cathode: $[e^- + H_2O + NiO(OH)(s) \longrightarrow Ni(OH)_2(s) + OH^-] \times 2$

Anode: $2OH^- + Fe(s) \longrightarrow Fe(OH)_2(s) + 2e^-$

Cell: $Fe(s) + 2NiO(OH)(s) + 2H_2O \longrightarrow 2Ni(OH)_2(s) + Fe(OH)_2(s)$

The cell is rechargeable and, unlike the lead storage cell, does not decrease in cell voltage as it is discharged.

The platinum metals   A word should be said about the six *platinum metals:* ruthenium, rhodium, and palladium in period 5, and osmium, iridium, and platinum in period 6. These are directly below the iron triad in the periodic table, but are very different in properties, being very noble, that is, unreactive. They occur together in nature, sometimes with gold, as uncombined elements. Many of the platinum metals are used in specialized alloys, and for protective plating on other metals. Some are useful as catalysts for various organic reactions such as hydrogenations.

The platinum metals are especially difficult to dissolve. Most will dissolve in aqua regia (HCl + HNO$_3$), forming chloro complexes such as $PtCl_6^{2-}$. Osmium and iridium resist even aqua regia but can be oxidized by heating in a molten KOH–KNO$_3$ mixture.

Copper Copper has a $3d^{10}4s^1$ configuration, but is little like the alkali metals in its properties. It is a familiar metal which is quite unreactive and is an excellent electrical conductor, second only to silver. It is used in alloys such as *brass* (with zinc) and *bronze* (with tin). Copper has a reduction potential above that of hydrogen:

$$Cu^{2+} + 2e^- \longrightarrow Cu(s) \qquad \mathcal{E}° = 0.34 \text{ V}$$

and so it will not dissolve in nonoxidizing acids. It is oxidized by HNO$_3$, however, reducing the nitrate to NO and/or NO$_2$, depending on concentration.

In its compounds copper shows two oxidation states, +1 and +2, with the +2 state being by far the more important.

Cu(I). In solution the *cuprous ion,* Cu$^+$, is easily reduced to the metal,

$$Cu^+ + e^- \longrightarrow Cu(s) \qquad \mathcal{E}° = 0.52 \text{ V}$$

or oxidized to the +2 state,

$$Cu^{2+} + e^- \longrightarrow Cu^+ \qquad \mathcal{E}° = 0.15 \text{ V}$$

so that the first reaction, above, can reverse the second; that is, Cu$^+$ can oxidize and reduce itself (disproportionate):

$$
\begin{array}{lll}
Cu^+ + e^- \longrightarrow Cu(s) & \mathcal{E}° = & 0.52 \text{ V} \\
\underline{Cu^+ \longrightarrow Cu^{2+} + e^-} & \underline{\mathcal{E}° = -0.15 \text{ V}} \\
2Cu^+ \longrightarrow Cu(s) + Cu^{2+} & \mathcal{E}° = & 0.37 \text{ V}
\end{array}
$$

This corresponds to an equilibrium constant equal to 2×10^6 at 25°C. The reaction can be reversed, however, in the presence of ligands which complex more strongly with Cu(I) than Cu(II). These include Cl$^-$ and CN$^-$. Insoluble cuprous compounds include CuCl and Cu$_2$O, which are stable in contact with water.

Cu(II). The *cupric ion* (d^9) forms a blue hexaaquo complex in which two of the six H$_2$O molecules are located farther away from the copper than are the other four, the structure being a tetragonally distorted octahedron. It is often formulated as $Cu(H_2O)_4^{2+}$ and classed as a square-planar complex. Addition of base to the cupric ion precipitates insoluble cupric hydroxide, which is somewhat amphoteric, dissolving slightly in concentrated OH$^-$ to form the *cuprite ion* $Cu(OH)_3^-$. Cu(OH)$_2$ dissolves readily in NH$_3$, forming deep blue $Cu(NH_3)_4^{2+}$, which is actually a tetragonally distorted $Cu(H_2O)_2(NH_3)_4^{2+}$ ion. Gradual addition of *acid* to this complex precipitates Cu(OH)$_2$ which then redissolves upon addition of excess acid.

Silver Just as the platinum metals are less reactive than the iron triad, so also are the reactivities of silver and gold less than that of copper. Silver is a soft, malleable metal with the highest known electrical and thermal conductivities. It is generally less reactive than copper, but can be dissolved in nitric acid, or in cyanide solutions in the presence of H_2O_2 or O_2, because of the formation of the stable $Ag(CN)_2^-$ complex.

In its compounds silver commonly exhibits only the $+1$ (*argentous*) state, although the $+2$ (*argentic*) and $+3$ states are known. When a strong base is added to the *silver ion*, Ag^+, dark brown *silver oxide* precipitates:

$$2Ag^+ + 2OH^- \longrightarrow Ag_2O(s) + H_2O$$

Addition of ammonia to Ag^+ will also precipitate Ag_2O but the solid redissolves upon addition of excess NH_3 to form the diamminesilver(I) ion, commonly called the silver-ammonia complex:

$$Ag_2O(s) + 2NH_3 + H_2O \longrightarrow Ag(NH_3)_2^+ + 2OH^-$$

Many silver salts are insoluble. These include the halides (except for AgF, which is soluble), AgCN, and Ag_2S. Silver salts of weak acids, such as Ag_2CO_3 and Ag_3PO_4, are insoluble in neutral or basic solutions, but soluble in acidic ones.

Most of the complex ions formed by Ag(I) are *linear*. These include $Ag(NH_3)_2^+$, $AgCl_2^+$, $Ag(CN)_2^-$, and $Ag(S_2O_3)_2^{3-}$.

SILVER AND THE PHOTOGRAPHIC PROCESS. Tons of silver are used each day in the manufacture of photographic films and papers. Black-and-white film is composed of a *backing*, or *base*, usually made of cellulose acetate plastic, uniformly coated with the *photographic emulsion*, tiny crystals of silver halide suspended in a thin layer of gelatin. When a photon of light, $h\nu$, strikes a crystal of AgBr, for example, it may be absorbed by a bromide ion. This causes the bromide ion to lose an electron,

$$h\nu + Br^- \longrightarrow Br + e^-$$

and the electron moves through the crystal and may be trapped at a defect, often an impurity center. The trapped electron attracts an interstitial Ag^+ ion which accepts the electron and becomes a silver atom:

$$Ag^+ + e^- \longrightarrow Ag$$

If several photons of light have struck the same crystal, the freed electrons may be trapped at the same impurity center and there they attract other silver ions. The end result is a cluster of a dozen or so silver atoms at the lattice defect. This cluster is believed to be part of the *latent image*, "latent" because it is so-far invisible.

Development of photographic film consists of treating it with a mild reducing agent, usually an organic compound such as hydroquinone or *para*-aminophenol.

Hydroquinone   *para*-Aminophenol

The developer penetrates the gelatin and if the concentrations, pH, and temperature are all just right, those AgBr crystals with sufficiently large clusters of silver atoms become totally reduced to silver metal,

$$e^- + AgBr(s) \longrightarrow Ag(s) + Br^-$$

while those crystals which have absorbed too few photons of light during exposure remain unreduced. Thus the latent image is transformed into a visible image, composed of grains of silver metal, where sufficient light has struck the film. The more light which originally struck the film, the more concentrated will be the assemblage of silver grains and the darker the image on the film.

After development the film is rinsed briefly in an acid stop bath, which changes the pH in the gelatin, stopping the redox reaction. Then the film is immersed in "hypo," a solution of sodium thiosulfate, $Na_2S_2O_3$. The *unreduced* AgBr crystals remaining in the gelatin are then dissolved,

$$AgBr(s) + 2S_2O_3^{2-} \longrightarrow Ag(S_2O_3)_2^{3-} + Br^-$$

and a final washing leaches $Ag(S_2O_3)_2^{3-}$ and other ions out of the emulsion. Drying reveals a dark image on the film wherever sufficient light struck it during exposure: a *negative* image. Printing or enlarging then produces a *positive* image on photographic paper by much the same process.

Color photography is much more complicated, but most processes rely upon light-sensitive silver salts to trap the photons of light in the emulsion. Color films may have as many as eight layers coated on the film base.

Zinc   Zinc is a somewhat soft, gray-silver metal with only a moderate melting point (419°C). It is reasonably reactive, but serves as a good protective coating for iron because it cathodically protects the iron and because it forms a self-protective layer of $Zn_2(OH)_2CO_3$ on its surface. Zinc is also used in various alloys and in dry-cell batteries. Intricately shaped articles such as automobile grills and trim have been made out of die-cast zinc and its alloys. (The chromium-plated product looks much like steel, but beware of the first little bump; zinc is not a strong metal.)

In its compounds the only important oxidation state is the +2 state, as the $(n-1)d^{10}$ configuration is hard to break into. The zinc ion, Zn^{2+}, is a tetrahedral tetraaquo ion, and its solution is acidic due to hydrolysis:

$$Zn^{2+} + H_2O \rightleftharpoons ZnOH^+ + H^+$$

Zinc hydroxide precipitates upon addition of base to the zinc ion:

$$Zn^{2+} + 2OH^- \longrightarrow Zn(OH)_2(s)$$

This is amphoteric and redissolves upon addition of excess OH^- to form *zincate* solutions containing species such as $Zn(OH)_3^-$ and $Zn(OH)_4^{2-}$.

Zinc forms a number of tetrahedral complex ions including $Zn(NH_3)_4^{2+}$, $Zn(CN)_4^{2-}$, and $Zn(C_2O_4)_2^{2-}$. Many zinc complexes are less stable than those of other transition metals, however.

Cadmium   Cadmium, a second-series transition metal, is much like zinc in physical and chemical properties. It is a gray-white metal which is somewhat softer and lower-melting than zinc. Its principal use is in plating iron, on which it can be deposited in a very smooth coat which, like zinc plating, is self-healing after it is scratched.

Cadmium exhibits only the $+2$ oxidation state in its important compounds. Cd^{2+} precipitates as $Cd(OH)_2$ in basic solutions. Unlike $Zn(OH)_2$, however, cadmium hydroxide is not amphoteric. $Cd(OH)_2$ will dissolve in NH_3 to form an ammine complex, however:

$$Cd(OH)_2(s) + 4NH_3 \longrightarrow Cd(NH_3)_4^{2+} + 2OH^-$$

Cadmium forms complex ions which are tetrahedral and much like those of zinc in many respects. Its halide complexes are more stable, however.

Cadmium is a bit unusual in that it forms numerous *weak salts,* compounds which dissociate incompletely in water. Conductivity measurements show the existence of species like $CdCl^+$ and $CdCl_2$ in an aqueous solution of cadmium chloride, for example. Compounds of cadmium are very toxic.

Mercury Mercury is a fascinating silvery liquid at room temperature (melting point $-39°C$). Its vapor pressure is low, about 1×10^{-3} mmHg at $20°C$, but this is high enough to be potentially dangerous because of mercury's great toxicity. Like many "heavy metals" mercury is a cumulative poison, and in the early days of chemistry many researchers experienced its effects after spending years working in unventilated laboratories in which spilled mercury lurked in the cracks and crevices of plank floors. Mercury is used in thermometers, barometers, switches, and mercury-vapor lamps, including fluorescent lamps. It is also used in alloys, called *amalgams,* which may be liquid or solid at room temperature. Dental amalgam is composed of mercury, silver, cadmium, tin, and copper.

Hg(I). Mercury forms a rather unexpected double ion, the *mercurous ion,* Hg_2^{2+}, which accounts for its showing a $+1$ oxidation state as well as the expected $+2$. (Zn_2^{2+} and Cd_2^{2+} have been observed, but only under special conditions and not in aqueous solution.) The mercurous ion is not paramagnetic, indicating that the $6s$ electron in each of two Hg^+ ions has been used in forming a single bond in Hg_2^{2+}.

The mercurous ion might be expected to disproportionate but does not, as can be seen from the following standard reduction potentials:

$$2Hg^{2+} + 2e^- \longrightarrow Hg_2^{2+} \qquad \mathcal{E}° = 0.92 \text{ V}$$

$$Hg_2^{2+} + 2e^- \longrightarrow 2Hg(l) \qquad \mathcal{E}° = 0.78 \text{ V}$$

Reversing the first half-reaction and adding it to the second gives us

$$
\begin{array}{ll}
Hg_2^{2+} \longrightarrow 2Hg^{2+} + 2e^- & \mathcal{E}° = -0.92 \text{ V} \\
2e^- + Hg_2^{2+} \longrightarrow 2Hg(l) & \mathcal{E}° = 0.78 \text{ V} \\
\hline
2Hg_2^{2+} \longrightarrow 2Hg^{2+} + 2Hg(l) & \mathcal{E}° = -0.14 \text{ V}
\end{array}
$$

This corresponds to $\Delta G°$ (for the reaction as written) equal to 27 kJ, and so for

$$Hg_2^{2+} \longrightarrow Hg^{2+} + Hg(l) \qquad \Delta G° = 14 \text{ kJ}$$

The corresponding equilibrium constant is

$$K = \frac{[Hg^{2+}]}{[Hg_2^{2+}]} = 4 \times 10^{-3}$$

which is small enough to ensure the stability of Hg_2^{2+} in solution. But addition of OH^- forces the $Hg(I) \longrightarrow Hg(0) + Hg(II)$ disproportionation to take place by forming a highly insoluble Hg(II) product, mercuric oxide:

$$Hg_2^{2+} + 2OH^- \longrightarrow Hg(l) + HgO(s) + H_2O$$

The most common Hg(I) compound is probably mercurous chloride, Hg_2Cl_2, a white, insoluble solid called *calomel*. It is like silver chloride in its insolubility, but turns black upon addition of NH_3. The reaction has been represented as

$$Hg_2Cl_2(s) + 2NH_3 \longrightarrow \underset{\text{White}}{HgNH_2Cl(s)} + \underset{\text{Black}}{Hg(l)} + NH_4^+ + Cl^-$$

but is apparently more complicated.

Hg(II). The *mercuric ion,* Hg^{2+}, is invariably formed when mercury is dissolved in a strong oxidizing agent; almost anything which can oxidize Hg(0) to Hg(I) can oxidize it to Hg(II):

$$Hg_2^{2+} + 2e^- \longrightarrow 2Hg(l) \qquad \mathcal{E}° = 0.79 \text{ V}$$

$$Hg^{2+} + 2e^- \longrightarrow Hg(l) \qquad \mathcal{E}° = 0.85 \text{ V}$$

The mercuric ion forms a chloride, $HgCl_2$, called *corrosive sublimate,* which is a *weak salt,* existing in solution as linear $HgCl_2$ molecules which are only slightly dissociated. Mercuric sulfide HgS, is a very insoluble salt ($K_{sp} = 10^{-54}$) which will not dissolve in hot concentrated HNO_3, but will in aqua regia, in which $HgCl_4^{2-}$ ions are formed and in which the sulfide is oxidized to sulfur (and perhaps some sulfate).

Mercury poses a serious threat in the environment today. During years of ignorance and lack of concern we have polluted some parts of the world with mercury rather badly. Mercury compounds in wastewater from industrial plants get partially converted to dimethyl mercury, CH_3—Hg—CH_3, which is soluble in fats and gets into the food chain, eventually becoming concentrated in fish. This is why fishing is now prohibited in certain inland waters and why some fish have all but disappeared from the market. In most of the world the discharge of mercury into rivers, lakes, and the ocean has been prohibited, but it will take a long time to clean up the world's water.

SUMMARY

In this chapter, which completes our survey of descriptive inorganic chemistry, we have discussed the transition metals, the elements which intervene between group IIA and group IIIA in the periodic table. These are generally *typical metals,* as far as physical properties are concerned, but have *reduction potentials* which vary from highly negative values for scandium and others to very positive values of the *noble metals. Densities* and *melting points* also vary widely, a result of the variation in number of $(n-1)d$ electrons and in the resulting bonding in the solid. A characteristic of the transition metals is the variability of their oxidation states in compounds.

In this chapter we discussed *bonding* and *structure* in complex ions. Most of the discussion of bonding was in terms of *crystal field theory,* a simplified version of the molecular-orbital-based *ligand field theory.* In crystal field theory the d-orbital energies are split by the field produced by the ligands in the complex. Although metal-ligand bonding is essentially covalent, the electrostatic field can be imagined as interacting differentially with the various d orbitals, so that the resulting orbital subsets, when populated with the d electrons, help account for the *magnetic and spectral properties* of complex ions. Ligands can be ordered in the *spectrochemical series,* in which they are listed in order of increasing field strength (ability to split the d-orbital energies).

The discussion of the *stereochemistry* of complex ions was based on the different *coordination numbers* which are observed. Coordination numbers of 2 (linear), 4 (tetrahedral and square planar), and 6 (octahedral) were described, and examples were given for each. *Stereoisomerism* was briefly discussed, and examples of *cis-trans* and *optical isomerism* were given.

Titanium is a light, hard metal which finds application in the aerospace industry. It is highly corrosion-resistant because of the formation of an impervious oxide-nitride coat. In its compounds titanium shows the +2, +3 (titanous), and +4 (titanic) states. *Vanadium* is denser than titanium but has similar properties. It is mostly used in alloy steels. In its compounds it exhibits the +2 (vanadous), +3 (vanadic), +4, and +5´ states. *Chromium* is another important corrosion-resistant metal which is used in alloys and in plating. It commonly exhibits the +2 (chromous), +3 (chromic), and +6 oxidation states in its compounds. Chromates and dichromates are important compounds of $Cr(VI)$. *Manganese* is a somewhat reactive metal used in alloy steels. In its compounds it commonly exhibits the +2 (manganous), +4, +7 states, with the +3 (manganic) and +6 states being less common. The permanganate ion MnO_4^- is an important oxidizing agent. *Iron* is an abundant and useful, if somewhat corrosion-prone metal. It is obtained by reduction from its ores in a blast furnace, followed by further purification. The corrosion of iron is a process in which the metal first dissolves by being oxidized to the +2 state and then is oxidized to the +3 state, in which it precipitates as hydrous ferric oxide, rust. Rusting can be prevented by coating the iron surface with a metal whose reduction potential is more negative than that of iron, an application of *cathodic protection.* In its compounds iron commonly exhibits the +2 (ferrous) and +3 (ferric) states. *Cobalt* is a hard, unreactive metal used mostly in alloys. As an aquo complex the +2 (cobaltous) state is stable, but the +3 (cobaltic) state is common in many complex ions with ligands other than water. *Nickel* is a hard metal used in alloys and plating. It commonly shows only the +2 (nickelous) state in compounds, although higher oxidation states are known. *Copper* is a fairly unreactive metal used in plating, alloys, and as an electrical conductor. Its chemistry is largely that of the +2 (cupric) state, although the +1 (cuprous) state is stable in certain solids and complexes. *Silver* is an unreactive metal with very high electrical and thermal conductivities. Its chemistry is largely that of the +1 (argentous) state. Much silver is used in the manufacture of photographic film and other light-sensitive materials. *Zinc* is a metal used in plating iron (galvanizing) and in alloys. Its chemistry is that of the +2 state. *Cadmium* resembles zinc in its properties, although it is much more toxic. It is used in plating, and in its compounds it shows only the +2 oxidation state. *Mercury,* a liquid at room temperature, shows the +1 (mercurous) and +2 (mercuric) states. The mercurous ion is an unusual double ion, Hg_2^{2+}. Alloys of mercury are called *amalgams.*

KEY TERMS

Actinoid (Sec. 22-1)
Amalgam (Sec. 22-6)
Blast furnace (Sec. 22-6)
Cathodic protection (Sec. 22-6)
Chelate (Sec. 22-5)
Chiral (Sec. 22-5)
cis- (Sec. 22-5)
Coordination number (Sec. 22-3)
Crystal-field theory (Sec. 22-4)
d-d transition (Sec. 22-4)
Domain (Sec. 22-6)
Enantiomer (Sec. 22-5)
fac- (Sec. 22-5)

Ferromagnetism (Sec. 22-6)
High-spin complex (Sec. 22-4)
Lanthanoid (Sec. 22-1)
Lanthanoid contraction (Sec. 22-6)
Ligand-field splitting (Sec. 22-4)
Low-spin complex (Sec. 22-4)
Magnetic balance (Sec. 22-4)
mer- (Sec. 22-5)
Noble metal (Sec. 22-2)
Optical isomer (Sec. 22-5)
Paramagnetism (Sec. 22-4)
Passivity (Sec. 22-6)
Polarimeter (Sec. 22-5)

Polydentate ligand (Sec. 22-5)
Spectrochemical series (Sec. 22-4)
Stereochemistry (Sec. 22-5)
Stereoisomer (Sec. 22-5)

Strong-field ligand (Sec. 22-4)
Tetragonal distortion (Sec. 22-6)
trans- (Sec. 22-5)
Weak-field ligand (Sec. 22-4)

QUESTIONS AND PROBLEMS

Transition metals: general

22-1 What are the transition metals? Why are they so-named? Why are there no transition metals in the second or third periods of the periodic table?

22-2 Justify the inclusion of the lanthanoids and actinoids among the transition metals.

22-3 Which elements included with the transition metals in this chapter would have been excluded if the transition metals had been defined as **(a)** those metals which as free elements have partly filled *d* subshells in their atoms **(b)** those whose atoms have partly filled *d* subshells in *either* their uncombined atoms *or* in compounds?

22-4 What is the lanthanoid contraction? What are its consequences? Why do we hear little about an actinoid contraction?

22-5 Contrast the properties of the alkali metals with those of the group IB elements Cu, Ag, and Au. How do you account for the differences?

22-6 Which of the first-series transition metals would you expect to be able to purify by electrolytically plating out of solution? Which not? Why not?

22-7 How can the $3d^5 4s^1$ electronic configuration of chromium be accounted for? Considering that this is its configuration, how do you account for the absence of a $+1$ oxidation state for chromium?

22-8 The reduction potential ($M^{2+} \longrightarrow M(s) + 2e^-$) for titanium is -1.63 V and for platinum is $+1.2$ V. Given this large difference, why is it that both metals resist corrosion so well?

22-9 Using only the periodic table, predict the *maximum* oxidation state to be expected for **(a)** $Ta(Z = 73)$ **(b)** $Tc(Z = 43)$ **(c)** $Ce(Z = 58)$ **(d)** $Pt(Z = 46)$.

22-10 Why is it that so many transition-metal compounds are **(a)** colored **(b)** paramagnetic?

Bonding in complex ions

22-11 Describe the formation of molecular orbitals in an octahedral complex.

22-12 Crystal-field theory is a modified electrostatic model. Since bonding in transition-metal complexes is clearly covalent, what justification is there for using crystal-field theory?

22-13 Describe the splitting of the *d*-orbital energies in **(a)** octahedral, **(b)** tetrahedral, and **(c)** square-planar complexes.

22-14 Predict the changes which would occur in the *d*-orbital splitting in an octahedral complex when **(a)** the complex is elongated by pulling a trans pair of ligands out from the metal ion **(b)** the complex is squashed by pushing in a trans pair of ligands.

22-15 Predict the number of unpaired electrons in each of the following complexes: **(a)** TiF_6^{3-} **(b)** $FeCl_6^{3-}$ **(c)** $Fe(CN)_6^{3-}$ **(d)** $Zn(NH_3)_4^{2+}$ **(e)** $Cu(NH_3)_4^{2+}$ **(f)** CoF_6^{3-} **(g)** $Co(NH_3)_6^{3+}$ **(h)** $Ni(CN)_4^{2-}$ (square planar) **(i)** $NiCl_4^{2-}$ (tetrahedral).

22-16 Why are different high- and low-spin complexes not observed for **(a)** d^3 octahedral complexes **(b)** d^8 square-planar complexes?

22-17 Predict the way the *d*-orbital splitting in a tetrahedral complex will change, if the complex is **(a)** squashed in the *z* direction **(b)** elongated in the *z* direction.

22-18 Why is it that d^8 ions often form square-planar complexes?

22-19 What *d*-orbital splitting would you expect for a complex which showed *cubic* coordination (eight ligands at the corners of a cube)?

22-20 Why is the coordination number of Cu(II) in its complexes sometimes given as 4 and sometimes as 6?

22-21 Certain ligands can π bond with metal ions. Which metal *d* orbitals are used in the formation of these MOs?

Stereochemistry

22-22 Why are cis and trans isomers not found in tetrahedral complexes?

22-23 Why is optical isomerism not observed in square-planar complexes?

22-24 What structural characteristic must a tetrahedral complex possess in order to be chiral?

22-25 Sketch all the isomers of the square-planar complexes MA_2BC and MABCD.

22-26 Sketch all the stereoisomers of each of the following octahedral complexes:
(a) $CoCl_2(NH_3)_4^+$ (b) $CoCl_3(NH_3)_3$
(c) $CrCl_2(en)_2^+$ (d) $Co(C_2O_4)(en)_2^+$.

22-27 Of the two octahedral complexes $Co(en)_3^{3+}$ and $Co(NH_3)_6^{3+}$, the chelate is much more stable, in spite of the fact that in each complex six nitrogen atoms are bonded to the cobalt. Explain. (*Hint:* Consider entropy effects.)

Nomenclature

22-28 Give the IUPAC name for each of the following: (a) $CrBr_2(NH_3)_4^+$ (b) $Fe(C_2O_4)_3^{3-}$
(c) $Cu(NH_3)_4^{2+}$ (d) $AgBr_2^-$ (e) $PtCl_2(NH_3)_2^{2+}$
(f) $Co(NCS)_4^{2-}$ (g) $Fe(CN)_5NO^{2-}$
(h) $Os(NO_2)_5^{2-}$ (i) $W(OH)Cl_5^{2-}$.

22-29 Write the formula for each of the following species:
(a) Octafluorozirconate(IV)
(b) Chloropentaaquotitanium(III)
(c) Trifluorotriaquovanadium(III)
(d) Octacyanotungstate(IV)
(e) Dichlorodithiocyanatodiaquochromium(III)
(f) Dichlorodiaquodiamminecobalt(III)
(g) Dithiocyanatobis(ethylenediamine)iron(III)
(h) Bromochlorodicyanonickelate(II)
(i) Oxotetrafluorochromate(V)

Descriptive chemistry

22-30 List (a) the highest and (b) the lowest oxidation states commonly exhibited by each of the following elements in its compounds: Ti, Cr, Mn, Fe, Cu.

22-31 Briefly account for each of the following: (a) $TiCl_4$ is a liquid at room temperature. (b) When powdered zinc is added to an acidified solution of ammonium metavanadate, the color of the solution undergoes several changes, ending up violet. (c) Chromic hydroxide is amphoteric, but chromous hydroxide is not. (d) Silver metal is insoluble in hydrogen peroxide or potassium cyanide, but dissolves in a solution containing both. (e) When MnO_4^- is reduced, the oxidation state of Mn tends to decrease more, the more acidic the solution. (f) Copper and iron water pipes should never be connected together without an electrically-insulating fitting. (g) Liquid mercury is much less poisonous than its compounds.

22-32 Which of the following metals would be expected to provide cathodic protection to iron: Mg, Co, Cr, Al, Cu?

22-33 Freshly precipitated zinc hydroxide will dissolve in either a solution of sodium hydroxide (to form the zincate ion) or one of ammonia. How could you show experimentally that when NH_3 is added, the zincate ion is *not* formed?

22-34 Refining of copper involves electrolyzing a solution of $CuSO_4$ using an impure copper anode and a pure copper cathode. Explain how impurities such as iron and silver are removed by this process.

22-35 Write a balanced equation for each of the following: (a) mercuric sulfide dissolves in aqua regia (b) ammonium dichromate thermally decomposes to form nitrogen and Cr_2O_3 (c) the chromous ion is oxidized by permanganate in acidic solution (d) a limited amount of NH_3 is added to a copper sulfate solution until a blue precipitate forms (e) the precipitate in part (d) is dissolved by adding an excess of NH_3 (f) the decomposition of a solution of cobaltic sulfate (g) the thermal dehydration of ferric hydroxide (h) the oxidation of Mn^{2+} to MnO_4^- by Pb_3O_4 (acidic solution).

22-36 Describe the operation of a blast furnace in smelting iron. Write equations for pertinent reactions.

22-37 Write an equation for each of the following *successive* chromium reactions: (a) A solution of chromous chloride is oxidized by oxygen. (b) A solution of NaOH is added to form a precipitate. (c) An excess of NaOH is added to dissolve the precipitate. (d) H_2O_2 is added in excess. (e) The excess H_2O_2 is decomposed by heating. (f) The solution is acidified. (g) More H_2O_2 is added. (h) Powdered zinc is added.

22-38 What is ferromagnetism? Which elements are ferromagnetic?

22-39 Why are zirconium and hafnium always found together in ores?

22-40 Suggest a laboratory method for distinguishing between each member of each of the following pairs of solutions: (a) $AgNO_3$ and KNO_3 (b) $Cr_2(SO_4)_3$ and $NiSO_4$ (c) $FeCl_3$ and K_2CrO_4 (d) $AgNO_3$ and $Hg_2(NO_3)_2$ (e) $K_3Fe(CN)_6$ and $K_4Fe(CN)_6$.

22-41 It is often said that "the cupric ion is blue." Why, then, is $CuSO_4$ a white powder?

22-42 How can the disproportionation of Cu(I) in solution be prevented?

22-43 Describe the role of (a) the developer and (b) the hypo in the processing of photographic film.

22-44 A series of different solutions is prepared each containing the mercurous ion and the mercuric ion in equilibrium with liquid mercury. The solutions are analyzed and the ratio of Hg(I) to Hg(II) is found to be constant. Show that this proves that the mercurous ion is diatomic.

22-45 Account for the fact that the hexaaquo complex of Co(II) is more stable than that of Co(III) but that the order is reversed for hexacyano complexes.

22-46 List all of the theoretically possible different compounds having the empirical formula $CoCr(NH_3)_6Cl_6$. Sketch the structure of each.

22-47 Zinc metal will dissolve in either strong base or strong acid, but iron will dissolve only in acids. Explain, writing appropriate equations.

22-48 Vitamin B_{12} is a cobalt complex in which a cyanide ligand and a large pentadentate organic ligand are coordinated to a cobalt atom. Its formula is $C_{63}H_{88}CoN_{14}O_{14}P$. If the recommended daily allowance of vitamin B_{12} is 5 micrograms (μg), how many cobalt atoms combined in vitamin B_{12} should be eaten per day?

23

ORGANIC CHEMISTRY

TO THE STUDENT

To early chemists *organic* compounds were those which were found in or produced by living organisms and hence were part of the life process. *Inorganic* compounds were all the rest. Further, the prevailing doctrine of *vitalism* maintained that there was an unbridgeable gap between the two, because organic compounds were infused with a "vital force," which made them inherently different from other compounds. But in 1828 Friedrich Wöhler found that heating crystals of the inorganic salt *ammonium cyanate* produced *urea*, known to be a component of *urine:*

$$NH_4OCN \longrightarrow \underset{\substack{\\ H_2N \qquad NH_2}}{\overset{\substack{O \\ \| \\ C}}{}}$$

Ammonium   Urea
cyanate

Thus Wöhler dealt a sharp blow to the vitalists, and today *organic* and *inorganic* are used in a different sense.

Unfortunately, it is impossible to come up with a definition of an organic compound which is completely satisfactory. It is sometimes said that organic compounds are compounds of carbon, but this definition includes the oxides of carbon, the carbonates, the cyanides, the carbides and other compounds which have traditionally been considered inorganic. It is better to say that organic compounds are those of carbon, hydrogen, and possibly other elements. Although this definition is not completely satisfactory to all chemists, it is generally accepted, and so we will use it here.

At present perhaps 20 organic compounds are known for each known inorganic compound, the total approaching 2 million. Why is carbon such a prolific compound producer? There are several reasons, but the main one is that carbon atoms form strong bonds with each other, and so *chains* of carbon atoms of any length are possible.   These chains, which serve as skeletons for organic molecules, may be simple and unbranched:

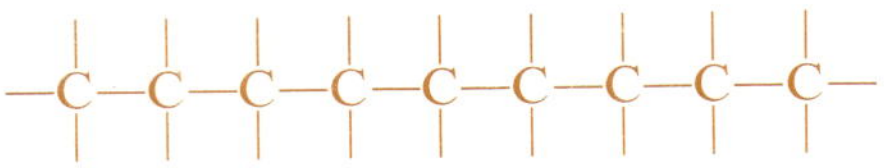

$$-\overset{|}{\underset{|}{C}}-\overset{|}{\underset{|}{C}}-\overset{|}{\underset{|}{C}}-\overset{|}{\underset{|}{C}}-\overset{|}{\underset{|}{C}}-\overset{|}{\underset{|}{C}}-\overset{|}{\underset{|}{C}}-\overset{|}{\underset{|}{C}}-\overset{|}{\underset{|}{C}}-$$

or they may be branched at any number of places:

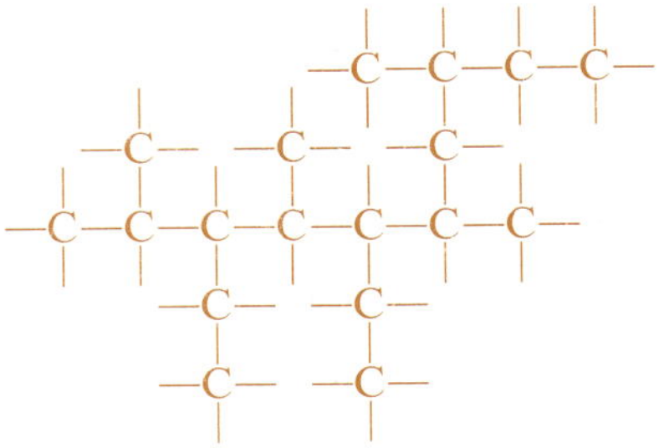

They sometimes even circle back on themselves to form *cyclic* structures:

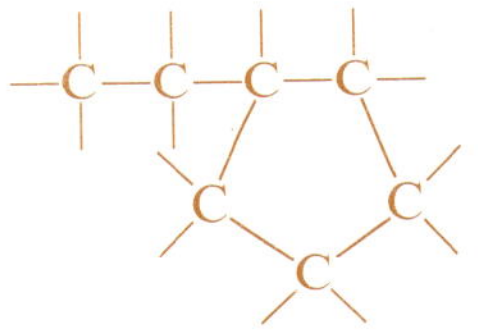

Thus the possibilities are limitless, accounting for an endless variety of organic compounds.

23-1
Saturated
hydrocarbons

Hydrocarbons are compounds which consist of carbon and hydrogen only. *Saturated hydrocarbons* are those in which all carbon-carbon bonds are single bonds.   They are so-named because they will not react with hydrogen, while *unsaturated hydrocarbons* (Sec. 23-2) can react with hydrogen and *become* saturated.

The alkanes

The simplest saturated hydrocarbon is methane, CH_4.   In the methane molecule a carbon atom uses four equivalent sp^3 hybrid orbitals to bond tetrahedrally to four hydrogen atoms.   Methane is the first in a series of saturated hydrocarbons called the *alkanes,* all of which have the general formula C_nH_{2n+2}.   Each carbon in an alkane bonds tetrahedrally to four other atoms, C or H.   The structural formulas of the first three alkanes, methane (CH_4), ethane (C_2H_6), and propane (C_3H_8), are shown in Fig. 23-1.   Also shown are space-filling and ball-and-stick models of these molecules.

NORMAL ALKANES.   Alkanes in which there is no chain branching are called *normal,* or *straight-chain, alkanes*.   These are all nonpolar compounds found in natural gas or petroleum.   They have different boiling points, as Table 23-1 shows for the first ten normal alkanes, and so they can be separated by fractional

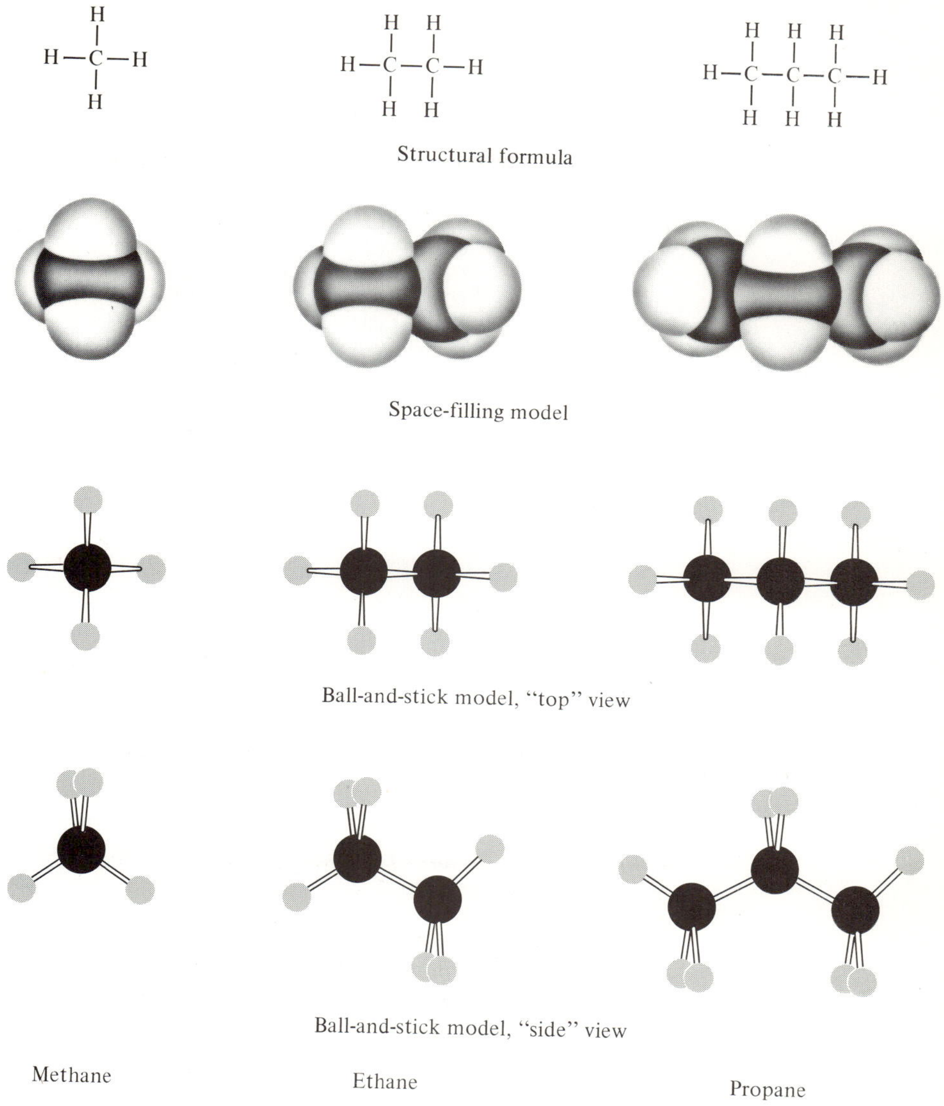

Structural formula

Space-filling model

Ball-and-stick model, "top" view

Ball-and-stick model, "side" view

Figure 23-1
The first three alkanes.

Methane   Ethane   Propane

distillation (Sec. 12-4). The increase in boiling point with number of carbon atoms is accounted for by increased London forces (Sec. 10-5) in the liquid.

Table 23-1 also shows the *condensed structural formulas* of the first 10 normal alkanes. Such a formula is derived from a complete structural formula but does not specifically show all the bonds. For example, for normal butane (C_4H_{10}) the complete structural formula is

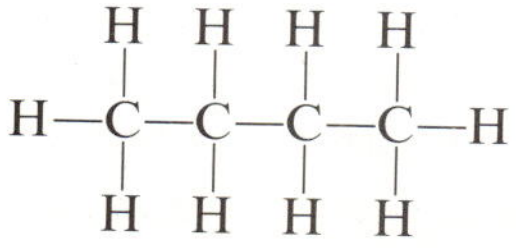

Table 23-1	Name	Molecular formula	Condensed structural formula	Normal melting point, °C	Normal boiling point, °C
The first ten normal alkanes	Methane	CH_4	CH_4	−182	−161
	Ethane	C_2H_6	CH_3CH_3	−183	−89
	Propane	C_3H_8	$CH_3CH_2CH_3$	−188	−42
	Butane	C_4H_{10}	$CH_3CH_2CH_2CH_3$	−138	−1
	Pentane	C_5H_{12}	$CH_3CH_2CH_2CH_2CH_3$	−130	36
	Hexane	C_6H_{14}	$CH_3CH_2CH_2CH_2CH_2CH_3$	−95	69
	Heptane	C_7H_{16}	$CH_3CH_2CH_2CH_2CH_2CH_2CH_3$	−91	98
	Octane	C_8H_{18}	$CH_3CH_2CH_2CH_2CH_2CH_2CH_2CH_3$	−57	126
	Nonane	C_9H_{20}	$CH_3CH_2CH_2CH_2CH_2CH_2CH_2CH_2CH_3$	−50	151
	Decane	$C_{10}H_{22}$	$CH_3CH_2CH_2CH_2CH_2CH_2CH_2CH_2CH_2CH_3$	−30	174

and the condensed structural formula is written as

$$CH_3-CH_2-CH_2-CH_3$$

or as

$$CH_3CH_2CH_2CH_3$$

Note that although the normal alkanes are called *straight-chain* alkanes, the carbon chain which is the backbone of each molecule is by no means straight. Rather, it is a zigzag chain, as can be seen from the "side" views of the ball-and-stick models in Fig. 23-1. The word *straight* means "unbranched."

Note also that the name of each normal alkane with more than four carbon atoms is taken from the Greek word for the number of carbons. (See Table 23-1.) Thus *oct*ane, C_8H_{18}, has *eight* carbon atoms in its molecule.

BRANCHED-CHAIN ALKANES. The molecule of a *branched-chain* alkane has, as the name implies, a chain of carbon atoms which is branched. If, for example, we replace one of the hydrogens bonded to the second carbon in a molecule of pentane,

by a methyl group, CH_3-, we obtain a molecule of the branched-chain alkane known as 2-methylpentane,

The name itself tells us that there is a methyl group on the second carbon of a five-carbon (pentane) chain. Note that a meth*yl* group is just a molecule of meth*ane*, CH_4, minus one of its hydrogen atoms. The condensed structural formula of 2-methylpentane is

$$CH_3\!-\!\underset{\underset{CH_3}{|}}{CH}\!-\!CH_2\!-\!CH_2\!-\!CH_3 \quad \text{or} \quad CH_3\underset{\underset{CH_3}{|}}{CH}CH_2CH_2CH_3$$

According to the IUPAC system branched chain alkanes are named by first identifying the longest straight carbon chain in the molecule and then using *locator numbers* to indicate the positions of side groups or chains. For example, the longest chain in the compound

$$\overset{1}{CH_3}\!-\!\underset{\underset{CH_3}{|}}{\overset{2}{CH}}\!-\!\overset{3}{CH_2}\!-\!\overset{4}{CH_3}$$

is the four carbon chain of *butane* (see Table 23-1). The methyl group can be seen to be on the *second* carbon, and so the locator number 2 is made part of the name 2-methylbutane. The carbons in a chain are numbered so that the locator number in the name of a compound is as small as possible. Thus the above compound is *not* named 3-methylbutane, as it would appear to be if the carbons were numbered starting from the other end of the chain.

ADDED COMMENT

Note that the carbons in a chain are always numbered starting from one end of the chain. Note, also, that the numbering does *not* depend on the way the structural formula is written. Thus the following all represent 2-methylbutane:

$$\overset{4}{CH_3}\!-\!\overset{3}{CH_2}\!-\!\underset{\underset{CH_3}{|}}{\overset{2}{CH}}\!-\!\overset{1}{CH_3} \qquad \underset{\underset{_4CH_3}{|}}{\overset{3}{CH_2}}\!-\!\underset{\underset{CH_3}{|}}{\overset{2}{CH}}\!-\!\overset{1}{CH_3}$$

$$\overset{4}{CH_3}\!-\!\overset{3}{CH_2}\!-\!\underset{\underset{CH_3}{|}}{\overset{2}{\overset{|}{\overset{1}{CH_3}}{CH}}} \qquad \overset{1}{CH_3}\!-\!\underset{\underset{_3CH_2}{|}}{\overset{2}{CH}}\!-\!CH_3$$

Hydrocarbon groups (such as the methyl group) which are derived from alkanes are called *alkyl* groups. Some of these are shown in Table 23-2 with their IUPAC and common names.

● **EXAMPLE 23-1**

Problem Give the IUPAC name for

$$\underset{\underset{CH_2-CH_3}{|}}{CH_2}\!-\!\underset{\underset{}{}}{CH}\!-\!CH_3 \;\; \overset{CH_3}{|}$$

Solution The name of a compound does not depend on how its structural formula happens to be

Table 23-2
Some alkyl groups

Group	IUPAC name	Common name	Comment
CH_3-	Methyl	Methyl	
CH_3CH_2-	Ethyl	Ethyl	
$CH_3CH_2CH_2-$	Propyl	*n*-Propyl	*n* means "normal" (straight-chain).
CH_3-CH- $\quad\quad CH_3$	1-Methylethyl	Isopropyl	*iso* means that there is a methyl group on the next to the last carbon.
$CH_3CH_2CH_2CH_2-$	Butyl	*n*-Butyl	
CH_3CH_2CH- $\quad\quad\quad CH_3$	1-Methylpropyl	*sec*-Butyl	*sec* (also abbreviated *s*) stands for "secondary," which means that the first carbon has *two* other carbons attached.
CH_3CHCH_2- $\quad\quad CH_3$	2-Methylpropyl	Isobutyl	
$\quad\quad CH_3$ CH_3C- $\quad\quad CH_3$	1,1-Dimethylethyl	*tert*-Butyl	*tert* (also abbreviated *t*) stands for "tertiary," which means that *three* other carbons are attached to the first carbon.
$CH_3CH_2CH_2CH_2CH_2-$	Pentyl	*n*-Amyl	

drawn. This is 2-methylpentane, again. The numbering of carbons in the main chain starts at the end which gives the lowest number (or numbers) in the name. ●

● **EXAMPLE 23-2**

Problem   Give the IUPAC name for

$$CH_3-CH_2-\underset{\underset{CH_2-CH_2-CH_3}{|}}{CH}-CH_2-CH_3$$

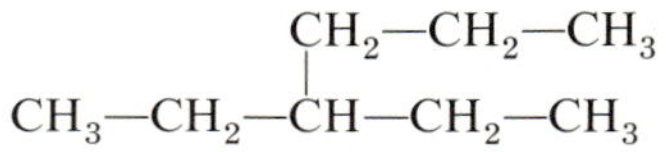

Solution   Be careful here! The longest chain has *six* carbons, not five. Therefore this is a molecule of 3-ethylhexane. (It is *not* named 3-propylpentane.) ●

● **EXAMPLE 23-3**

Problem   Give the IUPAC name for

$$CH_3-CH_2-CH_2-\underset{CH_3-CH-\underset{\underset{CH_2-CH_2-CH_3}{|}}{\underset{CH-CH_2-CH_3}{|}}CH_2-CH_3}{|}CH_2$$

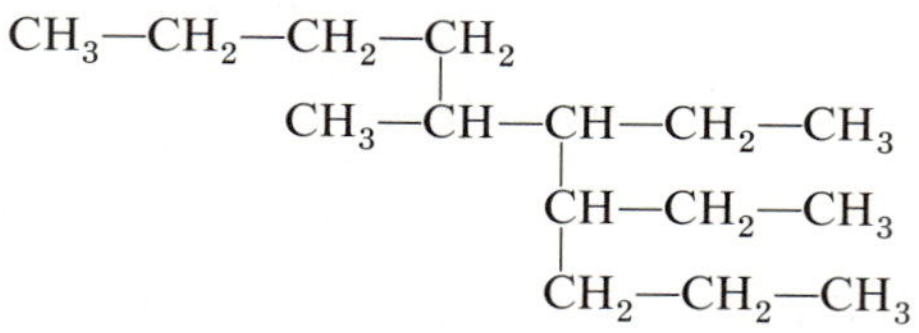

Solution   The longest chain has 10 carbons and is numbered from lower right (carbon 1) to upper left, as drawn above. The name of this molecule is, therefore, 4,5-diethyl-6-methyl-decane. ●

SKELETAL ISOMERISM. The alkanes show a kind of structural isomerism known as *skeletal*, or *chain, isomerism*. Two alkanes which have the same molecular formula but differ in the branching of their carbon chains are called *skeletal isomers*. Skeletal isomerism is possible if the molecule has four or more carbon atoms. The two isomers having the molecular formula C_4H_{10} are

$$CH_3-CH_2-CH_2-CH_3 \qquad \text{and} \qquad CH_3-\overset{\overset{\displaystyle CH_3}{|}}{CH}-CH_3$$

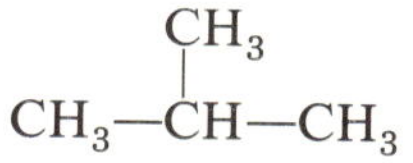

Butane   2-Methylpropane
(*n*-butane)   (isobutane)

It should be emphasized that these are *different* compounds. They have different normal boiling points, for example, $-0.5°C$ and $-12°C$, respectively.

The number of possible isomers increases sharply as the number of carbon atoms in an alkane increases. Thus there are three isomers with the molecular formula C_5H_{12}, five with C_6H_{14}, nine with C_7H_{16}, and 18 with C_8H_{18}. It has been calculated that 366,319 isomers with molecular formula $C_{20}H_{42}$ are possible!

● **EXAMPLE 23-4**

Problem   Indicate and name the skeletal isomers with molecular formula C_5H_{12}.

Solution

$$CH_3-CH_2-CH_2-CH_2-CH_3 \qquad CH_3-\overset{\overset{\displaystyle CH_3}{|}}{CH}-CH_2-CH_3 \qquad CH_3-\overset{\overset{\displaystyle CH_3}{|}}{\underset{\underset{\displaystyle CH_3}{|}}{C}}-CH_3$$

Pentane   2-Methylbutane 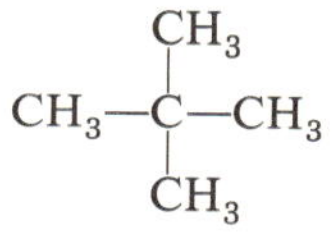   2,2-Dimethylpropane ●

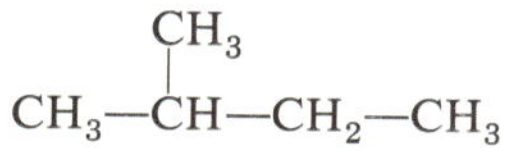

The cycloalkanes   A saturated hydrocarbon whose molecules have carbon-chain skeletons which are closed to form rings is called a *cycloalkane,* general formula C_nH_{2n}. Each carbon in a cycloalkane bonds tetrahedrally, or approximately tetrahedrally, to two other carbons and to two hydrogens. Table 23-3 shows the regular and simplified structural formulas of the first four cycloalkanes. Note that as with the alkanes their names indicate the number of carbon atoms in each molecule, but that the prefix *cyclo* is added.

Table 23-3 Some cycloalkanes

Molecular formula	Structural formula	Simplified structural formula	IUPAC name
C_3H_6	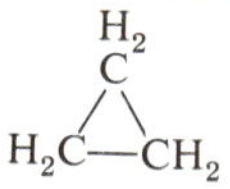	 △	Cyclopropane
C_4H_8	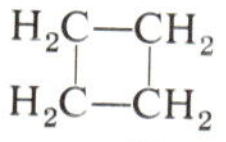	▢	Cyclobutane
C_5H_{10}	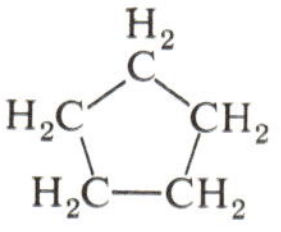	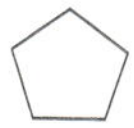 ⬠	Cyclopentane
C_6H_{12}	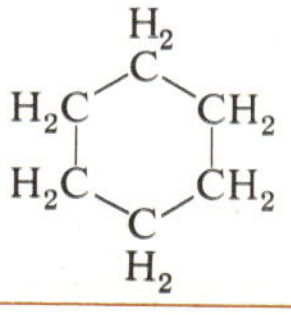	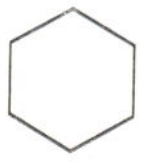 ⬡	Cyclohexane

Properties of the saturated hydrocarbons

The alkanes and cycloalkanes are nonpolar compounds which are used as solvents for other nonpolar substances, as raw materials for the synthesis of other organic compounds, and as fuels. They are generally less dense than water and have melting and boiling points which tend to increase with molecular weight, although isomers which have more compact molecules (with weaker London forces) are generally lower melting and boiling than those with long extended chains.

The alkanes are generally quite unreactive to acids, bases, oxidizing agents, and reducing agents. However, they will react with powerful oxidizing agents such as O_2 (in combustion) and Cl_2. The cycloalkanes are similarly unreactive, except that cyclopropane (C_3H_6) and cyclobutane (C_4H_8) tend to undergo reactions in which the carbon-ring skeleton is opened. In these compounds the bond angles in the ring are not close to the ideal tetrahedral angle, and so orbital overlap is not very efficient, decreasing stability.

The saturated hydrocarbons are found in natural gas and petroleum. Natural gas consists mostly of methane with some ethane and other low-boiling hydrocarbons, as well. The decomposition of animal and plant organisms in swamps and bogs produces methane, which has been called *marsh gas*. It is also produced by the decomposition of household garbage and farm waste materials, and technologies are now being developed for exploiting these potential sources of a valuable fuel.

Reactions of the saturated hydrocarbons

Hydrocarbons are generally rather unreactive. They will burn in an excess of air or oxygen to produce water and carbon dioxide:

$$CH_4(g) + 2O_2(g) \longrightarrow CO_2(g) + 2H_2O(g) \qquad \Delta H^\circ = -802 \text{ kJ mol}^{-1}$$

Limiting the amount of oxygen produces carbon monoxide and less heat:

$$CH_4(g) + \tfrac{3}{2}O_2(g) \longrightarrow CO(g) + 2H_2O(g) \qquad \Delta H^\circ = -519 \text{ kJ mol}^{-1}$$

Using even less oxygen produces carbon in the form of soot or *lamp black:*

$$CH_4(g) + O_2(g) \longrightarrow C(s) + 2H_2O(g) \qquad \Delta H^\circ = -409 \text{ kJ mol}^{-1}$$

The alkanes and cycloalkanes will react with the halogens to produce mixtures of various *halogenated hydrocarbons*. For example, methane reacts with chlorine gas either at elevated temperatures or in the presence of light to form CH_3Cl, CH_2Cl_2, $CHCl_3$, and CCl_4.

The cycloalkanes will undergo certain ring-opening reactions. An example is the hydrogenation of cyclopropane in the presence of a nickel catalyst:

$$\triangle \quad + H_2 \xrightarrow{\text{Ni}} CH_3CH_2CH_3$$

Cyclopropane   Propane

23-2
Unsaturated hydrocarbons

The *unsaturated hydrocarbons* are hydrocarbons whose molecules have multiple bonds. These include the *alkenes* and *alkynes.*

The alkenes

Molecules which have one carbon-carbon double bond are called *alkenes* and like the cycloalkanes, have the general formula C_nH_{2n}. The simplest alkene is *ethene*

(IUPAC name), also called *ethylene* (common name):

$$\begin{array}{ccc} H & & H \\ \diagdown & & \diagup \\ & C{=}C & \\ \diagup & & \diagdown \\ H & & H \end{array}$$

Ethene (ethylene)

In ethene each carbon uses its s, p_x, and p_y orbitals to form three sp^2 (trigonal) hybrid orbitals as is shown in Fig. 23-2a. Two of these three orbitals overlap with $1s$ orbitals of hydrogens to form C—H σ bonds. The other sp^2 orbital overlaps with the corresponding orbital of the other carbon atom to form a C—C σ bond. This completes the *sigma skeleton* of the molecule, shown in Fig. 23-2b. The two unhybridized p_z orbitals overlap side-to-side (Fig. 23-2c) to form a π bond (Fig. 23-2d), thus completing the carbon-carbon double bond. (For clarity the sigma skeleton is not shown in Figs. 23-2c and d.)

The carbon-carbon double bonding in all alkenes is similar to that in ethene. The names and structures of a few alkenes are given in Table 23-4.

The alkenes are named by taking the name of the corresponding alkane, substituting the suffix *-ene* for *-ane,* and, when necessary, using a number to locate

Figure 23-2
Bonding in ethylene. (*a*) **Atoms approach.** (*b*) σ **skeleton.** (*c*) **Overlap of carbon** p_z **orbitals.** (*d*) π_z **bond.**

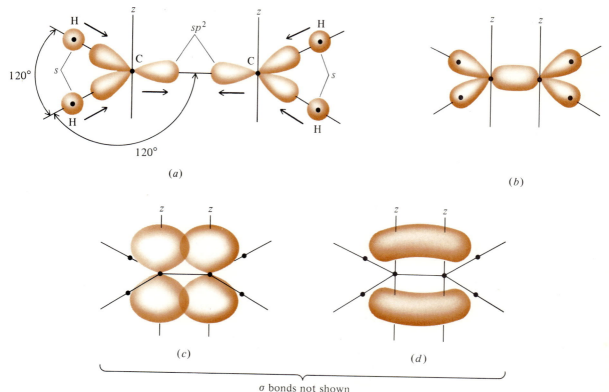

ORGANIC CHEMISTRY

Table 23-4 Some alkenes	Molecular formula	Structural formula	Name	Normal melting point, °C	Normal boiling point, °C
	C_2H_4	$CH_2\!\!=\!\!CH_2$	Ethene (IUPAC), ethylene (common)	−169	−104
	C_3H_6	$CH_2\!\!=\!\!CH\!\!-\!\!CH_3$	Propene (IUPAC), propylene (common)	−185	−48
	C_4H_8	$CH_2\!\!=\!\!CH\!\!-\!\!CH_2\!\!-\!\!CH_3$	1-Butene (IUPAC)	−130	−6
	C_4H_8	(cis structure)	cis-2-Butene (IUPAC)	−139	4
	C_4H_8	(trans structure)	trans-2-Butene (IUPAC)	−106	1
	C_4H_8	(2-methylpropene structure)	2-Methylpropene (IUPAC)	−140	−7

the *beginning* of the double bond. The carbons are numbered so that this number is as small as possible. Thus

$$CH_3\!\!-\!\!CH_2\!\!-\!\!CH_2\!\!-\!\!CH\!\!=\!\!CH_2$$

is 1-pentene, *not* 4-pentene.

● **EXAMPLE 23-5**

Problem   Name each of the following alkenes.

(a) $CH_3\!\!-\!\!\underset{\underset{CH_3}{|}}{CH}\!\!-\!\!CH\!\!=\!\!CH_2$ (b) $CH_3\!\!-\!\!\underset{\underset{CH_2-CH_3}{|}}{C}\!\!=\!\!CH\!\!-\!\!CH_3$

Solution   (a) 3-Methyl-1-butene. (b) 3-Methyl-2-pentene. (Note that the longest chain in this molecule is a *five*-carbon chain.) ●

Rotation about a double bond in an alkene is not possible because such a rotation destroys the overlap necessary to form the π bond. This means that cis-trans isomerism (Sec. 22-5) is possible. The cis (methyl groups adjacent) and trans (methyl groups opposite) forms of 2-butene are shown in Table 23-4.

The alkynes   Hydrocarbons with one *triple bond* per molecule are called *alkynes* and have the general formula C_nH_{2n-2}. The simplest alkyne is *ethyne* (IUPAC name), commonly called *acetylene*:

$$H\!\!-\!\!C\!\!\equiv\!\!C\!\!-\!\!H$$
Ethyne (acetylene)

The bonding in the alkynes involves the use of *sp* (linear) hybrid orbitals by each carbon to form a linear H—C—C—H sigma skeleton. If we assume the bond axis to be the x axis, then the unhybridized p_y and p_z orbitals of each carbon overlap respectively side-to-side to form *two* π bonds. The $\sigma + \pi_y + \pi_z$ bonding MOs constitute the triple bond.

Table 23-5 gives the names and structures of a few alkynes. The alkynes are named as the alkenes are, except that the suffix *-yne* is used, instead of *-ene*.

Table 23-5 Some alkynes

Molecular formula	Structural formula	Name	Normal melting point, °C	Normal boiling point, °C
C_2H_2	$HC{\equiv}CH$	Ethyne (IUPAC), acetylene (common)	-81	-84 (sublimes)
C_3H_4	$HC{\equiv}C-CH_3$	Propyne (IUPAC), methylacetylene (common)	-101	-23
C_4H_6	$HC{\equiv}C-CH_2-CH_3$	1-Butyne (IUPAC), ethylacetylene (common)	-126	8
C_4H_6	$CH_3-C{\equiv}C-CH_3$	2-Butyne (IUPAC), dimethylacetylene (common)	-32	27

Properties of the unsaturated hydrocarbons

The alkenes and alkynes are nonpolar compounds which, like the alkanes, have melting and boiling points which increase with molecular weight. Unlike the alkanes, however, these compounds are chemically quite reactive.

There are no important natural sources of unsaturated hydrocarbons. Alkenes are obtained industrially as one of the products of the refining of petroleum. Alkenes can be synthesized by means of *elimination reactions,* in which two atoms or groups on adjacent carbons in a saturated chain are simultaneously removed, leaving a double bond between the carbon atoms. Thus chloroethane, when heated with a solution of potassium hydroxide in ethanol, reacts to form ethene:

$$H-\underset{\underset{H}{|}}{\overset{\overset{H}{|}}{C}}-\underset{\underset{Cl}{|}}{\overset{\overset{H}{|}}{C}}-H + OH^- \longrightarrow \underset{H}{\overset{H}{>}}C=C\underset{H}{\overset{H}{<}} + Cl^- + H_2O$$

Chloroethane   Ethene

Ethyne (acetylene) can be prepared by the addition of water to calcium carbide:

$$CaC_2 + 2H_2O \longrightarrow HC{\equiv}CH + Ca^{2+} + 2OH^-$$

Reactions of the unsaturated hydrocarbons

An alkene generally owes its reactivity to the presence of its π bond. An *electrophile,* an atom or group of atoms which seeks electrons, can break this π bond. Thus bromine, Br_2, reacts with propene to form 1,2-dibromopropane:

$$Br_2 + CH_2{=}CH-CH_3 \longrightarrow \underset{\underset{Br}{|}}{CH_2}-\underset{\underset{Br}{|}}{CH}-CH_3$$

Propene   1,2-Dibromopropane

This reaction is called an *addition reaction,* and bromine is said to add to the double bond. Addition of hydrogen to an alkene yields an alkane:

$$H_2 + CH_2{=}CH-CH_3 \xrightarrow{\text{Ni}} CH_3-CH_2-CH_3$$
Propene   Propane

Hydrogen halides will also add to double bonds:

$$HCl + CH_2{=}CH-CH_3 \longrightarrow CH_3-\underset{\underset{Cl}{|}}{CH}-CH_3$$

Propene   2-Chloropropane

For such a reaction *Markovnikov's rule* states that the hydrogen adds to the carbon of the double bond which is already bonded to more hydrogen atoms. Thus the major product is 2-chloropropane, not 1-chloropropane.

Electrophilic addition to the triple bond of an alkyne is also possible, although a catalyst is frequently needed. Thus one mole of HBr adds to one mole of propyne to form 2-bromopropene:

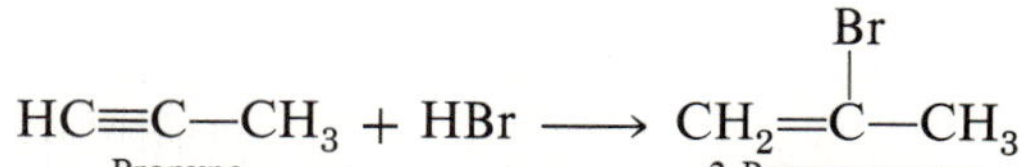

$$HC \equiv C-CH_3 + HBr \longrightarrow CH_2 = \underset{\underset{\displaystyle Br}{|}}{C} - CH_3$$

Propyne   2-Bromopropene

Addition of a second mole of HBr yields 2,2-dibromopropane:

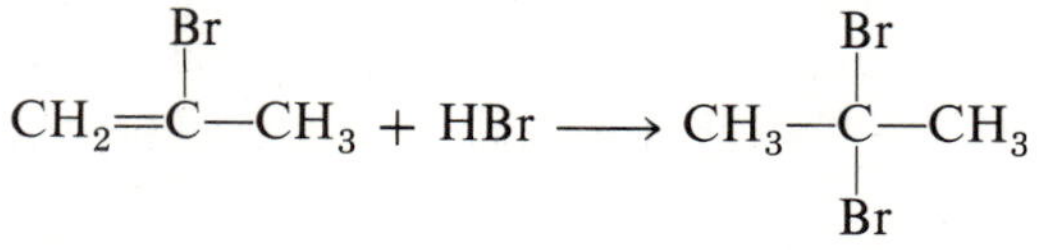

$$CH_2 = \underset{\underset{\displaystyle Br}{|}}{C} - CH_3 + HBr \longrightarrow CH_3 - \underset{\underset{\displaystyle Br}{|}}{\overset{\overset{\displaystyle Br}{|}}{C}} - CH_3$$

2-Bromopropene   2,2-Dibromopropane

Unsaturated hydrocarbons will burn in air or oxygen to form water and CO_2, CO, C, or a mixture of these, depending on the availability of O_2.

23-3
Aromatic
hydrocarbons

The *aromatic hydrocarbons* constitute a special type of unsaturated hydrocarbon.[1] The simplest example of an aromatic hydrocarbon is *benzene*.

Benzene

Benzene, C_6H_6, has a cyclic molecule which, according to valence-bond theory, is described as a resonance hybrid:

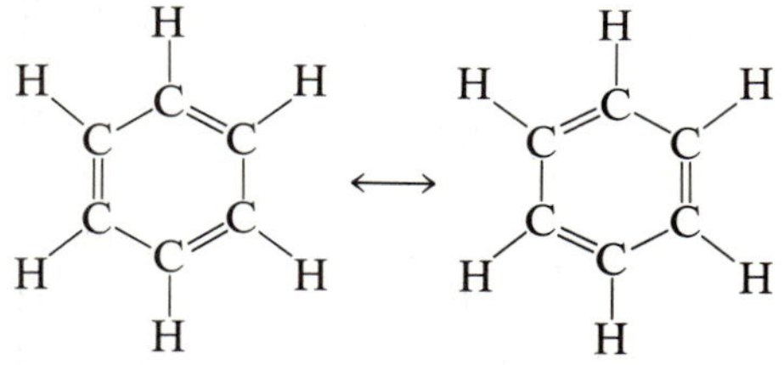

Since a resonance hybrid is a composite or average structure, each carbon-carbon bond is halfway between a single and a double bond.

Molecular-orbital theory describes the bonding in benzene as follows: Each carbon uses its s, p_x, and p_y orbitals to form three sp^2 hybrids. These are used to bond the carbon to one hydrogen and two other carbons, establishing the sigma skeleton of the molecule, shown in Fig. 23-3a. The remaining unhybridized p_z orbitals, one for each carbon, overlap side-to-side above and below the plane of the ring, as is shown in Fig. 23-3b. The result of the combination of the six p_z orbitals is three (bonding) π MOs shown in Fig. 23-3c, and three (antibonding) π^* MOs, not shown in the illustration. The bonding MOs are *delocalized* because they extend over several atoms. Now let us count electrons: there is a total of 30 valence electrons to be assigned. Twenty four of these go into the sigma skeleton

[1] The term *aromatic* was used originally because many of these compounds have pronounced odors.

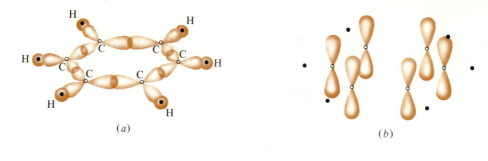

(a)

(b)

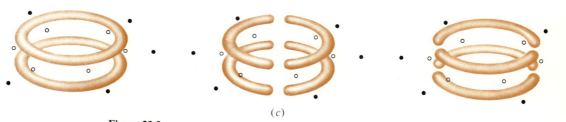

(c)

Figure 23-3
Bonding in benzene. (a) σ **skeleton.** (b) p_z **orbitals (lobes shrunk for clarity).** (c) **Delocalized bonding** π_z **orbitals.**

(two in each bond). The remaining six go into the three delocalized π orbitals. Since three filled orbitals for six bonds works out to one-half of a π bond per pair of carbon atoms, the bond order for each carbon-carbon bond is 1 (the σ bond) plus $\frac{1}{2}$ (the half-π bond), for a total of $\frac{3}{2}$. This is consistent with the description of the molecule made in terms of resonance (above).

The structural formula of benzene is often abbreviated as

and, because the π bonds are delocalized, it is often shown as

where the hexagon represents the sigma skeleton of the molecule, and the circle, the delocalized π electrons. All of the simpler aromatic hydrocarbons contain the benzene ring as part of their molecular structures.

Other aromatic hydrocarbons   The hydrogens on a benzene molecule may be replaced by alkyl groups. Replacement of one hydrogen by a methyl group gives *methylbenzene,* usually called *toluene:*

Methylbenzene (toluene)

Addition of two methyl groups to the benzene ring gives the possibility of three isomers. In the IUPAC system these are named with the numbers 1,2-; 1,3-; and 1,4-; respectively. A commonly used older system uses the prefixes *ortho-, meta-,* and *para-* (abbreviated *o-, m-,* and *p-*, respectively), together with the name *xylene*. The three isomeric dimethylbenzenes are

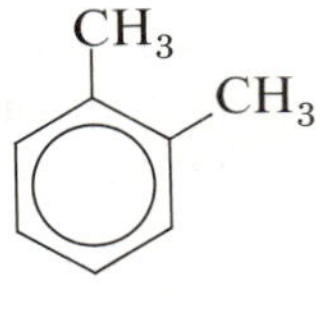

| 1,2-Dimethylbenzene | 1,3-Dimethylbenzene | 1,4-Dimethylbenzene |
| (*o*-xylene) | (*m*-xylene) | (*p*-xylene) |

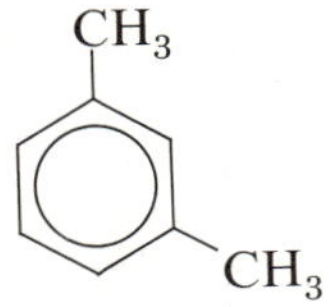

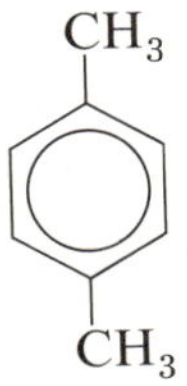

Another type of aromatic hydrocarbon has *condensed* benzene rings. The simplest example of such a compound is naphthalene, commonly used as mothballs:

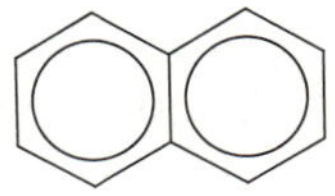

Another is 3,4-benzpyrene,

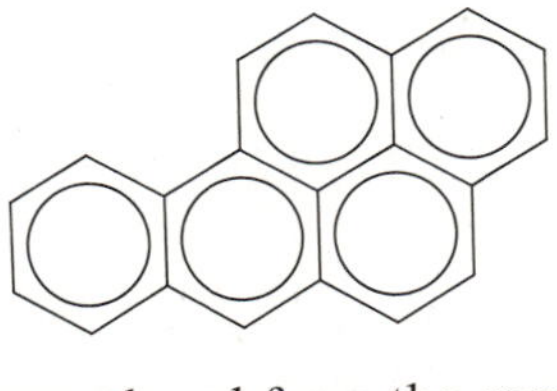

a potent carcinogen which is produced from the combustion of most plant materials, including wood, paper, and tobacco.

Sources and properties of the aromatic hydrocarbons

Mixtures of aromatic hydrocarbons can be obtained from coal tar and separated by fractional distillation. Today most of the benzene, toluene, and xylenes are obtained from alkane hydrocarbons in petroleum. Thus, hexane is heated with platinum as a catalyst at 400°C to produce benzene:

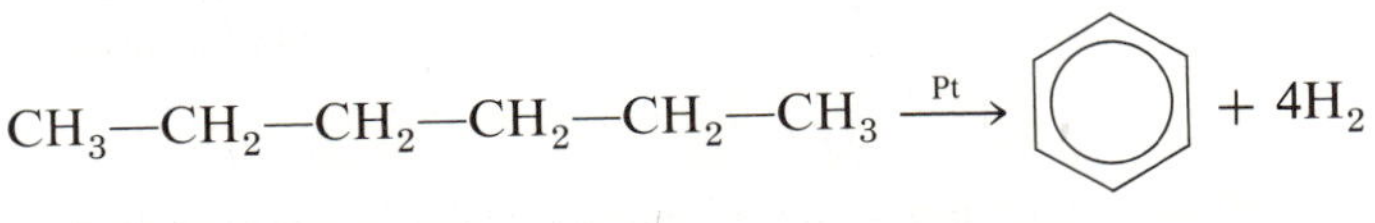

$$CH_3-CH_2-CH_2-CH_2-CH_2-CH_3 \xrightarrow{\text{Pt}} \quad + \quad 4H_2$$

Hexane   Benzene

This process is an example of *catalytic reforming* or *platforming*.

The aromatic hydrocarbons are compounds of low polarity and are toxic to varying degrees. Benzene is a volatile liquid (normal boiling point, 80°C) which is toxic and carcinogenic. Toluene and the xylenes are apparently not carcinogenic but are still quite toxic. Special care must be taken to avoid breathing the vapors of benzene and its derivatives.

The aromatic hydrocarbons, being highly unsaturated, might be expected to be quite reactive. In fact, they are not. For example, although Br_2 undergoes addition reactions with alkenes and alkynes, it will not undergo a similar reaction with benzene. It will, however, undergo a substitution reaction, but only if iron is present as a catalyst (see below). Benzene and other aromatic hydrocarbons burn in air with a very sooty, luminous flame.

Reactions of the aromatic hydrocarbons

In spite of their low reactivities, the aromatic hydrocarbons can be induced to undergo a variety of reactions if in each case the appropriate catalyst is present. Benzene can undergo a *substitution reaction* with bromine in the presence of an iron catalyst:

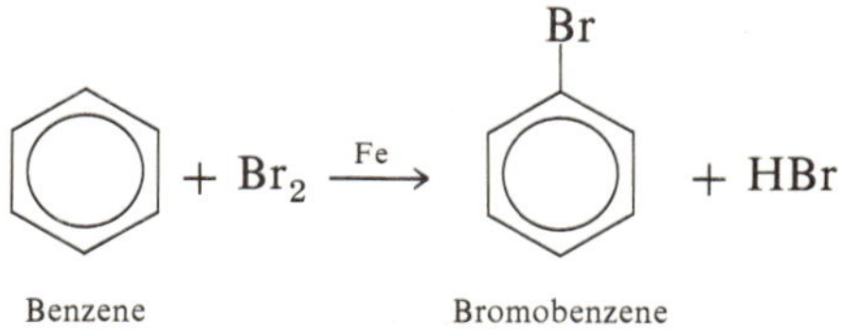

Benzene   Bromobenzene

Further reaction with bromine yields isomeric dibromobenzenes. These reactions are called *brominations*. *Chlorinations* are similar.

In a *nitration* reaction the nitro ($-NO_2$) group is substituted for a hydrogen on the benzene ring. With benzene the reaction is

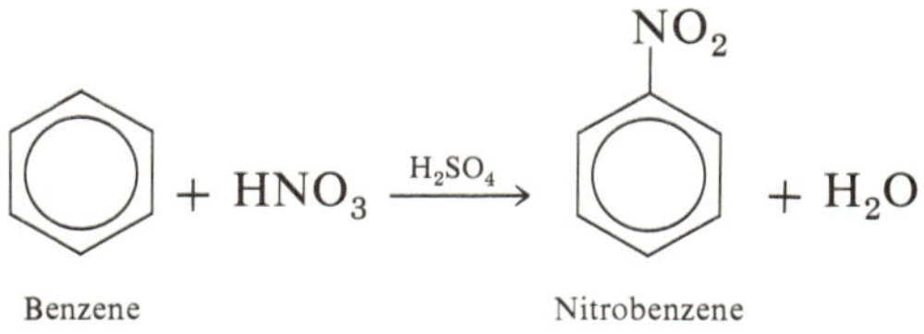

Benzene   Nitrobenzene

When toluene undergoes extensive nitration, the product is 2,4,6-trinitrotoluene, known as TNT.

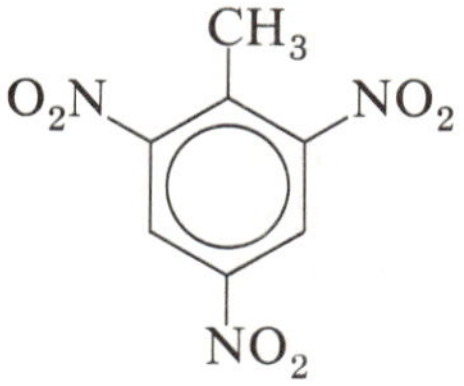

2,4,6-Trinitrotoluene

23-4 Functional groups

The molecules of many organic compounds can be considered to be hydrocarbons which have had one or more of their hydrogen atoms replaced by a new atom or group of atoms. These molecules are referred to as hydrocarbon *derivatives,* and

Functional group	General formula	Class of compound
—OH	ROH	Alcohol
—O—	ROR′	Ether*
$\overset{\text{O}}{\underset{\|}{-}}\overset{\|}{\text{C}}-\text{H}$	RCHO	Aldehyde†
$-\overset{\text{O}}{\overset{\|}{\text{C}}}-$	RCOR′	Ketone*
$-\overset{\text{O}}{\overset{\|}{\text{C}}}-\text{OH}$	RCOOH	Carboxylic acid†
$-\overset{\text{O}}{\overset{\|}{\text{C}}}-\text{O}-$	RCOOR′	Ester*,†
$\overset{\text{H}}{\underset{}{-}}\text{NH}$	RNH$_2$	Amine (primary)

Table 23-6
Some functional groups

*In these compounds the two hydrocarbon residues may be different.
†In these compounds the R bonded to the carbon may represent a hydrogen atom.

the replacing atoms or groups are called *functional groups*. Thus, if one hydrogen on a molecule of ethane,

$$CH_3CH_3$$

is replaced by an —OH functional group,

$$CH_3CH_2OH$$

the compound which results is *ethanol*. Each functional group gives a characteristic property or function to a molecule. The —OH group is called the *alcohol* functional group, and each organic molecule which incorporates it exhibits properties which are characteristic of *alcohols,* an important class of compounds. Thus the general formula for an alcohol can be written as ROH, where R represents the hydrocarbon part, or *residue,* of the molecule.

Table 23-6 shows some important functional groups.

23-5 Alcohols

In the molecule of an alcohol the —OH group has replaced a hydrogen atom of an alkane.

Nomenclature

The IUPAC name of an alcohol is obtained by replacing the -*e* at the end of the name of the parent alkane by the suffix -*ol*. Thus CH_3OH is methanol and CH_3CH_2OH is ethanol. Numbers are used when it is necessary to locate the position of the —OH group. For example, $CH_3CH_2CH_2CH_2OH$ is 1-butanol, and $CH_3CH_2CHOHCH_3$ is 2-butanol.

Alcohols are frequently given common names. This is done by adding the word *alcohol* to the name of the alkyl group to which the —OH is attached. Thus, CH_3OH is methyl alcohol, and $CH_3CH_2CH_2CH_2OH$ is *n*-butyl alcohol. Table 23-7 gives the names and structures of some alcohols.

Properties

The boiling points of the alcohols are considerably higher than those of their parent alkanes. (Compare Tables 23-1 and 23-7.) This is a result of the high

Table 23-7
Some alcohols

Condensed structural formula	IUPAC name	Common name	Normal melting point, °C	Normal boiling point, °C
CH_3OH	Methanol	Methyl alcohol	−98	65
CH_3CH_2OH	Ethanol	Ethyl alcohol	−114	78
$CH_3CH_2CH_2OH$	1-Propanol	*n*-Propyl alcohol	−127	97
$CH_3CHOHCH_3$	2-Propanol	Isopropyl alcohol	−90	82
$CH_3CH_2CH_2CH_2OH$	1-Butanol	*n*-Butyl alcohol	−90	117
$CH_3CH_2CHOHCH_3$	2-Butanol	*sec*-Butyl alcohol	−115	100
$(CH_3)_2CHCH_2OH$	2-Methyl-1-propanol	Isobutyl alcohol	−108	108
$(CH_3)_3COH$	2-Methyl-2-propanol	*tert*-Butyl alcohol	26	82

degree of association in the liquid brought about by hydrogen bonding (Sec. 12-3). Hydrogen bonding also accounts for the solubilities of the lower alcohols in water. Methane, ethane, and propane have low solubilities in water, but methanol, ethanol, and 1-propanol are infinitely soluble (they are completely miscible with water), a result of the strong alcohol-water hydrogen bonds in the solutions. But the solubilities of the alcohols decrease as the hydrocarbon chain length increases, as is shown in Table 23-8. When the hydrocarbon part of an alcohol is large, the formation of alcohol-water hydrogen bonds can not compensate for the water-water hydrogen bonds which must be broken to make room for the alcohol molecule during the dissolving process.

Methanol, CH_3OH, is sometimes called *wood alcohol* because it is one of the volatile products of the destructive distillation of wood (heating wood in the absence of air). Today most methanol is produced by the catalytic hydrogenation of carbon monoxide:

$$CO(g) + 2H_2(g) \longrightarrow CH_3OH(g)$$

Methanol is useful as a solvent and as a starting material for many organic syntheses. It is very toxic.

Ethanol, CH_3CH_2OH, also called *grain alcohol,* is produced in large quantities by the fermentation industries. The natural fermentation of carbohydrates in grains, fruits, and berries produces ethanol, which is separated from the reaction mixture by fractional distillation. The reaction is catalyzed by the enzyme *zymase* in yeast:

$$\underset{\text{Glucose}}{C_6H_{12}O_6} \xrightarrow{\text{zymase}} \underset{\text{Ethanol}}{2CH_3CH_2OH} + 2CO_2$$

Much ethanol goes into the production of alcoholic beverages. It is commonly called, simply, "alcohol," and is the least toxic of the alcohols. Much ethanol is

Table 23-8
Solubilities of some alcohols in water

Alcohol	Formula	Solubility at 25°C, M
Methanol	CH_3OH	∞
Ethanol	CH_3CH_2OH	∞
1-Propanol	$CH_3(CH_2)_2OH$	∞
1-Butanol	$CH_3(CH_2)_3OH$	1.1
1-Pentanol	$CH_3(CH_2)_4OH$	0.25
1-Hexanol	$CH_3(CH_2)_5OH$	0.01
1-Heptanol	$CH_3(CH_2)_6OH$	0.0008

also produced by the reaction of sulfuric acid with ethene,

$$CH_2{=}CH_2 + H_2SO_4 \longrightarrow CH_3CH_2OSO_3H$$

Ethene   Ethyl hydrogen sulfate

followed by reaction with water:

$$CH_3CH_2OSO_3H + H_2O \longrightarrow CH_3CH_2OH + HSO_4^- + H^+$$

Ethanol

Ethanol is a useful solvent and raw material for organic syntheses. It has been used for many years in Europe and more recently in the United States as an extender and octane booster for gasoline in a mixture known as *gasohol*.

Polyols are alcohols which contain more than one —OH group per molecule. An example of a *diol* is 1,2-ethanediol, also called *ethylene glycol,*

$$\begin{array}{cc} CH_2{-}CH_2 \\ | \quad\;\; | \\ OH \quad OH \end{array}$$

which is used in automotive antifreeze. Glycerol (glycerin) is a *triol,* 1,2,3-propanetriol,

$$\begin{array}{ccc} CH_2{-}CH{-}CH_2 \\ | \quad\;\; | \quad\;\; | \\ OH \quad OH \quad OH \end{array}$$

It is a byproduct from soap manufacture, is used in cosmetic and medicinal creams, ointments, and lotions, and is added to tobacco products to keep them moist.

Reactions   Alcohols can be oxidized by numerous oxidizing agents. Ethanol reacts in an acidified solution of potassium chromate to form a compound called an *aldehyde*. The balanced equation is

$$3CH_3{-}CH_2{-}OH + Cr_2O_7{}^{2-} + 8H^+ \longrightarrow 3CH_3{-}C{\overset{O}{\underset{H}{\big\langle}}} + 2Cr^{3+} + 7H_2O$$

Ethanol   Ethanal (acetaldehyde)

Organic reactions are often represented by abbreviated equations in which only the major organic reactants and products are shown and other reactants are written over the arrow. The oxidation of ethanol can thus be shown as

$$CH_3{-}CH_2{-}OH \xrightarrow{K_2Cr_2O_7} CH_3{-}C{\overset{O}{\underset{H}{\big\langle}}}$$

Ethanol   Ethanal

Notice that this oxidation accomplishes a *dehydrogenation,* removal of hydrogen atoms from the molecule. The product aldehyde can be further oxidized to form a carboxylic acid. (See Sec. 23-9.)

Ethanol is called a *primary alcohol,* because the —OH group is bonded to a carbon with at least two hydrogens attached. Primary alcohols upon oxidation yield aldehydes:

$$R-CH_2-OH \xrightarrow{\text{oxidation}} R-C{\overset{\displaystyle O}{\underset{\displaystyle H}{\Big\langle}}}$$

Primary alcohol   Aldehyde

A *secondary alcohol* has only one H atom attached to the carbon to which the —OH group is bonded. An example is 2-propanol, $CH_3CHOHCH_3$. Secondary alcohols can be oxidized to form *ketones:*

$$\overset{\displaystyle OH}{\underset{}{R-CH-R'}} \xrightarrow{\text{oxidation}} \overset{\displaystyle O}{\underset{}{R-C-R'}}$$

Secondary alcohol   Ketone

In a *tertiary alcohol* the —OH functional group is bonded to a carbon to which no H atoms are attached. Tertiary alcohols are not readily oxidized.

Alcohols react with hydrogen halides in reactions known as *nucleophilic substitutions.* We have seen that an electrophile is an electron-seeking atom or group. A *nucleophile* is an electron-rich atom or group, such as a Lewis base (Sec. 13-1), which seeks to share its electrons with a relatively positive nucleus. When methanol reacts with hydrogen bromide the products are methyl bromide and water:

$$\underset{\text{Methanol}}{CH_3-OH} + HBr \longrightarrow \underset{\text{Methyl bromide}}{CH_3-Br} + H_2O$$

The product, methyl bromide is an example of an *alkyl halide.* The reaction proceeds by the formation of an intermediate, $CH_3-OH_2{}^+$,

$$CH_3-OH + HBr \longrightarrow \left[CH_3-\overset{\displaystyle H}{\underset{}{O}}-H \right]^+ + Br^-$$

Then the Br^- nucleophile substitutes for the H_2O nucleophile to produce the product:

$$\left[CH_3-\overset{\displaystyle H}{\underset{}{O}}-H \right]^+ + Br^- \longrightarrow CH_3-Br + H_2O$$

At elevated temperatures and in the presence of sulfuric acid an alcohol will undergo an *elimination reaction* in which a molecule of water is eliminated and an alkene is formed:

$$\underset{\text{Ethanol}}{CH_3CH_2OH} \xrightarrow[200°C]{H_2SO_4} \underset{\text{Ethene}}{CH_2{=}CH_2} + H_2O$$

This is an example of a *dehydration.* When the reaction is run at lower temperatures an *intermolecular dehydration* occurs, as one molecule of water is lost by two alcohol molecules and an *ether* is formed.

$$CH_3CH_2OH + HOCH_2CH_3 \xrightarrow[135°C]{H_2SO_4} CH_3CH_2\!-\!O\!-\!CH_2CH_3$$

Ethyl ether

23-6
Ethers

Ethers are compounds of the general type R—O—R′, where the two hydrocarbon residues are not necessarily the same.

Nomenclature

Ethers are not usually named by using the IUPAC system. More often they are named by specifying the hydrocarbon groups which are bonded to the oxygen and adding the word *ether*. When the two groups are the same, their name is used, often without the prefix *di-*. For example, when the two hydrocarbon residues are methyl groups, the compound is named methyl ether:

$$CH_3\!-\!O\!-\!CH_3$$
Methyl ether

Table 23-9 lists several common ethers.

Properties

The boiling points of the ethers are much lower than those of alcohols of comparable molecular weight. They are about as low as those of comparable hydrocarbons. Ethers are more soluble in water than are comparable hydrocarbons, largely as a result of water-ether hydrogen bonding. Ethers are good solvents for a wide variety of organic compounds. Methyl ether, a gas at room temperature, has been used as a refrigerant. Ethyl ether is an extremely volatile liquid, which because of its flammability must be handled with great care. Its vapor is much more dense than air, and so it can flow from an open beaker along the surface of a laboratory bench to an open flame some distance away. Ethyl ether is the "ether" used as a general anesthetic. Methyl propyl ether is also used for this purpose and is less irritating to the nose, throat, and lungs.

Reactions

The ethers are quite unreactive. They will, of course, burn in air to form CO_2 and H_2O. An ether can be *cleaved* by reaction with a strong halogen acid. If an excess of acid is used, the products are organic halides.

$$\underset{\text{Methyl ethyl ether}}{CH_3\!-\!O\!-\!CH_2CH_3} + 2HI \longrightarrow \underset{\substack{\text{Methyl}\\\text{iodide}}}{CH_3I} + \underset{\substack{\text{Ethyl}\\\text{iodide}}}{CH_3CH_2I} + H_2O$$

It is dangerous to store opened bottles of ether for long periods of time, because they react with oxygen to form small quantities of highly explosive peroxides. Since these tend to become concentrated by evaporation of the ether, an old bottle still containing a small amount of ether is potentially very dangerous to handle.

Table 23-9
Some ethers

Condensed structural formula	IUPAC name	Common name	Normal melting point, °C	Normal boiling point, °C
CH_3OCH_3	Methoxymethane	Methyl ether	−142	−25
$CH_3CH_2OCH_2CH_3$	Ethoxyethane	Ethyl ether	−116	35
$CH_3CH_2OCH_3$	Methoxyethane	Ethyl methyl ether	—	11
$CH_3CH_2CH_2OCH_3$	1-Methoxypropane	Methyl propyl ether	—	39

23-7
Aldehydes

We have seen (Sec. 23-5) that the oxidation of a primary alcohol yields an aldelyde:

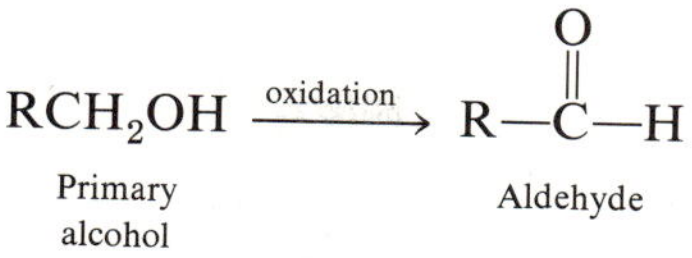

$$\underset{\substack{\text{Primary} \\ \text{alcohol}}}{RCH_2OH} \xrightarrow{\text{oxidation}} \underset{\text{Aldehyde}}{R\overset{\displaystyle O}{\overset{\|}{-C}}-H}$$

An aldehyde has one hydrocarbon residue and one hydrogen atom attached to the

$-\overset{\displaystyle O}{\overset{\|}{C}}-$, or *carbonyl* group. (*Exception:* Methanal, HCHO, commonly called *formaldehyde,* has two hydrogens bonded to the carbonyl carbon.)

Nomenclature

Using IUPAC nomenclature an aldehyde is named by substituting the suffix *-al* for *-e* at the end of the name of the parent hydrocarbon. Thus HCHO is methanal, CH_3CHO is ethanal, etc. Aldehydes are frequently known by their common names, all of which contain the suffix *-aldehyde,* and which are shown for some common aldehydes in Table 23-10.

Properties

The boiling points of the aldehydes are higher than those of the hydrocarbons and ethers of comparable molecular weight. This is a direct consequence of the higher polarity produced by the presence of the carbonyl group in an aldehyde. But since the carbonyl oxygen has no attached hydrogen atoms, hydrogen bonding is not possible in the liquid, and the boiling points of the aldehydes are lower than those of comparable alcohols. On the other hand, the solubilities of the aldehydes in water are comparable to those of the alcohols because of aldehyde-water hydrogen bonding in solution:

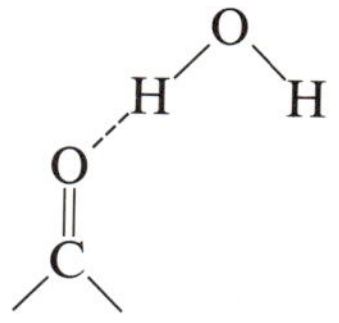

Although aldehydes with only a few carbons have odors which are somewhat unpleasant (acetaldehyde is present in photochemical smog), certain higher-molecular-weight aldehydes are used in the manufacture of perfumes and fragrances for cosmetics, soaps, etc. Vanillin (accent on the first syllable) is a component of natural vanilla flavoring and is the major constituent of artificial vanilla.

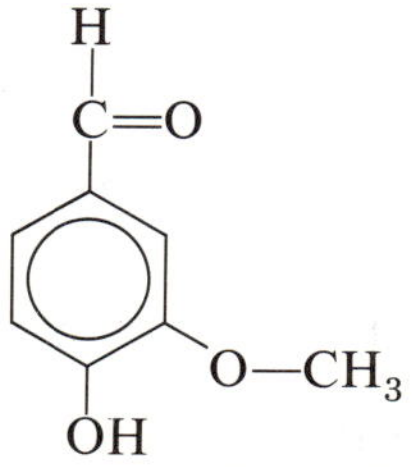

Vanillin

Table 23-10
Some aldehydes

Condensed structural formula	IUPAC name	Common name	Normal melting point, °C	Normal boiling point, °C
HCHO	Methanal	Formaldehyde	-117	-19
CH_3CHO	Ethanal	Acetaldehyde	-123	21
CH_3CH_2CHO	Propanal	Propionaldehyde	-81	49

Citral has a strong lemon-like odor and is used as a flavoring and in "citrusy" fragrances.

$$
\underset{\substack{\text{3,7-Dimethyl-2,6-octadienal}\\\text{(citral)}}}{\overset{\displaystyle CH_3}{\underset{\displaystyle CH_3}{C}}=CHCH_2CH_2-\overset{\displaystyle O}{\underset{\displaystyle CH_3}{C}}=CH\overset{\displaystyle\parallel}{C}-H}
$$

Reactions   The highly polar carbonyl group makes aldehydes quite reactive. As has been described in Sec. 23-5, alcohols can be *oxidized* to aldehydes using fairly mild conditions. The resulting aldehyde can be further oxidized to a *carboxylic acid*.

$$
CH_3-\overset{O}{\overset{\parallel}{C}}-H \xrightarrow{\text{KMnO}_4} CH_3-\overset{O}{\overset{\parallel}{C}}-OH
$$

$$
\begin{array}{cc}
\text{Ethanal} & \text{Ethanoic acid} \\
\text{(acetaldehyde)} & \text{(acetic acid)}
\end{array}
$$

The relative ease of oxidation of aldehydes can be used to distinguish them from another class of carbonyl compounds, the ketones (Sec. 23-8). One way of doing this is by using *Tollens' reagent,* a solution containing the diamminesilver(I) ion, $Ag(NH_3)_2{}^+$. When this ion oxidizes an aldehyde, it becomes reduced to silver metal, which is deposited on clean glass surfaces as a silver mirror. (Silver mirrors are produced commercially by this reaction.) Ketones will not undergo similar reactions.

$$
H-\overset{O}{\overset{\parallel}{C}}-H + 2Ag(NH_3)_2{}^+ + 2OH^- \longrightarrow
$$

Methanal
(formaldehyde)

$$
\left[H-\overset{O}{\overset{\parallel}{C}}-O \right]^- + 2Ag(s) + NH_4{}^+ + 3NH_3 + H_2O
$$

Methanoate
(formate) ion

Aldehydes will undergo *nucleophilic addition.* Hydrogen cyanide, for example, reacts with ethanal (acetaldehyde) to form a *cyanohydrin:*

$$
CH_3-\overset{O}{\overset{\parallel}{C}}-H \xrightarrow{\text{HCN}} CH_3-\overset{\displaystyle OH}{\underset{\substack{\displaystyle C\\\displaystyle\parallel\\\displaystyle N}}{C}}-H
$$

$$
\begin{array}{cc}
\text{Ethanal} & \text{A cyanohydrin} \\
\text{(acetaldehyde)} &
\end{array}
$$

The reaction takes place in two steps: First the cyanide ion, a nucleophile, is attracted to the carbonyl carbon:

$$CH_3-\overset{\overset{O}{\|}}{C}-H + [C{\equiv}N]^- \longrightarrow \left[CH_3-\overset{\overset{O}{\|}}{\underset{\underset{\underset{N}{\|\|}}{C}}{C}}-H \right]^-$$

This is followed by the addition of a proton to the oxygen:

$$\left[CH_3-\overset{\overset{O}{\|}}{\underset{\underset{\underset{N}{\|\|}}{C}}{C}}-H \right]^- + H^+ \longrightarrow CH_3-\overset{\overset{OH}{|}}{\underset{\underset{\underset{N}{\|\|}}{C}}{C}}-H$$

Cyanohydrins are useful intermediates in organic syntheses.

23-8 Ketones

As was described in Sec. 23-5, the oxidation of a secondary alcohol yields a ketone:

$$R-\overset{\overset{OH}{|}}{C}H-R \xrightarrow{\text{oxidation}} R-\overset{\overset{O}{\|}}{C}-R$$

Ketones have a carbonyl group bonded to two hydrocarbon residues, which may or may not be the same.

Nomenclature   Using IUPAC nomenclature a ketone is named by adding the suffix *-one* to the stem of the name of the parent alkane. A locator number for the carbonyl group is specified if more than four carbons are present. The common name of a ketone is composed of the names of the two groups bonded to the carbonyl group followed by the word *ketone*. Some examples are given in Table 23-11.

Properties   Because ketone molecules contain the carbonyl group, their volatilities and solubilities in water are similar to those of aldehydes of comparable molecular weight. Many ketones are found in nature as products or intermediates in plant and

Table 23-11
Some ketones

Condensed structural formula	IUPAC name	Common name	Normal melting point, °C	Normal boiling point, °C
CH_3COCH_3	Propanone	Dimethyl ketone (acetone)	-95	56
$CH_3CH_2COCH_3$	Butanone	Ethyl methyl ketone (MEK)	-86	80
$CH_3CH_2COCH_2CH_3$	3-Pentanone	Diethyl ketone	-42	102

706 ORGANIC CHEMISTRY

animal metabolism. Many have pleasant fragrances. *Camphor* is a ketone which is obtained from the camphor tree of the Orient and South America, and which is used for medicinal purposes.

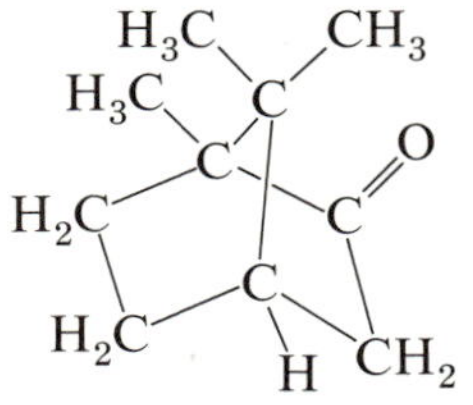

Camphor

Muscone is a ketone obtained from the musk deer in the Himalayas and used in perfume formulations.

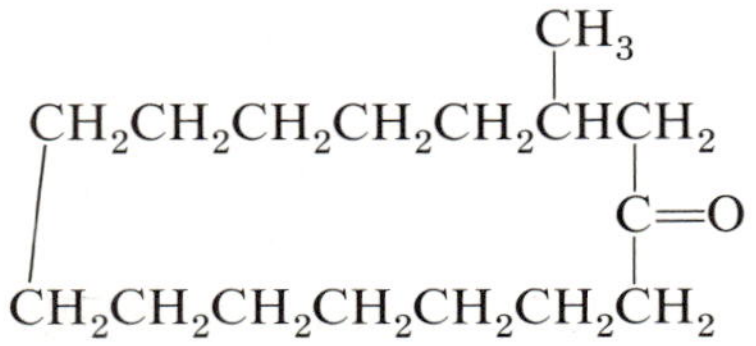

3-Methylcyclopentadecanone (muscone)

Reactions Ketones are not as easily oxidized as aldehydes and can be distinguished from them in the laboratory as described above (Sec. 23-7). Like the aldehydes, they will undergo *nucleophilic* addition. Addition of a *Grignard* (pronounced green'-yard) *reagent* to a ketone produces a tertiary alcohol. A Grignard reagent is prepared by reacting an organic halide with magnesium metal in the absence of water:

$$RBr + Mg \xrightarrow{ether} RMgBr$$

Alkyl halide   Grignard reagent

This reacts with a ketone

$$R'-\overset{O}{\underset{}{C}}-R'' + RMgBr \xrightarrow{ether} R'-\overset{OMgBr}{\underset{R}{C}}-R''$$

and water is added to form the tertiary alcohol

$$R'-\overset{OMgBr}{\underset{R}{C}}-R'' \xrightarrow{H_2O} R'-\overset{OH}{\underset{R}{C}}-R'' + Mg^{2+} + OH^- + Br^-$$

(Grignard reagents react with methanal to form a primary alcohol and with other aldehydes to form secondary alcohols.)

23-9
Carboxylic acids

We saw in Sec. 23-7 that oxidation of an aldehyde yields a carboxylic acid:

$$R-\overset{\overset{\displaystyle O}{\|}}{C}-H \xrightarrow{\text{oxidation}} R-\overset{\overset{\displaystyle O}{\|}}{C}-OH$$

Aldehyde   Carboxylic acid

These acids have the $-\overset{\overset{\displaystyle O}{\|}}{C}-OH$, functional group, called the *carboxyl* group.

Nomenclature

The IUPAC names of the carboxylic acids which are derived from alkanes are formed by adding the suffix *-oic* to the stem of the name of the parent alkane and adding the word *acid*. For example, CH_3COOH is named *ethanoic acid*. However, most of the carboxylic acids are widely known by common names. Thus ethanoic acid is almost universally called *acetic acid*. The IUPAC and common names of some carboxylic acids are given in Table 23-12. Note that oxalic acid is a *dicarboxylic acid* and is therefore diprotic.

Properties

In the liquid state the carboxylic acids are highly associated because of hydrogen bonding. This accounts for their comparatively high boiling points, higher even than those of alcohols of comparable molecular weight. In both liquid and gaseous states they are partially associated in dimers held together by two hydrogen bonds:

$$R-C\overset{\displaystyle O\text{---}H-O}{\underset{\displaystyle O-H\text{---}O}{}}C-R$$

The solubilities of the shorter-chain acids in water are high; one- through four-carbon acids are completely miscible with water. The longer-chain acids are less soluble because of the increased hydrocarbon portion of the molecule. The dissociation constants of most of the monocarboxylic acids are about 10^{-5}. The substitution of electronegative atoms for hydrogens on the alkyl chain withdraws

Table 23-12
Some carboxylic acids

Condensed structural formula	IUPAC name	Common name	Normal melting point, °C	Normal boiling point, °C	pK_a (25°C)
HCOOH	Methanoic acid	Formic acid	8	101	3.77
CH_3COOH	Ethanoic acid	Acetic acid	17	118	4.74
CH_3CH_2COOH	Propanoic acid	Propionic acid	−22	141	4.88
$CH_3CH_2CH_2COOH$	Butanoic acid	Butyric acid	−5	163	4.82
$CH_3(CH_2)_{16}COOH$	Octadecanoic acid	Stearic acid	69	383	4.85
HOOC—COOH	Ethanedioic acid	Oxalic acid	189	—	{1.27 4.28

electrons from the O—H bond making the acid stronger, as is shown by the following dissociation constants:

$$CH_3COOH \qquad K_a = 1.8 \times 10^{-5}$$

$$CH_2ClCOOH \qquad K_a = 1.4 \times 10^{-3}$$

$$CHCl_2COOH \qquad K_a = 5.6 \times 10^{-2}$$

$$CCl_3COOH \qquad K_a = 2.3 \times 10^{-1}$$

Carboxylic acids are found almost everywhere. Methanoic (formic) acid is present in the venom of red ants. Ethanoic (acetic) acid is the major component of vinegar. These have somewhat sharp, irritating odors, and as the alkyl chain lengthens, the acids develop even more unpleasant odors. Butanoic (butyric) acid smells like (and is present in) rancid butter. Pentanoic (valeric) acid has been described as smelling like old, dirty socks.

Lactic acid is *difunctional,* having both a carboxyl group and a hydroxyl group in its molecule:

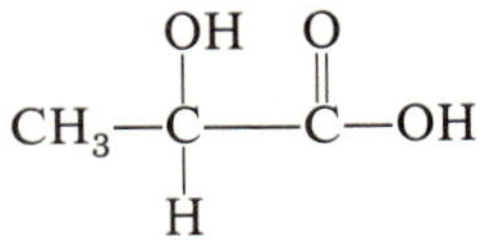

Lactic acid

It is found in sour milk and is a metabolic product of muscle activity. Note that lactic acid has an *asymmetric carbon atom* (Sec. 22-5). This means that lactic acid can exist as a pair of *enantiomers,* one of which, the (+) form, will rotate the plane of polarized light to the right, and the other of which, the (−) form, rotates it to the left. Only the (+) form is produced in muscle metabolism. (See Sec. 23-13.)

Reactions   Carboxylic acids can be *neutralized* by inorganic bases in aqueous solution:

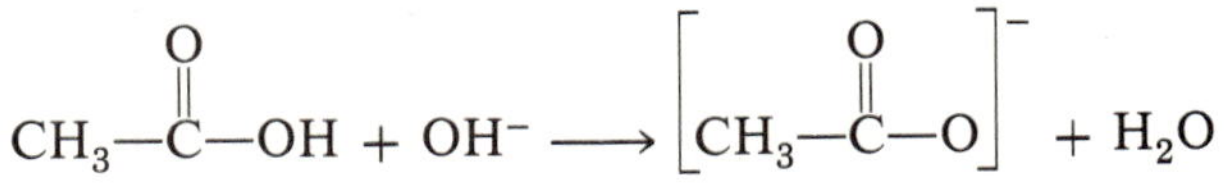

Ethanoic acid   Ethanoate ion
(acetic acid)   (acetate ion)

In such reactions the O—H bond of the carboxyl group is broken. The anion formed, a *carboxylate ion,* has a structure which can be described as a resonance hybrid:

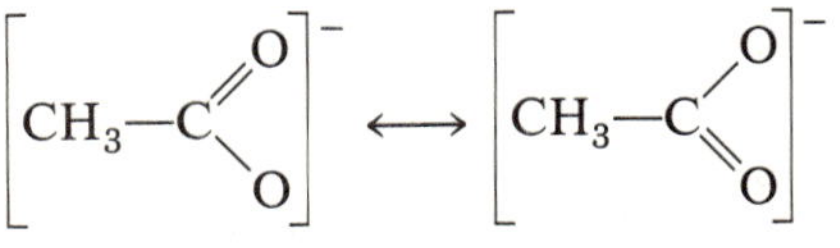

The C—O bond of the carboxyl group is broken in reactions known as *esterifications.* These are acid-catalyzed reactions of a carboxylic acid with an alcohol:

$$CH_3CH_2-\overset{\overset{\displaystyle O}{\|}}{C}-\boxed{OH + H}OCH_2CH_3 \xrightarrow{H^+} CH_3CH_2-\overset{\overset{\displaystyle O}{\|}}{C}-O-CH_2CH_3 + H_2O$$

Propanoic acid   Ethanol   Ethyl propanoate

(propionic acid)   (ethyl alcohol)   (ethyl propionate)

The product is called an *ester* (Sec. 23-10). Evidence that the C—O bond in the acid is broken in esterifications comes from experiments using isotopically labeled reactants. If some of the ethanol used in the above reaction contains ^{18}O, the atoms of this isotope are all found to end up in the product ester, not the water.

23-10
Esters

Esters have the general formula $R-\overset{\overset{\displaystyle O}{\|}}{C}-O-R'$ and can be synthesized, as has just been described, from the acid-catalyzed reaction of an alcohol with a carboxylic acid.

Nomenclature

Esters are named by specifying the *alkyl group* bonded to the oxygen followed by the name of the *carboxylate group*. Either IUPAC or common names may be used, but they should not be mixed. For example,

$$CH_3-\overset{\overset{\displaystyle O}{\|}}{C}-O-CH_2CH_2CH_3$$

Ethanoate   Propyl

(acetate)   (*n*-propyl)

group   group

is called either propyl ethanoate (IUPAC name) or *n*-propyl acetate (common name). A few esters are listed in Table 23-13.

Properties and uses

The lower-molecular-weight esters are soluble in water as a result of hydrogen bonding between water and the carbonyl oxygen. As molecular weight increases, the solubilities decrease, however.

Esters are found widely distributed in nature. Many are responsible for natural flowery and fruity odors and flavors. Pentyl ethanoate (*n*-amyl acetate) smells like bananas; 2-methylpropyl methanoate (isobutyl formate) smells and tastes like raspberries; and pentyl propanoate (*n*-amyl propionate) tastes like apricots.

Animal and vegetable fats are esters of fatty acids and the triol glycerol (Sec. 23-5). For example the triester formed from stearic acid and glycerol is called *glyceryl tristearate*, commonly known as *tristearin*:

Table 23-13 Some esters	Condensed structural formula	IUPAC name	Common name	Normal melting point, °C	Normal boiling point, °C
	$HCOOCH_3$	Methyl methanoate	Methyl formate	−99	32
	CH_3COOCH_3	Methyl ethanoate	Methyl acetate	−98	57
	$CH_3COOCH_2CH_3$	Ethyl ethanoate	Ethyl acetate	−84	77
	$CH_3CH_2COOCH_3$	Methyl propanoate	Methyl propionate	−88	80

$$CH_3(CH_2)_{16}-\overset{\overset{\displaystyle O}{\|}}{C}-O-CH_2$$

$$CH_3(CH_2)_{16}-\overset{\overset{\displaystyle O}{\|}}{C}-O-CH$$

$$CH_3(CH_2)_{16}-\overset{\overset{\displaystyle O}{\|}}{C}-O-CH_2$$

Glyceryl tristearate (tristearin)

The term *fat* is applied to a glyceryl ester which is solid or semisolid, the term *oil,* to a liquid. Fats and oils are often called *glycerides.* Most natural glycerides are derived from two or three different carboxylic acids. When these are relatively unsaturated, that is, when they have many carbon-carbon double bonds, the glyceride is an oil. When they are mostly saturated (few double bonds) it is a fat. Vegetable oils can be converted into semi-hardened vegetable shortenings by hydrogenation, which reduces the number of double bonds in the carboxylate parts of the glycerides. Thus, the *linoleate* part of a glyceride may be converted into a *stearate:*

$$CH_3(CH_2)_4-\overset{\overset{\displaystyle H}{|}}{C}=\overset{\overset{\displaystyle H}{|}}{C}-CH_2-\overset{\overset{\displaystyle H}{|}}{C}=\overset{\overset{\displaystyle H}{|}}{C}-(CH_2)_7-\overset{\overset{\displaystyle O}{\|}}{C}-O-\xrightarrow[\text{Ni}]{\text{H}_2}$$

Linoleate

$$CH_3(CH_2)_4-\overset{\overset{\displaystyle H}{|}}{\underset{\underset{\displaystyle H}{|}}{C}}-\overset{\overset{\displaystyle H}{|}}{\underset{\underset{\displaystyle H}{|}}{C}}-CH_2-\overset{\overset{\displaystyle H}{|}}{\underset{\underset{\displaystyle H}{|}}{C}}-\overset{\overset{\displaystyle H}{|}}{\underset{\underset{\displaystyle H}{|}}{C}}-(CH_2)_7-\overset{\overset{\displaystyle O}{\|}}{C}-O-$$

Stearate

Saturated glycerides have been implicated in the deterioration of the vascular systems of animals (atherosclerosis).

Saponification of glycerides

Glycerides undergo *base hydrolysis,* commonly known as *saponification,* and the product is a *soap.*

$$CH_3(CH_2)_{16}-\overset{\overset{\displaystyle O}{\|}}{C}-O-CH_2$$

$$CH_3(CH_2)_{16}-\overset{\overset{\displaystyle O}{\|}}{C}-O-CH + 3NaOH \longrightarrow 3NaOOC(CH_2)_{16}CH_3 + \begin{array}{l} HO-CH_2 \\ HO-CH \\ HO-CH_2 \end{array}$$

$$CH_3(CH_2)_{16}-\overset{\overset{\displaystyle O}{\|}}{C}-O-CH_2$$

Tristearin   Sodium stearate   Glycerol
(a fat)   (a soap)

Most common soaps are sodium or potassium salts. The carboxyl end of a soap anion, such as the stearate ion, is soluble in water and is said to be *hydrophilic.*

The long hydrocarbon tail of the ion is soluble in oils and is said to be *hydropho-bic*. This structure allows the soap anions to disperse small oil globules in water. The hydrocarbon tails of soap anions bury themselves in oil globules leaving the carboxyl ends at the surfaces of the globules. This prevents globules from sticking to each other and keeps the oil *emulsified*. Particles of dirt are often associated with oils, and so washing with soap is effective, because the dirt is removed with the emulsified oil. As we have already mentioned (Sec. 21-2) soap anions are precipitated by certain ions such as Ca^{2+} in hard water.

23-11 Amines

Amines are compounds which can be regarded as derivatives of ammonia. In a *primary amine,* $R-NH_2$, one hydrogen of ammonia has been replaced by a hydrocarbon group. Replacement of two or three hydrogens yields a *secondary*

$$amine,\ R-\overset{\displaystyle H}{\underset{\displaystyle |}{N}}-R',\ or\ a\ tertiary\ amine,\ R-\overset{\displaystyle R''}{\underset{\displaystyle |}{N}}-R',\ respectively.$$

Nomenclature

The IUPAC name of an amine is formed (according to the 1969 rules) by adding the suffix *-amine* to the name of the attached alkyl group. Thus, CH_3NH_2 is methylamine. Some of the simpler amines are listed in Table 23-14.

Properties

Because nitrogen is not as electronegative as oxygen, hydrogen bonding in liquid primary and secondary amines is weaker than in alcohols of comparable molecular weight, and so the boiling points of the amines tend to be lower than those of the alcohols. Tertiary amines have no hydrogens to contribute to hydrogen bonding, and so their boiling points are less than those of similar primary and secondary amines. All classes of amines can hydrogen bond with water, and so the lower-molecular-weight amines have high solubilities in water.

The amines have various odors. The lower amines smell something like ammonia, but as the hydrocarbon proportion of the molecule increases, the odors become quite putrid. Putrescine, $H_2N(CH_2)_4NH_2$, and cadaverine, $H_2N(CH_2)_5NH_2$, are appropriate names for two diamines.

Reactions

An amine has a lone pair of electrons on its nitrogen, and so amines react as bases much as ammonia does. In water the dissociation of methylamine can be written as

$$CH_3-\overset{\displaystyle H}{\underset{\displaystyle H}{\overset{\displaystyle |}{\underset{\displaystyle |}{N}}}}:\ +\ H_2O\ \rightleftharpoons\ \left[CH_3-\overset{\displaystyle H}{\underset{\displaystyle H}{\overset{\displaystyle |}{\underset{\displaystyle |}{N}}}}-H\right]^{+}\ +\ OH^{-}$$

Methylamine   Solution of
                              "methylammonium hydroxide"

The alkyl amines are generally slightly stronger bases than ammonia. Amines react with inorganic acids to form salts:

$$CH_3CH_2NH_2\ +\ HCl\ \longrightarrow\ CH_3CH_2NH_3Cl$$
Methylamine   Methylammonium chloride

Table 23-14
Some amines

Condensed structural formula	IUPAC name	Normal melting point, °C	Normal boiling point, °C	pK_b (25°C)
CH_3NH_2	Methylamine	−94	−6	3.36
$(CH_3)_2NH$	Dimethylamine	−96	7	3.28
$(CH_3)_3N$	Trimethylamine	−117	3	4.30
$CH_3CH_2NH_2$	Ethylamine	−81	17	3.33
$CH_3CH_2CH_2NH_2$	Propylamine	−83	48	3.29

The reaction is analogous to the reaction of ammonia with an acid:

$$NH_3(g) + HCl(g) \longrightarrow NH_4Cl(s)$$

Secondary amines react with nitrous acid or nitrites to form *N*-nitrosoamines:

$$\underset{\substack{\text{Secondary} \\ \text{amine}}}{R-\overset{\overset{\displaystyle R}{|}}{N}H} + HNO_2 \longrightarrow \underset{\text{N-Nitrosoamine}}{R-\overset{\overset{\displaystyle R}{|}}{N}-N=O} + H_2O$$

This is an interesting reaction which has been found to take place during the cooking of meat to which sodium nitrite has been added. (Sodium nitrite is added to sausage and lunch meat to preserve its pink color and to prevent the growth of deadly *Clostridium botulinum* bacteria.) During cooking, protein, which contains amino groups, reacts with the nitrite to form various *N*-nitrosoamines, many of which have been found to be carcinogenic to animals.

23-12 Synthetic organic polymers

A *polymer* is a very large molecule which may be thought of as being built up from many identical small molecules, *monomers,* linked together by covalent bonds. The artificial plastics, including fibers and films, are examples of synthetic polymers.

Addition polymerization

Some polymers are synthesized by a reaction known as *addition polymerization,* in which many small molecules add to each other to form the polymer. Thus the polymerization of ethene (ethylene) produces the plastic *polyethylene:*

$$n \; \underset{\substack{| \quad | \\ H \;\; H}}{\overset{\substack{H \;\; H \\ | \quad |}}{C=C}} \longrightarrow \left(\underset{\substack{| \quad | \\ H \;\; H}}{\overset{\substack{H \;\; H \\ | \quad |}}{C-C}} \right)_n$$

Many $-CH_2-CH_2-$ units are linked together, so that the resulting product has a very high molecular weight.

Many addition polymerizations are believed to proceed by a *free-radical mechanism.* A *free radical* is a molecule which is uncharged, but which is highly reactive because it has an unpaired valence electron. At elevated temperatures organic peroxides split into free radicals:

$$\underset{\text{Organic peroxide}}{R-O-O-R} \longrightarrow \underset{\text{Alkoxy free radical}}{2R-O \cdot}$$

The free radical combines with a molecule of a monomer, say, ethylene, to form a new free radical,

$$R-O\cdot \; + \; \underset{H\;\;H}{\overset{H\;\;H}{C=C}} \longrightarrow R-O-\underset{H\;\;H}{\overset{H\;\;H}{C-C}}\cdot$$

which adds to another monomer,

$$R-O-\underset{H\;\;H}{\overset{H\;\;H}{C-C}}\cdot \; + \; \underset{H\;\;H}{\overset{H\;\;H}{C=C}} \longrightarrow R-O-\underset{H\;\;H\;\;H\;\;H}{\overset{H\;\;H\;\;H\;\;H}{C-C-C-C}}\cdot$$

and so on, until a very large molecule has been built up. The polymerization stops when two free radicals react with each other. Table 23-15 shows some addition polymers which are found in some commercially important plastics.

Condensation polymerization   In a second type of polymerization, *condensation polymerization,* long molecules are built up when two different bifunctional monomers split out small molecules, such as water. Adipic acid, a dicarboxylic acid, reacts with hexamethylenediamine, for example:

$$\underset{\text{Adipic acid}}{HO-\overset{O}{\overset{\|}{C}}-(CH_2)_4-\overset{O}{\overset{\|}{C}}-\boxed{OH+H}}-\underset{\text{Hexamethylenediamine}}{\overset{H}{\overset{|}{N}}-(CH_2)_6-\overset{H}{\overset{|}{N}}-H} \longrightarrow$$

$$HO-\overset{O}{\overset{\|}{C}}-(CH_2)_4-\overset{O}{\overset{\|}{C}}-\overset{H}{\overset{|}{N}}-(CH_2)_6-\overset{H}{\overset{|}{N}}-H \; + \; H_2O$$

The process repeats many times until a long molecule of a *nylon* has been built

Tables 23-15
Some addition polymers

Monomer	Polymer formula	Name of polymer
$CH_2=CH_2$ (ethylene)	$+CH_2-CH_2+_n$	Polyethylene
$CH_2=CH$ $\quad\;\; CH_3$ (propylene)	$\left(CH_2-CH\right)_n$ $\qquad\;\; CH_3$	Polypropylene
$CH_2=CH$ $\quad\;\; Cl$ (vinyl chloride)	$\left(CH_2-CH\right)_n$ $\qquad\;\; Cl$	Polyvinyl chloride (PVC; vinyl)
$CF_2=CF_2$ (tetrafluoroethylene)	$+CF_2-CF_2+_n$	Polytetrafluoroethylene (Teflon)
$CH_2=CH$ $\quad\;\; C$ $\quad\;\; \|\|\|$ $\quad\;\; N$ (acrylonitrile)	$\left(CH_2-CH\right)_n$ $\qquad\;\; C$ $\qquad\;\; \|\|\|$ $\qquad\;\; N$	Polyacrylonitrile (Orlon; Acrilan)

up. This nylon is called Nylon 66, because both adipic acid and hexamethyl-enediamine are six-carbon molecules.

Another condensation polymer is the polyester Dacron (as a fiber) or Mylar (as a thin sheet or film) which is produced from ethylene glycol and terephthalic acid.

$$HO-CH_2CH_2-O\boxed{H + HO}-\overset{\overset{O}{\|}}{C}-\langle\bigcirc\rangle-\overset{\overset{O}{\|}}{C}-OH \longrightarrow$$

Ethylene glycol   Terephthalic acid

$$HO-CH_2CH_2-O-\overset{\overset{O}{\|}}{C}-\langle\bigcirc\rangle-\overset{\overset{O}{\|}}{C}-OH + H_2O$$

Again, the process continues until a large molecule has been built up.

23-13
Optical isomerism
in organic
compounds

In Sec. 22-5 we showed that a molecule is *chiral* if one of its atoms is *asymmetric,* that is, if it is bonded tetrahedrally to four different groups. This is a common occurrence in organic chemistry, since carbon atoms frequently use sp^3 (tetrahedral) hybrids for bonding. A pair of molecules of opposite chirality are called *enantiomers.* In the past these have been labeled *dextrorotatory, dextro* (Latin: "right"), or *d*, if the molecule rotates the plane of polarized light to the *right,* and *levorotatory, levo* (Latin: "left"), or *l*, if it rotates the plane of polarized light to the *left.* However, it has been recommended that these terms be dropped in favor of the symbols ($+$) and ($-$) for rotation to the right and left, respectively. Thus

$$\overset{COOH}{\underset{CH_3}{\overset{|}{H\text{---}\underset{|}{C}\text{---}OH}}}$$

is ($-$)-lactic acid. ("Wedge" bonds extend above the plane of the paper; "dotted" bonds below.)

Many molecules are known which have a single asymmetric carbon. The hydrocarbon 3-methylhexane is an example:

$$CH_3CH_2-\overset{\overset{H}{|}}{\underset{\underset{CH_3}{|}}{C}}-CH_2CH_2CH_3$$

3-Methylhexane

Since the third carbon has four different groups attached (a methyl group, an ethyl group, a propyl group, and a hydrogen atom) the above formula represents a pair of enantiomers. They can be shown schematically as

$$\overset{C_3H_7}{\underset{C_2H_5}{\diagdown}}\overset{H}{\underset{CH_3}{C}} \quad \text{and} \quad \overset{C_3H_7}{\underset{C_2H_5}{\diagdown}}\overset{H}{\underset{CH_3}{C}}$$

Enantiomers of 3-methylhexane

Other examples of chiral molecules are

$$CH_3-\underset{\underset{H}{|}}{\overset{\overset{Cl}{|}}{C}}-CH_2CH_3 \qquad CH_2=CH-\underset{\underset{CH_3}{|}}{\overset{\overset{H}{|}}{C}}-CH_2CH_3$$

2-Chlorobutane   3-Methyl-1-pentene

$$CH_3CH_2-\underset{\underset{H}{|}}{\overset{\overset{CH_3}{|}}{C}}-CH_2OH \qquad CH_3-\underset{\underset{NH_2}{|}}{\overset{\overset{H}{|}}{C}}-\overset{\overset{O}{\|}}{C}-OH$$

2-Methyl-1-butanol   Alanine (an amino acid)

Molecules with more than two asymmetric carbons can exist as more than two stereoisomers.

Absolute configurations of enantiomers   Two methods of specifying the absolute configuration of the four groups around an asymmetric carbon are in use today. One, the D, L *system*, makes use of *Fischer projections*. In a Fischer projection the asymmetric carbon is considered to be in the plane of the paper, and the carbon chain is arranged vertically on the paper's surface so that the most oxidized carbon, or the carbon with the lowest locator number (Sec. 23-1), is placed *above* the asymmetric carbon. Then the groups which have been drawn *above* and *below* the asymmetric carbon are considered to project *below* the plane of the paper, and the groups drawn at the *right* and *left, above* the plane. For example, the Fischer projection for the amino acid alanine

$$H_2N-\underset{\underset{CH_3}{|}}{\overset{\overset{COOH}{|}}{C}}-H$$

is a simplified representation of

$$\overset{\displaystyle O \diagdown \diagup OH}{\underset{\underset{CH_3}{\vdots}}{H_2N\blacktriangleright\overset{\overset{\textstyle C}{}}{C}\blacktriangleleft H}}$$

The compound is then labeled D (for *dextro*) if the functional group (—NH$_2$, in this case) is on the right, or L (*levo*) if it is on the left. The two enantiomeric alanines are, therefore,

$$H-\underset{\underset{CH_3}{|}}{\overset{\overset{COOH}{|}}{C}}-NH_2 \qquad H_2N-\underset{\underset{CH_3}{|}}{\overset{\overset{COOH}{|}}{C}}-H$$

D-Alanine   L-Alanine

The *R and S system* is the other method commonly used for specifying absolute configurations. Using this system a *priority* is first assigned to each group attached to the asymmetric carbon.   Priorities are based on the atomic number of the atom directly bonded to the asymmetric carbon, and the higher the atomic number, the higher the priority.   If two bonded atoms have the same atomic number, such as if they are both carbon atoms, then a higher priority is assigned to the atom which has higher-priority atoms bonded to it.   Thus in the molecule of alanine, above, the priorities of the four groups bonded to the asymmetric carbon are

$$-NH_2 \; > \; \overset{\displaystyle O \atop \|}{-C-OH} \; > \; -CH_3 \; > \; -H$$

| 4 | 3 | 2 | 1 |
| (Highest) | | | (Lowest) |

Next, mentally orient the molecule so that the atom or group of *lowest priority* is directly below the plane of the paper.   With D-alanine this looks like

$$\underset{\displaystyle CH_3}{\overset{\displaystyle COOH}{H-C-NH_2}} \xrightarrow{\text{rotate}} \underset{4}{H_2N}-\underset{\displaystyle CH_3}{\overset{\displaystyle {}^3COOH}{C}}{}_{2}$$

(Hydrogen hidden below central carbon)

Start at the highest priority group, —NH$_2$, in this case, and see if the priorities *decrease* going to the right (clockwise) or to the left (counterclockwise).   If they decrease going to the *right,* the configuration is labeled *R,* for *rectus* (Latin: "right"); if they decrease going to the left, it is labeled *S,* for *sinister* (Latin: "left").   The above configuration of alanine is thus the *R* configuration:

$$\underset{\displaystyle CH_3}{\overset{\displaystyle COOH}{H-C-NH_2}}$$

(*R*)-Alanine

23-14
Carbohydrates
and proteins

From a chemical point of view *life* can be considered to be the set of complex reaction systems found in organisms which, by tradition, we call *living.*   The study of such systems is the focus of *biochemistry.*   Although saying very much about biochemistry is inconsistent with the scope and purpose of this book, we will give brief examples of two classes of compounds which are biologically important.

Carbohydrates

The name *carbohydrate* reflects the fact that carbohydrates were once thought to be hydrates of carbon.   Thus, glucose, $C_6H_{12}O_6$, can be represented as $C_6 \cdot 6H_2O$.   But there are no waters of hydration in glucose, as we will see shortly.   Nevertheless, many (though not all) carbohydrates have the general formula $C_x(H_2O)_y$.   Carbohydrates are best defined as *aldehydes* or *ketones* which are also *polyols,* or their polymers.   The carbohydrates include the sugars, starches, cellulose, and other compounds and are found in all plant and animal organisms.

The simplest carbohydrates are the *monosaccharides,* of which all are sugars.   A common example is D-(+)-glucose:

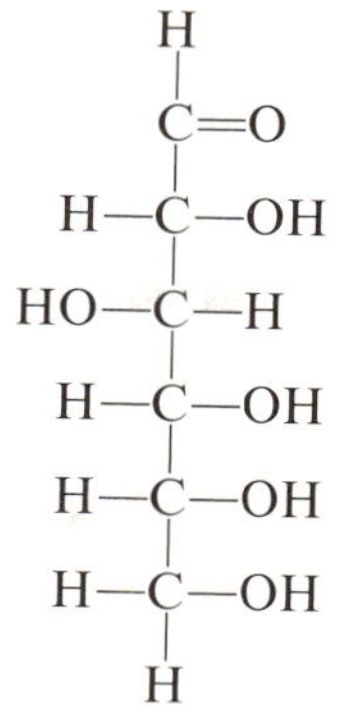

$$
\begin{array}{c}
\text{H} \\
| \\
\text{C}{=}\text{O} \\
| \\
\text{H}-\text{C}-\text{OH} \\
| \\
\text{HO}-\text{C}-\text{H} \\
| \\
\text{H}-\text{C}-\text{OH} \\
| \\
\text{H}-\text{C}-\text{OH} \\
| \\
\text{H}-\text{C}-\text{OH} \\
| \\
\text{H}
\end{array}
$$

D-(+)-Glucose

(The D isomer is the one in which the —OH group on the second carbon from the bottom is on the right in the above Fischer projection.) Glucose is classified as an *aldohexose, aldo-* because it is an aldehyde, and *-hex-* because it has six carbons. D-(+)-Glucose is found naturally in many places including honey and corn. It is also blood sugar and serves as a source of energy for much animal life. Because it rotates the plane of polarized light to the right, it has been called *dextrose*.

Another naturally occurring monosaccharide is D-(−)-fructose,

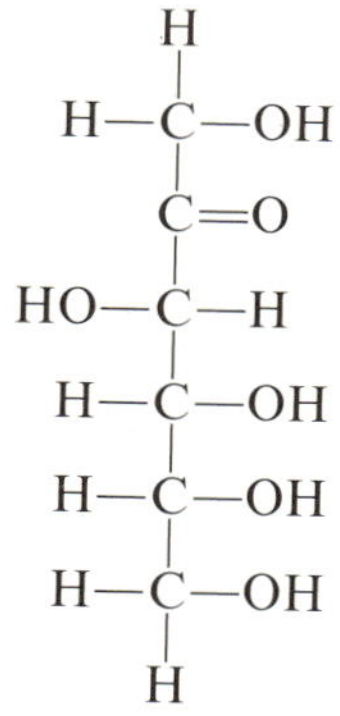

$$
\begin{array}{c}
\text{H} \\
| \\
\text{H}-\text{C}-\text{OH} \\
| \\
\text{C}{=}\text{O} \\
| \\
\text{HO}-\text{C}-\text{H} \\
| \\
\text{H}-\text{C}-\text{OH} \\
| \\
\text{H}-\text{C}-\text{OH} \\
| \\
\text{H}-\text{C}-\text{OH} \\
| \\
\text{H}
\end{array}
$$

D-(−)-Fructose

which is a *ketohexose*, because it has a ketone group in its structure. Fructose is found in grapes and in honey.

Strictly speaking, in solids the monosaccharides exist in *cyclic* forms. In solution the cyclic forms predominate but are in equilibrium with *acyclic* forms, such as those shown above. The cyclic forms of D-(+)-glucose and D-(−)-fructose are shown below:

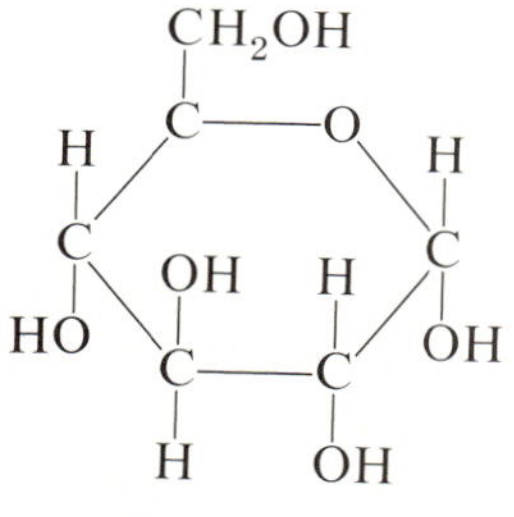

D-(+)-Glucose 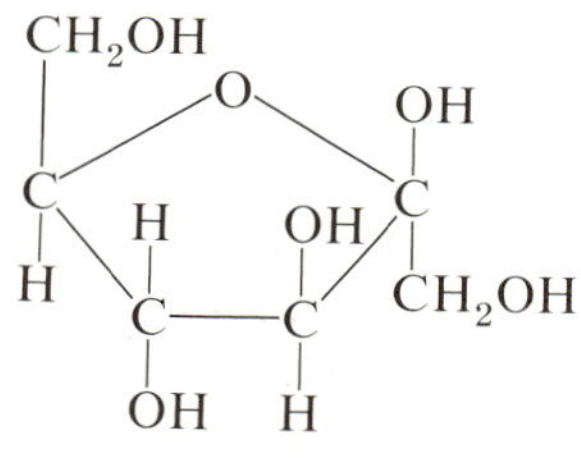   D-(−)-Fructose

Disaccharides consist of two monosaccharide units linked together. The most common disaccharide is *sucrose*. It may be thought of as a molecule of D-(+)-glucose which has been linked to one of D-(−)-fructose by splitting out a water molecule. Its structure is

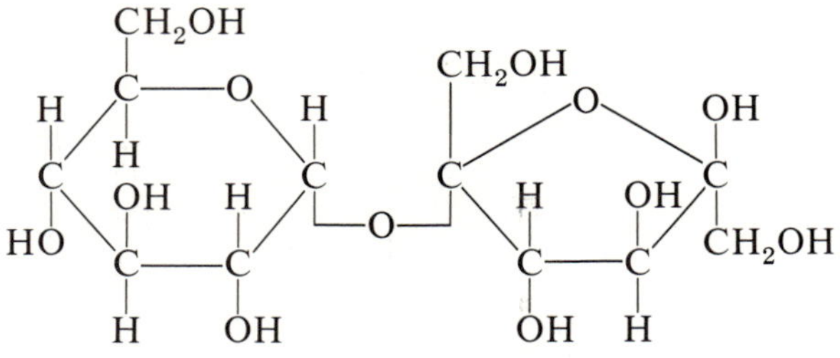

Sucrose

Sucrose is the most common sugar and is obtained commercially from sugar cane and sugar beets. It undergoes hydrolysis when catalyzed by acid or the enzyme *sucrase* to form its daughter molecules glucose and fructose. The mixture of the two is called *invert sugar* and is found in honey.

The linking of *many* monosaccharides produces *polysaccharides*. One form of starch, *amylose,* is composed of D-(+)-glucose units linked together as shown below:

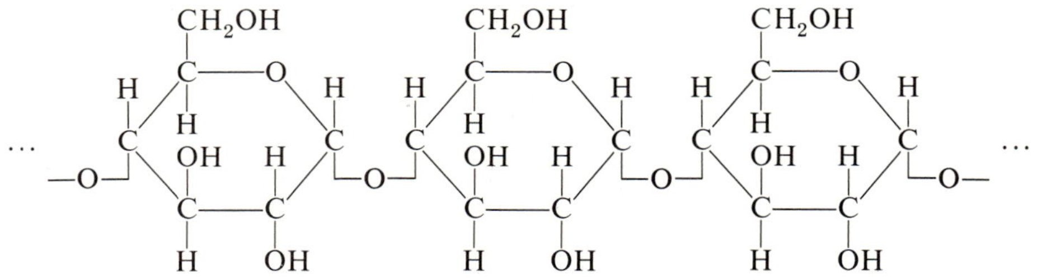

Amylose (starch)

This linkage is called an α *linkage*. Natural starches consist of mixtures of amylose and *amylopectin,* which is like amylose, except that the chains of glucose units are cross-linked via the side CH_2OH groups.

The most abundant carbohydrate, indeed the most abundant organic compound, is *cellulose*. It is a polysaccharide which has a structure just like that of amylose, except that the linkage between the glucose units is different:

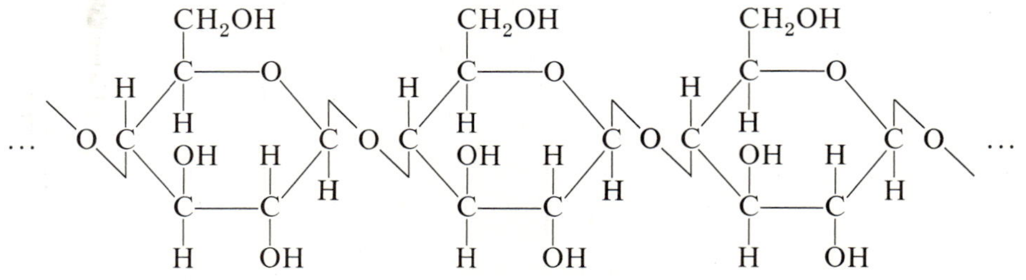

Cellulose

This linkage is called the *β linkage*. It is interesting that the slight difference between the way glucose units are linked in starch and in cellulose makes the difference between whether we can or cannot digest these substances. If our bodies possessed an enzyme capable of hydrolyzing cellulose, we could perhaps live on trees.

Proteins   *Proteins* are high-molecular-weight compounds consisting of many *amino acids* linked together. An amino acid is a bifunctional molecule having the amine and carboxylic acid functional groups. Some amino acids are

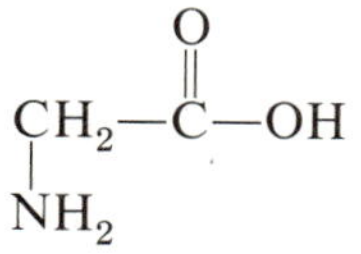

$$
\begin{array}{ccc}
\underset{\underset{\displaystyle NH_2}{|}}{CH_2}-\overset{\displaystyle O}{\overset{\|}{C}}-OH & CH_3\underset{\underset{\displaystyle NH_2}{|}}{CH}-\overset{\displaystyle O}{\overset{\|}{C}}-OH & CH_3\underset{\underset{\displaystyle CH_3}{|}}{CH}-\underset{\underset{\displaystyle NH_2}{|}}{CH}-\overset{\displaystyle O}{\overset{\|}{C}}-OH \\
\text{Glycine} & \text{Alanine} & \text{Valine}
\end{array}
$$

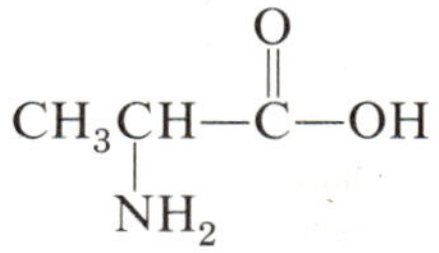

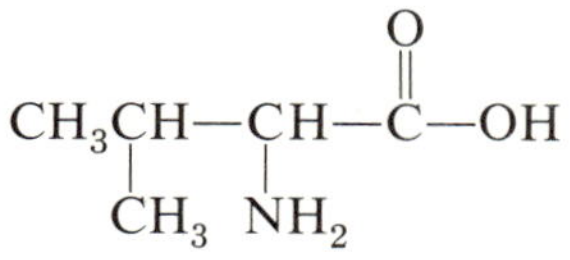

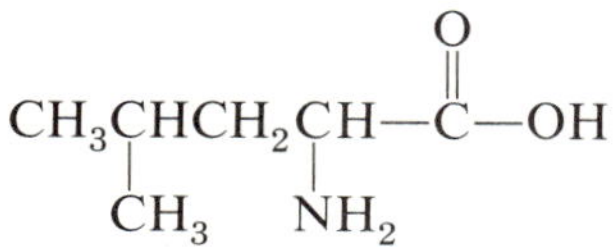

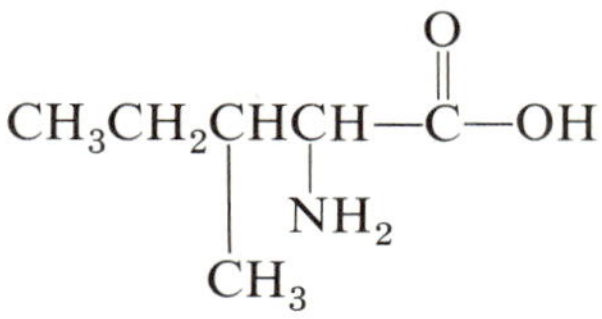

$$
\begin{array}{cc}
CH_3\underset{\underset{\displaystyle CH_3}{|}}{CH}CH_2\underset{\underset{\displaystyle NH_2}{|}}{CH}-\overset{\displaystyle O}{\overset{\|}{C}}-OH & CH_3CH_2\underset{\underset{\underset{\displaystyle CH_3}{|}}{\underset{\displaystyle NH_2}{|}}}{CH}CH-\overset{\displaystyle O}{\overset{\|}{C}}-OH \\
\text{Leucine} & \text{Isoleucine}
\end{array}
$$

Twenty different amino acids are found in the proteins of most animals. Some of these are called *essential* amino acids, because they cannot be synthesized by the animal's body and must be taken in as food. The number of essential amino acids varies from one animal to another. Eight amino acids are essential for the human, nine for the dog, and ten for the honeybee.

In each amino acid except for glycine the carbon atom to which the ammine group is attached is asymmetric, making possible the existence of enantiomers. Interestingly enough, only L-amino acids are found in proteins.

The amino acids in a protein are connected by linkages in which the amine group of one amino acid may be thought of as having reacted with the carboxyl of another to establish what is called a *peptide link:*

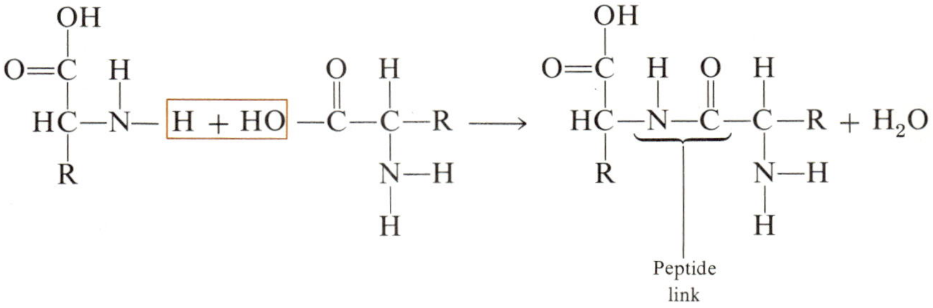

When a few amino acids are linked together this way, the product is called a *peptide*. Naturally occurring peptides include the *hormones*. *Insulin* is a typical peptide hormone. It consists of 51 amino acids in two chains which are bonded together by two disulfide, —S—S—, cross-linkages.

When many amino acids, say more than 60, are linked together, the molecule is classed as a protein. Some protein molecules have molecular weights in the millions. Each protein has its own specific sequence of amino acids. This sequence is known as the *primary structure* of the protein. The protein hemoglobin, found in red blood cells, consists of four chains of 146 amino acids each. The sequence in each chain is specific. The genetic disease sickle-cell anemia is a condition in which two of the hemoglobin chains have a *valine* amino acid unit where a *glutamic acid* unit is present in normal hemoglobin. This seemingly insignificant difference greatly reduces the ability of the red blood cells to carry oxygen, and the cells assume a sickle shape instead of a disk.

The *secondary structure* of a protein is the way in which the amino acid chain is folded or bent. The two principal secondary protein structures are the α and β structures. The α structure is a right-handed helix, a spiral turning in the direction of a right-hand screw. The protein chain is held in this position by hydrogen bonds between the amine of one amino acid and the carboxyl oxygen of an amino acid in the adjacent turn of the helix. Wool is an example of a protein with the α-helical structure.

The β secondary structure of a protein consists of several protein chains stretched out next to each other, adjacent chains being held together with hydrogen bonds. The result is a pleated or puckered sheet. The β-sheet structure is found in the protein of silk.

The *tertiary structure* of a protein is the three-dimensional shape into which a protein α-helix is bent. Specific proteins have specific tertiary structures which are maintained by hydrogen bonds, disulfide linkages, and electrostatic forces. Some proteins have a *quaternary structure*. This refers to the way several protein molecules are arranged in space to form a large unit. Hemoglobin's quaternary structure is the arrangement of its four bent α-helical chains.

SUMMARY

Our introduction to organic chemistry began with a discussion of the *hydrocarbons,* compounds composed of carbon and hydrogen only. *Saturated hydrocarbons* are those whose molecules contain no multiple bonds. The *alkanes* are saturated hydrocarbons which have the general formula C_nH_{2n+2}. They can be either *normal* (straight-chain) or *branched chain*. The *cycloalkanes* are saturated hydrocarbons with the general formula C_nH_{2n}. They have carbon chains which are closed to form rings.

Unsaturated hydrocarbons have multiple bonds. They include the *alkenes* (C_nH_{2n}), which have double bonds, and the *alkynes* (C_nH_{2n-2}), which have triple bonds. Cis-trans isomerism is possible in the alkenes, but not in the alkanes or alkynes. The unsaturated hydrocarbons undergo many *electrophilic addition* reactions in which the multiple bond is converted to one of lower order.

Aromatic hydrocarbons contain a special type of un-

saturation. The most common aromatic compounds contain the benzene ring in their molecules. The molecule of benzene, C_6H_6, has a six-membered ring of carbon atoms with each carbon bonded to one hydrogen. Three π bonds are delocalized over the ring and contribute to the stability of this structure. The aromatic hydrocarbons are quite unreactive but can be made to undergo catalytic substitution reactions in which one or more hydrogens are replaced by other atoms or groups.

Many organic compounds can be regarded as *derivatives* of hydrocarbons. These contain *functional groups* bonded to hydrocarbon residues. In this chapter we mentioned the *alcohols* (R—OH), the *ethers* (R—O—R′), the *aldehydes* (R—$\overset{\displaystyle O}{\overset{\|}{C}}$—H), the *ketones* (R—$\overset{\displaystyle O}{\overset{\|}{C}}$—R′), the *carboxylic acids* (R—$\overset{\displaystyle O}{\overset{\|}{C}}$—OH), the *es-*

$$O$$

ters $(R-\overset{O}{\underset{}{C}}-O-R')$, and the *amines* $(R-NH_2, R-\overset{R'}{\underset{}{N}}H$,

and $R-\overset{R'}{\underset{R''}{N}})$.   The compounds in each class can be given IUPAC or common names, and they undergo similar chemical reactions.

Polymers are very large molecules made up of repeating monomer units.   Many synthetic polymers are synthesized by *addition polymerization,* in which monomer molecules add to each other to produce long chains. *Condensation polymerization* involves the splitting out of a small molecule such as water, as when two different bifunctional molecules bond to each other.

Optical isomerism is shown by organic molecules which have an asymmetric carbon atom.   The symbols $(+)$ and $(-)$ are used to indicate rotation of the plane of polarized light clockwise and counterclockwise, re-

spectively. *Absolute configurations* of enantiomers can be specified by using either the D, L system, or the *R* and *S* system.

Carbohydrates and proteins are two biologically important classes of compounds.   *Carbohydrates* are polyols which are also aldehydes or ketones. The simplest carbohydrates are the *monosaccharides.* Two monosaccharides bonded together constitute a *disaccharide.* *Polysaccharides* are naturally occurring polymers of monosaccharides. These include starch and cellulose, which differ mainly in the linkage between the glucose monomers.

Proteins are also naturally occurring polymers. These are long-chain molecules composed of many *amino acids* bonded together by means of *peptide links.* Proteins occur in very complex forms and can be described in terms of their *primary, secondary, tertiary,* and often *quaternary* structures.

KEY TERMS

Addition polymerization (Sec. 23-12)
Addition reaction (Sec. 23-2)
Alcohol (Sec. 23-5)
Aldehyde (Sec. 23-7)
Alkane (Sec. 23-1)
Alkene (Sec. 23-2)
Alkyl group (Sec. 23-1)
Alkyne (Sec. 23-2)
Amine (Sec. 23-11)
Amino acid (Sec. 23-14)
Aromatic hydrocarbon (Sec. 23-3)
Branched chain (Sec. 23-1)
Carbohydrate (Sec. 23-14)
Carbonyl group (Sec. 23-7)
Carboxylate ion (Sec. 23-9)
Carboxylic acid (Sec. 23-9)
Chain isomerism (Sec. 23-1)
Condensation polymerization (Sec. 23-12)
Condensed structural formula (Sec. 23-1)
Cycloalkane (Sec. 23-1)
D, L system (Sec. 23-12)
Derivative (Sec. 23-4)
Disaccharide (Sec. 23-14)
Electrophile (Sec. 23-2)
Elimination reaction (Sec. 23-5)
Ester (Sec. 23-10)
Ether (Sec. 23-6)
Fat (Sec. 23-10)
Fischer projection (Sec. 23-13)
Free radical (Sec. 23-12)

Functional group (Sec. 23-4)
Glyceride (Sec. 23-10)
Grignard reagent (Sec. 23-8)
Halogenated hydrocarbon (Sec. 23-1)
Hydrocarbon (Sec. 23-1)
Hydrophilic (Sec. 23-10)
Hydrophobic (Sec. 23-10)
Intermolecular dehydration (Sec. 23-5)
Ketone (Sec. 23-8)
Locator number (Sec. 23-1)
Markovnikov's rule (Sec. 23-2)
Monomer (Sec. 23-12)
Monosaccharide (Sec. 23-14)
Normal (23-1)
Nucleophile (Secs. 23-5,7,8)
Oil (Sec. 23-10)
Polymer (Sec. 23-12)
Polyol (Sec. 23-5)
Polysaccharide (Sec. 23-14)
Primary alcohol (Sec. 23-5)
Primary amine (Sec. 23-11)
Primary protein structure (Sec. 23-14)
Protein (Sec. 23-14)
Quaternary protein structure (Sec. 23-14)
R and *S* system (Sec. 23-12)
Saturated (Sec. 23-1)
Saponification (Sec. 23-10)
Secondary alcohol (Sec. 23-5)
Secondary amine (Sec. 23-11)
Secondary protein structure (Sec. 23-14)

Skeletal isomerism (Sec. 23-1)
Straight chain (Sec. 23-1)
Substitution reaction (Sec. 23-3)
Tertiary alcohol (Sec. 23-5)

Tertiary amine (Sec. 23-11)
Tertiary protein structure (Sec. 23-14)
Unsaturated (Secs. 23-1,2)

QUESTIONS AND PROBLEMS

Carbon and organic compounds

23-1 Why are there so many compounds of carbon? Which element forms more? Why?

23-2 Is it possible to prove that there is no such thing as a "vital force"? Explain.

23-3 (a) Carbon is normally thought of as forming four covalent bonds. Can you think of a molecule in which carbon forms less than four bonds? (b) A carbon atom can use two of its p orbitals to form *methylene*, CH_2, a short-lived molecule. Why is methylene not a stable species? (c) Comment on the possibility that the bond in C_2 is a quadruple bond.

23-4 What is the origin of the word *organic*? How is it used in chemistry today?

23-5 Draw a Lewis structure for each of the following molecules: CH_4, C_2H_6, C_2H_4, C_2H_2, CH_4O, CH_5N.

Saturated hydrocarbons

23-6 Write a structural formula for each of the following alkanes: (a) 2-methylpentane (b) 2,3,5-trimethylheptane (c) 2-methyl-5-propyldecane (d) 3,3-dimethylhexane (e) 3,3,4,4,5,5-hexamethylnonane (f) 1,1,3-trimethylcyclohexane (g) cyclopentylcyclohexane.

23-7 Draw the structural formula for each isomer having the formula (a) C_4H_{10} (b) C_5H_{12} (c) C_6H_{14}.

23-8 What classes of compounds are represented by the general formula C_nH_{2n}? Draw the structure of each isomer of C_5H_{10}.

23-9 (a) Why is cis-trans isomerism not found in the alkanes? (b) Is it possible in the cycloalkanes? Explain.

23-10 Cyclohexane, C_6H_{12}, exists in two conformational isomers, or *conformers,* called the *chair form* and the *boat form.* Draw or construct models of these two forms.

23-11 Give the IUPAC name for each of the following:

(a) CH₃CHCH₂CH₃
      |
      CHCH₃
      |
      CH₂CH₂CH₃

(b) CH₂CH₂CH——CH₂
              | |
             CH₂ CH₂

(c) CH₃C——CHCH₃ with CH₃ above first C and CH₃, CH₃ below

$$CH_3C\text{---}CHCH_3$$

(d) CH₃CH₂CHCH₂CH₃
              |
         CH₃CH₂CHCH₂CH₃

23-12 In addition to cyclohexane there are 15 cyclic isomers having the formula C_6H_{12}. Indicate the structure of each.

23-13 Indicate the product of the ring-opening addition reaction of cyclopropane with (a) H_2 (b) Br_2 (c) HBr.

23-14 The reaction of 1 mol of Cl_2 with 1 mol of an alkane is a substitution reaction in which a Cl atom substitutes for an H atom and a molecule of HCl is formed. What are the products of the reaction of 1 mol of Cl_2 with 1 mol of (a) CH_4 (b) C_2H_6 (c) C_3H_8 (d) C_4H_{10}?

Unsaturated hydrocarbons

23-15 Describe the orbitals used by carbon for bonding in (a) C_2H_4 (b) C_2H_2 (c) CH_3CHCH_2.

23-16 (a) Explain in terms of orbital overlap why rotation around a C—C single bond is possible but rotation around a double bond does not occur. (b) Comment on the concept of rotation around a carbon-carbon triple bond.

23-17 Write a structural formula for each of the following: (a) 3-hexene (b) 1,4-pentadiene (c) 4,4-dimethyl-2-hexene (d) 2-pentyne (e) 3-penten-1-yne.

23-18 Derive a general formula for the *dienes,* hydrocarbons with two double bonds.

23-19 Give the IUPAC name for each of the following:

(a) CH₃C≡CCH₃

(b) CH₃CH₂C=CCH₃ (with H H above the double-bonded carbons)

(c) CH₃C≡CCH₂C≡CCH₃

(d) CH₃CHC=CH₂ (with CH₃ below the CH)

(e) CH₃C=C—CH₂CH₃ (with H above and H below the double bond)

23-20 *Allene* is propadiene. **(a)** Describe the orbitals used for bonding by each carbon in allene. **(b)** What is the geometrical relationship among the four hydrogen atoms in this molecule? **(c)** Draw the structure of each *dimethylallene*.

23-21 What is the product of the addition of 1 mol of HBr to 1 mol of **(a)** propene **(b)** acetylene **(c)** 4-methyl-2-pentyne **(d)** 1-bromopropene.

23-22 Why is it that cyclopentyne is nonexistent?

Aromatic hydrocarbons

23-23 Describe the bonding in benzene from the standpoint of **(a)** resonance theory **(b)** MO theory.

23-24 Draw the structural formula and give the IUPAC name for each isomer of each of the following:
(a) dimethylbenzene   **(b)** trimethylbenzene
(c) tetramethylbenzene.

23-25 Draw contributing resonance forms for

(a) Naphthalene,

(b) Anthracene,

(c) Phenanthrene,

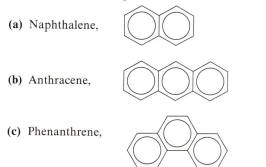

23-26 Show that the following is an impossible structure:

23-27 Write the structural formula for **(a)** *p*-dichlorobenzene (mothballs), **(b)** *m*-bromonitrobenzene, **(c)** 1,3,5-trinitrobenzene, **(d)** *o*-chlorotoluene, and **(e)** chloromethylbenzene.

Derivatives of hydrocarbons

23-28 Give the structures and IUPAC names of the isomeric, acyclic, saturated, five-carbon alcohols.

23-29 What kind of compound is formed from the oxidation of a **(a)** primary alcohol **(b)** secondary alcohol?

23-30 In the synthesis of aldehydes from alcohols, it is necessary to minimize the further oxidation of the product to a carboxylic acid. Suggest one way by which this might be done.

23-31 A mixture of methanol, ethanol, and sulfuric acid is reacted at 135°C. Name each of the ethers which is formed.

23-32 Draw a structural formula for each of the following: **(a)** 3-chloro-2-pentanone **(b)** cyclohexanol **(c)** triethylamine **(d)** trichloroacetic acid **(e)** 3-hexanone **(f)** isopropyl butyrate **(g)** 2,3-butanediol **(h)** cyclodecanone **(i)** 2-ethyl-3-hydroxyhexanal **(j)** 2-chloropropanoic acid **(k)** 2-amino-3-methylheptane.

23-33 Why is it that tertiary alcohols are not readily oxidized?

23-34 Draw the structures of the possible esters which could be formed from the reaction of glycerol with propanoic acid.

23-35 Is carbonic acid a carboxylic acid? What is its anhydride? What is the anhydride of formic acid?

23-36 Justify the classification of double and triple bonds as functional groups.

23-37 Explain why the normal boiling point of methyl ether is so much lower than that of its structural isomer, ethanol.

23-38 Show how an amino acid can react **(a)** as an acid **(b)** as a base **(c)** as both, simultaneously.

23-39 Two dicarboxylic acids are oxalic, HOOCCOOH, and malonic, HOOCCH$_2$COOH. Justify the fact that oxalic acid is the stronger of the two.

23-40 Why do most of the normal carboxylic acids have about the same dissociation constant?

23-41 Show that cis and trans isomers exist for 1,2-dichlorocyclopentane.

23-42 Describe the orbitals used for bonding by each atom in **(a)** formaldehyde **(b)** methylamine **(c)** acetic acid.

23-43 Classify each of the following as a primary, secondary, or tertiary amine:
(a) CH$_3$NHCH$_2$CH$_3$ **(b)** CH$_3$CH$_2$CH$_2$NH$_2$
(c) CH$_3$NHCH$_2$NH$_2$ **(d)** (CH$_3$)$_2$NCH$_2$CH$_3$.

23-44 List the following in order of decreasing acid dissociation constant:

CH$_3$CHClCOOH   CH$_3$CH$_2$COOH
CH$_3$CCl$_2$COOH   CH$_2$ClCH$_2$COOH

Polymers

23-45 Contrast addition polymerization with condensation polymerization, and give an example of each.

23-46 Why does commercial polyethylene consist of chains of various lengths?

23-47 1,3-Butadiene can undergo polymerization to form three different polymers. Draw the structure of each.

23-48 Lucite and Plexiglas are names for polymethylmethacrylate. If its monomer, methylmethacrylate is

$$CH_2=\overset{\underset{\displaystyle CH_3}{|}}{C}-\overset{\underset{\displaystyle O}{\|}}{C}-OCH_3,$$ what is the structure of the polymer?

Optical isomerism

23-49 What is the smallest **(a)** *hydrocarbon* and **(b)** *carbon compound* which can exhibit optical isomerism?

23-50 Draw Fischer-projection formulas for the two enantiomeric lactic acids (see p. 708). Label each as either D or L.

23-51 Indentify each isomer in Prob. 23-50 as being either *R* or *S*.

23-52 Determine whether L-alanine has the *R* or *S* configuration.

23-53 Give structural formulas and names for all the isomeric butanols, C_4H_9OH.

Carbohydrates

23-54 Define the following terms: monosaccharide, aldohexose, ketopentose.

23-55 Describe a simple laboratory test which could be used to distinguish D-glucose from D-fructose.

23-56 The simplest sugars are an aldotriose and a ketotriose. Draw a structure for each of these. Which one shows optical isomerism?

23-57 What are the similarities and differences between cellulose and starch?

23-58 Sucrose and glucose are often used in the manufacture of candy. Which sugar serves as a faster energy source for the body? Explain.

Proteins

23-59 Only one amino acid does not show optical isomerism. Which one is it? Why is it not enantiomeric?

23-60 In an electrolytic cell an amino acid migrates in one direction at a low pH and in the other direction at a high pH. Explain.

23-61 The C—N—C bond angle in the peptide link has been shown to be about 120°. Show how a resonance form having a C=N double bond can explain the observed bond angle.

23-62 Describe what is meant by the primary, secondary, tertiary, and quaternary structures of proteins.

23-63 Comment on structural similarities between nylon and the proteins.

23-64 A certain peptide was found to consist of six different amino acids. How many different amino acid sequences are possible for such a peptide?

23-65 If the average molecular weight of the amino acids is 115, approximately how many amino acid residues are present in a protein having a molecular weight of 700,000?

24

NUCLEAR PROCESSES

TO THE STUDENT

Until now, we have paid little attention to the atomic nucleus, except to point out how nuclear charge affects atomic properties such as electronegativity, ionization energy, and atomic radius. The nucleus itself remains largely unaffected during a chemical reaction. Buried deep in the center of the atom, it seems oblivious to all the tumult and change going on out in the extremities of the atom. But the atomic nucleus can indeed undergo change, and the study of these changes is the focus of this chapter.

24-1
Radioactivity

In 1895 Wilhelm Röntgen discovered the x rays which are emitted from the anode of a high-voltage cathode-ray tube. Less than one year later Antoine Henri Becquerel thought he had found a natural source of x rays: potassium uranyl sulfate, $K_2UO_2(SO_4)_2 \cdot 2H_2O$, but he later decided that the natural rays emanating from uranium compounds were different from Röntgen's x rays. It was Becquerel who coined the word *radioactivity* to describe the production of these rays.

Eventually three kinds of radioactive emissions were identified and characterized, and all were shown to be ejected by atomic nuclei undergoing changes in composition or structure. These were called *alpha, beta,* and *gamma rays*. Alpha (α) rays consist of a stream of particles (now called *alpha particles*) which are identical to ^4_2He nuclei. Beta (β) rays consist of a stream of electrons, often of very high energy, called *beta particles* and designated $_{-1}^0e$. Here the -1 subscript indicates the charge, and the superscript 0, the extremely small mass, of the electron. *Gamma* (γ) *rays* are not particles; they are electromagnetic radiation like x rays, but are generally of higher frequency, and thus, energy ($E = h\nu$).

Natural
radioactivity

In the absence of external influences many nuclides are permanently stable.[1] But some are not and undergo *radioactive decay,* also known as *nuclear disintegration.*

---

[1] A *nuclide* is a specific atomic species with a certain nuclear composition.

Such a process is represented by a nuclear equation in which the symbol, atomic number (Z), and mass number (A) of each particle are specified. The radioactivity detected by Becquerel was a result of the alpha decay of a uranium isotope, $^{238}_{92}U$. The nuclear equation for this process is written

$$^{238}_{92}U \longrightarrow {}^{234}_{90}Th + {}^{4}_{2}He$$

$$\underset{\text{nuclide}}{\underset{\text{Parent}}{}} \qquad \underset{\text{nuclide}}{\underset{\text{Daughter}}{}} \quad \underset{\text{particle}}{\underset{\text{Alpha}}{}}$$

The equation shows the *parent* uranium nuclide emitting an alpha particle and forming a *daughter* thorium nuclide. Note that the above equation shows conservation of nucleons and of charge. In other words, each side of the above equation shows a total of 238 nucleons and a total number of protons, or positive charge, of 92.

A daughter nuclide may itself be unstable. Thorium 234 undergoes beta decay:

$$^{234}_{90}Th \longrightarrow {}^{234}_{91}Pa + {}^{0}_{-1}e$$

$$\underset{\beta \text{ particle}}{}$$

In this case it can be seen that the total number of nucleons in the daughter protactinium nuclide is the same as in the parent thorium nuclide. But the daughter has one more proton than the parent. When a beta particle was emitted from the nucleus, a neutron was evidently converted into a proton.

Gamma emission is often not explicitly shown in nuclear equations, since neither mass nor charge change during this kind of emission. Gamma emission is very common, often accompanies other kinds of decay, and represents loss of energy when a nucleus drops from a higher to a lower energy state.

● EXAMPLE 24-1

Problem   Radon 219 decays by α emission. What daughter nuclide is formed?

Solution   Loss of an α particle means that the daughter nuclide must have two fewer protons and four fewer total nucleons. In other words, the nuclear equation is

$$^{219}_{86}Rn \longrightarrow {}^{215}_{84}(?) + {}^{4}_{2}He$$

Since 84 is the atomic number of polonium, the daughter nuclide is $^{215}_{84}Po$. ●

● EXAMPLE 24-2

Problem   When lead 210 decays, it forms bismuth 210. What kind of decay is this?

Solution   The nuclear equation is

$$^{210}_{82}Pb \longrightarrow {}^{210}_{83}Bi + (?)$$

In order to conserve charge and nucleons, we write

$$^{210}_{82}Pb \longrightarrow {}^{210}_{83}Bi + {}^{0}_{-1}(?)$$

Clearly the emitted particle is a β particle, $^{0}_{-1}e$. ●

Measuring   Many methods have been used for detecting radioactivity and observing its
radioactivity   properties. One of the oldest is the *Geiger-Müller counter,* shown in Fig. 24-1. Radiation enters the counter tube through the thin window at the end. It strikes

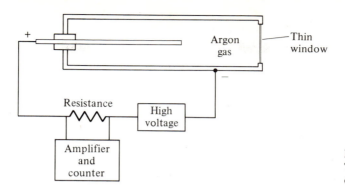

Figure 24-1
The Geiger-Müller
counter.

argon atoms inside the tube, ionizes them, and causes an electrical discharge between the central wire and the outer shell. This discharge is detected and each such event is counted electronically.

Even older than the Geiger-Müller counter is the use of phosphorescence for detecting high-energy radiation. Today *scintillation counters* can be regarded as descendants of the early phosphorescent screen. In these, a specially doped crystal of an alkali-metal halide gives off a flash of light when high-energy radiation hits it. This light is detected by a photomultiplier tube, an ultrasensitive photocell, which is in turn connected to an amplifier and counter.

Photographic emulsions have been used for many years for observing the tracks of high-energy particles. Used similarly are *cloud chambers* in which a particle's path is made visible by a trail of droplets of water or other liquid condensed from a supersaturated vapor. *Bubble chambers* reveal a particle's path by a trail of tiny bubbles in liquid hydrogen, and *spark chambers* by a trail of spark discharges between thin, oppositely charged electrodes. In the cloud, bubble, and spark chambers the tracks are usually photographed to provide a permanent record.

Radioactive decay *** series*** If a nucleus is unstable, it decays, and if the daughter nucleus is unstable, it also decays. This process continues until a stable nucleus is formed. The ordered sequence of unstable nuclei is called a *radioactive decay series*. There are several natural decay series. One, the uranium series, starts with $^{238}_{92}U$ and ends with the stable nuclide $^{206}_{82}Pb$. This series is shown in Fig. 24-2. As can be seen, each alpha emission reduces the mass number by 4 and the atomic number by 2, as an $^{4}_{2}He$ leaves the nucleus. Similarly, each beta emission leaves the mass number unchanged, but increases the atomic number by 1, as an $^{0}_{-1}e$ leaves the nucleus. The succession of nuclear disintegrations continues through all the intermediate nuclides, until $^{206}_{82}Pb$, which is stable, is produced. Note that at several places there is branching in the sequence. For example, there are two ways $^{218}_{84}Po$ can decay into $^{214}_{83}Bi$.

Other naturally ***occurring nuclear*** ***processes*** Many years after alpha, beta, and gamma emission processes were discovered a fourth type of natural decay was observed: *electron capture* (EC). In this, the nucleus captures an orbital electron:

$$^{40}_{19}K + ^{0}_{-1}e \xrightarrow{\text{EC}} {}^{40}_{18}Ar$$

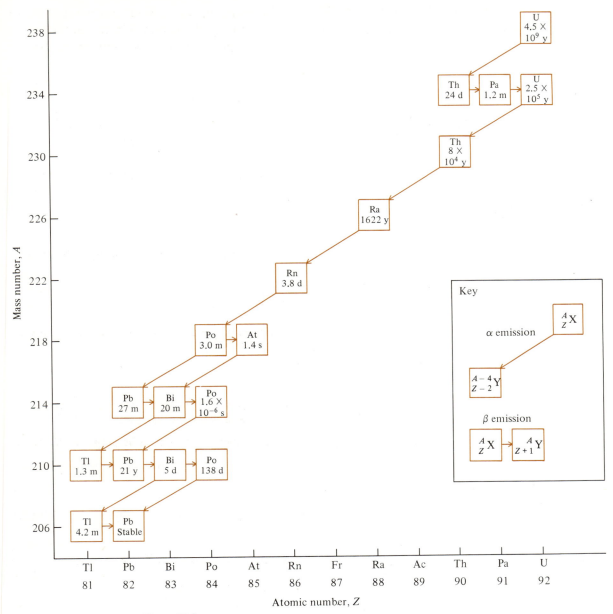

Figure 24-2
The uranium decay series. (Abbreviations: s, seconds; m, minutes; d, days; y, years.)

Note that the mass number is unchanged, but the atomic number is decreased by one. Often the electron captured is a K-shell electron, in which case the capture is called *K capture*. A fifth kind of natural nuclear process is *spontaneous fission*, which we will discuss in Sec. 24-5.

24-2
The kinetics of
nuclear decay

The rate of nuclear disintegration is proportional to the number of unstable nuclei present in a sample. Using the symbolism of chemical kinetics developed in Chap. 14 and letting N be the number of parent nuclei in a given sample, we can write

$$\text{Rate of decay} = -\frac{dN}{dt} = kN$$

where k is the rate constant and t is time. This equation describes a *first-order reaction;* nuclear decay does indeed follow first-order kinetics.

First-order
radioactive decay

In Sec. 14-2 we stated that by using calculus it is possible to take the above rate law and derive the relationship

$$\log N = -\frac{kt}{2.303} + \log N_0$$

where N in this case is the number of parent nuclei at time t, and N_0 is the number at $t = 0$. This can be rewritten as

$$\log \frac{N}{N_0} = -\frac{kt}{2.303}$$

This useful relationship gives the fraction of the unstable nuclei left, N/N_0, after a period of time, t.

● **EXAMPLE 24-3**

Problem

The rate constant for the α decay of $^{222}_{86}\text{Rn}$ is 0.18 day^{-1}. To what quantity will 4.5×10^{-5} g of this nuclide be reduced after a period of 8.5 days?

Solution

Since for any one nuclide the mass and number of atoms must be proportional, we can write

$$\log \frac{x}{x_0} = -\frac{0.18\ t}{2.303}$$

where x_0 and x are the masses of radon at the start and at time t, respectively. Substituting, we have

$$\log \frac{x}{4.5 \times 10^{-5}\ \text{g}} = -\frac{(0.18\ \text{day}^{-1})(8.5\ \text{days})}{2.303}$$

$$x = 9.7 \times 10^{-6}\ \text{g} \quad ●$$

The stability of a nuclide is measured by its half-life, $t_{\frac{1}{2}}$. As we saw in Chap. 14, this is the time necessary for one-half of a reactant to disappear. For any nuclear disintegration, then, we can write

$$\log \frac{(N_0/2)}{N_0} = -\frac{kt_{\frac{1}{2}}}{2.303}$$

or

$$t_{\frac{1}{2}} = \frac{2.303 \log 2}{k} = \frac{0.693}{k}$$

● **EXAMPLE 24-4**

Problem   From the data in Example 24-3 calculate the half-life of $^{222}_{86}$Rn.

Solution

$$t_{\frac{1}{2}} = \frac{0.693}{k}$$

$$= \frac{0.693}{0.18 \text{ day}^{-1}} = 3.8 \text{ days} \quad ●$$

Radiochemical dating   One of the uses of radioactive decay is in determining the ages of ancient relics, fossils, and rocks. For example, the $^{238}_{92}$U decay series diagrammed in Fig. 24-2 is used for *uranium dating*. Since the first step in the decay series has by far the longest half-life, the series is like a multistep chemical reaction in which the first step is rate-determining. In this case the number of lead atoms (in the final, stable product) is essentially equal to the number of uranium atoms decayed. In determining the age of rocks the numbers of uranium and lead atoms are obtained by analysis and the total equated to the number of uranium atoms at $t = 0$, that is, when the rocks were formed. Since the half-life of uranium 238 is known, the age of the rocks can be calculated from the ratio of the number of uranium atoms present to the original number. The oldest rocks yet found have an age of about 3×10^9 years, as determined by this method.

● **EXAMPLE 24-5**

Problem   A sample of rock is found to contain 1.3×10^{-5} g of uranium 238 and 3.4×10^{-6} g of lead 206. If the half-life of $^{238}_{92}$U is 4.5×10^9 years, how old is the rock?

Solution   We first need to find how many grams of uranium 238 have decayed.

$$^{238}_{92}\text{U decayed} = 3.4 \times 10^{-6} \text{ g Pb} \frac{238 \text{ g U mol}^{-1}}{206 \text{ g Pb mol}^{-1}} = 3.9 \times 10^{-6} \text{ g}$$

$$\text{Total } ^{238}_{92}\text{U at start} = 1.3 \times 10^{-5} \text{ g} + 3.9 \times 10^{-6} \text{ g} = 1.7 \times 10^{-5} \text{ g}$$

Let x = grams of uranium 238. Then

$$\log \frac{x}{x_0} = -\frac{kt}{2.303}$$

$$t = \frac{-2.303 \log (x/x_0)}{k}$$

From the half-life we can find the rate constant k:

$$k = \frac{0.693}{t} = \frac{0.693}{4.5 \times 10^9 \text{ years}} = 1.5 \times 10^{-10} \text{ year}^{-1}$$

So $$t = \frac{-2.303 \log (1.3 \times 10^{-5}/1.7 \times 10^{-5})}{1.5 \times 10^{-10} \text{ year}^{-1}} = 1.8 \times 10^9 \text{ years} \quad ●$$

Carbon dating is used for objects which once were part of living organisms. In the upper atmosphere neutrons from cosmic radiation bombard nitrogen 14 nuclei to form carbon 14:

$$^{14}_{7}\text{N} + ^{1}_{0}n \longrightarrow ^{14}_{6}\text{C} + ^{1}_{1}\text{H}$$

The carbon nucleus is unstable and decays to form $^{14}_{7}N$ through beta emission:

$$^{14}_{6}C \longrightarrow {}^{14}_{7}N + {}^{0}_{-1}e \qquad t_{\frac{1}{2}} = 5670 \text{ years}$$

It is generally assumed that the ratio of carbon 14 to carbon 12 has been constant for many thousands of years and that the $^{14}_{6}C$ gets oxidized to CO_2, absorbed by plants, which are then eaten by animals, which excrete the carbon 14, etc., so that the ratio of the two isotopes remains constant during the lifetime of the plant or animal. But after death, the carbon 14 continues decaying in the organism and is not replaced, so that the $^{14}_{6}C/^{12}_{6}C$ ratio begins to drop. From the observed ratio in an ancient piece of wood, bone, etc., the age of the object can be determined.

● **EXAMPLE 24-6**

Problem   An old wooden scraper was unearthed in an archaeological dig in Africa. Its $^{14}_{6}C/^{12}_{6}C$ ratio is only 0.714 of that in the atmosphere today. How old is the scraper?

Solution   From the half-life for the beta decay of carbon 14 we calculate the rate constant

$$k = \frac{0.693}{t_{\frac{1}{2}}}$$

$$= \frac{0.693}{5670 \text{ years}} = 1.22 \times 10^{-4} \text{ year}^{-1}$$

As in Example 24-5 we have

$$t = \frac{-2.303 \log x/x_0}{k}$$

$$= \frac{-2.303 \log 0.714}{1.22 \times 10^{-4} \text{ year}^{-1}} = 2760 \text{ years} \quad ●$$

24-3
Nuclear reactions   The natural emission of an alpha or beta particle by a nucleus transforms that nucleus into a new one with a different number of protons. Thus each such radioactive decay represents the *transmutation* of one element into another. Transmutation can also be accomplished artificially. Thus the dream of the ancient alchemists has been realized!

Transmutation   In 1919 Rutherford bombarded nitrogen 14 with alpha particles from the radio-active decay of radium. The product nuclides were oxygen 17, and the equation for this, the first recorded successful artificial transmutation, is

$$^{14}_{7}N + {}^{4}_{2}He \longrightarrow [{}^{18}_{9}F] \longrightarrow {}^{17}_{8}O + {}^{1}_{1}H$$

The highly unstable intermediate, a high-energy state of fluorine 18, is sometimes called a *compound nucleus*. Its half-life is less than 10^{-12} s, and it decays by emitting a proton to form stable oxygen 17.

In many cases the product of a nuclear bombardment reaction is unstable and undergoes subsequent radioactive decay. For example, when cobalt 59 is bombarded with a neutron, the following reaction takes place:

$$^{59}_{27}Co + {}^{1}_{0}n \longrightarrow [{}^{60}_{27}Co] \longrightarrow {}^{56}_{25}Mn + {}^{4}_{2}He$$

But the product manganese 56 is not stable, decaying with a half-life of 2.6 h to form iron 56, which is stable:

$$^{56}_{25}\text{Mn} \longrightarrow \,^{56}_{26}\text{Fe} + \,^{0}_{-1}e$$

This is an example of *induced,* or *artificial,* radioactivity.

Induced radioactivity illustrates several modes of decay not found in natural radioactivity. One of these is *neutron emission,* as is illustrated in one mode of decay of bromine 87.

$$^{87}_{35}\text{Br} \longrightarrow \,^{86}_{35}\text{Br} + \,^{1}_{0}n$$

Another decay mode is *beta-plus* (β^+) *decay,* also known as *positron emission.* Ordinary beta particles are more properly called *beta-minus* (β^-) particles in order to distinguish them from β^+ particles, which are positrons, $^{0}_{1}e$. A *positron* is a particle which has the mass of an electron, but a *positive* electrical charge. An example of β^+ decay is the disintegration of a nitrogen 13 nuclide:

$$^{13}_{7}\text{N} \longrightarrow \,^{13}_{6}\text{C} + \,^{0}_{1}e$$

Induced radioactivity and artificial transmutation are possible because of the development of high-energy particle accelerators such as the *cyclotron,* the *synchrotron,* and the *linear accelerator.* Figure 24-3 shows a schematic diagram of a cyclotron. The evacuated, hollow, metal electrodes, called *dees,* are located between the poles of a large magnet (not shown in the diagram), and a source of high-frequency alternating current is connected to the dees. Protons injected at the center travel an ever-widening spiral path. They gain considerable energy as they whirl around, and finally they collide with a target. In Batavia, Illinois, there is currently in operation a proton synchrotron which produces protons with energies in excess of 200 BeV (billion electronvolts), or about 1.9×10^{13} kJ mol^{-1}.

The *transuranium elements,* the elements following uranium in the periodic table, have been prepared by bombardment techniques. For example, neptunium

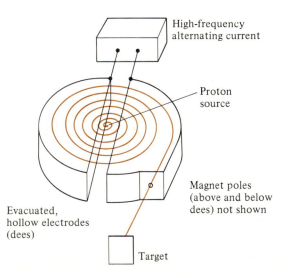

High-frequency
alternating current

Proton
source

Magnet poles
(above and below
dees) not shown

Evacuated,
hollow electrodes
(dees)

Target

**Figure 24-3
The cyclotron.**

($Z = 93$) has been synthesized by bombarding uranium 238 nuclei with deuterons, nuclei of hydrogen 2:

$$^{238}_{92}\text{U} + ^{2}_{1}\text{H} \longrightarrow ^{238}_{93}\text{Np} + 2\,^{1}_{0}n$$

The atoms of highest atomic number have been prepared by bombardment using comparatively massive particles such as $^{10}_{5}\text{B}$, $^{12}_{6}\text{C}$, and $^{14}_{7}\text{N}$. For instance, hahnium ($Z = 105$) was synthesized by bombarding californium 249 with nitrogen 15 nuclei:

$$^{249}_{98}\text{Cf} + ^{15}_{7}\text{N} \longrightarrow ^{260}_{105}\text{Ha} + 4\,^{1}_{0}n$$

**24-4
Nuclear stability**

Although an extensive discussion of the factors which contribute to stability and instability in nuclei is beyond the scope of this book, a few observations are in order. First, with the exception of $^{1}_{1}\text{H}$, all stable nuclei contain at least one neutron. Second, as the number of protons in the nucleus increases, the number of neutrons *per proton* increases in the stable nuclei. Apparently neutrons are necessary to keep a nucleus from flying apart as a result of proton-proton repulsions, and the more protons which are present in the nucleus, the greater the neutron/proton ratio must be for the nucleus to be stable. Third, when there are more than 83 protons in a nucleus, no number of neutrons will stabilize it. In the periodic table bismuth ($Z = 83$) is the last element having a stable isotope.

The belt of stability

Figure 24-4 shows a graph of the numbers of neutrons and of protons for all stable nuclei. These nuclei are found in a *belt of stability*, a region on the graph in which the neutron/proton ratio is close to 1 for the lighter nuclei, but in which the ratio increases as the number of protons increases. Thus in $^{6}_{3}\text{Li}$ the ratio is $1:1$, in $^{110}_{48}\text{Cd}$ it is $1.29:1$, and in $^{202}_{80}\text{Hg}$ it is $1.53:1$.

On this kind of neutron-proton plot unstable nuclei lie either above, below, or beyond the end of the belt of stability. Those lying above the belt have too high a neutron/proton ratio, those below, too low a ratio, and those beyond the belt simply have too many nucleons to be stable. Nuclei in these regions tend to undergo transformation into nuclei within the belt, or, at least, closer to it.

Nuclei lying *above* the belt of stability lower their neutron/proton ratios by β^- decay or less commonly, neutron emission. In the β^--*decay* process

$$^{133}_{54}\text{Xe} \longrightarrow ^{133}_{55}\text{Cs} + ^{0}_{-1}e$$

the daughter cesium nuclide is stable, but in some cases several successive decays are necessary before the nucleus gets into the belt of stability. For example, antimony 131 must undergo three consecutive β^- decays to form a stable nucleus:

$$^{131}_{51}\text{Sb} \longrightarrow ^{131}_{52}\text{Te} + ^{0}_{-1}e$$
$$\longrightarrow ^{131}_{53}\text{I} + ^{0}_{-1}e$$
$$\longrightarrow ^{131}_{54}\text{Xe} + ^{0}_{-1}e$$

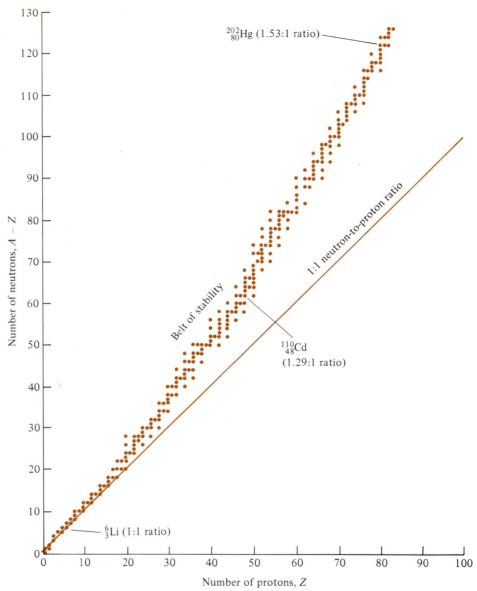

Figure 24-4
The belt of stability.

An occasionally observed process for lowering the neutron/proton ratio is *neutron emission.* An example is

$$\ce{^{90}_{36}Kr} \longrightarrow \ce{^{89}_{36}Kr} + \ce{^{1}_{0}}n$$

Nuclei lying *below* the belt of stability raise their neutron/proton ratios either by β^+ (positron) emission or by electron capture. An example of β^+ *emission* is

$$^{105}_{48}\text{Cd} \longrightarrow {}^{105}_{47}\text{Ag} + {}^{0}_{1}e$$

In *electron capture* (EC) a low-energy electron is captured by the nucleus

$$^{127}_{54}\text{Xe} + {}^{0}_{-1}e \xrightarrow{\text{EC}} {}^{127}_{53}\text{I}$$

and the energy lost in the process is often emitted as an x ray.

Sometimes it is not an unfavorable neutron/proton ratio which produces decay. Rather, the total number of nucleons may just be so large that the nuclear binding force is not strong enough to hold them together. In other words, the nuclide is *beyond* the end of the belt of stability. This situation often leads to alpha emission, because in this way the nucleus can get rid of two protons and two neutrons at the same time:

$$^{211}_{84}\text{Po} \longrightarrow {}^{207}_{82}\text{Pb} + {}^{4}_{2}\text{He}$$

If the daughter nuclide is far enough beyond the belt of stability, several successive emissions may occur, as in the uranium 238 decay series (Fig. 24-2).

Nuclear fission   Sometimes a nucleus which is far beyond the end of the belt of stability will split into two major pieces, rather than emitting a succession of alpha particles. This process, *nuclear fission,* is one of the modes, though not a common one, by which uranium 235 can spontaneously decay:

$$^{235}_{92}\text{U} \longrightarrow {}^{140}_{56}\text{Ba} + {}^{92}_{36}\text{Kr} + 3\,{}^{1}_{0}n$$

As can be seen from the equation, the uranium nucleus divides itself into two fragments, with three neutrons left over.

The fission of uranium 235 described above is a natural, *spontaneous* process. Fission can be *induced,* however, when a uranium 235 nucleus captures a slow, or *thermal,* neutron:

$$^{235}_{92}\text{U} + {}^{1}_{0}n \longrightarrow [{}^{236}_{92}\text{U}] \longrightarrow {}^{90}_{38}\text{Sr} + {}^{143}_{54}\text{Xe} + 3\,{}^{1}_{0}n$$

This equation represents only one of many ways the nucleus can split. Other ways uranium 235 can undergo fission include

$$^{235}_{92}\text{U} + {}^{1}_{0}n \longrightarrow [{}^{236}_{92}\text{U}] \longrightarrow {}^{94}_{36}\text{Kr} + {}^{139}_{56}\text{Ba} + 3\,{}^{1}_{0}n$$

and

$$^{235}_{92}\text{U} + {}^{1}_{0}n \longrightarrow [{}^{236}_{92}\text{U}] \longrightarrow {}^{90}_{38}\text{Sr} + {}^{144}_{54}\text{Xe} + 2\,{}^{1}_{0}n$$

Thus the fission fragments are variable. When many uranium 235 nuclides undergo fission, a collection of many different fragments distributed over a wide range of masses is produced. Figure 24-5 shows the yields of various products of the induced fission of uranium 235. Nuclear fission is the energy-producing process used in nuclear weapons and nuclear reactors (See Secs. 24-5 and 24-6).

24-5
Fission, fusion,
and nuclear
binding energy

We are all aware of the vast amounts of energy which can be obtained from nuclear processes. Where does this energy come from, and why is there so much of it? The answer is found in the Einstein relationship, $E = mc^2$ (Sec. 6-1), based on the idea that mass can be converted into energy, and vice versa. (Or, perhaps better, mass and energy are different, but interconvertible, manifestations of the same thing.)

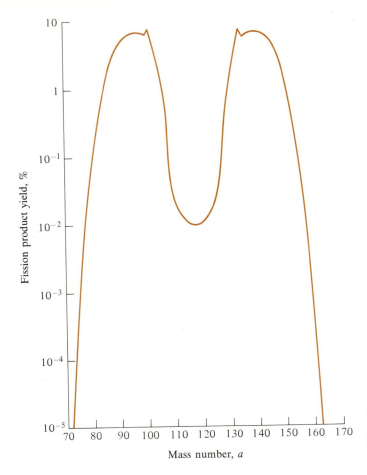

Figure 24-5
Product yields from the fission of uranium 235. (*Note: Vertical scale is logarithmic.*)

Mass and energy changes during nuclear fission

In nuclear fission there is a significant loss in mass. Consider the reaction:

$$^{235}_{92}U + ^{1}_{0}n \longrightarrow ^{94}_{38}Sr + ^{139}_{54}Xe + 3\,^{1}_{0}n$$

Let us compare the masses of the products with those of the reactants. From the following data

Particle	Mass, amu
$^{235}_{92}U$ atom	235.0439
$^{94}_{38}Sr$ atom	93.9154
$^{139}_{54}Xe$ atom	138.9178
Neutron	1.0087

we calculate the overall change in mass, Δm, during the above fission of one uranium 235 nuclide:

$$\Delta m = \Sigma(\text{mass})_{\text{products}} - \Sigma(\text{mass})_{\text{reactants}}$$

$$= [93.9154 + 138.9178 + 3(1.0087) - 235.0439 - 1.0087]\,\text{amu}$$

$$= -0.1933\,\text{amu}$$

The negative sign indicates that the system *loses* 0.1933 amu per atom of uranium. This corresponds to a loss of 0.1933 g mol^{-1}, or 1.933 × 10^{-4} kg mol^{-1}. The speed of light in a vacuum, c, is 2.998 × 10^8 m s^{-1}, and so we can calculate the energy which is equivalent to the observed loss in mass by using the Einstein relationship.

$$E = mc^2$$
$$= (1.933 \times 10^{-4} \text{ kg mol}^{-1})(2.998 \times 10^8 \text{ m s}^{-1})^2$$
$$= 1.737 \times 10^{13} \text{ kg m}^2 \text{ s}^{-2} \text{ mol}^{-1}$$

Since 1 kg m^2 s^{-2} is 1 J, this is 1.737 × 10^{13} J mol^{-1}, or, 1.737 × 10^{10} kJ mol^{-1}.

We can see that the amount of energy produced by the fission of one mole of uranium-235 atoms is colossal. It is greater by a factor of about 10^6 than the energy released in highly exothermic chemical processes. In nuclear fission about seven-eighths of this energy shows up as kinetic energy of the product particles and one-eighth as electromagnetic (radiant) energy.

Nuclear binding energy   The stability of a nucleus is measured by its *binding energy,* the energy released when it is formed from its component protons and neutrons. For example, consider the formation of a $^{57}_{27}$Co nucleus, mass = 56.9215 amu, from 27 protons and 30 neutrons. Since the mass of a proton is 1.00728 amu and that of a neutron is 1.00866 amu, the loss in mass which occurs when a cobalt-57 nucleus is formed is

$$\text{Mass of 27 protons} = 27(1.00728) = \quad 27.1966 \text{ amu}$$
$$\text{Mass of 30 neutrons} = 30(1.00866) = \quad \underline{30.2598 \text{ amu}}$$
$$57.4564 \text{ amu}$$

$$\text{Less the mass of } ^{57}_{27}\text{Co nucleus,} \quad \underline{-56.9215 \text{ amu}}$$
$$\text{Loss in mass} = \quad 0.5349 \text{ amu}$$

This is

$$0.5349 \text{ amu} \frac{1 \text{ g}}{6.022 \times 10^{23} \text{ amu}} = 8.882 \times 10^{-25} \text{ g}$$

$$= 8.882 \times 10^{-28} \text{ kg}$$

which corresponds to

$$E = mc^2$$

$$= (8.882 \times 10^{-28} \text{ kg})(2.998 \times 10^8 \text{ m s}^{-1})^2 = 7.983 \times 10^{-11} \text{ J}$$

Nuclear binding energies are usually expressed *per nucleon.* Since there are 57 nucleons in $^{57}_{27}$Co, the nuclear binding energy is

$$\frac{7.982 \times 10^{-11} \text{ J}}{57 \text{ nucleons}} = 1.401 \times 10^{-12} \text{ J nucleon}^{-1}$$

Figure 24-6 shows the way nuclear binding energy varies with mass number. This plot shows that when a uranium-235 nucleus undergoes fission to produce two lighter fragments with mass numbers closer to the middle of the curve, there is an increase in stability, as binding energy per nucleon increases. This increase is really the energy released when uranium undergoes fission.

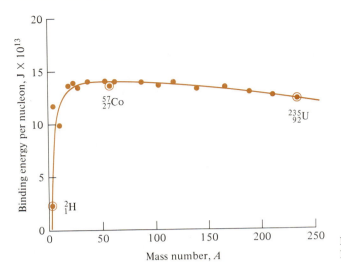

Figure 24-6
Nuclear binding energies.

Nuclear weapons and nuclear reactors

We have seen that when a nucleus undergoes fission, it splits into two fragments and several neutrons. If each of these neutrons is captured by other fissionable nuclei, the process continues, and the result is a *chain reaction* in which the sudden fission of many nuclei and the resulting production of huge quantities of energy produce a *nuclear explosion*. In the so-called atomic bomb (not a very descriptive name) a certain quantity, the *critical mass,* of a fissionable nuclide is suddenly assembled by the bomb's mechanism. If the mass is less than critical, too many neutrons escape and the chain reaction cannot be sustained. One way of triggering a bomb is to use a chemical explosion to blow two separate subcritical masses of fissionable material at each other, so that the critical mass is suddenly exceeded. Both uranium 235 and plutonium 239 have been used in nuclear weapons. Plutonium 239 is made by bombarding uranium 238, the most common isotope of uranium, with neutrons.

$$^{238}_{92}U + ^{1}_{0}n \longrightarrow ^{239}_{92}U$$

The uranium 239 then decays in two steps to form plutonium 239:

$$^{239}_{92}U \longrightarrow ^{239}_{93}Np + ^{0}_{-1}e$$

$$^{239}_{93}Np \longrightarrow ^{239}_{94}Pu + ^{0}_{-1}e$$

In a *nuclear reactor* no more than one of the neutrons emitted when a nucleus undergoes fission is allowed to be captured by another fissionable nucleus. In this way the reaction is kept under control. Fission goes on, but at a slower rate than in a bomb. The reactor is kept under control by adjusting the positions of control rods which are inserted among the nuclear fuel elements in the reactor. These control rods are usually made out of cadmium or boron, two elements which are highly efficient at absorbing neutrons. Figure 24-7 shows a schematic diagram of a nuclear reactor. Note that the reactor serves only as a source of heat used to boil water. Then, as in a conventionally fueled power plant, the steam drives a turbine-generator which produces electricity.

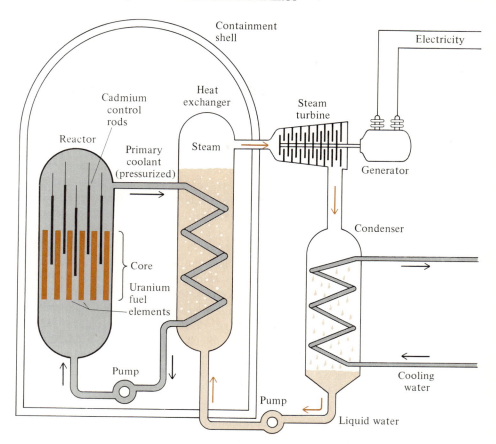

Figure 24-7
A nuclear reactor
(schematic).

Nuclear fusion　From the curve of nuclear binding energies (Fig. 24-6) it can be seen that conversion of very light nuclei (at the left of the curve) to heavier ones also results in an increase in binding energy per nucleon and so should release large amounts of energy, even larger than in fission. Such reactions are called *fusion reactions* because in them smaller nuclei fuse to form larger ones.

The source of energy in the sun is a series of reactions, the net result of which is the fusion of four protons to form a single $_2^4$He nucleus. The reaction undoubtedly takes place in steps. One possibility is

$$_1^1H + {}_1^1H \longrightarrow {}_1^2H + {}_1^0e$$
$$_1^2H + {}_1^1H \longrightarrow {}_2^3He$$
$$\underline{_2^3He + {}_1^1H \longrightarrow {}_2^4He + {}_1^0e}$$
$$\text{Net: } 4\,_1^1H \longrightarrow {}_2^4He + 2\,_1^0e$$

The only successful "practical" application of fusion reactions has been in so-called hydrogen, or thermonuclear, bombs. One problem with fusion reactions is getting them to start. In order for two light nuclei to fuse they must have extremely high energies so that the electron clouds in the atoms' extranuclear regions do not prevent nuclei from getting close to each other. What is needed is an ultrahigh temperature, about $10^8\,^\circ$C. In a thermonuclear weapon a fission reaction is used to provide the high energies needed to initiate fusion. In one device an ordinary fission bomb is surrounded by a shell of lithium deuteride. Neutrons from the fission reaction are captured by lithium nuclei,

$$\text{$^6_3$Li} + \text{$^1_0$}n \longrightarrow \text{$^4_2$He} + \text{$^3_1$H}$$

and under the high-energy conditions supplied by the fission reaction the tritium undergoes fusion with deuterium:

$$\text{$^3_1$H} + \text{$^2_1$H} \longrightarrow \text{$^4_2$He} + \text{$^1_0$}n$$

In a so-called fission-fusion-fission bomb a shell of $^{238}_{92}$U surrounds the fusion device. Normally this isotope of uranium will not undergo fission, but under energy-rich conditions it will. The "advantage" of this weapon is that $^{238}_{92}$U is relatively inexpensive because it is by far the most abundant isotope of this element.

Controlling nuclear fusion so it can be used for the production of usable power is a problem which has challenged scientists and engineers for over 20 years. Presumably reactions employing $^1_1$H and $^2_1$H would be employed, and although only about 0.015 percent of the hydrogen on earth is deuterium, there is so much water in the oceans that the total potentially available amount of this isotope of hydrogen is very large. But in addition to the problem of "igniting" a fusion reaction there is the difficulty of containing a reaction mixture at a temperature high enough to vaporize all matter. Currently experiments are being performed using a battery of high-power lasers all aimed at a very small sample of hydrogen. It is hoped that the lasers will concentrate enough energy to start the fusion reaction. Other experiments are being run in which ultrahigh-temperature mixtures are contained in a "magnetic bottle," an arrangement of intense magnetic fields which prevents escape of the reaction mixture.

24-6 Applications of radioactivity

Since the discovery of radioactivity by Becquerel in 1896, chemistry has played an active role in the application of nuclear processes to all fields of science, medicine, and technology. We have already mentioned radiochemical dating (Sec. 24-2), and we will conclude this chapter by describing a few more of these applications.

Radioactive tracers

Because radioactivity can be detected at low levels, very small amounts of radioactive materials can be used as *tracers* for following the progress of many kinds of processes. For example, some kinds of vascular disease can be diagnosed by injecting a small quantity of sodium chloride containing some sodium 24 into the bloodstream. This isotope of sodium is a β^- and γ emitter, and its progress through the arteries, capillaries, and veins can easily be followed. A similar technique is used in the petroleum industry: when a change is to be made in the type of oil being pumped through a pipeline, a small quantity of a radioactive nuclide is added at the time of changeover. Many miles away the eventual arrival of the new oil is signaled by the radioactive emission from the tracer.

Tracer studies have been useful in chemistry for many purposes. The rate of some *exchange reactions* has been studied by using tracers. For instance, the electron interchange between Fe^{3+} and Fe^{2+} can be followed by mixing radioactive $^{55}Fe^{3+}$ ions with nonradioactive Fe^{2+} ions. The rate of the exchange

$$^{55}Fe^{3+} + Fe^{2+} \longrightarrow {}^{55}Fe^{2+} + Fe^{3+}$$

is determined by taking periodic samples of the mixture, separating Fe^{2+} from Fe^{3+}, and determining how fast the radioactivity has decreased in the Fe(III) or increased in the Fe(II).

Tracer studies have been used to help elucidate structure. The Lewis structure for the thiosulfate ion is

$$\left[\begin{array}{c} \ddot{\underset{\cdot\cdot}{O}} \\ :\ddot{\underset{\cdot\cdot}{O}}:\ddot{S}:\ddot{\underset{\cdot\cdot}{S}}: \\ :\ddot{\underset{\cdot\cdot}{O}}: \end{array} \right]^{2-}$$

Support for this structure comes from evidence that the two sulfur atoms in the ion are not equivalent. When sulfur 35 (a β^- emitter) is dissolved in a solution containing the sulfite ion, the product is a thiosulfate ion in which one of the sulfur atoms, the one from the elemental sulfur, is *tagged* or *labeled*:

$$^{35}S(s) + SO_3^{2-} \longrightarrow {}^{35}SSO_3^{2-}$$

When this solution is acidified, elemental sulfur and sulfurous acid are produced. If the two sulfur atoms in the thiosulfate ion are equivalent, then each has a 50:50 chance of ending up in the elemental sulfur, and so half the radioactivity should be in the S(s) and half in the H_2SO_3. But the results are otherwise: all of the radioactivity ends up in the precipitated sulfur and none in solution. This is clear evidence that the two atoms of sulfur in $S_2O_3^{2-}$ are structurally nonequivalent and is consistent with the above Lewis structure.

Tracer studies have been of great use in establishing metabolic pathways in living organisms. The American chemist Melvin Calvin relied heavily on CO_2 tagged with carbon 14 (a β^- emitter) in determining the photosynthetic mechanism by which plants convert carbon dioxide and water to glucose and oxygen. By analyzing the leaves of plants exposed to CO_2 and sunlight for various periods of time, Calvin was able to break down the overall process,

$$6H_2O(l) + 6CO_2(g) \longrightarrow C_6H_{12}O_6 + 6O_2(g)$$

into a complex sequence of steps.

Analytical techniques   Radioactive isotopes have a number of applications in chemical analysis. One technique, known as *isotope dilution analysis,* is useful in analyzing for a component which is difficult to separate completely from a mixture. For example, suppose that it is desired to determine the amount of $NaNO_3$ which is present in a solution. First, a known quantity of additional $NaNO_3$ containing radioactive sodium 24 is dissolved in the solution. Then water is evaporated until some $NaNO_3$ has crystallized out of solution. The radioactivity of this solid is then measured, and if, for example, it is 3 percent of that of the added radioactive $NaNO_3$, then the ratio of original to added $NaNO_3$ in the crystals must be 97:3. But this must also be the ratio in the original solution, and so from the quantity of radioactively tagged $NaNO_3$ added, the quantity of $NaNO_3$ originally present can be determined.

Another analytical technique, *neutron activation analysis,* is useful for the nondestructive analysis of many substances. When a stable nuclide is bombarded with neutrons, a heavy isotope of the element is formed, as the stable nuclide captures a neutron:

$$_Z^A X + {}_0^1 n \longrightarrow {}_Z^{A+1} X$$

If the product nuclide lies above the belt of stability it will undergo β^- decay with a characteristic half-life:

$$_Z^{A+1} X \longrightarrow {}_{Z+1}^{A+1} Y + {}_{-1}^0 e$$

Even if the product nuclide is stable, it is generally in an excited state, a result of the energy it has received from the colliding neutrons. It thus emits one or more photons (γ rays) of *characteristic energy,* until it drops down to its ground state. Since half-lives and characteristic γ-emission energies are known for most nuclides, measurement of these quantities permits identification of the unknown substance. Neutron activation analysis is especially useful for analysis of very small samples and of samples in which the element being sought is at a very low concentration, often less than 10^{-4} percent.

Structure modification

High-energy radiation from unstable nuclides can produce important structural changes in matter. Irradiating polyethylene plastic with β^- or γ radiation dissociates some of the hydrogen atoms from the hydrocarbon chains (Sec. 23-12). These atoms form H_2 molecules, and adjacent chains form *cross-linkages* where the H atoms were. Cross-linking increases the strength and durability of plastics. Cobalt 60 is now used in the treatment of cancer. (Radium therapy is no longer common because of the high cost of this element.) $_{27}^{60}$Co is a β^- and γ emitter, and the radiation is focused on malignant tissue to destroy it. Small amounts of compounds containing iodine 131 are taken internally by patients with thyroid-gland cancer. The radioactive iodine concentrates in the thyroid and destroys tissue in the gland with minimal effect on the rest of the body.

Neutron diffraction

One of the most exciting techniques in the area of structure determination is *neutron diffraction.* This is analogous to x-ray diffraction (Sec. 10-2), but there is one big difference: x rays are strongly scattered only by high concentrations of electrons. Thus atoms such as hydrogen which have few electrons are difficult or impossible to locate in a crystal by the use of x-ray diffraction. On the other hand, neutrons are scattered by *nuclei* and can be used to locate the positions of the lighter elements, even hydrogen, in solid-state structures.

SUMMARY

We have concluded this book with a brief introduction to nuclear transformations. *Natural radioactivity* is exhibited by many naturally occurring nuclides. This is usually characterized by the emission of *alpha* (α) *particles, beta-minus* (β^-) *particles,* and *gamma* (γ) *rays.* An alpha particle is essentially a $_2^4$He nucleus, and as it is emitted, the mass number of the decaying nucleus de-

creases by 4, and its atomic number, by 2. A beta-minus particle is a high-energy electron, $_{-1}^0 e$, and as it is emitted, the atomic number of the decaying nucleus increases by 1, its mass number remaining unchanged. Gamma rays are high-energy electromagnetic radiation and are not accompanied by any change in either atomic or mass number of the decaying nucleus. Less common modes of natural radioactive decay are *electron capture* and *spontaneous fission*.

Stable nuclei may be made radioactive by bombarding them with high-energy particles such as alpha particles, neutrons, or electrons. This *induced radioactivity* is much like natural radioactivity, except that additional modes of decay are observed. These include neutron and positron (β^+) emission.

The stable nuclei form a *belt of stability* when they are plotted on a neutron-versus-proton plot. Unstable nuclei tend to emit particles which will either change their neutron/proton ratios or will reduce their total number of nucleons so that they are located closer to the belt of stability.

All radioactive decay follows first-order kinetics. Because first-order half-lives are constants characteristic of the reacting species, measurement of radioactivity or of decay products provides a method of determining the age of old objects. This method is called *radiochemical dating* and is based on the known half-lives of radioactive nuclides.

Nuclear binding energy is the energy released when a nucleus is assembled from its component nucleons. This energy is equivalent to the decrease in mass during this process, the two being related by the Einstein formula $E = mc^2$. During *nuclear fission,* heavy nuclei split to form lighter nuclei with higher binding energies. Neutrons are also released, and if these are captured by other fissionable nuclei, a *chain reaction* is initiated. During *nuclear fusion,* lighter nuclei merge to form heavier ones with higher binding energies. In both fission and fusion the loss in mass shows up as a very large emission of energy.

The many applications of radioactivity include the use of tracers to follow the physical motion of substances, to follow the path of atoms undergoing reaction, and to determine structure. Radioactive nuclides are used in analytical chemistry in *isotope dilution analysis* and *neutron activation analysis*. High-energy radiation produces fundamental structural changes in matter. Advantage is taken of this to modify the structures of substances like plastics and to destroy malignant tissue in living organisms. *Neutron diffraction* is a structure-determination technique which is similar to x-ray diffraction, but which provides information about the locations of light as well as heavy atoms in a structure.

KEY TERMS

Alpha particle (Sec. 24-1)
Belt of stability (Sec. 24-4)
Beta-minus particle (Secs. 24-1,3)
Beta-plus particle (Sec. 24-3)
Chain reaction (Sec. 24-5)
Compound nucleus (Sec. 24-3)
Critical mass (Sec. 24-5)
Cyclotron (Sec. 24-3)
Electron capture (Secs. 24-1,3)
Fission (Secs. 24-4,5)
Fusion (Sec. 24-5)
Gamma ray (Sec. 24-1)
Geiger-Müller counter (Sec. 24-2)
Induced radioactivity (Sec. 24-3)

Isotope dilution analysis (Sec. 24-6)
K capture (Sec. 24-1)
Natural radioactivity (Sec. 24-1)
Neutron activation analysis (Sec. 24-6)
Neutron diffraction (Sec. 24-6)
Neutron emission (Sec. 24-3)
Positron (Sec. 24-3)
Radioactivity (Sec. 24-1)
Radiochemical dating (Sec. 24-2)
Thermonuclear reaction (Sec. 24-5)
Tracer (Sec. 24-6)
Transmutation (Sec. 24-3)
Transuranium element (Sec. 24-3)

QUESTIONS AND PROBLEMS

Atomic nuclei

24-1 Distinguish between the terms *nuclide* and *isotope*.

24-2 How is it possible that (a) alpha particles, (b) beta particles, and (c) gamma rays can come from nuclei which did not have them to start with?

24-3 When a high-energy particle from an accelerator collides with a nucleus, a new particle, a *meson,* is often formed. Its mass is intermediate between that of a proton and that of an electron. If no protons or neutrons are destroyed, where did it come from?

24-4 Consider the neutron/proton ratio in an isolated neutron, and then write a nuclear equation showing how a neutron decays.

24-5 What is a *positron*? What happens to a nucleus when it emits a positron? A positron and an electron both vanish when they collide with each other. Where do they go?

Radioactivity

24-6 What are the kinds of naturally occurring radioactive decay? How do they differ from each other?

24-7 Describe the process of electron capture. How can this process be detected?

24-8 What can we conclude about the structure of the nucleus from the fact that gamma emission is a commonly observed mode of decay?

24-9 What particle is emitted during each of the following nuclear disintegrations?

(a) $^{223}_{88}Ra \longrightarrow ^{219}_{86}Rn + ?$

(b) $^{241}_{94}Pu \longrightarrow ^{241}_{95}Am + ?$

(c) $^{56}_{27}Co \longrightarrow ^{56}_{26}Fe + ?$

(d) $^{89}_{36}Kr \longrightarrow ^{88}_{36}Kr + ?$

24-10 Complete the following equations by indicating the daughter nucleus in each case:

(a) $^{49}_{24}Cr \longrightarrow ? + ^{0}_{-1}e$

(b) $^{126}_{53}I \longrightarrow ? + ^{0}_{+1}e$

(c) $^{214}_{83}Bi \longrightarrow ? + ^{4}_{2}He$

(d) $^{214}_{83}Bi + ^{0}_{-1}e \xrightarrow{EC} ?$

24-11 Write the nuclear equation for the emission of an alpha particle by (a) $^{178}_{79}Au$ (b) $^{227}_{90}Th$ (c) $^{257}_{104}Rf$.

24-12 Write the nuclear equation for the β^- emission of (a) $^{127}_{50}Sn$ (b) $^{10}_{4}Be$ (c) $^{56}_{25}Mn$.

24-13 Write the nuclear equation for electron capture by (a) $^{125}_{55}Cs$ (b) $^{195}_{82}Pb$ (c) $^{243}_{98}Cf$.

24-14 Write the nuclear equation for positron emission by (a) $^{203}_{83}Bi$ (b) $^{105}_{48}Cd$ (c) $^{70}_{33}As$.

24-15 Complete the following equations by indicating in each case the particle captured by the nuclide at the left:

(a) $^{59}_{27}Co + ? \longrightarrow ^{56}_{25}Mn + ^{4}_{2}He$

(b) $^{43}_{20}Ca + ? \longrightarrow ^{46}_{21}Sc + ^{1}_{1}H$

(c) $^{7}_{3}Li + ? \longrightarrow ^{7}_{4}Be + ^{1}_{0}n$

(d) $^{130}_{52}Te + ? \longrightarrow ^{130}_{53}I + 2\,^{1}_{0}n$

(e) $^{246}_{96}Cm + ? \longrightarrow ^{254}_{102}No + 5\,^{1}_{0}n$

24-16 How do a gamma ray and an x ray differ in (a) energy? (b) wavelength? (c) frequency?

24-17 In the thorium radioactive decay series the starting $^{232}_{90}Th$ nucleus emits the following particles in succession: α, β, β, α, α, α, β, α, α, β. What stable daughter nuclide is formed?

24-18 In the actinium decay series a total of seven alpha particles are emitted. If the respective starting and ending nuclides are $^{235}_{92}U$ and $^{207}_{82}Pb$, how many beta particles are emitted, assuming that no other particles are released?

Nuclear reaction kinetics

24-19 Thallium 202 decays by electron capture with a half-life of 12 days. If a 1.00-g sample of thallium 202 decays for a period of 36 days, how much thallium is left?

24-20 Half-lives for radioactive decay processes are found to be independent of the size of the sample. What does this tell you about the kinetics of these processes?

24-21 The rate constant for the beta decay of $^{118}_{48}Cd$ is 1.4×10^{-2} min^{-1}. What fraction of any sample of cadmium is left after 12 h?

24-22 The rate constant for the alpha decay of thorium 230 is 8.7×10^{-6} year^{-1}. What is the half-life of this isotope of thorium?

24-23 A sample of pitchblende, a uranium ore, was analyzed and found to contain 51.09 percent $^{238}_{92}U$ and 2.542 percent $^{206}_{82}Pb$. If the half-life of uranium 238 is 4.5×10^9 years, how old was the rock found to be?

24-24 Some fragments of bone found in an excavation in Egypt were found to have a $^{14}_{6}C$ radioactivity 0.629 times that in the bones of living animals. How old are the fragments?

24-25 A sample of granite is found to have a $^{40}_{19}K/^{40}_{18}Ar$ mass ratio of 0.592. If the half-life of $^{40}_{19}K$ is 1.28×10^9 years, what is the age of the rock?

Nuclear stability and nuclear energy

24-26 What is the belt of stability?

24-27 List two ways by which an unstable nuclide might approach the belt of stability if (a) its neutron/proton ratio were too high (b) its neutron/proton ratio were too low (c) the total number of nucleons present were too great.

24-28 Predict a likely mode of decay for each of the following nuclides: (a) $^{37}_{19}K$ (b) $^{129}_{56}Ba$ (c) $^{143}_{57}La$ (d) $^{255}_{101}Md$.

24-29 Nobelium undergoes spontaneous fission. Write nuclear equations showing three possible pairs of fission products. (Assume in each case that three neutrons are produced.)

24-30 Calculate the average binding energy per nucleon for (a) $^{6}_{3}Li$ (mass = 6.01512 amu) (b) $^{58}_{26}Fe$ (mass = 57.9333 amu) (c) $^{235}_{92}U$ (mass = 235.0439 amu).

24-31 All things being equal, which process produces more energy per gram of matter: fusion or fission?

24-32 In a fusion or fission reaction mass seems to be converted to energy. Why, then, do the numbers of protons and neutrons remain constant?

24-33 Comment on the statement: In an endothermic chemical reaction the total mass of the products is greater than the total mass of the reactants.

24-34 What problems are there in "igniting" a fission bomb; in controlling a nuclear fission reactor?

24-35 What problems are there in "igniting" a fusion bomb; in controlling a nuclear fusion reactor?

24-36 Estimate the energy (kJ mol^{-1}) released in each of the following processes:

(a) $^{238}_{92}U \longrightarrow ^{234}_{90}Th + ^{4}_{2}He$

(b) $^{1}_{1}H + ^{3}_{1}H \longrightarrow ^{4}_{2}He$

(c) $^{235}_{92}U + ^{1}_{0}n \longrightarrow ^{139}_{56}Ba + ^{94}_{36}Kr + 3\,^{1}_{0}n$

Applications

24-37 Describe an experiment which you could perform in order to determine whether the carbonyl-oxygen or the oxygen-ethyl bond breaks in the hydrolysis

$$CH_3\overset{\displaystyle O}{\overset{\|}{C}}-O-CH_2CH_3 + H_2O \xrightarrow{\;H^+\;}$$

$$CH_3\overset{\displaystyle O}{\overset{\|}{C}}-O-H + CH_3CH_2OH$$

24-38 Suppose that you wish to determine the volume of water in a small pond. You pour 55 gal of water which has been enriched with tritium ($^{3}_{1}H$) into the pond, mix well with a motorboat, and then take samples. If the radioactivity is 1.4×10^{-4} percent of that in the original radioactive water, how many gallons of water are in the pond?

24-39 Describe a radioactive tracer experiment which you could do to **(a)** determine the solubility product of lead bromide **(b)** determine the rate of solid-state diffusion of one metal into another **(c)** measure the vapor pressure of a substance of low volatility **(d)** determine the rate of wear of piston rings in an automobile engine.

24-40 The most stable isotope of astatine is $^{210}_{85}At$, with a half-life of only 8.3 h. Without using any more than a trace quantity of astatine, how could you show that this element is a halogen?

APPENDIX A
GLOSSARY OF IMPORTANT TERMS

absolute zero. The temperature at which all particles in a substance are in their lowest energy states: 0 K, or $-273.15°C$.

accuracy. The truthfulness or correctness of a number obtained by measurement.

acidic solution. An aqueous solution in which the concentration of hydrogen (hydronium) ions exceeds that of hydroxide ions.

acid. A substance which (1) produces hydrogen ions in aqueous solution (Arrhenius definition); (2) is a proton donor (Brønsted-Lowry); (3) is an electron-pair acceptor (Lewis); or (4) increases the concentration of dissolved cations related to the solvent (solvent system).

acid salt. A salt, the anion of which can serve as an acid by losing an H^+ (donating a proton).

actinoid. A member of the series of 14 elements following actinium in the periodic table; also called *actinide.*

activated complex. A short-lived combination formed by collision of reactant particles in an elementary process; also called *transition state.*

activation energy. The kinetic energy which reactant particles must have in order for their collision to result in the formation of an activated complex.

active site. A location on the surface of a heterogeneous catalyst or an enzyme at which reactant molecules can combine and react with a low required activation energy.

activity. A quantity which measures the apparent or effective concentration (or, for a gas, partial pressure) of a species and which takes into account interparticle interactions which produce nonideal behavior. At low concentrations (or pressures) activity is essentially equal to concentration (or pressure).

addition polymerization. A polymer-forming reaction in which monomers bond through addition reactions.

addition reaction. A reaction in which one molecule adds to another with no loss of atoms from either.

adiabatic change. A change which takes place without gain or loss of heat.

aerosol. A colloidal dispersion of a solid or liquid in a gas: a fog or a smoke.

alchemy. The chemical period from about 300 B.C. to A.D. 1500 during which a central goal was the transmutation of base metals into gold.

alcohol. An organic compound of the type R—OH.

aldehyde. An organic compound of the type $R-\overset{\overset{\displaystyle O}{\|}}{C}-H$.

alkali metal. A member of group IA in the periodic table.

alkaline-earth metal. A member of group IIA in the periodic table.

alkane. A saturated hydrocarbon with the general formula C_nH_{2n+2}.

alkene. An unsaturated hydrocarbon with the general formula C_nH_{2n}. Contains a C=C double bond.

alkyl group. A hydrocarbon group consisting of an alkane molecule less one of its hydrogen atoms.

alkyne. An unsaturated hydrocarbon with the general formula C_nH_{2n-2}. Contains a C≡C triple bond.

allotropes. Different forms of an uncombined element.

alloy. A combination of metals; may be a solid solution, a compound, or a mixture of these.

alpha (α) particle. A radioactive emission consisting of two protons and two neutrons as in a $^4_2$He nucleus.

amalgam. An alloy in which one component is mercury.

amine. An organic compound with the general formula

$$R-NH_2 \text{ (primary)}, \quad R-\underset{\underset{H}{|}}{N}-R' \text{ (secondary)}, \quad \text{or}$$

$$R-\underset{\underset{R''}{|}}{N}-R' \text{ (tertiary)}.$$

amino acid. A carboxylic acid with an amine ($-NH_2$) group on a noncarboxyl carbon.

ammine. The name given to ammonia, NH_3, when it serves as a ligand.

amorphous solid. A substance with the external appearance and characteristics of a solid, but with the irregular structure typical of a liquid; a highly supercooled liquid; a *glass*.

amphiprotism. The ability of a species either to gain or to lose a proton.

amphoterism. The ability of a substance to react either as an acid or as a base.

anion. A negatively charged ion.

anode. The positively charged electrode in a gas-discharge tube. In an electrochemical cell, the electrode at which oxidation occurs.

antibonding orbital. A molecular orbital in which electrons have higher energies than in the unbonded atoms; an orbital characterized by a region of low electron probability density between the bonded atoms, producing a destabilizing effect on the molecule.

antinode. A region or location of maximum disturbance in a standing wave.

antiparallel spins. Two spins which are in opposite directions.

aqua regia. A powerful oxidizing and complexing solvent mixture composed of about three parts concentrated HCl to one part concentrated HNO_3.

aquo complex. A complex in which water molecules serve as ligands.

aromatic hydrocarbon. A hydrocarbon usually containing at least one benzene ring in its structure.

Arrhenius plot. A plot of the logarithm of the specific rate constant for a reaction against the reciprocal of the absolute temperature; useful for determining the activation energy of the reaction.

atmosphere (standard atmosphere). A unit of pressure; 1 atm = 1.013×10^5 pascals (Pa) = 760 mmHg.

atom. The smallest particle of an element which shows the properties of the element.

atomic mass unit. A unit of mass; one atomic mass unit (amu) is defined as $\frac{1}{12}$ the mass of one $^{12}_6$C atom.

atomic number. The number of protons in the nucleus of an atom.

atomic weight. The weighted (by abundance) average of the masses of the isotopes of an element expressed in atomic mass units.

Aufbau procedure. The procedure of starting at the lowest of a set of energy levels and gradually adding electrons to progressively higher levels; "building-up" procedure.

autodissociation. Self-dissociation. The production of cations and anions by dissociation of solvent molecules without interaction with other species.

auto-oxidation. A reaction in which one substance acts simultaneously as an oxidizing agent and a reducing agent; also called *disproportionation*.

Avogadro's number. The number of atoms in exactly 12 g of $^{12}_6$C; 6.02×10^{23}. The number of units in one mole of units.

azimuthal quantum number, *l*. A quantum number which specifies a subshell for an electron in an atom.

base. A substance which (1) produces hydroxide ions in aqueous solution (Arrhenius definition), (2) is a proton acceptor (Brønsted-Lowry), (3) is an electron-pair donor, or (4) increases the concentration of anions related to the solvent (solvent system).

basic solution. An aqueous solution in which the concentration of hydroxide ions exceeds that of hydrogen (hydronium) ions.

battery. (1) A set of galvanic cells connected in series. (2) In common usage, a single commercial galvanic cell.

beta-minus (β^-) particle. A radioactive emission consisting of a high-energy electron, $^0_{-1}e$. Also known, simply, as a *beta particle*.

beta-plus (β^+) particle. A radioactive emission consisting of a positron, ^0_1e.

bimolecular process. An elementary process in which the activated complex is formed as a result of the collision of two particles.

boiling-point elevation. A colligative property: the increase in boiling point of a solvent brought about by the presence of a solute.

bond axis. A line passing through the nuclei of two bonded atoms.

bond distance. The distance between nuclei of two bonded atoms. Also known as *bond length*.

bonding orbital. A molecular orbital in which electrons have lower energies than they would in the unbonded atoms; an orbital characterized by a region of high electron probability density between the bonded atoms, leading to a stabilizing effect on the molecule.

bonding pair. A pair of electrons shared between two atoms and constituting a covalent bond (VB model).

boundary surface. A surface of constant electron probability density, Ψ^2.

buffer. A solution which contains moderate or high concentrations of a Brønsted-Lowry conjugate acid-base pair; a solution whose pH does not change greatly in response to added acids or bases.

bumping. Sudden, explosive boiling following the superheating of a liquid.

cage effect. The trapping of two reactant particles in liquid solution by a shell of solvent molecules.

calorie (cal). A unit of energy or work: 1 cal = 4.184 J.

carbohydrate. An aldehyde or ketone which is a polyol, or a polymer of these.

carbonyl group. The $-\overset{\overset{\text{O}}{\|}}{\text{C}}-$ structure.

carboxylate ion. The anion of a carboxylic acid.

carboxylic acid. An organic compound with the general formula $R-\overset{\overset{\text{O}}{\|}}{\text{C}}-OH$.

catalyst. A substance or species which is not consumed in a reaction, but whose presence provides an alternate, low-activation-energy mechanism for the reaction. A catalyst increases the rate of a reaction.

cathode. The negatively charged electrode in a gas-discharge tube. In an electrochemical cell, the electrode at which reduction occurs.

cathode ray. The stream of electrons emanating from the cathode in a discharge tube.

cathodic protection. Preventing the oxidative corrosion of a metal by forcing it to be a cathode.

cation. A positively charged ion.

centered cell. A unit cell which has entities (atoms, molecules, ions) at locations in addition to the cell corners. A nonprimitive cell.

chain branching. In a chain reaction, a step which produces more chain carriers than it consumes.

chain isomerism. Isomerism involving differences in skeletal chains of atoms in molecules.

chain reaction. A reaction type characterized by the formation of products of a later step (chain carriers) which are reactants for an earlier step.

chalcogen. A member of group VIA in the periodic table.

chelate. A complex in which the ligands are polydentate.

chemical change (reaction). A change in which one or more substances are transformed into one or more new substances.

chemical property. A property of a substance which can be described only by referring to a chemical reaction.

chemistry. The science of the compositions and structures of substances and of the changes which these undergo.

chirality. Handedness; *chiral* means "handed," as in right- or left-handed. A chiral molecule cannot be superimposed on its mirror image.

cis isomer. Any isomer in which two identical atoms or groups are adjacent to each other or on the same side of a structure.

classical mechanics. A system of mechanics developed before the inception of quantum ideas; useful for describing the behavior of objects and particles much larger than atoms.

clathrate. A "cage compound" in which atoms or molecules are trapped in a cage of covalently bonded atoms, but not directly bonded to them.

colligative property. A property of a solution which depends upon the concentration of solute particles, but not on their identities.

colloid. A dispersion of one phase in another in which the particles or units of the dispersed phase have at least one dimension which is larger than usual molecular dimensions, but too small to be observed visually.

color center. A kind of point defect in a crystal. (See Sec. 11-3.)

common-ion effect. The shifting of an ionic equilibrium due to the addition of an ion involved in the equilibrium; usually refers to the repression of dissociation of a weak electrolyte or the decrease in the solubility of an electrolyte brought about by the addition of an ion which is a dissociation product.

complex. An ion or, sometimes, molecule which consists of a central atom or ion surrounded by some peripheral atoms, ions, or molecules (ligands) bonded to it.

complexation. The formation of a complex.

compound. A pure substance composed of atoms of different elements, whose components cannot be separated by physical means.

condensation polymerization. A polymer-forming reaction in which a small molecule is split out when each pair of monomers bond together.

conjugate acid-base pair. An acid and the base formed by removal of a proton from the acid, or a base and the acid formed by the addition of a proton to the base (Brønsted-Lowry).

contour diagram. A closed curve which represents the intersection of a boundary surface with a plane passing through the nucleus or nuclei; a curve of constant probability density, Ψ^2, in such a plane.

coordinate covalent bond. A covalent bond in which both shared electrons appear to have been contributed by one atom.

coordination number. The number of atoms, ions, or molecules surrounding a central atom or ion in a complex or, sometimes, in a solid.

core. All of an atom except for its valence shell of electrons; also called the *kernel*.

coulomb (C). The SI derived unit of electrical charge; the quantity of charge passing through a conductor carrying a current of one ampere in one second.

counterion. An ion with a charge which is opposite in sign to that of some ion under consideration.

covalent bond. A bond consisting of a pair of electrons shared between the bonded atoms.

covalent solid. A solid in which atoms are bonded covalently to form a giant extended network.

critical mass. The mass of fissionable material necessary for the fission reaction to be self-sustaining.

critical point. The temperature and pressure above which the liquid and gaseous states become indistinguishable.

crystal lattice. The regular, repeating arrangement of particles (atoms, ions, molecules) in a crystal.

crystalline solid. A true solid, one with a regular internal structure.

cyclic. Having a ring-type structure in which a chain of atoms is closed on itself.

cycloalkane. A cyclic hydrocarbon with the general formula C_nH_{2n}.

decomposition potential. The minimum voltage which must be applied across a pair of inert electrodes immersed in a medium in order to electrolyze the medium.

defect. An internal irregularity or flaw in the structure of a crystal.

delocalized molecular orbital. An orbital which extends over more than two atoms in a molecule, ion, or larger aggregate.

density. The mass of a substance which occupies a unit of volume: density = mass/volume.

deuterium. Hydrogen 2; ^2_1H; "heavy hydrogen."

dielectric constant. A measure of the ease with which the particles of a substance may be polarized (distorted or oriented) by an electric field.

differentiating solvent. A solvent which will discriminate among the strengths of acids or bases, all of which are strong (completely dissociated) in water (Brønsted-Lowry).

diffraction grating. A set of closely spaced lines ruled or scribed on a mirror (reflection grating) or transparent piece of glass or plastic (transmission grating).

diffusion. The passage of one substance through another.

dipole. A polar molecule; a molecule in which the centers of positive and negative charge do not coincide.

dipole-dipole forces. Forces between polar molecules.

dipole moment. The product of the magnitude of the charge at one end of a dipole times the distance between the opposite charges.

diprotic acid. An acid with two available H^+, or two donatable protons.

dimer. A combination of two identical molecules bonded together.

dislocation. A structural defect in which the lattice planes in a crystal are incomplete or warped. Also known as a *line defect*.

dispersion forces. Weak forces between atoms or molecules due to momentary fluctuations in their electronic charge-cloud distributions. Also called *London forces*.

disproportionation. A reaction in which one substance acts simultaneously as an oxidizing agent and a reducing agent; auto-oxidation.

dissociation. The splitting apart of a molecule to form two fragments; the reaction of an electrolyte with a solvent to form ions. Sometimes called *ionization*.

dissociation constant, K_{diss}. The equilibrium constant for a dissociation equilibrium.

domain. A region in a ferromagnetic substance within which the magnetic moments of all the atoms are aligned.

doping. The addition of controlled small amounts of a foreign substance to an otherwise pure substance.

edge dislocation. A dislocation in which a layer of particles in the crystal is incomplete.

effusion. The passage of a substance through a small orifice.

electrochemical cell. Any device which converts electrical into chemical energy, or vice versa.

electrode potential. The voltage associated with a half-reaction written, by convention, as a reduction; a reduction potential.

electrolysis. The passage of an electric current through a medium to produce a chemical change.

electrolyte. A substance which produces ions when dissolved in a solvent.

electrolytic cell. An electrochemical cell in which electrical energy is used to produce chemical change; a cell in which electrolysis takes place.

electron. A (perhaps) fundamental subatomic particle with a very low mass and a unit negative electrical charge; found in the extranuclear region of an atom.

electron affinity. The quantity of energy released when a gaseous, isolated, ground-state atom (or, sometimes, ion) gains an electron.

electron capture. A mode of radioactive decay in which an electron from the extranuclear region, usually the K shell, is captured by a nucleus.

electron-deficient compound. A compound in which insufficient electrons are available to bond all the atoms with conventional (two-center) covalent bonds.

electron gas. The delocalized electrons in a metal.

electronegativity. The relative tendency of a bonded atom to attract electrons to itself.

electrophile. An atom or group of atoms which appears to seek electrons in its reactions.

element. A pure substance composed of atoms each having the same atomic number. An element cannot be chemically decomposed.

elementary process. A single step of a reaction mechanism.

elimination reaction. A reaction in which atoms or groups on adjacent atoms in a molecule are removed to leave a double or triple bond between the atoms.

empirical formula. A formula expressing the simplest whole-number ratio of atoms of each element in a compound. (Also called *simplest formula*.)

emulsion. A colloidal dispersion of a liquid in another liquid.

enantiomer. One of a pair of optical isomers.

encounter. The period of time during which two reactant particles are trapped by a solvent-molecule cage in a liquid-solution reaction.

endothermic reaction. A reaction which absorbs heat.

energy, E. The ability to do work.

enthalpy, H. A thermodynamic quantity which is useful for describing heat exchanges taking place under constant-pressure conditions. The enthalpy of a system is defined as the sum of its energy, E, and its pressure-volume product: $H = E + PV$.

enthalpy of formation, ΔH_f. The enthalpy change for a reaction in which a compound is formed from its uncombined elements.

entropy, S. A thermodynamic quantity which measures the degree of disorder or randomness in a system.

enzyme. A protein which serves as a biochemical catalyst.

equilibrium. A state in which opposing processes or reactions take place at the same rate so that no net change is observed.

equilibrium condition. The condition, that the mass-action expression equals the value of the equilibrium constant for a reaction, which is satisfied when the reacting system is at equilibrium.

equilibrium constant. The value of the mass-action expression for a reaction at equilibrium.

equivalence point. The state in a titration at which equal numbers of equivalents of reactants and products have been mixed.

equivalent, acid-base. The quantity of an acid (or base) which will furnish (or react with) one mole of H^+.

equivalent, redox. The quantity of an oxidizing agent (or a reducing agent) which will accept (or furnish) one mole of electrons.

equivalent weight. The mass in grams of one equivalent.

escaping tendency. The tendency shown by a substance to escape from its phase to another.

ester. An organic compound with the general formula

$$R-\overset{\displaystyle O}{\overset{\displaystyle \|}{C}}-O-R'.$$

ether. An organic compound with the general formula $R-O-R'$.

excited state. Any state higher in energy then the ground state.

exothermic reaction. A reaction which liberates heat.

extranuclear region. All of an atom except the nucleus.

facial (fac-) isomer. An isomer of an octahedral complex in which three adjacent octahedral positions are occupied by one kind of ligand.

faraday ($\mathscr{F}$). A unit of electrical charge: 1 faraday equals 9.65×10^4 coulombs.

fat. A solid or semisolid ester of the triol glycerol and fatty acids.

ferromagnetism. A strong attraction into a magnetic field.

fission. A nuclear process in which massive nuclei split to form lighter nuclei, several neutrons, and much energy.

fog. A colloidal dispersion of a liquid in a gas.

formal charge. A somewhat arbitrary but useful way of indicating the approximate electrical characteristic or charge of an atom. (See Sec. 8-4.)

formula unit. The group of atoms indicated by the formula of a substance.

formula weight. The sum of the masses of the atoms indicated in a formula, expressed in atomic mass units; the mass of one formula unit.

free energy, G. A thermodynamic quantity which measures the energy in a system which is available for doing work, other than work of expansion. The free energy of a system is defined as the difference between its enthalpy, H, and its temperature-entropy product: $G = H - TS$.

free energy of formation, ΔG_f. The free energy change for a reaction in which one mole of a compound is formed from its uncombined elements.

free expansion. The expansion of a substance, usually a gas, against no opposing pressure.

free radical. A molecule, often a reaction intermediate, with an unpaired electron.

freezing-point depression. The lowering of the freezing point of a solvent brought about by the presence of a solute.

frequency, ν. Number of vibrations, oscillations, or excursions per second; measured in hertz (Hz), cycles per second (cps), or reciprocal seconds (s^{-1}), all of which are equivalent.

frequency factor. The quantity preceding the exponential term in the Arrhenius equation (Sec. 14-3).

fuel cell. A galvanic cell in which reactants can be replaced continuously as energy is drawn from the cell.

functional group. A group of atoms in a molecule which cause the molecule to undergo a set of characteristic reactions.

fusion. A nuclear process in which lighter nuclei merge or fuse to form more massive nuclei, releasing much energy.

galvanic cell. An electrochemical cell which produces electrical energy as a result of a spontaneous chemical reaction.

gamma (γ) ray. High-energy electromagnetic radiation emitted from a nucleus.

gel. A solid-liquid colloid in which the solid phase is one- or two-dimensional and is continuous throughout the colloid.

glass. An amorphous solid.

ground state. The state of lowest energy of a particle.

group. A vertical column of elements in the periodic table, sometimes called a family of elements.

half-life. The period of time necessary for one-half of a reactant to be consumed in reaction or process.

half-reaction. An equation for an oxidation or a reduction which specifically indicates the electrons lost or gained.

halogen. A member of group VIIA in the periodic table.

heat. Energy which is in transit from a hot to a cold object because of the temperature difference.

heat capacity. The quantity of heat necessary to raise the temperature of a substance 1 K. The *molar heat capacity, C,* is the heat capacity per mole of substance.

heat of formation, ΔH_f. The change in enthalpy for a reaction in which one mole of a compound is formed from its uncombined elements (*enthalpy of formation*).

Heisenberg uncertainty principle. The product of the uncertainty in the position of a particle times that in its momentum is a constant. It is impossible to determine simultaneously both the position and the momentum of a particle exactly.

heterogeneous. Composed of two or more phases.

hole. The absence of an electron in a semiconductor.

homogeneous. Composed of a single phase.

homonuclear. Possessing nuclei of atoms of the same element.

Hund's rule. The rule which states that two electrons tend to remain unpaired and in separate orbitals of the same energy, rather than paired in the same orbital.

hybrid orbital. An atomic orbital formed by combining or mixing two or more ground-state atomic orbitals.

hydrate. A solid compound which incorporates water molecules in its crystal lattice.

hydration. The interaction of a solute with water; the clustering of water molecules around a solute particle.

hydride. A compound containing hydrogen in the -1 oxidation state.

hydrocarbon. A compound containing hydrogen and carbon only.

hydrogen bond. An attraction between a hydrogen atom bonded to an electronegative atom and an electronegative atom of another molecule.

hydrolysis. The reaction of an anion with water to form a weak acid and OH^-, or the reaction of a cation with water to form a weak base and H^+ (Arrhenius). Also, any reaction in which H_2O is split.

hydrolysis (hydrolytic) constant, K_h. The equilibrium constant for a hydrolysis equilibrium.

hydronium ion. The hydrated hydrogen ion; represented by H_3O^+. Also known as the *oxonium ion.*

hydrous oxides. Poorly characterized compounds formed from the combination of certain oxides with water or by precipitation from the addition of base to certain ions in aqueous solution.

iatrochemistry. The chemical period from about 1500 to 1650 during which alchemical principles were applied to medicine.

ideal gas. A gas whose behavior is described by the ideal-gas law, $PV = nRT$.

ideal solution. A solution which obeys Raoult's law (Sec. 12-4); a solution in which all species act independently of each other.

indicator. A conjugate acid-base pair in which at least one of the pair is highly colored.

inert pair. The pair of s electrons in an atom with a valence shell containing at least one p electron.

inertia. The resistance shown by all matter to any attempt to change its state of motion.

inhibitor. A substance which decreases the rate of a reaction.

insoluble. Of low solubility; slightly soluble; sparingly soluble.

interhalogen. A compound of two different halogens.

interstice. A space between objects, such as between atoms in a crystal.

inversion temperature. A temperature at which free expansion of a real gas produces neither heating nor cooling of the gas.

ion. An atom or covalently bonded group of atoms which carries an electrical charge.

ion current. The migration of ions between the electrodes of an electrochemical cell.

ion product. (1) The product of the concentrations of hydrogen (hydronium) ions and hydroxide ions in water. (2) The mass-action expression for a solubility equilibrium.

ionic atmosphere. The space around a given ion in solution, occupied largely by counterions.

ionic bond. The electrostatic attraction between ions of opposite electrical charge.

ionic solid. A solid composed of anions and cations in its lattice.

ionization. (1) Loss of an electron by an atom, molecule, or ion. (2) Dissociation of an electrolyte.

ionization energy. The energy used to remove an electron from a gaseous, isolated, ground-state atom (or, sometimes, ion). Also called *ionization potential*.

isolated system. A system which can exchange neither matter nor energy in any form with its surroundings.

isotope effect. The dependence of a property such as reaction rate on the mass number of an element.

isotopes. Atoms of an element having different numbers of neutrons in their nuclei, and hence, different mass numbers.

IUPAC. The International Union of Pure and Applied Chemistry.

joule (J). The SI derived unit of energy or work: $1 \text{ J} = 1 \text{ N m}$ (newton-meter). $1 \text{ J} = 0.2390 \text{ cal}$.

Kelvin temperature scale. An absolute temperature scale, in which the unit is the *kelvin* (K), defined as $\frac{1}{273.16}$ of the temperature difference between absolute zero (0 K) and the triple point of water. $\text{K} = {}^\circ\text{C} + 273$.

kernel. All of an atom except for its valence shell of electrons; also called the *core*.

ketone. An organic compound of the type $R\!-\!\overset{\displaystyle O}{\overset{\|}{C}}\!-\!R$.

kinetic energy. Energy associated with the motion of an object. An object of mass m moving at speed s has kinetic energy $\frac{1}{2}ms^2$.

lanthanoid. A member of the series of 14 elements following lanthanum in the periodic table. Also called *lanthanide*.

lanthanoid contraction. The gradual shrinkage of the atomic radii of the lanthanoids with increasing atomic number.

lattice energy. The energy necessary to separate one mole of a crystalline solid into a gaseous collection of the units at the lattice points. (*Exception:* In the case of a metallic solid a gaseous collection of *atoms* is formed.)

law (natural). A generalization which describes natural behavior.

Le Châtelier's principle. When a system at equilibrium experiences a stress, it will adjust, if possible, so as to minimize the effect of the stress.

Lewis structure. A method of indicating the assignment of valence electrons in an atom, molecule, or ion by representing them as dots placed around symbols, which represent the cores, or kernels, of the atoms.

ligand. An atom, molecule, or ion bonded to the central atom in a complex.

liquid-junction potential. A voltage produced across the junction between two dissimilar liquids.

London forces. Weak forces between atoms or molecules due to momentary fluctuations in their electronic charge-cloud distributions; also called *dispersion forces*.

lone pair. A pair of electrons which belongs to only one atom and hence is not shared (VB model).

magnetic quantum number, m_l. A quantum number which indicates the orbital occupied by an electron.

mass. A measure of quantity of matter.

mass-action expression, Q. The product of the concentrations or partial pressures (or, better, activities) of the products in a reaction, divided by those of the reactants. Each term is raised to an exponential power corresponding to the coefficient written before the corresponding substance or species in the balanced equation. Pure solids and liquids are omitted, as are substances present in large excess, and therefore almost constant concentration.

mass number. The total number of nucleons (protons and neutrons) in an atom.

matter. Anything which has real, physical existence; the stuff of which substances are made.

mechanism. The set of steps (elementary processes) which together comprise an overall reaction.

meridional (mer-) isomer. An isomer of an octahedral complex in which a plane contains three identical ligands and the central ion.

metal. An element which has high electrical and thermal conductivities, a characteristic luster, and a low ionization energy, electron affinity, and electronegativity.

metallic solid. A solid in which positive ions are bonded together by delocalized electrons.

metalloid. An element which has properties which are intermediate between those of a typical metal and those of a typical nonmetal. Also called a *semimetal.*

millimeter of mercury (mmHg). A unit of pressure: $1 \text{ mmHg} = \frac{1}{760} \text{ atm}$. Also known as *torr.*

mixture. A combination of two or more different substances which has a variable composition and can be separated by physical means.

molality (molal concentration; m). A concentration unit: the number of moles of solute per kilogram of solvent.

molar heat of fusion, ΔH_{fus}. The heat necessary to melt one mole of a substance.

molar heat of solution, ΔH_{sol}. The heat liberated (if negative) or absorbed (if positive) when one mole of solute dissolves in a solvent.

molar heat of vaporization, ΔH_{vap}. The heat necessary to vaporize one mole of a substance.

molar volume, V_m. The volume occupied by one mole of a substance. For an ideal gas, 22.4 liter mol^{-1}.

molarity (molar concentration; M). A concentration unit: number of moles of solute per liter of solution.

mole. Avogadro's number of particles or units.

mole fraction (X). A concentration unit: the number of moles of one component in a solution divided by the total number of moles of all components.

mole percent (mol%). Mole fraction multiplied by 100.

molecular formula. A formula expressing the actual number of atoms of each element in one molecule.

molecular orbital (MO). An electronic energy level in a molecule and the corresponding charge-cloud distribution in space.

molecular solid. A solid in which molecules are held together by dipole-dipole or London forces.

molecular weight. The sum of the masses (atomic weights) of the atoms in a molecule, expressed in atomic mass units; the mass of one molecule.

molecularity. The number of particles which collide to form the activated complex in an elementary process.

molecule. The smallest aggregate of atoms capable of acting as a unit and exhibiting the chemical properties of a substance. A combination of two or more atoms.

monomer. The repeating unit in a polymer, or the small molecule from which a polymer is synthesized.

n-type semiconductor. A semiconductor in which the charge carriers are weakly bound electrons.

net equation. An equation which shows only actual reactants at the left and only actual products at the right of the arrow.

neutral solution. An aqueous solution in which the concentrations of hydrogen and hydroxide ions are equal. A solution of pH = 7.0 at 25°C.

neutralization reaction. An acid-base reaction (Arrhenius).

neutron. A nucleon which carries no electrical charge; much more massive than an electron.

noble gas. A member of group 0 in the periodic table.

node. A region or location of minimum disturbance in a standing wave. A surface where Ψ^2 equals zero.

nonelectrolyte. A solute which does not dissociate into ions in solution.

nonmetal. An element with generally low electrical and thermal conductivities, dull luster, and a high ionization energy, electron affinity and electronegativity.

nonpolar covalent bond. A covalent bond in which the bonding electron pair is shared equally between atoms of identical electronegativity.

nonpolar molecule. A molecule in which the centers of positive and negative charge coincide.

normal boiling point. The boiling point of a substance at 1 atm pressure.

normal covalent bond. A covalent bond in which one of the shared electrons appears to have been contributed by the first bonded atom, the other, by the second.

normal freezing point. The freezing point of a substance at 1 atm pressure.

normal hydrocarbon. A hydrocarbon whose skeletal carbon chain is unbranched.

normality (N). A concentration unit: the number of equivalents of solute per liter of solution.

nucleon. A particle in the nucleus of an atom; a proton or a neutron.

nucleophile. An electron-rich atom or group of atoms which seeks to share its electrons with a relatively positive atom.

octahedral hole. A space in a close-packed structure which is bounded by six spheres located at the corners of an octahedron.

octet rule. A rule which states that a (valence-shell) ns^2np^6 configuration in an atom is an especially stable one.

oil. A liquid ester of the triol glycerol and fatty acids.

optical isomer. An isomer which will rotate the plane of polarized light.

orbital. A discrete electronic energy level; also, the spatial distribution of the electron probability density, Ψ^2, for such a level.

osmosis. The passage of solvent molecules through a semipermeable membrane from a solution of higher solvent cencentration (lower solute concentration) to one of lower solvent concentration (higher solute concentration).

osmotic pressure. A colligative property: the pressure which must be applied to a solution phase on one side of a semipermeable membrane in order to stop osmosis.

oxidation. The loss of electrons by a species or substance in a reaction.

oxidation number (oxidation state). A somewhat arbitrary but useful way of indicating the approximate electrical characteristic of an atom. (See Sec. 8-4.)

oxidation potential. A measure of the tendency of an oxidation half-reaction to occur; expressed as the voltage produced by a cell employing the half-reaction at its anode and using the standard hydrogen electrode as its cathode.

oxidation-reduction reaction. An electron-transfer reaction.

oxidizing agent. A substance or species which gains electrons in a reaction. An electron acceptor.

oxonium ion. Hydronium ion; H_3O^+.

p-type semiconductor. A semiconductor in which the charged carriers are weakly localized holes (missing electrons).

parallel spins. Spins which are in the same direction.

paramagnetism. A weak attraction into a magnetic field, a result of the presence of unpaired electrons in a substance.

partial pressure. The pressure which a gas in a mixture would exert on the walls of a container if no other gases were present.

percent by mass (mass percent). A concentration unit: 100 times the mass of one component divided by the total mass of the solution.

period. A horizontal series (row) of elements in the periodic table.

pH. The negative common logarithm of the concentration (actually, activity) of the hydrogen (hydronium) ions in an aqueous solution.

phase. A physically distinct region with a uniform set of properties throughout.

photon. A quantum of electromagnetic energy: $E_{photon} = h\nu$.

physical change. A change in which no new substances are formed.

physical property. A property which can be described without referring to a chemical reaction.

pi (π) bond. A covalent bond in which the electron charge cloud of the shared pair of electrons is located in two regions on opposite sides of the bond axis.

pK. The negative common logarithm of an equilibrium constant.

pnicogen. A member of group VA in the periodic table.

polar covalent bond. A covalent bond in which the bonding electron pair is not shared equally but is drawn closer to the more electronegative atom.

polar molecule. A molecule in which the centers of positive and negative charge do not coincide; a dipole.

polarizability. The ease by which the particles of a substance may be distorted or oriented by an electric field.

polydentate ligand. A polyatomic ligand with more than one lone pair of electrons which can simultaneously bond to the central ion in a complex.

polymer. A long-chain molecule composed of many repeating units (monomers).

polymorphism. The ability of a substance to crystallize in different structures.

polyol. An organic molecule with more than one alcohol (—OH) functional group.

polyprotic acid. An acid which has more than one available H^+, or can donate more than one proton.

positron. A particle with the same mass as an electron and carrying a charge of the same magnitude as the electronic charge, but opposite in sign (positive).

potential energy. Energy associated with the position or configuration of an object.

precipitation. The formation of a condensed phase (solid or liquid) during a reaction.

precision. The degree of exactness or sharpness with which a number obtained by measurement is expressed.

pressure. The force exerted on a unit of surface area.

primitive cell. A unit cell which has entities (atoms, molecules, ions) only at the corners of the cell.

principal quantum number, n. A quantum number which specifies a shell for an electron in an atom.

probability density, Ψ^2. The probability of finding an electron in a small element of volume; the square of the wave function for an electron; the density of the electronic charge cloud.

product. A substance formed in a chemical reaction.

protein. A polymeric combination of amino acids.

protium. Ordinary hydrogen, ^1_1H.

proton. A nucleon carrying a positive charge equal in magnitude to that on the electron, but which is much more massive than the electron.

quantization of energy. The restriction of the energy of a system to certain, specific, discrete amounts.

quantum mechanics. The branch of physics which describes the behavior of small particles by assigning wavelike properties to them. Also known as *wave mechanics*.

quantum number. A number used to describe the state of an electron.

radioactivity. The decay or decomposition of nuclei of atoms.

rate-determining step. The slowest step in a sequential reaction mechanism.

rate law. The algebraic statement of the dependence of the rate of a reaction upon the concentrations of various substances, generally reactants.

reactant. A substance consumed in a chemical reaction.

reaction order. The exponent on a concentration term in a simple rate law (order with respect to one component); or the sum of all such exponents (overall order).

reaction rate. The time rate of change of concentration (or, sometimes, quantity) of a reactant or product in a reaction.

redox reaction. An oxidation-reduction, or electron transfer, reaction.

reducing agent. A species or substance which loses electrons in a reaction.

reduction. The gain of electrons by a species or substance during a reaction.

reduction potential. A measure of the tendency of a reduction half-reaction to occur, expressed as the voltage produced by a cell employing the half-reaction at its cathode and using the standard hydrogen electrode as its anode.

refractory. Having a very high melting point.

representative element. A member of one of the main, or A, groups in the periodic table.

resonance hybrid. A structure which cannot be represented by a single valence-bond Lewis structure but rather is shown as a combination, or average, of two or more structures.

salt. A compound formed from the positive ions of an Arrhenius base and the negative ions of an Arrhenius acid.

saponification. The basic hydrolysis of a fat or oil to form glycerol and a soap.

saturated hydrocarbon. A hydrocarbon containing no multiple bonds.

saturated solution. A solution which is or can be at equilibrium with excess solute.

screw dislocation. A dislocation in which layers of particles in a crystal are warped around a screw axis.

self-dissociation. The production of cations and anions by dissociation of solvent molecules without interaction with other species.

semiconductor. A substance whose electrical conductivity increases with increasing temperature.

semipermeable membrane. A membrane which will permit passage of some components of a solution but which will stop others.

shell. A principal set of electron energies in an atom; designated by K, L, M, N, . . . , or by n (principal quantum number) = 1, 2, 3, 4, ...

sigma (σ) bond. A covalent bond in which the electron charge cloud of the shared pair is centered on and symmetrical around the bond axis.

smoke. A colloidal dispersion of a solid in a gas.

solubility. The concentration of a solute in a saturated solution; the maximum amount of solute which can be dissolved in a solution by simply adding it to a solvent at constant temperature.

solubility product, K_{sp}. The equilibrium constant for an ionic solubility equilibrium. (Also called the *solubility product constant*.)

solute. A component of a solution present at a concentration which is low relative to that of the solvent.

solution. A homogeneous mixture.

solvation. The interaction of a solute with a solvent; the surrounding of solute particles by solvent particles.

solvent. A component of a solution present at relatively high concentration.

space lattice. A regular, repeating array of points in space.

specific rate constant. The constant of proportionality in a rate law.

spectrochemical series. A listing of ligands in order of their ability to cause *d*-orbital splitting in a complex.

spin quantum number, m_s. A quantum number which specifies the spin of an electron.

spontaneous change. A change which is thermodynamically probable; a possible change.

standard hydrogen electrode (s.h.e.). An electrode consisting of a piece of platinum (coated with platinum black) in contact with hydrogen gas at 1 atm and with a solution in which the concentration of hydrogen ions is 1 *M*. (Actually, the hydrogen gas and hydrogen ions are at unit *activity*.) The voltage of the s.h.e. is defined as zero.

standard state. A reference state for specifying thermodynamic quantities, usually defined as the most stable form of the substance at 1 atm pressure. For a solute the standard state is the ideal, 1 *M* solution.

stereochemistry. The study of the spatial geometries of molecules and polyatomic ions.

stereoisomers. Molecules or polyatomic ions with the same atoms and the same bonds but differing in the geometrical orientations of the atoms and bonds.

steric factor. The fraction of the collisions in an elementary process in which the colliding particles have the proper geometrical orientation with respect to each other to produce the activated complex.

steric number. The sum of the number of bonds and lone pairs around a bonded atom. (VSEPR theory.)

STP. Standard temperature (0°C) and pressure (1 atm).

strength (of an electrolyte). The extent or degree of dissociation of an electrolyte in solution.

strong (electrolyte). Completely, or almost completely, dissociated in solution.

structural formula. A drawing showing which atoms are bonded to each other in a molecule or polyatomic ion.

sublimation. Direct conversion from solid to gas.

subshell. A subset of electron energies in an atom; designated by *s*, *p*, *d*, *f*, . . . , or by *l* (azimuthal quantum number) = 0,1,2,3, ...

substitution reaction. A reaction in which one atom or group of atoms is substituted for another in a molecule.

substrate. The substance or species which bonds and then reacts at an active site in heterogeneous or enzyme catalysis.

supercooling. The cooling of a liquid below its freezing point without freezing taking place.

superheating. The heating of a liquid above its boiling point without boiling taking place.

supersaturated solution. An unstable solution in which the concentration of solute is greater than its solubility.

surface tension. A measure of the energy necessary to increase the surface area of a liquid.

system. A portion of the universe under observation or consideration.

temperature. The property of a substance which determines the direction of heat flow into or out of the substance; heat flows from a substance of higher temperature to one of lower temperature. The temperature of a substance measures the average kinetic energy of its particles.

termolecular process. An elementary process in which the activated complex is formed from the simultaneous collision of three particles.

theory. A proposed explanation or justification of observed behavior made in terms of a model.

thermochemical equation. A chemical equation which includes an indication of the heat liberated or absorbed during the reaction.

three-center bond. A bond consisting of an electron pair shared among, and bonding, three atoms; found in electron-deficient compounds.

titrant. The substance slowly added during a titration.

titration. The slow addition of a solution of one reactant to one of a second reactant until the equivalence point is signaled by an indicator color change or other method.

torr. A unit of pressure: 1 torr = 1 mmHg = $\frac{1}{760}$ atm.

trans isomer. Any isomer in which two identical groups are located on opposite sides of a structure.

transition element. A member of one of the B groups, which intervene between group IIA and IIIA in the periodic table.

transmutation. The transformation of one element into another.

tritium. Hydrogen 3, $^3_1$H.

Tyndall effect. The scattering of a beam of light by a colloid.

unimolecular process. An elementary process in which the probability of collisional deactivation of the activated complex greatly exceeds the probability of its decomposing to form products.

unit cell. A small portion of a crystal lattice which can be used to generate the entire lattice by moving the cell distances equal to its edge lengths in directions parallel to its edges.

unsaturated hydrocarbon. A hydrocarbon with one or more multiple bonds.

unsaturated solution. A solution in which the concentration of solute is less than its solubility.

vacancy. A lattice point in a crystal at which a particle is missing.

valence shell. The shell of electrons with the highest principal quantum number n in an atom.

van der Waals forces. Weak forces between atoms or molecules; include London and dipole-dipole forces.

vapor pressure. The pressure of a gas when it is in equilibrium with its liquid.

vapor-pressure lowering. A colligative property: the decrease in solvent vapor pressure brought about by the presence of a solute.

viscosity. The resistance to flow shown by a fluid.

VSEPR theory. Valence-shell electron-pair repulsion theory (Sec. 9-1).

wave equation. A mathematical equation describing the motion of a wave.

wave function. The mathematical relation which solves a wave equation. Each correctly obtained electron wave function corresponds to a discrete energy state and spatial distribution of an electron in an atom or molecule.

wave mechanics. The branch of physics which describes the behavior of small particles by assigning wavelike properties to them. Also, known as *quantum mechanics*.

wavelength, λ. The distance between successive crests (or other corresponding points) on a wave.

weak (electrolyte). Only partially dissociated.

weight. The force of gravitational attraction between an object and (usually) the earth.

work. The product of the distance which an object is moved times the force opposing the motion.

x-ray diffraction. The diffraction of x rays as by the atoms in a crystal.

zero-point energy. The energy of a solid at absolute zero due to its residual nuclear, electronic, atomic, and molecular motion. The lowest energy state of a substance.

APPENDIX B
UNITS, CONSTANTS, AND CONVERSION EQUATIONS

B-1 UNITS

SI units

The version of the metric system of units which has been approved by the International Union of Pure and Applied Chemistry (IUPAC) and other international bodies is the *Système International d'Unités,* or SI. Seven *base units* serve as the foundation for this system:

Physical quantity	Unit	Symbol
Length	Meter	m
Mass	Kilogram	kg
Time	Second	s
Electric current	Ampere	A
Temperature	Kelvin	K
Luminous intensity	Candela	cd
Amount of substance	Mole	mol

In addition to these, two *SI supplementary units* have been approved. These are the *radian* (rad) for measuring plane angles and the *steradian* (sr) for measuring solid angles.

Many other units are derived from the SI base units. Some of these *SI derived units* have no special names. These include

Physical quantity	Unit	Symbol
Area	Square meter	m^2
Volume	Cubic meter	m^3
Velocity	Meter per second	$m\,s^{-1}$
Acceleration	Meter per second squared	$m\,s^{-2}$
Density	Kilogram per cubic meter	$kg\,m^{-3}$

Other SI derived units have special names. These include

Physical quantity	Unit	Symbol	Definition
Force	Newton	N	$kg\,m\,s^{-2}$
Energy	Joule	J	$kg\,m^2\,s^{-2}$
Pressure	Pascal	Pa	$kg\,m^{-1}\,s^{-2} = N\,m^{-2}$
Electrical charge	Coulomb	C	$A\,s$
Electrical potential difference	Volt	V	$J\,A^{-1}\,s^{-1} = J\,C^{-1}$
Frequency	Hertz	Hz	s^{-1}

Many older, non-SI units have been defined in terms of SI units and are to be retained because they are traditional and useful. These include

Physical quantity	Unit	Symbol	Definition
Time	Minute	min	60 s
	Hour	h	3600 s
	Day	d	86,400 s
Plane angle	Degree	°	$(\pi/180)$ rad
	Minute	′	$(\pi/10,800)$ rad
	Second	″	$(\pi/648,000)$ rad
Volume	Liter	l or L	$10^{-3}\,m^3$
Temperature	Degree Celsius	°C	$K - 273.15$

It has been recommended that the use of certain units be gradually phased out. The most important of these for our purposes is the *standard atmosphere* (atm) which is defined as 101,325 Pa. In addition, it has been recommended that use of the pressure unit *millimeter of mercury* (mmHg), essentially the same as the torr (no abbreviation), be abandoned immediately ($1 \text{ mmHg} = \frac{1}{760} \text{ atm}$). The atmosphere and millimeter of mercury (torr) will probably disappear from use eventually, but they are of such a convenient size that they continue to be commonly used. On the other hand, the calorie (cal), a unit of energy, appears to be rapidly disappearing from use ($1 \text{ cal} = 4.184 \text{ J}$).

Metric prefixes

The following metric prefixes for multiple and submultiple units may be applied to both SI and non-SI units.

Multiple or submultiple	Prefix	Symbol
10^{18}	exa	E
10^{15}	peta	P
10^{12}	tera	T
10^{9}	giga	G
10^{6}	mega	M
10^{3}	kilo	k
10^{2}	hecto	h
10^{1}	deka	da
10^{-1}	deci	d
10^{-2}	centi	c
10^{-3}	milli	m
10^{-6}	micro	μ
10^{-9}	nano	n
10^{-12}	pico	p
10^{-15}	femto	f
10^{-18}	atto	a

B-2   PHYSICAL CONSTANTS

Constant	Symbol	Value
Atomic mass unit	amu	1.660566×10^{-24} g
Avogadro's number	N	6.02204×10^{23} mol^{-1}
Electron charge	e	1.602189×10^{-19} C
Electron mass	m_e	$\begin{cases} 9.10953 \times 10^{-28} \text{ g} \\ 5.48579 \times 10^{-4} \text{ amu} \end{cases}$
Faraday's constant	$\mathscr{F}$	9.64846×10^{4} C mol^{-1}
Gas constant	R	$\begin{cases} 8.2054 \times 10^{-2} \\ \text{liter atm K}^{-1} \text{ mol}^{-1} \\ 8.3144 \text{ J K}^{-1} \text{ mol}^{-1} \end{cases}$
Molar volume of ideal gas at STP	V_m	22.4138 liter mol^{-1}
Neutron mass	m_n	$\begin{cases} 1.674954 \times 10^{-24} \text{ g} \\ 1.00866 \text{ amu} \end{cases}$
Planck's constant	h	6.62618×10^{-34} J s
Proton mass	m_p	$\begin{cases} 1.672648 \times 10^{-24} \text{ g} \\ 1.00728 \text{ amu} \end{cases}$
Speed of light in vacuum	c	2.99792458×10^{8} m s^{-1}

B-3   CONVERSION EQUATIONS

Conversion	Equation
Calories–joules	$1 \text{ cal} = 4.184 \text{ J}$
Electronvolts–joules	$1 \text{ eV} = 1.602 \times 10^{-19} \text{ J}$
Liter-atmospheres–joules	$1 \text{ liter atm} = 101.3 \text{ J}$
Ergs–joules	$1 \text{ erg} = 10^{-7} \text{ J}$
Ångstroms–meters–nanometers	$1 \text{ Å} = 10^{-10} \text{ m} = 10^{-1} \text{ nm}$
Inches–centimeters	$1 \text{ in} = 2.540 \text{ cm}$
Miles–kilometers	$1 \text{ mi} = 1.609 \text{ km}$
Pounds–kilograms	$1 \text{ lb} = 0.4536 \text{ kg}$
Ounces–grams	$1 \text{ oz} = 28.35 \text{ g}$
Quarts–liters	$1 \text{ qt} = 0.9464 \text{ liter}$
Atmospheres–pascals	$1 \text{ atm} = 1.013 \times 10^{5} \text{ Pa}$

APPENDIX C
CHEMICAL NOMENCLATURE

C-1   COMMON NAMES

Many substances have been known for years by their common names. In fact, these names for some substances are much better known than their systematic names. Thus *water* is almost never called *hydrogen oxide*. Other common names are used in specialized applications. Thus *sodium thiosulfate* is photographers' *hypo*, and the minerologist knows *zinc sulfide* as *sphalerite*. Some common names must be used with caution. *Alcohol*, for instance, is a common name for a specific compound and is also a name for a general class of compounds. The following list shows both common and systematic names for a number of compounds:

Formula	Common name	Systematic name
NH_3	Ammonia	Hydrogen nitride
Al_2O_3	Alumina	Aluminum oxide
CH_3CH_2OH	Alcohol; grain alcohol	Ethanol
CH_3OH	Wood alcohol	Methanol
$NaOH$	Lye; caustic soda	Sodium hydroxide
KOH	Caustic potash	Potassium hydroxide
$NaCl$	Salt; table salt	Sodium chloride

C-2   SYSTEMATIC INORGANIC NOMENCLATURE

(See also Chaps. 20 to 22.)

Elements

The English names and the symbols of the elements are listed inside the front cover of this book. In order to identify a specific molecular allotrope, the following prefixes are used to indicate numbers of atoms:

mono	1	penta	5	nona or ennea	9
di	2	hexa	6	deca	10
tri	3	hepta	7	undeca or hendeca	11
tetra	4	octa	8	dodeca	12

Some examples of the use of these prefixes are given below:

Formula	Systematic name	Common name
O_2	Dioxygen	Oxygen
O_3	Trioxygen	Ozone
P_4	Tetraphosphorus	White phosphorus
S_8	Octasulfur	Sulfur
H	Monohydrogen	Atomic hydrogen

Cations

MONATOMIC CATIONS. When an element forms only one monatomic cation, the name of the cation is the same as that of the element. Some examples are

K^+	potassium ion
Ba^{2+}	barium ion
Al^{3+}	aluminum ion

When an element forms two monatomic cations (representing different oxidation states), the ions are differentiated from each other by means of either the *ous-ic* system or the *Stock* system. In the *ous-ic* system the suffix *-ous* indicates the *lower* of two oxidation states, and the suffix *-ic*, the *higher*. These are added to a root formed by dropping *-um* or *-ium* from the English or Latin name of the element. (If the element name does not end in *-um* or *-ium*, sometimes the last syllable is dropped, and sometimes the root is the element name itself.)

The *ous-ic* system does not actually indicate oxidation states, only that an element is in the higher or lower of two states. The *Stock* system avoids ambiguity by clearly specifying oxidation numbers. These are indicated by a Roman numeral written in parentheses immediately after the name of the element. Some examples of cations and their *ous-ic* and Stock names are given below:

Formula	Stock-system name	ous-ic name
Cu^+	Copper(I) ion	Cuprous ion
Cu^{2+}	Copper(II) ion	Cupric ion
Fe^{2+}	Iron(II) ion	Ferrous ion
Fe^{3+}	Iron(III) ion	Ferric ion
Mn^{2+}	Manganese(II) ion	Manganous ion
Mn^{3+}	Manganese(III) ion	Manganic ion
Sn^{2+}	Tin(II) ion	Stannous ion
Sn^{4+}	Tin(IV) ion	Stannic ion
Co^{2+}	Cobalt(II) ion	Cobaltous ion
Co^{3+}	Cobalt(III) ion	Cobaltic ion

POLYATOMIC CATIONS. Polyatomic cations are given various names, most of which should be memorized. Note that the *-yl* suffix in the name of a cation indicates an *oxocation*. Some examples include

NH_4^+	ammonium ion
H_3O^+	hydronium (oxonium) ion
UO_2^+	uranyl(V) ion; dioxouranium(V) ion
UO_2^{2+}	uranyl(VI) ion; dioxouranium(VI) ion
NO^+	nitrosyl ion
Hg_2^{2+}	mercurous ion

The IUPAC name for H_3O^+ is the *oxonium* ion, but *hydronium* ion is still in common use in the United States.

Anions

MONATOMIC ANIONS. The names of monatomic anions are obtained by adding the suffix *-ide* to the root, usually the first syllable, of the element's name.

Examples:

F^-	fluoride ion
Cl^-	chloride ion
Br^-	bromide ion
I^-	iodide ion
H^-	hydride ion
O^{2-}	oxide ion
S^{2-}	sulfide ion
N^{3-}	nitride ion
P^{3-}	phosphide ion

A few *polyatomic* anions are also named with an *-ide* suffix. These include

OH^-	hydroxide ion
CN^-	cyanide ion
S_2^{2-}	disulfide ion
S_x^{2-}	polysulfide ion
O_2^{2-}	peroxide ion
O_2^-	superoxide ion
HO_2^-	hydrogen peroxide ion

OXOANIONS. Oxoanions are binary (two-element) anions in which one element is oxygen. These can be named as complex ions (see below), but traditional names are more often used. When the central, or principal, atom commonly occurs in only one oxidation state as an oxoanion, the suffix *-ate* is used.

CO_3^{2-}	carbonate ion
SiO_4^{4-}	silicate ion

The suffixes *-ite* and *-ate* are used to distinguish between two oxoanions with the same central or principal atom in two different respective oxidation states. The lower state is indicated by *-ite*, the higher by *-ate*.

NO_2^-	nitrite ion
SO_3^{2-}	sulfite ion
HPO_3^{2-}	phosphite ion
NO_3^-	nitrate ion
SO_4^{2-}	sulfate ion
PO_4^{3-}	phosphate ion

Up to four different oxidation states can be differentiated in oxoanions by using the prefix *hypo-* to indicate a state lower than the *-ite* state and the prefix *per-* to indicate one higher than the *-ate* state. Some examples of oxoanions of bromine are

Br oxidation state	Formula	Name
+1	BrO^-	Hypobromite ion
+3	BrO_2^-	Bromite ion
+5	BrO_3^-	Bromate ion
+7	BrO_4^-	Perbromate ion

ANIONS WITH ACIDIC HYDROGENS. Anions which may be regarded as resulting from the incomplete neutralization of an acid are named with the word *hydrogen*.

HSO_4^-	hydrogen sulfate ion
HCO_3^-	hydrogen carbonate ion
$H_2PO_4^-$	dihydrogen phosphate ion
HS^-	hydrogen sulfide ion

When such anions contain only one hydrogen and are derived from a diprotic acid, they have traditionally been named with a *bi-* prefix.

HSO_4^-	bisulfate ion
HSO_3^-	bisulfite ion
HCO_3^-	bicarbonate ion
HS^-	bisulfide ion

Binary compounds

SALTS. Binary salts are named by indicating the name of the cation followed by that of the anion. Prefixes (*di*, *tri*, etc.) are not used (except in some common names). Some examples are

NaCl	sodium chloride
$FeCl_2$	iron(II) chloride; ferrous chloride
$FeBr_3$	iron(III) bromide; ferrous bromide
Na_3N	sodium nitride
K_2S	potassium sulfide
CaI	calcium iodide
$AlCl_3$	aluminum chloride
Ca_3P_2	calcium phosphide

OXIDES. Oxides are named as if they were salts.

Na_2O	sodium oxide
SnO	tin(II) oxide; stannous oxide
SnO_2	tin(IV) oxide; stannic oxide
Fe_2O_3	iron(III) oxide; ferric oxide

BINARY ACIDS. Compounds which are binary Arrhenius acids are named by using the word *hydrogen* followed by the anion name.

HCl	hydrogen chloride
HBr	hydrogen bromide
H_2S	hydrogen sulfide
H_2Se	hydrogen selenide

Binary acids are also named by using the prefix *hydro-*, the root of the anion name, the suffix *-ic*, and the word *acid*.

HCl	hydrochloric acid
HBr	hydrobromic acid

H_2S	hydrosulfuric acid
H_2Se	hydroselenic acid

Note: HCN is named as if it were a binary acid: hydrogen cyanide or hydrocyanic acid.

BINARY COMPOUNDS OF NONMETALS, OTHER THAN ACIDS. Compounds containing two nonmetals are named much as salts are, with the more electropositive element indicated first. For these compounds the prefixes *di-*, *tri-*, *tetra-*, etc., are used as necessary. (See *elements*, above.) The prefix *mono-* is often, though not always, omitted. Some examples of these compounds are

SO_2	sulfur dioxide
SO_3	sulfur trioxide
PCl_3	phosphorus trichloride
PCl_5	phosphorus pentachloride
CCl_4	carbon tetrachloride
N_2O_5	dinitrogen pentaoxide
CO	carbon monoxide
CO_2	carbon dioxide

Ternary compounds

Most common ternary (three-element) inorganic compounds are oxoacids or oxosalts.

OXOSALTS. As in the case of binary salts, oxosalts are named by indicating the cation first, followed by the anion.

$CaSO_4$	calcium sulfate
$NaNO_3$	sodium nitrate
NH_4NO_2	ammonium nitrite
Na_2SO_4	sodium sulfate
Na_3PO_4	sodium phosphate
$Al_2(SO_4)_3$	aluminum sulfate
$CuSO_4$	copper(II) sulfate; cupric sulfate
Cu_2SO_4	copper(I) sulfate; cuprous sulfate

OXOACIDS. Oxoacids can be named as hydrogen compounds, but they seldom are. (H_2SO_4 is rarely called hydrogen sulfate.) Instead, they are named by dropping the *-ite* or *-ate* from the anion name and adding *-ous acid* or *-ic acid*, respectively. Thus we have

Anion		Corresponding acid	
Formula	Name	Formula	Name
ClO^-	Hypochlorite ion	HClO	Hypochlorous acid
ClO_2^-	Chlorite ion	$HClO_2$	Chlorous acid
ClO_3^-	Chlorate ion	$HClO_3$	Chloric acid
ClO_4^-	Perchlorate ion	$HClO_4$	Perchloric acid

HYDROXIDES. Hydroxides are named as if they were salts.

NaOH	sodium hydroxide
$Ca(OH)_2$	calcium hydroxide
$Al(OH)_3$	aluminum hydroxide
$Cu(OH)_2$	copper(II) hydroxide; cupric hydroxide

Complexes

We will consider a complex to be a central, or principal atom or ion bonded to one or more ligands, which may be atoms, ions, or molecules. (See Chap. 22.) The *formula* of a complex shows the central atom first, followed by the ligands, written with anionic ligands preceding neutral and cationic ligands. Examples include

$NiF_6{}^{2-}$	$Ag(S_2O_3)_2{}^{3-}$	$CoCl_4(NH_3)_2{}^-$
$Al(OH)(H_2O)_5{}^{2+}$	$Cr(NH_3)_6{}^{3+}$	$Co(NCS)(NH_3)_5{}^{2+}$

The formula of a complete compound which contains a complex ion is often written with square brackets to indicate the complex.

$K_2[Pt(NO_2)_4]$	$Na_3[Co(NO_2)_6]$
$Na_3[Ag(S_2O_3)_2]$	$K[Co(CN)(CO)_2(NO)]$

NAMES OF COMPLEXES. Complexes are named by specifying first the ligands and then the central, or principal, atom. (Note that this is the reverse of the order in a formula.) In the case of cationic or neutral complexes no suffix is added, but the suffix *-ate* is added to indicate an anion. The oxidation state of the central, or principal, atom is indicated by using a Roman numeral according to the Stock system. (The Arabic 0 is used for zero, and a minus sign is used with a Roman numeral in the case of a negative oxidation state.)

$Cr(H_2O)_6{}^{3+}$	hexaaquochromium(III) ion
$CrCl_6{}^{3-}$	hexachlorochromate(III) ion
$Ni(CO)_4$	tetracarbonylnickel(0)
$Fe(CO)_4{}^{2-}$	tetracarbonylferrate(-II) ion

ANIONIC LIGANDS. The suffix in an anion name is changed when the ion serves as a ligand:

-ide	usually becomes	*-o.*
-ate	becomes	*-ato.*
-ite	becomes	*-ito.*

In a few cases the *-ide* suffix of an anion is replaced by *-ido*.

Formula	Name as an anion	Name as a ligand
Cl^-	Chloride	Chloro
Br^-	Bromide	Bromo
OH^-	Hydroxide	Hydroxo
CN^-	Cyanide	Cyano
O^{2-}	Oxide	Oxo
H^-	Hydride	Hydrido
$CO_3{}^{2-}$	Carbonate	Carbonato
$SO_4{}^{2-}$	Sulfate	Sulfato
$SO_3{}^{2-}$	Sulfite	Sulfito
$HSO_3{}^-$	Hydrogen sulfate	Hydrogen sulfito
$C_2O_4{}^{2-}$	Oxalate	Oxalato
OCN^-	Cyanate	Cyanato

NEUTRAL LIGANDS. Neutral ligands are named as the neutral molecule. There are, however, some important exceptions. These include H_2O, named the *aquo* ligand; NH_3, *ammine;* CO, *carbonyl;* NO, *nitrosyl.*

THIO LIGANDS. A ligand in which one or more sulfurs have replaced oxygens is named with the prefix *thio-*. Thus we have SCN^-, the *thiocyanato* ligand; $S_2O_3{}^{2-}$, *thiosulfato;* and $CS_3{}^{2-}$, *trithiocarbonato.*

NUMBERS OF LIGANDS IN A COMPLEX. The prefixes *di-, tri-, tetra-, penta-, hexa-,* etc., are used to indicate the numbers of ligands in a complex. When the name of a ligand itself contains such a prefix, *bis-* (twice), *tris-* (thrice), *tetrakis-* (four times), etc., are used.

$Fe(CN)_5NO^{2-}$	pentacyanonitrosylferrate(III) ion
$AuCl_4{}^-$	tetrachloroaurate(III) ion
$CrCl_2(H_2O)_4{}^+$	dichlorotetraaquochromium(III) ion
$CoCO_3(NH_3)_5{}^+$	carbonatopentaamminecobalt(III) ion
$CoCl_2en_2{}^+$	dichlorobis(ethylenediamine) cobalt(III) ion

COMPLEX IONIC SALTS. As usual, the name of the cation precedes that of the anion in a salt.

$Na_3[Ag(S_2O_3)_2]$	sodium dithiosulfatoargentate(I)
$[Al(OH)(H_2O)_5]Cl_2$	hydroxopentaaquoaluminum(III) chloride
$[Coen_3]_2(SO_4)_3$	tri(ethylenediamine)cobalt(III) sulfate

C-3 SYSTEMATIC ORGANIC NOMENCLATURE

(See also Chap. 23.)

The hydrocarbons

NORMAL ALKANES. The names of most of the normal (straight-chain; continuous-chain; unbranched-chain) alkanes are derived from the Greek word for the number of carbon atoms present, followed by the suffix -ane. (The first four normal alkanes have special names: methane, ethane, propane, and butane, respectively.)

NORMAL ALKYL GROUPS. The name of the hydrocarbon group formed by the removal of one hydrogen from a terminal atom of a normal alkane is obtained by changing the suffix -ane to -yl. The resulting group is called a *normal alkyl group*.

No. of C atoms	Alkane Name	Alkane Formula	Alkyl group Name	Alkyl group Formula
1	Methane	CH_4	Methyl	CH_3-
2	Ethane	CH_3CH_3	Ethyl	CH_3CH_2-
3	Propane	$CH_3CH_2CH_3$	Propyl	$CH_3CH_2CH_2-$
4	Butane	$CH_3(CH_2)_2CH_3$	Butyl	$CH_3(CH_2)_2CH_2-$
5	Pentane	$CH_3(CH_2)_3CH_3$	Pentyl	$CH_3(CH_2)_3CH_2-$
6	Hexane	$CH_3(CH_2)_4CH_3$	Hexyl	$CH_3(CH_2)_4CH_2-$
7	Heptane	$CH_3(CH_2)_5CH_3$	Heptyl	$CH_3(CH_2)_5CH_2-$
8	Octane	$CH_3(CH_2)_6CH_3$	Octyl	$CH_3(CH_2)_6CH_2-$
9	Nonane	$CH_3(CH_2)_7CH_3$	Nonyl	$CH_3(CH_2)_7CH_2-$
10	Decane	$CH_3(CH_2)_8CH_3$	Decyl	$CH_3(CH_2)_8CH_2-$

BRANCHED-CHAIN ALKANES. Branched-chain alkanes are named by adding the name(s) of substituent hydrocarbon group(s) as a prefix (or prefixes) to the name of the longest continuous hydrocarbon chain. The position of each substituent group is indicated with a locator number, and the longest chain is numbered from the end that gives the lowest locator numbers to the substituent groups.

$$CH_3CHCH_2CH_3$$
2-Methylbutane

$$CH_3CH_2CH_2CHCH_3$$
2-Methylpentane

The prefixes di-, tri-, tetra-, etc., are used to indicate the presence of more than one substituent.

$$CH_3CCH_2CH_3$$
2,2-Dimethylbutane

$$CH_3CHCH_2CCH_3$$
2,2,4-Trimethylpentane

BRANCHED-CHAIN ALKYL GROUPS. IUPAC names (and common names, in parentheses) are given below for several important hydrocarbon groups.

$$CH_3CHCH_3$$
1-Methylethyl
(isopropyl)

$$CH_3CHCH_2CH_3$$
1-Methylpropyl
(sec-butyl)

$$CH_3CCH_3$$
1,1-Dimethylethyl
(tert-butyl)

$$CH_3CHCH_2-$$
2-Methylpropyl
(isobutyl)

When different hydrocarbon groups are attached to the longest hydrocarbon chain, they are listed in alphabetical order. (The prefixes di-, tri-, etc., are ignored.)

$$CH_3C—CHCH_2CH_3$$
3-Ethyl-2,2-dimethylpentane

ALKENES AND ALKYNES. Alkenes, hydrocarbons with carbon-carbon double bonds, are named by replacing the -ane suffix on the name of the corresponding alkane with -ene. (If two double bonds are present, the new suffix is -adiene; if three, -atriene, etc.) Alkynes, hydrocarbons with carbon-carbon triple bonds, are named by replacing the -ane suffix on the name of the corresponding alkane with -yne. (If two double bonds are present, the new suffix is -adiyne; if three, -atriyne.) The longest hydrocarbon chain is numbered from the end that gives the lowest locator number to the carbon at the beginning of the multiple bond(s).

$$CH_3CH_2CH=CH_2$$
1-Butene

$$CH_3C≡CCH_3$$
2-Butyne

$$CH_2=CHCH=CH_2$$
1,3-Butadiene

$$CH≡CCH=CHCH_3$$
3-Penten-1-yne

CYCLOALKANES. Cycloalkanes are named by adding the prefix cyclo- to the name of the alkane with the same number of carbon atoms.

$$H_2C—CH_2 \quad | \quad C \quad H_2$$
Cyclopropane

$$H_2C—CH_2 \quad | \quad | \quad H_2C—CH_2$$
Cyclobutane

Cyclopentane

AROMATIC HYDROCARBONS. The names of many aromatic hydrocarbons should simply be memorized. Examples of these compounds include

Benzene   Methylbenzene   Ethylbenzene
(toluene)

Styrene   Cumene   Naphthalene

When two hydrocarbon substituents are present on a benzene ring their positions are indicated by using either locator numbers or the prefixes o- (for *ortho*-), m- (for *meta*-), or p- (for *para*-).

1,2-Dimethylbenzene   1,3-Dimethylbenzene   1,4-Dimethylbenzene
(o-xylene)   (m-xylene)   (p-xylene)

Derivatives of hydrocarbons

Some hydrocarbon derivatives are named with a *prefix* indicating the functional group plus the name of the parent hydrocarbon. Functional groups named by prefix include —F (*fluoro*-), —Cl (*chloro*-), —Br (*bromo*-), —I (*iodo*-), and —NO$_2$ (*nitro*-).

CH_3F CH_2Cl_2 CH_3CH_2Br

Fluoromethane   Dichloromethane   Bromoethane

CH_3NO_2 $CH_3CH_2CH_2I$ $CH_3\overset{\displaystyle I}{C}HCH_2$

Nitromethane   1-Iodopropane   2-Iodopropane

Other derivatives of hydrocarbons are most commonly named by using a *suffix* to indicate the functional group. In these the *-e* of the parent hydrocarbon is replaced with the appropriate suffix. Some classes of compounds which are named this way are

Class	Functional group	Suffix
Alcohols	—OH	*-ol*
Aldehydes	$-\overset{\displaystyle O}{\overset{\|}{C}}-H$	*-al*
Ketones	$-\overset{\displaystyle O}{\overset{\|}{C}}-$	*-one*
Carboxylic acids	$-\overset{\displaystyle O}{\overset{\|}{C}}-OH$	*-oic acid*

Some examples of the use of these suffixes follow. (Common names are given in parentheses.)

CH_3-OH $CH_3-\overset{\displaystyle O}{\overset{\|}{C}}-H$ $CH_3-\overset{\displaystyle O}{\overset{\|}{C}}-CH_3$

Methanol   Ethanal   Propanone
(methyl alcohol)   (acetaldehyde)   (acetone)

$CH_3-\overset{\displaystyle O}{\overset{\|}{C}}-OH$ $CH_3CH_2CH_2OH$ $CH_3\overset{\displaystyle OH}{C}HCH_3$

Ethanoic acid   1-Propanol   2-Propanol
(acetic acid)   (*n*-propyl alcohol)   (isopropyl alcohol)

Amines carry the —NH$_2$ functional group. · Primary amines, R—NH$_2$, are most commonly named by adding the suffix *-amine* to the name of the bonded alkyl group.

CH_3-NH_2 $CH_3CH_2-NH_2$

Methylamine   Ethylamine

Secondary amines, R—NH—R′, and tertiary amines,
$\overset{\displaystyle R''}{\overset{\|}{}}$
R—N—R′, are named as *N*-substituted derivatives of the amine that has the longest carbon chain. (The *N*-prefix shows that the smaller group is attached to nitrogen.)

$CH_3CH_2CH_2-\overset{\displaystyle H}{\overset{\|}{N}}-CH_3$ $CH_3CH_2-\overset{\displaystyle CH_3}{\overset{\|}{N}}-CH_3$

N-Methylpropylamine   *N*,*N*-Dimethylethylamine

Ethers carry the —O— functional group and are most commonly named by specifying the alkyl groups present in each molecule. When both groups are the same, the *di*- prefix is often omitted.

CH_3-O-CH_3 $CH_3-O-CH_2CH_3$

Methyl ether   Ethyl methyl ether

Esters are derivatives of carboxylic acids and have the general formula $R-\overset{\displaystyle O}{\overset{\|}{C}}-O-R'$. They are named by specifying the alkyl group attached to the oxygen and then changing the *-ic acid* of the carboxylic acid to *-ate*.

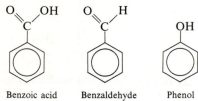

Ethyl ethanoate
(ethyl acetate)

Methyl propanoate
(methyl propionate)

Aromatic derivatives are usually named nonsystematically or semisystematically.

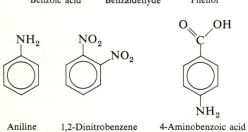

Benzoic acid   Benzaldehyde   Phenol

Aniline   1,2-Dinitrobenzene   4-Aminobenzoic acid
(*o*-dinitrobenzene)   (*p*-aminobenzoic acid)

APPENDIX D
MATHEMATICAL OPERATIONS

D-1 GRAPHS OF LINEAR EQUATIONS

An equation of the form

$$y = mx + b$$

where b is any constant and m is any nonzero constant, is called a *linear equation*. When values of x and the corresponding values of y which are related by the above equation are plotted on a graph using two-dimensional cartesian (rectangular) coordinates, the result is a set of points which fall on a straight line. Furthermore the *slope* of the straight line is the value of m in the above equation.

When data points (x, y) are plotted and the points fall on a straight line, it is evidence that x and y are related by the linear equation $y = mx + b$. The slope m of the line can be determined from the graph by choosing any two points (x_1, y_1) and (x_2, y_2) on the line and calculating m from the relationship

$$m = \frac{y_2 - y_1}{x_2 - x_1}$$

The constant b in the equation $y = mx + b$ is the y *intercept*, the place where the line crosses the y, or vertical, axis. The x intercept is $-b/m$.

D-2 QUADRATIC EQUATIONS

An equation of the form

$$ax^2 + bx + c = 0$$

where b and c are constants and a is any nonzero

constant, is called a *quadratic equation*. Each such equation has two roots,

$$x = \frac{-b \pm \sqrt{b^2 - 4ac}}{2a}$$

where the "plus" of the "plus-or-minus" sign is used in one root, and the "minus" in the other. This relationship is called the *quadratic formula*.

In order to solve the equation

$$\frac{(x)(x)}{0.2 - x} = 1.6 \times 10^{-2}$$

we first rearrange it so that it is in the form $ax^2 + bx + c = 0$.

$$x^2 + 1.6 \times 10^{-2}x - 3.2 \times 10^{-3} = 0$$

In this equation $a = 1$, $b = 1.6 \times 10^{-2}$, and $c = -3.2 \times 10^{-3}$. Substituting in the quadratic formula, we have

$$x = \frac{-1.6 \times 10^{-2} \pm \sqrt{(1.6 \times 10^{-2})^2 - 4(1)(-3.2 \times 10^{-3})}}{2(1)}$$

from which we get

$$x = \frac{-1.6 \times 10^{-2} + 0.114}{2} = 4.9 \times 10^2$$

or $\quad x = \dfrac{-1.6 \times 10^{-2} - 0.114}{2} = -6.5 \times 10^{-2}$

Although each of these two roots solves the original equation, if that equation describes a real, physical situation, only one of the roots may give a solution which is meaningful. (An example of a root which is not physically meaningful is one which corresponds to a negative concentration.)

D-3 LOGARITHMS

The logarithm of a number N is the power (or exponent) x to which a base number b must be raised in order to yield the number. In other words, if

$$b^x = N$$

then $$x = \log_b N$$

where $\log_b N$ is read "logarithm of N to the base b."

There are two bases for logarithms which are especially important. These are 10, the base of the *common logarithms,* and e, the base of the *natural logarithms.* (e is an irrational number, $2.71828 \cdots$.) Logarithms to the base 10 are often represented by the symbol *log* and logarithms to the base e, by *ln*. In other words,

Common logarithm of $N = \log_{10} N = \log N$

Natural logarithm of $N = \log_e N = \ln N$

The common and natural logarithms are related by the simple equation

$$\ln N = 2.303 \log N$$

Common logarithms of numbers may be found from tables (such as the one following) or from a calculator with the logarithm function. As an example of the use of the table consider the following example:

$$N = 7.31 \times 10^{-17}$$

$$\log N = ?$$

Our procedure is to find the logarithm of 7.31 and then *add* the logarithm of 10^{-17}. From the table we find that to three significant figures the logarithm of 7.31 is 0.864. The logarithm of 10^{-17} is simply the exponent itself, -17. Therefore,

$$\log (7.31 \times 10^{-17}) = \log 7.31 + \log 10^{-17}$$
$$= 0.864 + (-17) = -16.136$$

Finding the number whose logarithm is known is called finding the *antilogarithm.* Suppose that we wish to find the number whose logarithm is 3.921, that is, antilog 3.921.

$$\log N = 3.921$$

First we write the logarithm as the sum of an integer and a positive decimal fraction less than 1.

$$\log N = 3 + 0.921$$

The number whose logarithm equals 3 is 10^3, and (from the table) the number whose logarithm equals 0.921 is 8.34. Therefore, $N = 10^3 \times 8.34 = 8.34 \times 10^3$. If the logarithm is negative, essentially the same procedure is followed.

$$\log N = -8.192$$
$$= -9 + 0.808$$
$$N = 10^{-9} \times 6.43 = 6.43 \times 10^{-9}$$

Natural logarithms are handled in much the same way. Tables of natural logarithms are not as easy to find as those of common logarithms, but conversion from one to the other is easy using the equation $\ln N = 2.303 \log N$. Alternatively, many pocket calculators have the natural logarithm function.

Common logarithms

N	0	1	2	3	4	5	6	7	8	9
10	0000	0043	0086	0128	0170	0212	0253	0294	0334	0374
11	0414	0453	0492	0531	0569	0607	0645	0682	0719	0755
12	0792	0823	0864	0899	0934	0969	1004	1038	1072	1106
13	1139	1173	1206	1239	1271	1303	1335	1367	1399	1430
14	1461	1492	1523	1553	1584	1614	1644	1673	1703	1732
15	1761	1790	1818	1847	1875	1903	1931	1959	1987	2014
16	2041	2068	2095	2122	2148	2175	2201	2227	2253	2279
17	2304	2330	2355	2380	2405	2430	2455	2480	2504	2529
18	2553	2577	2601	2625	2648	2672	2695	2718	2742	2765
19	2788	2810	2833	2856	2878	2900	2923	2945	2967	2989
20	3010	3032	3054	3075	3096	3118	3139	3160	3181	3201
21	3222	3243	3263	3284	3304	3324	3345	3365	3385	3404
22	3424	3444	3464	3483	3502	3522	3541	3560	3579	3598
23	3617	3636	3655	3674	3692	3711	3729	3747	3766	3784
24	3802	3820	3838	3856	3874	3892	3909	3927	3945	3962
25	3979	3997	4014	4031	4048	4065	4082	4099	4116	4133
26	4150	4166	4183	4200	4216	4232	4249	4265	4281	4298
27	4314	4330	4346	4362	4378	4393	4409	4425	4440	4456
28	4472	4487	4502	4518	4533	4548	4564	4579	4594	4609
29	4624	4639	4654	4669	4683	4698	4713	4728	4742	4757
30	4771	4786	4800	4814	4829	4843	4857	4871	4886	4900
31	4914	4928	4942	4955	4969	4983	4997	5011	5024	5038
32	5051	5065	5079	5092	5105	5119	5132	5145	5159	5172
33	5185	5198	5211	5224	5237	5250	5263	5276	5289	5302
34	5315	5328	5340	5353	5366	5378	5391	5403	5416	5428
35	5441	5453	5465	5478	5490	5502	5514	5527	5539	5551
36	5563	5575	5587	5599	5611	5623	5635	5647	5658	5670
37	5682	5694	5705	5717	5729	5740	5752	5763	5775	5786
38	5798	5809	5821	5832	5843	5855	5866	5877	5888	5899
39	5911	5922	5933	5944	5955	5966	5977	5988	5999	6010
40	6021	6031	6042	6053	6064	6075	6085	6096	6107	6117
41	6128	6138	6149	6160	6170	6180	6191	6201	6212	6222
42	6232	6243	6253	6263	6274	6284	6294	6304	6314	6325
43	6335	6345	6355	6365	6375	6385	6395	6405	6415	6425
44	6435	6444	6454	6464	6474	6484	6493	6503	6513	6522
45	6532	6542	6551	6561	6571	6580	6590	6599	6609	6618
46	6628	6637	6646	6656	6665	6675	6684	6693	6702	6712
47	6721	6730	6739	6749	6758	6767	6776	6785	6794	6803
48	6812	6821	6830	6839	6848	6857	6866	6875	6884	6893
49	6902	6911	6920	6928	6937	6946	6955	6964	6972	6981
N	0	1	2	3	4	5	6	7	8	9

(*Continued on next page*)

Common logarithms (*Continued*)

N	0	1	2	3	4	5	6	7	8	9
50	6990	6998	7007	7016	7024	7033	7042	7050	7059	7067
51	7076	7084	7093	7101	7110	7118	7126	7135	7143	7152
52	7160	7168	7177	7185	7193	7202	7210	7218	7226	7235
53	7243	7251	7259	7267	7275	7284	7292	7300	7308	7316
54	7324	7332	7340	7348	7356	7364	7372	7380	7388	7396
55	7404	7412	7419	7427	7435	7443	7451	7459	7466	7474
56	7482	7490	7497	7505	7513	7520	7528	7536	7543	7551
57	7559	7566	7574	7582	7589	7597	7604	7612	7619	7627
58	7634	7642	7649	7657	7664	7672	7679	7686	7694	7701
59	7709	7716	7723	7731	7738	7745	7752	7760	7767	7774
60	7782	7789	7796	7803	7810	7818	7825	7832	7839	7846
61	7853	7860	7868	7875	7882	7889	7896	7903	7910	7917
62	7924	7931	7938	7945	7952	7959	7966	7973	7980	7987
63	7993	8000	8007	8014	8021	8028	8035	8041	8048	8055
64	8062	8069	8075	8082	8089	8096	8102	8109	8116	8122
65	8129	8136	8142	8149	8156	8162	8169	8176	8182	8189
66	8195	8202	8209	8215	8222	8228	8235	8241	8248	8254
67	8261	8267	8274	8280	8287	8293	8299	8306	8312	8319
68	8325	8331	8338	8344	8351	8357	8363	8370	8376	8382
69	8388	8395	8401	8407	8414	8420	8426	8432	8439	8445
70	8451	8457	8463	8470	8476	8482	8488	8494	8500	8506
71	8513	8519	8525	8531	8537	8543	8549	8555	8561	8567
72	8573	8579	8585	8591	8597	8603	8609	8615	8621	8627
73	8633	8639	8645	8651	8657	8663	8669	8675	8681	8686
74	8692	8698	8704	8710	8716	8722	8727	8733	8739	8745
75	8751	8756	8762	8768	8774	8779	8785	8791	8797	8802
76	8808	8814	8820	8825	8831	8837	8842	8848	8854	8859
77	8865	8871	8876	8882	8887	8893	8899	8904	8910	8915
78	8921	8927	8932	8938	8943	8949	8954	8960	8965	8971
79	8976	8982	8987	8993	8998	9004	9009	9015	9020	9025
80	9031	9036	9042	9047	9053	9058	9063	9069	9074	9079
81	9085	9090	9096	9101	9106	9112	9117	9122	9128	9133
82	9138	9143	9149	9154	9159	9165	9170	9175	9180	9186
83	9191	9196	9201	9206	9212	9217	9222	9227	9232	9238
84	9243	9248	9253	9258	9263	9269	9274	9279	9284	9289
85	9294	9299	9304	9309	9315	9320	9325	9330	9335	9340
86	9345	9350	9355	9360	9365	9370	9375	9380	9385	9390
87	9395	9400	9405	9410	9415	9420	9425	9430	9435	9440
88	9445	9450	9455	9460	9465	9469	9474	9479	9484	9489
89	9494	9499	9504	9509	9513	9518	9523	9528	9533	9538
90	9542	9547	9552	9557	9562	9566	9571	9576	9581	9586
91	9590	9595	9600	9605	9609	9614	9619	9624	9628	9633
92	9638	9643	9647	9652	9657	9661	9666	9671	9675	9680
93	9685	9689	9694	9699	9703	9708	9713	9717	9722	9727
94	9731	9736	9741	9745	9750	9754	9759	9763	9768	9773
95	9777	9782	9786	9791	9795	9800	9805	9809	9814	9818
96	9823	9827	9832	9836	9841	9845	9850	9854	9859	9863
97	9868	9872	9877	9881	9886	9890	9894	9899	9903	9908
98	9912	9917	9921	9926	9930	9934	9939	9943	9948	9952
99	9956	9961	9965	9969	9974	9978	9983	9987	9991	9996
N	0	1	2	3	4	5	6	7	8	9

APPENDIX E
KINETIC-MOLECULAR THEORY AND THE IDEAL GAS

The ideal-gas law can be derived from the assumptions of kinetic-molecular theory as follows: Consider a cubical container, edge-length e, and the cartesian coordinate axes shown in the accompanying diagram. The container is oriented so that each wall is parallel to two axes. (Thus there are two xy walls, two yz walls, and two xz walls.) Consider next that a molecule in the container has a velocity u in the direction shown.[1] This

[1] u is a vector, having a direction and a magnitude. The magnitude is often called the *speed, s.*

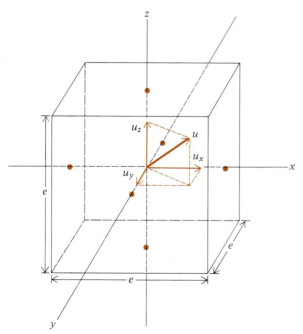

velocity can be resolved into three components, u_x, u_y, and u_z, parallel to the x, y, and z axes, respectively. The relationship among these component velocities is obtainable from geometrical considerations:

$$u^2 = u_x{}^2 + u_y{}^2 + u_z{}^2$$

Now we consider u_x separately by imagining that there is a molecule moving parallel to the x axis with a velocity u_x. The molecule collides elastically with the right-hand yz wall, rebounds, traverses the distance e to the left-hand yz wall, collides and rebounds again, and continues in this way bouncing back and forth indefinitely.

The *momentum* of a moving object is defined as the product of its mass, m, times its velocity, u. *Before* colliding with the right-hand yz wall, the momentum of the molecule is mu_x. *After* the collision the molecule is now moving in the opposite direction, and so its velocity is now $-u_x$, and its momentum, $-mu_x$. Thus for a single collision with the right-hand yz wall the *change* in momentum of the molecule is

$$mu_x - (-mu_x) = 2mu_x$$

Now we determine how many of these collisions take place per second. Between collisions with the right-hand yz wall the molecule travels a distance $2e$. Thus the time necessary for each such collision to occur is

$$\frac{2e}{u_x}$$

and the number of collisions per second is the reciprocal of this, or

$$\frac{u_x}{2e}$$

Therefore the change of momentum per second is

$$2mu_x \frac{u_x}{2e} = \frac{mu_x{}^2}{e}$$

But this change is for the molecule hitting *one* yz wall only. Considering both yz walls, the *total* change of momentum per second is

$$\frac{2mu_x{}^2}{e}$$

Similar results are reached for collisions with the xy and xz walls. Thus the *total* change of momentum per second experienced by a molecule colliding with *all* walls of the container is

$$\frac{2mu_x{}^2}{e} + \frac{2mu_y{}^2}{e} + \frac{2mu_z{}^2}{e} = \frac{2m}{e}(u_x{}^2 + u_y{}^2 + u_z{}^2)$$

$$= \frac{2mu^2}{e}$$

Furthermore, if there are q molecules in the container, the grand total momentum change becomes

$$\frac{2qm\overline{u^2}}{e}$$

where $\overline{u^2}$ is the *average* value of the velocities squared of the molecules.

We learn from physics that the change in momentum experienced by a molecule when it collides with a wall equals the force exerted by the molecule on the wall. Thus the *total force* exerted outward on *all* walls is

$$F = \frac{2qm\overline{u^2}}{e}$$

Pressure is force per unit area, where in this case the area is the total surface area of the cube, $6e^2$. Therefore,

$$P = \frac{F}{A} = \frac{2qm\overline{u^2}/e}{6e^2} = \frac{qm\overline{u^2}}{3e^3}$$

$$= \frac{qm\overline{u^2}}{3V}$$

where V is the volume of the cube.

Rearranging this we get

$$PV = \tfrac{1}{3}qm\overline{u^2}$$

We have assumed in this derivation that the container is a cube, but it can be shown that exactly the same result holds true for a container of *any* shape. The relationship is thus perfectly general.

Now, a little more juggling produces:

$$PV = \tfrac{2}{3}q(\tfrac{1}{2}m\overline{u^2}) = \tfrac{2}{3}q\overline{E_k}$$

where $\overline{E_k}$ is the average of the kinetic energies of the molecules.

According to kinetic-molecular theory, the average kinetic energy is proportional to the absolute temperature:

$$\overline{E_k} = cT$$

where c is a constant of proportionality. Substituting, we get

$$PV = \tfrac{2}{3}qcT$$

and since the number of molecules q is equal to the product of the number of moles n times Avogadro's number N_0,

$$PV = \tfrac{2}{3}nN_0cT = n(\tfrac{2}{3}N_0c)T$$

which is just the ideal-gas law,

$$PV = nRT$$

in which $R = \tfrac{2}{3}N_0c$.

It should be noted that one of the justifications for assuming the proportionality between average kinetic energy and absolute temperature is that doing so permits this derivation to be made.

APPENDIX F
VAPOR PRESSURE OF WATER

Temperature, °C	Pressure, mmHg
−10	2.15
−5	3.16
0	4.58
1	4.93
2	5.29
3	5.69
4	6.10
5	6.54
6	7.01
7	7.51
8	8.04
9	8.61
10	9.21
11	9.84
12	10.52
13	11.23
14	11.99
15	12.79
16	13.63
17	14.53
18	15.48
19	16.48
20	17.54
21	18.65
22	19.83

Temperature, °C	Pressure, mmHg
23	21.07
24	22.39
25	23.76
26	25.21
27	26.74
28	28.35
29	30.04
30	31.82
35	42.18
40	55.32
45	71.88
50	92.51
55	118.04
60	149.38
65	187.54
70	233.7
75	289.1
80	355.1
85	433.6
90	525.76
95	633.90
100	760.00
105	906.07
110	1074.56
115	1267.98

APPENDIX G
SELECTED THERMODYNAMIC PROPERTIES AT 25°C

Standard heats (enthalpies) of formation, ΔH_f°; standard absolute entropies, S°; and standard free energies of formation, ΔG_f°

Substance	ΔH_f°, kJ mol^{-1}	S°, J K^{-1} mol^{-1}	ΔG_f°, kJ mol^{-1}
$H_2(g)$	0	130.6	0
GROUP IA			
$Li(s)$	0	28.0	0
$LiCl(s)$	−408.8	55.2	−383.7
$Li_2O(s)$	−595.8	37.9	−560.2
$Na(s)$	0	51.0	0
$NaCl(s)$	−412.1	72.4	−384.0
$Na_2O(s)$	−415.9	72.8	−376.6
$NaNO_3(s)$	−466.7	116.3	−365.9
$K(s)$	0	63.6	0
$K_2O(s)$	−361.5	94.1	−318.8
$KCl(s)$	−435.9	82.7	−408.3
GROUP IIA			
$Mg(s)$	0	32.5	0
$MgCl_2(s)$	−641.8	89.5	−592.3
$MgO(s)$	−601.8	26.8	−569.6
$Ca(s)$	0	41.6	0
$CaO(s)$	−635.1	39.7	−604.2
$CaCl_2(s)$	−795.8	104.6	−748.1
$Ca_3(PO_4)_2(s)$	−4137	236.0	−3899.
GROUP IIIA			
$B(s)$	0	6.5	0
$B_2O_3(s)$	−1264	54.0	−1184
$Al(s)$	0	28.3	0
$AlCl_3(s)$	−705.6	109.3	−630.1
$Al_2O_3(s)$	−1676	51.0	−1576

Substance	ΔH_f°, kJ mol^{-1}	S°, J K^{-1} mol^{-1}	ΔG_f°, kJ mol^{-1}
GROUP IVA			
C(diamond)	1.90	2.38	2.87
C(graphite)	0	5.74	0
$CH_4(g)$	−74.8	187.9	−50.8
$CH_3OH(l)$	−239.0	126.3	−166.3
$C_2H_2(g)$	226.8	200.8	209.2
$C_2H_4(g)$	52.3	219.5	68.1
$C_2H_6(g)$	−84.6	229.5	−32.9
$C_2H_5OH(l)$	−277.6	160.7	−174.8
$CO(g)$	−110.5	197.6	−137.2
$CO_2(g)$	−393.5	213.6	−394.4
$CCl_4(l)$	−139.5	214.4	−68.7
$HCN(g)$	130.5	201.8	120.1
$Si(s)$	0	18.7	0
SiO_2(quartz)	−859.4	41.8	−805.0
GROUP VA			
$N_2(g)$	0	191.5	0
$NH_3(g)$	−46.1	192.3	−16.5
$NH_4Cl(s)$	−314.4	94.6	−202.7
$NO(g)$	90.4	210.6	86.7
$NO_2(g)$	33.8	240.4	51.8
$N_2O(g)$	81.5	220.0	103.6
P_4(white)	0	44.0	0
$P_4O_{10}(s)$	−2940	228.9	−2675

(Continued on next page)

Standard heats (enthalpies) of formation, ΔH_f°; standard absolute entropies, S°; and standard free energies of formation, ΔG_f° (Continued)

Substance	ΔH_f°, kJ mol^{-1}	S°, J K^{-1} mol^{-1}	ΔG_f°, kJ mol^{-1}	Substance	ΔH_f°, kJ mol^{-1}	S°, J K^{-1} mol^{-1}	ΔG_f°, kJ mol^{-1}
GROUP VIA				$Kr(g)$	0	164.0	0
$O_2(g)$	0	205.1	0	$Xe(g)$	0	169.6	0
$O_3(g)$	143	237.6	163.2	$Rn(g)$	0	176.2	0
$H_2O(g)$	−241.8	188.7	−228.6				
$H_2O(l)$	−285.8	69.9	−237.2				
$H_2O_2(l)$	−187.6	92	−120.4	TRANSITION ELEMENTS			
S_8(rhombic)	0	31.9	0	$Mn(s)$	0	31.8	0
S_8(monoclinic)	0.30	32.6	0.10	$MnCl_2(s)$	−481.3	118.2	−440.5
$SO_2(g)$	−296.8	248.1	−300.2	$MnO_2(s)$	−519.6	53.1	−466.1
$SO_3(g)$	−395.7	256.6	−371.1	$Fe(s)$	0	27.2	0
$H_2S(g)$	−20.6	205.7	−33.6	$FeCl_2(s)$	−341.8	117.9	−302.3
$H_2SO_4(l)$	−814.0	156.9	−690.1	$FeCl_3(s)$	−399.5	142.3	−334.1
				$FeO(s)$	−272.0	60.8	−251.5
GROUP VIIA				$Fe_2O_3(s)$	−824	90.0	−741.0
$F_2(g)$	0	203.3	0	$Fe_3O_4(s)$	−1121	146.4	−1014
$HF(g)$	−268.6	173.5	−270.7	$Cu(s)$	0	33.3	0
$Cl_2(g)$	0	222.9	0	$CuCl(s)$	−137.2	86.2	−119.9
$HCl(g)$	−92.3	186.8	−95.3	$CuCl_2(s)$	−205.9	108.1	−161.9
$Br_2(g)$	0	152.3	0	$CuO(s)$	−155.2	43.5	−127.2
$HBr(g)$	−36.2	198.5	−53.2	$Cu_2O(s)$	−166.7	100.8	−146.4
$I_2(s)$	0	116.7	0	$CuSO_4(s)$	−769.9	113.4	−661.9
$HI(g)$	25.9	206.3	1.72	$CuSO_4 \cdot 5H_2O$	−2278	305.4	−1880
				$Hg(l)$	0	77.4	0
GROUP 0				$HgCl_2(s)$	−230.1	144	−185.8
$He(g)$	0	126.1	0	$Hg_2Cl_2(s)$	−264.9	195.8	−210.7
$Ne(g)$	0	144.1	0	HgO(red)	−90.7	72.0	−58.5
$Ar(g)$	0	154.7	0	HgO(yellow)	−90.2	73.2	−58.4

APPENDIX H
EQUILIBRIUM CONSTANTS AT 25°C

H-1 DISSOCIATION CONSTANTS OF WEAK ACIDS

Formula	Name	K_a
H_3BO_3	Boric acid	6.0×10^{-10}
HCN	Hydrogen cyanide (hydrocyanic acid)	4.0×10^{-10}
$HC_2H_3O_2$	Acetic acid	1.8×10^{-5}
H_2CO_3 ($H_2O + CO_2$)	Carbonic acid	4.2×10^{-7} (K_1) 5.6×10^{-11} (K_2)
$H_2C_2O_4$	Oxalic acid	5.4×10^{-2} (K_1) 5.0×10^{-5} (K_2)
$H_2C_6H_6O_6$	Ascorbic acid	5.0×10^{-5} (K_1) 1.5×10^{-12} (K_2)
HNO_2	Nitrous acid	5.0×10^{-4}
H_3PO_3	Phosphorous acid	1.6×10^{-2} (K_1) 7×10^{-7} (K_2)
H_3PO_4	Phosphoric acid	7.6×10^{-3} (K_1) 6.3×10^{-8} (K_2) 4.4×10^{-13} (K_3)
H_2S	Hydrogen sulfide (hydrosulfuric acid)	1.1×10^{-7} (K_1) 1.0×10^{-14} (K_2)
H_2SO_3 ($H_2O + SO_2$)	Sulfurous acid	1.3×10^{-2} (K_1) 6.3×10^{-8} (K_2)
HSO_4^-	Hydrogen sulfate ion (bisulfate ion)	1.2×10^{-2}
HF	Hydrofluoric acid (hydrogen fluoride)	6.7×10^{-4}
HOCl	Hypochlorous acid	3.2×10^{-8}
$HClO_2$	Chlorous acid	1.1×10^{-2}

H-2 DISSOCIATION CONSTANTS OF WEAK BASES

Formula	Name	K_b
NH_3	Ammonia	1.8×10^{-5}
NH_2OH	Hydroxylamine	9.1×10^{-9}
CH_3NH_2	Methylamine	4.4×10^{-4}
$C_{10}H_{14}N_2$	Nicotine	7.4×10^{-7} (K_1) 1.4×10^{-11} (K_2)
PH_3	Phosphine	1×10^{-14}

H-3   SOLUBILITY PRODUCTS

Compound	K_{sp}
BROMIDES	
AgBr	5.0×10^{-13}
Hg_2Br_2	9×10^{-17}
$PbBr_2$	4.0×10^{-5}
CARBONATES	
$MgCO_3$	2.1×10^{-5}
$CaCO_3$	4.7×10^{-9}
$SrCO_3$	7.0×10^{-10}
$BaCO_3$	1.6×10^{-9}
Ag_2CO_3	8.2×10^{-12}
CHLORIDES	
AgCl	1.7×10^{-10}
Hg_2Cl_2	1.1×10^{-18}
$PbCl_2$	1.6×10^{-5}
CHROMATES	
$CaCrO_4$	7.1×10^{-4}
$SrCrO_4$	3.6×10^{-5}
$BaCrO_4$	8.5×10^{-11}
Ag_2CrO_4	1.9×10^{-12}
$PbCrO_4$	2.0×10^{-16}
FLUORIDES	
MgF_2	7.9×10^{-8}
CaF_2	1.7×10^{-10}
SrF_2	2.5×10^{-9}
BaF_2	1.7×10^{-6}
PbF_2	4×10^{-8}
HYDROXIDES	
$Mg(OH)_2$	8.9×10^{-12}
$Ca(OH)_2$	1.3×10^{-6}
$Sr(OH)_2$	3.2×10^{-4}
$Ba(OH)_2$	5.0×10^{-3}
$Cr(OH)_3$	6.7×10^{-31}
$Mn(OH)_2$	2×10^{-13}
$Fe(OH)_2$	2×10^{-15}
$Fe(OH)_3$	6×10^{-38}
$Co(OH)_2$	2.5×10^{-16}
$Ni(OH)_2$	1.6×10^{-16}
$Cu(OH)_2$	1.6×10^{-19}
$Zn(OH)_2$	5×10^{-17}
$Cd(OH)_2$	2.8×10^{-14}
$Al(OH)_3$	5×10^{-33}
$Pb(OH)_2$	4×10^{-15}
$Sn(OH)_2$	3×10^{-27}

Compound	K_{sp}
IODIDES	
AgI	8.5×10^{-17}
Hg_2I_2	4.5×10^{-29}
HgI_2	2.5×10^{-26}
PbI_2	8.3×10^{-9}
OXALATES	
CaC_2O_4	1.3×10^{-9}
SrC_2O_4	5.6×10^{-8}
BaC_2O_4	1.5×10^{-8}
PbC_2O_4	8.3×10^{-12}
PHOSPHATES	
$Ca_3(PO_4)_2$	1.3×10^{-32}
$Sr_3(PO_4)_2$	1×10^{-31}
$Ba_3(PO_4)_2$	6.0×10^{-39}
Ag_3PO_4	1×10^{-21}
$Pb_3(PO_4)_2$	8.0×10^{-43}
SULFATES	
$CaSO_4$	2.4×10^{-5}
$SrSO_4$	7.6×10^{-7}
$BaSO_4$	1.5×10^{-9}
$PbSO_4$	1.3×10^{-8}
Ag_2SO_4	1.7×10^{-5}
SULFIDES	
MnS	7×10^{-16}
FeS	4×10^{-19}
CoS	5×10^{-22}
NiS	3×10^{-21}
CuS	8×10^{-37}
Ag_2S	5.5×10^{-51}
ZnS	1.2×10^{-23}
CdS	1.0×10^{-28}
HgS	1.6×10^{-54}
PbS	7×10^{-29}
SnS	1×10^{-26}
SnS_2	1×10^{-70}
Bi_2S_3	1×10^{-96}
SULFITES	
Ag_2SO_3	5×10^{-14}
$BaSO_3$	6×10^{-5}

APPENDIX I
STANDARD REDUCTION POTENTIALS AT 25°C

Half-reaction	$\mathcal{E}°$, V
$e^- + Li^+ \rightleftharpoons Li(s)$	-3.05
$e^- + K^+ \rightleftharpoons K(s)$	-2.93
$2e^- + Ba^{2+} \rightleftharpoons Ba(s)$	-2.91
$2e^- + Ca^{2+} \rightleftharpoons Ca(s)$	-2.76
$e^- + Na^+ \rightleftharpoons Na(s)$	-2.71
$2e^- + Mg^{2+} \rightleftharpoons Mg(s)$	-2.38
$e^- + \frac{1}{2}H_2(g) \rightleftharpoons H^-$	-2.25
$3e^- + Al^{3+} \rightleftharpoons Al(s)$	-1.67
$2e^- + V^{2+} \rightleftharpoons V(s)$	-1.19
$2e^- + Mn^{2+} \rightleftharpoons Mn(s)$	-1.18
$2e^- + Zn^{2+} \rightleftharpoons Zn(s)$	-0.76
$3e^- + Cr^{3+} \rightleftharpoons Cr(s)$	-0.74
$3e^- + Ga^{3+} \rightleftharpoons Ga(s)$	-0.53
$2e^- + Fe^{2+} \rightleftharpoons Fe(s)$	-0.44
$e^- + Cr^{3+} \rightleftharpoons Cr^{2+}$	-0.41
$2e^- + Cd^{2+} \rightleftharpoons Cd(s)$	-0.40
$2e^- + PbI_2(s) \rightleftharpoons Pb(s) + 2I^-$	-0.37
$2e^- + PbSO_4(s) \rightleftharpoons Pb(s) + SO_4^{2-}$	-0.36
$2e^- + Co^{2+} \rightleftharpoons Co(s)$	-0.28
$e^- + V^{3+} \rightleftharpoons V^{2+}$	-0.26
$2e^- + Ni^{2+} \rightleftharpoons Ni(s)$	-0.25
$2e^- + Sn^{2+} \rightleftharpoons Sn(s)$	-0.14
$2e^- + Pb^{2+} \rightleftharpoons Pb(s)$	-0.13
$2e^- + 2H^+ \rightleftharpoons H_2(g)$	$0.$
$e^- + AgBr(s) \rightleftharpoons Ag(s) + Br^-$	$+0.07$
$2e^- + S(s) + 2H^+ \rightleftharpoons H_2S(aq)$	$+0.14$
$2e^- + Sn^{4+} \rightleftharpoons Sn^{2+}$	$+0.15$
$e^- + Cu^{2+} \rightleftharpoons Cu^+$	$+0.15$
$2e^- + SO_4^{2-} + 4H^+ \rightleftharpoons H_2SO_3 + H_2O$	$+0.17$
$e^- + AgCl(s) \rightleftharpoons Ag(s) + Cl^-$	$+0.22$
$2e^- + Hg_2Cl_2(s) \rightleftharpoons 2Hg(l) + 2Cl^-$	$+0.27$
$2e^- + Cu^{2+} \rightleftharpoons Cu(s)$	$+0.34$
$e^- + VO^{2+} + 2H^+ \rightleftharpoons V^{3+} + H_2O$	$+0.36$
$e^- + Cu^+ \rightleftharpoons Cu(s)$	$+0.52$
$2e^- + I_2(s) \rightleftharpoons 2I^-$	$+0.54$
$e^- + MnO_4^- \rightleftharpoons MnO_4^{2-}$	$+0.56$
$2e^- + O_2(g) + 2H^+ \rightleftharpoons H_2O_2(aq)$	$+0.68$
$e^- + Fe^{3+} \rightleftharpoons Fe^{2+}$	$+0.77$
$2e^- + Hg_2^{2+} \rightleftharpoons 2Hg(l)$	$+0.78$
$e^- + Ag^+ \rightleftharpoons Ag(s)$	$+0.80$
$2e^- + 2NO_3^- + 4H^+ \rightleftharpoons N_2O_4(g) + 2H_2O$	$+0.80$
$2e^- + 2Hg^{2+} \rightleftharpoons Hg_2^{2+}$	$+0.92$
$3e^- + NO_3^- + 4H^+ \rightleftharpoons NO(g) + 2H_2O$	$+0.96$
$2e^- + V(OH)_4^+ + 2H^+ \rightleftharpoons VO^{2+} + 3H_2O$	$+1.00$
$2e^- + Br_2(l) \rightleftharpoons 2Br^-$	$+1.09$
$2e^- + Br_2(aq) \rightleftharpoons 2Br^-$	$+1.06$
$4e^- + O_2(g) + 4H^+ \rightleftharpoons 2H_2O$	$+1.23$
$2e^- + MnO_2(s) + 4H^+ \rightleftharpoons Mn^{2+} + 2H_2O$	$+1.23$
$6e^- + Cr_2O_7^{2-} + 14H^+ \rightleftharpoons 2Cr^{3+} + 7H_2O$	$+1.33$
$2e^- + Cl_2(g) \rightleftharpoons 2Cl^-$	$+1.36$
$2e^- + PbO_2(s) + 4H^+ \rightleftharpoons Pb^{2+} + 2H_2O$	$+1.46$
$3e^- + Au^{3+} \rightleftharpoons Au(s)$	$+1.50$
$e^- + Mn^{3+} \rightleftharpoons Mn^{2+}$	$+1.51$
$5e^- + MnO_4^- + 8H^+ \rightleftharpoons Mn^{2+} + 4H_2O$	$+1.51$
$e^- + Au^+ \rightleftharpoons Au(s)$	$+1.69$
$2e^- + H_2O_2 + 2H^+ \rightleftharpoons 2H_2O$	$+1.78$
$2e^- + XeO_3(s) + 6H^+ \rightleftharpoons Xe(g) + 3H_2O$	$+1.8$
$e^- + Co^{3+} \rightleftharpoons Co^{2+}$	$+1.81$
$e^- + Ag^{2+} \rightleftharpoons Ag^+$	$+1.98$
$2e^- + O_3(g) + 2H^+ \rightleftharpoons O_2(g) + H_2O$	$+2.07$
$2e^- + F_2(g) \rightleftharpoons 2F^-$	$+2.87$
$2e^- + F_2(g) + 2H^+ \rightleftharpoons 2HF(aq)$	$+3.06$

APPENDIX J
ANSWERS TO SELECTED NUMERICAL PROBLEMS

Chapter 1

1-17(a) 1.004×10^2; (b) 4.3×10^{-3}; (c) 1.0×10^9; (d) 4.00×10^5; (e) 1.56×10^5 **1-18**(a) 3; (c) 5; (e) 3; (g) 2; (i) 2; (k) 3 **1-19**(a) 45.3292 g; (b) 46.526 g; (c) 73.98 g; (d) 112 g; (e) 2.00×10^2 g **1-21**(a) 1.50×10^2; (c) 8.8; (e) 2.4×10^2; (g) 6.6×10^{15}; (i) 6.91×10^{-8}; (k) 4.1×10^2; (m) 1.6×10^{-2}; (o) 5.8×10^{11} **1-25**(a) 1 m; 10^3 times larger; (c) equal **1-26**(a) 2.65×10^{-1} g; (c) 8.14×10^6 g; **1-28**(a) $\dfrac{1 \text{ ft}}{12 \text{ in}}$; (c) $\dfrac{1 \text{ cm}}{10 \text{ mm}}$ **1-29**(a) $\dfrac{1.06 \text{ qt}}{1 \text{ liter}}$ or $\dfrac{1 \text{ qt}}{0.946 \text{ liter}}$; (c) $\dfrac{1.61 \text{ km}}{1 \text{ mi}}$ or $\dfrac{1 \text{ km}}{0.621 \text{ mi}}$; (e) $\dfrac{10^9 \text{ mm}^3}{1 \text{ m}^3}$ **1-30**(a) 4.76×10^2 m; (b) 47.6 mm; (c) 1.87 in; (d) 0.156 ft **1-33** \$0.19/tomato, or, 19¢/tomato

Chapter 2

2-4(a) 0.684 g cm^{-3} **2-7** 0.437 g cm^{-3} **2-9** 144 g **2-16**(a) 1; (c) 3; (e) 3 **2-21**(a) 100.0 g; (c) 22.6 g **2-22** 14.5% Br **2-23**(a) 2.052 g **2-25** 50% S; 50% O **2-26** 40% S; 60% O **2-35**(a) 298 K; (c) 374 K **2-36**(a) 24°C; (c) 163°C

Chapter 3

3-1(a) 35.5 amu; (b) 108 amu; (c) 6.41×10^3 amu; (d) 2.41×10^{25} amu **3-3** 20.2 amu **3-5**(a) 3.16×10^{-23} g; (b) 0.663 g; (c) 7.89×10^{-3} mol; (d) 4.75×10^{21} atoms **3-7**(a) 35.5 g; (b) 127 g; (c) 5.62 g; (d) 0.342 g; (e) 70.9 g; (f) 33.3 g **3-10**(a) 0.719 mol; (b) 7.67 mol; (c) 0.458 mol; (d) 4.20×10^{-2} mol; (e) 3.93×10^{-2} mol **3-13** 47.5 amu **3-15** Na_2SO_4 **3-18** $FeC_{10}H_{10}$ **3-21**(a) 0.12 mol; (b) 3.0×10^{-2} mol

3-22(a) 31.3 g; (b) 27.1 g **3-23** $HgC_2N_2O_2$ **3-25** 80.207% C, 9.617% H, 10.176% O **3-28** $C_9H_8O_4$ **3-30**(a) C_2H_3; (b) 0.0999 g; (c) C_4H_6 **3-35**(a) 2; (b) 28; (c) 2; (d) 76; (e) 2; (f) 14; (g) 38 **3-36**(a) 0.100 mol; (b) 0.200 mol; (c) 0.346 mol; (d) 0.913 mol **3-37**(a) 22.0 g; (b) 229 g; (c) 50.1 g; (d) 12.3 g **3-40** 1×10^{-2} mol O atoms

Chapter 4

4-2 10.3 m (33.7 ft) **4-3** 5.32 atm **4-5** 81.5 ml **4-10**(a) H_2 diffuses 3.98 times as fast **4-12** 40.9 liters **4-14** 7.48 ml **4-16**(a) 0.700 liter; (b) 0.536 liter **4-19**(a) 0.406 atm; (b) 0.956 atm **4-22** 0.800 atm **4-24**(a) 402 ml; (b) 420 ml **4-27** 0.591 g liter^{-1} **4-29** 1.45 g liter^{-1} **4-30** 30.0 amu **4-34**(a) 7.96 times as many H_2 molecules; (b) P_{H_2} is 7.96 times as high; (c) s_{H_2} is 2.82 times as great **4-38** $C_2H_4Cl_2$ **4-40** 1.64 liters **4-41** 8.43 liters **4-43** 0.605 liter

Chapter 5

5-16 35.46 amu **5-18** $2.0 \times 10^1\%$ ^{10}B, $8.0 \times 10^1\%$ ^{11}B **5-24** 16 in s^{-1} **5-25** 455 nm

Chapter 6

6-2 15 nm

Chapter 7

7-25 112 kJ mol^{-1}

Chapter 10

10-8 0.115 nm **10-17** 0.362 nm **10-21** 9.03 g cm^{-3} **10-25** 0.0750 nm **10-53** 1.06×10^3 mmHg **10-55** 64°C **10-57** 32.9 kJ mol^{-1} **10-62** 46.8

Chapter 11

11-7(a) 4.37×10^2 atm; (b) 1.00×10^3 atm **11-9**(a) 214°C; (b) 218°C **11-36** 72.3 kJ mol^{-1}

Chapter 12

12-6 7.3×10^4 amu **12-16**(a) $1.71 \times 10^{-2} M$ **12-17**(a) 0.187 m **12-18**(a) 0.663 **12-19**(a) 52.0 mass %; (b) 5.40 mol %; (c) 1.88 M; (d) 3.16 m **12-23**(a) 1.10 mass %; (b) 1.80×10^{-3} **12-26** 2.8 M **12-30** 1.29×10^{-4} **12-39** 69.7 atm **12-42**(a) −0.49°C; (b) 100.13°C **12-44** 141 amu **12-48** 1.96×10^5 **12-49**(b) 39.7 mmHg **12-53** 0.690 M Na$^+$, 0.138 M Cl$^-$, 0.828 M K$^+$, 0.690 M SO$_4{}^{2-}$ **12-55** −0.28°C **12-64** 0.69 M H$^+$, 2.07 M Cl$^-$, 1.38 M Na$^+$

Chapter 13

13-33(a) 3.38×10^{-2} mol; (b) 3.38×10^{-2} equiv **13-36** 0.110 mol **13-37** 714 ml **13-41** 0.264 N **13-45**(a) 0.125 M; (b) 0.0833 M; (c) 0.0625 M **13-50** 284 g equiv^{-1} **13-52**(a) 0.400 M; (b) 0.363 M; (c) 0.230 M; (d) 0.124 M

Chapter 14

14-5 $7.80 \times 10^{-3} M$ s^{-1} **14-7** 51.0 mmHg **14-8**(c) $6.32 \times 10^{-3} M$ s^{-1} **14-9**(a) $6.07 \times 10^{-3} M$ min^{-1} **14-15**(b) $1.08 \times 10^{-4} M^{-2}$ s^{-1}; (c) $4.31 \times 10^{-7} M$ s^{-1} **14-18**(b) $5.24 \times 10^{-4} M^{-2}$ s^{-1} **14-21** $k = 2.65 \times 10^{-2}$ min^{-1} **14-25** 5.0×10^1 min **14-27** 6.5×10^3 s **14-29**(a) 0.659 M; (b) 7.97 h **14-40** 34°C **14-45** 1.1×10^2 kJ mol^{-1}

Chapter 15

15-22 8.60×10^{14} **15-25** 4.7×10^2 **15-28** −185 kJ **15-30** 20.1 mol liter^{-1} **15-33**(a) 1.89 M; (b) 1.89 M; (c) 0.115 M; (d) 0.115 M **15-35** [CO$_2$] = 0.378 M; [H$_2$O] = 0.022 M; [CO] = 0.022 M, [H$_2$] = 0.378 M **15-37** $4.4 \times 10^{-12} M$ **15-40** $4.5 \times 10^{-17} M$ **15-42** $9.1 \times 10^{-4} M$ **15-44** 6.6×10^{-2} atm **15-49** 0.1 atm

Chapter 16

16-1(a) 0.00; (b) 3.80 **16-2**(a) $2.4 \times 10^{-5} M$; (b) $1.1 \times 10^{-8} M$ **16-3**(a) $1.5 \times 10^{-11} M$; (b) $7.4 \times 10^{-5} M$ **16-4**(a) 12.04; (b) 10.56 **16-5**(a) 0.59; (b) 11.15 **16-6** 5.3 ml **16-10** $4.9 \times 10^{-5} M$

16-12 4.97 **16-14**(a) 2.55 **16-15**(a) 2.47 **16-16**(a) 0.72% **16-18** 6.4×10^{-7} **16-20** $7.9 \times 10^{-4} M$ **16-22** 0.16 M **16-23**(a) 11.52 **16-27**(a) 9.07; (c) 4.93 **16-28**(a) 10.10 **16-29**(a) $4.7 \times 10^{-3}\%$ **16-30** 5.0×10^{-4} **16-36** 5.19 **16-39**(a) 12.91; (b) 12.52 (c) 11.72; (d) 10.70; (e) 7.00 **16-44**(a) 4.74; (b) 9.3 **16-46** 0.065 mol **16-47** 0.04 mol **16-49**(a) 2.54; (b) 4.82; (c) 4.7 **16-51**(a) 2.00; (b) 7.00; (c) 12.00; (d) 10.70 **16-57** [H$^+$] = $8.4 \times 10^{-2} M$, [H$_2$PO$_4{}^-$] = $8.4 \times 10^{-2} M$, [HPO$_4{}^{2-}$] = $6.3 \times 10^{-8} M$, [PO$_4{}^{3-}$] = $3.3 \times 10^{-19} M$

Chapter 17

17-1 8.5×10^{-11} **17-3** 7.9×10^{-8} **17-5** 1.7×10^{-14} **17-6** 1.2×10^{-13} **17-8** $3.5 \times 10^{-12} M$ **17-9** $4.8 \times 10^{-23} M$ **17-13** $1.6 \times 10^{-4} M$ **17-15** $4.8 \times 10^{-5} M$ **17-22** $9 \times 10^{-9} M$ **17-24** 3.4 M **17-28** 1.18 **17-32** $2.2 \times 10^{-7} M$

Chapter 18

18-2 $w = 81$ J, $\Delta E = -685$ **18-8**(a) 9.60 liter atm; (b) −9.60 liter atm **18-10** 3.48 kJ **18-12**(a) 1.13 kJ; (b) 2.04 kJ **18-14** $\Delta H = 14.0$ kJ, $\Delta E = 13.3$ kJ **18-17** 6.009 kJ mol^{-1} **18-19** -1.2996×10^3 kJ mol^{-1} **18-22** −562.0 kJ **18-25** −465.5 kJ **18-27** 2.92 kJ **18-29** 29.0°C **18-35** 11.2 J K^{-1} **18-37** 7.5×10^2°C **18-42** 175.9 J K^{-1} **18-48** −1235.2 kJ mol^{-1} **18-54** 7×10^{24} **18-60** 54.23 kJ mol^{-1} **18-62** 1.72 kJ mol^{-1} less

Chapter 19

19-11 745 ml **19-12** 1.40×10^4 **19-14** 1.54 g **19-16** 18.7 h **19-19** 6.6×10^{-3} A **19-28**(a) −0.41 V; (b) −0.34 V **19-29**(a) 1.83 V; (b) 0.29 V **19-33**(a) 1.30 V; (b) 3.18 V **19-34** 1.14 V **19-36** 0.38 V **19-38** 1.00 V **19-40** $2 \times 10^{-6} M$ **19-42** 8.61 **19-44**(a) -1.06×10^3 kJ; (b) −56 kJ **19-45**(a) -1.5×10^2 kJ; (b) +66 kJ **19-46**(a) −251 kJ; (b) −614 kJ **19-47**(a) −413 kJ **19-49**(a) 4×10^{-81}

Chapter 24

24-19 0.125 g **24-22** 8.0×10^4 years **24-30**(a) 8.142×10^{-13} J nucleon^{-1}

INDEX

INDEX

Units, constants, and conversion equations (See also Appendix B)

SI base units

Physical quantity	Unit	Symbol
Length	Meter	m
Mass	Kilogram	kg
Time	Second	s
Electric current	Ampere	A
Temperature	Kelvin	K
Luminous intensity	Candela	cd
Amount of substance	Mole	mol

Some SI derived units

Physical quantity	Unit	Symbol	Definition
Force	Newton	N	$m\,kg\,s^{-2}$
Energy	Joule	J	$m^2\,kg\,s^{-2}$
Pressure	Pascal	Pa	$m^{-1}\,kg\,s^{-2} = N\,m^2$
Electrical charge	Coulomb	C	$A\,s$
Electrical potential difference	Volt	V	$J\,A^{-1}\,s^{-1}$
Frequency	Hertz	Hz	s^{-1}

Metric prefixes

Multiple	10^{18}	10^{15}	10^{12}	10^9	10^6	10^3	10^2	10^1
Prefix	exa	peta	tera	giga	mega	kilo	hecto	deka
Symbol	E	P	T	G	M	K	h	da
Submultiple	10^{-1}	10^{-2}	10^{-3}	10^{-6}	10^{-9}	10^{-12}	10^{-15}	10^{-18}
Prefix	deci	centi	milli	micro	nano	pico	femto	atto
Symbol	d	c	m	μ	n	p	f	a